FEDERAL GOVERNMENT CONSTRUCTION CONTRACTS

SECOND EDITION

MICHAEL A. BRANCA
AARON P. SILBERMAN
JOHN S. VENTO

ADRIAN L. BASTIANELLI III
ANDREW D. NESS
JOSEPH D. WEST

EDITORS

FORUM ON THE
CONSTRUCTION
INDUSTRY

AMERICAN BAR ASSOCIATION
Defending Liberty
Pursuing Justice

Printed in the United States of America.

14 13 12 11 10 5 4 3 2 1

Library of Congress Cataloging-in-Publication Data

Federal government construction contracts / edited by Michael A. Branca.—2nd ed.
   p. cm.
   Includes index.
   ISBN-13: 978-1-60442-751-6
   ISBN-10: 1-60442-751-5
   1. Government purchasing—Law and legislation—United States. 2. Letting of contracts—United States. 3. Construction contracts—United States. I. Branca, Michael A.
   KF865.F44 2010
   346.7302'3—dc22

                                                                                    2009050403

Discounts are available for books ordered in bulk. Special consideration is given to state bars, CLE programs, and other bar-related organizations. Inquire at Book Publishing, ABA Publishing, American Bar Association, 321 North Clark Street, Chicago, Illinois 60654-7598.

www.ababooks.org

CONTENTS

ABOUT THE EDITORS

Michael A. Branca is a partner in the Washington, D.C. office of Peckar & Abramson, P.C. He is a member of the Board of Governors of the Construction and Public Contract Law Section of the Virginia State Bar and is a past Chair of the Government Contract Section of the Federal Bar Association. Mr. Branca received his B.A. degree from the University of Virginia and his J.D. degree from George Mason University. He regularly counsels his clients with respect to all aspects of government contract law including bid protests on the front end of projects and claims on the back end. He also handles contractor integrity and compliance issues and federal and state false claims act matters. Mr. Branca is the co-author of the Thomson West treatise *Virginia Construction Law* and also writes and lectures regularly on construction and government contract topics.

Aaron P. Silberman is a shareholder at Rogers Joseph O'Donnell in San Francisco, California. He specializes in government contracts and construction counseling and litigation involving federal, state, and local public projects. His work includes evaluating, negotiating, and drafting solicitations, proposals, and bids; negotiating and evaluating contract terms; litigating bid protests; investigating and litigating claims for delay, disruption, payment, defects, and false claims; and advising and representing clients about contract terminations, debarments, bond requirements, liens and payment issues, and disabled access issues. Mr. Silberman has spoken and written on false claims, terminations, payment issues, BIM, green building, cardinal change claims, disabled access, and licensing. He chairs the Specialty Trade Contractors & Suppliers Division of the ABA Forum on the Construction Industry, co-chairs of the ABA Public Contract Law Section's Construction Division, and serves on the Board of Directors of the Bay Area District of AGC-California.

John S. Vento is a shareholder in the Tampa office of Trenam Kemker where he chairs the firm's practice group on Government Contracting and Construction Law. Prior to joining the firm in 1981, Mr. Vento was a Judge Advocate on active duty with the U.S. Air Force and retired from the Reserve in January 2005 in the rank of Colonel. While in the Reserve, he served as an Adjunct Professor of Law at the Air Force's Judge Advocate General's School, which published the handbook he authored on contingency contracting and fiscal law. Mr. Vento frequently speaks and writes on government contract law. He received his B.A., cum laude, from the University of Pittsburgh; a J.D., cum laude, from Duquesne University School of Law; and an LL.M. in International Law from the University of Michigan Law School, where he was a Michigan Scholar for highest academic achievement. While on academic leave of absence from the faculty of the U.S. Air Force Academy, Mr. Vento was a National Endowment for the Humanities Fellow at Yale Law School.

Adrian L. Bastianelli, III is the co-managing partner of the Washington, D.C. office of Peckar & Abramson, P.C. Mr. Bastianelli is Chair of the Forum on the Construction Industry, a fellow of the American College of Construction Lawyers, past president of the Washington Building Congress, and a past editor of *The Construction Lawyer* and the *Journal of the American College of Construction Lawyers*. Mr. Bastianelli received his BSCE degree from Purdue University and his J.D. degree from the University of Louisville. He worked as an engineer on claims for the Corps of Engineers, served as a law clerk for the U.S. Court of Claims, and has been in the private practice of law specializing in construction and government law for more than 35 years. Mr. Bastianelli also has an active ADR practice, having served as a mediator, arbitrator, and DRB member on over 300 disputes, including many involving the federal government. Mr. Bastianelli writes and lectures regularly on construction, government contract, and ADR topics.

Andrew D. Ness is a partner in the Washington, D.C. office of Howrey LLP, and co-chair of its 45 lawyer construction practice group. He assists both owners and contractors with respect to troubled projects, solving complex construction and design-related problems without need for formal dispute resolution whenever practicable. He has served as lead counsel for numerous large construction disputes that were resolved in federal and state courts, boards of contract appeals, and via arbitration. He has resolved disputes on many government projects, varying from GSA buildings to DOE radioactive waste processing facilities to Corps of Engineers dams and waterways. Mr. Ness is repeatedly recognized in both *Chambers, USA* and *Best Lawyers in America*, and is also in *Washington, D.C. Superlawyers* (Top 50 Washington, D.C. Lawyers 2008). He is a Fellow of the American College of Construction Lawyers and currently Publications Chair of the ABA Forum on the Construction Industry.

Joseph D. West is the co-chair of the Government and Commercial Contracts Practice Group of Gibson, Dunn & Crutcher LLP, an international law firm with offices in fifteen cities worldwide. He is resident in the firm's Washington, D.C. office. Mr. West graduated magna cum laude from Villanova University in 1971 with a Bachelor of Civil Engineering Degree and was elected to Tau Beta Pi (National Engineering Honorary) and Chi Epsilon (National Civil Engineering Honorary). In 1977, he graduated from the George Washington

University Law School, with high honors, where he was elected to the Order of the Coif and was an editor of *The George Washington Law Review*. He is also a registered professional engineer in the State of Virginia. Mr. West has written and lectured extensively on various government, construction, and commercial contract issues. He is currently a member of the American Bar Association Forum on the Construction Industry (member of the Forum's Governing Committee since 2008; chair of the Forum's Steering Committee of Division 1 (Dispute Avoidance & Resolution) from 2002 to 2004), as well as the Public Contract Law and Litigation Sections (currently co-chair of the Public Contract Law Section's Debarment and Suspension Committee and former vice-chair of the Section's Pro Bono and Acquisition Reform and Emerging Issues Committees). He is the former co-chair of the Steering Committee of the D.C. Bar Government Contracts and Litigation Section and a former member of the Board of Governors of the Virginia State Bar Construction and Public Contract Law Section, and he currently serves on the Advisory Board of The West Group's The Government Contractor. In November 2004, Mr. West was named by *The Legal Times* as one of Washington's Top Twelve Federal Procurement Attorneys, and in September 2006, he was selected by *The Washington Business Journal* as the city's Top Government Contracts Lawyer.

ABOUT THE AUTHORS

Robert S. Brams is a partner with the law firm of Patton Boggs LLP in Washington, D.C. He has over 20 years of experience in negotiating the various complex contractual arrangements associated with the development of major engineering and construction projects and in mediating, arbitrating, and litigating disputes arising on many of the largest infrastructure and construction projects in the world. Mr. Brams regularly assists owners, developers, architects, engineers, and contractors on a broad range of project delivery, procurement, and risk management issues associated with both public and private sector projects. Mr. Brams' practice also focuses on pre- and post-award bid protests on federal and state projects. Mr. Brams started his career as a judicial law clerk to the Honorable Christine Miller, U.S. Court of Federal Claims. He serves on various industry boards, including the Advisory Board for Thompson West's publication, *The Construction Contractor*.

Christyne K. Brennan is an associate in the Government and Commercial Contracts practice in the Washington, D.C. office of Gibson, Dunn & Crutcher LLP. Ms. Brennan is Managing Editor of the American Bar Association's *Public Contract Law Journal*, Vice-Chair of the Contract Claims and Disputes Resolution Committee of the Section of Public Contract Law, and a member of the Forum on the Construction Industry. Ms. Brennan received her B.A. from Furman University and earned a J.D. with honors from Georgetown University, where she served as Administrative Editor of the *Georgetown Journal of Law and Public Policy*. Ms. Brennan has written and lectured on various government and construction contract issues.

Michael J. Carrato is a Unit Counsel for ALSTOM, a global leader in the world of power generation and rail infrastructure. Prior to joining Alstom, Mr. Carrato was an Associate in the Government Contracts and Construction Groups at Patton Boggs LLP in Washington, D.C. In his current capacity with ALSTOM, Mr. Carrato oversees complex construction and product manufacturing transactions and manages the resolution of disputes. He holds a B.S. in Civil Engineering from Virginia Tech and a J.D. from Seton Hall University School of Law.

Robert K. Cox is a senior partner of Watt, Tieder, Hoffar & Fitzgerald, LLP, working in the McLean, Virginia, office. Since 1976, he has focused his legal practice on construction, government contracts, and surety law issues. Mr. Cox has authored numerous articles on construction and surety law issues and frequently teaches construction law and public contract courses. He was a member of the Board of Governors of the Virginia State Bar Section of Construction Law and Public Contracts, and was Chair of the Section 1999–2000. Chambers USA has included Mr. Cox in its list of Top Construction Lawyers in Virginia 2003–2009. He is also listed in *Best Lawyers in America 2007, 2008, 2010*; *International Who's Who of Business Lawyers*, Construction Section 2005–2009; *Virginia Super Lawyers 2006–2009*, and *Washington, D.C. Super Lawyers 2007–2009*. Mr. Cox is a graduate of Cornell University and an honors graduate of George Washington University Law School.

David H. Corkum is a Partner at Donovan Hatem LLP. He is intimately familiar with every aspect of the planning, design, and construction of heavy civil and underground construction projects. His practice includes the representation of architects, engineers, and construction managers involved in the construction industry, with a special emphasis on underground projects. He also assists project owners in developing procurement strategies and drafting contract documents for large construction programs. Mr. Corkum received his B.S. from Northeastern University and his J.D. from Suffolk University.

W. Stephen Dale is a partner with the law firm Smith Pachter McWhorter, PLC where his practice focuses on construction law, government contracts, and commercial litigation. Smith Pachter McWhorter PLC's construction practice encompasses all phases of the contracting process, from contract bidding and negotiation through contract administration and performance to claims and litigation, including terminations. Its clients have included a wide variety of public and private owners, general and specialty contractors, EPC contractors, architects and engineers, equipment manufacturers and sureties. They are a mix of Engineering News Record 500 firms, as well as small and medium size organizations, including minority- and women-owned businesses. Mr. Dale's experience includes counseling national, international, and local construction and engineering firms in all forms of construction issues from contract administration to claims and litigation, on federal, state, and private construction projects. His practice has also included counseling construction clients on Public-Private-Partnerships for major infrastructure improvements. Mr. Dale has advised on a broad range of projects, including rail transit, highway, mining, environmental remediation, wastewater treatment, military installations, and navigational locks. Mr. Dale completed both his undergraduate and legal education at the University of North Carolina at Chapel Hill, where he served as the Executive Editor of the *North Carolina Journal of International Law and Commercial Regulation*. He currently serves on the Board of Governors of the American Society of Civil Engineers Construction Institute and the ASCE Committee on Professional Practice.

Daniel J. Donohue is a shareholder in the Tysons Corner, Virginia, office of Akerman Senterfitt LLP. His law practice concentrates on litigation of disputes arising from federal, state and municipal construction contracts and contracts for supplies and services. He advises client on government contracts compliance issues and tries government contracts disputes in the Federal Agency Boards of Contract Appeals and Court of Federal Claims. Mr. Donohue is a member of the adjunct faculty of George Mason University Law School, where he teaches Appellate Writing and Legal Drafting. He is a 1975 graduate of the College of the Holy Cross in Worcester, Massachuetts, and received his J.D. from the Catholic University of America, Washington, D.C.

Denise E. Farris is the managing member of The Farris Law Firm LLC in Kansas City, Missouri. Ms. Farris is past Chair of Division 10 of the American Bar Association Forum on the Construction Industry. She served as past Chair of the Missouri Bar and Kansas City Metropolitan Bar Association Construction Law Committees and was recently named "Best of the Bar Construction Law" by the *Kansas City Business Journal*. Ms. Farris currently represents the national organization of Women Impacting Public Policy on an advisory council to the U.S. Senate Small Business Committee, where she was offered congressional testimony on various aspects of affirmative action legislation and rules. She has enjoyed more than 20 years of practice in commercial construction law with a special emphasis in government contracting, constitutional parameters of affirmative action programs, and federal schedule or IDIQ contracts, and served as the first female representative to Kansas City, Missouri, and the Unified Government Wyandotte County/Kansas City, Kansas Fairness in Contract Boards.

Cheryl A. Feeley is an associate in Holland & Knight's Litigation Section in Washington, D.C. Ms. Feeley's construction litigation practice focuses on representing general contractors on major construction disputes. Her experience includes bid protests, claims for equitable adjustment, qui tam actions under the False Claims Act, Buy American Act compliance issues, and general government contracts advice. Ms. Feeley is a 2008 graduate of George Washington University Law School.

Donald G. Gavin is a shareholder in the Tysons Corner, Virginia, office of Akerman Senterfitt, LLP. He was a founding Shareholder of Wickwire Gavin, P.C., a past National Chair of the Section of Public Contract Law of the American Bar Association, and a past Member of the Board of Governors of the Bar Association for the U.S. Court of Federal Claims. Early in his career, Mr. Gavin served as a Captain in the United States Army Judge Advocate General

Corps, principally as a procurement legal advisor to the Army Corps of Engineers. Mr. Gavin has provided representation to owners, contractors, public entities, design professionals, and sureties with issues involving public contracts, construction, suretyship, and grants. His experience includes disputes resolution responsibility related to public buildings and infrastructure; major petroleum, gas, chemical, wastewater, power production facilities; and supplies in South America, Europe, Africa, and Asia, as well as within the United States, both offshore and land-based. Mr. Gavin also acts as an arbitrator and mediator for disputes resolution around the world. Currently, he is a Fellow of the American College of Construction Lawyers, a Member of the Chartered Institute of Arbitrators of London, and a Certified International Mediator. Mr. Gavin is a graduate of the Wharton School of the University of Pennsylvania, and received his J.D. from the University of Pennsylvania Law School in 1967. In 1972, he received his LL.M. from the National Law Center of George Washington University in Government Procurement Law.

Elizabeth M. Gill is a senior associate with the law firm of Patton Boggs LLP specializing in federal government contracting. She counsels clients on a variety of procurement matters concerning contract preparation, negotiation, and performance. Ms. Gill litigates contract actions in federal courts, the Civilian Board of Contract Appeals, and the Government Accountability Office. She routinely works with companies to prepare and implement compliance and training programs and counsels small businesses on the intricacies of contracting with the federal government. Ms. Gill is a graduate of the University of Richmond and the American University Washington College of Law and is a member of the Public Contract Law Section of the ABA, the Board of Contract Appeals Bar Association, and the Federal Bar Association. Ms. Gill writes and speaks frequently on public contract law issues.

Michael Guiffré is a partner at Patton Boggs LLP and concentrates his practice on government and commercial contract litigation and arbitration. A graduate of Georgetown University Law Center, Mr. Guiffré has over 15 years of experience representing companies, foundations, individuals, and government agencies before federal and state courts and administrative boards in cases involving claims, default terminations, subcontractor disputes, and bid protests. Mr. Guiffré also counsels and represents clients regarding the full spectrum of government contract matters, including contract administration, claims, procurement fraud, and suspension and debarment matters. Prior to joining Patton Boggs, Mr. Guiffré served as an officer and an attorney in the Army, where he obtained considerable trial and appellate experience. Working in the Office of the Chief Trial Attorney, he managed complex contract litigation and fraud investigations. He was lead counsel in dozens of appeals, and he prosecuted numerous criminal offenses.

Dirk D. Haire is a partner at the Washington, D.C. office of Smith, Currie & Hancock LLP. Mr. Haire currently sits on the Associated General Contractors of America (AGC) and Maryland Chapter AGC Boards of Directors and serves on multiple national committees and councils, including the AGC of America Federal Acquisition Regulation Committee. He practices in the areas of construction law and government contracts, with a focus on federal construction. Mr. Haire represents contractors and owners in virtually all phases of both the public and private sectors of the construction industry. He is a frequent lecturer on matters relating to federal construction and recently presented for AGC's comprehensive webinar series examining the procurement, contract administration, reporting, and disputes phases of federal construction contracting. A graduate of George Washington University Law School, Mr. Haire was a senior director at AGC of America before entering private practice.

Larry D. Harris is a shareholder in the Washington, D.C. office of Greenberg Traurig LLP. Mr. Harris' experience in government contracts and construction matters includes counseling, transactions, and litigation before boards of contract appeals, in state and federal courts, and arbitration tribunals in cases involving changes, acceleration, delay, disruption, defective work, and termination. He has counseled domestic and international contractors, subcontractors, design professionals, and private and public owners in matters regarding contract formation, joint venturing, bidding, contract performance, minority business participation, and dispute resolution under supply, service, aerospace, and construction contracts. Mr. Harris is a past Chair of the ABA Forum on the Construction Industry (1998–1999) and is a Fellow of the American College of Construction Lawyers, Board of Governors (1999–2002).

David J. Hatem is a Founding Partner of the multi-practice law firm, Donovan Hatem LLP. He leads the firm's Professional Practices Group, which represents engineers, architects, and construction management professionals. Mr. Hatem has been practicing for over 30 years and is nationally recognized for his expertise in representing engineers and architects. He frequently lectures on issues of professional liability for design and construction management professionals, risk management, and design-build procurement issues, and has authored numerous related articles. He edited *Subsurface Conditions: Risk Management for Design and Construction Management Professionals, Design-Build Subsurface Projects*, now undergoing a second edition, and is co-editor and chapter contributor for *Megaprojects: Challenges and Recommended Practices* (American Council of Engineering Companies 2009). Mr. Hatem has served as ACEC/Massachusetts Counsel since 1988 and was the recipient of the 2008 American Council of Engineering Companies of Massachusetts Distinguished Service Award. He was also selected for the sixth consecutive year for inclusion in *The Best Lawyers in America 2009* in the fields of Construction

Law and Professional Malpractice. He currently teaches Legal Aspects of Engineering at Tufts University. Mr. Hatem received his B.A. and J.D. from Boston University.

J. Todd Henry is a partner in the Seattle, Washington, office of Oles Morrison Rinker & Baker, LLP. His litigation practice focuses on the representation of parties in construction-related disputes. A fourth generation contractor, Mr. Henry spent nearly 20 years in construction management before receiving his J.D. cum laude from Seattle University School of Law, where he received the additional honors of election to the Order of the Barristers and Phi Delta Phi. Mr. Henry is a frequent lecturer on construction law issues to industry groups as well as an instructor on litigation skills to lawyers and law students. In addition to having published many articles on construction law, Mr. Henry has been a contributing author and editor of the annual *Aspen Construction Law Update* since 2003.

David Innis is an attorney in the Office of Counsel of the United States Army Corps of Engineers, Kansas City District. Mr. Innis served as a member of the Council of the American Bar Association Public Contracts Law Section. He currently serves as co-chair of its State and Local Procurement Division and of its Teleconference Educational Programs Committee. All opinions are exclusively those of the author and not any instrumentality of the Department of Defense.

Geoffrey T. Keating is counsel to contractors, engineers, public and private owners, and sureties. His work has encompassed all phases of construction contracting with particular emphasis on public works and international infrastructure. Representative projects include military installations, dams, subways, bridges, and power plants. Overseas work includes U.S. Embassy, USAID, and other civil projects in Egypt, Saudi Arabia, Iraq, Afghanistan, Brazil, Russia, Italy, and Mexico. Mr. Keating has litigated many complex construction disputes in courts, boards of contract appeals, and arbitration. He also serves as an arbitrator. In 2006–2007 he consulted on public procurement regulations for the new Iraqi government and the government of Vietnam. He is a fellow of the American College of Construction Lawyers and on the West Group Construction Advisory Board. Mr. Keating is Of Counsel to McManus & Darden, LLP, Washington, D.C.

Thomas J. Kelleher, Jr. is a senior partner of Smith, Currie & Hancock LLP. His practice areas are construction law and government contracts. He received an A.B. from Harvard University, 1965, cum laude and a J.D. from the University of Virginia, 1968. Mr. Kelleher is co-author of *Inspection Under Fixed Price Construction Contracts*, Briefing Papers, No. 766, Federal Publications, Inc., December 1976; *Preparing and Settling Construction Claims*, Construction

Briefings, No. 8312, Federal Publications, Inc., December 1983; *Construction Litigation: Practice Guide with Forms*, Aspen Law & Business; and *Development in Federal Construction Contracts*, Wiley Construction Law Update, Wiley Law Publications (1992–1999). He is also creator and editor of his firm's construction law newsletter, *Common Sense Contracting*; co-editor of *Common Sense Construction Law*, Fourth Edition; and co-editor of *Federal Government Construction Contracts*. Mr. Kelleher is a member of the American College of Construction Lawyers; the American Bar Association, Public Contract Law Section; Associated General Contractors of America, Inc.; and Corps of Engineers, Naval Facilities Engineering Command, Governmental Affairs, and Contract Documents Committees. He was also Chair of the Federal Acquisition Regulation Committee (1999–2003).

Lori Ann Lange is an attorney in the Washington, D.C. office of Peckar & Abramson, P.C., where she specializes in counseling government contractors in all aspects of contracting with the federal government from proposal preparation to claims resolution. Ms. Lange has written extensively on government contracts and construction law issues and is a former revision author of McBride & Touhey's *Government Contracts*. She also is a contributing author to *Construction Checklists, State-by-State Guide to Construction Contracts and Claims, State-by-State Guide to Architect, Engineer, and Contractor Licensing, Construction Checklists*, and *Construction Claims Deskbook*. Ms. Lange co-authored *Construction Briefing Papers on Recovering Delay Damages for Home Office Overhead and Innovative Financing for Infrastructure Project*. She received her B.A. and J.D. degrees from The George Washington University.

Adam K. Lasky is an associate in the Seattle office of Oles Morrison Rinker & Baker LLP. Mr. Lasky concentrates his practice on construction litigation and government contracts. He has represented contractors involved in government contract disputes before both state and federal courts, boards of appeal, and administrative agencies. Mr. Lasky received his J.D., cum laude, from the University of Minnesota Law School, where he was a Trogner Scholar. During law school, he served as a judicial extern for Hon. Janet N. Poston, Fourth Judicial District of Minnesota. He has also studied law in Ireland at the University College Dublin, and holds a B.A. in economics and political science from the University of Washington.

Michael C. Loulakis is the President/CEO of Capital Project Strategies, LLC, a specialized consulting firm that provides strategic procurement, contracting, and risk management advice to clients who develop and construct capital projects. Prior to forming the firm in 2007, Mr. Loulakis practiced law for almost 30 years with Wickwire Gavin, P.C. He has extensive and diverse experience in the design-build process, having represented clients on design-build projects around the world in a variety of sectors. He writes and speaks extensively on design-build and authored several highly regarded design-build publications, including *Design-Build for the Public Sector* (Aspen Law Publications, 2003) and *Design-Build: Planning through Development* (McGraw-Hill, 2001). Mr. Loulakis is an active member of the Design-Build Institute of America and played a lead role in the development of DBIA's model contracts. He graduated magna cum laude from Tufts University with a B.S.C.E. and received his J.D. from Boston University School of Law.

Lauren P. McLaughlin is a founding partner of the Tysons Corner, Virginia, office of BrigliaMcLaughlin, PLLC. Ms. McLaughlin received her B.A., Phi Beta Kappa, magna cum laude, from Trinity University in Washington, D.C., and her J.D. degree from Catholic University, Columbus School of Law. She has been in the private practice of law specializing in construction and government law for more than 10 years. Ms. McLaughlin is a regular contributor to the American Society of Civil Engineering's (ASCE) monthly publication where she co-authors articles addressing legal issues facing design professionals.

Tamara M. McNulty, LEED AP, is a partner in the Washington, D.C. office of Duane Morris LLP. Ms. McNulty is a trial lawyer and practices primarily in the areas of construction law, government contracts, and surety. She received both her B.A. and her J.D. from the University of Wisconsin–Madison. Ms. McNulty is a member of the ABA's Forum on the Construction Industry, the Tort and Trial Insurance Practice Section, the Public Contracts Division, the Virginia Bar Association's Construction Law section, and the National Bond Claims Association. She is on the Board of Directors of the Washington Building Congress. She is an instructor of advanced trial skills for the National Institute of Trial Advocacy. Ms. McNulty is the author of numerous articles and chapters of books on construction law and government contracting and has written two books on construction law, *Virginia Construction Law* and *Maryland Construction Law*.

Angeline R. Nelson is an associate in the Washington, D.C. office of Seeger, Faughnan, Mendicino P.C. Ms. Nelson received her J.D. from The George Washington University Law School and her B.A. in International Relations from Michigan State University. Her practice at Quagliano & Seeger involves government contract law as well as commercial and construction litigation. While in law school, Ms. Nelson focused primarily on government contract law and wrote her thesis on the use of cascading Set-Asides.

James F. Nagle is a nationally known expert in government contracts and construction law who received his Bachelor's degree from Georgetown University School of Foreign Service; his J.D. from Rutgers; and his LL.M. and S.J.D. in government contracts from the National Law Center, George Washington University. He practices with Oles Morrison Rinker & Baker in Seattle and chairs its Federal Practice Group. Besides litigation, he is a frequent expert witness, arbitrator, and teacher on federal contracts. Mr. Nagle has written six books on federal contracting: Nash, Cibinic and Nagle, *Administration of Government Contracts*, 4th ed. (George Washington University Press 2006); Whelan and Nagle, *Cases and Materials on Federal Government Contracts*, 3rd ed. (Foundation Press 2007), the leading law school textbook on the subject; *Federal Procurement Regulations: Policy, Practice and Procedures*, (American Bar Association Press [ABA] 1987); *How to Review a Federal Contract and Research Federal Contract Law*, 2nd ed. (ABA Press 2000); *Federal Construction Contracting* (Wiley Law Publications 1992); and *The History of Government Contracting*, 2nd ed. (George Washington University Press 1999). He has also co-authored and co-edited *Washington Building Contracts and Construction Law* (Butterworth (now Aspen) Publishers 1994).

Krista L. Pages is a Senior Contracts Specialist for Infrastructure with International Relief and Development, a government contractor in Arlington, Virginia. She is a past member of the Forum Governing Committee and past Division 10 Chair. Prior to joining IRD, Ms. Pages was Of Counsel with Winston & Strawn, LLP from 2001 to 2009 and a partner at Smith, Pachter, McWhorter and D'Ambrosio from 1989 to 2001. She specializes in government contracts and construction law and is the co-editor of *False Claims in Construction Contracts* (ABA 2007).

Bryan R. Phillips, AIA, is an attorney and licensed architect practicing in the Litigation Department, focusing his practice in the area of construction law. Prior to entering private practice, Mr. Phillips was a project architect at the international firms of Skidmore, Owings & Merrill and Ellerbe Becket, Inc. His architectural practice has included work on a variety of building types, including Class A office buildings, embassies, and courthouses. He represents project owners, developers, and general contractors in all phases of construction and on a wide variety of building types. With over 20 years of experience in the design and construction industry, Mr. Phillips provides a unique insight to his clients from contract negotiation to project close-out. The primary focus of his practice is negotiating and drafting construction agreements, professional design services agreements, design-build agreements, and construction management agreements. He has mediated, arbitrated, and litigated all manner of construction disputes including building defects and delay claims.

Lawrence M. Prosen, a partner with K&L Gates in Washington, D.C., concentrates his practice in the areas of construction litigation, government contracts and architects/engineers errors and omissions, and real estate and corporate transactional work. He has represented some of the largest construction contractors in the U.S. in both domestic and foreign projects ranging in value to over $200 million. He has represented clients in all aspects of bench and jury trials, as well as administrative appeals at various levels of county, state, and federal tribunals, arbitrations, and mediations. He has extensive experience in protests, having represented parties in protests at the federal, state, and local levels of government. Mr. Prosen has worked on real estate projects and transactions involving the sale, purchase, leasing, and construction of commercial, office, industrial, residential, and institutional buildings valued at up to and exceeding $250 million.

Steven L. Reed, a partner, in the Washington, D.C., firm of Smith, Currie & Hancock LLP, was an Administrative Judge, Armed Services Board of Contract Appeals, until retirement from federal service and was formerly with the Corps of Engineers Board of Contract Appeals. Mr. Reed received his B.S. magna cum laude from the University of Georgia and his J.D. degree from the University of Georgia School of Law. He worked as a contracts trial attorney and litigation counsel for the Corps of Engineers prior to joining the Boards of Contract Appeals. Since 1981, his federal government service and private practice have focused on federal government contracts, particularly construction contracts. Mr. Reed counsels and represents federal government contractors in dispute avoidance and resolution. He also has acted as a neutral in more than 50 ADR proceedings, both as a mediator and arbitrator, assisting with more than 220 disputed matters, most involving federal government contract disputes.

Joel S. Rubinstein, a partner with K&L Gates in Washington, D.C., has 35 years of experience in government contracts and construction law matters. He concentrates his practice in construction law, including bid issues, small business contracting, changes, claims, and other contract disputes before various administrative boards of contract appeals, state and federal courts, and arbitration tribunals, and has participated in actions involving prime contractors, subcontractors sureties regarding payment, and performance bond issues. Mr. Rubinstein has also been involved in Davis-Bacon Act cases and Walsh-Healey Act cases, as well as defense of OSHA citations. As a frequent author and lecturer, he addresses construction and government contracts issues of bid protests, contract payment and performance problems, small and minority business matters, drug-free workplace, federal supply schedule matters and minority certification.

Laurence Schor concentrates his practice on all phases of construction and government contracts law and has represented large and small business clients in various courts and forums across the United States. He served as the Assistant General Counsel for NASA Support in the office of the General Counsel, United States Army Corps of Engineers, and as an attorney at the Marshall Space Flight Center for NASA. He is now with Asmar, Schor & McKenna, PLLC. He has authored articles, book chapters, and course manuals on government contracts and construction topics and lectures regularly for professional groups on issues arising in these areas. He is an active arbitrator and mediator. Mr. Schor is a member of the Bars of the District of Columbia, Texas, and Maryland, as well as a member of the American Bar Association, Public Contract Law Section, where he chaired the Construction Committee and also served as Budget and Finance Officer on the governing Council. He is a member and lecturer for the ABA Forum Committee on the Construction Industry, and is also a member of the Federal Bar Association. He holds a Bachelor's degree in Business Administration from Southern Methodist University, a law degree from the University of Texas, Austin, School of Law, and a Master of Laws degree from George Washington University with an emphasis in Government Procurement. He is a founding member, served on the Board of Governors, and was the president (2001–2002) of the American College of Construction Lawyers. Mr. Schor has been recognized in *Best Lawyers in America* (2005–2009), *Who's Who Legal, The International Who's Who of Business Lawyers,* and *Super Lawyers for the Washington, D.C.* metropolitan area, and is listed in *Who's Who in America, Who's Who in American Law,* and *Who's Who in the World.*

Stephen M. Seeger is a partner at the law firm of Seeger, Faughnan, Mendicino P.C., where he spends most of his time trying to keep the firm's government contracts and construction industry clients out of trouble. The balance of his work life is devoted to talking and writing about government contracts and construction law. Before his world was so limited, Mr. Seeger attended and graduated from Lehigh University and George Washington University Law School.

Stephen B. Shapiro is a partner in the Washington, D.C. office of Holland & Knight LLP. He is a board member of the Associated General Contractors of Metropolitan Washington, D.C. and serves on the NAVFAC, Corps of Engineers, Federal Acquisition Regulation, and Project Delivery Committees of the Associated General Contractors of America. Mr. Shapiro is also a member of the American Bar Association's Public Law Section and Forum on the Construction Industry. He received his B.A. from University of Maryland and his J.D. degree from George Washington University Law School. He has been in the private practice of law specializing in construction and government contracts law for more than 22 years. Mr. Shapiro represents public contractors and government entities in public contract matters throughout the United States.

Richard F. Smith is Senior Counsel to the Vienna, Virginia, firm of Smith Pachter McWhorter, PLC. Mr. Smith is a fellow in the American College of Construction Lawyers and a fellow in the College of Commercial Arbitrators. Twice he has been Chair of the Construction and Public Contracts Section of the Virginia State Bar, and has been on the Boards of the American College of Construction Lawyers, the Construction and Public Contracts Section of the Virginia Bar Association, and the ABA Forum on the Construction Industry. He received his B.S. from Wake Forest University, his L.L.B. from the University of Virginia, and his LL.M. in Government Procurement from George Washington University. Mr. Smith has been in private practice for more than 35 years and has an active ADR practice, having served as a mediator, arbitrator, project neutral, and DRB member on over 150 disputes. He is the editor of the *Virginia CLE Construction Law Deskbook*.

Donald A. Tobin is a partner in the Washington, D.C. office of Peckar and Abramson, PC. He has over 35 years of experience in litigation and general counseling in both construction and government contracting. He has represented government contractors and construction contractors in the negotiation and preparation of contract and subcontract documents. He has assisted in the preparation and negotiation of requests for equitable adjustment, with an emphasis on the pricing of extra work, delay and inefficiency claims. Mr. Tobin has litigated government contract cases at the Federal Boards of Contract Appeals and the U.S. Court of Federal Claims, and he has appeared often at the Government Accountability Office on bid protest matters. Mr. Tobin began his career as an attorney with the Naval Air Systems Command, specializing in government contracts. He is a graduate of the College of the Holy Cross (B.A. Mathematics 1969) and Boston College Law School (J.D. 1973 cum laude).

Daniel E. Toomey is a senior counsel in the Construction Group of the Washington, D.C. office of Duane Morris, LLP, and a mediator for the McCammon Group. Mr. Toomey clerked on the D.C. Court of Appeals and was an Assistant U.S. Attorney for the District of Columbia. He previously served as president of the AUSA Association. Mr. Toomey is a graduate of the Georgetown University Law Center where he was an Adjunct Professor of Construction and Criminal Trial Practice Law for 20 years. He was a 2007 recipient of the Paul R. Dean Award from his alma mater for his distinguished service to Georgetown. Mr. Toomey is currently listed by Chambers and by the Virginia, District of Columbia, and Corporate Counsel editions of *Super Lawyer* magazine in the area of Construction Litigation. In 2000, he was elected as a Fellow in the American College of Trial Lawyers. Mr. Toomey has written often for the *Construction Lawyer* and other construction publications and co-authored numerous chapters of books on construction and related topics. He has lectured on these topics throughout the United States.

Brian P. Waagner is a shareholder in the Tysons Corner, Virginia, office of Akerman Senterfitt, LLP. His law practice focuses on the representation of construction contractors and subcontractors seeking to resolve disputes on federal contracts. In addition to advising clients on contract administration issues and litigating affirmative claims in court proceedings, arbitration, and mediation, Mr. Waagner defends clients in cases alleging violations of the False Claims Act. Mr. Waagner began his law practice with Wickwire Gavin, P.C. He received his law degree from Cornell Law School.

Kenneth B. Walton is a Partner at Donovan Hatem LLP and focuses his practice on the defense of professionals, including engineers, architects and surveyors. He represents clients in federal and state courts, in both jury and nonjury cases, in Massachusetts, Connecticut, and New Hampshire, and has been admitted pro hac vice in a number of other states. Mr. Walton has handled litigation matters for design professionals arising out of a number of major construction projects. He has handled hundreds of mediations, arbitrations and trials for design professionals in cases across the country. Mr. Walton received his B.A. from St. Lawrence University and his J.D. from New England School of Law. After graduating from law school, he served as a law clerk to the Honorable Frank H. Freedman in the U.S. District Court for the District of Massachusetts.

PREFACE

The U.S. government has been and, for the foreseeable future, will continue to be the largest purchaser of construction services in the world. This is especially the case with the passage of the American Recovery and Reinvestment Act of 2009. The government has developed detailed regulations governing almost every aspect of construction contracting from procurement of the construction services to final payment under the contract. There also is a set of standard contract clauses that provide for a reasonable allocation of risk between the government and the contractor. Regulations governing compliance and the prevention of fraud, waste, and abuse have also moved into the forefront.

Disputes arising under federal construction contracts are brought to the Court of Federal Claims and the boards of contract appeals. These forums hear contract disputes without a jury and generally issue reasoned decisions discussing the basis for the decision and applicable law. As a result, there is a large body of construction case law involving federal government contract claims.

State courts around the country often rely on federal government contract law, particularly where no law on a subject has evolved in the state. State common law generally mimics the federal common law in the area of construction law. Most disputes on private construction contracts are resolved using the principles established by federal government contract law. Therefore, lawyers handling private construction disputes need a basic understanding of federal government contract law or the ability to access it quickly.

Most construction lawyers will handle federal government contract cases at some point in their careers. While construction lawyers do not need to have detailed knowledge of government contracts to handle these cases, they do need a starting point. The myriad of federal government contract regulations, clauses, and case law can present problems and pitfalls for nongovernment contract practitioners. Understanding the principles and operation of these rules and regulations can be difficult task without a road map.

There are several excellent treatises on federal government contract law, the classic being the Nash and Cibinic series. These treatises, however, do not focus on construction cases. There also are many excellent construction law books, including many published by the Forum on the Construction Industry of the American Bar Association. Although federal government construction law represents the largest segment of the construction law in the United States, there were no books dedicated solely to federal government construction law until the First Edition of this book. The Forum decided to fill this gap by publishing a book that addresses all aspects of federal government construction contract law. This Second Edition is both updated and infused with new topics to include architect-engineering contracts and federal grants.

This Second Edition is written by many of the top experts in government contracts and construction law. Its goal, as with the First Edition, is to provide the occasional construction lawyer, consultant, and contractor with the basic knowledge of federal government construction contracting regulations and case law. It also is designed to provide the experienced government contract practitioner with a sophisticated analysis of the issues and a ready source of case law.

—Michael A. Branca

ACKNOWLEDGMENTS

The editors of the Second Edition extend their sincere thanks and appreciation to the editors of the First Edition: Adrian Bastianelli, Andrew Ness, and Joe West. Their efforts on the First Edition, and the guidance, support, and advice provided over the past 18 months, made possible the successful development and publication of the Second Edition.

Michael A. Branca
Aaron P. Silberman
John S. Vento

CHAPTER I

Introduction

JAMES NAGLE

I. Introduction

Contracting with the federal government is like dancing with a gorilla. With very few exceptions, you dance the way the gorilla wants you to dance and if you don't do what the gorilla wants, things can get very ugly very quickly.

The U.S. government spends over $400 billion a year on its contracting. This is spread over supplies, services, construction, research, and development. Even if the prime contract is not a construction contract, very often it will have construction aspects that will result in the award of a construction subcontract. A construction subcontract under a supply or services contract still must adhere to many of the same federal contracting requirements plus those that apply solely to federal construction contracting.

Contractors must never forget that, in conducting business with the government, they are dealing not only with an enormously large and wealthy contracting party, but also with a sovereign entity. For that reason, things that would be considered perfectly acceptable in private contracting might be criminal acts in federal contracting. One prime example is gratuities. In private contracting it is routine, even expected and encouraged, to take customers out to lunch, or to send them gifts. In federal contracting, at the very least, such actions would be considered impermissible gratuities if not outright bribes. Puffing up one's qualifications in order to win the contract might be considered commonplace in private contracting, but in federal contracting, it could constitute a false statement under 18 U.S.C. § 1001.

Contractors should be aware that, when they get involved in government contracting, the government wears many hats. Quite often the government will clearly and undeniably delay or make more expensive the costs of performing the contract, but the contractor's remedies are severely limited. In one case, for example, the contractor was performing work in the Washington Navy Yard in the District of Columbia. The contractor had a schedule to meet under a fixed-price contract. Unfortunately, the Secret Service had ordered the

1

contractor off the site and sealed the area because President George H.W. Bush would be visiting the Navy Yard, where the construction was occurring. Neither the contractor nor the agency had authority to refuse the Secret Service's order. The contractor thus left the site and returned after the presidential visit. During the interim, the contractor still had to pay its workers and still had to incur the costs of owning or renting equipment.[1] The contractor submitted a claim to the contracting officer and it was denied.

The contracting officer did not dispute that the delay was excusable, because it was beyond the control and without the fault or negligence of the contractor. Therefore, the contractor was entitled to a time extension under the Excusable Delay clause of the contract, FAR 52.249-10(c). The problem was that this was an act of the government in its sovereign capacity, not its contractual capacity. Accordingly, it did not fall under the Changes clause of the contract, nor any of the other remedy-granting clauses. The contractor received a time extension but no money. This same result occurs if the contractor's performance under its fixed-price contract is made more expensive or longer because of new regulations issued by the Environmental Protection Agency or any other department or agency.

It is important, therefore, that people in the construction industry have an appreciation of the governing rules, the government officials they are most likely to deal with, and the forums to which grievances may be taken.[2]

A. Governing Law

Federal contracts are normally subject only to federal procurement law comprising federal contracting statutes and regulations, as interpreted by the U.S. Court of Appeals for the Federal Circuit, the U.S. Court of Federal Claims, the Boards of Contract Appeals, and the U.S. Comptroller General. Sometimes it will be necessary to review state laws such as those regarding warranties under a state's Uniform Commercial Code.

While federal law does not typically apply to subcontracts, it can be adopted by the prime and the subcontractor so that both the prime contract with the government and the subcontract are governed by the same body of law.[3] This does not, however, give a federal court subject-matter jurisdiction over a dispute between the prime and subcontractor.

B. Statutes

Procurement statutes are spread throughout the U.S. Code. The main statutes are in Title 41, Public Contracts, but if you deal with military agencies such as the Army Corps of Engineers, many of those statutes are included in Title 10 dealing with the Armed Forces. The two main statutes involved in federal contracting are the Armed Services Procurement Act of 1947[4] and the Federal Property and Administrative Services Act of 1949.[5] These old statutes are still very current. Whenever Congress issues a new government contracts statute, it amends those two 1940s-era statutes. For example, the Truth in Negotiations Act,[6] which requires the submission of cost or pricing data on large-dollar-value contracts, and which is often called the Defective Pricing Statute, is codified in both Title 10 and Title 41. Similarly, the Competition in Contracting Act, which mandates full and open competition in government contracts, is codified in 10 U.S.C. § 2304 and 41 U.S.C. § 253. The Federal Acquisition Streamlining Act,[7] which created a major overhaul of federal contracting in the mid-1990s, is codified at numerous places throughout the U.S. Code.

Besides these comprehensive procurement statutes, Congress will often enact specific requirements in the public laws that authorize programs or appropriate monies for the particular federal agency. For example, the budget for the Army Corps of Engineers or the General Services Administration may well contain paragraphs stating that no money of the funds appropriated under that public law can be used except in a particular method that Congress wants to impose.

C. Regulations

Obviously, federal contractors must comply with the applicable statutes, executive orders, and regulations. Fortunately, the federal government, unlike many other large owners, has published a publicly available compilation of the rules that must be followed. This compilation is known as the Federal Acquisition Regulation (FAR).

The FAR is at Title 48 of the Code of Federal Regulations, Chapter 1. Actually, Title 48 in its entirety is often referred to as the FARS—the Federal Acquisition Regulatory System—because while the FAR itself is Chapter 1, many agencies have issued their supplements to the FAR and these are scattered throughout the rest of Title 48. Chapter 2 of Title 48, for example, contains the Defense Federal Acquisition Regulatory Supplement—the DFARS. Chapter 9 contains the Department of Energy Acquisition Regulatory Supplement—the DEARS. The supplements are discussed in more detail below.

The FAR came into existence on April 1, 1984. It was an attempt by the government to make uniform the contracting practices of the federal agencies so that, for example, the contract procedures and the contract itself for the Agriculture Department would not be radically different from a contract for the Department of the Army. In fact, the contracts would look amazingly similar.

The FAR comprises 53 different parts that are chronologically designed to walk the reader through the contracting process from initial planning to termination. Its 53 parts are grouped into eight subchapters. Set out below are those parts that seem particularly important for construction contractors.

Subchapter A is General. The first part, the Introduction, deals with how the FAR is put together. Part 2, Definitions, defines such important individuals as the contracting officer and agency head, plus terms that permeate the process. Part 3 is Improper Business Practices and Personal Conflicts of Interest. In this part, guidance is given on such issues as anti-kickback rules, buying in, etc. Part 4, Administrative Matters, explains how the contracting office is supposed to keep its contract files and how long the parties must keep records after the contract is complete.

Subchapter B is Competition and Acquisition Planning. Part 5, Publicizing Contract Actions, explains how the government goes about notifying prospective bidders and offerors of its planned contracts. Part 6 is Competition Requirements. The government will normally try to achieve full and open competition on all of its buys. This part explains how that is achieved, what the exceptions are, and how they can be accomplished.

Part 9 is Contractor Qualifications. It is here where the government explains the concept of contractor responsibility and also discusses one of the ultimate sanctions that it has against the contractor, i.e., debarment and suspension. Part 11 is Describing Agency Needs. In this part, the government describes its method of specification writing, which includes functional, performance, design, and brand name or equal specifications.

Part 12, Commercial Items Acquisition, deserves special mention. Government officials now have a way to cut through much of the traditional red tape that permeated government contracting. In the mid-1990s, Congress discovered that many contractors avoided doing business with the federal government because of all the rules, regulations, and standards imposed. Congress thus enacted the Federal Acquisition Streamlining Act, which put a greater emphasis on buying commercially. It expanded the definition of what was a commercial item to include services.

While the definition does not use the word "construction," many agencies and many contracting officers view certain construction work, such as painting or roofing, as services traditionally sold in the commercial marketplace. If a contracting officer attempts to procure construction services as a commercial item, he does not need the volumes of rules and regulations that are associated with standard contracting. Instead, the agency may use FAR Part 12, which provides for a simplified process involving very few standard clauses and specific provisions. These are stripped down to the essentials: for example, the Changes clause becomes the Changes sentence.

Subchapter C is Contracting Methods and Contract Types. Historically, Part 14, Sealed Bidding, was the most important part. This is the traditional method by which the government would award construction contracts. However, the importance of Part 14 has diminished in recent years as the government moves more and more to awarding contracts by negotiation through the Request for Proposal (RFP) process. Competitive negotiation is addressed in Part 15. Parts 16 and 17 deal with types of contracts and special contracting methods. This is where the reader should go to learn the details for a firm-fixed-price contract, a cost-plus-incentive-fee contract, an options contract, a requirements contract, or an indefinite-delivery indefinite-quantity (IDIQ) contract.

Subchapter D is Socio-Economic Programs. Part 19 focuses on small business programs including small business subcontracting, small disadvantaged business contracting, and HUBZone (Historically Underutilized Business Zone) contracting. Part 22 is Application of Labor Laws to Government Acquisitions. The most important subparts here are those dealing with the Davis-Bacon Act, Subpart 22.4, and the subpart dealing with equal employment opportunities, Subpart 22.8.

Included within Subchapter D is Part 25, Foreign Acquisition, which deals with the Buy American Act. Subpart 25.2, Buy American Act—Construction Materials, applies to contracts for the construction, alteration, or repair of any public building or public work in the U.S.. It is essential that contractors become familiar with this subpart. Contracting for the federal government often requires certifications that the contractor will comply with the Buy American Act and use only domestic material in the performance of the contract. Contractors often overlook this requirement and bid jobs expecting to be able to use cheaper foreign material. Once they have been awarded a contract at a set price, they are then horrified to learn that their cheaper substitutes will not be accepted. Instead, they must use more expensive domestic materials. Worse yet, they have already installed the foreign materials and are required to rip them out or to substantially reduce their contract price and possibly face criminal prosecution.

Subchapter E is General Contracting Requirements. The most important part in this subchapter for construction contractors is Part 28, Bonds and Insurance. This part describes the regulations regarding the Miller Act requirement for performance and payment bonds and also bid bonds. Contractors must study this part regarding these requirements and such substitutes as individual sureties, letters of credit, etc.

In Part 31, Contract Cost Principles and Procedures, the government describes the costs that are allowable on cost-reimbursement contracts or on modifications for many types of fixed-price contracts. Often a contractor incurs costs that are required in business, and recognized by its bank and by the Internal Revenue Service, yet discover that the contracting agency will not pay for them. Prime examples are interest on borrowings and advertising.

In Part 32, Contract Financing, the government describes its process for progress payments and for the requirements of the Prompt Payment Act. This part also explains its invoicing and debt-collection process.

Part 33 is Protest, Disputes, and Appeals. This vital part explains how and where disappointed bidders may file protests regarding the award of a contract. It is also this part that describes the disputes process in which a contractor that feels it is entitled to more money or more time from the government may pursue its claim.

Subchapter F is Special Categories of Contracting. The part here that is most important is Part 36, Construction and Architect-Engineer Contracts. This part, only about 20 pages, should be carefully reviewed by anyone engaged in construction contracting with the federal government. It discusses particular requirements for forming the construction contract, and then explains in detail the two-phase design-build procedures in Subpart 36.3. Subpart 36.5 describes the unique clauses that are required or suggested in government construction contracts. Many of these will be familiar to contractors, such as the Differing Site Conditions clause or the Site Investigation and Conditions Affecting the Work clause. Others, however, may require special scrutiny, such as the Material and Workmanship clause, or the Permits and Responsibilities clause.

Subpart 36.6 specifically deals with architect-engineer services and contains a variety of clauses that require review. Subpart 36.7 deals with standard and optional forms for contracting for construction, architect-engineer services and dismantling, demolition, or removal of improvements.

Subchapter G, Contract Management, deals with the administration of an awarded contract and should be carefully reviewed by anyone embarking on the performance of a government contract. The part that should be thoroughly examined is Part 43 dealing with Contract Modifications. This part covers how the government modifies its contract on the basis of any of the modification clauses, such as the Changes clause, the Differing Site Conditions clause, or any of the Government Delay of Work clauses.

Part 44 deals with subcontracting and addresses how a government prime contractor should subcontract. Depending on the size and type of the prime contract, the prime contractor may have to get the contracting officer's approval before awarding subcontracts. Moreover, the contracting officer may want to review the subcontract form itself or the subcontracting method the prime plans to use. The contracting officer also may need to know if there is sufficient competition, if all the appropriate clauses flowed down, or if the prime is subcontracting too much of the work.

Part 45, Government Property, imposes requirements on the contractor to properly identify, preserve, protect, account for, and dispose of that property. These requirements may apply where the government has furnished property such as equipment for the contractor to use during construction, or if the contractor has bought property that either temporarily during the performance of the contract or permanently at the end of the contract is to be government property.

Part 46, Quality Assurance, explains how the government will inspect and accept the work of the contractor; it also specifies any warranties that will be required. Part 49 deals with termination of contracts and explains how the government will terminate contracts for convenience or for default.

Subchapter H is Clauses and Forms. Part 52 is Solicitation Provisions and Contract Clauses. Provisions essentially apply only before the contract is awarded; clauses may apply both before and after the contract is awarded. The FAR is logically set out. All the provisions and clauses are in FAR 52.2. The next two digits will refer back to the part dealing with the substance of the clause. For example, since FAR Part 49 deals with terminations, all of the government's termination clauses will be in Section 52.249. That number will then be followed by a dash and another number, which will identify the specific applicable clause. For example, 52.249-10 is the default clause for fixed-price construction contracts.

Section 52.301 is the matrix. This lists all the solicitation provisions and contract clauses that will be required (R), required when applicable (A), or optional (O), in all the various types of contracts the government awards. Readers of this book should specifically reference those clauses that are required under FP CON (Fixed-Price Construction) contracts or CR CON (Cost-Reimbursement Construction) contracts. They may also want to review the clauses under T&M (Time and Material) or LH (Labor Hour) contracts or IND DEL (Indefinite-Delivery) contracts.

This matrix has particular importance because of a doctrine in federal contracting known as the *Christian* doctrine. The *Christian* doctrine came out of the case *G.L. Christian & Associates v. United States*,[8] in which the government awarded a construction contract but neglected to include a Termination for Convenience clause even though it was required by the applicable regulation at the time. The court ruled that if a clause was required to be in the contract by regulation having the force and effect of law and represented a fundamental procurement policy, it would be incorporated into the contract by operation of law despite its physical omission from the contract. The *Christian* doctrine is discussed in detail in Chapter 7.

FAR 53 is a list of all the standard forms used in government contracting. They are all contained in FAR 53.301. FAR 53.301 will be followed by a dash and then the clause number. For example, a contractor who wanted to locate Standard Form 1442 (Standard Form 1442, Solicitation, Offer and Award (Construction, Alteration or Repair)) would go to FAR 53.301-1442. The form is attached as Appendix A.[9]

Many agencies have issued their own supplements to the FAR. These supplements are not intended to contradict or overrule the FAR, unless a statute unique to that agency authorizes such special requirements. Rather, the supplements are designed to implement agency-specific forms, clauses, or procedures. As an example of the multiple layers of regulations that may apply to a single contract, if you have a contract with the Army you should review not only the FAR and DFARS, but also the AFARS, the Army Federal Acquisition Regulation Supplement. All of the agencies will follow the same numbering system as the FAR. Thus, if you wanted to see if there are any special Navy requirements on construction contracts, you would look at Part 36 of the Navy supplement.

D. Individuals

The contractor will deal with a variety of individuals. The most important one is the contracting officer, who is the person who speaks for the government in its contracting capacity. Contracting officers are specifically delegated authority to bind the government contractually. This binding authority, often called the contracting officer's "warrant," is a Standard Form 1402, Certificate of Appointment. The Standard Form 1402 may list limits on the contracting officer's authority. For example, an individual may have a contracting officer's warrant but it is limited to $5 million, so he is not authorized to bind the government on any contracts over $5 million. Contractors are on notice of any limit on the contracting officer's warrant. It behooves a contractor to make sure that it is dealing with an individual, even a contracting officer, who is acting within the scope of his authority.

There are various types of contracting officers. The procuring contracting officer (PCO) is the main contracting officer. When the FAR mentions "contracting officer," it refers to the PCO. PCOs award the contract and retain their authority throughout the process. They may delegate some authority to an administrative contracting officer (ACO) (some agencies, especially the Navy, refer to this individual as an ROICC, resident officer in charge of construction). This individual might be the daily interface at the construction site with the contractor and he speaks on behalf of the PCO as long as he acts in accordance with the authority delegated by the PCO. A termination contracting officer (TCO) will be appointed if a termination for convenience or default becomes necessary. These matters are very complex and often a specially trained contracting officer will be appointed to oversee the process.

Besides the contracting officer, there are a variety of other officials with whom the contractor will come in contact. It is important to understand their roles and the powers they possess and do not possess. The contracting officer's representative (COR), sometimes called the contracting officer's technical representative (COTR), is designated by the contracting officer as his representative in certain technical areas. For example, the inspector or the government engineer may be specifically designated the authority to approve submittals or accept work. Normally, the contracting officer will send a letter designating a named individual as his representative, but the letter will emphasize that the representative does not have the right to change the contract in any way that entitles the contractor to more money or more time. However, as will be discussed elsewhere in this book, the COR is frequently the focal point when the issue of constructive changes occurs.

To foster the role that small businesses play in federal contracting, each contracting agency, at its procurement offices, has officials known as Small and Disadvantaged Business Utilization Specialists (SADBUs) who will work with the contracting officer and the Small Business Administration (SBA) to ensure that contracts are properly set aside for small businesses or that subcontracting opportunities are identified for small businesses.

E. Forums for Dispute Resolution

When contractors litigate with the federal government, they are suing the sovereign. One cannot sue the sovereign unless the sovereign has waived its sovereign immunity. On federal contracts, the government has waived its immunity under the Contract Disputes Act,[10] although the waiver is limited and actions may only be brought in the Agency Board of Contract Appeals or U.S. Court of Federal Claims.

The highest court that deals specifically with government contracts is the U.S. Court of Appeals for the Federal Circuit. The only appeal from that court is to the U.S. Supreme Court, which rarely will take a government contract case. Normally, there is no constitutional issue involved nor is there any split between the circuits because all contract matters will eventually be funneled through the Court of Appeals for the Federal Circuit. The U.S. Court of Appeals for the Federal Circuit came into existence in 1982. It succeeded the U.S. Court of Customs and Patent Appeals and the U.S. Court of Claims. The U.S. Court of Claims had been in existence for approximately 120 years. It was an Article III court. Its decisions have been adopted by the U.S. Court of Appeals for the Federal Circuit and are binding upon the U.S. Court of Federal Claims and the Boards of Contract Appeals.

The lineage of the U.S. Court of Federal Claims is more confusing. It is not an Article III court, but an Article I court. The old U.S. Court of Claims, the Article III court, had two levels, the judges' level, which was composed of the Article III judges, and a commissioners' level. The commissioners actually heard the cases and issued a recommended decision that the judges were free to adopt or reject. When the Court of Appeals for the Federal Circuit was created in 1982, the judges of the U.S. Court of Claims were transferred and became the judges of the U.S. Court of Appeals for the Federal Circuit. Basically, the commissioners of the old U.S. Court of Claims became the judges of what was originally called the U.S. Claims Court.

The U.S. Claims Court was eventually renamed the U.S. Court of Federal Claims to avoid confusion with the old U.S. Court of Claims, the Article III court. Obviously, however, many people still confuse the two.

That explanation is still probably as clear as mud, but the result is this: the present day U.S. Court of Federal Claims is not an Article III court, and has never been an Article III court. The judges of the Court of Federal Claims are appointed by the President and confirmed by the Senate and sit for 15 years. They are not employees of the individual agencies. The court's decisions are not binding on the Boards of Contract Appeals (nor are the Boards of Contract Appeals decisions binding on the Court of Federal Claims), but both the Boards and the Court of Federal Claims are bound by decisions of the Court of Appeals for the Federal Circuit and its predecessor court, the U.S. Court of Claims.

Protests occur quite often. Because it is dealing with public monies, the government has a formalized protest process to ensure that federal money is spent in accordance with the law. Protests will be covered in more detail in Chapter 6, but the protest forums are the agency itself, the U.S. Government Accountability Office (GAO), and the Court of Federal Claims. The

GAO, headed by the Comptroller General of the U.S., is the primary forum for protests.

F. Agencies

Besides the contracting agency itself, there are other federal departments, agencies, and offices that substantially affect federal contracting.

The federal government actively encourages the use of its contracting process to foster a variety of socioeconomic goals. One of the most important of these goals is the use of small businesses in federal contracting, as either prime contractors or subcontractors. The agency tasked to further this goal is the SBA. The SBA intersects with the federal contracting process through FAR Part 19. Contracting officers frequently will set aside contracts specifically for small businesses, which are defined under the North American Industry Classification System (NAICS), compiled by the Department of Commerce. If there is an issue of whether a business is small under this classification, the SBA will make the determination.

Federal contractors will deal with two other important agencies from the Department of Labor (DoL), the Occupational Safety and Health Administration, which issues safety and health standards, and the Wage and Hour Division, which issues wage determinations under the Davis-Bacon Act. The Davis-Bacon Act (which will be discussed in Chapter 12) affects how, and how much, the contractor pays its employees.

Another DOL office that government contractors should be aware of is the Office of Federal Contracts Compliance Programs (OFCCP). This is the office charged with ensuring that the contractor not only does not discriminate but also engages in affirmative action. The requirements of this program are set forth in FAR 22.8. The OFCCP will conduct investigations to ensure that the contractor is eligible for continued government contracts.

II. Conclusion

This is obviously a brief overview of an enormously complicated subject. Ideally, we hope it has served to introduce the reader to the people, the rules, the concepts, the forums, and the entities that the remaining chapters will deal with in detail.

Notes

1. Mergentime Corp., ENG BCA No. 5765, 92-2 BCA ¶ 25,007.

2. For a more detailed explanation of this subject, see James F. Nagle, How to Review a Federal Contract and Research Federal Contract Law (2d ed. 2000).

3. *See id.*

4. 10 U.S.C. §§ 2301–2314.

5. 41 U.S.C. §§ 251–260.

6. Pub. L. No. 87-653, 10 U.S.C. § 2306, 41 U.S.C. § 254.

7. Pub. L. No. 103-355.

8. 160 Ct. Cl. 1, 312 F.2d 418, *reh'g denied*, 160 Ct. Cl. 58, 320 F.2d 345, *cert. denied*, 375 U.S. 954 (1963).

9. See Nagle, *supra* note 2, for a detailed breakdown as to what information is required to be placed in each block on this form and where else in the contract such information will relate.

10. 41 U.S.C. §§ 601 *et seq.*

CHAPTER 2

Sealed Bidding

J. TODD HENRY
JAMES F. NAGLE

I. Introduction

For the first century and a half of U.S. government contracting, sealed bidding was the method by which the federal government contracted for construction work. It is the method with which most experienced government contractors are familiar. They understand this process, but often do not understand the government's newfound preference for negotiated procurements. Those contractors prefer the sealed bidding method because it is totally objective. The low, responsive, and responsible bidder wins, and that procurement method allows for no use of subjective criteria that may inject favoritism into the process.

While negotiated procurements under Federal Acquisition Regulation (FAR) 15 are now commonly seen in government construction contracting, sealed bidding must be utilized if four conditions are met:

1. Time permits the solicitation, submission, and evaluation of sealed bids. Normally the solicitation must be on the street for at least 30 days prior to the bid date, and there must be sufficient time for the government to evaluate the sealed bids;

2. The award will be made on the basis of price and other price-related factors. This means that the lowest evaluated price bid will receive the award;

3. It is not necessary to conduct discussions with the responding offerors about their bids. This is a critical distinction from negotiated procurements. In sealed bidding the contractor lives and dies by its bid. If it omits something from the bid, there will be no opportunity to correct it by negotiation or discussion. Offering parties have but one opportunity to prevail, and that is on bid day. Additionally, the bid must agree with and respond to all the material aspects of the solicitation. Otherwise, it will be summarily rejected as nonresponsive; and

4. There is a reasonable expectation of receiving more than one sealed bid. In negotiated procurements, the government has extensive rights

13

to conduct a pre-award audit of the contractor's proposal and examine the contractor's books. In sealed bidding, there is no such right. Consequently, it is critical that competition be expected from multiple bidders, because it is this expectation that ensures that bidders work to keep their prices low.

The FAR allows the contracting officer to convert an invitation for bids into a request for proposals, allowing the agency to then require the contractor to submit cost or pricing data so the government can be assured that it is paying a reasonable price.[1] If this occurs, the Truth in Negotiations Act (TINA), the defective pricing statute, will apply to the formation of the contract. Absent a sealed bid procurement converted to a request for proposals, a contractor is generally not required to disclose its costs, pricing data, or other components of its bid, including overhead, general and administrative expenses, or profit rate. Contractors should be aware, however, that even a contract awarded by the sealed bidding method may have to comply with the TINA if the contract is modified, and especially if the modification involves a price adjustment greater than $650,000.[2]

Because there is no right to a pre-award audit, only firm-fixed-price contracts or fixed-price contracts with economic price-adjustment clauses can be used under the sealed bidding method. This does not mean that sealed bid contracts avoid audits. If the contract is likely to exceed $650,000, the contracting officer is required, by FAR 14.201-7(a), to insert the "Audit and Records—Sealed Bidding" clause into the agreement. This clause assures the right of the government to audit the "accuracy, completeness, and currency of the cost or pricing data" for any modification to the contract. It also requires the contractor to keep records available for three years after final payments.[3]

Telegraphic bids and mailgrams may be an authorized means of bid submission. If so, they will be specifically identified in the solicitation.[4] Similarly, facsimile bids may also be authorized pursuant to FAR 14.202-7, and electronic bids per FAR 14.202-8.

Ultimately, sealed bidding is a process of government procurement designed both to promote fairness to all those interested in securing federal projects and to promote competition. Its aim is to ensure that taxpayers receive the best possible value in an award for public works.

This chapter covers the requirements for sealed bidding, and discusses how determinations and interpretations of both "responsibility" and "responsiveness" are to be made. Finally, a brief discussion of what constitutes a bid mistake, and the methods by which such occurrences can be corrected, will be addressed.

II. Sealed Bidding

Many federal statutes and regulations either allow or require that agencies desiring to procure services or materials by contract do so through a process of sealed bidding.[5] Sealed bidding is the process in which proposing contractors submit bids comprised generally of fixed prices, which are publicly opened and then evaluated for both the "responsibility" and "responsiveness" of the bidders by the agency awarding the contract. Most sealed bid procurements are required to be awarded to the "lowest responsive, responsible" bidder by statute or regulation.

FAR Part 14 outlines the sealed bid methodology as including the following five steps: (a) preparation of the invitation for bids (IFB); (b) publicizing the invitation for bids; (c) submission of bids; (d) evaluation of bids; and (e) contract award.[6]

Federal courts have held that "the purpose of such a system [the sealed bid process] is to ensure free and equal competitive bidding."[7] The FAR makes clear that a sealed bid award, while considering a bidder's qualifications and conformance to the bid request, is primarily focused on achieving the best value for the agency, "the responsible bidder whose bid is responsive to the terms of the invitation for bids and is most advantageous to the Government, considering only price."[8]

In *Weeks Marine Inc. v. United States*,[9] the Court of Federal Claims ruled that the U.S. Army Corps of Engineers had violated applicable requirements that certain projects be procured using the sealed bid method. The South Atlantic Division of the Corps (SAD) sought to let a contract for certain dredging work on a negotiated basis, under a new SAD Acquisition Plan that provided that a multiaward task contract using an indefinite-delivery indefinite-quantity (IDIQ) contracting approach was the preferred procurement method for that type of work.

In upholding a disappointed bidder's protest, the Court of Federal Claims ruled that SAD failed to show that any criteria other than price was the ultimate basis of the award, and therefore no justification existed that the sealed bidding method should not have been employed. The court cited a complete lack of analysis in the Acquisition Plan of why sealed bidding was not the proper method for the procurement, as well as a lengthy track record of the Corps' successful use of the sealed bidding method in similar procurements.

Absent factors that exclude a bidder's proposal from the competition on grounds of responsiveness or responsibility (see below), consideration of a bidder's proposal in a sealed bid procurement is limited under the FAR to price alone, and to those items within a bid that can be equated to price. FAR 14.201-8 sets out the factors, including the costs of transportation and taxes, that must be evaluated by an awarding agency in determining which bid is, in fact, the lowest.

III. Responsiveness

A sealed bidding procurement requires strict adherence to the instructions included in the invitation for bid. For example, failure to provide any portion of a bid allowing an evaluation of price (i.e., the FAR 14.201-8 factors) is a ground upon which a bid can be declared nonresponsive. This requirement is designed to further the twin aims of sealed bidding: fairness and value. "The requirement that a bid be responsive is designed to avoid unfairness to other contractors who submitted a sealed bid on the understanding that they must comply with all of the specifications and conditions in the invitation for bids, and who could have made a better proposal if they imposed conditions upon or variances from the contractual terms the government had specified."[10]

In *Atlas Iron & Machine Works, Inc. v. Secretary of the Air Force*,[11] a bidder omitted from its bid the required citation of the transportation weight of equipment and materials required for the construction of a radar installation in Greenland. When the bid was deemed nonresponsive for this omission, the bidder sought an injunction to prevent the Air Force from awarding the contract to another bidder. In determining that the Air Force's action declaring the bid nonresponsive was correct, the court held that "obviously the public has an interest in the awarding of these contracts to the lowest responsive and responsible bidder. But the public has just as strong, if not stronger, interest in preserving the objective fairness of the competitive bidding system, and that interest might well be jeopardized by a rule allowing, or compelling, contracting officers to simulate a bidder's thought processes and to award the contract accordingly."[12]

For sealed bidding purposes, the FAR defines responsiveness in terms of the bid's "materiality" and "conformance" with the invitation for bid, and reinforces the sealed bidding system's focus on fairness: "To be considered for award, a bid must comply in all material respects with the invitation for bids. Such compliance enables all bidders to stand on equal footing and maintains the integrity of the sealed bidding system."[13] "Responsiveness involves a determination of whether a bidder has unequivocally offered to provide supplies or services in conformity with all the material terms and conditions of the IFB."[14]

In *Qualicon Corp.*,[15] the Navy denied a bid protest by the second low bidder on a chiller replacement project. The Navy determined that a notation made on the outside of the deemed low bidder's sealed envelope was a deductive addendum to its bid inside, and reduced the bid price on the bid form accordingly, making it the low bid. The protester asserted that failure to have all the pricing within the sealed envelope violated the invitation for bid, and rendered the bid nonresponsive. The Comptroller General, in upholding denial of the protest, stated that "responsiveness concerns whether a bid constitutes an offer to perform, without exception, the exact thing called for in the invitation. Since [the low bidder's] bid complied with all of the IFB's material provisions, it was responsive."[16]

A bidder's responsiveness will also be judged on its technical compliance with the requirements of the solicitation. In *Luther Construction*,[17] the protest of a bidder deemed nonresponsive was upheld. Luther was the low bidder on a Bureau of Indian Affairs project, the solicitation for which required that a certain percentage of the overall work be performed directly by the successful bidder. Luther signed the bid form without alteration, yet the agency determined that Luther did not intend to conform to this requirement and declared its bid nonresponsive. Overturning that determination, the Comptroller General held that compliance with an invitation for bid is adequate in and of itself to establish *prima facie* responsiveness. The Comptroller General noted that "by signing and submitting the bid documents without taking any exceptions therein, Luther offered to perform the work in conformity with all material terms and conditions of the solicitation, including the retention of work requirement. Thus, Luther's bid was responsive on its face, and should not have been rejected as nonresponsive."[18]

The concept of conformity with all material terms of the IFB includes agreeing to not only the terms and conditions of the invitation for bids, but often also the precise requirements for inclusions in such bid. Responsiveness can hinge on the proper completion of the bid form documents, and the requirements for the inclusion of other documents required by the IFB.[19]

In *Firth Construction Co. v. United States*,[20] a disappointed bidder successfully challenged the award of a contract by the Corps of Engineers to lay railroad track at Fort Campbell, Kentucky. The challenge claimed that the low bidder's proposal was nonresponsive based on its omission of the required bond commitment and statement of the period of bid validity. The GAO upheld the award on the ground that the low bidder had incorporated the missing terms "by reference" to another document submitted after bid time. In striking down the award on appeal, the Court of Federal Claims said that responsiveness "is determined at the time of bid opening. Accordingly, a bid that is nonresponsive on opening may not be made responsive by subsequent submissions or communications."[21]

Similarly, in *Interstate Rock Products, Inc. v. United States*,[22] the Court of Federal Claims held that the omission of the required penal sum on a bidder's bid bond rendered the entire bid nonresponsive. The court rejected the bidder's argument that the omission was not "material." Instead, the court observed that a part of the IFB was compliance with the bid bond requirement, and that the penal sum needed to be included to assure the bond was enforceable against the surety. The failure to include an unambiguous obligation on the part of the surety was adequate to substantiate the determination of nonresponsiveness. The court observed that "a surety's obligation must be objectively discernable from the bidding documents so that the extent and character of its liability is clearly ascertainable."[23] Echoing the *Firth Construction* holding, the court agreed that allowing the omission to be corrected post-bid was unacceptable because doing so "would place the surety in a position to disavow its obligation, thus compromising the integrity of the sealed bidding system by permitting the bidder to decide after bid opening whether or not to make its bid acceptable."[24]

Nonconformance with an IFB's requirements that specific documents be included with a bid has been held time and again to be a "material" deviation not excusable as a "minor deviation" allowed by 41 C.F.R. § 1-2.405. The omission of required documents brings into question the bidder's willingness to be bound by the terms of the IFB, and if allowed to stand, creates the kind of uneven playing field the sealed bidding system seeks to eliminate.

In determining whether an omission can be ignored, the Comptroller General has stated "the general rule is that where a bidder fails to return with his bid all of the documents which were part of the invitation, the bid must be submitted in such form that acceptance would create a valid and binding contract requiring the bidder to perform in accordance with all of the material terms and conditions of the invitation."[25] For example, a bond form failing to manifest the surety's required conditions and levels of indemnity, as in *Interstate Rock,* cannot meet the criterion that no exception to an IFB's requirements may in any way affect a bidder's intent to be unconditionally bound to the terms of that IFB. It is on the basis of this reasoning that many determinations of nonresponsiveness have been made.[26]

Despite the general rule that strict compliance with the invitation for bid is required, some deviation has been deemed allowable. The FAR provides that minor informalities or irregularities in a bid may be corrected by the bidder or waived by the awarding agency. A deviation that does not materially affect the price, delivery time, quality, or quantity of the item solicited is considered a minor waivable informality.[27] Those four elements—price, delivery time, quality, and quantity, sometimes referred to as the "PDQQ rule"—are the essential elements of any invitation for bid. The litmus test in any analysis of the materiality of a deviation in a sealed bid is fairness to the other bidders and value to the taxpayer.

If the awarding agency has any confusion as to what the basis of the bidder's proposal is, that bid normally must be rejected.[28] However, in cases in which a deviation occurred, but did nothing to violate the PDQQ rule, protests by unsuccessful bidders have been denied. In *Northeast Construction v. Romney,* the Court of Appeals for the Federal Circuit dealt with an instance in which a low bidder's proposal was deemed nonresponsive by HUD for its failure to have completed a bid form setting out certain goals for minority employment. Though Northeast Construction had signed the appendix in the appropriate spots, it had not filled in blanks in the form that specifically set out the bidder's goals. Despite the fact that Northeast Construction orally indicated its willingness to meet the goals at the bid opening (and confirmed that commitment in writing two days later), HUD deemed the bid to be nonresponsive for failure to complete the appendix. The Comptroller General likewise determined the bid to be nonresponsive.

Northeast Construction brought an action to enjoin HUD from awarding to the second low bidder, and was granted a preliminary injunction preventing the award by the district court. The Federal Circuit reversed, indicating that contrary to the district court's conclusion, Northeast Construction's error was not a "minor informality or irregularity" per 41 C.F.R. § 1-2.405, which would allow the contracting officer to permit the bidder to correct its bid or, alternately, to waive the deficiency.

Rather, the court determined that the failure to complete the appendix was a matter of "responsiveness." The important distinction was the fact that the goals included in the project were mandated by Congress, and therefore this was not information that the government agency could deem to be "only a formality." Therefore, this did not fall within the kinds of bid omissions correctable under 41 C.F.R. § 1-2.405.[29]

Where there exists no question of the bidder's intent to be bound by the terms of the IFB, and a violation of the PDQQ rule is not threatened by a deviation, the awarding agency has discretion to ignore that deviation.[30] For the same reason that the government may not force a bidder to accept a contract deviating materially from that upon which the bid was made, in a competitive sealed bidding environment the government may not fairly accept a proposal deviating to such a degree that other bidders are disadvantaged. Any bidder that protests an agency's discretionary award of a contract to the low bidder, based on deviation from the requirements of the IFB, bears the burden of demonstrating that there was "no reasonable basis" on which the award was made.[31]

IV. Responsibility

While "responsiveness" refers to a bidder's willingness to perform in a manner precisely as requested by the government, "responsibility" involves a determination of the bidder's ability to perform as promised. This is the least objective element of the sealed bidding process because contracting agencies have broad discretion to determine the responsibility of prospective contractors.[32]

FAR 14.408 requires an agency to determine a bidder's responsibility prior to making an award to that bidder. The FAR policies and definitions regarding "responsibility" are found in FAR Part 9.1. FAR 9.103(b), in describing those policy considerations, states that "no purchase or award shall be made unless the contracting officer makes an affirmative determination of responsibility. In the absence of information clearly indicating that the prospective contractor is responsible, the contracting officer shall make a determination of nonresponsibility."

FAR 9.104-1 sets out the standards for the determination of responsibility. For a prospective contractor to be determined responsible, it must be found that the bidder:

a. Has adequate financial resources to perform the contract, or the ability to obtain them. See 9.104-3(a);

b. Is able to comply with the required or proposed delivery or performance schedule, taking into consideration all existing commercial and governmental business commitments;

c. Has a satisfactory performance record. See 9.104-3(b) and subpart 42.15. A prospective contractor shall not be determined responsible or nonresponsible solely on the basis of a lack of relevant performance history, except as provided in 9.104-2;

    d.   Has a satisfactory record of integrity and business ethics;.

    e.   Has the necessary organization, experience, accounting and operational controls, and technical skills, or the ability to obtain them (including, as appropriate, such elements as production control procedures, property control systems, quality assurance measures, and safety programs applicable to materials to be produced or services to be performed by the prospective contractor and subcontractors. See 9.104-3(a);

    f.   Has the necessary production, construction, and technical equipment and facilities, or the ability to obtain them. See 9.104-3(a); and

    g.   Is otherwise qualified and eligible to receive an award under applicable laws and regulations.

FAR 9.105 requires the contracting agency to obtain information from prospective contractors in order to demonstrate compliance with the standards of FAR 9.104. It also lists the kinds of information to be solicited, and the sources where the contracting agency should seek that information.

Much like determinations of responsiveness, responsibility evaluations are reviewed on a "rationality" basis. Determinations of "nonresponsibility" require the contracting agency to state its rationale for so finding, but no such rationale is required from an agency making a discretionary determination of responsibility.[33]

The underlying purpose of the responsibility requirement is to assure that the government makes an award in the best interest of the taxpayers. "The contracting agency must consider all relevant factors, such as the low bidder's reliability and honesty, in addition to the amount of the bid in order to determine whether a contract would be advantageous to the government."[34]

Unlike the responsiveness analysis, which is focused on the bidder's agreement to the technical requirements of the IFB, a responsibility analysis considers a broad array of factors, many of which have little to do with the prospective contractor's bid amount.

As noted by the Claims Court, "responsibility addresses the performance capability of a bidder, and normally involves an inquiry into the potential contractor's financial resources, experience, management, past performance, place of performance, and integrity."[35] The federal procurement system has long provided that "the term 'responsible' means something more than pecuniary ability; it includes also judgment, skill, ability, capacity and integrity."[36]

An awarding agency is free to analyze a potential contractor's experience and skill in the specific type of procurement in its responsibility determination. An agency also has the ability to augment the FAR factors for determining responsibility by additional criteria tailored to the particular procurement so long as the criteria are specific and objective and the bidders are made aware of them in the bid solicitation.[37]

In *Maintenance Engineers v. United States*,[38] the Court of Federal Claims held that a procurement responsibility analysis method employed by the

Navy, which awarded points based upon a proposing contractor's experience and the size and complexity of previous contracts, was not administered in a wrongful manner or otherwise improper. As the court noted: "An agency has discretion to determine the scope of the offeror's performance history to be considered provided all proposals are evaluated consistently."[39]

The bid process in *Maintenance Engineers,* while technically a FAR Part 15 negotiated procurement, provides important lessons to sealed bidders regarding the evaluation of responsibility on the basis of "experience." The "experience" criterion was viewed in two ways by the Navy and the court, considering both direct experience in the kind of work involved and experience with the kind and size of contract to be awarded. The protest involved the award of landscape and maintenance contracts for 11 Navy facilities in and around San Diego, California. The solicitation consolidated seven previous agreements, two of which had been held by the plaintiff. The Navy's RFP required bidders to provide their proposals in two parts, one a fixed-price proposal and the other a "non-price" technical and management proposal.

According to the terms of the RFP, specific standards for evaluation were established, with review accomplished by a Technical Evaluation Board (TEB) in accordance with those standards. The plaintiff submitted the lowest bid, and received the second best evaluation of the eight reviewed by the TEB. The only proposing contractor receiving a higher score also provided a bid $1.2 million higher than the plaintiff's. Ultimately, the Navy chose, based partly on the comparatively lower amounts of the plaintiff's past contracts, to award the contract to the second bidder. The plaintiff's action claimed, among other things, that the award was improperly based on evaluation criteria not identified in the RFP, and that the Navy improperly weighted its experience analysis based on prior contract prices, rather than for the stated purpose of evaluating prior contracts to assess the bidder's analogous experience.

Commenting on this process, the *Maintenance Engineers* court observed that "the wide discretion afforded contracting officers extends to a broad range of procurement functions, including the determination of what constitutes an advantage over other proposals."[40] The court also noted precedent allowing "an agency [to] give more weight to one contract over another if it is more relevant to an offeror's future performance on the solicited contract."[41]

Further, the court discounted the plaintiff's assertion that an "undisclosed evaluation factor" was improperly employed by the Navy. "An agency's consideration of whether the contractor's past performance included work on contracts of a similar size and complexity is a proper consideration in determining the degree to which the past contracts are relevant to the present evaluation."[42]

Finally, the court dismissed the plaintiff's argument that the past performance criteria was improperly used as indicative of a bidder's ability to perform contracts of the size to be awarded, rather than as a method for determining a bidder's experience in similar work. Citing the requirement of *Seattle Security Services*[43] that "even handed" evaluation of all proposals is required, the court held that "the Navy's actions in the present case are entirely consistent with the holding of *Seattle Security Services*. The Navy followed a consistent procedure and employed consistent criteria in evaluating each of the bidders' experience/past performance."

Evaluations of responsibility can also hinge more directly on a prospective contractor's financial ability to undertake the contract being procured, focusing directly on the bidder's financial strength and track record as determinative of its ability to perform. Additionally, where a disappointed bidder has an administrative avenue for review of a nonresponsibility assessment, there is no guarantee that additional grounds to support the determination of nonresponsibility will not be found. Even in cases where the initial assessment was flawed, grounds for nonresponsibility overlooked by the contracting agency's initial assessment may be discovered in the supplemental assessment and utilized as a reason to deny award.

In *C&G Excavating v. United States*,[44] the plaintiff was the apparent low bidder on a dredging project to be awarded by the Corps of Engineers. After bid opening, the Corps performed a responsibility review, determined that the plaintiff was not in compliance with a technical aspect of the project requirements, and declared the bidder nonresponsible on a "capacity" basis. Since the project was a "set aside" for award to qualifying small businesses, the plaintiff had the option to pursue a Certificate of Competency (COC) from the Small Business Administration (SBA).

The plaintiff applied for a COC, and the SBA began its own, independent assessment of the plaintiff's responsibility. The SBA's review included visits to the plaintiff's place of business, and a thorough review of its financial data. That review disclosed significant financial concerns about the plaintiff, including that the plaintiff was undercapitalized, had failed to compile a financial statement for a recent business year, and had outstanding tax issues with the IRS. Based upon the shaky financial condition of the plaintiff, the SBA refused to issue a COC.

The plaintiff nevertheless sought injunctive relief to prohibit the Corps from awarding the contract to another bidder, asserting that in its assessment of the plaintiff's responsibility, the SBA should have been limited to reviewing only the grounds that the contracting officer cited as her reason for nonresponsibility. In holding that the SBA had acted rationally in making its own determination of nonresponsibility on other grounds, the Claims Court observed that "the SBA in conducting a COC review may review all elements of responsibility and is not limited to the deficiencies cited by the contracting officer."[45]

Under the Small Business Act,[46] Congress granted the SBA the authority to make pre-award responsibility determinations in regard to the issuance of COCs. Both the Small Business Act and FAR 9.104-1 allow agencies considering a prospective contractor's responsibility to inquire into "all elements of responsibility, including, but not limited to . . . capacity, credit."[47] The court interpreted this authority as not limiting a subsequent COC investigation to merely the basis for nonresponsibility relied on by the contracting officer. The court again noted that the core concern regarding responsibility is that the government ultimately award contracts to those providing the best overall value. "In light of this authority, the SBA would be remiss if it did not evaluate all factors comprising the responsibility determination."[48]

Ultimately, whether determined on the basis of financial wherewithal, capacity, or previous performance, or even on reputation for integrity, a determination of responsibility often plays as important a role in a sealed bidding process as does the lowest bid price. Awareness of the evaluation criteria included in the IFB, or otherwise to be employed by the awarding agency, and how a bidder's qualifications honestly compare against such criteria are among the many factors bidders must consider when choosing contracts on which to bid, and how best to structure responses to IFBs.

V. Bid Mistake

FAR 14.407 addresses how awarding agencies are to deal with mistakes made by prospective contractors in a sealed bidding process. FAR 14.407-1 addresses the general objective of the government in dealing with mistakes made by bidders, requiring an examination by the contracting officer of all bids for obvious mistakes, and notice to the bidder of such discovery.

The FAR includes procedures to be employed when dealing with identified pre-bid errors as well as errors not identified until after award. Clerical or computational errors often present themselves as bids significantly out of the range of other bids received. In the case of a clerical error, 14.107-2(a) requires the contracting officer to obtain a pre-award "verification" of the bid intended by the bidder. In cases where the mistake is not obvious, however, no such duty may exist.

Chris Berg, Inc. v. United States[49] is a pre-FAR case illustrating the government's duty to verify a bid containing an obvious mistake, and to allow that mistake to be corrected pre-bid when clerical in nature. In *Berg*, the bidder made a computational error that resulted in its submitted bid price for a Navy project being significantly low. Suspecting an error, the Navy asked for the price to be confirmed. In doing so, the bidder discovered its errors and sought to reform its pricing, which apparently would have had no effect on its being the low bidder. The Navy refused to allow the reformation, providing the bidder only the options of either accepting the contract at its bid price or rescinding its bid.

The bidder chose to execute the contract for its bid price, while simultaneously reserving its rights to a contract modification correcting the bid mistake. The applicable pre-FAR regulation allowed a bidder to either rescind a mistaken bid or, if proof was available that correction would not displace lower bids (a fairness issue), the bid could be reformed.

Ruling in favor of the bidder's right to reform its bid, the court characterized the Navy's actions as violative of its own procurement regulations,[50] and
held that "the award of the contract to plaintiff at the bid price, with knowledge of its mistake and over its protest, was a clear-cut violation of the law."[51]

Hunt Construction Co. v. United States[52] contrasts with *Berg*, demonstrating
that the government need not provide an opportunity to all bidders to reform
their bids based on its knowledge of one bidder's mistake. Hunt was the successful bidder on a Department of Veterans Affairs project. It brought suit to
collect taxes it claimed were not included in its bid, but were required to be
paid by it in connection with the project. The solicitation included a notice that
all applicable taxes were to be included in the bidders' pricing. However, it
also included a clause stating that any available exemptions from taxes should
be sought by the bidders.

During the course of the project, Hunt requested that the government designate it as a government "purchasing agent," apparently in the hope that governmental exemption from certain taxes would then accrue to the contractor
as a result. The government, feeling the designation not in its best interest,
declined to do so, and Hunt was left with a tax burden not accounted for in its
bid pricing.

One of Hunt's assertions was that under FAR 14.407-1, the government was
obligated to verify its bid before the award, because it had constructive notice,
through an exclusion of taxes in another bidder's proposal, that an ambiguity regarding taxes existed in the IFB. Rejecting this argument, the Court of
Appeals noted that Hunt was arguing not that a mistake in its *own* bid should
have prompted verification but that a mistake in *another* bid should have triggered this action. The court said that "the government, when confronted with
one bidder's unreasonable interpretation of the contract, does not have an
obligation to verify *other* bids to ensure that no other bids incorporated the
same unreasonable contract interpretation."[53]

The *Hunt Construction* holding is consistent with the general rule that the
government's duty to warn a bidder of problems in a bid is limited to situations in which the awarding agency knew or should have known that a bid
(a) included an obvious typographical error; (b) included an obvious computational error; or (c) was based on an obvious misreading of the project
specifications.[54]

This concept is also well illustrated by *J. Rose Corp. v. United States*,[55] in
which a contractor asserted that its unit price bid for excavation was so low as
to have required a verification of its bid by the Corps of Engineers. While the
contractor's unit price for "unclassified" excavation was significantly lower
than the government's estimate, the contractor's overall bid was less than 2
percent lower than the government's estimate for the project. Despite the fact
that the next lowest bidder's price was nearly $100,000 higher, the Court of
Claims determined that "under all the circumstances presented at trial, plaintiff's bid was not so low as to necessarily alert the contracting officer that the
bid was the result of plaintiff's mistake. . . . [N]othing in the bid, aside from

the price differential, could have alerted the contracting officer to the possibility of an error."[56]

In sum, the burden remains squarely on bidders to assure themselves that
their pricing is accurately presented. Without sufficient and apparent reason
for the need for verification, a contracting agency has no obligation to assure
itself that every bid is without error. "When a mistake is suspected, the contracting officer must seek verification of a contractor's bid; yet, government
officials are not required to speculate as to the bases of the contractor's bid."[57]

VI. Conclusion

The sealed bidding procurement process as set out in the FAR focuses squarely
and consistently on fairness. While certainly designed to promote a competitive environment from which the taxpayer theoretically receives the best value,
the system is designed expressly to ensure fairness by and between bidders,
and on the part of the government in awarding its contracts. Any action that
upsets the level playing field will likely be rejected by the courts.

Through the key measures of responsiveness and responsibility, the system's aim of fairness extends both to protecting unbiased competition among
bidders and against unreasonable harm to an honestly mistaken bidder. The
sealed bid process continues to thrive as a key method of government construction contracting. Understanding its requirements, its limitations, and its
ultimate risks therefore remains important to any entity attempting to procure government work.

Notes

1. FAR 14.404-1(f).
2. FAR 15.403-4.
3. FAR 52.214-26.
4. FAR 14.202-2.
5. *E.g.*, 10 U.S.C. § 2305(b)(3) (1999) (addressing sealed bidding procurement
by the Departments of Defense and Transportation, as well as the Army, Navy,
Air Force, and NASA).
6. FAR 14.101(a)–(e).
7. Minor Metals v. United States, 38 Ct. Cl. 16, 21 (1977).
8. FAR 14.103-2(d).
9. 79 Fed. Cl. 22 (2007).
10. Toyo Menka Kaisha, Ltd. v. United States, 597 F.2d 1371, 1377, 220 Ct. Cl.
210 (1979).
11. 429 F. Supp. 11 (1977).
12. *Id.* at 13.
13. FAR 14.301(a).
14. Gardner Zemke Co., B-238334, Apr. 5, 1990, 90-1 CPD ¶ 372.
15. B-237288, Feb. 7, 1990, 90-1 CPD ¶ 158.

16. *Id.*, citing Cent. Mech. Constr., Inc., B-220594, Dec. 31, 1985, 85-2 CPD ¶ 730.

17. B-241719, Jan. 28, 1991, 91-1 CPD ¶ 76.

18. *Id.*

19. *See* FAR 14.404-2.

20. 36 Fed. Cl. 268 (1996).

21. *Id.* at 275, citing Cent. States Bridge Co., B-219559, Aug. 9, 1985, 85-2 CPD ¶ 154, STEPHEN W. FELDMAN, GOVERNMENT CONTRACT AWARDS § 27:15 (1995).

22. 50 Fed. Cl. 349 (2001).

23. *Id.* at 364, citing Allen County Builders Supply, 64 Comp. Gen. 505 (1985), 85-1 CPD at ¶ 507.

24. *Id.*

25. Leasco Info. Prods., Inc., 53 Comp. Gen. 932, 940 (1972).

26. *E.g.*, Prof'l Bldg. Concepts, Inc. v. City of Cent. Falls, 974 F.2d 1 (1st Cir. 1992) (submission of noncertified corporate check, contrary to bid instructions, warranted a finding of material noncompliance with IFB and rejection of bid); Honeywell, Inc. v. United States, 16 Cl. Ct. 173 (1989) (ambiguity of bidder's identity at bid time as a corporation or joint venture determined to be a "material defect," warranting rejection of the bid); McMaster Constr., Inc. v. United States, 23 Cl. Ct. 679 (1991) (bidder's failure to submit, as required by the IFB, a signed certificate of procurement integrity, rendered the bid nonresponsive).

27. 30 Comp. Gen. 179 (1950).

28. *E.g.*, Atlas Iron & Mach. Works, 429 F. Supp. 11 (1977).

29. 485 F.2d 752, 761 (1973).

30. *E.g.*, Croman Corp. v. United States, 31 Fed. Cl. 741 (1994).

31. *See* Cont'l Bus Enters. v. United States, 452 F.2d 1016, 1021 (Ct. Cl. 1971).

32. *E.g.*, Trilon Educ. Corp. v. United States, 578 F.2d 1356 (Ct. Cl. 1978).

33. *E.g.*, Marwais Steel Co. v. Dep't of Air Force, 871 F. Supp. 1448 (D.D.C. 1994).

34. Joseph Constr. Co. v. Veterans Admin., 595 F. Supp. 448, 451 (N.D. Ill. 1988).

35. Blount, Inc. v. United States, 22 Cl. Ct. 221, 227 (1990), referencing FAR 9.105-1.

36. O'Brien v. Carney, 6 F. Supp. 761, 762 (D. Mass., 1934), citing Williams v. Topeka, 85 Kan. 857 (1911).

37. *E.g.*, News Printing Co. v. United States, 46 Fed. Cl. 740 (2000).

38. 50 Fed. Cl. 399 (2001).

39. *Id.* at 423, citing Seattle Sec. Servs. v. United States, 45 Fed. Cl. 560 (2000).

40. *Id.* at 412.

41. *Id.* at 415, citing Forestry Surveys & Data v. United States 44 Fed. Cl. 493, 499 (1999).

42. *Id.*

43. Seattle Security Services v. United States, 45 Fed. Cl. 560 (2000).

44. 32 Fed. Cl. 231 (1994).

45. *Id.* at 248.

46. 15 U.S.C. § 637(b) (1988).

47. *Id.* § 637(b)(7)(A).

48. 32 Fed. Cl. at 241, citing Honeywell, Inc. v. United States, 870 F.2d 644, 647 (Fed. Cir. 1989); Logicon, Inc. v. United States, 22 Cl. Ct. 776, 786 (1991).

49. 426 F.2d 314 (Ct. Cl. 1970).

50. *Id.* at 317, paraphrasing G.L. Christian & Assocs. v. United States, 312 F.2d 418, 160 Ct. Cl. (1963), *reh'g denied,* 320 F.2d 345, 160 Ct. Cl. 58, *cert. denied,* 375 U.S. 954, 84 S. Ct. 444 (1963).

51. *Id.* at 318.

52. 381 F.3d 1369 (Fed. Cir. 2002).

53. *Id.* at 1376 (emphasis in original).

54. *See* CTA, Inc. v. United States, 44 Fed. Cl. 684 (1999).

55. 1 Cl. Ct. 231 (1982).

56. *Id.* at 231–32, citing Armed Services Procurement Regulation (ASPR) 2-406.1.

57. Troise v. United States, 21 Cl. Ct. 48, 63 (1990), citing BCM v. United States, 2 Cl. Ct. 602 (1983).

CHAPTER 3

Competitive Negotiation

ROBERT S. BRAMS
MICHAEL GUIFFRE
ELIZABETH M. GILL
MICHAEL J. CARRATO

I. Introduction

A. What Is Competitive Negotiation?

Use of competitive negotiations provides the government with flexibility in defining its requirements, evaluating proposals, exchanging information with offerors, and awarding contracts. That flexibility, in turn, facilitates the government's objective of securing the "best value" for the goods and services it purchases.

Historically, the government procured construction services using sealed bidding procedures. Sealed bidding is a procurement process that affords little flexibility to either the government or the contractor. In sealed bidding, the government issues its solicitation (known as an Invitation for Bids, or IFB) stating its requirements and the contractor's obligations clearly, accurately, and completely. Contractors respond to the IFB by submitting their prices and any other information required by the IFB. The parties are not permitted to engage in any give-and-take (i.e., negotiations), nor are bidders entitled to revise their bids, unless certain conditions exist (e.g., bid mistakes). Award of a contract is based solely on price and price-related factors.[1]

There have been several changes to the statutes and regulations governing federal contracting over the past several decades, including the addition of a competitive negotiation regime for construction services. Federal Acquisition Regulation (FAR) 15.000 defines "competitive negotiation" as a "contract

The authors would like to recognize Rodney Grandon and William Slade, who coauthored this chapter in the first edition of this book. Section V, "Past Performance Information," was adapted by Michael Branca from Chapter 7 of the first edition, which was authored by Joseph West and Robert Wagman.

awarded using other than sealed bidding procedures. . . ." The use of competitive negotiations in federal construction contracting was greatly expanded in 1996 based on a statutory preference for design-build contracting, under which the agency awards a single contract for project design and construction using competitive negotiation.[2] The government now primarily relies on competitive negotiations to procure its construction requirements. However, the government still must use sealed bidding to acquire construction services if (1) time permits; (2) award will be made on the basis of price and price-related factors; (3) discussions are not necessary; and (4) there is a reasonable expectation of receiving more than one bid.[3] But where the use of sealed bids is not appropriate, or where construction contracts are to be performed outside the United States, its possessions, or Puerto Rico, government contracting officers (COs) may request competitive proposals.[4] The decision to engage in competitive negotiations largely is a matter falling within the business judgment of each CO. And the CO's business judgment to use competitive negotiations will be upheld so long as there is reasonable justification for not using sealed bidding procedures.[5]

B. Acquisition Reform

The Competition in Contracting Act of 1984 (CICA)[6] expanded the use of competitively negotiated procurements by eliminating the long-standing preference for sealed bidding. The significant changes to FAR Part 15 in the late 1990s also greatly expanded the discretion vested in federal agencies to negotiate and award contracts. As it relates to the FAR Part 15 changes, the government sought to

> infuse innovative techniques into the source selection process, simplify the process, incorporate changes in pricing and unsolicited proposal policy, and facilitate the acquisition of best value products and services. The rewrite emphasizes the use of effective and efficient acquisition methods and eliminates unnecessary burdens imposed on industry.[7]

In addition to the emphasis on best value contracting, the revised regulations affected competitively negotiated construction contracts by placing a greater emphasis on design-build contracting, expanded communication opportunities between agencies and offerors, and the use of past performance information in making award decisions.

1. Best Value Concept

Obtaining the "best value" is the government's objective when it competes and awards negotiated contracts. FAR 2.101 defines "best value" as "the expected outcome of an acquisition that, in the Government's estimation, provides the greatest overall benefit in response to the requirement." As discussed in greater detail below, the best-value concept envisions a flexible contracting process in which the government may exercise reasonable discretion to secure the best product or service at the most advantageous price, even if securing the "best value" results in the government paying a higher price for superior products or services.

2. Design-Build Contracts Must Be Negotiated

Prior to the acquisition system revisions in the 1990s, the government relied extensively on design-bid-build construction contracting. Under design-bid-build procedures, the government awarded two contracts—one for project design and another for construction. In 1996, Congress greatly expanded the use of design-build contracting procedures, and new regulations followed shortly thereafter.[8] Design-build contracting combines the design and construction phases of a project into a single contract.[9] Notwithstanding the discretion normally afforded COs on whether to use competitive negotiations, the government is required to use competitive negotiation procedures when awarding design-build contracts.[10]

3. Open Communications

The changes associated with the FAR Part 15 revisions encouraged and expanded communication between the government and contractors. FAR Part 15 is crafted to encourage the government to begin communicating with potential offerors well before it releases the solicitation for a project, primarily to gain a better understanding of its requirement and industry capabilities. While developing this information, the procurement procedures contemplate that the government will be able to sharpen and refine its acquisition planning and procurement strategies, resulting in the purchase of better products and services at lower prices. The FAR now also gives contracting personnel greater freedom to exchange information with competitors after the solicitation has been issued.

4. Use of Past Performance Information

Acquisition reform has imposed on agencies the obligation to consider past performance as an evaluation factor in almost all procurements as part of their best-value analyses. Past performance information means "relevant information, for future source selection purposes, regarding a contractor's actions under previously awarded contracts."[11] It includes "the contractor's record of conforming to contract requirements and to standards of good workmanship; the contractor's record of forecasting and controlling costs; the contractor's adherence to contract schedules, including the administrative aspects of performance; the contractor's history of reasonable and cooperative behavior and commitment to customer satisfaction; and generally, the contractor's businesslike concern for the interest of the customer."[12]

II. Source Selection Processes and Techniques

A. Best Value Continuum

FAR Part 15 governs the competitive negotiation process.[13] FAR Part 15 advises that "[w]hen contracting in a competitive environment, the procedures of [FAR Part 15] are intended to minimize the complexity of the solicitation, the evaluation, and the source selection decision, while maintaining a process designed to foster an impartial and comprehensive evaluation of the offerors' proposals, leading to selection of the proposal representing the best value to the Government."[14]

While what constitutes the best value will vary greatly in each procurement, much depends on the relative importance of cost or price in the acquisition as compared to other evaluation factors.[15] Where the government can clearly define and state its need, and where the risk of unsuccessful performance by the contractor is minimal, cost or price is likely to be more important than other evaluation factors. Conversely, where the government is not able to clearly define or state its need, where the requirement is developmental or technically demanding, or where the risk of unsuccessful performance is high, the government is likely to elevate technical considerations or past performance over cost or price factors.[16]

Within the context of securing the "best value," contracting activities can use "one or a combination of source selection approaches."[17] Most often, the agency will engage in a cost or price versus technical trade-off process.[18] The agency has the discretion to award a contract to the "lowest price technically acceptable" offeror when the agency expects the best value to result from such an award.[19]

B. Trade-off Process

Best-value procurement procedures permit agencies to engage in cost or price/technical trade-offs in evaluating and awarding contracts. In the interest of simplicity, the following discussion will refer only to price/technical trade-offs (price is defined as "cost plus any applicable fee or profit"[20]) although either cost or price can be the subject of the trade-off process. Price/technical trade-offs mean that agencies are permitted to award contracts to a higher-priced offeror, so long as the agency reasonably determines that the price/technical trade-off justifies the price premium. Likewise, agencies may award contracts to other than the highest-rated technical offeror, so long as the agency reasonably determines that the technical superiority of the offer does not warrant the increased price. As discussed below, the discretion granted to the agency in the price/technical trade-off process is generally broad but depends greatly on the evaluation factors and subfactors identified in the solicitation.[21]

When it is necessary or desirable for the agency to engage in a price/technical trade-off analysis, FAR 15.101-1 requires the following:

- All evaluation factors and significant subfactors that will affect contract award and their relative value shall be clearly stated in the solicitation.

- The solicitation shall state whether all evaluation factors other than cost or price, when combined, are significantly more important than, approximately equal to, or significantly less important than cost or price.

Negotiated best value procurements vest agency source selection officials with broad discretion over the conduct of the procurement (e.g., selecting the evaluation factors and subfactors), proposal evaluation, and award.[22] That discretion, however, is not unlimited. In exercising this broad discretion, agencies are subject to the tests of rationality and consistency in applying the requirements set forth in the solicitation.[23] In other words, the agency's award decision must be *rationally related* to the requirements and evaluation criteria announced in the solicitation, and the agency must *consistently* evaluate each offeror's proposal in accordance with the requirements and criteria called out in the solicitation.

Even if the agency determines that a particular offer presents the best overall value, the agency may not accept the offer if that offer fails to satisfy the mandatory minimum requirements called out in the solicitation.[24] For example, in *Mangi Environmental Group, Inc.*,[25] the agency attempted to waive what it apparently regarded as a minor technical flaw in the awardee's offer. The Court of Federal Claims (COFC) viewed the matter differently. Concluding that the awardee's proposal failed to conform to the material terms and conditions of the solicitation, the court ruled that the agency could not award a contract to an offeror whose proposal was technically unacceptable. The lesson for the agency was clear: best-value procurements do not vest source selection officials with the discretion to overlook the technical requirements called out in the solicitation. Instead, if a proposal departs from the stated requirements called out in the solicitation, the agency may only alter those requirements by amending the solicitation.[26]

Offerors and agency evaluators alike must pay careful attention to the specifics of the solicitation. In *A&D Fire Protection, Inc.*,[27] the Government Accountability Office (GAO) concluded that the agency improperly rejected an offeror's proposal. The solicitation stated that the government would base the award for construction services on cost and technical considerations. The solicitation specified that price, however, was the most important factor—all technical factors combined were approximately equal to price. Notwithstanding this evaluation scheme, the agency rejected the protester's technically acceptable, lower-priced proposal. The GAO concluded that the agency failed to give sufficient weight in the evaluation to the protester's lower-priced offer. In sustaining the protest, the GAO directed the agency to perform its price/technical trade-off analysis in accordance with the terms of the solicitation.

In conducting a price/technical trade-off analysis, the agency may not simply rely on the fact that an offeror's prices are fair and reasonable (although the agency must make such a determination).[28] Rather, the agency must evaluate price in accordance with the factors specified in the solicitation.[29] Even where technical evaluation factors outweigh price, as part of the cost/technical trade-off, the agency must make a value determination that the higher-rated, higher-priced proposal presents a better value to the government than a lower-rated, lower-priced proposal.

While the best-value price/technical trade-off vests federal agencies with considerable discretion and flexibility in the contracting process, agencies must ensure that their concept of what constitutes "best value" is clearly stated in the work requirements and standards set forth in the solicitation. Just as important, agencies must ensure that the evaluation factors and subfactors included in the solicitation clearly communicate to the offerors the relative importance of price versus the technical elements of their proposals.

C. Special Procedures for Cost-Reimbursement Contracts for Construction

There is a strong preference within the federal government to award contracts, including construction contracts, on a firm fixed-price basis (meaning that price is stated in the contract or ascertainable from the contract). Agencies may use a cost-reimbursement contract (such as a time and materials contract) to acquire construction services "only when uncertainties involved in contract performance do not permit costs to be estimated with sufficient accuracy to use any type of fixed-price contract."[30] Furthermore, the contractor's accounting system must be adequate for determining costs applicable to the contract, and the government must sufficiently monitor contractor performance to "provide reasonable assurance that efficient methods and effective cost controls are used."[31] Once an agency elects to award a cost-reimbursement contract, the agency must compete and award the contract in accordance with FAR Part 15.[32]

D. Lowest Price Technically Acceptable Source Selection Process

Even when the government is engaging in competitive negotiations, it nevertheless may elect to award to the lowest-priced technically acceptable offeror without engaging in a price/technical trade-off. Awarding to the lowest-priced offeror is appropriate when the government expects that it will secure the best-value products or services from the lowest-priced technically acceptable offer. If the government plans to award on this basis, the solicitation must inform all offerors that the agency will award the contract to the lowest-priced proposal that meets or exceeds the acceptability standards for non-cost factors.[33]

When using this procedure, price/technical trade-offs are not permitted.[34] The agency may consider past performance but only to determine the acceptability of the offerors' relevant performance history.[35] The agency may evaluate other non-cost/non-price factors only to determine the adequacy of the proposal, and it may not rank the proposals based on non-cost/non-price factors. Exchanges between the agency and the offerors are permitted, including clarifications, communications before the establishment of the competitive range, and negotiations.[36] The agency also may give offerors the opportunity to correct technically unacceptable offers, although whether to do so rests in the sound discretion of the agency. However, price is the only permissible discriminator between otherwise acceptable proposals.[37] By electing to award a contract based on these procedures, the government significantly limits the discretion and flexibility it has in a best-value procurement.

III. *Solicitation and Receipt of Proposals and Information*

A. Distribution of Advance Notices and Solicitations

COs must publicize proposed contract actions in order to increase competition, broaden industry participation in meeting government requirements, and assist small business concerns in obtaining contracts and subcontracts.[38] Advance notices and solicitations must be distributed to reach as many prospective offerors as practicable.[39] If the FAR requires notice of a proposed contract action, COs may transmit the notices to the government-wide point of entry (GPE), which may be accessed via the Internet at http://www .fedbizopps.gov.[40] With limited exceptions (e.g., when disclosure would compromise national security), the CO must synopsize in the GPE all proposed contract actions expected to exceed $25,000 at least 15 days before issuance of a solicitation.[41] When a solicitation is synopsized in the GPE, the CO must also make the solicitation available through the GPE.[42]

Unless the requirement is waived by the head of the contracting activity or designee, the CO is required to send a presolicitation notice to prospective offerors on any construction requirement when the proposed contract is expected to equal or exceed $100,000.[43] Notices may also be used if the contract is expected to fall below that threshold.[44] The agency must issue the notice sufficiently in advance of the solicitation to stimulate interest in the procurement from the greatest number of prospective bidders.[45]

B. Inspection of Work Site and Examination of Data

When time and circumstances permit, the CO should make arrangements for prospective offerors to inspect the work site and to examine all data available to the government that may assist potential offerors to understand the required work.[46] Useful data may include boring samples, original boring logs, and records and plans of previous construction.[47] The solicitation should advise offerors of the time and place for the site inspection and data examination.[48]

It is important for an offeror to avail itself of the opportunity to conduct a site visit and examine the government's data. If a contractor suffers increased performance costs as a result of a condition that it could have discovered by a reasonable site inspection, the contractor cannot recover its increased costs from the government but instead must continue to perform at the contract price.[49]

C. Request for Proposals—The Solicitation

1. Describing Agency Needs

In a negotiated procurement, the agency issues a Request for Proposal (RFP) or solicitation to communicate its requirements and solicit responses from potential offerors. The RFP must describe the government's requirements in a manner designed to promote full and open competition.[50] At a minimum, the RFP must describe (1) the government's requirements; (2) anticipated terms and conditions that will apply to the contract; (3) information required to be in the offeror's proposal; and (4) factors and significant subfactors that will be used to evaluate the proposal and their relative importance.[51] In describing agency needs, the agency may include restrictive provisions or conditions only to the extent necessary to satisfy the needs of the agency or as authorized by law.[52]

An agency cannot achieve full and open competition unless the requirements set forth in the solicitation are sufficiently definite to permit the preparation and evaluation of offers on an equal basis. In fulfilling this obligation, an agency must state its requirements in terms of functions to be performed, required performance, and/or essential physical characteristics.[53] The party responsible for any risk associated with the accuracy and suitability of the stated requirements depends greatly on the type of specification used. Typically, the agency will express its requirements by using a combination of types of specifications to identify its needs.[54]

a. Design Specifications/Construction Drawings

Design specifications are properly used when the government's technical requirements are known, understood, and capable of being clearly communicated to potential offerors. Design specifications generally detail precise measurements, tolerances, materials, in-process and finished product tests, quality control, inspection requirements, and other specific information.[55]

By requiring contractors to meet detailed design specifications, the government assumes responsibility for design and related omissions, errors, and deficiencies in the specifications and drawings,[56] as the government warrants the adequacy and completeness of the specifications provided to the construction contractor.

b. Performance Specifications

Performance specifications provide the offeror with the fewest restrictions in responding to the agency's needs because the contractor can choose the methods, materials, and equipment to meet the performance criteria. This flexibility frequently fosters greater participation from the contracting community in the procurement, thereby promoting full and open competition. Performance specifications set forth the agency's needs and operational characteristics desired for the item being procured. No detailed information as provided in design specifications is needed, and the essential consideration is whether the contractor is capable of meeting the stated performance requirements.[57]

If the government uses a performance specification, the contractor accepts general responsibility for the design, engineering, and attainment of the performance requirements.[58] Accordingly, performance specifications shift the burden of potential design defects from the government to the contractor and require a higher degree of sophistication on the part of the contractor in terms of the ability to perform design work.[59]

c. Brand Name/Proprietary Specifications

Agencies generally cannot specify a particular manufacturer's product. However, specifications often establish a "brand name or equal" requirement by stating a need for specific commercial brand names of materials and equipment. Government specifications may refer to equipment, material, articles, or patented processes by trade name, make, or catalog number. Such references are regarded as establishing a standard of quality and are not to be construed as limiting competition.[60] The contractor may, at its option, use any equipment, material, article, or process that, in the judgment of the CO, is equal to that named in the specifications.[61] By utilizing a "brand name or equal" requirement, the government generally assumes responsibility for proper performance of the specified item.

2. Pre-Award Problems Relating to Specifications
a. Ambiguous Specifications

Specifications that are subject to two or more reasonable interpretations are ambiguous and require cancellation or amendment of the solicitation.[62] Further, if an ambiguous specification is not corrected or clarified prior to contract award, the contract is vulnerable to price revision during performance based on a constructive change theory.[63]

As a result of the many specifications, drawings, schedules, and contract terms contained in construction contracts, an offeror frequently will encounter conflicts among the contract documents. To deal with potential ambiguities and disputes arising from such a situation, most negotiated federal contracts include an order-of-precedence clause, which establishes priorities among the different sections of the contract.[64]

b. Unduly Restrictive Specifications

Unduly restrictive specifications (e.g., specifications that unnecessarily contain geographical limitations or that unnecessarily specify a need for a particular product) interfere with the agency's ability to secure full and open competition. Thus, an agency may include restrictive provisions or conditions only to the extent necessary to satisfy the needs of the agency or as authorized by law.[65]

3. Evaluation Factors and Significant Subfactors

The agency has the discretion to determine the evaluation factors and significant subfactors that it will use to make a source selection (contract award) decision. These factors and subfactors should be tailored to meet the needs of each acquisition.[66] Evaluation factors and significant subfactors must represent the key areas of importance and emphasis to be considered in the source selection decision and must support meaningful comparison and discrimination between and among proposals.[67]

a. Required Evaluation Factors

In every source selection decision, the agency is required to evaluate the price or cost of the product or service to be acquired.[68] The agency also is required to evaluate "quality" factors, such as past performance, compliance with solicitation requirements, technical excellence, management capability, personnel qualifications, and prior experience.[69] Absent a specific determination that past performance is not an appropriate evaluation factor, past performance must be evaluated in all negotiated competitive acquisitions that are expected to exceed the simplified acquisition threshold.[70] Agencies also must evaluate the extent to which small disadvantaged business concerns will participate in the performance of unrestricted acquisitions expected to exceed $1 million in the case of construction contracts.[71]

b. Relative Importance of Evaluation Factors

The contracting agency determines the relative importance it assigns to each evaluation factor and subfactor.[72] The agency also determines the scoring procedures to be applied by the evaluation teams (e.g., color or adjectival ratings, numerical weights, and ordinal rankings).[73]

c. Notice of the Basis for Award and the Evaluation Factors

The RFP must clearly state all of the factors and significant subfactors that will affect the contract award decision, as well as their relative importance.[74] Agencies are not required to include in the solicitation the specific rating method that will be utilized to make the award decision.[75] However, agencies must disclose the basis upon which they intend to make the award decision (e.g., price/technical trade-off or lowest-price technically acceptable offer).[76] In addition to the factors and subfactors and their relative importance, the RFP must also disclose, at a minimum, whether all of the non-cost- or non-price-related factors, when combined, are (1) significantly more important than cost or price; (2) approximately equal in importance to cost or price; or (3) significantly less important than cost or price.[77] Whenever past performance information is to be used in the evaluation, the "general approach" for evaluating this information also must be described.[78]

Failure to specifically state factors and subfactors, or to identify their relative importance, will not undermine a solicitation and award so long as the solicitation fairly advises the offerors that the factor or subfactor would be considered. If the RFP does not state the relative importance of the listed evaluation factors, it is presumed that all factors are of equal importance.[79]

D. Two-Phase Design-Build Procedures

Design-build contracting combines the design and construction of the facility into a single contract, as opposed to the traditional approach of design-bid-build contracting. The design-build process can both reduce the time required to complete design and construction and shield the government from contractor claims based on design defects.[80]

Two-phase design-build contracting is appropriate when (1) three or more offerors are anticipated; (2) design work must be performed by offerors to prepare price or cost proposals, and offerors will incur substantial expense to prepare offers; and (3) the CO has considered the following criteria: the extent to which the project requirements have been adequately defined, the time constraints for delivery of the project, the capability and experience of potential contractors, the suitability of the project for use of the two-phase selection method, the capability of the agency to manage the two-phase selection process, and any other criteria established by the head of the contracting activity.[81]

Each procurement is divided into two phases, and the agency may issue one solicitation covering both phases or two solicitations in sequence.[82] Phase One is a qualification process intended to narrow the potential field of competition. Phase One identifies the scope of the work and seeks from potential offerors general information, including (1) the offerors' general technical approach to performing the contract; (2) the offerors' technical qualifications, experience, capability to perform, and relevant past performance; and (3) any other appropriate information.[83] The agency cannot solicit information concerning cost- or price-related factors during Phase One.[84] Phase One also identifies the Phase Two evaluation factors and states the maximum number of offerors that will be selected to submit Phase Two proposals.[85] Offerors selected to compete in Phase Two submit technical and price proposals, which are separately evaluated in accordance with the competitive negotiation procedures set forth in FAR Part 15.[86]

In design-build procurements, the government RFP will include performance, rather than design, specifications. The contractor is responsible for preparing a detailed design based on general statements of the agency's needs and for building in accordance with its design. As with other performance

specification procurements, the allocation of risk for defective design specifications or construction drawings shifts markedly to the contractor. Even
though the contractor's design is subject to agency approvals, the contractor
assumes responsibility both for the design specifications and construction
drawings and for the completed construction.[87]

E. Submission, Modification, Revision, and Withdrawal of Proposals

Proposals and any revisions must reach the government office designated in
the solicitation by the date and time specified in the solicitation.[88] If no time
is stated in the solicitation, the time for receipt of proposals is 4:30 p.m. local
time for the designated government office.[89] Late proposals generally will not
be considered for award.[90]

There are limited exceptions to the "late is late" rule. The agency may consider an otherwise late offer or revised offer, provided that the proposal or
revision is received before an award is made, the CO determines that accepting the late proposal would not unduly delay the procurement, and one of the
following conditions is met:

(1) the proposal was transmitted through an electronic commerce method
      authorized by the solicitation and was received at the initial point of
      entry to the government infrastructure not later than 5:00 p.m. one
      working day prior to the date specified for receipt of proposals;
(2) there is acceptable evidence to establish that the proposal was received
      at the government installation designated for receipt of proposals and
      was under the government's control prior to the time set for receipt of
      proposals; or
(3) the proposal was the only proposal received.[91]

In most situations, the agency must promptly notify any offeror if its
proposal is late, and it must inform the offeror whether its proposal will be
considered.[92] Finally, a late revision to an otherwise successful proposal that
makes its terms more favorable to the government may be accepted, regardless of when it is received.[93]

An offeror may withdraw its proposal by providing written notice of withdrawal to the CO any time before award.[94]

F. Amendment and Cancellation of the Solicitation

If the government significantly changes its requirements or terms and conditions after the solicitation is issued, the agency must amend the solicitation.[95]
If an amendment is issued before the deadline for receipt of proposals, the
amendment must be issued to all parties receiving the solicitation.[96] However,
if an amendment is issued after the agency establishes the competitive range,
the agency is obligated to issue the amendment only to those offerors that are
included in the competitive range.[97]

If the CO determines that a proposed amendment is so substantial as to exceed what prospective offerors reasonably could have expected, such that additional sources likely would have submitted offers had the substance of the amendment been known to them, the CO must cancel the original solicitation and issue a new one regardless of the stage of the acquisition.[98]

IV. Source Selection

A. Source Selection Objective

Subpart 15.3 sets forth the policies and procedures that govern selection of a source or sources in competitively negotiated acquisitions.[99] The government's objective with regard to source selection is to select the proposal that represents the best value to the government.[100] The discussion below addresses the responsibilities of agencies with respect to source selection; the process used in evaluating and, where applicable, revising proposals; and the final source selection decision.

B. Source Selection Responsibilities

The FAR places various responsibilities upon the agency official designated as the source selection authority, or SSA. The SSA normally will be the CO unless the agency head appoints another individual for a particular acquisition or group of acquisitions.[101] While the ultimate responsibility for source selection rests with the agency head, the SSA is the official who is directly responsible for most of the tasks associated with the source selection process.[102]

The SSA is required to establish an evaluation team that is tailored to the particular acquisition and includes appropriate staffing of contracting, logistics, technical, and other resources to ensure a comprehensive evaluation of offers.[103]

The SSA also is responsible for approving the source selection strategy or acquisition plan, if applicable, before the solicitation is released.[104] With regard to the solicitation itself, the SSA must ensure that there is consistency among the solicitation requirements, notices to offerors, proposal preparation instructions, evaluation factors and subfactors, solicitation provisions or contract clauses, and data requirements.[105]

Once proposals are submitted in response to a solicitation, it is the SSA's responsibility to ensure that the proposals are evaluated solely on the factors and subfactors contained in the solicitation.[106] After proposals are evaluated, the SSA selects the source or sources whose proposal represents the best value to the government.[107] In making the selection decision, the SSA also must consider any recommendations provided by applicable advisory boards or panels, if any such boards or panels exist.[108]

As noted above, the CO is the person routinely designated to act as the SSA. Even where this is not the case, the CO will serve as the focal point for any inquiries from offerors or prospective offerors and will control any applicable exchanges in accordance with FAR 15.306 (discussed in detail below).[109] In addition, it is the CO who is responsible for actually awarding the contract.[110]

C. Proposal Evaluation

1. Compliance with the Factors in the RFP

Proposal evaluation involves an assessment of two things: (1) the proposal itself, and (2) the offeror's ability to perform the prospective contract successfully.[111] While agencies have substantial discretion to determine which proposal represents the best value for the government, they must evaluate proposals and assess their relative qualities based *solely* on the factors and subfactors specified in the solicitation.[112] In *GlassLock, Inc.*,[113] the Comptroller General found an award to be improper where the solicitation had advised offerors that the agency would perform a price/technical trade-off that emphasized technical excellence but the agency never performed the trade-off and instead awarded the contract to the lowest-priced technically acceptable offer.

Agencies also are prohibited from making an award based on a proposal that ultimately fails to conform to a material term or condition of the solicitation.[114] In determining the acceptability of a proposal, an agency may not accept at face value the proposer's promise to meet a material requirement where the evaluators know or should know of significant countervailing evidence that raises doubt concerning whether the requirement will be met. In *Carson Helicopter Service, Inc.*,[115] for example, the Comptroller General found an award to be improper where evidence provided to the evaluators cast doubt on the awardee's ability to meet a material requirement of the solicitation, even though the proposal, on its face, indicated compliance. In another case, *Metcalf Construction Co., Inc.*,[116] an agency's rejection of a proposal was upheld where the offeror exceeded a line-item cost limitation in the solicitation, even though the total cost proposed was under the prescribed budget ceiling.

Agencies have the authority to reject all proposals if it is in the best interest of the government to do so, such as in cases where none of the proposals received appears to be sufficient to fulfill the agency's needs or all proposals exceed the funds authorized for the project.[117]

2. Agency Assessment of Proposals

The FAR provides no guidance regarding exactly how agencies should perform the task of assessing proposals—only the guiding principle that whatever process is chosen must be conducted with integrity and fairness.[118] However, the FAR does require that evaluators make their assessments based on the factors and subfactors contained in the RFP, which, as discussed above, must include price or cost, quality, and, in most cases, past performance.[119] Agencies often assign different teams to evaluate different aspects of the proposals, such as one team to evaluate cost or price and a different team to evaluate technical requirements or other non-cost factors. Because cost information is often relevant to assessing an offeror's understanding of, and ability to perform, the technical requirements of a prospective contract, the FAR specifically allows the government to provide such information to members of the technical evaluation team, even where a separate team is established specifically to evaluate cost or price.[120]

Regardless of the actual procedures used to evaluate proposals, agencies must document the relative strengths, deficiencies, significant weaknesses, and risks supporting the evaluations in the contract file.[121]

3. *Scoring/Ranking of Proposals*

In conducting evaluations, agencies are free to use any rating method or combination of methods that they wish, so long as the chosen method fairly differentiates proposals based on their relative merit.[122] Methods that are commonly used include adjectival ratings, numerical weightings, color-coded ratings, and ordinal rankings. Adjectival rating systems, as the name suggests, utilize adjectives to describe whatever criteria are being ranked, such as "excellent," "good," "acceptable," or "unacceptable." Numerical weighting systems typically assign a given number of possible points for each factor or subfactor being scored, with the maximum points for each criterion corresponding to its relative importance in the overall proposal evaluation scheme. Color-coded systems employ various colors instead of adjectives or numbers to describe the relative perceived merits of measured criteria. And ordinal rankings—less commonly used than the other methods described—refers to ranking proposals from best to worst without the aid of more detailed scoring systems using adjectives, colors, numbers, or other such devices.

Regardless of which scoring method an agency decides to use, the scores or ratings themselves do not control the source selection decision but, rather, are only guides to intelligent decision making.[123] The source selection decision must be supported by documentation of the relative differences between proposals, as well as the reasons for the decision itself.[124]

4. *Price Reasonableness and Cost Realism*

The government must evaluate price or cost in every procurement.[125] Normally, adequate competition establishes price reasonableness.[126] Accordingly, when an agency contracts on a firm-fixed price or fixed-price with economic price adjustment basis, comparison of the offered prices should satisfy the government's requirement to perform a price analysis, and a cost analysis need not be performed.[127] However, with cost-reimbursement contracts, cost realism analyses are mandatory.[128] A cost realism analysis determines what the government should realistically expect to pay, the offeror's understanding of the work, and the offeror's ability to perform the contract.[129] COs are required to document their cost or price evaluations.[130]

5. *Contract Pricing—Cost or Pricing Data Considerations*

It has long been government policy that procuring activities must ensure that the prices they pay for goods and services are "fair and reasonable."[131] Within this context, the government has developed special procedures relating to price negotiation in construction contracting.[132] Under these procedures, for each construction project anticipated to cost $100,000 or more, the government must prepare an independent estimate of construction costs.[133] It is within the CO's discretion to require an independent estimate if the anticipated work may fall below the $100,000 threshold.[134] The government subsequently uses this independent estimate to evaluate the reasonableness of the offerors' prices. If an element of an offeror's proposed cost "differs significantly from the Government estimate," the agency is obligated to "request the offeror to submit cost information concerning that element (e.g., wage rates or fringe benefits, significant materials, equipment allowances, and subcontractor costs)."[135] Conversely, when a proposed price is "significantly lower than the government estimate, the contracting officer shall make sure both the offeror and the government estimator completely understand the scope of the work."[136] Either way, the government must reconcile the offered prices with its independent estimate.

One of the tools available to the government to ensure that it is paying a fair and reasonable price is "cost or pricing data." FAR 15.401 very broadly defines cost or pricing data as "all facts that . . . prudent buyers and sellers would reasonably expect to affect price negotiations significantly." The definition of cost or pricing data in FAR 2.101(b) includes more than mere historical accounting data; it includes:

> all facts that can be reasonably expected to contribute to the soundness of estimates of future costs and to the validity of determinations of costs already incurred. They also include such factors as: vendor quotations; nonrecurring costs; information on changes in production methods and in production or purchasing volume; data supporting projections of business objectives and related operations costs; unit-cost trends such as those associated with labor efficiency; make-or-buy decisions; estimated resources to attain business goals; and information on management decisions that could have a significant bearing on costs.

In establishing price reasonableness, however, contracting activities shall not obtain more information than is necessary.[137] This mandate takes on critical importance in the context of most negotiated procurements, because the applicable authority limits the government from requiring the submission of cost or pricing data where prices are based on "adequate price competition."[138] A price is based on adequate price competition if "[t]wo or more responsible offerors, competing independently, submit priced offers that satisfy the Government's expressed requirement," provided the resulting award will be made to the offeror whose proposal represents the best value "where price is a substantial factor in source selection," and there is "no finding that the price of the otherwise successful offeror is unreasonable."[139] Thus, in most competitively awarded construction contracts, offerors should not have to submit cost or pricing data.

Furthermore, before the government seeks any cost or pricing information from the offeror, it must first attempt to determine price reasonableness/cost realism from information that is either publicly available or available within the government.[140] If satisfactory information is not available from these sources, the government must turn next to sources other than the offeror (e.g., market analyses).[141] Lastly, and only when no other sources are available to support the reasonableness of the offeror's prices or realism of the offeror's costs, the government may seek information from the offeror.[142]

On those rare occasions where it is necessary for an offeror to submit cost or pricing data, the FAR outlines certain procedural requirements that must be met.[143] In the past, offerors submitting cost or pricing data used a Standard Form (SF) 1411, Contract Pricing Proposal Cover Sheet. This form, however, is obsolete. Instead, COs may (1) require offerors to submit cost or pricing data in the format specified in FAR 15.408, Table 15-2; (2) specify a reasonable alternative format; or (3) allow offerors to use their own format.[144]

Offerors must exercise care to ensure that the data is provided to the proper official, generally the CO or the CO's authorized representative.[145] When in doubt, offerors should seek written guidance from the CO. Offerors must ensure that they have provided *all* cost or pricing data in existence as of the date of the price agreement. This requirement imposes a burden on the offeror to update its information up to the point at which the government and the offeror reach agreement on price. To meet this obligation, many offerors conduct a "sweep" (that is, a thorough review of their records followed by a final production or disclosure of information to the government) of cost or pricing data just before the completion of the final agreement on price.

Finally, offerors required to submit cost or pricing data must also execute and provide a Certificate of Current Cost or Pricing Data, certifying that the data is accurate, current, and complete as of the date of agreement on price, using the format found at FAR 15.406-2(a).[146] Offerors must submit the certificate as soon as practicable after the date the parties conclude negotiations and agree on contract price.[147] An offeror's failure to properly certify its cost or pricing data will not relieve it of liability for "defective" cost or pricing data.

"Defective" cost or pricing data is that data which is subsequently discovered to have been inaccurate, incomplete, or noncurrent as of the date of final agreement on price (or an earlier date agreed upon by the parties).[148] The submission of defective cost or pricing data can result in serious problems for the offeror. Specifically, the government is entitled to an adjustment in the contract price, to include profit or fee, if it relied on the defective cost or pricing data. To enforce this right, the government is vested with substantial audit rights.[149] Thus, when an offeror must submit cost or pricing data, the offeror must exercise extreme caution to ensure that its data is accurate, complete, and current.

The requirement to submit cost or pricing data presents very serious data collection and control challenges for most contractors. The requirement also imposes considerable burdens on the government's contracting officials. In the current contracting environment, particularly given the FAR's direction that COs should not obtain more information than is necessary to determine price reasonableness or cost realism, it clearly is in the best interests of the contracting community and the government to cooperate in developing the appropriate level of information necessary to support award decisions. Accordingly, the government and the contractor should take reasonable steps to avoid the submission of cost or pricing data whenever possible and appropriate.

6. Evaluation of Past Performance Information

Past performance information is considered to be an important indicator of an offeror's ability to perform the contract successfully.[150] Consideration of past performance information is mandatory in competitively negotiated procurements and is separate from the responsibility determination required under FAR 9.1 (discussed in section 7, below). Agencies are afforded broad discretion in evaluating an offeror's past performance. Agencies may obtain past performance information directly from offerors or through various data collection systems. The government captures and maintains past performance information using an SF 1420 performance evaluation report for all construction contracts of $550,000 or more and for those over $10,000 that the government terminated for default.[151] FAR 15.305(a)(2)(iv) mandates that agencies may not evaluate either favorably or unfavorably offerors that lack relevant past performance history. If the agency establishes a competitive range, agencies must communicate with offerors concerning adverse past performance information to which an offeror has not had a prior opportunity to respond.[152]

The subject of past performance information is addressed in greater detail in Section V, *infra*.

7. Responsibility Determination

COs are required to make affirmative responsibility determinations as a prerequisite to the award of any contract.[153] In the absence of information that clearly indicates that a prospective contractor is responsible, COs are required to make a determination of nonresponsibility.[154] Prospective contractors must affirmatively demonstrate their responsibility and, when necessary, the responsibility of any proposed subcontractors as well.[155] If the prospective contractor is a small business concern, the CO is required to follow the Certificates of Competency and Determinations of Responsibility guidelines set forth under FAR Subpart 19.6.[156]

Current responsibility standards under FAR Part 9 are divided into two categories—general standards and special standards.[157] General standards apply to all contractors regardless of whether the solicitation specifically mentions responsibility, and relate to whether a contractor has the general ability to perform the contract adequately. In order to meet the general standards, a prospective contractor must (a) have adequate financial resources or the ability to obtain them; (b) be able to meet the required contract schedule; (c) have a satisfactory performance record;[158] (d) have a satisfactory record of integrity and business ethics; (e) have the necessary organizational and operational skills, or the ability to obtain them; (f) have the necessary equipment and facilities, or the ability to obtain them; and (g) be otherwise eligible and qualified to receive an award under applicable laws and regulations.[159]

Special standards, also known as "definitive performance criteria," are used by agencies when the general standards are deemed to be inadequate for a particular job. Special standards allow the government to prescribe more detailed and stringent minimum standards when it is deemed necessary for a particular contract or class of acquisitions.[160] In order to ensure fairness, special standards are required to be identified as such in the solicitation, must apply to all offerors, and cannot be waived by the CO.[161]

D. Clarifications and Award without Discussions

FAR 15.306(a) addresses the rights of agencies to make awards on the basis of initial proposals without engaging in discussions and to use limited exchanges known as "clarifications" either to resolve minor or clerical errors or to clarify certain aspects of proposals.[162] In order to make an award on the basis of initial proposals, agencies must provide express notice in the solicitation that the government intends to evaluate proposals and make the award without discussions.[163] Once a solicitation contains such a notice, if the agency decides that discussions are nonetheless needed, it must document the rationale for conducting discussions in the contract file.[164]

Agencies have broad discretion in determining whether to seek clarification from a particular offeror with respect to conflicting or adverse information.[165] An agency may choose to seek clarification where, for example, there is a question concerning the relevance of an offeror's past performance information or where there exists adverse past performance information to which the offeror has not previously had an opportunity to respond.[166] But an offeror's right to have an opportunity to clarify adverse past performance information may be limited to situations where there is a clear question about the validity of such information, such as where there is an obvious inconsistency between a reference's narrative comments and the actual ratings it gives an offeror. In *A.G. Cullen Construction Inc.*,[167] for example, the GAO found that the CO had reasonably exercised his discretion by not seeking any clarification regarding a reference that had rated the protester as "marginal" on one subfactor. The GAO rejected the protester's argument that the agency was obligated to provide an opportunity to respond to the marginal rating, finding that there was nothing on the face of the reference that would create any concern about its validity.[168]

When an agency chooses to seek additional information from an offeror, the context in which the information exchange takes place may be important in establishing whether it qualifies as a clarification, discussion, or other communication. For example, in *Information Technology & Applications Corp. v. United States*,[169] the COFC addressed the issue of whether an evaluation notice sent by the agency, which sought additional information regarding an offeror's past performance, constituted a discussion or a clarification. The court concluded that the exchange did not fit neatly into either category but instead fell more appropriately within the definition of a "communication" under FAR 15.306(b). In denying the protest, the court determined that the communication was a permissible pre-discussion exchange.[170] It should be noted, however, that the decisions to date that have addressed the new "clarifications" rule under the FAR Part 15 rewrite have been inconsistent.[171] At least one COFC decision has indicated that "clarifications" may occur even after discussions have taken place.[172]

E. Communications with Offerors before the Competitive Range Is Established

As noted above, "communications" are another type of exchange between the government and certain offerors that may occur after receipt of initial proposals but before the establishment of a competitive range.[173] The government conducts such communications with eligible offerors for the limited purpose of addressing issues that must be explored to determine whether a proposal should be placed within the competitive range.[174] Only offerors whose past performance information is the determining factor preventing them from being placed in the competitive range or whose exclusion or inclusion from the competitive range remains uncertain are eligible to receive such communications from the government.[175] In the case of the former—offerors whose past performance information is preventing their inclusion in the competitive range—the government is required to hold communications and must address any adverse past performance information to which the offeror has not had a prior opportunity to respond.[176] In the case of the latter—where inclusion in the competitive range is still uncertain—the government has broad discretion whether or not to conduct communications.[177]

Communications may be conducted to (a) enhance an agency's understanding of proposals; (b) address ambiguities and allow reasonable interpretation of a proposal; (c) address perceived deficiencies, errors, weaknesses, or omissions in a proposal; or (d) facilitate the evaluation process.[178] However, communications between the government and an offeror do not necessarily provide an opportunity for the offeror to revise its proposal. In *Firearms Training Systems, Inc.*,[179] the agency decided to delay its competitive range determination and allow six of seven offerors to demonstrate their proposed systems for a computer-operated simulator used to train military personnel. After the demonstrations, the government informed one offeror that it would not be included in the competitive range. The offeror filed a pre-award protest at the COFC contending that, by holding the demonstrations, the agency had already included it in a competitive range and was obligated to enter discussions to identify any perceived weaknesses in its proposal. The court denied the protest, concluding that the agency had properly exercised its "broad discretion" to determine the appropriate level of communication with the offerors, as well as the timing on when to make the competitive range determination.[180]

F. Competitive Range

If an agency determines that discussions are necessary, it must narrow the field of offerors by establishing what is known as the "competitive range." It is permissible for the government to narrow the competitive range to just one offeror.[181] The government may narrow the competitive range to include only the "most highly rated" proposals.[182] The CO has additional discretion to further reduce the competitive range to "the greatest number that will permit an efficient competition among the most highly rated proposals."[183] However, in order for a CO to be able to reduce the competitive range in the interest of efficiency, the solicitation must contain the requisite FAR notice that advises offerors that this limitation may be imposed.[184]

The determination of whether a proposal will be included in the competitive range is a matter within the reasonable discretion of the procuring agency.[185] The GAO has stated that in reviewing an agency's evaluation of proposals and subsequent competitive range determinations, it will not make a de novo evaluation to determine the relative merits of the proposals or their acceptability.[186] Instead, the GAO, like the COFC, will only examine the agency record to determine whether the documented evaluation was fair, reasonable, and consistent with the evaluation criteria set forth in the solicitation.[187]

Although the discretion afforded agencies regarding competitive range decisions certainly is broad, it is not unlimited. Judgments exercised by procuring agencies regarding which proposals to include in a competitive range must be made in a relatively equal, consistent manner. For example, an agency cannot exclude a proposal if the strengths and weaknesses of that proposal are substantially similar to those found in proposals that are included in the competitive range.[188] In *Columbia Research Corp.*,[189] for example, the GAO found that the agency had improperly eliminated the protester's proposal from the competitive range where there was no material distinction from a technical standpoint between the protester's proposal and the proposals that had been included in the competitive range.

In addition to the requirement that the government make reasonable distinctions between proposals to justify excluding a proposal from the competitive range, agencies must always consider the relative cost or price of proposals in making their competitive range determinations. Cost or price to the government is a mandatory evaluation factor in every RFP.[190] In *Meridian Management Corp.*,[191] the GAO sustained a protest where the agency had failed to consider the protester's proposed price before deciding to exclude its proposal from the competitive range. The protester in that case had offered a price that was considerably lower than that of one of the competing proposals that had been included in the competitive range—notwithstanding a technical evaluation score that was very close to the technical score given to the protester's proposal.[192]

G. Exchanges with Offerors After Establishment of the Competitive Range

Once the competitive range has been established, the FAR provides for agencies to engage in full-scale bargaining or robust exchanges, known as "negotiations," with those offerors that remain in the competition. "Negotiations" are defined as "exchanges . . . between the Government and offerors that are undertaken with the intent of allowing the offeror to revise its proposal" and may include "bargaining."[193] "Bargaining" is defined to include "persuasion, alteration of assumptions and positions, give-and-take, and may apply to price, schedule, technical requirements, type of contract, or other terms of a proposed contract."[194] In a competitive acquisition, negotiations take place after the establishment of the competitive range and are called "discussions."[195]

1. Meaningful Discussions

The CO must conduct discussions with each offeror included in the competitive range and must tailor those discussions to fit each offeror's proposal.[196] The primary objective of discussions is to "maximize the Government's ability to obtain best value, based on the requirement and the evaluation factors set forth in the solicitation."[197] Consistent with the government's objective to obtain best value and with the overarching requirement that it conduct its business with "integrity, fairness, and openness,"[198] COs are required to make their discussions with qualified offerors "meaningful." In order to be meaningful, discussions must give offerors sufficient information regarding perceived weaknesses in their proposals to provide them with a reasonable opportunity to address those areas of weakness that could have a competitive impact.[199] The requirement for meaningful discussions is addressed at FAR 15.306(d), which provides guidance to COs with regard to their discussions of deficiencies, significant weaknesses, and other aspects of offerors' proposals.

2. Discussions of Deficiencies, Significant Weaknesses, and Other Aspects of the Proposal

As noted above, once the competitive range is established, COs must conduct meaningful discussions with each offeror still being considered for award. In order to qualify as meaningful, these discussions must, at a minimum, indicate any deficiencies or significant weaknesses related to the offeror's proposal, as well as any adverse past performance information to which the offeror has not yet had an opportunity to respond.[200] The term "deficiency" is defined as "a material failure of a proposal to meet a Government requirement or a combination of significant weaknesses in a proposal that increases the risk of unsuccessful contract performance to an unacceptable level."[201] The term "significant weakness" is defined as a flaw in the proposal that "appreciably increases the risk of unsuccessful contract performance."[202]

While discussions are mandatory with all of the "most highly rated offerors," offerors that were originally included in the competitive range may be eliminated after discussions have begun if the CO determines that the offeror no longer remains among the "most highly rated" group of offerors, regardless of whether all material aspects of the proposal have been discussed or the offeror has been given an opportunity to submit a revised proposal.[203] The mandatory discussion rule, found at FAR 15.306(d), makes clear that COs are *not* required to discuss every area in which a proposal could be improved. Rather, COs have considerable discretion regarding what to discuss beyond the mandatory topics of deficiencies, significant weaknesses, and adverse past performance to which the offeror has not previously been given an opportunity to respond.[204]

Although the FAR provides significant discretion to the government with regard to the scope and extent of discussions, it also specifically encourages the CO to discuss "other aspects" of an offeror's proposal that could, in the CO's opinion, be altered or explained in a way that might materially enhance the offeror's potential for award.[205] As an example of the bargaining that may occur during discussions, the government may, in situations where the solicitation stated that evaluation credit would be given for technical solutions exceeding any mandatory minimums, negotiate with offerors for increased performance beyond any mandatory minimums.[206] On the other hand, the government may suggest to offerors that have exceeded mandatory minimums (in ways that are not integral to the design) that their proposals would be more competitive if the excesses were removed and the offered prices decreased.[207]

Although the FAR bestows considerable discretion upon COs with regard to conducting discussions with the most highly rated offerors being considered for award, clear limitations apply. FAR 15.306(e) expressly prohibits government personnel involved in an acquisition from engaging in conduct that (1) favors one offeror over another; (2) reveals an offeror's technical solution or any information that would compromise an offeror's intellectual property; (3) reveals an offeror's price without permission; (4) reveals the names of individuals providing reference information about an offeror's past performance; or (5) knowingly furnishes source selection information in violation of the Procurement Integrity Act[208] and implementing regulations. With respect to price discussions, COs are permitted to inform an offeror that its price is considered to be too high or too low and to reveal the analysis that supports that conclusion.[209]

There are a number of protest decisions that also serve to illustrate the limits imposed upon agency discretion. For example, in *Chemonics International, Inc.*,[210] an agency gave detailed advice to the awardee during discussions regarding the importance of increasing its proposed level of effort for certain key personnel, but it failed to give similar advice to the protester. The

GAO sustained the protest, finding that the agency had conducted unequal and misleading discussions with the protester in violation of the FAR 15.306(e) prohibition on conducting discussions in a manner that favors one offeror over another. In a similar case, *Dynacs Engineering Co. Inc. v. United States*,[211] the COFC sustained a protest where an agency conducted a second round of discussions with an awardee that addressed weaknesses that had been discussed during the first round of negotiations, but did not afford a similar opportunity to the protester.[212]

Agencies also run afoul of the rules governing discussions if they fail to provide enough information to offerors. In *Cotton & Company, LLP*,[213] an agency's "obliqueness" in its oral and written communications with the protester during discussions was found to be misleading and in violation of the agency's duty to provide meaningful discussions.[214] In that case, the agency's chief negotiator recognized that an offeror had not understood the agency's concerns over what the agency viewed as clear deficiencies in the offeror's proposal. Rather than making its concerns clear, the agency decided it had no duty to challenge the protester's approach in light of what it believed was sufficient information within the offeror's possession to allow making the necessary revisions. After the agency selected another proposal, the offeror protested. The GAO sustained the protest, concluding that by remaining silent, or offering only oblique suggestions regarding the perceived deficiencies, the agency had failed to meet its duties to conduct meaningful discussions and avoid misleading an offeror.[215] Although the GAO noted that an agency is not required to "spoon-feed" an offeror, it also warned that "negotiators need to keep in mind that their reaction, or failure to react, to an offeror's position during oral discussions risks misleading an offeror," which can provide sufficient basis for a protest.[216]

H. Oral Presentations

Agencies have long used oral presentations to gain a better understanding of offerors' proposals. The FAR specifically permits agencies to require oral presentations as a substitute for, or supplement to, written proposal information.[217] Oral presentations often help to streamline the source selection process. In deciding what information to obtain through oral presentations, agencies are encouraged to consider (1) the government's ability to evaluate the information presented; (2) the need to incorporate information into resulting contracts; (3) the overall efficiency of the procurement process; and (4) the impact on small businesses and other potential competitors.[218] When oral presentations are required, the agency must include in the solicitation (1) a description of the type of information to be presented; (2) the qualifications of the presenters; (3) other required or limited media; (4) the location, time and date for presentations; (5) the restrictions governing the time permitted for each oral presentation; and (6) the scope of any exchanges that may occur during presentations.[219]

The government is required to maintain careful records of the information presented during oral presentations, as the impact of the presentation on the award decision may not otherwise be apparent.[220] The GAO has sustained protests in cases where an agency has failed to maintain adequate records of oral presentations that would allow the agency to establish the relative strengths and weaknesses of proposals.[221]

Agencies are permitted to ask questions during oral presentations.[222] Agencies may also engage in discussions during oral presentations but, if they do so, must fully comply with the discussion and proposal revision rules set forth under FAR 15.306 and 15.307.[223]

I. Proposal Revisions

Historically, revisions made to proposals were known as best and final offers, or BAFOs. Proposal revisions are still allowed by the FAR but only from offerors that remain in the competitive range and only as the result of negotiations.[224] If an offeror's proposal is eliminated from the competitive range, the government is not permitted to accept or consider further revisions to that proposal.[225] For those offerors remaining in the competitive range, the CO may request or allow proposal revisions to clarify or document understandings reached during negotiations.[226] Furthermore, at the conclusion of discussions, all offerors still in the competitive range must be given the opportunity to submit final proposal revisions.[227] The CO is required to establish a common cutoff date for receipt of final proposal revisions and must advise all eligible offerors both that the final proposal revisions must be in writing and that the government intends to make an award without obtaining any further revisions.[228]

J. Source Selection Decision

The SSA is required to exercise independent judgment in making the source selection decision.[229] This decision must be based on a comparative assessment of proposals against all source selection criteria contained in the solicitation, and it must be documented.[230] Although the documentation itself is not required to *quantify* the trade-offs that led to the selection decision, it must nonetheless include the rationale for any business judgments or trade-offs that were made by the SSA, including the benefits associated with any additional costs.[231]

The GAO has stated that, in reviewing source selection decisions, it will "examine the supporting record to determine whether the decision was reasonable, consistent with the stated evaluation criteria, and adequately documented."[232] In *Johnson Controls World Services, Inc.,*[233] for example, the GAO sustained a protest where the contemporaneous source selection statement did not document any substantive consideration as to whether a proposal rated technically superior to the awardee's proposal might have represented a better value to the government than the lower-rated, lower-priced proposal actually selected for award. The record only contained the SSA's general statement in the record that "no discernible benefits" offset the "significant advantage" of the lowest cost/price offered by the awardee. The GAO found that, given the evaluation scheme, which made mission suitability and past performance combined approximately equal to cost/price, the SSA's statement was insufficient and fell "far short" of the requirement under FAR 15.308 to document the rationale justifying cost/technical trade-off decisions.[234]

V. Past Performance Information

Over the last 10 years, past performance has become the single most important non-price evaluation factor in federal government procurement. In many procurements, past performance and price are the only two evaluation factors. The Office of Federal Procurement Policy (OFPP) encourages agencies to make past performance information (PPI) an essential factor in the evaluation process and to weigh it to ensure that "significant consideration" is given to past performance. According to OFPP, this means that past performance should comprise at least 25 percent of non-cost (or non-price) factors.[235]

Agencies enjoy enormous discretion both in assessing a contractor's current performance and in using past performance information in making source selection decisions. As a result, the definition and evaluation of past performance can vary tremendously from agency to agency, and even from one contract to the next, based on what information is considered and how it is collected.

Further, the government and contractors often have vastly different perspectives of past performance information and its impact on contract procurement and administration. The government sees the use of past performance information as both "critical for source selection and essential to ensure enhanced performance on existing contracts."[236] To contractors, it is at a minimum an administrative burden, if not a necessary evil, in both its use for source selection and performance.

A. What Is Past Performance Information?

The FAR defines "past performance information" as follows:

> Past performance information is relevant information, for future source selection purposes, regarding a contractor's actions under previously awarded contracts. It includes, for example, the contractor's record of conforming to contract requirements and to standards of good workmanship; the contractor's record of forecasting and controlling costs; the contractor's adherence to contract schedules, including the administrative aspects of performance; the contractor's history of reasonable and cooperative behavior and commitment to customer

satisfaction; the contractor's record of integrity and business ethics, and generally, the contractor's business-like concern for the interest of the customer.[237]

Several of these factors, particularly a contractor's "reasonable and cooperative behavior" and its "business-like concern for the interest of the customer," are so vague that they could be given almost any meaning by a CO or SSA.

B. Past Performance, Experience, and Responsibility

1. *Different Terms, Different Meanings*

Although "past performance," "experience," and "responsibility" are three distinct terms of art in government contract law, the lines separating these terms are often blurred. As a result, these terms may be misused in the source selection process and therefore evaluated incorrectly.

Experience is objective. One can tell with certainty how long a company has been in the business of producing a product or performing a service, how many contracts that company has performed for the government, and the total dollar value of those contracts. Experience as an indicator of ability is very straightforward; if a company has done something for a lengthy period of time, it will likely be able to do it again.

The Department of Defense (DoD) has articulated the clear and important distinction between past performance and experience. Experience reflects *whether* contractors have performed similar work before, whereas past performance reflects *how well* contractors have performed the work.[238] An even more significant distinction is that past performance is defined by the solicitation. While a company's experience is fixed and will not be subject to interpretation, its past performance will vary from one solicitation to the next.

For example, in *Foundation Health Federal Services, Inc.*,[239] the GAO held that an award to an offeror was improper where the solicitation stated that offerors' past performance scores would be based on government experience, but the agency based its past performance evaluation on the offeror's commercial experience.

Responsibility determinations are intended to address the threshold question of whether an offeror has the present capability to perform a particular contract based upon financial resources, operational controls, technical skills, quality assurance, and performance history.[240] A past performance evaluation, on the other hand, is a comparative evaluation of how well a contractor performed in the past that is used as an indicator of how well the offeror will perform in the future. This comparative assessment of past performance information is separate from the responsibility determination required under FAR Subpart 9.1.[241]

Historically, *responsibility* has been a post hoc evaluation applied only to the prospective awardee. As a matter of administrative convenience, if a contractor was not in line for award, there was no need for the agency to expend the time and resources determining if that contractor was capable of performing the contract. This has changed in part,[242] through the increased use of contracting by negotiation and by CICA, which statutorily removed the preference for sealed bidding and instead established an absolute preference for competition.[243] Agencies now must make a preliminary determination of an offeror's ability to perform a contract before being able to determine which offeror represents the best value to the government. According to OFPP, "[i]f a contractor's past performance record passes the responsibility determination, then the record should be compared to the other responsible offerors to determine the offeror that provides the best value to the government."[244]

2. The Intersection of Past Performance Evaluations and Responsibility Determinations

Both past performance and responsibility share a common factor: they are based, at least in part, on a contractor's experience. In addition, since performance history is part of a responsibility determination, the evaluations are often very similar. Although the evaluation of past performance information may be the basis for both a nonresponsibility determination and an offeror being excluded from the competitive range (both of which have the identical practical effect), case law suggests that courts may hold responsibility determinations to a higher level of scrutiny.

For example, in *Ryan Co. v. Dalton*, the U.S. District Court for the District of Columbia held that a responsibility determination based on inaccurate or incomplete past performance information lacked a reasonable basis, and the agency's nonresponsibility determination was reversed.[245] In that case, the court looked at the information reviewed by the Navy and noted that information in the Construction Contractor Appraisal Support System (CCASS) was erroneous because it had misreported a satisfactory evaluation as "unsatisfactory" and listed two other contracts as incomplete because they had not been entered into the system. The court also looked at the comments actively solicited by the Navy through telephone interviews and noted that the negative comments did not correspond to the written SF 1420 evaluations. In light of these findings, the court held that the Navy's reliance on inaccurate and incomplete information displayed a lack of a rational basis, especially where the CO had the means to verify the information by requesting copies of the SF 1420s but had failed to do so.

The GAO, on the other hand, has often held that past performance may be based on the agency's reasonable perception at the time.[246] It has upheld cases in which agencies evaluate past performance with less than complete information about an offeror's prior experience. Further, once an agency receives past performance information from someone with specific knowledge about the prior contract, the GAO has stated that the agency has no further obligation to verify information contained in past performance references.[247]

3. Small Business Contractors

The relationship between past performance and responsibility has also arisen in the context of small business concerns.[248] Protesters in several cases have argued that performance history is part of a responsibility determination and, therefore, under the exclusive jurisdiction of the Small Business Administration (SBA). The GAO held in each case that nothing in the Small Business Act prohibits an agency from making a *comparative evaluation* of a small business offeror's past performance without referral to the SBA for a Certificate of Competency.[249] A *comparative evaluation* means that competing proposals will be rated on a scale relative to each other, rather than on a pass/fail basis.[250] OFPP guidance also recognizes this distinction:

> Using past performance as an evaluation factor to rank an otherwise responsible contractor for award of a contract is not . . . part of the responsibility determination. Evaluation factor rankings are not subject to the Small Business Administration's Certificate of Competency (COC) ratings.[251]

Accordingly, agencies may properly make past performance an element of evaluation of offerors, including small businesses, as long as it is used in a comparative fashion to draw distinctions between offerors.[252]

4. Objective Experience Evaluations

The COFC has held that a request by an agency creates a minimum requirement applicable to all offerors. In *Chas. H. Tompkins Co. v. United States*,[253] the agency required bidders to submit past performance information for at least five projects completed within the last five years within the same size and scope range as the subject procurement. The "size and scope" range was defined as within 10 percent of the bid price offered for the project. The low bidder submitted 24 references, none of which were within the 10 percent requirement. The agency argued that its request for specific past performance information was merely a preference for the types of contracts it wanted to review and not a mandatory requirement. The GAO upheld the agency's determination; however, the COFC reversed the decision based on the plain language of the solicitation. The court held that, even though the agency may not have intended to create a minimum requirement, the agency's unexpressed intent plays no role in interpreting the solicitation.

C. Opportunity to Respond to or Challenge Past Performance Evaluations

One of the major issues associated with past performance evaluations is that a contractor's only recourse to dispute evaluations rendered on current/completed contracts (known as the contractor's "report card") is a challenge within the contracting agency to one level above the CO.[254] The ultimate conclusion on performance evaluations is a decision of the contracting agency.[255]

In general, contractors must be given the opportunity to respond to negative or adverse past performance information before that information may be considered in a subsequent procurement. Contractors must be provided a copy of passive evaluations and have a minimum of 30 days to submit rebuttal information.[256] Additionally, in negotiated procurements, the FAR requires that a contractor be given the opportunity to respond to adverse past performance information gathered on an ad hoc basis during source selection to which it has not had a prior opportunity to respond if (1) the past performance information is the determining factor preventing the offeror's inclusion in the competitive range, or (2) the past performance is a significant weakness or deficiency that could be altered or explained to materially enhance the offeror's potential for award.[257]

Opportunities for responding to adverse past performance information are not always express. For example, the GAO has held that when a solicitation announces that past performance evaluations will be based solely on passive information contained in a database, contractors are on notice. If incomplete or adverse information is contained in the database, the contractor has an affirmative duty to contact the agency and notify it of the discrepancy. If the contractor waits until after the due date for proposals to protest that its score is too low, that protest ground will be dismissed as untimely.[258] Similarly, the GAO has held that a protest that the agency should have used a different past performance database than that set forth in the solicitation was untimely filed.[259] The GAO also held, in *Rohmann Services, Inc.*,[260] that an agency did not need to provide an offeror the opportunity to respond to adverse information where the protester was the incumbent and, thus, the agency was directly aware of its performance history. In that case, the GAO also held that, if the solicitation states that the contract may be awarded without discussions (under FAR 15.306(a)(2), which states that offerors may be given the opportunity to clarify certain aspects of proposals), this does not create an agency duty to discuss past performance.

The COFC has recently held that a contractor's allegation that the CO's final performance evaluation was factually and legally insupportable qualified as a non-monetary Contract Disputes Act of 1978 (CDA)[261] claim upon which the contractor could bring suit.[262]

D. Relevance of Past Performance

The term "relevant" represents a major disconnect between PPI theory and its actual use in source selection. The concept of relevance gives broad discretion to an agency as to how that information is used, if at all, and it is often successfully used by agencies to justify complete disregard of past contracts or reevaluations of performance assessments during source selection, because it allows agencies to limit the information that is considered and determine how that information is weighed.

Relevance is litigated more than any other aspect of past performance because there is no express definition of it in statute or regulation. For example, the DoD guide on past performance suggests that relevant past performance is "[i]nformation that has a logical connection with the matter under consideration and applicable time span."[263] Nevertheless, an agency's determination as to what is relevant will be afforded almost absolute deference.

1. Limitation on Information Agencies Will Consider

Procuring agencies have almost complete discretion to determine what past performance information is relevant to a particular award decision. For example, an agency could decide that a single project is the most relevant example of an offeror's past performance and could make that single project the entire basis for the past performance rating on that procurement. Therefore, even if a company has received the highest past performance rating possible on all of its other appraisals, its lower score on the one project deemed relevant by the CO could be its downfall.

In addition, there is no requirement that a reasonable past performance determination consider all of the references submitted by a contractor.[264] The GAO has held on several occasions that, since no statute or regulation requires an agency to consider all information, there is no legally enforceable duty to consider all references. As a result, an agency may base evaluations on only what it considers to be the most relevant information. An example of this is a case involving an Air Force contract to design/build a software maintenance facility.[265] The protester submitted 15 references for current or past contracts. The evaluation committee deemed relevant two projects performed at the same base, out of the 15 that the contractor submitted, because they were the protester's only federal design/build projects, and the committee based its PPI evaluation solely on information from a telephone conversation with one individual at the base who served as a reference for both projects. The GAO stated that the agency's evaluation was reasonable because there was no legal or RFP requirement that the agency check all references listed in an offeror's proposal.

2. Using Relevance to Modify Existing Evaluations

Relevance can also be used to downgrade an offeror's past performance rating based on a different size or scope of previous contracts, regardless of the fact that references rated the offeror highly. For example, in a Veterans Affairs procurement for the distribution of pharmaceuticals, the protester received the same past performance rating (highly acceptable) as the awardee. The protester claimed that its rating should have been higher (exceptional) because all of its references provided outstanding recommendations. The agency claimed that, although the protester's references all reported consistently good performance, they all represented much smaller volumes than contemplated for the instant procurement. The GAO found that the language of the RFP, which stated that each offeror "will be evaluated on the depth, breadth, and relevance of its experience," clearly permitted the agency to consider any relevant aspects, including size and complexity. As a result, the GAO concluded that the agency's evaluation was reasonable and consistent with the solicitation.[266]

In another case, involving a contract with the Navy to manage and operate a health club at the Naval Intelligence Office, the protester argued that it should have been rated as outstanding since its references listed its past performance as "outstanding" or "neutral." The CO decided that the "outstanding" projects were smaller than the instant procurement, and therefore neutral, and translated the references to "acceptable." The GAO held that this was a mere disagreement between the protester and the CO, which does not make the agency's evaluation unreasonable.[267]

The GAO has even upheld an agency's evaluation where the agency rated the protester's past performance as "fair" (which was the second-lowest rating on a five-point evaluation scale) despite the fact that each of the protester's prior contract references rated it as "good or "very good." In *Clean Venture, Inc.*,[268] the agency felt that each of the protester's past contracts was simpler in scope than the present contract and downgraded the protester's evaluation. While the GAO had consistently upheld not considering an offeror's past performance that the agency did not consider relevant, that case appears to be the first instance where a protester's favorable rating was actually rated less than neutral, that is, three out of five, based on an agency's relevance determination.

In another case, the Army rated the past performance of both the awardee and the protester as "good." The protester objected to receiving the same score because there were deficiencies in the awardee's past projects. The GAO held that the evaluation was reasonable because it considered the awardee's past performance problems and the corrective actions it had taken. In addition, the GAO found that, because the protester's past projects were all considered smaller projects, they were considered not relevant, even though all had positive past performance information, and the agency was reasonable in rating the protester's past performance as only "good."[269]

While an agency may weigh more complex prior contracts more heavily in the evaluation process, there is no requirement that it do so.[270] In fact, in one case the GAO held that it was reasonable to give more weight to an offeror's less complex contracts when evaluating performance history because the contract in question called for less complex service. In that case, the protester's past performance was rated only as "satisfactory" despite the fact that it had received excellent ratings on several of its past contracts. The agency argued that success on the offeror's more complex contracts did not translate to success on less complex contracts and therefore disregarded the offeror's more complex contracts, all of which were rated as outstanding (the highest possible rating) by previous COs.[271]

3. *Agency Failure to Use Readily Available Information*

There is some information that is considered too close at hand not to be considered by a procuring agency. In some limited instances, the GAO has held that it is simply unreasonable not to consider information that the agency knows about. The GAO stated that this narrow exception generally applies to cases in which contracts involved the same services with the same procuring activity or where the activity possessed information personally known to the evaluators.[272]

For example, the GAO sustained a protest where the Navy evaluated the protester's past performance without considering its current contract at the site and awarded the contract to another offeror. The protester filed an initial protest, and the Navy decided to conduct a post-protest reevaluation before the GAO ruled on the protest. In its reevaluation, the Navy considered the relevant project but did not change the protester's performance rating. The protester filed a subsequent protest, and the GAO held that it was unreasonable for the Navy not to consider information that was within the personal knowledge of the evaluators and that the reevaluation did not cure this defect. The GAO sustained the subsequent protest, finding that the Navy failed to conduct meaningful discussions because the information contained adverse PPI on which the protester was not given the opportunity to comment. The GAO also held that it was unreasonable for the Navy not to consider the government's termination of one of the awardee's prior contracts, and it stated that the Navy failed to comply with the past performance evaluation criteria of the solicitation when it gave nothing more than cursory consideration to whether one of the awardee's contracts was similar to the one at issue.[273]

The GAO also sustained a protest in which the protester argued that its past performance evaluation was flawed because the Department of Energy (DOE) failed to consider another DOE project in which the protester was the incumbent and that involved the exact type of service that was being procured. (The RFP even listed the protester's project as an example of what was being procured.) The protester's proposal included 15 pages of information about the work it was performing on the project. The GAO held that, while there is no legal requirement to consider all past projects, it was patently unfair to the protester not to consider information so close at hand.[274]

This case should be contrasted, however, with a case involving another DOE procurement. There, the protester argued that it was unreasonable for it to receive a neutral past performance evaluation because the COs responsible for the protester's other DOE contracts did not return questionnaires. The DOE argued that, since there is no obligation to prepare past performance evaluations before the end of the contract, the COs were under no obligation to complete the questionnaires. The GAO agreed, holding that the agency acted reasonably because there is no legal requirement that all past performance references be included for a valid review.[275]

4. Age of Past Performance Information

In several cases, the GAO has held that more recent contract performance is more relevant and, therefore, appropriately given more weight in the evaluation process.[276] For example, "[a]n agency may also place more significance on recent performance problems."[277] Conversely, the GAO has stated that "[a]n agency may reasonably give less weight to older performance problems where the contractor's subsequent performance has been good."[278]

5. Performance History of Predecessors, Parent/Subsidiary Companies, Subcontractors, and Individual Employees

Several cases have been decided generally holding that agencies have discretion to attribute a subcontractor's performance history to an offeror. For example, the GAO denied a protest where the Navy limited the credit given to the protester's subcontractor because it appeared that the subcontractor would be performing a minimal amount of work on the contract.[279] However, in another case, the GAO upheld the agency's evaluation giving the awardee credit for the work performed by a subcontractor, even though that subcontractor was not part of the instant procurement. The GAO held that, since the prime contractor is ultimately responsible for contract performance, it was reasonable to attribute the subcontractor's past performance to the awardee.[280] The GAO has also held that it was reasonable to downgrade an offeror who relied too much on a subcontractor's experience. Even though the protester argued that performance history should not be segregated, the GAO held that the agency is free to decide the relevance it should give to subcontractor experience.[281]

A similar issue is how an agency may treat an offeror's past performance as a subcontractor. In *Acepex Management Corp.*,[282] for example, the GAO denied a protest where the agency had downgraded the protester's past performance rating because the protester's experience was gained as a subcontractor or as part of a joint venture, rather than as a prime contractor. The GAO held that, given the inherent differences between acting as a prime and a sub, it was reasonable for the agency to look less favorably on the protester's performance history as a subcontractor.

Another issue is the relevance of the past performance of an offeror's parent, subsidiaries, or other related companies, rather than that of the offeror itself. In *Universal Building Maintenance, Inc.*,[283] for example, the GAO sustained a protest where the agency improperly considered past performance of another subsidiary of the awardee's parent company. The GAO stated that, in determining whether one company's performance should be attributed to another, the agency must consider the nature and extent of the relationship between the two companies. Particularly, the GAO stated that agencies should consider factors such as management, facilities, workforce, or other resources shared by the companies that may have an effect on contract performance. In that case, the GAO sustained the protest because the agency favorably evaluated the past performance information for another subsidiary company without first evaluating the relationship between the two companies. The GAO cases hold that it is improper for the agency to attribute the resources of the affiliated entity to the offeror unless the proposal clearly demonstrates that the resources of the affiliated entity will be relied upon and that there will be *meaningful involvement* of that entity in contract performance.[284] The COFC has also recently adopted the *meaningful involvement* requirement.[285]

In addition to attributing a subcontractor's performance history to an offeror, other cases have considered whether an individual employee's performance history may be considered by a procuring agency. Generally, agencies have discretion in determining whether or not to consider individual employees' performance histories. For example, in one case, a protester argued that it was improper for the agency to disregard the past performance of potential employees that it intended to hire if awarded the contract, while downgrading it based on the negative performance history of its special projects manager. The GAO noted that it was reasonable for the agency not to consider the potential hires' performance histories because the protester did not include any evidence establishing the employees' willingness to work for it. The GAO also held that, because the protester's special projects manager had a significant role in its proposal, it was reasonable for the agency to downgrade its proposal based on its employee's poor performance history.[286]

6. Consideration of Contracts Involved in Litigation

One of the most contentious issues surrounding past performance information is the relevance of a contractor's performance on a project in which a claim or litigation is pending. From the contractor's perspective, this scenario often leads to unfair evaluations, especially if the reason for negative performance evaluation is the subject matter of the litigation. The government's perspective is that, if agencies were not permitted to consider contracts subject to litigation, it would encourage litigation simply to avoid consideration of valid information.

The GAO has generally taken the position that it lacks jurisdiction to consider matters that are pending before a court. Under this view, the GAO will not evaluate such matters because, if it were to determine that a performance evaluation was unreasonable because it was based on a contract subject to a claim, it would be deciding the merits of the claim.

For example, in *Oahu Tree Experts*,[287] an incumbent protester was given a "poor" past performance rating on its evaluation for its follow-on contract. Even though the protester provided five references, its past performance rating was based solely on its report card from the current project. During the evaluation, the protester filed an action in U.S. District Court for injunctive relief, alleging that the low evaluation report was in retaliation for reporting the reviewing official for ethical violations. The protester argued that the agency knew that the evaluation was biased and vindictive and did not accurately reflect its performance. Even though the instant procurement was not mentioned in the District Court proceeding, the GAO stated the subject matter of the procurement was at issue. The GAO held that it would have to decide if there was in fact bias to determine if the rating was unreasonable, and it lacked jurisdiction to do so because the issue was before the District Court.

A similar case involved a protester that was removed from the competitive range after it received an unacceptable past performance rating. The protester argued that the Department of Agriculture unfairly considered a previous contract in which it had entered into an agreement to settle claims filed before the Agriculture Board of Contract Appeals (AGBCA) for improper suspension and termination of the predecessor contract. The protester argued that the settlement agreement required the parties to act in good faith, which therefore precluded the agency from using past performance information from the disputed contract in downgrading its score. The GAO denied the protest, holding that nothing in the agreement required the agency "to disregard the protester's performance under the terminated contract."[288]

7. Using Claims, Protest, and Dispute Resolution Histories to Evaluate Past Performance

One of the private sector's greatest criticisms about the use of past performance is the perception that COs use PPI evaluations as a way to punish contractors for filing claims. The GAO has held that the use of claims history as part of a past performance evaluation during source selection may be unreasonable. In *AmClyde*,[289] a 1999 bid protest, a contractor argued that it was improper for the Navy to cite claims history as a justification for lowering the contractor's past performance rating. The GAO denied the protest on this ground because it found no basis to conclude that the protester would have been entitled to a materially better rating even if it had eliminated claims history as an area of concern. In a footnote, the GAO cautioned that "agencies should not lower a firm's past performance claim based solely on its having filed claims" and that firms should not be penalized for pursuing statutorily created remedies.[290]

In another case, the GAO sustained a protest where the only reason for the protester's past performance rating being downgraded was its claims history. In that case, the Navy requested information about "any and all claims submitted, reason for claim and disposition" for the past 15 years. The Navy SSA downgraded the protester's past performance score, finding that the protester's nine claims filed over the previous 15 years raised questions about its "cooperation/responsiveness with regards to customer satisfaction." The GAO did not hold that evaluating claims history under past performance was per se unreasonable but did hold that it was unreasonable in the case before it. Citing *AmClyde*, the GAO stated that, absent evidence of abuse of the contract disputes process, an offeror should not have its past performance evaluation lowered based solely on filing claims. The GAO sustained the protest, finding that the agency acted unreasonably because the protester had prevailed on eight of its nine claims, with the ninth still pending, and had been evaluated as satisfactory to outstanding on all its previous contracts.[291]

The GAO has also questioned an agency's negative reference to the protester's prior agency level protest that questioned certain solicitation provisions. In *SOS Interpreting, Ltd.*,[292] the GAO advised that "we question the propriety of the SSA's considering SOS's [agency level] protest in evaluating its proposal under this RFP, in the absence of some evidence of abuse of the bid protest process."

Closely related to considering a contractor's claims or protest history in evaluating past performance is the consideration of a contractor's willingness to engage in alternative dispute resolution (ADR) in lieu of filing claims, which is often tracked by agencies.

OFPP has denounced the use of claims history, prior protests, or willingness to use ADR in evaluating a contractor's past performance. On April 1, 2002, OFPP administrator Angela Styles issued a memorandum to all Agency Senior Procurement Executives providing, in relevant part, that "[c]ontractors may not be given 'downgraded' past performance evaluations for availing themselves of their rights by filing protests and claims or for not deciding to use ADR."

VI. Pre-Award, Award, and Post-Award Notifications and Debriefings

In the mid-1990s, federal procurement practices were changed to expand the opportunity for information exchanges between the government and unsuccessful offerors.[293] These changes require the government to provide timely notification to unsuccessful offerors (those eliminated from the competition, as well as those not selected for award) regarding the reasons for the government's decisions. In addition, if requested by an unsuccessful offeror, the government must also provide a debriefing in which the government is obligated to provide additional details regarding the procurement.

A. Notices to Unsuccessful Offerors

1. Pre-Award Notification

FAR 15.503(a)(1) requires COs to provide prompt written notice to offerors excluded or eliminated from the competitive range. The written notice must state the basis for the government's determination to exclude or eliminate the offeror from the competition. The notice also must inform the unsuccessful offeror that the government will not consider any further proposal revisions from the unsuccessful offeror.[294]

2. Post-Award Notification

The government must notify unsuccessful offerors within three days of the date of award of the government's decision to make award to another offeror.[295] The post-award notice must include, at a minimum, (i) a statement of the total number of offerors solicited; (ii) the number of proposals received;

(iii) the name and address of the offeror receiving an award (or each offeror, in the event of multiple awards); (iv) the items, quantities, and any stated prices for each award; and (v) in general terms, the reason(s) the unsuccessful offeror's proposal was not accepted. In no event, however, shall the government identify or discuss any other offeror's cost breakdown, profit, overhead rates, trade secrets, manufacturing processes and techniques, or any other confidential business information that an offeror disclosed in the competition.[296]

B. Debriefings

1. Pre-Award Debriefings

An offeror excluded from the competitive range, or otherwise eliminated from consideration for award, may request a pre-award debriefing.[297] The unsuccessful offeror is not entitled to the pre-award debriefing unless it submits a written request for a debriefing to the CO within three days of the date it receives the required pre-award notice, although the government may agree to provide a debriefing notwithstanding the lack of a timely written request.[298] Once the government receives a timely written request for a debriefing, the CO *must* make every effort to conduct the debriefing as soon as practicable.[299] The CO may delay the debriefing until after award if the CO concludes that delaying the debriefing is in the government's best interest.[300] Likewise, an offeror may request that the CO delay the debriefing until after contract award.[301] If the debriefing is delayed until after award, the debriefing shall also include all information required in a post-award debriefing (*see* FAR 15.506(d); section 2, below). Offerors are cautioned, however, that requesting a delay in the debriefing will not toll the timing requirements for pursuing protests.[302]

Pre-award debriefings may be conducted orally, in writing, or by any other method acceptable to the CO. At a minimum, FAR 15.505(e) establishes that the pre-award debriefings shall include (i) the agency's evaluation of significant elements of the offeror's proposal; (ii) a summary of the agency's rationale for excluding the offeror; and (iii) reasonable responses to the offeror's relevant questions. Pre-award debriefings must not disclose (i) the number of offerors; (ii) the identity of other offerors; (iii) the content of other offerors' proposals; (iv) the ranking of the other offerors; (v) the evaluation of other offerors; or (vi) any information prohibited by FAR 15.506(e), which mandates that the debriefing shall not include point-by-point comparisons of the debriefed offeror's proposal with those of the other offerors.[303] Furthermore, the debriefing shall not reveal trade secrets, privileged or confidential manufacturing processes or techniques, privileged or confidential commercial or financial information, the names of individuals providing information about an offeror's past performance, or any other information exempt from release under the Freedom of Information Act (FOIA).[304]

2. Post-Award Debriefings

Following notice of the award decision, an unsuccessful offeror may request a post-award debriefing. As with pre-award debriefings, the offeror must submit a written request for debriefing within three days of the date it receives the notice.[305] To the maximum extent practicable, the CO must conduct the post-award debriefing within five days of the date the agency receives a timely request.[306] The government may accommodate untimely requests. However, the agency is not obligated to do so, nor does the debriefing serve to extend the deadlines for filing protests or suspending contract performance.[307]

FAR 15.506(d) establishes that, at a minimum, post-award debriefings shall include (i) the agency's evaluation of the deficiencies and significant weaknesses in the offeror's proposal; (ii) the overall ratings of the debriefed offeror and the successful offeror; (iii) the overall rankings of the offerors; (iv) a summary of the rationale for the award decision; (v) the make and model of any commercial items the successful offeror will deliver; and (vi) reasonable responses to the offeror's relevant questions. Post-award debriefings shall not include a point-by-point comparison of the debriefed offeror's proposal with any other offeror's proposal, nor is the government permitted to provide a copy of any information exempt from release under the FOIA.[308] Information excluded from release under the FOIA includes trade secrets, privileged or confidential manufacturing processes or techniques, privileged or confidential commercial or financial information, and the names of individuals providing information about an offeror's past performance.[309]

In both pre-award and post-award debriefings, the government should strive to provide enough information to the unsuccessful offerors to help them understand the rationale underlying the government's decisions. By doing so, the government frequently avoids protests, many of which are based on an unsuccessful offeror's incomplete or inaccurate understanding of the government's decisions.

VII. Mistakes

Unlike procurements conducted under the rigid sealed bidding procedures of FAR Part 14, in competitive negotiations the offeror and the government are likely to discover mistakes in the offeror's proposal during the give-and-take of the pre-award exchanges. To the extent mistakes are discovered during the pre-award phases of a negotiated procurement, the offeror has several options, including withdrawing its proposal,[310] or, if given the opportunity, revising its proposal.[311] Occasionally, however, the parties will discover mistakes included in the contractor's proposal only after the government has awarded the contract (and after the mistaken information has become part of the awarded contract). When this occurs in a negotiated procurement, FAR 15.508 directs the parties to process the mistake in accordance with the procedures for mistakes in bids at FAR 14.407-4.

FAR 14.407-4(a) provides that when a mistake is discovered after award, the parties may correct the mistake, if doing so would be favorable to the government without changing the essential requirements of the specifications or contract requirements.[312] Furthermore, FAR 14.407-4(b) authorizes the government to (i) rescind the contract; (ii) reform the contract to delete the items involving the mistake, or to increase the contract price if the increased contract price does not exceed that of the next lowest evaluated proposal; or (iii) if the evidence does not warrant rescission or reformation of the contract, to make no change to the contract as awarded.

The government may rescind or reform a contract only on the basis of clear and convincing evidence that a mistake was made.[313] Also, before the government may rescind or reform a contract, it must be clear that the mistake was mutual (that is, both the government and the offeror failed to recognize the mistake), or, if unilateral, that the mistake was so apparent as to have charged the CO with notice of the probability of mistake.[314]

Because of the many concerns introduced into the contracting process by either rescission or reformation based on mistake, as well as the increased likelihood of a protest, FAR 14.407-4(e) establishes procedures that the government must follow to rescind or reform a contract based on mistake. Fundamental to this process is the requirement that the CO verify by clear and convincing evidence that a mistake has been made and that, under the tests set forth above, the contractor is entitled to a rescission or modification.[315]

The GAO has repeatedly affirmed this evidentiary obligation. In *Gulf-Atlantic Constructors, Inc.*,[316] the GAO denied a protest challenging the agency's decision to allow the awardee to upwardly adjust its low bid. In doing so, it noted that a bidder seeking an upward correction of its contract price must submit clear and convincing evidence that a mistake was made and of the intended price—conditions met by the contractor in *Gulf-Atlantic* and reasonably confirmed by the government. The GAO also noted, however, that the contractor is not required to establish the exact amount of its corrected price, provided there is clear and convincing evidence that the amount of the intended price would fall within a "narrow range of uncertainty and would remain low after correction."[317]

To support an upward adjustment in price, the contractor must submit written statements and other pertinent information that clearly support the intended price, provided that there does not exist any contrary evidence.[318] Other pertinent information includes such things as the contractor's copy of its proposal, data and work papers used by the contractor to prepare its proposal, subcontractor's and supplier's quotations, and published price lists.[319] Whether the available evidence satisfies the clear and convincing standard mandated by the FAR is a question of fact that will rise or fall based on the specific circumstances of each case.[320]

To the extent the government refuses to reform or rescind the contract based on the alleged mistake, and the contractor continues to insist that it is entitled to rescission or reformation, the contractor must seek relief from the COFC or the appropriate board of contract appeals under the dispute process mandated by the CDA.[321]

Notes

1. FAR 14.101.
2. FAR 36.303. For further discussion of federal design-build contracting, see Chapter 5.
3. FAR 6.401, 36.103.
4. *Id.* For further discussion of federal construction contracting outside the United States, see Chapter 25.
5. *See* Enviroclean Sys., B-278261, Dec. 24, 1997, 97-2 CPD ¶ 172 (the agency reasonably determined that discussion might be required); Specialized Contract Serv., Inc., B-257321, Sept. 2, 1994, 94-2 CPD ¶ 90 (need to evaluate more than price supported decision to engage in competitive negotiations).
6. 41 U.S.C. § 253.
7. 62 Fed. Reg. 51,224 (Sept. 30, 1997).
8. 10 U.S.C. § 2305a; 41 U.S.C. § 253m; FAR subpt. 36.3.
9. FAR 36.303.
10. *Id.*
11. FAR 42.1501.
12. *Id.*
13. FAR 36.214.
14. FAR 15.002(b).
15. FAR 15.101.
16. *Id.*
17. *Id.*
18. FAR 15.101-1.
19. FAR 15.101-2.
20. FAR 15.401.
21. FAR 15.101-1.
22. Tessada & Assocs., Inc., B-293942, July 15, 2004, 2004 CPD ¶ 170.
23. *Id.*
24. Mangi Envtl. Group, Inc. v. United States, 47 Fed. Cl. 10 (2000) (agency does not have discretion in a best value procurement to award to an offeror whose proposal is not technically acceptable under the terms of the solicitation).
25. *Id.*
26. FAR 15.206(d) ("If a proposal of interest to the Government involves a departure from the stated requirements, the contracting officer shall amend the solicitation, provided this can be done without revealing to the other offerors the alternate solution proposed or any other information that is entitled to protection (see [FAR] 15.207(b) and 15.306(e)").
27. A&D Fire Prot., Inc., B-288852, Dec. 12, 2001, 2001 CPD ¶ 201.
28. FAR 15.402(a) (COs must "[p]urchase supplies and services from responsible sources at fair and reasonable prices").

29. Beacon Auto Parts, B-287483, June 13, 2001, 2001 CPD ¶ 116 (agency failed to consider whether the awardee's technically superior proposal was worth the price premium).

30. FAR 36.215, 16.301-2.

31. FAR 16.301-3.

32. *See* FAR 16.301-1 (because the contract cannot be awarded on the basis of price, as opposed to cost, sealed bidding procedures are inappropriate); FAR 6.401.

33. FAR 15.101-2.

34. FAR 15.101-2(b)(2).

35. FAR 15.101-2(b)(1).

36. FAR 15.101-2(b)(4).

37. FAR 15.101-2(b)(3).

38. FAR 5.002.

39. FAR 5.204; 36.211.

40. FAR 5.003; FAR 5.201(d)(1). The Comptroller General has determined that issuance of a solicitation only in electronic format does not unduly restrict competition. NuWestern USA Constructors, Inc., No. B-275514, Feb. 27, 1997, 97-1 CPD ¶ 90.

41. FAR 5.101(a)(1); 5.201(a), (b)(1)(i); 5.202; 5.203.

42. FAR 5.102(a).

43. FAR 36.213-2.

44. *Id.*

45. FAR 36.213-2, 36.204.

46. FAR 36.210.

47. *Id.*

48. *Id.*

49. FAR 52.236-3; *see also* Weeks Dredging & Contracting, Inc. v. United States, 13 Cl. Ct. 193 (1987); Avisco, Inc., ENG BCA No. 5802, 93-3 BCA ¶ 26,172; Signal Contracting, Inc., ASBCA No. 44963, 93-2 BCA ¶ 25,877; Fred Burgos Constr. Co., ASBCA No. 41395, 91-2 BCA ¶ 23,706. For further discussion of site inspections and the effect of a contractor's failure to inspect on its right to an equitable adjustment for increased costs, see chapters 10 and 15, respectively.

50. FAR 11.002(a)(1)(i).

51. FAR 15.203(a).

52. FAR 11.002(a)(1)(ii).

53. FAR 11.002(a)(2)(i).

54. For further discussion of defective government specifications, see Chapter 18.

55. *See* Apollo Sheet Metal, Inc. v. United States, 44 Fed. Cl. 210 (1999); Blake Constr. Co. v. United States, 987 F.2d 743 (Fed. Cir. 1993); Monitor Plastics Co., ASBCA No. 14447, 72-2 BCA ¶ 9626.

56. *See Apollo*, 44 Fed. Cl. 210; Neal & Co. v. United States, 19 Cl. Ct. 463 (1990).

57. *See* Blake Constr., 987 F.2d 743; *Monitor*, 72-2 BCA ¶ 9626.

58. *See Apollo*, 44 Fed. Cl. 210; Blake Constr., 987 F.2d 743 (Fed. Cir. 1993).

59. *See Apollo*, 44 Fed. Cl. 210; *Monitor*, 72-2 BCA ¶ 9626.

60. FAR 52.236-5(a).

61. *Id.*

62. *See* RMS Indus., B-248678, Aug. 14, 1992, 92-2 CPD ¶ 109; Flow Tech., Inc., B-228281, Dec. 29, 1987, 87-2 CPD ¶ 633.

63. See Chapter 8.

64. *See, e.g.,* FAR 52.215-8, 52.236-21.

65. FAR 11.002(a)(2).

66. FAR 15.304(a).

67. FAR 15.304(b).

68. FAR 15.304(c)(1); 10 U.S.C. § 2305(a)(3)(A)(ii); 41 U.S.C. § 253a(c)(1)(B); *see also* Spectron, Inc., B-172261, 51 Comp. Gen. 153 (1971).

69. FAR 15.304(c)(2); 10 U.S.C. § 2305(a)(3)(A)(i); 41 U.S.C. § 253a(c)(1)(A).

70. FAR 15.304(c)(3)(i).

71. FAR 15.304(c)(4).

72. 10 U.S.C. § 2305(a)(3); 41 U.S.C. § 253a(c).

73. FAR 15.305(a).

74. FAR 15.304(d); 10 U.S.C. § 2305(a)(2)(A)(i); 41 U.S.C. § 253a(b)(1)(A); *see also* Qual-Med, Inc., B-254397.13, July 20, 1994, 94-2 CPD ¶ 33.

75. FAR 15.304(d); *see also* D.N. Am., Inc., B-292557, Sept. 25, 2003, 2003 CPD ¶ 188, n.6; ABB Power Generation, Inc., B-272681, Oct. 25, 1996, 96-2 CPD ¶ 183.

76. FAR 15.101-1; FAR 15.101-2.

77. FAR 15.304(e).

78. FAR 15.304(d).

79. *See* Hyperbaric Tech., Inc., B-293047.2, Feb. 11, 2004, 2004 CPD ¶ 87.

80. See Chapter 5.

81. FAR 36.301(b)(3).

82. FAR 36.303.

83. FAR 36.303-1(a).

84. FAR 36.303-1(a)(2)(iii).

85. FAR 36.303-1(a)(3)–(4).

86. FAR 36.303-2(b); *see also* FAR 15.305.

87. See Chapter 18.

88. FAR 15.208(a).

89. *Id.*

90. FAR 15.208(b)(1).

91. *Id.*

92. FAR 15.208(f).

93. FAR 15.208(b)(2).

94. FAR 15.208(e).

95. FAR 15.206(a); *see also* Northrop Grumman, B-295526, Mar. 16, 2005, 2005 CPD ¶ 45; EP Prod., Inc. v. United States, 63 Fed. Cl. 220, 224–25 (2005).

96. FAR 15.206(b).

97. FAR 15.206(c).

98. FAR 15.206(e).

99. FAR 15.300; *see* FAC 97-2, 62 Fed. Reg. 51,224 (Sept. 30, 1997).

100. FAR 15.302.

101. FAR 15.303(a). *See* Beneco Enters., Inc., B-283512.3, July 10, 2000, 2000 CPD ¶ 176 (GAO recommended that the agency appoint a new SSA in light of the existing SSA's apparent misrepresentation on the record and repeated failure to act reasonably in evaluating past performance information).

102. FAR 15.303(b).

103. FAR 15.303(b)(1).

104. FAR 15.303(b)(2).

105. FAR 15.303(b)(3).

106. FAR 15.303(b)(4).

107. FAR 15.303(b)(6).

108. FAR 15.303(b)(5).

109. FAR 15.303(c)(1), (2).

110. FAR 15.303(c)(3).

111. FAR 15.305(a).

112. FAR 15.305(a); *see* United Architecture & Eng'g Inc., 46 Fed. Cl. 56, 63 (2000), *aff'd*, 251 F.3d 170 (Fed. Cir. 2000).

113. B-299931, Oct. 10, 2007, 2007 CPD ¶ 216; *see also* Marquette Med. Sys., Inc., B-277827.5, Apr. 29, 1999, 99-1 CPD ¶ 90 (award improper where agency deviated from stated selection criteria).

114. *See* Wiltex, Inc., B-297234.2, Dec. 27, 2005, 2006 CPD ¶ 13; Special Operations Group, Inc., B-287013, Mar. 30, 2001, 2001 CPD ¶ 73.

115. B-299720, July 30, 2007, 2007 CPD ¶ 142.

116. B-289100, Jan. 14, 2002, 2002 CPD ¶ 31.

117. FAR 15.305(b).

118. *See* FAR 1.102-2(c).

119. *See* FAR 15.304.

120. FAR 15.305(a)(4).

121. FAR 15.305(a); *see also* Honeywell Tech. Solutions, Inc., Wyle Labs., B-292354, B-292388, Sept. 2, 2003, 2005 CPD ¶ 107.

122. FAR 15.305(a).

123. *See* Wackenhut Servs., Inc., B-400240, Sept. 10, 2008, 2008 CPD ¶ 184.

124. *See* Fedcar Co., Ltd., B-310980, Mar. 25, 2008, 2008 CPD ¶ 70.

125. FAR 15.304(c)(1).

126. FAR 15.305(a)(1); *see also* FAR 15.403-1(c)(1).

127. FAR 15.305(a)(1).

128. *Id.*

129. *Id.*

130. *Id.; see also* Rockwell Elec. Commerce Corp., B-286201, Dec. 14, 2000, 2001 CPD ¶ 65 (agency improperly failed to evaluate third-party costs included in proposal as required by the solicitation).

131. FAR 15.402(a).

132. FAR 36.214.

133. FAR 36.203.

134. *Id.*

135. FAR 36.214(b)(1).

136. FAR 36.214(b)(2).

137. FAR 15.402(a).

138. FAR 15.402(a)(1).

139. FAR 15.403-1(c).

140. FAR 15.402(a).

141. *Id*.

142. *Id*.

143. The requirement to submit cost or pricing data also extends to subcontractors. FAR 15.403-4 requires the submission of cost or pricing data prior to the "award of a subcontract at any tier, if the contractor and each higher-tier subcontractor have been required to furnish cost or pricing data."

144. FAR 15.403-5.

145. 10 U.S.C. § 2306a(a)(3); 41 U.S.C. § 254b(a)(3).

146. *See* 10 U.S.C. § 2306a(a)(2); 41 U.S.C. § 254b(a)(2) (offerors submitting cost or pricing data must certify that the data is accurate, current, and complete).

147. FAR 15.406-2(a).

148. FAR 15.407-1(b)(1).

149. 10 U.S.C. § 2313(a)(2), (e)–(f); 41 U.S.C. § 254d(a)(2), (e)–(f); and FAR 52.215-2 (contracting agency's right to audit); 10 U.S.C. § 2313(b) and 41 U.S.C. § 254d(b) (DCAA empowered to subpoena records); 10 U.S.C. § 2313(c), (e)–(f) and 41 U.S.C. § 254d(c), (e)–(f) (Comptroller General's right to audit).

150. *See* FAR 15.305(a)(2)(i).

151. FAR 36.201.

152. FAR 15.306(b)(1)(i).

153. FAR 9.103(b).

154. *Id*.

155. FAR 9.103(c).

156. FAR 9.103(b).

157. *See* FAR 9.104.

158. A prospective contractor cannot be determined responsible or nonresponsible based solely on the lack of relevant performance history, except as provided under FAR 9.104-2. *See* FAR 9.104-1(c).

159. FAR 9.104-1.

160. FAR 9.104-2(a).

161. FAR 9.104-2; *see* John C. Grimberg Co., Inc. v. United States, 185 F.3d 1297 (Fed. Cir. 1999).

162. FAR 15.306(a)(1), (2); *see* 10 U.S.C. § 2305(b)(4)(A)(ii); 41 U.S.C. § 253b(d)(1)(b).

163. FAR 15.306(a)(3).

164. *Id*.; *see* FAR 52.215-1 (Instructions to Offerors—Competitive Acquisitions).

165. *See* U.S. Constructors, Inc., B-282776, July 21, 1999, 99-2 CPD ¶ 14; Inland Servs. Corp., B-282272, June 21, 1999, 99-1 CPD ¶ 113 (CO not required to provide offeror with opportunity to comment on adverse past performance information).

166. *See* FAR 15.306(a)(2).

167. B-284049.2, Feb. 22, 2000, 2000 CPD ¶ 45.

168. *Id*.; *see also* NMS Mgmt., Inc., B-286335, Nov. 24, 2000, 2000 CPD ¶ 197.

169. 51 Fed. Cl. 340 (2001).

170. *Id.*

171. See *The Nash & Cibinic Report,* June 2000, for a more in-depth discussion of this topic.

172. *See* Antarctic Support Assocs. v. United States, 46 Fed. Cl. 145 (2000) (telephone call from CO to offeror seeking revised information after proposal revision had been submitted was held to be a "clarification" rather than a discussion). *Cf.* Dubinsky v. United States, 43 Fed. Cl. 243 (1999) (holding that the term "clarifications" has no application to exchanges occurring after discussions).

173. FAR 15.306(b).

174. FAR 15.306(b)(3).

175. FAR 15.306(b)(1).

176. FAR 15.306(b)(1)(i).

177. FAR 15.306(b)(1)(ii).

178. FAR 15.306(b)(2), (3).

179. 41 Fed. Cl. 743 (1998).

180. *Id.* at 747.

181. *See, e.g.,* SDS Petroleum Prods., Inc., B-280430, Sept. 1, 1998, 98-2 CPD ¶ 59.

182. FAR 15.306(c)(1).

183. FAR 15.306(c)(2).

184. *Id.*; Matrix Gen., Inc., B-282192, June 10, 1999, 99-1 CPD ¶ 108.

185. *Matrix,* 99-1 CPD ¶ 108.

186. *Id.*

187. *Id.*; Ervin & Assocs., Inc., B-280993, Dec. 17, 1998, 98-2 CPD ¶ 151 at 3; *see, e.g.,* Impresa Construzioni Geom. Domenico Garufi v. United States, 238 F.3d 1324, 1340 (Fed. Cir. 2001).

188. Nations, Inc., B-280048, Aug. 24, 1998, 99-2 CPD ¶ 94 at 4–5.

189. B-284157, Feb. 28, 2000, 2000 CPD ¶ 158.

190. 41 U.S.C. § 253a(c)(1)(B); FAR 15.304(c)(1).

191. B-285127, July 19, 2000, 2000 CPD ¶ 121.

192. *Id.*

193. FAR 15.306(d).

194. *Id.*

195. *Id.*

196. FAR 15.306(d)(1).

197. FAR 15.306(d)(2).

198. FAR 1.102(b)(3).

199. Dynacs Eng'g Co., Inc. v. United States, 48 Fed. Cl. 124, *recon. denied,* 48 Fed. Cl. 240 (2000).

200. FAR 15.306(d)(3); *see* Tiger Truck, LLC, B-400685, Jan. 14, 2009 (finding that discussions cannot be meaningful if a vendor is not advised of the significant weaknesses or deficiencies that must be addressed in order for its quotation to be in line for award).

201. FAR 15.001.

202. *Id.*

203. FAR 15.306(d)(5).

204. *See* 66 Fed. Reg. 65,439.

205. FAR 15.306(d)(3).

206. FAR 15.306(d)(4).
207. *Id.*
208. 41 U.S.C. § 423(h)(1)(2).
209. FAR 15.306(e)(3).
210. B-282555, July 23, 1999, 99-2 CPD ¶ 61.
211. 48 Fed. Cl. 124, *recon. denied*, 48 Fed. Cl. 240 (2000).
212. *Id.* at 136.
213. B-282808, Aug. 30, 1999, 99-2 CPD ¶ 48.
214. *Id.* at p. 8.
215. *Id.*
216. *Id.; see also* Bank of Am., B-287608 *et al.*, July 26, 2001, 2001 CPD ¶ 137.
217. FAR 15.102.
218. FAR 15.102(c).
219. FAR 15.102(d).
220. FAR 15.102(e).
221. *E.g.*, J&J Maint., Inc., B-284708.2 *et al.*, June 5, 2000, 2000 CPD ¶ 106; Checchi & Co. Consulting, Inc., B-285777, Oct. 10, 2000, 2001 CPD ¶ 132.
222. FAR 15.102(d)(6).
223. FAR 15.102(g).
224. FAR 15.001, 15.307.
225. FAR 15.307(a).
226. FAR 15.307(b).
227. *Id.*
228. *Id.*
229. FAR 15.308.
230. *Id.*
231. *Id.*
232. Johnson Controls World Servs., Inc., B-289942, May 24, 2002, 2002 CPD ¶ 88 (citing AIU North Am., Inc., B-283743.2, Feb. 16, 2000, 2000 CPD ¶ 39 at p. 7).
233. *Id.*
234. *Id.* at p. 4.
235. *See* Best Practices for Collecting and Using Current and Past Performance Information 12 (Office of Federal Procurement Policy, May 2000), *available at* https://www.acquisition.gov/comp/seven_steps/library/OFPPbp-collecting.pdf (hereinafter OFPP Guide).
236. *See* A Guide to Collection and Use of Past Performance Information 1 (Office of the Under Secretary of Defense for Acquisition, Technology and Logistics, May 2003), *available at* http://www.acq.osd.mil/dpap/Docs/PPI_Guide_2003_final.pdf (hereinafter, DoD Guide); *see also* foreword of Deidra A. Lee, (Former) Administrator, OFPP, in OFPP Guide, *supra* note 235, at 2.
237. FAR 42.1501.
238. DoD Guide, *supra* note 236, at 7.
239. B-278189.3, Feb. 4, 1998, 98-2 CPD ¶ 51.
240. FAR 9.104.
241. FAR 15.305(a)(2).
242. *Id.*
243. *See* S. Rep. No. 97-655, at 14 (1982).

244. *See* OFPP GUIDE, *supra* note 235, at 11.

245. Civ. Action No. 96-2803 (D.D.C. 1997).

246. *See, e.g.,* Buckeye Park Servs., Inc., B-282082, June 1, 1999, 99-1 CPD ¶ 97.

247. *See* JGB Enters., Inc., B-291432, Dec. 9, 2002, 2002 CPD ¶ 213; Black & Veatch Special Projects Corp., B-279492.2, June 26, 1998, 98-1 CPD ¶ 173.

248. For further discussion of the government's small business program, see Chapter 12.

249. Xerxe Group, Inc. B-280180.2 *et al.*, Sept. 28, 1998, 98-2 CPD ¶ 80.

250. J. Womack Enters., Inc., B-299344, Apr. 4, 2007, 2007 CPD ¶ 69.

251. *See* OFPP GUIDE, *supra* note 235, at 11.

252. *See* Goode Constr., Inc., B-288655 *et al.*, Oct. 19, 2001, 2001 CPD ¶ 186; *see also Womack*, 2007 CPD ¶ 69.

253. 43 Fed. Cl. 716 (1999).

254. *See* FAR 42.1503.

255. *Id.*

256. FAR 42.1503(b).

257. FAR 15.306.

258. Dayton-Granger, Inc., Reconsideration, B-279553.3, Oct. 2, 1998, 98-2 CPD ¶ 90.

259. Midwest Metals, B-299805, July 17, 2007, 2007 CPD ¶ 131.

260. B-280154.2, Nov. 16, 1998, 98-2 CPD ¶ 134; *accord,* TLT Constr. Corp., B-286226, Nov. 7, 2000, 2000 CPD ¶ 179.

261. 41 U.S.C. § 601.

262. Todd Constr., Inc. v. United States, 85 Fed. Cl. 34 (2008). For further discussion of claims, see Chapter 15.

263. *See* DoD GUIDE, *supra* note 236, app. A.

264. Gulf Group, Inc., B-287697, B-287697.2, July 24, 2001, 2001 CPD ¶ 135.

265. Black & Veatch Special Projects Corp., B-279492.2, June 26, 1998, 98-1 CPD ¶ 173.

266. Walsh Distribution, Inc., Walsh Dohmen Se., B-281904; B-281904.2, Apr. 29, 1999, 99-1 CPD ¶ 92.

267. Hard Bodies, Inc., B-279543, June 23, 1998, 98-1 CPD ¶ 172.

268. B-284176, Mar. 6, 2000, 2000 CPD ¶ 47.

269. Protection Total/Magnum Sec., B-278129.4, May 12, 1998, 98-1 CPD ¶ 137.

270. Oceaneering Int'l, Inc., B-287325, June 5, 2001, 2001 CPD ¶ 95.

271. Ti-Hu, Inc., B-284360, Mar. 31, 2000, 2000 CPD ¶ 62.

272. TRW, Inc., B-282162, B-282162.2, June 9, 1999, 99-2 CPD ¶ 12.

273. GTS Duratek, Inc., B-280511.2, B-280511.3, Oct. 19, 1998, 98-2 CPD ¶ 130.

274. SCIENTECH, Inc., B-277805.2, Jan. 20, 1998, 98-1 CPD ¶ 33.

275. Advanced Data Concepts, Inc., B-277801.4, June 1, 1998, 98-1 CPD 145; *see also* Beck's Spray Serv., Inc., B-299599, June 18, 2007, 2007 CPD ¶ 113.

276. Chemical Demilitarization Assocs., B-277700, Nov. 13, 1997, 98-1 CPD ¶ 171.

277. *Id.*

278. *Id.* (citing E. Huttenbauer & Sons, Inc., B-257778, B-257779, Nov. 8, 1994, 94-2 CPD ¶ 206).

279. Xeno Technix, Inc., B-278738, Mar. 11, 1998, 98-1 CPD ¶ 110.

280. Battele Mem'l Inst., B-278673, Feb. 27, 1998, 98-1 CPD ¶ 107.

281. Oceanometrics, Inc., B-278647.2, June 9, 1998, 98-1 CPD ¶ 159; *accord*, Strategic Res., Inc., B-287398, B-287398.2, June 18, 2001, 2001 CPD ¶ 131.

282. B-279173.5, July 22, 1998, 98-2 CPD ¶ 128.

283. Universal Building Maintenance, Inc., B-282456, July 15, 1999, 99-2 CPD ¶ 32.

284. *Id.*; Ecompex, Inc., B-292865.4 *et al.*, June 18, 2004, 2004 CPD ¶ 149; Perini/Jones, Joint Venture, B-285906, Nov. 1, 2000, 2002 CPD ¶ 68.

285. Femme Comp Inc., 83 Fed. Cl. 704 (2008).

286. Lynwood Mach. & Eng'g, Inc. B-287652, Aug. 2, 2001, 2001 CPD ¶ 138; *see also* Wilson Beret Co., B-289685, Apr. 9, 2002, 2002 CPD ¶ 206.

287. B-282247, Mar. 31, 1999, 99-1 CPD ¶ 69.

288. Wilderness Mountain Catering, B-280767.2, Dec. 28, 1998, 99-1 CPD ¶ 4; *see also* KELO, Inc., B-284601.2, June 7, 2000, 2000 CPD ¶ 110 (GAO "will not infer a condition to a settlement that is not clearly set out in the language of the settlement agreement").

289. AmClyde Engineered Prods. Co., Inc., B-282271, June 21, 1999, 99-2 CPD ¶ 5.

290. *Id.* at n.5.

291. Nova Group, Inc., B-282947, Sept. 15, 1999, 99-2 CPD ¶ 56.

292. B-293026 *et al.*, Jan. 20, 2004, 2005 CPD ¶ 26.

293. 10 U.S.C. § 2305(b)(5); 41 U.S.C. § 253b(e); FAR 15.5.

294. In addition to the information required by FAR 15.503(a)(1), FAR 15.503(a)(2) requires the government to provide additional information to unsuccessful offerors if the government is completing its requirement through one of the many small business programs (*see* FAR Part 19). This additional information includes, when available, the name and address of the apparently successful offeror.

295. FAR 15.503(b).

296. *Id.*

297. FAR 15.505.

298. FAR 15.505(a)(3).

299. FAR 15.505(b).

300. *Id.*; *see also* Global Eng'g & Const. Joint Venture, B-275999, Feb. 19, 1997, 97-1 CPD ¶ 77 (GAO declines to review the CO's decision to delay the debriefing).

301. FAR 15.505(a)(2).

302. *Id.* For further discussion of bid protests, see Chapter 6.

303. FAR 15.505(f).

304. *Id.* For further discussion of FOIA (5 U.S.C. § 552), see Chapter 7.

305. FAR 15.506(a)(1).

306. FAR 15.506(a)(2).

307. FAR 15.506(a)(4) (government accommodation of an untimely request for debriefing or request to delay the debriefing does not automatically extend the deadlines for filing protests). *See also* FAR 33.104(c) (government must stay the procurement when it receives notice of a protest from GAO within 10 days after contract award or within five days after a debriefing date offered to the protester for any debriefing that is required by FAR 15.505 or 15.506); 4 C.F.R. § 21.2(a)(2) (GAO rules mandate that protests shall be filed within 10 days after "the basis of

the protest is known or should have been known (whichever is earlier), with the exception of protests challenging a procurement conducted on the basis of competitive proposals under which a debriefing is requested and when requested, is required. In such cases, with respect to any protest basis which is known or should have been known either before or as a result of the debriefing, the initial protest shall not be filed before the debriefing date offered to the protester, but shall be filed not later than 10 days after the date on which the debriefing is held.").

308. FAR 15.506(e).

309. *Id.*

310. *See* FAR 15.208(e) (proposals may be withdrawn any time before award).

311. *See* FAR 15.307.

312. The government also may make *de minimus* changes to the contract, so long as such changes are minor and have no impact on the relative standings of the offerors. *See* Gulf-Atl. Constr., Inc., B-289032, Jan. 4, 2002, 2002 CPD ¶ 2 (pricing discrepancy of $652.80 on a bid totaling over $5 million regarded as *de minimus*).

313. FAR 14.407-4(c).

314. *Id.*

315. *See* FAR 14.407-4; *Gulf-Atl.*, 2002 CPD ¶ 2.

316. 2002 CPD ¶ 2.

317. *Id.* at 4 (citing C Constr. Co., Inc., B-253198.2, Sept. 30, 1993, 93-2 CPD ¶ 198).

318. FAR 14.407-4(e); *see also Gulf-Atl.*, 2002 CPD ¶ 2; *C Constr.*, 93-2 CPD ¶ 198.

319. FAR 14.407-4(e)(1).

320. *Gulf-Atl.*, 2002 CPD ¶ 2 at 4 (citing *C Constr.*, 93-2 CPD ¶ 198).

321. 41 U.S.C. § 601; FAR 14.407-4; *see* FAR subpt. 33.2.

CHAPTER 4

Architect-Engineer Contracting

DAVID S. HATEM
KENNETH B. WALTON
DAVID H. CORKUM

I. Introduction

The federal government spends billions of dollars every year for architectural and engineering services on construction projects involving federal buildings and other public works. While long ago the government used design-build as its delivery model for design and construction services, as of enactment of the Armed Services Procurement Act of 1947 (the Procurement Act) and the Federal Property and Administrative Services Act of 1949 (the Property Act), construction services could only be procured through competitive bidding. The government did not solicit competitive bids from architects and engineers, however, because it was thought that responding to advertising for competitive bids for professional services was unethical, and conflicted with professional standards established by architectural and engineering professional associations.

Both the Procurement and Property Acts permitted the procurement of design services according to a negotiated process, and ensured price reasonableness of architectural or engineering services by mandating a maximum fee of "6 percent of the estimated cost of the project."[1] This separation—procuring design services through negotiation and construction services through competitive bidding—resulted in the complete separation of design from construction services. Today, however, the pendulum has swung back toward design-build, the use of which is broadly authorized by Federal Acquisition Regulation (FAR) subpart 36.3.

We would like to thank the other Donovan Hatem LLP attorneys who contributed to this chapter, including Patricia B. Gary, Peter Lenart, Matthew Tuller, Gwen Weisberg, Joshua S. Wernig, Sue Yoakum, Steven Wojtasinski, Brian C. Newberry, Sa'adiyah Masoud, Douglas M. Marrano, Jordan S. Rattray, Meghan McNamara, and William D. Gillis, Jr.

By carving out an exception for design services from the general rule requiring price competition, the government recognized that price-based procurement policies are not necessarily well suited for selecting design professionals, because an architect requires special taste, skill, and technical learning to perform its job. This exception was formalized in 1972, when the Property Act was amended by the Brooks Architect-Engineers Act (Brooks Act), clarifying federal policy regarding procurement of architectural and engineering services. The Brooks Act requires that all architect-engineer (A-E) services be procured through a negotiated procedure that focuses on the "demonstrated competence and qualification" of each design professional, and mandates a selection process for federal design contracts where price is not a selection factor.

As initially enacted, the Brooks Act required the government to "negotiate contracts for architectural and engineering services on the basis of demonstrated competence and qualification for the type of professional services required."[2] In 2002, the Brooks Act was revised and recodified at 40 U.S.C. §§ 1101–1105. [3] In its present form, the Brooks Act sets forth the following evaluation and selection procedures:

> (c) **Evaluation.** For each proposed project, the agency head shall evaluate current statements of qualifications and performance data on file with the agency, together with statements submitted by other firms regarding the proposed project. The agency head shall conduct discussions with at least 3 firms to consider anticipated concepts and compare alternative methods for furnishing services.
>
> (d) **Selection.** From the firms with which discussions have been conducted, the agency head shall select, in order of preference, at least 3 firms that the agency head considers most highly qualified to provide the services required. Selection shall be based on criteria established and published by the agency head.

Many states have followed the federal practice by enacting "Baby Brooks Acts."[4] In effect, the combination of the 1972 Brooks Act and the underlying Property and Procurement Acts operated as a statutory bar to other methods of procurement and delivery such as design-build,[5] although as previously noted this is no longer the case.

FAR subpart 36.6 prescribes "policies and procedures applicable to the acquisition of architect-engineer services." In addition to specifying the process of acquiring A-E services under the Brooks Act, FAR subpart 36.6 also establishes policies applicable to the design of federal projects. For example, FAR 36.601-3 provides that for "facility design contracts, the statement of work shall require that the architect-engineer specify . . . use of the maximum practicable amount of recovered materials . . . . Where appropriate, the statement of work shall also require the architect-engineer to consider energy conservation, pollution prevention, and waste reduction to the maximum extent practicable in developing the construction design specifications."[6]

II. The Architect-Engineer (A-E) Selection Process

The selection of a design professional for any public construction project can be an involved undertaking, and it is generally most complex for federal government projects. While the selection process requires a design firm to comply with detailed requirements and protocols, the government typically selects its design services contractors based upon the experience and qualifications of the design team. As in the private sector, design teams that have excelled on earlier projects for a government agency will have an advantage over newcomers.

Design teams are generally ranked on the technical merits of their proposal during the course of a selection process. Beyond the technical component, procurement decisions focus on firm qualifications, and in the final negotiation step fees can play a role.

A. Lack of Price Competition

The Brooks Act compels federal agencies to utilize qualifications-based selection (QBS) in contracting for professional design services.[7] The QBS method is also endorsed by most major design professional and builder associations, the American Bar Association, and the American Public Works Association.

B. Insight into the QBS Process

The QBS process is intended to be uncomplicated and nonconfrontational. If the government determines that a design team it has worked with previously is qualified to work on a pending project, it may work directly with that design team from the outset to attempt to reach an agreement for a project. If no design team with a prior relationship is viewed as qualified, a scope of work is compiled, which may include the need for specialized skill sets and experience.

Design teams deemed qualified by the government may be interviewed. Key factors often include the experience of specific team members intended to work on the project, a company's overall workload, and the ability to work well with the government agency.

Concurrently, the government will supply the design team candidates with more specific information on the project's scope. This is an opportunity for a design team to demonstrate its abilities by helping the government to refine the scope, to offer design themes and practical suggestions beneficial to the project, and to help the government identify requirements that may not yet have been contemplated or explored fully.

C. The Design Professional Contract and Scope of Services

After the selection process has been completed, the contracting officer (CO) must negotiate the terms of the A-E contract.[8] The negotiations begin with the submittal of a proposal from the prospective A-E contractor outlining the terms (including fees) and proposed design methods.[9]

Negotiating a contract for A-E services involves the development of an accurate description of the scope of services and the effect of applicable provisions of the FAR, such as the use of renewable or recoverable resources, energy efficiency, pollution prevention, standard of care, and choice of law.

1. Scope of Services

The design professional's scope of services determines its responsibilities and its potential liability. How the parties define and negotiate that scope of services, including the construction cost limitation, can be central to later disputes. The starting point for the definition of the scope of services should be the publicly announced requirements that are issued by all government agencies prior to the negotiation of A-E contracts.[10] That announcement should set forth all of the requirements for the services, including preliminary technical requirements, expected scope of the work, whether award of a fixed-price contract is anticipated, and express notice that the proposed project must be designed within funding limits. As discussed below, the pre-contract funding limits and related requirements are critical to determining the design professional's potential liability or entitlement to compensation for redesign services when costs exceed the original funding limitations.

The FAR contains specific definitions of such terms as "design," "plans and specifications," and "record drawings,"[11] and these definitions will govern the contract. For example, the term "design" means "defining the construction requirement (including the functional relationships and technical systems to be used . . .), producing the technical specifications and drawings, and preparing the construction cost estimate."[12]

Construction cost estimates play an important role in government contracting. The government uses those estimates to determine if the costs of the project will meet funding limitations and, in some circumstances, to establish the compensation of the design professional.

Indeed, where a cost-plus-fixed-fee contract is used because the project scope is broad and undefined, 41 U.S.C. § 254(b) mandates that design fees will be capped at 6 percent of the original construction estimate furnished by the government.[13] 10 U.S.C. § 2306(d) is a similar limitation applicable to fixed price A-E contracts. Therefore, the A-E needs to be aware of this limitation when developing its proposal, as the scope of the design services should be developed to fit within this limitation.

2. Funding Limits and Fixed Price Contracts

As noted, the government normally requires a design professional to design the project so that construction costs will not exceed a fixed amount, defined as the estimated construction contract price or funding limitation. The required FAR language then states that additional design services necessitated by the construction bids exceeding the estimated construction contract price shall be performed by the A-E at "no increase in price."[14] As such, this limitation effectively requires redesign work to be done as part of the original scope of services.

The sole exception to this provision occurs when the unfavorable bids are "the result of conditions beyond [the design professional's] reasonable control," such as unexpected fluctuations in commodities pricing. The onerous nature of this provision has led to the dual requirements that the applicable funding limitation must be negotiated between the design professional and the government, and that the amount of the funding limitation must be set forth in the contract.[15] The government is required to provide its original estimate and supporting data to the design professional, as a basis for negotiating the agreed funding limitation. The failure to establish a timely preconstruction estimate may void the contract for architectural or design services.[16]

Similarly, when a contractor enters into a cost-plus-fixed-fee contract that includes design services, 41 U.S.C. § 254(b) caps the fee for the design services portion at 6 percent of the estimated construction costs. The government has no authority to exceed the 6 percent cap, and the design professional that performs work under a contract that exceeds the 6 percent limitation does so at its own peril.[17] Design services fees that exceed the cap are not recoverable.

The design professional should also be aware of required FAR provisions such as choice of law,[18] dispute resolution,[19] and differing site conditions,[20] which are required provisions and are not negotiable.

3. Standard of Care

The professional standard of care is one of the most important concepts involved in assessing the potential legal exposure of architects and engineers on any project, and government work is no exception. The legal standard for professional services, including services provided by design professionals, does not require perfection.[21] In *Klein v. Catalano*,[22] a seminal case on the standard of care, the Massachusetts Supreme Judicial Court held that the standard of care required of a design professional is not perfection or even satisfactory results, but only the exercise of skill and judgment reasonably expected from similarly situated professionals:

> As a general rule, an architect's efficiency in preparing plans and specifications is tested by the rule of ordinary and reasonable skill usually exercised by one of that profession. . . . In the absence of a special agreement he does not imply or guarantee a perfect plan or satisfactory result. Architects, doctors, engineers, attorneys, and others deal in somewhat inexact sciences and are continually called upon to exercise their skilled judgment in order to anticipate and provide for random factors which are incapable of precise measurement. The indeterminable nature of these factors makes it impossible

for professional service people to gauge them with complete accuracy in every instance. . . . Because of the inescapable possibility of error which inheres in these services, the law has traditionally required, not perfect results, but rather the exercise of that skill and judgment which can be reasonably expected from similarly situated professionals.[23]

The Court of Federal Claims has similarly defined the standard of care owed by a design professional in fulfilling the terms of a government contract. In *C.H. Guernsey & Co. v. United States*,[24] the court determined that the architect did not breach its professional duty of care or the implied warranty under its contract simply because its design was costly or difficult to implement.

By some assessments, cost overruns on a construction project attributable to design professional errors and omissions may fall within 3 to 5 percent of total construction costs and still be within the acceptable professional standard of care.[25] This range may be higher or lower depending upon the complexity and schedule for an individual project. For example, design-build and fast-track projects typically have a higher level of allowable errors and omissions—6 to 8 percent—due to nontraditional, higher risk procedures used on those projects.

III. Particular Issues in A-E Contracts

A. Site Investigation and Geotechnical Reporting

It is impossible to eliminate all uncertainty regarding subsurface conditions of a particular site. Uncertainty means risk, and the uncertainties associated with providing site investigative and geotechnical services, whether with the government or a private entity, present risks that the design professional must learn to manage.[26]

Defective design claims frequently involve allegations of inadequate site investigations, inadequate surveying services, and/or improper or imprecise reporting of the results of a site investigation. Inadequate site investigation can result in the discovery of concealed and/or buried conditions that should have been identified, and that affect the design and/or constructability of the project. Inadequate surveying services may result in defective plans showing existing elevations higher or lower than actual conditions, resulting in problems such as inaccurate estimates of fill and excavation quantities. Improper or imprecise reporting can lead to confusion within the contract documents that causes the contractor to rely upon a misunderstanding of what was intended, again potentially leading to a defective design claim. These issues have the potential for large losses that may be attributed to the design professional.

Conducting a thorough site investigation is often in tension with budgetary requirements, and the design professional needs to work closely with the government in developing a scope of services for the entire project that enables collecting adequate site information for design and cost estimating. Uncertain site conditions can lead to increased design costs. Cutting corners on the subsurface investigation increases the likelihood of differing site condition claims, and potentially to a need to redesign the project during construction as unexpected conditions are encountered. Cost savings associated with cutting corners during the site investigation phase are almost always outweighed by resulting cost increases and claims.

Proper management of uncertainties associated with site investigations should begin at the earliest stages of a project. The government and design professional should agree on an unambiguous and clear scope of services that defines the responsibilities of the design professional throughout the life cycle of the project. Potential need for construction phase design modifications should be considered to address differing site conditions and/or means and methods of construction not assumed in the original design.

Proper planning and execution of site investigation plans is critical to minimizing uncertainties and avoiding potential exposure. The design professional must approach the site investigation with an aim of foreseeing potential problems and preventing future disputes. This is not an easy task, since site investigations rely primarily on subjective interpretations, often based on limited soil tests and investigations. Publications such as the American Society of Testing and Materials (ASTM) Standard D420-93, *Site Characterization for Engineering, Design, and Construction Purposes*, provide guidance to the design professional for the development and implementation of site investigation plans. In addition, the design professional must consider site- and project-specific considerations in order to properly tailor the investigation.

The first step in a site investigation is performing due diligence on the site, including researching the area and identifying any potential issues that may affect the project design. This can include preliminary site visits, inquiries of local landowners, research at municipal offices, review of geologic maps and soil surveys, and reports of subsurface investigations nearby. The level of due diligence required should be project-specific and is dictated by standard industry practices.

Next, the design professional will design a site investigation plan that is consistent with a reasonable level of accuracy in characterizing the site. The site investigation may include soil borings, installation of monitoring wells, test pits, geophysical surveying (e.g., ground-penetrating radar), surface surveys, sampling, and laboratory testing. The spacing and depth of the investigations should be determined by the geologic complexity and the particular design requirements.

Subsurface investigations are more art than science. Proper training and experience of field personnel is key to proper identification and representation of soil conditions and absolutely critical in managing the risks associated with site investigations. Subsurface investigations require a large degree of skill, judgment, experience, competence, and intuition to conduct and interpret.

Accurate and consistent soil classification and recording of observations in the field log are absolutely necessary.

Finally, the site investigation report interprets and reports the results. The report should detail all locations of borings, test pits, and samples; describe the investigative procedures; provide all logs; classify soils; and locate groundwater. It should also provide proper disclaimers disclosing the facts considered and the basis for the information provided and qualify the opinions and recommendations provided in the geotechnical report.

B. Building Information Modeling and Integrated Project Delivery in Government Contracts

The government is committed to utilizing recently developed tools such as Building Information Modeling (BIM) and is experimenting with integrated project delivery (IPD) concepts in its quest to build more quickly and efficiently.

Both the General Services Administration (GSA) and the Army Corps of Engineers (Corps) have committed to the use of BIM by architects, engineers, and contractors working on their projects. The GSA has published two key documents on BIM: *GSA Building Information Modeling Guide Series 01—Overview* and *GSA Building Information Modeling Guide Series 02—Spatial Program Validation*, and is specifying the use of BIM in some contracts.[27] The Corps has specific BIM requirements in certain of its contracts as well, which specify BIM hardware and software, and drawing requirements for BIM models.

The GSA is exploring use of BIM in spatial program validation, 4-D phasing, laser scanning, energy and sustainability, circulation and security validation, and building elements.[28]

GSA is focusing on the use of BIM by architects and engineers, as well as contractors and subcontractors. The 4-D aspects of BIM include using the BIM model to simulate the project construction schedule and sequencing, to allow the contractor to evaluate and optimize the construction sequence and virtually plan the construction before it starts.

Many large-scale private projects utilize BIM in order to better understand design, cost, and schedule very early in the design. BIM is very effective in its clash-detection ability, whereby the various design disciplines run conflict checks to understand spatial conflicts in the design. These same conflict checks can be run on the shop drawings when produced in BIM.

In addition, the government recognizes that BIM can be very helpful as a facility management tool. This requires the use of BIM and a BIM facility management program in the documentation of the construction markup set and preparing as-built documents that reflect information that is required for the government to manage the completed building. For example, the BIM facility management model should have detailed information about the products used and installed in the building, which can give the government specific information relating to objects within the model such as light fixtures and enable better predictions of bulb life. GSA clearly wants to take advantage of these elements of BIM.

IPD is a project delivery method that envisions the owner's sharing the risk to deliver the project on time and on budget. In view of this assumption of greater owner risk, it is unclear if the government will implement a true IPD project delivery method. But it is clear that agencies such as the Corps are exploring advanced project delivery methods. For example, the Corps is utilizing both Multiple Award Task Order Contracting (MATOC) and Early Contractor Involvement (ECI) on some projects. ECI uses a general contractor to work with the design team to develop cost and schedule estimates during the design phase of the project, rather than waiting until design completion, and is being implemented on selected large-scale construction projects.

IV. Guidelines for the Government Cost Estimate

As previously noted, the government's estimate of construction costs is very significant to the design professional in that it typically forms the basis for the funding limitation incorporated into the design services contract. As such, it is important for designers to understand the applicable requirements for arriving at this cost estimate. The government cost estimate also is critical to determining whether sufficient public funds are available to pursue the project.

The FAR, however, provides little guidance on preparation of government cost estimates. FAR 36.203 requires the government estimate to be prepared as though the government were competing for the award of the contract. The final cost estimate is then used to determine fair and reasonable prices and completeness of the offering at the time of bid. No guidelines, however, are referenced or provided to implement these requirements.

A. DOT Guidance

The Department of Transportation (DOT) provides guidance to assist state transportation agencies in developing reliable engineer's estimates to ensure a balanced assessment of construction bids, and to provide accurate estimates for the state agency financial planners. While not applicable to the federal government itself, these guidelines are illustrative of appropriate practices for preparing government construction cost estimates.

The DOT guidelines briefly describe the "Actual Cost Method," the alternative "Historical Data Approach," and a combined option that uses historical bid data in conjunction with actual cost data. DOT also notes the importance of credibility of the estimate within the bid process, while acknowledging that estimating is not an exact science. It cites a range of being within plus or minus 10 percent of low bid for 50 percent of the projects as a reasonable goal. DOT also provides a review guide for assessing an agency's procedures for developing the engineer's estimate.[29]

B. The 2007 GSA Project Cost Estimating Guide[30]

Prepared as a collaborative effort among the GSA Office of Chief Architect, its regional Public Buildings Service cost experts, and an outside consultant, this guide seeks to establish comprehensive cost-estimating and cost-management criteria for GSA's Public Buildings Service. It provides a matrix of estimating formats, project cost and building cost worksheets, tracking sheets, lists of useful data tables and cost calculation specifications, contingency guidelines, and sample figures derived from two broadly used estimating system tools—the Construction Specifications Institute (CSI) Masterformat and the GSA Uniformat design development reports.

Using standard estimating formats, the Guide provides advice on how to (1) ensure a uniform cost-control framework; (2) define the proper level of detail for setting expectations for the estimate; and (3) create a scope of work checklist. The Guide also includes a standardized historical database.

The GSA Guide also provides useful guidance on contingency amounts to be included in cost estimates, with the recommendation to start at 10 percent for site and design in the planning stage, and reducing the contingency amount thereafter in each estimate to cover incomplete design, unforeseen or unpredictable conditions, and other uncertainties. A useful table is provided to guide the contingency development plan, and a construction contingency of 7 percent is recommended for new construction and 10 percent for renovations, subject to a project-by-project determination.[31]

For preliminary concepts and concept design, the Guide notes that the estimator should calculate quantities for appropriate systems or apply parameters to each building area, including use of unit prices that combine labor and materials. Backup worksheets must support detailed estimates covering all cost-sensitive project data and document all major assumptions. The A-E is required to provide the government estimator advance copies of required concept estimates and other documentation to allow for preparation of the estimate.[32]

For the A-E contract itself, note also that FAR 36.605 requires an Independent Government Estimate for A-E services related to construction that will exceed $100,000. Per FAR 36.605, this estimate is used in negotiating a fair and reasonable fee with the selected A-E in the final step of the Brooks Act procurement process.

V. Construction Phase Services

A. Construction Phase Services by the Design Professional

The level of construction phase services is generally articulated in the designer's agreement with the government. The government does not use standard form contract documents such as American Institute of Architects (AIA) or Engineers Joint Contract Documents Committee (EJCDC) form documents in its contracting, and the FAR standard form clauses do not address in detail defining the scope of the design professional's construction phase responsibilities. Absent clear contractual language, design professionals must rely on analogies to common law interpretations of their role.

While it is the construction contractor's responsibility to perform the work in accordance with contract documents, the central issue regarding site visits during construction by the design professional is whether the design professional should have detected and either reported or prevented a construction defect that later comes to light. At common law, a key decision regarding the design professional's potential liability for its site observations, as compared to a more extensive clerk-of-the-works type of continuous presence and inspection services, is *Watson, Watson, Rutland/Architects, Inc. v. Montgomery County Board of Education.*[33] In that case, the owner had declined to exercise a contractual option whereby the design professional would provide clerk-of-the-works inspection services, but still attempted to hold the designer liable for not detecting contractor defects that led to a leaking roof. The *Watson* court found for the architect, but also noted that a design professional has a duty to report known deficiencies to the owner and cannot not turn a blind eye to contractor failings that might affect the finished project. This remains the basic standard to which the design professional will be held in making site visits.

Another aspect of the construction phase where design professionals may be exposed to professional liability claims from the government is if they fail to timely review a contractor's submittal and that failing results in contractor delay claims paid by the government. By requiring that the contractor submit a schedule of submittals, a design professional can avoid any delay allegations caused by contractor deviations from that schedule. However, government construction contracts do not require such a schedule by their standard terms, although the designer can request that the government add such a requirement. In any event, it is good practice for the design professional to maintain a log documenting dates of receipt and review return of contractor submittals.

B. Construction Management Services

The government sometimes retains the services of a construction management professional to assist its administration of complex projects. Indeed, the use of independent construction managers has expanded enough over the last decade that a new Special Item Number (SIN) 871-7, Construction Management (CM), has been added to the professional engineering services schedules in order to simplify and expedite the retention of these services. CM services are not meant to displace the construction phase services of the lead design professional. Rather, agencies utilizing CMs are essentially outsourcing some of the agency's role in administration and management of the project.

The role of a CM on a federal project is that of an advisor to the government, much like what would be known in the private sector as an "agency" CM. Decision-making power remains with the government, which may or may not rely on the CM's recommendation and advice. The CM cannot have had a stake in the design process or any financial involvement with the construction contractor. As such, the CM is supposed to provide the government with unbiased advice in the best interest of the project. Government agencies have found CM services to be valuable on both design-bid-build and design-build projects, and can be used at any phase within the development of the project. The menu of possible Construction Manager/Program Manager services anticipated under SIN 871-7 includes the following:

- Programmatic/planning services, including planning feasibility studies, economic studies, environmental site assessments, site surveys, or preparation of budgets and cost estimates, all of which are devised to support the government agency's decision making in scoping and planning the design and construction of the project. There may or may not be a conceptual design for the project developed by others at this point in the planning.
- Design phase services, including tasks such as design management, technical review of designs, code compliance reviews, constructability reviews, value engineering services, cost estimating and analysis, cost control monitoring, energy studies, site investigations including subsurface characterization and identification of hazardous material, site surveys, and general liaison services with the lead design engineer or architect and construction contractor, if selected at this point in time.
- Procurement phase services, including assistance to the CO in answering bid questions, attending, participating in, and organizing site visits and pre-bid conferences, and preparing and issuing solicitation documents in performing bid analyses.
- Construction phase services, including all tasks necessary to establish an organization capable of monitoring contractors' construction activities and administering the contract on behalf of the government agency. The CM's role will be defined by the consultant's agreement with the government agency. It is important that the scope of services be clearly articulated in that agreement and that the construction contractor's contract acknowledges the status of the CM.
- Commissioning services, including assistance for tasks like start-up planning and testing, and performance testing.
- Testing services, including specialty quality control, concrete and soil testing, and equipment installation tests.
- Claim analysis services, including analysis of contractor claims and negotiating assistance.
- Post-construction services, including outfitting, furnishing or planning the use of the facility, commissioning, and move-in and setup.

To the extent that the required services do not require performance by a registered or licensed architect or engineer (notwithstanding that architect-engineers also perform those services), the procurement process specified in the Brooks Act and FAR subpart 36.6 do not apply. This would be the case for some, but certainly not all, of the services contemplated by SIN 871-7, so a case-by-case analysis is required. When the required services are not "architect-engineer services" as defined in FAR 36.601-3 and -4, then they are procured by the normal methods of sealed bidding or contracting by negotiation (governed by FAR Parts 14 and 15, respectively).

VI. Government Claims Against the Design Professional—Cost Recovery Programs

Also of great interest to A-E contractors are the standards utilized by the government when determining whether to pursue a claim and cost recovery action against a design professional. In general, the causes of building performance problems, such as a leaking roof or windows, can be grouped into four categories: (1) design deficiencies, whether in architectural design or designation of improper materials; (2) material deficiencies (inadequate performance even when properly installed); (3) construction deficiencies (poor quality or substandard workmanship); and (4) subsurface/geotechnical problems, such as expansive soil conditions.[34] When a design deficiency has been identified, this may then trigger a cost recovery effort by the agency involved. Cost recovery actions may be asserted against all members of the project design team in an effort to recover costs from a potentially responsible party, and to mitigate increased costs and allocate risk on a project.[35] Cost recovery procedures have in the past been developed on a case-by-case basis, guided by contract language.[36] Recently, however, some government agencies have developed formal cost recovery processes.[37] A cost recovery procedure is the process by which the government's cost recovery action will be conducted and resolved, whether short of litigation or requiring litigation.[38]

Simply stated, a cost recovery action is a claim, which can be presented either in a notice letter or in the form of a CO's final decision. Typically, the agency asserts a cost recovery claim against the design professional in charge of the project, with whom the government has privity. The lead designer then has the option of passing the claim through to its subconsultants as indemnification claims.[39] In 2002, the Federal Highway Administration (FHWA) promulgated regulations requiring all state and local departments of transportation that are responsible for procuring engineering and design services to implement written procedures to determine "the extent to which the consultant, who is responsible for the professional quality, technical accuracy, and coordination of services, may be reasonably liable for costs resulting from errors or deficiencies in design furnished under its contract."[40] As a result, state transportation departments have been implementing formal cost recovery procedures.[41] In March 2009, the American Association of State Highway and Transportation Officials (AASHTO) published a study of 12 such procedures implemented and the process recommendations of the American Counsel of Engineering Companies (ACEC) (AASHTO study).[42]

The AASHTO study points out that the purpose of a cost recovery procedure is to develop a method of addressing errors and omissions fairly while keeping the project on track through (1) early notification of a design problem; (2) participation of the consultant in developing potential solutions; (3) agency analysis while maintaining communication with the consultant; (4) analysis of the benefit of potential cost recovery; (5) multiple levels of negotiation, review, decision, and appeal; (6) alternative dispute resolution (ADR) procedures; and (7) documentation of key events, decisions, and communications.[43] These steps are similar whether governed by a formal process implemented by an agency or by the contract between the parties.

The table below summarizes the steps of the cost recovery procedure recommended by the AASHTO study.[44]

Step	Procedure	Actions Taken
1	Discovery	An issue or problem is discovered and an initial review takes place. An initial determination of cause is made (e.g., did the issue arise out of a design deficiency or out of a construction error?).
2	Initial notification	Consultant is notified of issue identified and responds to the notice.
3	Investigation and decision on liability	Agency and consultant discuss the issue. Agency investigates the likelihood of an error and/or omission giving rise to the issue and whether the consultant was negligent. Other factors are considered when assessing potential liability. A damages review is conducted. The cost-effectiveness of pursuing the issue is examined. Legal principles are considered before deciding to proceed.
4	Notification of decision	Agency decides whether to pursue a cost recovery action against the consultant and informs the consultant of its decision.
5	Review of decision	The decision is reviewed with the assistance of outside technical experts, if needed. A panel makes a decision regarding liability and potential damages. Decision is communicated to the consultant.
6	Alternative dispute resolution	Parties engage in an ADR procedure (e.g., negotiations, mediation, or arbitration).
7	Recovery and collection	If the parties resolve the matter through ADR, restitution is made and releases are executed.
8	Litigation	If the parties do not resolve the matter through ADR, the agency may initiate litigation against the consultant.

To avoid ambiguity, the terms relevant to the cost recovery procedure should be clearly defined.[45] The primary goals of a cost recovery procedure should be to ensure that the project program requirements are met and to ensure the quality of the completed project. An effective procedure will allow for identification, discussion, and resolution of issues as they arise during construction in order to keep the project on track. In contrast, a procedure implemented ineffectively will hinder open dialogue during construction and will foster a culture of finger-pointing and refusal to take responsibility on all sides. The AASHTO study, while not directly applicable to the federal government, nevertheless is indicative of the principles generally followed by federal agencies in determining whether to pursue cost recovery from a design professional.

VII. Conclusion

Architect-engineer contracts enjoy an almost unique exemption from the strong governmental preference for price competition in awarding government contracts. Nevertheless, the contracting process for design services is highly regulated by statutory and FAR provisions. 41 U.S.C. § 254(b) and 10 U.S.C. § 2306(d) continue to cap the fee the government may pay for design services to a fee "not in excess of 6% of the estimated cost, exclusive of fees, . . . of the project to which such fee is applicable . . . in contracts for architectural or engineering services relating to any public works or utility project." Additionally, the government must prepare an independent estimate of the cost of the design services for use in negotiating a "fair and reasonable" fee not exceeding this cap, once the A-E firm has been selected for contract award. Moreover, the government must establish an estimate of construction costs and agree with the A-E on such an estimate to insert into the contract as the applicable funding limitation. The A-E will then be responsible for modifying the project design at no added cost to the government if the construction bids received are above this limitation, absent causes beyond the A-E's reasonable control.

Notes

1. *Id.* at 476. 41 U.S.C.A. § 254(b); 10 U.S.C. § 2306(d).
2. 40 U.S.C. § 542.
3. 40 U.S.C. § 1101 (formerly 40 U.S.C. § 542).
4. Construction Law (William Allensworth, Ross J. Altman, Allen Overcash & Carol J. Patterson eds., 2009).
5. Fluor Enters., Inc. v. United States, 64 Fed. Cl. 461, 483.
6. 48 C.F.R. § 36.601-3.
7. 40 U.S.C. §§ 1101–1105.
8. FAR 36.606.
9. FAR 36.606(b).

10. FAR 36.601-1.

11. FAR 36.102.

12. *Id*.

13. 41 U.S.C. § 254(b).

14. FAR 52.236-22.

15. FAR 36.609-1(b), 52.236-22.

16. Fluor Enters., Inc. v. United States, 64 Fed. Cl. 461, 491–92 (2005).

17. *Id*. at 492–95. For further discussion of CO authority issues, see Chapter 7.

18. FAR 52.233-4.

19. FAR 52.233-1.

20. FAR 52.236-2.

21. "A design professional must exercise reasonable care, technical skill and ability, and diligence, as is ordinarily required of similar designers in the locality in preparing plans and specifications." P.L. BRUNER & P. J. O'CONNOR, JR., BRUNER & O'CONNOR ON CONSTRUCTION LAW § 17:40, 631–32 (2002). "In the absence of an express agreement, 'an architect, like a physician or lawyer, does not guaranty, imply, or warrant a perfect plan, or favorable or satisfactory results. It follows that an architect's work can be inaccurate or imperfect without being an actionable deviation from the standard of care.'" 6 C.J.S. *Architects* § 12 (2004).

22. 386 Mass. 701, 437 N.E.2d 514 (1982).

23. *Id*. at 718.

24. 65 Fed. Cl. 582 (2005).

25. *See* WALLER S. POAGE, THE BUILDING PROFESSIONAL'S GUIDE TO CONSTRUCTION DOCUMENTS 40 (2000) (referencing National Research Council and Construction Industry Institute).

26. For further discussion of site condition issues generally, see Chapter 9.

27. GSA BIM Guide Overview, GSA BIM Guide Series 01, *available at* http://www.gsa.gov/bim; GSA BIM Guide for Spatial Program Validation, GSA BIM Guide Series 02, *available at* http://www.gsa.gov/bim.

28. The GSA is publishing the following BIM Guide Series: (1) Series 01—BIM Overview, (2) Series 02—Spatial Program Validation, and (3) Series 03—3D Imaging. Under consideration and in formulation are: (1) Series 04—4D Phasing, (2) Series 05—Energy Performance, (3) Series 06—Circulation and Security Validation, and (4) Series 07—Building Elements.

29. The attachment consists of 20 interrogatories or inquiries intended to expose the areas of research and procedures that attest to the state agency's quality of procurement systems, including a wide range of issues such as the existence of state law regarding release of estimates, establishing competitive bid procedures, methods for establishing unit prices, personnel assigned to the process, pricing data collection efforts, and ground rules for variant or insufficient bids.

30. P-120 Project Estimating Requirements for the Public Building Service, *available at* http://www.wbdg.org/ccb/gsaman/p120.pdf.

31. P-120 Project Estimating, Section 1.4 at 12–17.

32. P-120 Project Estimating, Section 2.2 at 29–31.

33. 559 So. 2d 168 (1990).

34. Consultants in Risk Management, Construction Defect, http://www
.c-risk.com/Construction_Risk/CR_CDs_01.htm.

35. David J. Hatem, *Errors/Omissions Cost Recovery Claims Against Design and
Construction Management Professionals*, CA/T PROF. LIABILITY REP. 1 (June 1996).

36. MICHAEL J. MARKOW, BEST PRACTICES IN THE MANAGEMENT OF DESIGN
ERRORS AND OMISSIONS at 1 (2009), *available at* http://www.transportation.org/
sites/design/docs/NCHRP%20Project%2020-7,%20Task%20225,%20Best%20Prac
tices%20in%20the%20Management%20of%20Design%20Errors%20and%20Omis
sions,%20Final%20Report%20March%202009.pdf (report prepared for the American Association of State Highway and Transportation Officials (AASHTO) Standing Committee on Highways).

37. *Id.*

38. ROBERT CERASOLI, OFF. OF THE INSPECTOR GEN., COMMONW. OF MASS., A
REVIEW OF THE CENTRAL ARTERY/TUNNEL PROJECT COST RECOVERY PROGRAM
(2000), *available at* http://www.mass.gov/ig/publ/catcrrpt.pdf.

39. The decision to assert a claim, including a pass-through claim, should be
carefully considered. There are a lot of factors to be considered before a claim is
passed through to a subconsultant (e.g., whether a joint defense approach would
benefit both members of the design team, what the ongoing business relationship
between the parties is, and what the effect of the claim would be).

40. 23 C.F.R. § 172.9(a)(b). The term "consultant" is defined in 23 C.F.R. § 172.3.

41. MARKOW, *supra* note 36.

42. The study offers a thorough discussion of cost recovery procedures of
state DOTs. The objectives of the study are stated as (1) to identify and describe
policies and practices now in place in transportation agencies nationwide that
effectively manage design errors and omissions; (2) to provide AASHTO's Preconstruction Engineering Management Technical Committee source material and a
point of departure to develop and publish recommended practices for AASHTO's
membership; and (3) to assist state DOTs, other transportation agencies, and the
private sector in understanding and adopting management and quality-based
practices that minimize and mitigate design errors and omissions.

43. *Id.* at 2–3.

44. MARKOW, *supra* note 36, at 98–100.

45. MARKOW, *supra* note 36, at 95.

CHAPTER 5

Alternate Delivery Systems: Design-Build, Construction Management, and IDIQ Task Order Contracts

MICHAEL C. LOULAKIS
LAUREN P. McLAUGHLIN
DONALD A. TOBIN

I. Introduction

This chapter examines three alternate delivery systems: design-build, construction management, and indefinite-delivery indefinite-quantity (IDIQ) contracts. Design-build contracts are utilized when a government agency has identified a specific need and desires one contract to provide both design and construction services to fulfill that need. Construction management contracts are also used when a specific need has been identified, but two separate contracts for design and construction services are desired. IDIQ contracts, by contrast, are used when an agency cannot identify the exact time and/or quantity of future construction projects. The agency awards an umbrella contract to one or more construction contractors; then, as the need arises, the agency places task orders for specific projects.

II. Design-Build

Fueled by legislative procurement reform and agency reevaluation of acquisition policy, the use of design-build by federal agencies has grown at a rapid pace. Federal agencies, like other public and private sector owners, have tired of the construction cost overruns, delays, and project acrimony associated

---

Michael C. Loulakis and Lauren P. McLaughlin have authored the design-build section of this chapter. Donald A. Tobin has authored the sections on construction management and indefinite delivery/task orders.

with the design-bid-build method of project delivery. The one attribute that differentiates design-build from other project delivery systems—the direct contractual relationship between the designer and contractor—has helped create what many federal agencies perceive as a "better mousetrap" in terms of project delivery.[1]

Despite the strong growth in the federal design-build market, best practices are still being developed and refined. Even though procurement and contracting for design-build services have been eased by legislative reform, many agencies still find these processes challenging. Agency personnel also find it difficult to understand the different roles they play in the execution of the project, particularly how much control they have the right to exercise.

The design-build section of this chapter will first review the historical evolution of federal law as it pertains to construction project delivery and the creation of today's federal design-build market. It will then examine how some agencies have used design-build on recent projects. Finally, it will review some of the more challenging issues agencies have with successfully executing the procurement and contracting of a design-build project.

A. Historical Overview of Federal Project Delivery

Many construction practitioners still think of public sector construction in terms of design-bid-build. Yet, through most of our country's history, virtually every major infrastructure project was delivered either through design-build or design-build-operate.[2] Therefore, it is critical to understand how the government came to shift away from an integrated project delivery system like design-build in favor of a segmented, sequential delivery system like design-bid-build.

Beginning in the early to mid-1800s, the construction industry began to separate the design and construction processes.[3] Much of this can be attributed to the impact of the industrial revolution, which caused productivity and technical efficiency to grow dramatically, in part through the systematic application of scientific knowledge to the manufacturing process. This placed new and increased demands on designers and builders. Because of the relative complexity of the new industrial facilities, design expertise and specialization were required of designers, but not to the same degree from builders.

Further creating a "separationist" philosophy was the growth of professional organizations in Europe and the United States due to increasing need for specialization. These organizations—including the American Society of Civil Engineers (ASCE) and the American Institute of Architects (AIA)—enabled their members to regulate practice and discuss and advance their knowledge base. It also enabled their members to distinguish themselves from craftsmen and trade groups, thereby increasing their social stature.[4]

In 1893, Congress permitted the Secretary of the Treasury, who controlled appropriations for the construction of federal buildings, to separately procure design and construction services.[5] A little over two decades later, Congress provided the first cash grants to state highway departments to improve rural post roads, conditioned on the advance federal approval of complete plans and specifications for each project.[6] This effectively began the concept of design-bid-build for federal construction projects.

The design-bid-build process picked up momentum in the early to mid-1900s, with the passage of the Public Buildings Act in 1926, which mandated that design be completed before construction.[7] This was followed by the Armed Services Procurement Act of 1947 (ASPA)[8] and the Federal Property and Administrative Services Act of 1949 (FPASA).[9] Both mandated that the federal government procure goods and services through a competitive process, using advertising to ensure full and free competition. Interestingly, however, the acts carved out an exception for design services, permitting the award of architectural and engineering contracts for the design of public works without negotiation or prior advertising. As a result, agencies that opted for non-competitive design procurement were nevertheless forced by this legislation to use competition for construction services. This induced many agencies to use a design-bid-build approach.[10]

Notwithstanding the legislation leading agencies to choose design-bid-build, design-build remained an option for the government. This was largely because the ASPA and FPASA only permitted, but did not mandate, the selection of design services through noncompetitive means. In fact, prior to 1972, there was no legislative requirement that complete plans and specifications be prepared for projects as a condition for entering a price competition for the construction component of federal projects.[11] The government continued to have the ability to solicit proposals to design, construct, finance, maintain, and operate specific infrastructure projects based on schematic designs only.[12] However, it appears that few projects were actually being delivered through a design-build process during the 1900s, perhaps because of the preference by agencies in using price competition as a basis for selecting construction contractors.

The federal construction project delivery environment changed dramatically with the 1972 passage of the Brooks Architect-Engineers Act (Brooks Act). The Brooks Act mandated that design professionals be selected based on qualifications, with the best-qualified firm negotiating with the government to reach a fair and reasonable contract price.[13] Once the Brooks Act was enacted, agencies were precluded from using design-build, design-build-operate, or other systems that combined design and construction services, since the procurement requirements for design and construction services had different standards.[14] This effectively "sealed the deal" for design-bid-build to be virtually the only method for delivering federal construction projects.

The Federal Acquisition Regulation (FAR),[15] which currently embodies the government's procurement policy, reinforced the requirement for design-bid-build of construction projects. The FAR mandates that construction services be procured on the basis of sealed, firm, fixed-price bids. It also contains regulations that implement the Brooks Act selection processes for design professionals.[16] These are accomplished through FAR 36.103, which states:

> Methods of Contracting. (a) Contracting officers shall acquire construction using sealed bid procedures . . . except that sealed bidding need not be used for construction contracts to be performed outside the United States, its possessions or Puerto Rico. (b) Contracting officers shall acquire architect-engineer services by negotiation, and select sources in accordance with applicable law.

The above-referenced legislation, reinforced by the government's tendencies to secure "competition" for construction contractors by an open, lowest-price bid philosophy, resulted in design-bid-build becoming virtually the only method of delivering federal projects.

Some believed that the Competition in Contracting Act of 1984 (CICA)[17] would assist in the effort to broaden the government's construction procurement philosophy and options. While CICA did put the sealed-bid method of contractor selection on par with competitive negotiation, it still expressed a general preference for competitive sealed bid procedures. Moreover, the government largely used CICA for defense contracts, rather than construction projects.

A major problem with CICA was the manner in which the government conducted the negotiation process. It was applied by federal agencies to allow even marginal proposals to be considered as being in the "competitive range," in the name of promoting full and open competition. This resulted in proposers spending substantial sums of money to compete on a project where there was, for all practical purposes, no short-listing. Although legislative initiatives were made in the early 1990s to facilitate short-listing (including the Federal Acquisition Streamlining Act of 1994),[18] none had any material effect on the construction industry.[19]

B. Early Use of Design-Build in the Federal Sector

Despite the restrictive legislation and the pragmatic effect of the government's disinterest in integrated construction project delivery, some special legislation was enacted in the mid- to late 1900s that allowed the use of design-build. For example, legislation enacted in the 1940s enabled the Naval Facilities Engineering Command to use design-build on housing projects. NASA began using design-build in 1962, followed by HUD's use of turnkey design-build for housing projects in 1968, and the Military Construction Authorization Act of 1986. The 1986 Act permitted each branch of the military to use design-build on three pilot projects, with a one-step turnkey selection procedure.[20]

As indicated above, several agencies used design-build under CICA, even though its use was sparse. One of the first reported government uses of design-build occurred in 1989 when the General Services Administration (GSA) began using the process for small office building projects, believing it to be the best process for these simple projects. GSA procured these contracts by first providing a request for proposals (RFP) that included performance specifications and conceptual drawings, then evaluating the proposers' qualifications and preliminary price proposals, and finally awarding through a competitive negotiation process.[21]

In 1991, GSA issued its *Design-Build Request for Proposal Guide* to enable it to continue using design-build.[22] One of GSA's more notable design-build projects was Foley Square, which it awarded in April 1991 and involved the design-build of a federal office building and U.S. courthouse in lower Manhattan.[23] While other agencies followed suit, the use of design-build was limited and often confined to special projects.[24]

C. Design-Build Legislation in the 1990s

For the reasons discussed above, the use of design-build in the federal sector—and indeed in state and local public sectors as well—was pretty much dead in the water until the mid-1990s. However, two major legislative efforts revived the process, creating the current robust federal environment for design-build services. As discussed below, these two legislative efforts involved the passage of the Clinger-Cohen Act and the rewrite of FAR Part 15.

1. The Clinger-Cohen Act

In 1994, major design and construction industry associations led by Preston Haskell, Chairman of the Design-Build Institute of America (DBIA) and one of DBIA's founders, formed a coalition to work with the GSA and the Army Corps of Engineers to build consensus on the adoption of legislation allowing federal agencies to have broader discretion to consider design-build, using a procurement approach that would be regarded as positive by both government and industry.[25] The result of these efforts was the codification of a two-phase procurement process, passed in February 1996, known as the Clinger-Cohen Act.[26]

The Clinger-Cohen Act did not require federal agencies to use design-build. Rather, it allowed them to do so when the head of the agency determines that a project is appropriate for the use of design-build.[27] The Act gives guidance as to the factors an agency needs to consider in determining when design-build may be appropriate.[28] These factors include:

- The contracting officer (CO) must anticipate that three or more offers for the contract will be received;
- Design work must be performed before a proposer can develop a price or cost proposal for the contract; and
- Proposers will incur significant expense in preparing the offer.

Other factors that the CO should consider include information such as "[t]he extent to which the project requirements have been adequately defined; [t]he time constraints for delivery of the project; [t]he capability and experience of potential [proposers]; [t]he suitability of the project for use of the two-phase selection procedures; [and] [t]he capability of the agency to manage the two-phase selection [procedures]."[29]

If the determination favors design-build, then the Act defines the two-phase selection procedure for procuring a design-build contract.[30]

Much of the Clinger-Cohen Act was developed to avoid the pitfalls that befell agencies and proposers under design-build competitions awarded under CICA. As a result, the essence of the two-phase process is to balance the interests of those proposing to be design-builders (which do not want to spend substantial sums of money competing for a project that will be open to many proposers) and the government (which needs some form of competition). The two-phase approach was also intended to alleviate concerns that the design-build procurement would be based entirely on price and would not adequately factor in a contractor's technical qualifications.

Phase One of the Act involves the submission of statements of qualifications demonstrating each proposer's specialized experience and technical competence. The agency must include in the solicitation a scope of work statement with a sufficiently detailed description of the project to put the offerors on notice of the information the agency needs to evaluate the proposals.[31] It is permissible for (and in fact typically expected that) the agency will contract with an outside firm to prepare this statement.[32] If the agency decides to contract for the preparation of this statement, it must do so in accordance with the Brooks Act.[33]

The Phase One solicitation must set forth the factors under which the proposals will be evaluated, including (1) experience and technical competence; (2) capability to perform; and (3) past performance of the proposed design-build team, including the design and construction members of the team.[34] In addition, the solicitation must explain the relative importance of each factor, and the agency must evaluate the factors in accordance with the solicitation. The Act specifically precludes cost from being considered in Phase One, and the proposal cannot contain any such cost or price information.[35]

The agency reviews the proposals submitted in response to the Phase One solicitation and selects a short list of proposers that will be permitted to participate in Phase Two. The maximum number of offerors selected to proceed to Phase Two may not exceed five, unless the agency determines that allowing more is in its best interest and consistent with the two-phase selection procedure.[36]

A second round of proposals for Phase Two is submitted by the short-listed firms, with the proposals including (1) the technical submission for the proposal, including design concepts and/or proposed solutions to the requirements addressed in the scope of work; and (2) the evaluation factors and subfactors, including pricing information, that will be considered when the agency evaluates the proposals.[37]

Following the passage of the Clinger-Cohen Act, the FAR was amended to incorporate and add detail to the Act's design-build procedures. C.F.R. sections 36.102 through 36.104 and sections 36.300 through 36.303 outline the process of soliciting and evaluating design-build proposals. The design-build regulations were first published for comment on August 7, 1996, and became final on January 1, 1997.[38]

The impact of the Clinger-Cohen Act on federal construction has been substantial. First, it created a great deal of interest in design-build and, for the first time, gave the government a process to eliminate marginal proposals and teams. It also gave comfort to government contractors that they could, at very low expense, submit proposals to become part of the short list and, if they made the short list, that there would be a limited number of competitors— justifying the more substantial investment in time and money to respond to the RFP. This two-phase process not only created a surge of interest by federal agencies in design-build but also became the catalyst for many state and local governments to adopt similar two-phase legislation, thereby increasing the use of design-build in these sectors as well.[39]

2. FAR Part 15 Rewrite

The FAR contains guidelines for procuring design-build services and integrated finance design-build-operate services by means other than the two-phase selection process developed through the Clinger-Cohen Act. Procedures that are used today, and were used prior to the passage of the two-phase legislation, are generally set forth in FAR Part 15, entitled "Contracting by Negotiation." FAR Part 15 contains, among other things, the protocol for source selection, proposals, evaluations, and award. It describes concepts such as "competitive range," "best and final offers," and "best value." It is also important to remember that the Clinger-Cohen Act itself does not identify how the government is to conduct the evaluations of Phase Two proposals. Rather, this is set forth in FAR Part 15.[40]

D. Agency Use of Design-Build in the New Millennium

Legislative activities in the 1990s (such as the Clinger-Cohen Act and the modifications to FAR Part 15) helped spur federal agencies to start using design-build.[41] At present, the reliance on design-build continues to grow at an ever-increasing rate, as federal agencies have turned to the design-build delivery method for some of its most important projects triggered by national and international events (e.g., the country's military operations in Iraq and Afghanistan, Hurricane Katrina, and the 2008 economic downturn and stimulus recovery of 2009).

While it is beyond the scope of this chapter to review the nature of the projects and the types of procurement approaches used on a detailed, agency-by-agency basis, it is fair to say that design-build is being used across the board by federal agencies. What follows is a representative sample of the different types of projects that are selected for design-build contracting.[42]

1. Army Corps of Engineers (Corps)

The Corps has been a major user and advocate of design-build over the past decade. While the number of projects is too numerous to mention, there are a few excellent examples of recent design-build projects.

Consider a recent Corps design-build project—the Red Wing lock and dam on the Mississippi River. The Corps determined that the lock had the second-highest risk of failure of any U.S. navigation project. Under the economic stimulus package of 2009, $70 million has been designated for safety improvements to the lock and dam, with design-build being the vehicle for accomplishing the Corps' accelerated project delivery goals (completion is set for September 2011).[43]

The Corps also elected to use a fast-track design-build project for the construction of infrastructure for a 6,000-person training base and brigade facility for the newly formed Afghan National Army, near Kabul. While the original contract provided for new water supply, wastewater treatment, underground electrical systems, power supply, dining hall, and barracks, later contract modifications allowed for operations, maintenance, and training for the new utilities, as well as extending the underground electrical distribution system.[44]

The Corps also engaged a U.S. design-build contracting firm under two IDIQ contracts regarding the United States Central Command's need for Task Force—Restore Iraq Electricity (TF-RIE) and military base construction projects for the Afghan National Army in Mazar-e-Sharif and Gardez.[45]

2. Bureau of Prisons[46]

Shortly after the Clinger-Cohen Act was passed, the Bureau of Prisons (BOP) began its transition from a sealed bid procurement process environment to a negotiated best value contract selection procedure, using design-build.[47] Most of its projects built through the late 1990s and early 2000s were delivered through design-build. One recent example of a successful design-build project by BOP is the U.S. Penitentiary and Federal Prison Camp in Tucson, Arizona. The penitentiary is a high-security institution, while the prison camp is a minimum-security facility for 128 inmates. The total project consisted of more than 25 structures, including 6 interconnected general inmate housing units. While it was expected to take 945 calendar days, the work was completed in only 925 days—after 6 months of design work to meet the desert region's unique architectural and energy requirements.[48]

Other notable and ongoing BOP projects include a high-security penitentiary and federal prison camp in Yazoo City, Mississippi (anticipated to cost more than $150 million and take 1,000 days to complete)[49] and a new medium-security federal correctional institution located in Berlin, New Hampshire, with an anticipated construction value of $240 million.[50]

3. Department of Defense (DoD)[51]

As with the Corps, the DoD has used design-build on a wide number of its national and domestic bases. It found the process a preferred way to overcome the many challenges of its reconstruction efforts in Iraq. For example, in March 2004, the DoD issued an IDIQ contract for the electrical transmission and distribution networks in the southern region of Iraq and for the construction of new substations.[52]

The DoD has also long looked to the private sector and privatization as a means for rapidly addressing two significant problems concerning military housing: (1) deteriorating conditions in DoD-owned housing and (2) a lack of affordable private housing. As part of the National Defense Authorization Act for Fiscal Year 1996, Congress enacted a series of authorities as part of a pilot program called the Military Housing Privatization Initiative (MHPI) that allow DoD to work with the private sector to build, renovate, operate, and maintain military family housing. The stated goals of the MHPI are to obtain private capital to leverage federal funds, make efficient use of limited federal resources, and use an array of private sector initiatives and ingenuity to build and renovate military family housing more quickly and at a lower cost than by traditional means.

As of mid-2009, DoD had successfully awarded over $12 billion in project development costs for its initiative to design-build 62 projects (132,000 housing units).[53] Over 40 projects are in either the planning or the solicitation stage with over $10 billion in development costs projected for almost 60,000 additional housing units.[54] Each of the recent programs used design-build, including the privatization efforts at Lackland Air Force Base (Texas), Fort Lewis Air Force Base (Washington), and Fort Drum (New York).[55]

4. Federal Highway Administration (FHWA)

Over the past two decades, FHWA has been a proverbial "trailblazer" for its use of design-build delivery on transportation projects. At this point, the FHWA is no longer experimenting with design-build, as it now has the tools and the framework in place to continue its widespread growth.

The birth of FHWA's use of the design-build delivery method began in 1990. The FHWA's Special Experimental Project #14 (SEP-14) established a Transportation Research Board task force to evaluate innovative contracting practices, including design-build.[56] Between 1990 and 2002, transportation agencies in 32 states had proposed approximately $14 billion for design-build contracting on 300 projects under SEP-14. As of 2009, there are still approximately 54 active projects in FHWA's SEP-14.

Based on the success of SEP-14, Congress enacted the Transportation Efficiency Act of the 21st Century (TEA-21) in 1998, allowing federal aid funding to be used for design-build highway projects contracted for by state departments of transportation. To qualify, the design-build projects were required to be larger than $50 million.[57]

TEA-21 did two important things to promote the use of design-build by FHWA. First, it mandated that FHWA implement a final rule allowing for design-build contracting (the Final Rule was enacted in 2003).[58] Second, it required that a comprehensive national study be conducted on the effectiveness of the design-build contracting method. The final report, *Design-Build Effectiveness Study*, was issued in January 2006, confirming widely held beliefs that (1) the ability to reduce the overall duration of the project development is enhanced by using design-build; (2) greater cost efficiencies were more likely to occur on design-build projects; and (3) project quality was not minimized or hampered from use of design-build.[59]

Due in large measure to SEP-14's success, SEP-15 (Explore Alternative and Innovative Approaches to the Overall Project Development Process)[60] was introduced in 2004 to allow experimental public-private partnerships (PPPs) in highway projects and encourage the involvement and assistance of private-sector teams with project planning, development, environmental requirements, construction, project finance, and operations.[61] SEP-15 anticipates the use of design-build but is largely driven by financing and operations and maintenance. Some of the largest "design-build" projects in the highway sector are being developed through SEP-15.

In 2005, the Safe, Accountable, Flexible, Efficient Transportation Equity Act: A Legacy for Users (SAFETEA-LU) was signed into law under President Bush, guaranteeing $244.1 billion for funding for highways and public transportation. SAFETEA-LU increased the flexibility states have to use design-build contracting by eliminating the TEA-21 requirement on the size of contracts qualifying for design-build ($50 million floor). This law directed FHWA to permit transportation agencies to proceed with design-build contracts *pending* approval by the National Environmental Policy Act (NEPA), rather than waiting for the issuance of such approval by a third-party federal bureaucracy. As a result of this measure, subsequent modifications were made to FHWA's Final Rule on Design-Build in August 2007,[62] consistent with the mandates of SAFETEA-LU to encourage more widespread use of design-build. SAFETEA-LU was set to expire on October 1, 2009, and it was anticipated that the next six-year period replacement bill would be signed into law prior to SAFETEA-LU's expiration. However, other major pieces of legislation considered during the 111th Congress, such as the economic stimulus recovery and healthcare reform, appeared to have taken precedence over transportation-funding issues and related legislation. As a result, the current Congress authorized highway and transit programs only through December 18, 2009, and anticipated passage of the SAFETEA-LU replacement bill sometime in 2010.

All indications, however, underscore and confirm the benefits of the design-build method for highway and transportation projects. In 2005, the National Cooperative Highway Research Program, in conjunction with the American Association of State Highway and Transportation Officials (AASHTO) issued a report, *Design-Build Environmental Compliance Process and Level of Detail: Eight Case Studies.*[63] While the report focused on the integration of environmental processes by state agencies using design-build contracts, it also confirmed that with respect to those eight case studies, the DBIA's reported benefits of design-build are "right on target" concerning quality, cost savings, time savings, sole point of responsibility, and improved risk management.

5. Internal Revenue Service (IRS)

The IRS's new design-build Kansas City Campus center was recently recognized by the AIA as a Top Green Project for its $370 million Leadership in Energy and Environmental Design (LEED)–certified complex. The 1.14-million-square-foot facility was designed with a major energy performance goal to accommodate IRS's peak seasons, when more space is needed, without operating a mostly empty building year-round. Additionally, the design intent for this project was to increase productivity by providing daylight for as many employees as possible and by providing good indoor air quality. Because of budget concerns, the project team included design-build teams for the mechanical, electrical, and plumbing systems. The entire project, from pre-construction work to dedication, was completed on time and under budget.[64]

6. U.S. Census Bureau

The Census Bureau reportedly saved an entire year of project delivery time by using the design-build delivery method to complete its new $300+ million headquarters office complex in Maryland. To ensure the headquarters would be fully operational in time to prepare its 2010 decennial census, strict adherence to its October 2006 completion date was critical. The Bureau determined that only a design-build method could offer an on-time completion and, possibly, a year's savings in construction project costs. For this award-winning building, the conceptual design was developed by GSA, and the design-builder delivered the remainder of the design and construction in just 39 months.[65]

E. Design-Build Challenges

There are a variety of challenges to the use of design-build in the public sector, particularly at the federal level. Some of these challenges are a result of the relative novelty of design-build to many agencies and COs, and the fact that best practices are still being developed.

1. Bridging[66]

Most federal agencies hire an independent design professional, sometimes called a "design criteria consultant," to assist in developing the program and preliminary design for the RFP. Some have suggested that the agency is well served by having the design criteria consultant advance the level of design through design development—generally 30 to 35 percent of the total design effort. This concept of advancing the design to this level of detail has come to be known as "bridging."[67]

For many years, the bridging concept has been the subject of heated debate within the design-build industry. Its advocates claim that the process combines the best of what design-bid-build and design-build have to offer. The owner is able to work with a design professional who is acting in the owner's interest to develop a design that meets the owner's needs. Once the design is developed, the completion of the design can be accomplished by the design-build team, which becomes the designer of record. Proponents claim that this levels the playing field for proposers, since it ensures that proposers are bidding "apples to apples." They also tout the benefit of the owner knowing in advance of receiving proposals that it will have a baseline design that meets its aesthetic and technical objectives.

Those who oppose bridging argue that this process loses many design-build benefits. First, the time required to perform to this level of design can be substantial, which may lessen some of the time benefits of design-build. They also note that the advancement of the design to the 30 to 35 percent level effectively forecloses the design-build team from using its ingenuity in developing a cost-effective design, which impacts price. Finally, the issue of who has responsibility for defects in the 30 to 35 percent design is unclear, which means that the owner may lose its benefit of having single point of responsibility liability protection.

This latter point has been well demonstrated by case law. One of the leading cases in this area is *M.A. Mortenson Co.*[68] This case involved a design-build contract awarded by the Corps of Engineers to Mortenson for a medical clinic replacement facility at Kirtland Air Force Base, New Mexico. The solicitation contained design documents that were approximately 35 percent complete, with the solicitation informing proposers that such documents expressed the minimum requirements for the project. The Corps' design criteria informed all proposers that "[t]hese requirements may be used to prepare the proposals."[69] The design documents furnished by the Corps contained a number of options for structural systems, including calculations for these systems.

In originally pricing the work, Mortenson's estimators did a take-off of the structural concrete and rebar quantities indicated in the solicitation design documents. While the final design was similar to that shown in the solicitation documents and was approved by the owner, the design evolution process required the building to "expand" somewhat from what was depicted on the RFP documents. Mortenson ultimately submitted a request for equitable adjustment based on the increased quantities of concrete and rebar associated with building to the final design. The Corps rejected the claim, believing that Mortenson assumed the risk of any cost growth due to these quantities because of the fixed-price nature of the design-build contract.

The board agreed with Mortenson, finding that, while the solicitation did not require that the proposers use the information in the drawings, it also did not indicate that "the information was to be used at the proposer's risk."[70] The board held that (1) the design-builder acted reasonably in relying on the technical information provided by the government; and (2) the changes provision applied to the changes in the structural concrete and rebar. It specifically found that the doctrine of *contra proferentum* (construing the ambiguity against the drafter) applied and that the government had warranted the adequacy of information on the solicitation design documents. The ruling in *Mortenson* remains good law today.

Another federal case, *Donahue Electric, Inc.*,[71] addressed whether the government or design-builder was responsible for problems arising out of the government's partial design. The issues in *Donahue* arose during construction of the Department of Veterans Affairs (VA) Ambulatory Care Center in Las

Vegas, Nevada. After the VA engaged an engineering firm, HCE, to develop the scope of work and 50 percent design drawings, it pursued a competitive design-build procurement based on the design drawings from HCE. Each bidder was expected to use those drawings to complete the design and construct the project. However, bidders were also advised that the HCE design was informational only and that bidders were encouraged to use their own design.

HCE's design specified that the design-builder was to install a VA-furnished sterilizer unit manufactured by Steris. The bridged design also listed a Parker B-3 as the steam boiler that should be furnished, which was a 7HP boiler. Donahue was awarded the design-build contract. After some preliminary design work, Donahue determined that the 7HP boiler would not meet the requirements of the Steris sterilizer unit and so notified the VA. It was ultimately agreed that Donahue would use a 25HP boiler to power the Steris sterilizer the VA required.

The change in boiler type prompted Donahue to submit a change order for the additional costs, which was rejected by the VA on grounds that Donahue had no right to rely upon the bridged drawings for its bid. The VA argued that Donahue had total design responsibility and that it should have ignored the HCE drawings to the extent there were any inconsistencies. The dispute prompted Donahue to file a claim.

The Veterans Affairs Board of Contract Appeals (VABCA) agreed with Donahue that the change in boiler size was a compensable change and that it was entitled to the difference in cost between the 7HP and 25HP boiler. The board noted that if the VA had stated in its bid, "install the Steris sterilizer and a boiler to operate it," Donahue would have been responsible for the costs of whatever boiler it procured. However, the VA was prescriptive in identifying the Parker B-3 boiler and therefore bore the risk when the specified boiler was not adequate.

In its ruling, the board underscored the *Mortenson* principle that risks of mistake in owner-furnished design documents are not transferred to the design-builder. In other words, while it is the design-builder's responsibility to determine the cost to design and construct the project based on conceptual drawings, to the extent the specifications include specific requirements the owner will be deemed to have warranted the accuracy of those specifications.

Other cases that have considered this issue have reached the same basic conclusions—that a government owner warrants the accuracy and reasonableness for bidding of the design it provides to design-builders in the RFP.[72] The "implied warranty of specifications" doctrine, commonly known as the *Spearin* doctrine, has been a fundamental tenet of construction and government contract law since the early 1900s and does not go away simply because design-build is used.[73]

2. Performance vs. Design Specifications[74]

There is a fair amount of case law that has addressed some of the challenges relative to performance specification deficiencies in the federal arena—particularly when the agency's performance specifications conflict with its design specifications. Some design-builders believe that the design-build process gives them the flexibility to ignore the design specification if they can still achieve the performance specification. This is not a correct assumption, as is made evident by several cases.

The project in *FSEC, Inc.*[75] involved the construction of a new abrasive blast and paint spray facility for the Naval Construction Battalion Center in California. The dispute between the government and the contractor involved the ventilation system in two abrasive rooms. The invitation for bids (IFB) contained detailed technical specifications, including drawings for the ventilation system that required two exhaust fans and two dust collectors for each abrasive room. The air handling capacity was specified and produced a cross-draft ventilation of 100 feet per minute (FPM). The IFB also required that the contractor comply with ANSI (American National Standards Institute) requirements for ventilation. Amendments to the IFB added the following language to the summary of the work:

> This is a design-build project. Some aspects of the design have not been completed, and it will be the responsibility of the contractor to complete design details and submit the design for approval by the Government.[76]

These amendments added design submittal language requiring the contractor to use the information shown in the IFB to prepare its design drawings. The amendments also identified which portions of the specifications were prescriptive and which were performance. The final amendment to the contract clarified that the abrasive blast medium was steel.

FSEC, a design-builder specializing in abrasive blast facilities, interpreted the specifications as giving it the flexibility to design a ventilation system that would meet the performance specifications. It concluded that four fans and dust collectors were unnecessary because the 100 FPM cross-draft ventilation they produced exceeded the ANSI requirement of 60 FPM for steel media. FSEC was the low bidder, basing its price on two fans and dust collectors. When it submitted a design reflecting this, the government rejected the design and directed FSEC to supply four fans and dust collectors. FSEC objected and filed a claim.

The Armed Services Board of Contract Appeals (ASBCA) agreed with the government and specifically rejected FSEC's argument that because the contract was design-build, FSEC was responsible for providing a design that would achieve the required performance specifications, regardless of whether design specifications were met. The decision noted that the contract very clearly contained a mixture of both performance and prescriptive specifications and it was incumbent upon the design-builder to comply with both.

The Board also was persuaded by testimony from the government that it wanted the ventilation system design to be prescriptive "to insure that the end result would meet applicable air pollution standards . . . [and not] leave it to chance for design-build contractors to design it."[77] The Board held that the use of the words "design-build" in Amendment 002 to the contract did not require a conclusion that FSEC was responsible for the preparation of the entire design as the government contended and that the prescriptive/design and performance labels alone do not create, limit, or remove a contractor's obligations. Rather, the Board found that the obligations imposed by the specifications are determined by the extent to which a particular specification is either performance or prescriptive.[78]

FSEC contended that the amendment specifying a steel abrasive blast media created an ambiguity, since ANSI required only a cross-draft ventilation rate of 60 FPM. According to the Board, this interpretation failed to recognize that ANSI ventilation rates were determined by both the blast media and the coating being blasted. This facility was blasting toxic coatings, which actually had an ANSI-specified ventilation rate of 100 FPM. Although FSEC complained that the specifications never expressly identified that the paint to be removed was toxic, the Board concluded that the specifications were so specific as to the 100 FPM ventilation requirement that FSEC was on notice of what was required.

The *FSEC* case reinforces the fact that the designation of a project as "design-build" does not give the design-builder carte blanche to design what it wants so long as performance specifications are met. The government has the right to dictate aspects of the design, and, if it does so, the design-builder is obligated to incorporate such criteria into its final design. Agencies should note, however, that this is a two-way street. Consistent with the *Spearin* doctrine mentioned in the preceding section, the government is generally responsible if what it specifies through a design specification precludes the design-builder from achieving expected performance.[79]

This two-way street is evident by another recent federal case, *White v. Edsall Construction Co.*[80] The project involved the Army's construction of an aircraft storage hangar for the Montana National Guard's helicopter fleet. The Army supplied schematic drawings to the bidders stating that the contractor should verify all information, including drawings depicting two tilt-up canopy doors. One drawing indicated the estimated weights and showed three "pick" lifting points of the doors. Another drawing reflected the truss details showing how the weight of the door would be distributed.

When the door subcontractor began performing its work, it determined that four lifting points were needed, not three. The Army approved the revised lifting design, but it rejected the sub's claim for added costs for the use of four lifting points. The ASBCA rejected the Army's position, noting that the government impliedly warrants the adequacy of the detailed design information it provides the design-builder.

On appeal to the U.S. Court of Appeals for the Federal Circuit, the government argued that, although the contractor did not have a duty to uncover hidden errors, the schematic drawings disclaimer clearly required the contractor to verify the three-lift-point design before bidding. The court rejected the government's claim and held that the disclaimer did not shift the risk of design flaws to the contractor and that the design-builder had no obligation to ferret out the subtle flaw before bidding.

The *Edsall* case is an important one for design-builders because it marks the first time the Federal Circuit addressed implied warranty of specifications in a design-build contract. The lesson from this case is that while specific disclaimers might be enough to put the design-builder on notice, ordinary disclaimers are not enough to shift the risk to the contractor for errors contained in government-furnished specifications.

Another recent design-build case involving liability for incorrect performance specifications is *Appeal of Lovering Johnson, Inc.*[81] The disputes in *Lovering* arose out of the Navy's desire to design and construct a housing office and community center in Glenview, Illinois. The specifications called for a storm drainage system that would be capable of handling a 10-year storm and indicated that the storm drains should have as a minimum 12-inch pipes. Due to high flow rates on the site, the design-builder (LJI) ultimately had to use 72-inch pipes and made a claim against the Navy for the cost difference.

LJI argued that the Navy's drawings "misled" it into presuming the Navy had already completed the storm drainage design work when it depicted the sizes of the pipes. The ASBCA held that, due to the preliminary nature of the Navy's drawings, no reasonable design-builder could have assumed that it was relieved of its design responsibilities for assessing the actual drainage design. The solicitation's drawing package clearly indicated that any information contained was required to be verified prior to development of a final design by the design-builder.

Moreover, the Board found that LJI would have determined the likelihood of high storm water flows had it performed an adequate site investigation. Such an investigation would have revealed the presence of twin 60-inch culverts and potentially huge flows from off-site water sources. The Board concluded that LJI's site investigation was insufficient and that high flow rates were reasonably implied from the discernible site conditions and drawings.[82]

In *Appeal of Strand Hunt*,[83] the disputes arose from the design-build contract award to Strand Hunt Construction (SHC) by the Corps of Engineers for the construction of a Joint Security Forces Complex in Alaska. The Corps' RFP for the project set forth certain performance specifications relative to the windows. Specifically, the windows were required to be both blast-resistant and capable of withstanding arctic conditions.

Although SHC's interim design submissions confirmed that it would meet the listed thermal and blast mitigation specifications of the windows, SHC had difficulties procuring the windows. SHC submitted a certified claim for the encountered delays and extra costs, alleging that the government provided a faulty window specification. SHC argued that the government's RFP was defective because no window existed at the time of its proposal that could meet the arctic and blast mitigation window requirements.

The ASBCA rejected the claim, stating that SHC as design-builder should have ensured that its proposal and design specifications mirrored the RFP requirements. "SHC was obligated to not just say that it would meet [performance] requirements, but also to be sure it could actually do so." The Board found that SHC should bear the risk for its faulty assumption that a "ready-made" window existed that met its budget and the RFP requirements. The Board also found evidence that windows meeting the performance specifications could have been manufactured had SHC allowed adequate time for manufacture or performed additional investigation prior to submitting its proposal.

3. *Conflict of Interest Affecting Design Professionals*

One of the commonly asked questions in the public sector design-build process is whether architects and engineers are conflicted in participating on the at-risk design-build team. Although several regulations have a bearing on this subject, there is little direct case law that explains these regulations in the context of design-build.

FAR Subpart 9.502 states that an "organizational conflict of interest" may result when there are factors that create "an actual or potential conflict of interest on an instant contract, or when the nature of the work to be performed on the instant contract creates an actual or potential conflict of interest on a future acquisition."[84] Consequently, a design professional who participates as the owner's consultant on such issues as project feasibility and planning may be deemed to violate this conflict-of-interest requirement.

Other regulations that have application to this issue are FAR 36.209 and 36.606(c), each of which may prevent architects and engineers from bidding on the projects they design. These regulations state that: "No contract for the construction of a project shall be awarded to the firm that designed the project or its subsidiaries or affiliates, except with the approval of the head of the agency or authorized representative."[85]

One case that considered the designer's potential conflict of interest in the context of a federal design-build project is *In the Matter of SSR Engineers, Inc.*[86] That case involved a protest by SSR of the Navy's determination to exclude SSR from participating in a design-build project for changes to the electrical distribution system at a naval air base in Biloxi, Mississippi. The basis for the Navy's position was that SSR had performed an engineering services contract for the Navy in 1996 to develop a long-range comprehensive master plan relative to a portion of this system. The Navy contended that Volume 1 of the three-volume master plan prepared by SSR under this services contract was being used as the statement of work for the design-build project. Volume 1 also contained the cost calculations that were the basis for the Navy's budget estimates for the project. Volumes 2 and 3 also contained information that was relevant to the design-build project.

The Navy's argument was based on FAR 9.505-2, which requires contracting officials to avoid, neutralize, or mitigate potential significant organizational conflicts of interest. The regulation states:

> If a contractor prepares, or assists in preparing, a work statement to be used in competitively acquiring a system or services—or provides material leading directly, predictably, and without delay to such a work statement—that contractor may not supply the system, major components of the system or the services [with certain exceptions].[87]

This regulation had been used in previous cases to preclude a firm that participated in the writing of a work statement from participating in the procurement of the services associated with the work statement.

SSR argued that the Subpart 9.5 conflict of interest provisions did not apply to architectural or construction services issued under FAR Part 36 (the regulations that deal with construction and A-E contracts). SSR concluded that "if the intent of the FAR was to preclude the firm developing the scope [of work] from participating on a design-build team the exclusions would be discussed or referenced in FAR Part 36."[88] The Comptroller General flatly rejected this contention, noting that FAR Part 36 does not specifically reference conflict-of-interest provisions and does not render Subpart 9.5 inapplicable to A-E contracts.

SSR also complained that it did not obtain a competitive advantage based on its prior work. It acknowledged that several contractors had expressed a desire to have SSR on their teams, but felt that this was based on SSR's design capabilities, not on any competitive advantage. SSR also argued that (1) contractors would not be relying upon SSR to establish pricing for the work, so that its cost estimating under the 1996 contract was not relevant, and (2) its fees were so small in relation to the total design-build contract price that any so-called competitive advantage would be insignificant. The Comptroller General rejected each of these premises, stating:

> The responsibility for determining whether a firm has a conflict of interest and to what extent a firm should be excluded from competition rests with the procuring agency, and we will not overturn such a determination unless it is shown to be unreasonable.[89]

Based on the facts, the Comptroller General concluded that because "SSR prepared material leading directly to the statement of work and prepared cost estimates which established the ceiling for the agency's budgeting of costs," the Navy's decision to exclude SSR was reasonable.

4. Impact of Design-Build on Public Sector Design Professionals

Another challenge for public sector design-build concerns a perception that this delivery method could result in major staff cutbacks for design professionals within public agencies. To evaluate whether this claim had any credence, a 2007 study was conducted through a survey of state departments of transportation and design-build RFPs by the USC Keston Institute for Public Finance and Infrastructure Policy.[90] The question posed was: "What is the impact on the state department of transportation professional workforce when the state authorizes it to deliver infrastructure projects utilizing design-build project delivery?" The conclusions underscored that there is no plausible basis for the perception that design-build negatively impacts design professionals employed by public agencies.

First, the study established that implementing design-build contracting does not shift professional engineering jobs from state agencies to the public sector. In other words, states with design-build experience did not reduce their engineering workforce as a result of implementing design-build. Other conclusions were that using design-build contracting "does not significantly reduce the use of the traditional design-bid-build method" and that implementing design-build requires a more competent and experienced workforce.[91]

5. Evaluation Factors

There are an increasing number of federal cases that address the issues associated with how an agency evaluates design-bid proposers. To date, the cases have stood for the proposition that an agency has broad discretion in evaluating proposals.[92]

For example, *SKE International, Inc.*[93] involved protests by SKE International over the award of an IDIQ job order contract for mechanical-electrical-plumbing (MEP) modernization projects at Fort Leavenworth. The RFP stated that the proposals would be evaluated for "best value" based on four factors (past performance, corporate experience, management approach, and price). The four factors were weighted on a descending scale of importance, with price being less important than the technical factors.

Five offerors submitted proposals and the source selection evaluation board (SEB) found that the technical benefits of one of the offerors outweighed any benefit from SKE's lower price. SKE's competitor was awarded the contract, and SKE filed a protest claiming that the evaluation was flawed because a proper evaluation would have resulted in its proposal being rated equal to or better than its competitor.

Specifically, SKE objected to the manner in which the SEB evaluated SKE's past performance as "satisfactory" under the timeliness subfactor. Of the eight references for SKE, four responded that the firm had completed its work "substantially ahead of schedule," while four responded that it had completed its work "on schedule, with minor delays." SKE believed the SEB should have rated its timeliness as "very good" based on the four ahead-of-schedule ratings.

The Comptroller General decided that SKE's arguments were meritless, noting that under the SEB's evaluation standards, a "very good" rating would have required "a majority of responses in the 'no time delays' category." Instead, the array of responses received from SKE's references squarely placed it in the "satisfactory" rating, and, as a result, the Comptroller endorsed the SEB's ranking.

As a second ground for protesting the award, SKE argued that the SEB improperly found a weakness in its offer concerning the corporate experience factor because the offer apparently did not give enough detail regarding completion dates and the results of task orders under the project examples. SKE challenged the protest by arguing it should have been rated "excellent" on its corporate experience factor instead of "very good." Specifically, SKE relied on the agency's response to a presolicitation question in which the agency stated that the data and descriptions provided should be overall contract data.

Again, the Comptroller General found that the agency could have reasonably considered SKE's failure to provide task order details as a proposal weakness. "While the agency confirmed that the offerors needed to identify the task orders, its response cannot reasonably be read as mandating a simple listing of task orders or as otherwise limiting the information offerors were to provide in their proposals." In fact, the Comptroller General noted that the RFP specifically states that additional detail would be rated higher. Finally, the protest was denied on the grounds that the SEB was correct in assigning a weakness to SKE's proposal for its failure to show who the Quality Control Manager (QCM) would report at the corporate level. The RFP specifically stated that the organizational chart was to show the relationship between the contract teams and the offeror's larger organization.

Like the ruling in *SKE International*, a protest was also denied *In the Matter of ICON Consulting Group*.[94] This case concerned the protester's allegations that certain corrective action by an agency unfairly excluded it from competing for the two awards reserved for 8(a) firms. The dispute arose from the issuance of an RFP by the Department of the Air Force for design-build construction solutions for facilities at Hill Air Force Base in northern Utah. The RFP provided it would award on a best value basis up to five contracts, with up to two of the awards reserved for 8(a) business concerns. The offers would be evaluated in two phases. Under the first phase, offers from 8(a) contractors would be evaluated separately, and up to two awards could be made. All remaining offers would be considered under the second phase. The RFP specifically stated, however, that the government was not required to award any minimum number of contracts.

Nine proposals were received and found to be in the competitive range. After discussions, ICON was rated unacceptable because its proposal did not contain structural calculations. The agency made four awards, two of which were to 8(a) contractors. ICON filed an agency-level protest, asserting that the Air Force did not effectively communicate the requirement for structural calculations and had improperly rejected ICON's proposal on those grounds. The Air Force denied the protest and said that the requirement for structural calculations was indeed clarified during discussions. Ultimately, the Air Force took corrective action by amending the solicitation to restate the need for structural calculations, but it only reopened discussions for the purposes of awarding one additional contract. The other four awards were still in place.

ICON filed another protest asserting that the agency's remedy had deprived it of the opportunity to compete in the first phase evaluation of 8(a) contractors. The Comptroller General cited the broad discretion agencies have to take corrective action where the agency seeks to ensure fair and impartial consideration. Specifically, the decision stated: "In our view, the corrective action is well within the discretion afforded to contracting agencies in these circumstances. . . . [T]he solicitation did not guarantee ICON, or any other 8(a) firm, that the agency would, in fact, make two 8(a) awards."

In *J.A. Jones/IBC Joint Venture*, the Jones/IBC Joint Venture (Jones) and Black Construction Company (Black) protested award of a design-build contract by the Navy to Dick Pacific Construction (Dick) for a project in Guam. The RFP included a number of mandatory requirements and indicated that the contract would be awarded to the offeror who provided the best technical and price value. The RFP stated that the agency intended to award the contract "without conducting discussions" and that the initial proposals should contain the offeror's best terms.

Three of the four offers were evaluated by the technical evaluation board (TEB) as "unacceptable but susceptible to becoming acceptable." Only Dick's proposal received a rating of "highly acceptable" and was ranked first on technical merit. It also had the highest price. Although the TEB recommended to the source selection board (SSB) that discussions be conducted with each offeror, the SSB decided that enough information was available to determine best value and recommended award of the contract to Dick without discussions.

After reviewing the reports of the TEB and SSB, the Source Selection Authority awarded the contract to Dick without discussions. It based its decision, in part, on the fact that Jones's proposal required significant redesign to achieve conformance with the RFP and was unacceptable. It also noted that Black's proposal, although containing several minor design weaknesses that would have to be corrected, did not compare as favorably to Dick's proposal, which contained features that exceeded the RFP requirements and justified its higher price.

Jones acknowledged that its proposal was deficient because it (1) failed to plant the minimum number of royal palm trees, and (2) provided main sewer lines with smaller diameters than specified in the RFP. Jones argued, however, that the Navy should have informed it of the perceived weaknesses in its proposal and allowed it to clarify the Navy's concerns. The Comptroller General disagreed, stating:

> Where award will be made without discussions, agencies may have limited exchanges with an offeror for the purpose of clarifying certain aspects of its proposal or to resolve minor or clerical errors. Such limited exchanges do not constitute discussions. . . . Discussions occur when an offeror is given an opportunity to revise or modify its proposal, or when information requested from and provided by an offeror is essential for determining the acceptability of its proposal.[95]

Because Jones's deficiencies were related to minimum RFP requirements, allowing Jones to provide information would have constituted "discussions" and required that the agency hold discussions with the other offerors. The decision of the Comptroller General further stated:

> There generally is no requirement that an agency hold discussions where, as here, the RFP advises that the agency intends to make award without discussions. . . . The burden was on Jones to submit an initial proposal containing sufficient information to demonstrate its technical merits and to show that all of the RFP's required design specifications were met.[96]

Finally, the Comptroller General rejected Jones's argument that the Navy should have awarded Jones the contract and then allowed it to modify its design specifications as necessary. This would have created potential problems since "the agency would not have been able to enforce all of the RFP's mandatory requirements on Jones if it awarded a contract to the firm based upon its nonconforming proposal."[97]

As to Black's protest, the Comptroller General noted several deficiencies related to the trees, ceilings, and water fountains. Even though they were minor deficiencies, Black bore the risk that its proposal would be rejected as unacceptable. As a result, this protest also was denied.

III. Construction Management

Under the design-build system, however, the owner solicits proposals based upon a statement of requirements, and then contracts with a single contractor to design and construct the project under one contract.[98]

A. Construction Management Generally

The construction management project delivery system differs from the traditional design-bid-build and design-build systems because, in addition to contracts for design and construction, the owner enters into an additional contract with the construction manager (CM). In most cases, the CM is a general contracting construction firm and is retained early in the design phase. The CM is expected to provide management skills to the owner throughout the project, from early design through completion of construction.[99] As an additional member of the design team, the CM may provide significant cost, schedule, constructability, and serviceability input to the design.

Construction management generally follows two models: (1) CM as Agent or (2) CM at Risk.[100] Under the agency approach, the CM contracts with the owner on a fee-for-service basis to provide a variety of services, such as construction scheduling, coordination, and technical services (including quality assurance and testing). The owner still retains separate contracts with a designer and a contractor. Under the risk approach, the CM assumes a role more similar to a general contractor, providing the owner with a lump sum or guaranteed maximum price (GMP) for the project.[101]

1. CM as Agent

Under the CM as Agent approach (also known as "pure construction management" or "CM as Advisor"), the CM acts as an agent of the owner, providing advice and services. The owner enters into a single contract for consulting services with the CM, while directly contracting with trade contractors and suppliers. The CM, who is paid on a fee-for-service basis, acts on the owner's behalf in managing and coordinating the trade contracts in the best interests of the owner. The owner acts as its own general contractor, with the assistance of the CM. The owner retains the contracting risks inherent in each of the trade contracts.[102]

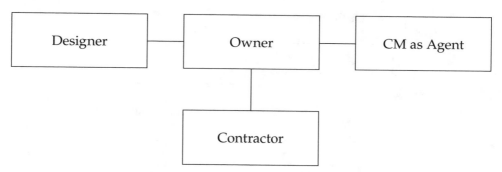

In some circumstances, the owner may decide to retain a general contractor, rather than contract directly with specialty subcontractors. In that case, the CM would oversee the efforts of the general contractor.[103]

The CM as Agent does not bear entrepreneurial risks for costs, timeliness, or quality of construction. Acting solely as the owner's agent, the CM functions as an advisor and/or consultant to the owner to assist with execution of the project, helping the owner complete a project that is properly constructed, on time, and on budget. The CM as Agent does not design the project and is not a party to the construction contract. It does not have control over and is not in charge of the construction means, methods, techniques, sequences, or procedures; those remain the responsibility of the construction contractor.[104]

A CM as Agent will typically provide some or all the following types of services: project planning, design overview and management, project estimating, assistance in the procurement process, contract administration support during construction, administration of submittals, commissioning and start-up support, testing services, claims support, and contract close-out support.[105]

2. CM at Risk

The CM at Risk form of construction management (sometimes referred to as "CM as Constructor") is similar in many ways to the traditional design-bid-build system. Under this approach, the CM provides advisory professional management services to the owner prior to construction by offering schedule, budget, and constructability advice during the project planning phase. However, at the completion of the design phase, the CM at Risk essentially assumes the role of a general contractor by guaranteeing the completion of the project for a lump sum guaranteed maximum price (GMP). The CM at Risk enters into multiple subcontracts with the specialty trade subcontractors and assumes responsibility, vis-à-vis the owner, for their performance. This is a hybrid approach, because the owner is dealing with a firm that functions as both CM and general contractor. The CM at Risk therefore assumes far greater entrepreneurial risk than a CM as Agent.[106]

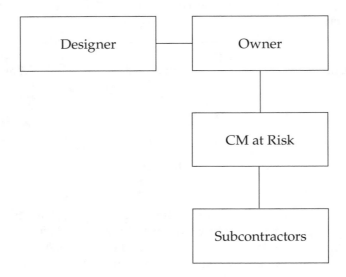

In addition to providing the owner with the benefit of preconstruction services during the planning phases of the project, the CM at Risk approach allows project construction to begin prior to design completion. In many instances, the CM at Risk enters into a guaranteed maximum price or lump sum contract prior to completion of design. This allows the CM at Risk to enter into subcontracts for portions of the work, while the design of unrelated portions continues.[107]

Under the CM at Risk approach, the owner enters into a single construction contract with the CM, who contracts directly with the specialty trades.[108] A disadvantage inherent in the CM at Risk system is the relationships among the owner, designer, and CM once construction begins. The CM's role shifts from professional advisor to the contractual (at risk) role of a general contractor. At that time, the interests of the parties become similar to those involved in the traditional design-bid-build system, and adversarial positions may develop. The CM at Risk may be inclined to take positions that are in its best interest and not that of the owner. Disputes over the construction quality and completeness of the design, schedule impacts, and delay responsibility may develop.[109]

B. CM and the Federal Government

The government has been using construction management contracts since the 1960s.[110] However, the 1990s and particularly the 2000s have seen significant growth in the government's use of these contracts, particularly by the GSA and the Army Corps.[111] These agencies utilize the construction management approach to supplement their in-house capabilities. By using CM contracts (both agency and at risk), the government can successfully complete more construction projects than it could with only in-house construction personnel.[112]

Neither FAR nor the agency supplements to the FAR discuss the use of construction management contracts or provide guidance regarding if and when such contracts should be utilized.[113] Moreover, the decisions of the Boards of Contract Appeals, the Court of Federal Claims (COFC), and the Federal Circuit provide little, if any, guidance on the government's use of construction management contracts.[114] As a result, agencies have developed their own approaches to the use of such contracts and prepared work statements or specialized contract provisions on a case-by-case basis.[115] They have used a variety of pricing methods, including but not limited to GMP, fixed price, fixed-price incentive, and shared savings contracts.

GSA has used numerous different types of construction management contracts in the construction of federal buildings and courthouses throughout the United States utilizing both CM as Agent and CM at Risk approaches. Typically, GSA will enter into a contract with the CM as Agent for the design phase of the project.[116] When the design phase reaches the 75 percent completion stage, GSA and the CM will negotiate a contract for the construction of the project, with the CM at Risk. The at-risk construction contract can be a GMP contract, a fixed-price contract, a fixed-price incentive contract with a not-to-exceed price, or some other variation of GMP and fixed-price contracts.

GSA has used IDIQ contracts to purchase construction management services on large-scale, complex construction projects. Typically, a GSA regional office will conduct a best value competition among interested construction firms and award a number of IDIQ contracts. Then, GSA competes individual task orders among the awardees. The task orders, which are at-risk, typically cover a wide range of design and construction services that are often performed by CMs or general contractors.[117]

C. The Impact of the American Recovery and Reinvestment Act of 2009

As discussed in this chapter, the use of alternative project delivery methods such as design-build, task order, and construction management contracts has been steadily increasing since the 1990s. A contributing factor to this increase has been the government's recent emphasis on privatization (i.e., using commercial entities whenever possible) and budgetary "downsizing," both of which have resulted in a smaller federal workforce. Also, agencies have lost a significant number of experienced procurement and construction personnel due to the number of post-WWII baby boomer retirements.[118] As a result, they have turned to CM contracts to obtain the necessary expertise.

On February 17, 2009, President Obama signed the American Recovery and Reinvestment Act of 2009 (ARRA) into law.[119] ARRA is widely considered to be the most sweeping economic recovery package ($787 billion) in American history. ARRA included $130 billion for state and federal construction spending: transportation ($49.3 billion), energy ($30.6 billion), water and environment ($20 billion), buildings ($13.4 billion), housing ($ 9.6 billion), and defense/veterans ($7.8 billion). ARRA includes approximately $5.5 billion for GSA construction projects and up to $8 billion for construction projects managed by the Corps of Engineers. Most of the funds must be obligated by September 30, 2010.

Federal and state governments will therefore be expected to award and manage a significantly greater volume of construction projects for several years beginning in mid-2009. As discussed above, this likely will require additional procurement and construction expertise.[120] Also, contracts will need to be awarded expeditiously. It is unlikely that the federal government can utilize the traditional design-bid-build construction model to perform the ARRA-required construction projects, given the scarcity of federal acquisition employees and the ARRA's time constraints.[121] In all probability, the government will have to utilize alternative project delivery methods.[122] It will need to retain more construction managers (as Agent) to assist in the planning and management of ARRA construction projects. Also, the CM at Risk (CM as Constructor) contracting share probably will increase significantly above its current levels. In summary, in order to successfully award, manage, and complete the ambitious ARRA construction program, the government will need to utilize alternative delivery methods, such as design-build, CM, and IDIQ task order contracts.

IV. Indefinite-Delivery and Task Order Contracts

The indefinite-delivery contract is a procurement mechanism designed to streamline the federal government contract competitive bidding process. It allows the government to place multiple orders over a period of time under a single umbrella contract. Some government agencies have made indefinite-delivery contracts a preferred method of contracting. For example, the Army Corps is dramatically increasing the use of regional multiple-award task order contracts to purchase design-build construction services. This is part of an overall Army strategy to maximize its expenditures, reduce the time to complete construction projects, and standardize common-use buildings. This section examines the different types of indefinite-delivery contracts and their unique characteristics and risks.

A. Purpose of Indefinite-Delivery Contracts

The FAR defines a "delivery order" contract as an indefinite "delivery contract for supplies."[123] Under a delivery order contract, the government issues orders for specified supplies during the contract period.[124] Similarly, the FAR defines a "task order" contract as an indefinite "delivery contract for services." Under a task order contract, the government issues orders for specified services during the contract period.[125] The government typically issues orders for design-build and construction services under task order contracts.[126]

Indefinite-delivery contracts are used to acquire supplies and/or services when the exact time and/or exact quantity of future deliveries is unknown.[127] Indefinite-delivery contracts include (a) definite quantity contracts, (b) requirements contracts, and (c) IDIQ contracts.[128] Generally speaking, these contracts provide the government with flexibility in both quantity and delivery scheduling, and they permit the government to defer ordering supplies or services until the actual need materializes.[129]

B. Definite Quantity Contracts

Definite quantity contracts provide for "the delivery of a definite quantity of specific supplies or services for a fixed period, with deliveries or performance to be scheduled at designated locations upon order."[130] The government uses this contracting mechanism when it can readily identify the specific supply or service to be purchased and the quantity. The government will issue subsequent orders establishing the delivery time(s) and place(s).[131]

C. Requirements Contracts

Under a requirements contract, a government agency agrees to purchase *all* of its actual requirements for specified supplies or services from one contractor during a specified contract period.[132] Requirements contracts are appropriate for acquiring supplies or services when the government anticipates recurring requirements, but it cannot predetermine the precise quantities of supplies or services that designated government activities will need during a definite period.[133]

1. Minimum and Maximum Quantities

FAR 16.503(a)(2) provides that a requirements contract should, whenever feasible, establish a maximum limit on the government's obligation to order and the contractor's obligation to deliver. The contract may also include specific maximum or minimum quantities that the government may order under each individual delivery order and the maximum it may order during a specified period of time.[134] The government is not obligated to purchase a particular quantity of items unless a minimum quantity is specified in the contract.[135]

2. Estimated Quantities

The FAR requires the CO to include a realistic estimated total quantity in the solicitation and resulting contract.[136] The CO may derive the estimated quantity from records of previous requirements and consumption or by other means, and the estimated quantity should be based upon the most current information available.[137] The CO, however, is not required to examine information that is not reasonably available.[138]

In requirements contracts, the contractor generally assumes the risks associated with a variance between the estimated quantity and the amount actually purchased.[139] There is no guarantee that the government actually will order the estimated quantity.[140]

The government must exercise reasonable care when it prepares the estimate of the quantity to be ordered under the contract.[141] Failure to exercise reasonable care in preparing the estimate will entitle the contractor to an equitable adjustment.[142] To receive an equitable adjustment, a contractor must prove by a preponderance of the evidence that the estimate was (1) inadequately or negligently prepared; (2) not in good faith; or (3) grossly or unreasonably inadequate at the time it was made.[143] The contractor also must prove that it relied on the estimate in preparing its bid.[144] In determining if an estimate was prepared negligently, the court or board will look at the reasonableness of the government's actions in relation to the information available to it, as well as the agency's reasons for including or not including certain information.[145] An order that is substantially less than the estimated quantity alone does not establish that the estimate was prepared negligently or without due care.[146]

3. Requirement to Order All Supplies/Services from the Contractor

Because the government agrees to purchase *all* of its specified supplies or services needs from the contractor,[147] the Federal Circuit has held that the government's promise of exclusivity is a material element of a requirements contract.[148] Thus, the government breaches the contract if it orders the specified supplies or services from a different source.[149]

D. IDIQ Contracts

IDIQ contracts provide the government with more flexibility than definite quantity and requirements contracts. The IDIQ contract "provides for an indefinite quantity, within stated limits, of supplies or services during a fixed period. The Government places orders for individual requirements."[150] The government typically uses an IDIQ contract when (a) the government cannot predetermine, above a specified minimum, the precise quantities of supplies or services that will be required during the contract period; and (b) it is inadvisable for the government to commit itself for more than a minimum quantity.[151]

1. Distinction Between IDIQ and Requirements Contracts

Under a requirements contract, the government agency must purchase *all* of its requirements for the supplies or services covered under the contract from the contractor.[152] Under an IDIQ contract, however, the government only agrees to order a *stated minimum* quantity.[153] Once the government purchases the minimum quantity, it is not obligated to purchase any additional supplies or services.

The COFC explains this distinction as follows:

> An indefinite quantities contract is a contract under which the buyer agrees to purchase and the seller agrees to supply whatever quantity of goods the buyer chooses to purchase from the seller. It differs from a requirements contract in that under a requirements contract the buyer agrees to purchase all his requirements from the seller. Under an indefinite quantities contract, even if the buyer has requirements, he is not obligated to purchase from the seller. In an indefinite quantities contract, without more, the buyer's promise is illusory and the contract unenforceable against the seller. To make such contract enforceable, the buyer must agree to purchase from the seller at least a guaranteed minimum quantity of goods or services. If the contract contains such a minimum quantity clause, the buyer is required to purchase at least this minimum amount, but this is the extent of his legal obligation. He can purchase more if he chooses to but is under no obligation to do so.[154]

2. Minimum and Maximum Quantities

The minimum quantity required under an IDIQ contract provides the necessary consideration to bind both parties.[155] As a general rule, a solicitation without a minimum quantity is unenforceable due to lack of consideration.[156] However, the COFC held in 2002 that a contract without a minimum value was enforceable, since the minimum value could be "inferred" from the solicitation.[157] The minimum quantity "must be more than a nominal quantity, but it should not exceed the amount that the Government is fairly certain to order."[158] There are no bright-line rules explaining what is "nominal." Although the agency must obligate the funds to cover the minimum quantity at the time of award,[159] there is no obligation to order the minimum quantity at the time of the award.[160]

The FAR requires that the maximum quantity should be "reasonable" and "based on market research, trends on recent contracts for similar supplies or services, survey[s] of potential users, or any other rational basis."[161] In a 1995 bid protest, the General Services Board of Contract Appeals (GSBCA) found that the government lacked "a reasonable basis" for its maximum estimated quantity.[162] The Board observed that the estimate "was not reached because of any established budgetary limitations or because of any true analysis of requirements, but because of the expedience of avoiding a level of review by GSA."

Bidders tend to view the maximum estimates as valid predictions upon which they can rely. The government's only obligation, however, is to order the minimum dollar threshold while not exceeding the stated contract maximum.[163]

3. Accuracy of Estimated Quantities

The government is generally not held accountable for inaccurate quantity estimates contained in IDIQ solicitations.[164] In the leading case on this issue, *Dot Systems, Inc. v. United States*,[165] the contractor sought an equitable adjustment based on the government's failure to order more than 10 percent of the solicitation's stated estimated quantities. The court focused on the allocation of risk under the contract, stating that "[t]he general rule is clear: a contractor cannot sign a contract which allocates the risk to it and then four years later come to this court, having lost its gamble, and insist that the risk be placed on the government."[166] The court emphasized that the solicitation expressly stated that the estimates were not guarantees and that the contractor in an IDIQ contract "cannot expect the kind of accuracy in estimation that it can in a requirements or fixed price contract . . . [and] the government cannot be held to the negligence standard for requirements contracts . . . ."[167] Courts are reluctant to enforce estimates in IDIQ contracts, even when it appears that the government knew, or should have known, that the estimates may have been erroneous.[168]

4. Government Preparation of IDIQ Solicitations

FAR Subpart 16.5, "Indefinite-Delivery Contracts," establishes a strong governmental preference for making multiple awards under indefinite-quantity contracts.[169] Towards this end, the FAR requires that COs "to the maximum extent practicable . . . give preference to making multiple awards of indefinite-quantity contracts under a single solicitation for the same or similar supplies or services to two or more sources."[170] COs should consider the following factors when determining the number of contracts to be awarded:

(1) The scope and complexity of the contract requirement.
(2) The expected duration and frequency of task or delivery orders.
(3) The mix of resources a contractor must have to perform expected task or delivery order requirements.
(4) The ability to maintain competition among the awardees throughout the contracts' period of performance.[171]

COs should not make multiple awards if there is only one contractor that can provide the required supplies or services, if a single award will provide more favorable terms and conditions, or if multiple awards otherwise are not in the best interests of the government.[172] COs must document their decision whether or not to award multiple contracts.[173] The head of the contracting agency must provide written approval of the award of any task or delivery order contract to a single source, if the estimated amount exceeds $100 million.[174]

Pursuant to FAR 16.504(a)(4), a solicitation for an indefinite-quantity contract must include:

(1) a defined contract period, including options;
(2) the minimum and maximum total quantities;
(3) a statement of work that sufficiently describes the work, so that prospective offerors can decide whether or not to bid;
(4) the government agencies authorized to place orders, and procedures for oral orders, if authorized;
(5) the agency task and delivery order ombudsman; and
(6) the procedures that the government will use to place orders.

When multiple awards may be made, the solicitation must identify the procedures and selection criteria "that the government will use . . . to provide awardees a fair opportunity to be considered for each order."[175]

5. Orders under IDIQ Contracts

FAR 16.505(a) sets forth general guidelines that apply to all IDIQ orders. Individual orders must be within the contract's scope, must be issued during the contract's period of performance, and must be issued within the contract's maximum value.[176] IDIQ orders must contain the following information: (a) contract and order number; (b) date; (c) contract item number, price, or estimated cost; (d) delivery or performance schedule and location; (e) packaging/shipping instructions; (f) appropriation and accounting data; and (g) payment information.[177]

6. Orders under Multiple Award Contracts

Each awardee under a multiple delivery order or multiple task order contract must be provided with a fair opportunity to be considered for each order with a value exceeding $3,000 (with certain exceptions).[178] The CO is given broad discretion to develop placement procedures, with the goal of keeping submission requirements to a minimum.[179] Along these lines, the CO may use streamlined procedures, including oral presentations.[180] The FAR specifically states that the Part 6 (sealed bid) competition requirements and the Subpart 15.3 (competitive negotiation) policies do not apply to the IDIQ ordering process.[181] However, the CO must develop placement procedures that will provide each awardee a fair opportunity to compete for each order.[182] In preparing task order placement procedures, COs must consider, among other things, the awardees' past performance under earlier orders issued under the contract, the potential impact of an additional award upon other orders placed with a contractor, minimum ordering requirements, and the time necessary for awardees to respond to potential orders.[183] In addition, COs must consider price or cost under each order in the selection decision.[184]

In the event that the delivery or task order contract does not establish prices for supplies or services to be purchased under an individual order, the CO must establish the prices by using the policies and procedures of FAR 15.4, which details if and when cost and pricing data is necessary.[185] While the government is not required to use formal evaluation or scoring plans in the order selection process, it must prepare documentation for the file explaining the rationale for the selection and price of each order.[186]

7. Additional Procedures for Orders Exceeding $5 Million

The National Defense Authorization Act for Fiscal Year 2008 imposed additional competition requirements for delivery orders above $5 million.[187] These requirements, codified in FAR 16.505(b)(1)(iii), are intended to ensure that all awardees have a fair opportunity to be considered for orders exceeding $5 million that are issued under the contract. For orders above $5 million, the FAR now requires that awardees be provided with:

(A) A notice of the task or delivery order that includes a clear statement of the agency's requirements;

(B) A reasonable response period;

(C) Disclosure of the significant factors and subfactors, including cost or price, that the agency expects to consider in evaluating proposals, and their relative importance;

(D) Where award is made on a best value basis, a written statement documenting the basis for award and the relative importance of quality and price or cost factors; and

(E) An opportunity for a postaward debriefing . . . .[188]

Most significantly, the government is required to inform potential awardees of the factors and methods by which proposals will be evaluated.

For orders exceeding $5 million, the CO is required to follow the FAR Part 15 procedures for notifying unsuccessful awardees and conducting debriefings.[189] The agency head is required to appoint a task order and delivery order ombudsman, who must be a senior agency official independent of the CO. The ombudsman's role is to ensure that awardees are afforded a fair opportunity to be considered for orders, consistent with the procedures set out in the awardee's contract.[190]

8. Modifications to Delivery Orders

The IDIQ contract was designed specifically to reduce the number of change orders in the contracting process. Thus, while a CO can modify a delivery order, such modifications are limited. Modifications are subject to the following criteria:

1. the contract must permit changes, and the change must be in accordance with the procedures of the contract;
2. unilateral changes must be within the scope of the contract;[191] and
3. changes to the contract must allow for an equitable adjustment.[192]

When determining whether a change order extends beyond the original scope of the contract, courts will consider the contract's express language.[193] They will also look to whether there is a material difference between the delivery order and the contract.[194] Courts may identify material differences by (1) reviewing the circumstances attending the procurement that was conducted; (2) examining any changes in the type of work, performance period, and costs between the contract as awarded and as modified by the task order; and (3) considering whether the original contract solicitation adequately advised offerors of the potential for the type of task order issued. The overall inquiry is whether the modification is of a nature that potential offerors would have reasonably anticipated.[195]

9. Terminations for Convenience

Government contracts contain a standard termination for convenience clause, which permits the government to terminate an IDIQ contract even if it has not purchased the minimum quantity.[196] The ASBCA addressed this issue in *Hermes Consolidated, Inc. d/b/a Wyoming Refining Co.*,[197] in which the Defense Energy Support Center (DESC) partially terminated two contracts for convenience, thus purchasing less than the stated minimum quantities of jet fuel. The DESC argued that the contractor was only entitled to the price for the delivered fuel plus "reasonable charges" under the termination for convenience clause, but the contractor claimed that it was entitled to the contract price for the unordered fuel required by the contracts' minimum purchase obligations. The Board sided with the DESC, holding that the agency could reduce the minimum quantities under the termination for convenience clause.[198]

In *International Data Products Corp. v. United States*,[199] the government awarded the contractor a five-year, fixed-price IDIQ contract for computer systems and services with an estimated value of $100 million and a maximum value of $739 million. Midway through the contract, after purchasing $35 million in equipment, the government terminated the contract for convenience. Because the contract contained a broad on-site warranty and service provision, the government required the contractor to provide these services for the equipment already on-site after the termination. The contractor then filed a claim for the cost of providing these services. COFC held not only that the government had the right to demand the warranty and on-site services but also that the contractor could not recover for such costs. It ruled that the termination for convenience clause precluded recovery of amounts above the contract price. The court's rationale was that the contract price (the $35 million that the government had paid for the ordered equipment) constituted payment in full and, thus, the contractor had no further rights to recovery. The result would likely have been different under a simple fixed-price contract for $100 million, because in such a case the contractor could have recovered its additional costs under a termination clause. Thus, an IDIQ contract may result in an increased risk to the contractor.

10. Protests

Both the award of IDIQ contracts and the award of individual task orders may be protested.[200] Prior to May 2008, protests against the issuance or proposed issuance of task or delivery orders were permitted only if the protest alleged that the order was beyond the scope, period, or maximum value of the underlying multiple award contract.[201] Section 843 of the National Defense Appropriations Act of 2008 expanded the bid protest jurisdiction of the GAO to allow protests against the issuance or proposed issuance of such task or delivery orders exceeding $10 million.[202] A GAO protest in such cases is based upon an agency's failure to follow the procedures or ground rules that the agency established for a task order competition, but only if the task or delivery order has a value above $10 million. The GAO lacks jurisdiction to review task orders between $5 and $10 million, even though FAR 16.505(b)(1)(iii) competition requirements apply to orders in that dollar range.

The first reported decision under GAO's new task and delivery order bid protest jurisdiction is *Triple Canopy, Inc.*[203] In that case, the protester asserted that the awardee of a task order was ineligible for award and that the agency's evaluation system was flawed. The Army argued that the GAO's new jurisdiction was limited to protests challenging an agency's failure to inform offerors about new competitive procedures for task or delivery order awards and that it did not extend to those challenging an agency's failure to follow its procedures. The GAO rejected this argument, holding that its bid protest jurisdiction over task orders "extend[s] to protests asserting that an agency's award decision failed to reasonably reflect the ground rules established for the task order competition."

In future cases, the GAO will likely provide further guidance as to what constitutes a "fair opportunity" to compete. The FAR provisions of 16.505 governing delivery and task order competitions are much less detailed than the competitive negotiations provisions of FAR Part 15. The requirements for task order competitions emphasize the broad discretion of the CO and encourage streamlining in ways not contemplated under FAR Part 15. Future cases will require the GAO to balance these competing policies.

Notes

1. *See generally* MICHAEL C. LOULAKIS, DESIGN-BUILD FOR THE PUBLIC SECTOR (2003) [hereinafter LOULAKIS, PUBLIC SECTOR]; ROBERT CUSHMAN & MICHAEL C. LOULAKIS, DESIGN-BUILD CONTRACTING HANDBOOK (Aspen Law and Business, 2d

ed. 2001); Michael C. Loulakis, Design-Build Lessons Learned (A/E/C Training Technologies 1995–2004).

2. John B. Miller, Principles of Public and Private Infrastructure Delivery (2000) [hereinafter Miller, Principles] contains one of the most comprehensive, well-written accounts of the history of the federal government's construction project delivery processes. Readers should particularly note Chapter 3 of this treatise. *See also* John B. Miller, Case Studies in Infrastructure Delivery (2002).

3. *See* Cushman & Loulakis, *supra* note 1, at ch. 1.

4. *See* Michael C. Loulakis, Construction Project Delivery Systems: Evaluating the Owner's Alternatives (1999) (http://www.aectraining.com/ConstructProjectDelivSys.html).

5. Miller, Principles, *supra* note 2, at 115.

6. *Id.*

7. Public Buildings Act, 44 Stat. 630 (1926). This Act required prior preparation of plans and specifications by federal employees of the Treasury Department before construction could begin, but it permitted the Secretary of the Treasury to hire architects or engineers to assist these federal employees. *See* Miller, Principles, *supra* note 2, at 116.

8. Armed Services Procurement Act of 1947 (ASPA), 62 Stat. 21 (1948). This Act covered procurement for all armed forces.

9. Federal Property and Administrative Services Act of 1949 (FPASA), 63 Stat. 377 (1949). This Act covered procurement for all civilian agencies.

10. The legislative history of both the ASPA and FPASA did not statutorily require that the construction aspect of a project be based on price alone or that competition occur only after a single design had been produced. The "competition" referred to in these statutes was left to the agency. *See* Miller, Principles, *supra* note 2, at 157 nn.51 & 156.

11. *See* Miller, Principles, *supra* note 2, at 158, n.156.

12. *Id.*

13. 40 U.S.C. §§ 1101–1104 (2003). By recent count, 44 states have the same requirement, under statutes generally referred to as "Little Brooks Acts." *See* http://www.acec.org/programs/qbs.htm. The Brooks Act specified that the compensation to be paid to the successful design firm could not exceed 6 percent of the estimated construction cost of the project. For further discussion of federal A-E contracts, see Chapter 4.

14. *See* Miller, Principles, *supra* note 2, at 117.

15. 48 C.F.R. § 1.000–1.707.

16. FAR subpt. 36.6.

17. Pub. L. No. 98-369, 98 Stat. 1175 (1984).

18. Pub. L. No. 101-355, 108 Stat. 3243 (1994).

19. *See* Loulakis, Public Sector, *supra* note 1, at ch. 1.

20. *Id.* at ch. 2.

21. *See* L.D. Harris, C. Emery, & S.H. Pope, *Federal Design-Build Contracting and Challenges to Contract Award*, Changing Trends in Project Delivery: The Move to Design-Build (ABA Forum on the Construction Industry, Apr. 26–29, 1995).

22. *Id.* at 5.

23. *Id.* at 7.

24. Even the Corps of Engineers used a bit of design-build during the late 1980s and early 1990s. The Corps issued *Design-Build Instructions* for military construction projects in late 1994.

25. The coalition of interest groups included AIA, ASCE, ABC (Associated Builders and Contractors), ACEC (American Council of Engineering Companies), AGC (Associated General Contractors of America), NSPE (National Society of Professional Engineers), and DBIA.

26. Pub. L. No. 103-355, 108 Stat. 3243 (1994); Pub. L. No. 104-106 § 4001, 110 Stat. 679 (1996).

27. 41 U.S.C. § 253m(a) (2003).

28. *Id.*

29. 41 U.S.C. § 253m(b) (2003).

30. 41 U.S.C. § 253m(c) (2003). *See generally* Cushman & Loulakis, supra note 1, at ch. 9.

31. *See* 41 U.S.C. § 253m(c)(1) (2003).

32. *Id.*

33. 40 U.S.C. 1101–1104 (2003); *see also* 41 U.S.C. § 253m(c)(1) (2003).

34. 41 U.S.C. § 253(c)(3) (2003).

35. *Id.*

36. 41 U.S.C. § 235m(d) (2003).

37. 41 U.S.C. § 235m(c)(4) (2003).

38. Federal Acquisition Regulation; Two-Phase Design Build Selection Procedures, 61 Fed. Reg. 69,288 (Dec. 31, 1997); 61 Fed. Reg. 41,212 (Aug. 7, 1996).

39. *See* Loulakis, Public Sector, *supra* note 1, at Ch. 1.

40. For further discussion of negotiated procurements, see Chapter 3.

41. Those interested in a detailed review of how federal agencies are using design-build should review Loulakis, *Design-Build for the Public Sector, supra* note 1. This book contains several chapters that specifically address the use of design-build in several discrete industry sectors.

42. In March 2000, the DBIA published its *Guide to the Federal Design-Build Marketplace*. DBIA's guide briefly recaps the history of design-build in the federal market and contains some examples of federal design-build projects. Particularly noteworthy are the appendices in the guide, which contain surveys of how various agencies have approached the design-build process, as well as the procurement processes they have used. The guide is updated on a periodic basis.

43. David Shaffer, *Red Wing Lock and Dam to Get $70 Million Upgrade*, Star Tribune (Minneapolis-St. Paul), May 26, 2009, *available at* http://www.startribune.com/46022472.html?elr=KArksDyycyUtyycyUiD3aPc:_Yyc:aUU.

44. Federal Programs—Department of Defense, U.S. Army Corps of Engineers Reconstruction Projects: Design/Build First Brigade, Pol-E-Charkhi, Afghanistan, http://www.perini.org/pmsi/federal_defense_body.htm (last visited May 20, 2009).

45. *Id.*

46. *See* Loulakis, Public Sector, *supra* note 1, at Ch. 8.

47. *Id.*

48. U.S. Penitentiary and Federal Prison Camp, http://www.hsmm.aecom .com/MarketsAndServices/61/58/index.html (last visited May 20, 2009).

49. Solicitation Number X00-0576, Design-Build Construction of a USP to be located in Yazoo County, near Yazoo, MS, http://www.fbo.gov (last visited May 20, 2009).

50. Press Release, Jacobs Eng'g Group, Jacobs Receives Contract from Federal Bureau of Prisons for Federal Correctional Institution, Berlin, New Hampshire (Feb. 10, 2009), *available at* http://www.redorbit.com/modules/news/tools .php?tool=print&id=1636752.

51. *See* Loulakis, Public Sector, *supra* note 1, at Ch. 13.

52. Federal Programs—Department of Defense, Iraq Project and Contracting Office: Electrical Transmission & Distribution, Southern Region, Iraq, http:// www.perini.org/pmsi/federal_defense_body.htm (last visited May 20, 2009).

53. The Department of Defense enlisted Novogradac & Company, LLP, as a Transaction Advisor and Portfolio/Asset Manager for the DoD's Military Housing Privatization program. The CPA firm publishes these and other statistics on its Website, www.novoco.com/govalgroup/mhpi/index.php (last visited March 2009).

54. *Id.*

55. Office of the Deputy Under Secretary of Defense Installation and Environment, Military Housing Privatization, http://www.acq.osd.mil/housing/ projawarded.htm (last visited May 21, 2009).

56. *Innovative Contracting Practices*, FHWA Special Experimental Project No. 14 (SEP-14), 66 Fed. Reg. 53,288 (Oct. 19, 2001).

57. For projects on which Intelligent Transportation Systems (ITS) are being installed, the threshold minimum size is $5 million.

58. 23 C.F.R. pt. 636.

59. Federal Highway Administration, Design-Build Effectiveness Study as Required by TEA-21, Section 1307(f), Final Report (Jan. 2006), *available at* http://www.fhwa.dot.gov/reports/designbuild/designbuild.htm.

60. Department of Transportation, Federal Highway Administration, New Special Experimental Project (SEP-15) to Explore Alternative and Innovative Approaches to the Overall Project Development Process; Information, 69 Fed. Reg. 59,983 (Oct. 6, 2004).

61. Kevin Sheys, "SEP-15" for Transit?, Address at Partnerships in Transit, a program of the National Council for Public-Private Partnerships) (May 30, 2008), *available at* http://www.ncppp.org/Publications/TransitDenver_0806/Roundtable Handout_080612.pdf.

62. Design-Build Contracting Final Rule, 72 Fed. Reg. 45,329 (Aug. 14, 2007) (to be codified at 23 C.F.R. pts. 630, 635, 636).

63. Standing Comm. on the Env't, Am. Ass'n of State Highway and Transp. Officials, Design-Build Environment Compliance Process and Level of Detail: Eight Case Studies (2005). The projects highlighted in the report were: Davis Dam in Mohave County, Arizona; Transportation Expansion (T-REX) Multi-Modal Project, Denver, Colorado; Interstate 95, St. John's County, Florida; Widening of I-4, Orange County, Florida; U.S. 113 Dualization, Worcester

County, Maryland; U.S. 64 Knightdale Bypass, Wake County, North Carolina; State Highway 130 Toll Project, Austin, Texas; New Tacoma Narrows Suspension Bridge, Tacoma, Washington.

64. *New IRS Kansas City Campus Transforms Downtown Cityscape*, MOD. BUILDER, Nov.–Dec. 2006 at 2–5.

65. Jim L. Whitaker, AIA, DBIA, *Counting on Design-Build: US Census Bureau Uses Design-Build for New Headquarters to Ensure Readiness for 2010 Census–Saving a Year in Project Delivery Time*, DESIGN-BUILD DATELINE, May 2007, *available at* http://www.dbia.org/pubs/dateline/archives/2007/05-07/Features/.

66. *See* LOULAKIS, PUBLIC SECTOR, *supra* note 1, at Ch. 1.

67. The bridging concept is believed to have been conceived by George Heery, FAIA, formerly of Heery International, now a principal of The Brookwood Group in Atlanta, Georgia.

68. ASBCA No. 39978, 93-3 BCA ¶ 26,189.

69. *Id.*

70. *Id.*

71. Donahue Electric, Inc., VABCA No. 6618, 03-1 BCA ¶ 32,129.

72. For other cases on this point, see MICHAEL C. LOULAKIS, DESIGN-BUILD LESSONS LEARNED, A/E/C TRAINING TECHNOLOGIES, L.L.C. (published annually, 1995–2004).

73. For further discussion of defective specifications, see Chapter 18.

74. *See* LOULAKIS, PUBLIC SECTOR, *supra* note 1, at Ch. 1.

75. ASBCA No. 49509, 99-2 BCA ¶ 30,512.

76. *Id.*

77. *Id.*

78. *Id.*

79. For a broader discussion of this case, see Loulakis, *supra* note 72.

80. 296 F.3d 1081 (Fed. Cir. 2002), *aff'g* Edsall Constr. Co., Inc., ASBCA No. 51787, 01-2 BCA ¶ 31,425.

81. ASBCA No. 53902, 05-2 BCA ¶ 33,126.

82. For further discussion of site investigations, see Chapter 10.

83. ASBCA No. 55671, 2008 WL 2231484 (May 22, 2008).

84. 48 C.F.R. § 9.502. For further discussion of OCIs, see Chapter 7.

85. FAR 36.209. Some agencies have made specific allowances in their regulations to address this subject. As an example, the U.S. Department of Agriculture's regulations permit the head of the contracting activity "to approve the award of a contract to construct a project, in whole or in part, to the firm (inclusive of its subsidiaries or affiliates) that designed the project." 48 C.F.R. § 436.209.

86. No. B-282244, June 18, 1999, 99-2 CPD ¶ 27.

87. FAR 9.505-2(b) (1).

88. 99-2 CPD ¶ 27 at 3 (*citing* LW Planning Group, No. B-215539, Nov. 14, 1984, 84-2 CPD ¶ 531 at 4).

89. *Id.* at 4.

90. Douglas D. Gransberg, Ph.D., PE & Keith R. Molenaar, Ph.D., The Impacts of Design-Build on the Public Workforce (April 2007), (Research Paper 07-01, USC Keston Institute for Public Finance and Infrastructure Policy) (*available at* http://www.usc.edu/schools/sppd/keston/pdf/20070413-design-build.pdf).

91. *Id.*

92. For other cases on this point, see Loulakis, *supra* note 72.

93. SKE Int'l, Inc., B-311383, B-311383-2, June 5, 2008, 2008 CPD ¶ 111.

94. ICON Consulting Group, B-310431.2, Jan. 30, 2008, 2008 CPD ¶ 38.

95. *Id.* at 5.

96. *Id.*

97. *Id.*

98. Ronald Cilensek, *The Role of Agency Construction Management in Alternate Project Delivery*, CM ADVISOR, Nov.–Dec. 2007, at 12. LOULAKIS, WICKWIRE BERRY & DRISCOLL, CONSTRUCTION MANAGEMENT LAW AND PRACTICE (John Wiley & Sons, Inc. 1995), at 11–14; Construction Management Association of America (CMAA) *Selecting the Best Delivery Method for Your Project*, http://cmaa.net.org/nde/283.

99. LOULAKIS, *supra* note 98, at 14; ROBERT F. CUSHMAN & JAMES T. MYERS, CONSTRUCTION LAW HANDBOOK, at 348–49 (1999).

100. The American Institute of Architects, the Associated General Contractors of America, and the CMAA have drafted standard form construction management agreements. *See* 2 PHILIP L. BRUNER & PATRICK J O'CONNOR, BRUNER & O'CONNOR ON CONSTRUCTION LAW §§ 6:62–6:64 (2008).

101. Cilensek, *supra* note 98, at 13; CUSHMAN, *supra* note 1, at 350–51; BRUNER, *supra* note 100, § 6:57; CMAA, What Is Construction Management?, http://cmaanet .org/node/112.

102. BRUNER, *supra* note 100, §§ 6:58, 6:67.

103. Cilensek, *supra* note 98.

104. LOULAKIS, *supra* note 98, at 14–15; BRUNER, *supra* note 100, § 6:58.

105. BRUNER, *supra* note 100, § 6:61; CUSHMAN, *supra* note 99, at 348.

106. BRUNER, *supra* note 100, § 6:59; CUSHMAN, *supra* note 99, at 350–51; Cilensek, *supra* note 98, at 13.

107. BRUNER, *supra* note 100, § 6:59; CUSHMAN, *supra* note 99, at 350–51; Cilensek, *supra* note 98, at 13; CMAA, *supra* note 101.

108. BRUNER, *supra* note 100, § 6:59; CUSHMAN, *supra* note 99, at 350–51; Cilensek, *supra* note 98, at 13; CMAA, *supra* note 101.

109. CMAA, *supra* note 101.

110. LOULAKIS, *supra* note 98, at 2–4 (Supp. 1998); Turner Constr. Co., ASBCA Nos. 25447, et seq., 90-2 BCA ¶ 22,649; G.L. Cory, Inc., GSBCA No. 4474, 78-1 BCA ¶ 185, *recon. denied*, 78-2 BCA ¶ 13,333.

111. For example, in 2009, GSA estimated that it would use the following delivery methods for its capital projects: traditional design-bid-build (50–60 percent), design-build (5–10 percent), and CM at risk (30–40 percent). Larry J. Smith, *Construction Management for Owners*, CM ADVISOR, Sept.–Oct. 2008, at 10.

112. *Id.*

113. FAR Part 16 ("Types of Contracts"), Part 17 ("Special Contracting Methods"), Part 36 ("Construction and Architect—Engineer Contracts") and Part 52 ("Solicitation Provisions and Contract Clauses") do not address CM contracts. Nor do the applicable FAR agency supplements.

114. However, the Boards of Contract Appeals have decided disputes that arose under CM contracts. For example, the Boards have concluded that a CM at risk may recover for a differing site condition. Whiting Turner/A.L. Johnson Joint

Venture, GSBCA No. 15401, 02-1 BCA ¶ 31,708; *Turner Constr.*, 90-2 BCA ¶ 22,649. Similarly, in *G.L. Cory*, 78-1 BCA ¶ 13,185, the Board applied standard termination provisions to a convenience termination under a CM contract.

115. A sample GSA solicitation for an IDIQ contract requiring CM services can be found at http://cmaanet.org/user_images/idiq_example.pdf.

116. Contractors may obtain descriptions of GSA and USACE solicitations for CM services (both agency and at risk) at the Federal Business Opportunities Web site, http://www.FedBizOpps.gov. For example, in 2008, GSA issued Solicitation GS-04P-08-Ex-C-0119 for a new federal courthouse in Tuscaloosa, Alabama. The solicitation anticipated the award to a CM as Constructor, for design phase services and construction phase services, at a firm fixed price. Similarly, in 2009, the GSA Great Lakes Region issued Solicitation GS-05P-09-GB-0015 for a CM as Constructor for large-scale construction projects. The solicitation anticipated the award of IDIQ contracts to multiple contractors. Task orders would be issued on a competitive basis.

117. *Id.*

118. Ralph Nash, *Rebuilding the Federal Acquisition Workforce: Some Good Input*, 22 Nash & Cibinic Rep. 23 (2008); Ralph Nash, *The Acquisition Workforce: Technical Competence Is Required*, 21 Nash & Cibinic Rep. 59 (2007); Ralph Nash, *"Improving" the Workforce: Can It Be Done*, 19 Nash & Cibinic Rep. 44 (2005).

119. Pub. L. No. 111-5, 123 Stat. 115.

120. *See Top Management Challenges Facing the Department of Transportation: Hearing Before the H. Subcomm. on Transp., Hous. & Urban Dev., & Related Agencies*, 111th Cong. 3 (2009) (statement of Calvin L. Scovel III, Inspector Gen., U.S. Dep't of Transp.), *available at* http://www.oig.dot.gov/StreamFile?file=/data/pdfdocs/cc2007021.pdf ("DOT must ensure that it has sufficient personnel with relevant expertise to meet the increased workload and accelerated timeframes associated with overseeing stimulus spending. . . . Our work has shown that DOT faces substantial challenges in developing and maintaining a competent acquisition workforce to support its mission.").

121. *See* Matthew Weigelt, *Stimulus Funds Bring Acquisition Showdown*, Fed. Computer Week, Mar. 27, 2009, *available at* http://www.fcw.com/Articles/2009/03/30/Acquisition-stimulus-cover-story.aspx.

122. To assist with selecting and implementing alternative delivery methods, government agencies may choose to employ acquisition support services such as the GSA Office of Assisted Acquisition Services or the National Business Center's Acquisition Services Directorate. Because these organizations receive fees for managing agencies' procurement, they are functionally similar to CMs. *See generally* Acquisition Services Directorate—Home, http://www.aqd.nbc.gov/index.asp; GSA—Office of Assisted Acquisition Services, http://www.gsa.gov/aas.

123. FAR 16.501-1.

124. *Id.*

125. *Id.*

126. Tyler Constr. Corp. v. United States, 83 Fed. Cl. 94 (2008), *aff'd*, 2009 WL 1796702 (Fed. Cir. June 25, 2009); Vernon Edwards, *When Acquisition Strategy Radically Changes the Market: The Army Corps of Engineers Use of IDIQ Contracts*, 22 Nash & Cibinic Rep. ¶ 66 (2008).

127. FAR 16.501-2(a).

128. *Id.*

129. FAR 16.501-2(b)(1)–(2).

130. FAR 16.502(a).

131. *Id.*

132. FAR 16.503(a).

133. FAR 16.503(b)(1).

134. FAR 16.503(a)(2).

135. Network Med. Servs., DOTBCA No. 4059, 2000-2 BCA ¶ 31,068; Robertson & Penn, Inc. d/b/a Cusset Laundry, Inc., ASBCA No. 55625, 08-2 BCA ¶ 33,951 ("When using a requirements-type contract, the government is required to order its requirements from the contractor but is under no obligation to actually have requirements as long as the absence of requirements is in good faith").

136. FAR 16.503(a)(1); Fed. Group, Inc. v. United States, 67 Fed. Cl. 87, 97–101 (2005); Datalect Computer Servs., Ltd. v. United States, 40 Fed. Cl. 28 (1997), *aff'd in part, vacated in part and remanded,* 215 F.3d 1344 (Fed. Cir. 1999) (Table), *cert. denied,* 529 U.S. 1037 (2000).

137. FAR.16.503(a)(1).

138. *Robertson & Penn,* 08-2 BCA ¶ 33,419; Medart, Inc., GSBCA No. 8939, 91-2 BCA ¶ 23,741.

139. Technical Assistance Int'l, Inc. v. United States, 150 F.3d 1369 (Fed. Cir. 1998); Fed. Group, Inc. v. United States, 67 Fed. Cl. 87, 97 (2005); Cmty. Res. for Justice v. U.S. Dep't of Justice Fed. Bureau, DOTCAB No. 4541, 06-2 BCA ¶ 33,419; Bannum, Inc. v. Dep't of Justice, Fed. Bureau of Prisons, DOTCAB No. 4450, 05-2 BCA ¶ 33,049.

140. *Cmty. Res.,* 06-2 BCA ¶ 33,419; *accord Medart,* 91-2 BCA ¶ 23,741; *see also* Shader Contractors, Inc. v. United States, 276 F.2d 1 (Ct. Cl. 1960).

141. *See generally* Medart v. Austin, 967 F.2d 579 (Fed. Cir. 1992); Engineered Demolition, Inc. v. United States, 70 Fed. Cl. 580, 592 (2006); *Robertson & Penn,* 08-2 BCA ¶ 33,951; S.P.L. Spare Parts Logistics, Inc., ASBCA Nos. 51118, 51384, 02-2 BCA ¶ 31,982.

142. Hi-Shear Tech. Corp. v. United States, 356 F.3d 1372, 1378–80 (Fed. Cir. 2004); *S.P.L.,* 02-2 BCA ¶ 31,982.

143. Rumsfeld v. Applied Cos., 325 F.3d 1328, 1335 (Fed. Cir. 2003) (quoting Clearwater Forest Indus., Inc. v. United States, 650 F.2d 233, 239 (Ct. Cl. 1981)); Cardiometrix, DOTBCA No. 3047, 98-2 BCA ¶ 29,901 (citing *Clearwater,* 650 F.2d 233).

144. HKH Capitol Hotel Corp., ASBCA No. 47575, 98-1 BCA ¶ 29,548 (citing Craft Mach. Works, Inc v. United States, 20 Cl. Ct. 355 (1990), *rev'd on other grounds,* 926 F.2d 1110 (Fed. Cir. 1991)).

145. Fairfax Opportunities Unlimited, Inc., AGBCA No. 96-178-1, 98-1 BCA ¶ 29,556; *cf.* Fed. Group, Inc. v. United States, 67 Fed. Cl. 87, 98 (2005) ("The common threads in the cases where courts have found estimates to be unreasonable are actual knowledge that the estimates were inaccurate or failure to take reasonable efforts to confirm doubtful estimates").

146. Fed. Group, Inc. v. United States, 67 Fed. Cl. 87, 98–101 (2005); United Mgmt., Inc. v. Dep't of the Treasury, GSBCA No. 13515-TD, 97-2 BCA ¶ 29,262 (citing Crown Laundry & Dry Cleaners, Inc. v. United States, 29 Fed. Cl. 506 (1993), MDP Constr., Inc., ASBCA No. 50603, 97-2 BCA ¶ 29,211).

147. Int'l Data Prods. Corp. v. United States, 64 Fed. Cl. 642, 648 (2005) (citations omitted); Inland Container, Inc. v. United States, 512 F.2d 1073 (Ct. Cl. 1975); Ready Mix Concrete Co. v. United States, 158 F. Supp. 571 (Ct. Cl. 1958); Ace-Fed. Reporters, Inc. v. Gen. Servs. Admin., GSBCA Nos. 13298 et al., 99-1 BCA ¶ 30,139.

148. Coyle's Pest Control, Inc. v. Cuomo, 154 F.3d 1302, 1305 (Fed. Cir. 1998); Modern Sys. Tech. v. United States, 979 F.2d 200, 205 (Fed. Cir. 1992).

149. Hi-Shear Tech. Corp., 356 F.3d at 1378–79; Applied Cos., 325 F.3d at 1339; Medart v. Austin, 967 F.2d 579 (Fed. Cir. 1992); MDP Constr., Inc., ASBCA No. 49527, 96-2 BCA ¶ 28,525; S&W Tire Servs., Inc., GSBCA No. 6376, 82-2 BCA ¶ 16,048.

150. FAR 16.504(a).

151. FAR 16.504(b) provides that an IDIQ contract should be used only when a recurring need is anticipated.

152. FAR 16.503(a).

153. FAR 16.504(a)(1); *see also* Int'l Data Prods. Corp. v. United States, 64 Fed. Cl. 642, 646–47 (2005); Rice Lake Contracting, Inc. v. United States, 33 Fed. Cl. 144 (1995) (holding that government not required to order all of its needs from one contractor under IDIQ contract).

154. Schweiger Constr. Co. v. United States, 49 Fed. Cl. 188, 194 (2001) (citations omitted).

155. FAR 16.504(a)(1)–(2).

156. Willard, Sutherland & Co. v. United States, 262 U.S. 489, 493 (1923); Coyle's Pest Control, Inc. v. Cuomo, 154 F.3d 1302, 1302 (Fed. Cir. 1998); Modern Sys. Tech. Corp. v. United States, 24 Ct. Cl. 360, 366 (1991); Mason v. United States, 615 F.2d 1343, 1350 (Ct. Cl. 1980) (concluding that to be enforceable a contract must state an ascertainable minimum quantity); *see also, e.g.,* So. Def. Sys., Inc., ASBCA No. 54045, 07-1 BCA ¶ 33, 536.

157. *See* Howell v. United States, 51 Fed. Cl. 516, 523 (2002) (although no minimum quantity stated in solicitation, court could supply the missing terms into the contract).

158. FAR 16.504(a)(2).

159. Interagency Agreements—Obligation of Funds Under an Indefinite Delivery, Indefinite Quantity Contract, Comp. Gen. B-308969, May 31, 2007, 07-1 CPD ¶ 120.

160. *See* Greenlee Constr., Inc. v. Gen. Servs. Admin., CBCA No. 416, 07-1 BCA ¶ 33,514; Mac's Cleaning & Repair Serv., ASBCA No. 49652, 97-1 BCA ¶ 28,748.

161. FAR 16.504(a)(1).

162. Dynamic Decisions, Inc. v. Dep't of Health and Human Res., GSBCA No. 27731, 95-2 BCA ¶ 27,732.

163. FAR 16.504(a)(2); *see also* Willard, Sutherland & Co. v. United States, 262 U.S. 489, 493 (1923) (contract without a minimum quantity term is unenforceable for lack of consideration and mutuality); Int'l Data Prods. Corp. v. United States, 64 Fed. Cl. 642, 646 (2005) (purchase of minimum extinguishes government obligation); Rice Lake Contracting, Inc. v. United States, 33 Fed. Cl. 144, 154 (1995) (government obligation was only to satisfy the stated minimum in the contract); Mason v. United States, 615 F.2d 1343, 1348–50 (Ct. Cl. 1980) (government was

under no obligation to order all construction services from its IDIQ contractor where the government ordered the minimum quantity); Coastal States Petrochem. Co. v. United States, 559 F.2d 1, 1 (Ct. Cl. 1977) (only obligation owed by government to contractor was to satisfy the minimum purchase requirement); Crown Laundry & Dry Cleaners, Inc., ASBCA No. 39982, 90-3 BCA ¶ 22,993, *aff'd*, 935 F.2d 281 (Fed. Cir. 1991) (courts will not examine the reasonableness of estimates in indefinite quantity contracts).

164. Schweiger Constr. Co. v. United States, 49 Fed. Cl. 188, 198 (2001) (citing *Crown Laundry*, 90-3 BCA ¶ 22,993); C.F.S. Air Cargo, Inc., ASBCA No. 40694, 91-2 BCA ¶ 23,985, *aff'd*, 944 F.2d 913 (Fed. Cir. 1992).

165. 231 Ct. Cl. 765 (1982).

166. *Id.*

167. *Id.* at 768; *accord C.F.S. Air Cargo*, 91-2 BCA ¶ 23,985 (public policy and essential fairness do not mandate government accountability for negligent preparation of estimates in IDIQ contract because such negligence is immaterial in light of the government's legal obligation to order only the guaranteed minimum); Deterline Corp., ASBCA No. 33090, 88-3 BCA ¶ 21,132; *see also* Travel Ctr. v. Gen. Servs. Admin., GSBCA No. 14057, 98-1 BCA ¶ 29,536, *rev'd*, 236 F.3d 1316 (Fed. Cir. 2001) (government owes no duty to provide realistic estimates in IDIQ contract once minimum amount under contract is ordered); Applied Devices Corp. v. United States, 591 F.2d 635, 640–41 (Ct. Cl. 1979) (requiring realistic estimates requirements to be read into the contract under the *Christian* doctrine); Schweiger Constr., 49 Fed. Cl. at 196–98 (government not liable for negligently prepared estimates in IDIQ contracts lacking evidence of bad faith); RJO Enters., Inc., ASBCA No. 50981, 03-1 BCA ¶ 32,137 ("Regardless of the accuracy of the estimates delineated in the solicitation, based on the language of the solicitation for the IDIQ contract, [contractor] could not have had a reasonable expectation that any of the government's needs beyond the minimum contract price would necessarily be satisfied under this contract.") (quoting Travel Centre v. Barram, 236 F.3d 1316, 1319 (Fed. Cir. 2001)); *Crown Laundry*, 90-3 BCA ¶ 22,993 (courts will not examine the reasonableness of estimated quantities in IDIQ context where government ordered the minimum required under the contract); *Deterline*, 89-3 BCA ¶ 22,069 (contractor bears risks of inaccurate estimates in IDIQ context so long as government orders the minimum dollar amount specified in contract); Art Anderson Assocs., ASBCA No. 27807, 84-1 BCA ¶ 17,225 (purpose of IDIQ contract is for use when government cannot provide estimated needs except in terms of minimum and maximum thresholds).

168. *Travel Centre*, 236 F.3d at 1319.

169. FAR 16.500(a).

170. FAR 16.504(c)(1)(i).

171. FAR 16.504(c)(1)(ii)(A).

172. FAR 16.504(c)(1)(ii)(B).

173. FAR 16.504(c)(1)(ii)(C).

174. FAR 16.504(c)(1)(ii)(D).

175. FAR 16.504(a)(4)(iv).

176. FAR 16.505(a)(2).

177. FAR 16.505(a)(6).

178. FAR 16.505(b)(1)(i). The exceptions to the fair opportunity process include: (a) an agency need that is so urgent that providing a fair opportunity to all award-ees would result in unacceptable delays; (b) supplies or services that are so unique or specialized that only one awardee can provide them in a timely manner; (c) an order that must be issued on a sole-source basis in the interest of efficiency and economy, because it is a logical follow-on to an order already issued under the contract and awardees were provided a fair opportunity to compete for the exist-ing order; and (d) a sole-source order that is necessary to comply with a minimum guarantee. FAR 16.505(b)(2)(i)–(iv).

179. FAR 16.505(b)(1)(ii).

180. *Id.*

181. *Id.*

182. FAR 16.505(b)(1)(ii)(A).

183. FAR 16.505(b)(1)(iv).

184. FAR 16.505(b)(1)(ii)(E).

185. FAR 16.505(b)(3).

186. FAR 15.505(b)(1)(iv)(B), (b)(5).

187. 10 U.S.C. § 2304(d); 41 U.S.C. § 253j(d); Vernon Edwards, *Taming the Task Order Contract: Congress Tries Again*, 22 NASH & CIBINIC REP. ¶ 31 (2008).

188. FAR 16.505(b)(1)(iii).

189. FAR 16.505(b)(4).

190. FAR 15.505(b)(6).

191. Ostensibly to avoid problems that arise where the change constitutes a cardinal change. For cases addressing cardinal change within an IDIQ or related contract context, see Int'l Data Prods. Corp. v. United States, 70 Fed. Cl. 387 (2006); HDM Corp. v. United States, 69 Fed. Cl. 243 (2005); Northrop Grumman Corp., 50 Fed. Cl. 443 (2001); Melrose Assocs., L.P. v. United States, 47 Fed. Cl. 595 (2000); Green Mgmt. Corp. v. United States, 42 Fed. Cl. 411 (1998).

192. Vernon Edwards, *Task Order Contracting: Understanding and Using Task Order Contracts,* GOVERNMENT CONTRACTING TRAINING AND CONSULTING 83 (Jan. 1999).

193. *See, e.g.,* Fire Sec. Sys., Inc., ASBCA No. 53498, 02-1 BCA ¶ 31,806 (court's objective assessment of contract language dispositive of whether change order fell within the scope); Brown & Root Servs., ASBCA No. 44020, 95-2 BCA ¶ 27,860 (scope of work for individual delivery order to be determined according to the operating document as defined within the contract instrument); *cf.* AT&T Commc'ns, Inc. v. Wiley, Inc., 1 F.3d 1201, 1205 (Fed. Cir. 1993).

194. Anteon Corp., B-293523, Mar. 29, 2004, 2004 CPD ¶ 51; Ervin & Assocs., Inc. B-279083, B-279219, 1998 U.S. Comp. Gen. LEXIS 194 (Apr. 30, 1998).

195. Morris Corp., B-400336, Oct. 15, 2008, 2008 CPD ¶ 204; Neal R. Gross & Co., B-237434, Feb. 23, 1990, 90-1 CPD ¶ 212.

196. For further discussion of convenience terminations, see Chapter 13.

197. Hermes Consol., Inc., ASBCA No. 52308, 02-1 BCA ¶ 31,767; IMS Eng'rs-Architects, P.C., ASBCA No. 53471, 06-1 BCA ¶ 33,231.

198. *See also* ACE-Fed. Reporters, Inc. v. Gen. Servs. Admin., GSBCA No. 13298, 1998 WL 1674436 (Oct. 30, 1998).

199. Int'l Data Prods. Corp. v. United States, 64 Fed. Cl. 642 (2005); Ralph Nash, *Fixed Price IDIQ Contracts: High Risk Ventures*, 21 Nash & Cibinic Rep. ¶ 43 (2007).

200. *See, e.g.*, Global Solutions Network, Inc., B-401230, June 26, 2009, 2009 WL 1856507 (IDIQ); Brooks Range Contract Servs., Inc., B-401231, June 23, 2009, 2009 WL 1770793 (task order); Northrop Grumman Info. Tech., Inc., B-401198, B-401198.2, June 2, 2009, 2009 WL 1622384 (task order); Triple Canopy, Inc., B-310566.9, B-400437.4, Mar. 25, 2009, 2009 CPD ¶ 62 (task order); Med. Staffing Joint Venture, LLC, B-400705.2, B-400705.3, Mar. 13, 2009, 2009 CPD ¶ 71 (IDIQ); Advanced Sci. Applications, Inc., B-400312.2, Feb. 5, 2009, 2009 CPD ¶ 41 (IDIQ).

201. 41 U.S.C. § 253j(e); FAR 16.505(a)(9)(i).

202. National Defense Appropriations Act of 2008, Pub. L. No. 110-181, 122 Stat. 239; FAR 16.505(a)(9)(B). Note that the GAO's jurisdiction over delivery or task order protests exceeding $10 million expires on May 27, 2011, unless otherwise extended. FAR 16.505(a)(9)(B)(2); *see also* 10 U.S.C. §§ 2304a(d), c(d); 41 U.S.C. §§ 253h(d), 253j(d). The COFC does not have jurisdiction over bid protests against task or delivery orders above $10 million.

203. B-310566.4, Oct. 30, 2008, 2008 CPD ¶ 207, Delex Sys., Inc., B-400403, Oct. 8, 2008, 2008 CPD ¶ 181 (SBA's Rule of Two applies to individually competed task orders under multiple award contracts).

CHAPTER 6

Federal Bid Protests

JAMES F. NAGLE
ADAM K. LASKY

I. Introduction

This chapter discusses the background and history of the federal bid protest forums, and provides a procedural guide for bringing a federal bid protest. Currently, an eligible bidder/proposer may choose to file a protest challenging a federal contract award, and the procedure by which the contract offers were solicited, in its choice of three forums: (1) the agency whose procurement procedures are being challenged; (2) the Government Accountability Office (GAO);[1] or, (3) the Court of Federal Claims (COFC). Agency actions taken pursuant to GAO decisions may be reviewed by the COFC. In some circumstances a protest may also be brought in a federal district court, although the extent of district court jurisdiction is both tenuous and unclear.[2] Formerly, the General Services Board of Contract Appeals (GSBCA) had jurisdiction to hear bid protests concerning certain procurements of information technology,[3] but the GSBCA's jurisdiction was eliminated by Congress in 1996.[4]

II. History and Background of Bid Protest Forums

A. Agency-Level Protests

Although agency-level bid protests have taken place for many years, it was not until the mid-1990s that any government-wide regulations were enacted to regulate such protests.[5] The current system was initiated by President Clinton in 1995, when, in an effort to "reduce litigation and increase cooperation between the Government and industry in the procurement process,"[6] he issued Executive Order 12979.[7] By this order, federal agencies were required to proscribe bid protest procedures that (a) require all parties to use their "best efforts" to resolve the matter with agency contracting officers;

(b) wherever possible to make available forums, such as alternative dispute resolution and mediation, that provide for inexpensive and expeditious resolutions to bid protests; (c) allow actual or prospective bidders or offerors whose direct economic interests would be affected by the award or failure to award the contract to request a review of any decision by a contracting officer at a level above that contracting officer; and (d) except in emergencies, to stay contract award or performance while a timely protest is pending before the agency.[8]

B. GAO Protests

Created by the Budget and Accounting Act of 1921,[9] the GAO became the first external forum for federal bid protests.[10] The GAO was established as an independent governmental agency under the control and direction of the Comptroller General for the United States.[11] Even though the statutes giving the GAO jurisdiction to hear bid protests were not enacted until the mid-1980s, the GAO has been hearing bid protests since the 1920s.[12]

C. Federal Court Protests

The Court of Claims was created in 1885 to adjudicate private claims against the federal government.[13] In 1887, Congress expanded the court's jurisdiction with the passage of the Tucker Act, allowing the court to adjudicate all claims against the government except tort, equitable, and admiralty claims.[14]

The origin of judicial bid protests can be traced back to the 1950s, when the Court of Claims held that a bidder for a government contract enters into an implied contract under which the government promises to consider its bid fairly and honestly.[15] However, prior to 1970, very few bid protests were heard by federal courts[16] because the protesting party had "no standing to sue because he had no 'right' to a government contract which could be invaded by improper governmental action," and the federal procurement agency often had sovereign immunity protection.[17] This changed with the 1970 landmark decision of *Scanwell Laboratories, Inc. v. Schaffer*.[18] In *Scanwell*, the United States Court of Appeals for the District of Columbia Circuit (D.C. Circuit) held that the Administrative Procedure Act (APA)[19] gave bidders standing to challenge agency action.[20] The *Scanwell* decision embraced "the basic presumption of judicial review to one suffering legal wrong because of agency action,"[21] and held that "one who has a prospective beneficial relationship has standing to challenge the illegal grant of a contract to another."[22] Other federal circuits subsequently confirmed that disappointed bidders had standing to protest a government procurement award in district court.[23]

In 1982, the COFC gained the power to grant injunctive relief in pre-award bid protests.[24] The Administrative Dispute Resolution Act of 1996 (ADRA)[25] significantly amended the COFC's bid protest jurisdiction.[26] The ADRA gave the COFC and federal district courts concurrent jurisdiction to adjudicate all pre- and post-award federal bid protests brought by an "interested party,"[27] including the right to "award any relief that the court considers proper, including declaratory and injunctive relief except that any monetary relief shall be limited to bid preparation and proposal costs."[28] The ADRA contained a sunset provision on concurrent jurisdiction, so that district courts' ADRA jurisdiction over bid protests would lapse on January 1, 2001, in the absence of any act of Congress to extend that jurisdiction.[29]

When Congress did not act to extend the district courts' jurisdiction, a schism arose among the federal courts as to whether the district courts retained jurisdiction to adjudicate bid protests pursuant to *Scanwell*.[30] The majority of courts hold that the COFC became the exclusive judicial forum for the adjudication of federal bid protest disputes as of January 1, 2001.[31] However, some courts have held that district courts retain *Scanwell* jurisdiction under the APA to adjudicate bid protests, so long as they are not initiated by "interested parties."[32] The basis for this argument is that the sunset provision only affects actions described in the ADRA, namely protests brought by an "interested party."[33] An "interested party" for purposes of the ADRA is "an actual or prospective bidder or offeror whose direct economic interest would be affected by the award of the contract or by failure to award the contract."[34] Therefore, one court has reasoned:

> the [ADRA] did not affect the district court's ability to hear cases challenging the government's contract procurement process so long as the case is brought by someone other than an actual or potential bidder. The district court retains subject matter jurisdiction over cases brought by non-bidders under 28 U.S.C. § 1331 and the waiver of sovereign immunity in the Administrative Procedure Act.[35]

Recently, a minority of courts have even held that district courts have jurisdiction under the APA to adjudicate bid protests brought by "interested parties."[36]

Today, the general jurisdiction of the COFC, as outlined in 28 U.S.C. § 1491, commonly referred to as the Tucker Act, covers most suits against the federal government, including bid protests.[37] The COFC acts as a finder of both fact and law, and adjudicates disputes without a jury.[38]

III. *Procedural Guide to Federal Bid Protests*

A. Agency-Level Protests

The first option available to a disappointed bidder in a federal procurement is to file its bid protest with the procuring agency. This option is the least expensive and most informal. Agency-level protests are governed by FAR 33.103, in conjunction with each procurement agency's individual regulations.[39]

1. Standing to Protest

Any "interested party" may bring an agency-level bid protest.[40]

2. Filing the Protest

Protests filed with the procuring agency should be directed to either the contracting officer or the official "at a level above the contracting officer" designated by the agency to independently review the protest.[41] Depending on the agency, "independent review" is available either as an alternative to consideration by the contracting officer or as an appeal of the contracting officer's decision on the protest.[42] The protest must be concise and logical,[43] and must include all of the following information:

(i)   Name, address, and fax and telephone numbers of the protester.
(ii)   Solicitation or contract number.
(iii)   Detailed statement of the legal and factual grounds for the protest, to include a description of resulting prejudice to the protester.
(iv)   Copies of relevant documents.
(v)   Request for a ruling by the agency.
(vi)   Statement as to the form of relief requested.
(vii)   All information establishing that the protester is an interested party for the purpose of filing a protest.
(viii)   All information establishing the timeliness of the protest.[44]

Failure to provide any of the above information may be grounds for dismissal of the protest.[45]

3. Timeliness of Protest

A party should act expeditiously when filing an agency-level protest. Failure to file in a timely manner can limit the relief available or eliminate the agency as a viable forum for the protest. The time limitations for when a protest may be filed depend in part on what aspect of the procurement is being protested. Where a protest is based on alleged improprieties in a solicitation, the protest must be filed before bid opening or the closing date for receipt of proposals.[46] In all other cases, protests must be filed no later than 10 calendar days after the basis of protest is known or should have been known, whichever is earlier.[47] The procuring agency is not required to consider protests that are not timely filed. However, the agency may still consider the merits of an untimely protest if "good cause" is shown, or the "protest raises issues significant to the agency's acquisition system."[48]

Pursuing an agency protest does not extend the timeliness requirements for obtaining a stay of award at the GAO.[49] However, some agencies' regulations create a "voluntary suspension period" between the time the agency denies the protest and the time the protester subsequently files its protest with the GAO.[50]

4. Actions of Agency upon Receiving a Timely Protest
a. Pre-award Protests—Withholding of Award

If a pre-award protest is filed in a timely manner, the contracting officer may not award the contract in question until the pending agency protest is resolved.[51] The contracting officer must inform the bidding parties of the pending protest and, if appropriate, request the bidders to extend the time for acceptance of their bids to avoid the need for resolicitation.[52] However, the contract may be awarded while the protest is pending if an agency official at a level above the contracting officer makes a determination "in writing" that the "contract award is justified . . . for urgent and compelling reasons or is determined . . . to be in the best interest of the Government."[53]

b. Post-award Protests—Suspension of Performance

"Upon receipt of a protest within 10 days after contract award or within 5 days after a debriefing date offered to the protester . . . , whichever is later, the contracting officer shall immediately suspend performance, pending resolution of the protest within the agency, including any review by an independent higher level official. . . ."[54]

Just as in pre-award protests, the "urgent and compelling reasons" or "best interest of the government" exceptions can be invoked to continue performance of the contract while the agency-level protest is pending.[55]

5. Agency Decision

The FAR requires agencies to make their "best efforts to resolve agency protests within 35 days after the protest is filed."[56] "Agency protest decisions shall be well-reasoned, and explain the agency position. The protest decision shall be provided to the protester using a method that provides evidence of receipt."[57]

In agency-level protests, the burden is on the protester to show that "a solicitation, proposed award, or award does not comply with the requirements of law or regulation."[58] Should the contracting officer or reviewing officer determine that there is noncompliance, the head of the agency may take any action that could be recommended by the Comptroller General had the protest been filed with the GAO,[59] and may award the protester its costs associated with filing and pursuing the successful protest.[60] Furthermore, the agency may "[r]equire the awardee to reimburse the Government's costs . . . where a postaward protest is sustained as the result of an awardee's intentional or negligent misstatement, misrepresentation, or miscertification."[61]

6. Review of Agency-Level Protests

If a party's agency-level bid protest is unsuccessful, the protester may begin the protest anew by filing a timely protest with the GAO or COFC. However, the GAO is only a viable forum for a subsequent protest if the GAO protest is filed "within 10 days of knowledge of initial adverse agency action."[62]

B. GAO Protests

"An interested party wishing to protest [a federal government procurement] is encouraged to seek resolution within the agency (see 33.103) before filing a protest with the GAO, but may protest to the GAO in accordance with GAO regulations (4 CFR part 21)."[63] The procedures for GAO protests are outlined at 4 C.F.R. Part 21,[64] and require strict compliance or else the GAO will not consider the protest.[65]

1. Standing to Protest

"An economically interested party may protest to GAO the proposed awarding of a government contract."[66] Bid protest regulations define an "interested party" as any "actual or prospective bidder or offeror whose direct economic interest would be affected by the award of a contract or by the failure to award a contract."[67] This includes any party who is a disappointed bidder on the federal procurement contract that is the subject of the protest.[68]

The protester has the burden of setting forth all information establishing that it is an interested party for the purpose of filing a protest.[69]

An interested party may protest any of the following to the GAO:

a solicitation or other request by a Federal agency for offers for a contract for the procurement of property or services; the cancellation of such a solicitation or other request; an award or proposed award of such a contract; and a termination of such a contract, if the protest alleges that the termination was based on improprieties in the award of the contract.[70]

2. Procedures

Protests must be in writing[71] and delivered to the GAO by hand, mail, commercial carrier, facsimile, or e-mail,[72] and must include:

(1) the name, street address, electronic mail address, and telephone and facsimile numbers of the protester;
(2) the signature of the protester or its representative;
(3) the identity of the agency and the solicitation and/or contract number;
(4) a detailed statement of the legal and factual grounds of the protest, including copies of relevant documents;
(5) all information establishing that the protester is an interested party for the purpose of filing a protest;
(6) all information establishing the timeliness of the protest;
(7) a specific request for a ruling by the Comptroller General of the United States, and
(8) a statement of the form of relief requested.[73]

Failure to comply with any of the above requirements may be grounds for dismissal of the protest.[74]

The protester must also furnish the contracting agency whose decision is being challenged with a copy of the GAO protest, including all attachments, within one day of filing the protest with the GAO.[75]

A protest may also include a request for a protective order, for production by the agency of specific documents relevant to the protest, and/or for a hearing.[76]

3. Timeliness of Protest

Protests based upon alleged improprieties in a solicitation must be filed before bid opening or the time established for receipt of the proposals, unless the alleged impropriety is not apparent before that time.[77] With the exception of negotiated procurement protests, all other bid protests must be filed no later than 10 calendar days after the basis of the protest is known or should have been known, whichever is earlier.[78]

Where the protest is initially filed with the contracting agency, special timeliness rules apply. In those cases, any subsequent protest to the GAO must be filed not later than 10 days after the protester learns of the "initial adverse agency action."[79] Additionally, if the agency-level protest is untimely filed, any subsequent protest to the GAO is also untimely.[80]

"A document is *filed* on a particular day when it is received by GAO by 5:30 p.m., Eastern Time, on that day."[81] The burden is on the protester to include all information establishing timeliness at the time the protest is filed.[82] "Protests untimely on their face may be dismissed."[83] "Because bid protests may delay the procurement of needed goods and services, GAO, except under limited circumstances, strictly enforces the timeliness requirements."[84]

4. GAO Actions upon Receiving Protest
a. Notice to Parties

Upon receiving a protest, unless the protest is summarily dismissed,[85] the GAO must give notice of the impending protest to the contracting agency by telephone within one day after the protest is filed, and must promptly send the protester and the agency a written acknowledgment that the protest has been received.[86] Upon receiving this notice, the agency must give all potential "intervenors" notice of the protest, and provide them with copies of the protest submissions.[87]

b. Intervention

Other interested parties may be permitted by the GAO to participate in the protest as "intervenors."[88] GAO regulations define an "intervenor" as "an awardee if the award has been made or, if no award has been made, all bidders or offerors who appear to have a substantial prospect of receiving an award if the protest is denied."[89]

An interested party wishing to intervene under the circumstances[90] should give notice to the GAO and the other parties of its intent to intervene,[91] and then contact the GAO to learn whether it will be permitted to intervene.[92] The potential intervenor, or its representative, should also enter a notice of appearance to the GAO to ensure that all communications in relation to the protest are promptly received.[93]

c. Summary Dismissal

If the agency and/or any intervenor discovers a reason why summary dismissal would be appropriate, it should file a request for dismissal as soon as practicable.[94] When a request is filed, the GAO will generally permit the protester to file a brief in opposition to the dismissal request, and the GAO will thereafter promptly address the dismissal request.[95]

Summary dismissal may be appropriate at any time that the GAO has information to determine that the protest is deficient on procedural or jurisdictional grounds.[96]

If the GAO grants the request for summary dismissal, either in whole or in part, the agency is not required to prepare a report in response to the protest or in response to those grounds of the protest that were dismissed.[97]

d. Automatic Stay

After the agency has received telephonic notice of the protest from the GAO, the agency may not award the contract, or, if the contract has already been awarded, must suspend performance of the contract.[98] The automatic stay provision in post-award protests is only triggered if the procuring agency receives notice of a protest from the GAO within 10 days after contract award or within five days after a debriefing date, whichever is later.[99] However, due to subtleties in GAO regulations it is extremely important that the protest be filed at least one full day in advance of the deadline to ensure that the automatic stay provision in post-award protests is triggered.[100]

An exception to the automatic stay provision exists when the head of the agency authorizes the award "upon a written finding that urgent and compelling circumstances which significantly affect interests of the United States will not permit waiting for the decision of the Comptroller General [on the protest,]"[101] and the Comptroller General is given notice of this finding.[102]

While the GAO will not review an agency's decision to override the automatic stay provision, some federal courts will review the agency's decision to determine if it is arbitrary, capricious, an abuse of discretion, or otherwise clearly and prejudicially in violation of law or regulation.[103]

5. Discovery
a. Agency Report

Once the agency receives telephone notice of the protest, it has 30 days to provide the GAO with a complete written report responding to the protest.[104] The report must include the contracting officer's statement of the relevant facts, including a best estimate of the contract value, a memorandum of law, copies of all relevant documents (or portions of documents) not previously produced, and a list of the aforementioned documents.[105] Although the report is to be simultaneously provided to the protester and any intervenors, "[t]he agency may omit documents, or portions of documents, from the copy of the report provided to the parties if the omitted information is protected and a party receiving the report is not represented by counsel admitted under a protective order."[106] "Protected" material includes "proprietary, confidential, or source-selection-sensitive material, as well as other information the release of which could result in a competitive advantage" to one of the parties bidding on the contract.[107]

b. Requests for Documents

In the protest filing, the protester may request the agency to produce specific documents that the protester can show are relevant to the protest.[108] After the agency report is filed, the protester may file a request for any additional relevant documents, but such a request is permitted only if made within two days of the protester having actual or constructive notice of the document's existence or relevance, whichever is earlier.[109] If the protester objects to the agency's withholding of any requested documents, the GAO must decide whether the agency is required to produce the withheld documents, or portions of documents, and whether this should be done under a protective order.[110]

c. Protective Orders

The purpose of a "protective order" is to ensure that the protester, through its counsel, learns the relevant facts when parts of the record are deemed protected.[111] The GAO views it as the responsibility of the protester's counsel in the first instance to request a protective order and to submit timely applications for admission to access protected material under the order.[112]

A protective order may be justified if a relevant document contains "protected" material.[113] If no protective order is issued, the agency may withhold from the parties those portions of the agency report that would ordinarily be subject to a protective order.[114]

After a protective order has been issued, only counsel to the parties and/or their consultants may apply for admission to access material under the protective order.[115] For this reason, it is crucial that the protester be represented by counsel in the bid protest process. If the protester is not represented by counsel, issuing a protective order serves no useful purpose since the protester cannot apply for access to the protected material.[116] In such instances, any portions of the record that the GAO determines cannot be released without a protective order will not be released at all if the protester refuses to obtain counsel.[117]

Generally, other parties to the protest have two days to object to an application for admission under a protective order.[118] If there is no objection, the GAO will generally admit the applicant under the protective order.[119]

If an applicant is granted access to protected material, the order prohibits disclosure of any protected information to those not admitted under the protective order.[120] This creates the unusual circumstance where the protester's attorney or consultants, who can be granted access to protected information, are prohibited from disclosing that information to their client.[121]

If the terms of the protective order are violated, both counsel and client are subject to a variety of sanctions, including dismissal of the protest.[122]

Absent express prior written authorization from the GAO, material to which parties gain access under a protective order may only be used in the protest proceedings for which the protest was issued.[123] However, "GAO has generally permitted the use of protected material in the filing of federal lawsuits and before other administrative tribunals where the party seeking to use such material establishes that the material will be safeguarded."[124]

d. Protester Comments on Agency Report

After receipt of the agency report, the protester has 10 days to submit its comments on the report to the GAO. If the protester does not submit comments within the 10-day period, the GAO will dismiss the protest.[125] Comments consisting solely of general statements requesting that the GAO review the protest on the existing record generally are not sufficient to rebut the agency report.[126]

In its comments, a protester may not introduce new grounds for protest that could have been raised in its initial protest submission.[127] However, the protester can raise new grounds of protest if the new grounds were first discovered upon receipt of the agency report and the protester raises these supplemental issues within 10 working days of its receipt of the agency report.[128] Following the comment period, neither the agency nor any other party may submit additional statements for the record without GAO permission.[129]

6. Hearings

At the request of a party or on its own initiative, the GAO may conduct a hearing in connection with a protest.[130] A protester requesting a hearing should do so in the initial protest filing, setting forth the reasons why a hearing is necessary to resolve the protest.[131] Because of the increased cost and burden associated with a hearing, the GAO holds hearings rarely, and only when necessary.[132] If the GAO grants a hearing, it usually holds a prehearing conference to resolve procedural issues.[133]

The GAO hearing is presided over by the GAO attorney assigned to the protest. Parties must submit a list of expected attendees to the GAO at least one day before the hearing, and the presiding GAO attorney may restrict access to the hearing to prevent the improper disclosure of protected information.[134]

If a witness whose attendance has been requested by the GAO fails to attend the hearing or fails to answer a relevant question, the GAO may infer that the witness's testimony would have been unfavorable to the party for whom the witness would have testified.[135]

Within five days after the hearing, parties should submit comments to the GAO.[136] If the protester fails to submit any comments, the protest is dismissed.[137]

7. GAO Decision

Unless the GAO finds the protest appropriate for fast-tracking under the "express option," it must issue a decision on the protest within 100 days after the protest is filed.[138] If the GAO chooses the "express option," a decision will be issued within 65 days after the protest is filed.[139] The GAO also has the option of using, where appropriate, "flexible alternative procedures to promptly and fairly resolve a protest."[140] Once signed by the presiding GAO attorney, a copy of the decision is generally available on the GAO's Web site within 24 hours, and is distributed to the parties.[141] If the decision contains protected information, it will only be distributed to the agency and individuals admitted under the protective order, and, if possible, a redacted version will be made available to the public.[142]

If the GAO determines that the agency's procurement activities did not comply with statute or regulation, and such noncompliance prejudiced the protester, the GAO will sustain the protest, and recommend such remedial action as it "determines necessary to promote compliance."[143] GAO decisions are recommendations only, and are not binding upon the procurement agency or any of the other parties to the protest.[144] However, the statutory language giving the GAO jurisdiction to review procurement decisions indicates "that Congress contemplated and intended that procurement agencies normally would follow the Comptroller General's recommendation."[145]

8. Costs

Generally, if a protest is sustained, the GAO will recommend that the agency reimburse the protester's costs incurred in filing and pursuing the protest,[146] including limited attorney, consultant, and expert witness fees.[147] If the protest is sustained, but the protester is deprived of an opportunity to compete for the contract at issue, then the GAO will likely award the protester its bid and proposal preparation costs.[148] But "even where an offeror has been wrongfully denied award of a contract, there is no legal basis for allowing recovery of lost profits."[149]

If the protest is denied, or closed after the agency takes corrective action prior to the GAO's final ruling, the protester may still be awarded its protest costs if the GAO determines that "the agency unduly delayed taking corrective action in the face of a clearly meritorious protest."[150]

If the GAO recommends that the protester be awarded costs, the protester must file with the GAO a detailed claim for costs, certifying the time expended and costs incurred in pursuing the protest, within 60 days of the GAO's decision to award costs.[151] The claim must be supported by adequate documentation.[152] Absent a "compelling reason beyond the control of the protester [that] prevented the protester from timely filing the claim," failure to file an adequately substantiated cost claim within the 60-day window will result in the forfeiture of the protester's right to recover costs.[153]

9. Review of GAO Decisions
a. Request for Reconsideration

Any party involved in the bid protest, including intervenors, may request reconsideration of the GAO decision.[154] A party must file the request "not later than 10 days after the basis for reconsideration is known or should have been known, whichever is earlier," and the request must "contain a detailed statement of the factual and legal grounds upon which reversal or modification is deemed warranted, specifying any errors of law made or information not previously considered."[155] Unlike the initial protest, a request for reconsideration will not result in an automatic stay of contract award or performance.[156]

b. Appeal to the COFC

Where the protester fails to obtain its desired relief from the GAO, or where the GAO's decision to sustain the protest and grant relief to the protester is not implemented by the procuring agency, the protester can seek relief in the COFC. The subject of the COFC's review is the agency decision, not the GAO recommendation.[157] Despite the fact that GAO decisions are not binding on the COFC, "the [COFC] recognizes GAO's longstanding expertise in the bid protest area and accords its decisions due regard."[158]

C. COFC Protests

1. Rules of Procedure

The Rules of the Court of Federal Claims (RCFC) incorporate the Federal Rules of Civil Procedure applicable to civil actions tried by a district court sitting without a jury, to the extent appropriate.[159] Appendix C of the RCFC, entitled "Procedure in Procurement Protest Cases Pursuant to 28 U.S.C. § 1491(B)," acts to supplement the RCFC, and provides step-by-step guidance for filing a bid protest action in the COFC.[160] Appendix C, along with the RCFC, should be referenced by any party filing a bid protest with the COFC.

At least 24 hours before filing a protest with the COFC, the protester must provide pre-filing notice to the court, the Department of Justice (DOJ), the procuring agency's contracting officer, and the awardee.[161] Failure to provide pre-filing notice is not grounds for dismissal, but is likely to delay the initial processing of the case.[162] As soon as practicably possible after the complaint is filed, the COFC will schedule an initial status conference with the parties to address relevant procedural and evidentiary issues,[163] including requests for temporary or preliminary injunctive relief,[164] and motions for protective orders.[165]

If the protester believes that its complaint, or material filed with the complaint, contains confidential or proprietary information, and wishes to protect that information from public view, the protester must file a motion under seal for leave to file the complaint, together with the complaint, requesting that the COFC allow the action to be filed and maintained under seal.[166]

a. Timeliness of Protest

In most cases there is no deadline, other than the applicable statute of limitations, for bringing a bid protest in the COFC.[167] However, the availability of injunctive relief is significantly diminished if the protest is not brought in a timely manner.[168] Additionally, the doctrine of waiver acts to bar any protest to the COFC premised upon patent errors in a solicitation, if the protester did not first raise these errors to the agency "before the close of the bidding process."[169] The waiver rule has been held not to apply in cases where the protester lacked knowledge of the alleged defect in the solicitation until after the close of bidding.[170]

b. Intervention

When a bid protest action is brought in the COFC, other interested parties, specifically the awardee, may be permitted to intervene pursuant to RCFC 24(a). On timely motion, the COFC must permit anyone to intervene "who claims an interest relating to the property or transaction that is the subject of the action, and is so situated that disposing of the action may as a practical matter impair or impede the movant's ability to protect its interest, unless existing parties adequately represent that interest."[171] "In considering a motion to intervene under RCFC 24(a), the [COFC] must construe the rule's requirements in favor of intervention."[172]

c. Jurisdiction to Review Agency's Override of Automatic Stay

When a bid protest is pending in the GAO, and the procurement agency overrides the automatic stay of the procurement, the protester may seek immediate review from the COFC of the agency's override decision.[173] To obtain interlocutory review, a protester should immediately file a complaint in the COFC seeking a temporary restraining order, preliminary and permanent injunctions, and declaratory relief, alleging that the override of the automatic stay is arbitrary, capricious, an abuse of discretion and/or a violation of regulation or procedure.[174] In such cases, the COFC has jurisdiction to review the agency's decision to override the stay, but the merits of the protest remain before the GAO.[175] The COFC standard of review is whether the agency's override decision was "arbitrary, capricious, an abuse of discretion, or otherwise not in accordance with law."[176] The "ultimate standard of review is a narrow one" and the COFC "is not empowered to substitute its judgment for that of the agency."[177]

2. COFC Scope and Standard of Review after GAO Decisions
a. Scope of Review

Generally, where the COFC reviews a procurement following a GAO decision on the same procurement, "it is the agency's decision, not the decision

of GAO, that is the subject of judicial review."[178] In such cases, the GAO decision "serves as a recommendation that becomes a part of the administrative
record."[179] Additionally, "in view of the expertise of GAO in procurement matters," the COFC may rely upon GAO decisions for "general guidance to the
extent it is reasonable and persuasive in light of the administrative record."[180]
However, where the procuring agency changed its conduct in response to the
GAO decision, the COFC reviews the propriety of GAO's decision, as well as
the decision of the agency.[181]

The scope of the COFC's review is generally confined to the administrative record, i.e., to the record before the decision maker when the final award
decision was made.[182] "In limited circumstances, a court may grant a party's
request to supplement the administrative record."[183] "Supplementation of the
administrative record is appropriate where the record is insufficient for the
court to render a decision."[184] The Federal Circuit has "stated that supplementation may be allowed if the court needs more information: (1) to understand if the agency 'possessed or obtained information sufficient to decide'
the issue or (2) to understand on 'what basis' the agency made a decision."[185]
Courts have also allowed supplementation "when the issues are so complex
or technical that the supplementary evidence is 'evidence without which the
court cannot fully understand the issues.'"[186] Other grounds enumerated for
supplementation include: the failure of the agency to consider factors relevant to its final decision; consideration of evidence by the agency that it
failed to include in the record; evidence arising after agency action shows
whether the decision was correct or not; cases in which the agency is sued
for a failure to take action; cases arising under the National Environmental
Policy Act; and cases where relief is at issue, especially at the preliminary
injunction stage.[187]

b. Standard of Review

The Federal Circuit recently described the COFC's standard of review in bid
protest cases as follows:

> In the exercise of its bid protest jurisdiction, the court reviews agency
> action "pursuant to the standards set forth in section 706 of title 5" of
> the Administrative Procedure Act ("APA"), 5 U.S.C. §§ 551–59, 701-06.
> 28 U.S.C. § 1491(b)(4). Section 706 of the APA provides, in relevant part,
> that a "reviewing court shall . . . hold unlawful and set aside agency
> action, findings, and conclusions found to be . . . arbitrary, capricious,
> an abuse of discretion, or otherwise not in accordance with law." 5
> U.S.C. § 706 (2006). Under these standards, a reviewing court may set
> aside a procurement action if "(1) the procurement official's decision
> lacked a rational basis; or (2) the procurement procedure involved
> a violation of regulation or procedure." *Impresa Construzioni Geom.
> Domenico Garufi v. United States*, 238 F.3d 1324, 1332 (Fed. Cir. 2001).[188]

When a bid protest is brought on the basis that the procurement official's decision lacked a rational basis, a court reviews the procurement "to determine whether the contracting agency provided a coherent and reasonable explanation of its exercise of discretion, and the disappointed bidder bears a heavy burden of showing that the award decision had no rational basis."[189] An agency decision lacks a rational basis where "the agency entirely failed to consider an important aspect of the problem, offered an explanation for its decision that runs counter to the evidence before the agency, or is so implausible that it could not be ascribed to a difference in view or the product of agency expertise."[190] It is not the province of the court to determine whether the agency's decision was "correct;" rather, the court's focus is on "whether the contracting agency provided a coherent and reasonable explanation of its exercise of discretion."[191] "Despite this highly deferential standard, 'the [COFC] must still conduct a careful review to satisfy itself that the agency's decision is founded on a rational basis.'"[192]

When a bid protest is brought on the basis that the procurement procedure involved a violation of regulation or procedure, the disappointed bidder must show a "clear and prejudicial violation of applicable statutes or regulations."[193] This requires the protester to demonstrate error that is both significant and prejudicial.[194] In order to demonstrate significant prejudice, the protester must show that there was a "substantial chance" it would have received the contract award if the alleged errors in the bidding process were corrected.[195]

3. COFC Decisions and Relief

The COFC has discretion to award "any relief that the court considers proper, including declaratory and injunctive relief, except that any monetary relief shall be limited to bid preparation and proposal costs."[196]

a. Injunctive Relief

Unlike agency or GAO protests, a protest to the COFC does not trigger an automatic stay of the procurement. To stay performance or award of a procurement pending the COFC's resolution of the protest, the protester must seek a temporary restraining order or a preliminary injunction.[197] It is the COFC's practice to expedite protest cases and to conduct hearings on motions for preliminary injunctions at the earliest practicable time. Thus, when a plaintiff seeks a preliminary injunction it may not need to request a temporary restraining order.[198] To obtain injunctive relief, the burden of proof is on the protester to establish entitlement by a preponderance of the evidence.[199]

In order to obtain a temporary restraining order or preliminary injunction, the protester has the burden of establishing entitlement to extraordinary relief based on the following factors:

(1) immediate and irreparable injury to the movant;
(2) the movant's likelihood of success on the merits;

(3) the public interest; and,

(4) the balance of hardship on all the parties.[200]

To determine if a permanent injunction is warranted, the court must consider whether:

(1) the plaintiff has succeeded on the merits;

(2) the plaintiff will suffer irreparable harm if the court withholds injunctive relief;

(3) the balance of hardships to the respective parties favors the grant of injunctive relief; and,

(4) the public interest is served by a grant of injunctive relief.[201]

b. Monetary Relief

A losing bidder on a government contract "may recover the costs of preparing its unsuccessful proposal if it can establish that the Government's consideration of the proposals submitted was arbitrary or capricious."[202] Courts have conclusively held that expectation damages, such as lost profits, are not recoverable.[203] Beyond bid preparation costs, the COFC may in some cases award attorney's fees and other related expenses if the protest is sustained.[204]

4. Appeal of COFC Decisions

The Federal Circuit has jurisdiction over an appeal from a final decision of the COFC pursuant to 28 U.S.C. § 1295(a)(3).[205] On appeal, the Federal Circuit reviews the COFC's "determination on the legal issue of the government's conduct, in a grant of judgment upon the administrative record, without deference."[206] Thus, the Federal Circuit applies the "'arbitrary and capricious' standard of [5 U.S.C. § 706] de novo, conducting the same analysis as the Court of Federal Claims."[207] The Federal Circuit has stated:

> A party may establish an abuse of discretion "by showing that the court made a clear error of judgment in weighing the relevant factors or exercised its discretion based on an error of law or clearly erroneous fact finding." A clear error of judgment in weighing the relevant factors or a clearly erroneous fact finding is present when a reviewing court is left with a "definite and firm conviction" that a clear error of judgment or a mistake has been committed.[208]

However, in the absence of "clear error," the Federal Circuit defers to the COFC's factual determinations, including findings regarding prejudice to the protester.[209]

Notes

1. Originally titled the General Accounting Office, GAO was renamed the Government Accountability Office in 2004. GAO Human Capital Reform Act of

2004, Pub. L. No. 108-271, § 8, 118 Stat. 811, 814 (2004). The name was changed to more accurately reflect GAO's true role. *See* James F. Nagle & Bryan A. Kelly, *Federal Forums for Government Contracts*, 2 J. Am. Coll. Constr. L. 189, 204 (2008).

2. *See infra* section II.C.

3. John R. Tolle & James D. Duffy, Jr., Briefing Papers #87-4: GSBCA Bid Protests 1, 1–3 (Mar. 1987) (citing Pub. L. No. 98-369, § 2713 (codified at 40 U.S.C. § 759(h))).

4. Pub. L. No. 104-106, § 5101, 110 Stat. 186 (1996).

5. John Cibinic, Jr. & Ralph C. Nash, Jr., Formation of Government Contracts 1484 (3d ed. 1998); Eric A. Troff, *The United States Agency-Level Bid Protest Mechanism: A Model for Bid Challenge Procedures in Developing Nations*, 57 A.F. L. Rev. 113, 144 (2005).

6. *Up Front: Clinton Tells Agencies to Take a Greater Role in Handling Protests*, 37 Gov't Contractor ¶ 554 (Nov. 1, 1995).

7. Exec. Order No. 12,979, 60 Fed. Reg. 5517 (Oct. 25, 1995).

8. *Id.*

9. Pub. L. No. 67-13, 42 Stat. 20 (1921).

10. The GAO derives its authority to resolve bid protests from the language in section 305 of the Budget and Accounting Act of 1921. Pub. L. No. 67-13, 42 Stat. 20, 24 (1921) ("All claims and demands whatever by the Government of the United States or against it . . . shall be settled and adjusted in the General Accounting Office."); *see* 31 U.S.C. §§ 3702, 3526. Prior to 1921, this authority was vested in the Accounting Office of the Treasury Department. *See* Globe Indem. Co. v. United States, 291 U.S. 476, 479–80 (1934).

11. Pub. L. No. 67-13, 42 Stat. 20, 23 (1921).

12. Section of Pub. Contract Law, Am. Bar Ass'n, Comments Regarding U.S. General Accounting Office Study of Concurrent Protest Jurisdiction 7 n.4 (1999), *available at* http://www.ogc.doc.gov/ogc/contracts/cld/papers/scanwell.pdf [hereinafter Comments re GAO Study], *printed in* U.S. Gen. Accounting Office, GAO/GGD/OGC-00-72, Bid Protests: Characteristics of Cases Filed in Federal Courts app. viii, at 58 (2000); Office of Gen. Counsel, U.S. Gov't Accountability Office, GAO-06-797SP, Bid Protests at GAO: A Descriptive Guide 5 (8th ed. 2006) [hereinafter Descriptive Guide].

13. *See* Court of Fed. Claims Bar Ass'n, Deskbook for Practitioners 1 (5th ed. 2008). Prior to 1855, private claims against the federal government were submitted directly to Congress by petition. *See* Nagle & Kelley, *supra* note 1, at 194.

14. *See* Deskbook for Practitioners, *supra* note 13, at 3.

15. *See* Comments re GAO Study, *supra* note 12, at 7 (citing Heyer Prods. Co. v. United States, 135 Ct. Cl. 63, 69 (1956)).

16. GAO, Bid Protests, *supra* note 12, at 5.

17. Richard E. Speidel, *Judicial and Administrative Review of Government Contract Awards*, 37 Law & Contemp. Probs. 63, 74 (1972).

18. Scanwell Labs., Inc. v. Shaffer, 424 F.2d 859 (D.C. Cir. 1970).

19. 5 U.S.C. § 702.

20. *See* Formation of Government Contracts, *supra* note 5, at 1561.

21. *Scanwell*, 424 F.2d at 866.

22. *Id.* at 870.

23. City of Albuquerque v. U.S. Dep't of the Interior, 379 F.3d 901, 908 (10th Cir. 2004). Although the vast majority of bid protests in the district courts pursuant to *Scanwell* were brought by disappointed bidders, some district courts granted standing to non-bidders. *Id.* (citations omitted).

24. Federal Courts Improvement Act of 1982, Pub. L. No. 97-164, § 133, 96 Stat. 25 (codified at 28 U.S.C. § 1491(b) until amended by Administrative Dispute Resolution Act of 1996, Pub. L. No. 104-320, § 12, 110 Stat. 3870); *see* FORMATION OF GOVERNMENT CONTRACTS, *supra* note 5, at 1536.

25. Pub. L. No. 104-320, § 12, 110 Stat. 3870.

26. 28 U.S.C. § 1491(b).

27. ADRA § 12, 28 U.S.C. § 1491(b)(1).

28. ADRA § 12, 28 U.S.C. § 1491(b)(2).

29. *See* ADRA § 12(d) (sunset provision).

30. Even before January 2001, some in the legal community recognized that it was unclear whether district courts would retain jurisdiction of bid protests under the *Scanwell* doctrine after the ADRA sunset provision came into effect. *See* COMMENTS RE GAO STUDY, *supra* note 12, at 6 n.2.

31. *See* ADRA § 12(d) (sunset provision); *see also* Emery Worldwide Airlines, Inc. v. United States, 264 F.3d 1071, 1079–80 (Fed. Cir. 2001) ("the Court of Federal Claims is the only judicial forum to bring any governmental contract procurement protest"); DESKBOOK FOR PRACTITIONERS, *supra* note 13, at 24 (same); Goodwill Indus. Servs. Corp. v. Comm. for Purchase from People Who Are Blind or Severely Disabled, 378 F. Supp. 2d 1290, 1297 (D. Colo. 2005) (sunset provision of the ADRA "confers exclusive jurisdiction of the COFC to hear government contract procurements protests by interested parties"); Fire-Trol Holdings, LLC v. U.S. Dep't of Agric. Forest Serv., No. CV-03-2039-PHX-JAT, 2004 WL 5066232, at *5–6 (D. Ariz. Aug. 13, 2004) (District Courts lack jurisdiction to hear any procurement protest brought by an actual or prospective bidder).

32. *See generally* City of Albuquerque v. U.S. Dep't of the Interior, 379 F.3d 901 (10th Cir. 2004); *see also* Nat'l Treasury Employees Union v. IRS, No. Civ.A. 04-CV-0820, 2006 WL 416161, at *3 (D.D.C. Feb. 22, 2006) (holding that the ADRA did not deprive the court of jurisdiction over parties that are not actual or prospective bidders or offerors); Inlandboatmen's Union of Pac., Marine Div., ILWU v. Mainella, No. C 06-2152-CW, 2006 WL 2583678, at *2 (N.D. Cal. Sept. 7, 2006) (District Court had subject-matter jurisdiction over government contract procurement protest brought by a non-bidding party).

33. *See City of Albuquerque*, 379 F.3d at 911.

34. Am. Fed'n of Gov't Employees Local 1482 v. United States, 258 F.3d 1294, 1302 (Fed. Cir. 2001) (adopting the definition of "interested party" set forth in the Competition in Contracting Act, 31 U.S.C. § 3551(2)); *City of Albuquerque*, 379 F.3d at 910 (adopting Federal Circuit's interpretation of the term "interested party"); DESKBOOK FOR PRACTITIONERS, *supra* note 13, at 25.

35. *See City of Albuquerque*, 379 F.3d at 911 (noting that the court expresses no view on whether *Scanwell* doctrine and APA would allow an interested party to bring a protest in district court).

36. *See, e.g.*, Am. Cargo Transp., Inc. v. Natsios, 429 F. Supp. 2d 139, 146 (D.D.C. 2006). Am. Cargo Transp., Inc. v. United States, No. C05-393JLR, 2007 WL 3326683,

at *3 (W.D. Wash. Nov. 05, 2007) ("Under the law governing procurement decisions, a disappointed bidder may challenge a government contract award under the APA."). At least one court, prior to the sunset of concurrent jurisdiction under the ADRA, held that, based on the *Scanwell* doctrine, district courts had subject matter jurisdiction to entertain a bid protest by a disappointed bidder (i.e., an "interested party"). *See* Iceland S.S. Co., Ltd.-Eimskip v. U.S. Dep't of Army, 201 F.3d 451, 453 (D.C. Cir. 2000).

37. DESKBOOK FOR PRACTITIONERS, *supra* note 13, at 7–8.

38. *Id.* at 4.

39. *See, e.g.,* 48 C.F.R. §§ 333.102, 333.103 (HHSAR); 48 C.F.R. § 433.103 (Agric. Dep't FAR); 48 C.F.R. § 533.103 (GSAR); 48 C.F.R. § 633.103 (DOSAR); 48 C.F.R. §§ 733.103–70, 733.103–71 (AIDAR); 48 C.F.R. § 833.103 (VAAR); 48 C.F.R. § 933.103 (DEAR); 48 C.F.R. § 1333.103 (CAR); 48 C.F.R. § 1533.103–70 (EPAAR); 48 C.F.R. § 1833.103 (NASA FAR); 48 C.F.R. § 2433.103 (HUDAR); 48 C.F.R. § 2933.103 (DOLAR); 48 C.F.R. § 3433.103 (EDAR).

40. FAR 33.103(d)(4). The Court of Federal Claims has limited the definition of "interested parties" to "actual or prospective bidders or offerors whose direct economic interest would be affected by the award of the contract or by failure to award the contract." Galen Med. Assocs., Inc. v. United States, 56 Fed. Cl. 104, 108 (2003), *aff'd* 369 F.3d 1324 (Fed. Cir. 2004) (quoting Am. Fed'n of Gov't Employees v. United States, 258 F.3d 1294, 1299–1302 (Fed. Cir. 2001)). "This standard weeds out protestors that do not submit proposals, withdraw from the procurement, or finish lower than second after evaluation." *Id.* (citing Impresa Construzioni Geom. Domenico Garufi v. United States, 238 F.3d 1324, 1334 (Fed. Cir. 2001)); *see also* Dismas Charities, Inc. v. United States, 75 Fed. Cl. 59, 61–62 (2007) ("Dismas submitted a Final Proposal Revision that did not conform to the solicitation requirements. As a result, it did not have a substantial chance of contract award and cannot be an 'interested party' for purposes of this Court's bid protest jurisdiction."). *But see* Dyonyx, L.P. v. United States, 83 Fed. Cl. 460, 467–69 (2008) (holding that standing is based on whether the protester was an actual or prospective bidder whose direct economic interest would be affected by the award of the contract to another party, not on whether protester's bid was found to be "compliant").

41. FAR 33.103(d)(3)–(4). The agency official designated to conduct independent review "need not be within the contracting officer's supervisory chain." FAR 33.103(d)(4). When possible, this official should not have had any "previous personal involvement in the procurement." *Id.*

42. FAR 33.103(d)(4).

43. FAR 33.103(d)(1).

44. FAR 33.103(d)(2).

45. FAR 33.103(d)(1) (requires that the protester "substantially comply" with the requirements of FAR 33.103(d)(2)).

46. FAR 33.101(d)(4).

47. *Id.*

48. *Id.*

49. *See infra* section III.A.4.d.

50. FAR 33.101(f)(4).

51. FAR 33.101(f)(1).

52. FAR 33.101(f)(2) (Contracting officer is only required to notify parties who are "eligible for award of the contract.").

53. FAR 33.101(f)(1) ("Such justification or determination shall be approved at a level above the contracting officer, or by another official pursuant to agency procedures."). "In the event of failure to obtain such extension of offers, consideration should be given to proceeding with award pursuant to paragraph (f)(1) of this section." FAR 33.101(f)(2). If the agency determines that the "urgency" exception applies, it is not required to give the protester an opportunity to supplement its protest prior to award. Applications Research Corp. v. Naval Air Dev. Ctr., 752 F. Supp. 660, 685 (E.D. Pa. 1990).

54. FAR 33.101(f)(3). The automatic stay provision contained in section (f)(3) applies regardless of whether the protest is lodged with the contracting officer or with the agency official designated to provide independent review. ES-KO, Inc. v. United States, 44 Fed. Cl. 429, 434–35 (1999).

55. FAR 33.101(f)(3).

56. FAR 33.101(g).

57. FAR 33.101(h).

58. *See* FAR 33.102(b); *see also* Grumman Data Sys. Corp. v. Dalton, 88 F.3d 990, 1000 (Fed. Cir. 1996) ("A protestor bears the burden of proving error in the procurement process sufficient to justify relief.").

59. FAR 33.102(b)(1). *See infra* section III.B.7.

60. FAR 33.102(b)(2).

61. FAR 33.102(b)(3).

62. FAR 33.103(d)(4); *see also* 4 C.F.R. § 21.2(a)(3).

63. 48 C.F.R. § 33.102(e).

64. 4 C.F.R. §§ 21.0–.14.

65. For example, protests not filed within the time limits set forth in 4 C.F.R § 21.2, or that lack a detailed statement of the legal and factual grounds of protest as required by 4 C.F.R § 21.1(c)(4), or that fail to clearly state legally sufficient grounds of protest as required by 4 C.F.R § 21.1(f), shall be dismissed. 4 C.F.R. §§ 21.5(e)–(f).

66. Stay, Inc. v. Cheney, 940 F.2d 1457, 1460 (11th Cir. 1991) (citing 31 U.S.C. §§ 3551–3556; 4 C.F.R. § 21.1(a)).

67. 4 C.F.R. § 21.0(a)(1).

68. *Stay, Inc.*, 940 F.2d at 1460.

69. Total Procurement Servs., Inc., B-272891, et. al., 96-2 CPD ¶ 92, 1996 WL 491790, at *3 (C.G. Aug. 29, 1996) (citing 4 C.F.R. § 21.1(c)(5)).

70. 4 C.F.R. § 21.1(a).

71. 4 C.F.R. § 21.1(b).

72. 4 C.F.R. § 21.0(f).

73. 4 C.F.R. § 21.1(c).

74. 4 C.F.R. § 21.1(i).

75. 4 C.F.R. § 21.1(e).

76. *See* DESCRIPTIVE GUIDE, *supra* note 12, at 9.

77. 4 C.F.R. § 21.2(a)(1).

78. 4 C.F.R. § 21.2(a).

79. 4 C.F.R. § 21.2(a)(3). Deciding when adverse agency action occurs is straightforward when the protester receives oral or written notice that the agency is denying the agency-level protest. *Protesters should keep in mind, however, that GAO views as adverse agency action any action that makes clear that the agency is denying the agency-level protest.* Examples of adverse agency action include the agency's proceeding with bid opening or the receipt of proposals, the rejection of a bid or proposal, or the award of a contract despite the agency-level protest. Firms that have filed an agency-level protest and are considering filing a subsequent protest with GAO should be alert to any possible agency action that could be viewed as indicating that the agency is denying the agency-level protest. *See* Descriptive Guide, *supra* note 12, at 13 (citing 4 C.F.R. § 21.1(d)).

80. 4 C.F.R. § 21.2(a)(3).

81. 4 C.F.R. § 21.0(f) (emphasis added).

82. 4 C.F.R. § 21.2(b).

83. *Id.*

84. *See* Descriptive Guide, *supra* note 12, at 11; *see also* 4 C.F.R. § 21.2(c) ("GAO, for good cause shown, or where it determines that a protest raises issues significant to the procurement system, may consider an untimely protest.").

85. If the GAO determines that the protest is lacking in any of the basic elements required by 4 C.F.R. § 21.1(c), the GAO may summarily dismiss the protest and is not required to provide the agency any notice. *See, e.g.,* N.M. State Univ., B-230669, B-230669.2, 88-1 CPD ¶ 523, 1988 WL 227202, at *4 (C.G. June 2, 1988).

86. 4 C.F.R. § 21.3(a).

87. *Id.*

88. *See* Descriptive Guide, *supra* note 12, at 19 (citing 4 C.F.R. § 21.0(b)).

89. 4 C.F.R. § 21.0(b)(1). However, for protests filed by an interested party regarding a public-private competition conducted under Office of Management and Budget Circular A-76 regarding an activity or function of a federal agency performed by more than 65 full-time equivalent employees of the federal agency, the representative of the majority of affected employees and the agency tender official may also be intervenors. Descriptive Guide, *supra* note 12, at 19; 4 C.F.R. § 21.0(b)(2).

90. *See* 4 C.F.R. §§ 21.0(b)(1)–(2).

91. *See* Descriptive Guide, *supra* note 12, at 19–20 ("The notice of intervention can be a brief letter that includes the name, address, and telephone and fax numbers of the intervenor or its representative, if any, and advises GAO and all other parties of the intervenor's status.").

92. *See id.* at 19–20.

93. *See id.* at 20 (notice should be delivered to GAO, the protester, and the procurement agency, and should contain the intervenor's name, address, telephone number, fax number, and e-mail address).

94. 4 C.F.R. § 21.3(b).

95. *See* Descriptive Guide, *supra* note 12, at 21.

96. 4 C.F.R. § 21.5; *see also* 4 C.F.R. §§ 21.1(i), 21.2(b), and 21.11.

97. 4 C.F.R. § 21.5; *see* Descriptive Guide, *supra* note 12, at 21.

98. 31 U.S.C. § 3553(c)–(d).

99. 48 C.F.R. § 33.104(c).

100. Once GAO receives a protest, it has one day to give the agency telephonic notice. 4 C.F.R. § 21.3(a). Therefore, if a protest is filed with GAO 10 days after the contract is awarded, the procuring agency may not receive notice from GAO until the eleventh day, and the automatic stay provision will not be triggered.

101. 31 U.S.C. § 3553(c)(2), (d)(3)(C). For a contract that has already been awarded, the agency may substitute this with a finding that "performance of the contract is in the best interests of the United States[.]" 31 U.S.C. § 3553(d)(3)(C)(i)(I).

102. 31 U.S.C. § 3553(c)(2), (d)(3)(C).

103. 3 Steven W. Feldman, Government Contract Awards: Negotiation and Sealed Bidding § 31:3 (2008).

104. 4 C.F.R. § 21.3(c). If GAO determines that fast-tracking the case is appropriate, then the agency has 20 days to issue its report. 4 C.F.R. § 21.10(d).

105. 4 C.F.R. § 21.3(d).

106. See Descriptive Guide, supra note 12, at 28 (citing 4 C.F.R. § 21.3(e)).

107. 4 C.F.R. § 21.4(a).

108. 4 C.F.R. § 21.1(d)(2).

109. 4 C.F.R. § 21.3(g). The agency must produce the requested documents, or explain why it is not required to do so, within two days of such a request. Id. GAO may grant the protester leave to make requests for documents outside the two-day window. Id.

110. 4 C.F.R. § 21.3(h).

111. See Descriptive Guide, supra note 12, at 25; 4 C.F.R. § 21.4(a).

112. See Descriptive Guide, supra note 12, at 25; 4 C.F.R. § 21.4(a) (GAO can issue an order on its own initiative).

113. 4 C.F.R. § 21.4(a). Any protective order shall include procedures for application for access to protected information, identification and safeguarding of that information, and submission of redacted copies of documents omitting protected information. Id.

114. 4 C.F.R. § 21.4(b). Though in these cases the discoverability of information is left to the judgment of the agency, GAO reviews in camera all information not released to the parties. Id.

115. 4 C.F.R. § 21.4(c).

116. Vistron, Inc., B-277497, Oct. 17, 1997, 97-2 CPD ¶ 107, at 3 n.2; Descriptive Guide, supra note 12, at 25.

117. Am. Indian Law Ctr., Inc., B-254322, Dec. 9, 1993, 94-1 CPD ¶ 165, at 1 n.1. However, the agency still must provide the protester with documents adequate to inform the protester of the basis of the agency's position. 4 C.F.R. § 21.3(e).

118. 4 C.F.R. § 21.4(c).

119. See Descriptive Guide, supra note 12, at 23.

120. See 4 C.F.R. § 21.4(c); Descriptive Guide, supra note 12, at 22.

121. See generally Network Sec. Techs., Inc., B-290741.2, Nov. 13, 2002, 2002 CPD ¶ 193 (GAO provides notice that, in a future case, it may impose the sanction of dismissal where protester's attorney discloses protected information to client); PWC Logistics Servs. Co. KSC(c), B-310559, Jan. 11, 2008, 2008 CPD ¶ 25 (GAO dis-

missed bid protest after protester's attorney admitted under protective order revealed protected information to protester).

122. 4 C.F.R. § 21.4(d). While the Bid Protest regulations did not explicitly permit dismissal as a sanction prior to June 2008, *see* 73 FR 111 32427-30 (June 9, 2008), GAO had acknowledged this sanction and applied it in prior cases. MICHAEL GOLDEN, MANAGING ASSOC. GEN. COUNSEL, GAO, NOTICE REGARDING CHANGES TO PROTECTIVE ORDER 1-2 (Apr. 7, 2008), http://www.gao.gov/legal/notice_protectiveorder04072008.pdf; *see also Network Sec. Techs.*, 2002 CPD ¶ 193 (GAO provides notice that, in a future case, it may impose the sanction of dismissal where protester's attorney discloses protected information to client); *PWC Logistics*, 2008 CPD ¶ 25 (GAO dismissed bid protest after protester's attorney admitted under protective order revealed protected information to protester).

123. OFFICE OF GEN. COUNSEL, U.S. GOV'T ACCOUNTABILITY OFFICE, GAO-06-716SP, GUIDE TO GAO PROTECTIVE ORDERS 12 (May 2006).

124. GUIDE TO GAO PROTECTIVE ORDERS, *supra* note 123, at 12.

125. 4 C.F.R. § 21.3(i). On a case-by-case basis, GAO may modify the time period for comments. *Id.* If the express option is used, the parties have five days to submit comments. 4 C.F.R. § 21.10(d)(2).

126. *See* DESCRIPTIVE GUIDE, *supra* note 12, at 30 ("protests are rarely sustained where the protester does not file substantive comments on the report").

127. Martin Warehousing & Distribution, Inc., B-270651, B-270651.2, Apr. 25, 2006, 96-1 CPD ¶ 205, at 4 n.2; Ahern & Assocs., Inc., B-254907, B-254907.4, Mar. 31, 1994, 94-1 CPD ¶ 236, at 5.

128. 4 C.F.R. § 21.2(a)(2); Anteon Corp., B-293523, B-293523.2, Mar. 29, 2004, 2004 CPD ¶ 51, at 4 n.6 (allowing supplemental protest issues that were first discovered from agency report, and raised within 10 days of receipt of agency report); Planning & Dev. Collaborative Int'l, B-299041, Jan. 24, 2007, 2007 CPD ¶ 28, at 11–12 (supplemental protest issues first discovered from agency report were untimely because they were raised in protester's comments filed more than 10 days after receipt of agency report; extension for filing comments does not extend to raising protest issues); Gen. Elec. Aerospace Electronic Sys., B-250514, 93-1 CPD ¶ 101, 1993 WL 35220, at *6 (C.G. Feb. 4, 1993) (supplemental protest issues first discovered from agency report, but not raised within 10 working days of receipt of the report, were dismissed as untimely).

129. 4 C.F.R. § 21.3(j).

130. 4 C.F.R. § 21.7(a).

131. 4 C.F.R. §§ 21.1(d)(3), 21.7(a).

132. *See* DESCRIPTIVE GUIDE, *supra* note 12, at 31.

133. 4 C.F.R. § 21.7(b); *see* DESCRIPTIVE GUIDE, *supra* note 12, at 32.

134. *See* 4 C.F.R. § 21.7(d); DESCRIPTIVE GUIDE, *supra* note 12, at 32.

135. 4 C.F.R. § 21.7(f); Dep't of Commerce—Recon., B-277260, B-277260.4, July 31, 1998, 98-2 CPD ¶ 35, at 3; Du & Assocs., Inc., B-280283.3, Dec. 22, 1998, 98-2 CPD ¶ 156, at 6.

136. 4 C.F.R. § 21.7(g). These comments are in addition to those comments submitted after the agency report. *See* DESCRIPTIVE GUIDE, *supra* note 12, at 33.

137. 4 C.F.R. § 21.7(g).

138. 4 C.F.R. § 21.9.

139. 4 C.F.R. §§ 21.9(b), 21.10. Requests for the express option shall explain in writing why the case is suitable for resolution within 65 days, and must be received by GAO not later than five days after the protest or supplemental/ amended protest is filed. 4 C.F.R. § 21.10; *see, e.g.*, B&S Transp., Inc., B-299144, Jan. 22, 2007, 2007 CPD ¶ 16, at 1 n.1 (granted agency's request to use the express option, where agency contended that fast-tracking would allow it to meet its deadlines in the Army's Base Realignment and Closure plan); AshBritt, Inc., B-297889, B-297889.2, Mar. 20, 2006, 2006 CPD ¶ 48, at 6 n.9 (express option used pursuant to agency request).

140. 4 C.F.R. § 21.10(e).

141. 4 C.F.R. § 21.12(b); *see* DESCRIPTIVE GUIDE, *supra* note 12, at 37.

142. 4 C.F.R. § 21.12(a).

143. 4 C.F.R. § 21.8(a); *see also* Centech Group, Inc. v. United States, 554 F.3d 1029, 1039 (Fed. Cir. 2009) ("Pursuant to 31 U.S.C. § 3554(b)(1), GAO is required to recommend that an agency take specific corrective action if an award does not comply with a statute or regulation, including terminating the contract and awarding a contract consistent with the requirements of the statute and regulations." (citing Honeywell, Inc. v. United States, 870 F.2d 644, 648 (Fed. Cir. 1989)). "In determining the appropriate recommendation(s), GAO shall . . . , consider all circumstances surrounding the procurement or proposed procurement including the seriousness of the procurement deficiency, the degree of prejudice to other parties or to the integrity of the competitive procurement system, the good faith of the parties, the extent of performance, the cost to the government, the urgency of the procurement, and the impact of the recommendation(s) on the contracting agency's mission." 4 C.F.R. § 21.8(b).

144. *Honeywell*, 870 F.2d at 648; The Centech Group, Inc. v. United States, 78 Fed. Cl. 496, 507 (2007) ("Because the Comptroller General may only 'recommend' a remedy upon finding a procurement violation, GAO's rulings do not legally bind the parties to a bid protest." (citing 31 U.S.C. § 3554(b), (c))); Advanced Sys. Dev., Inc. v. United States, 72 Fed. Cl. 25, 30 (2006) ("A GAO decision adverse to an agency is only a recommendation—the GAO has no enforcement powers."); Cubic Applications, Inc. v. United States, 37 Fed. Cl. 339, 341 (1997) ("Neither the agency nor this court is bound by the determination of the GAO.").

145. *Honeywell*, 870 F.2d at 648; *see also Centech*, 554 F.3d at 1039 ("a procurement agency's decision to follow [GAO's] recommendation even though that recommendation differed from the contracting officer's initial decision was proper unless [GAO's] decision itself was irrational." (quoting *Honeywell*, 870 F.2d at 648)). The head of the procuring agency must report to the Comptroller General if the agency has not fully implemented the Comptroller General's recommendations within 60 days. 31 U.S.C. § 3554(b)(3) (agency must report failures to implement GAO recommendation within five days after the end of the 60-day period for implementation). The Comptroller General must report annually to Congress each instance of agency noncompliance. *See* 31 U.S.C. § 3554(e)(2); *Centech*, 78 Fed. Cl. at 506 n.19.

146. 4 C.F.R. § 21.8(d)(1); *see, e.g.*, Doyon-Am. Mech., JV; NAJV, LLC, B-310003, B-310003.2, Nov. 15, 2007, 2008 CPD ¶ 50, at 5; R & G Food Serv., Inc., d/b/a Port-A-Pit Catering, B-296435.4, B-296435.9, Sept. 15, 2005, 2005 CPD ¶ 194, at 8.

147. 4 C.F.R. § 21.8(d)(1). Although 31 U.S.C.A. § 3554(c)(2)(B) caps costs for attorney's fees at $150 per hour, GAO has repeatedly declined to impose a strict cap on fees. GAO has generally allowed for increased fees if the protester requests an upward adjustment and presents a basis upon which the adjustment should be calculated, such as an increase in the cost of living. *See, e.g.,* EBSCO Publ'g, Inc.—Costs, B-298918.4, May 7, 2007, 2007 CPD ¶ 90, at 2–3 (granting attorney's fees at $197 per hour, as increased by the change in Consumer Price Index for All Urban Consumers between 2000 and 2007). Reimbursement of protest costs associated with the use of consultants or expert witnesses is limited to the highest rate of pay for expert witnesses paid by the federal government pursuant to 5 U.S.C. § 3109 and 5 C.F.R. § 304.105. Dept. of the Army; ITT Fed. Servs. Int'l Corp.—Costs, B-296783.4; B-296783.5, Apr. 26, 2006, 2006 CPD ¶ 72, at 3–5 (citing 48 C.F.R. § 33.104). As of January 1, 2009, the maximum fee is the daily rate for a GS-15 step 10 federal employee, $489.13 per day. *See* 5 C.F.R. § 304.105 (equation for daily rate); U.S. Office of Personnel Management, Salary Table 2009-GS, *available at* http://www.opm.gov/oca/09tables/html/gs.asp (last visited Dec. 27, 2008) (listing 2009 annual wage of GS-15 step 10 federal employee at $127,604).

148. *See* 4 C.F.R. § 21.8(d)(2); *see, e.g.,* Advanced Tech. Sys., Inc., B-296493.6, Oct. 6, 2006, 2006 CPD ¶ 151, at 10; Aberdeen Tech. Servs.—Modification of Recommendation, B-283727.3, Aug. 22, 2001, 2001 CPD ¶ 146, at 2.

149. Al Long Ford, B-297807, Apr. 12, 2006, 2006 CPD ¶ 68, at 11 n.12 (citing Firebird Constr. Corp.—Recon., B-246182.2, May 27, 1992, 92-1 CPD ¶ 473, at 2).

150. Major Contracting Servs., Inc., B-400737.2, Dec. 17, 2008, 2008 CPD ¶ 230, at 3 (citing 4 C.F.R. § 21.8(e)); Alaska Mech., Inc.-Costs, B-289139.2, Mar. 6, 2002, 2002 CPD ¶ 56, at 1. A request for these costs must be filed within 15 days of the protester having some notice that GAO has closed the protest based on the agency's decision to take corrective action. 4 C.F.R. § 21.8(e).

151. 4 C.F.R. § 21.8(f)(1).

152. *See, e.g.,* Malco Plastics, B-219886, B-219886.3, 86-2 CPD ¶ 193, 1986 WL 63904, at *2 (C.G. Aug. 18, 1986) (denying protester's request for attorney's fees where protester failed to provide adequate documentation of these fees).

153. Keeton Corrs., Inc.—Costs, B-293348.3, Oct. 25, 2004, 2005 CPD ¶ 44, at 2–3.

154. 4 C.F.R. § 21.14(a); Stay, Inc. v. Cheney, 940 F.2d 1457, 1460 (11th Cir. 1991) ("Any interested party is also entitled to seek reconsideration of the GAO's decision . . . .").

155. 4 C.F.R. § 21.14.

156. 4 C.F.R. § 21.14(c).

157. *See infra* notes 178–181 and accompanying text.

158. Idea Int'l, Inc. v. United States, 74 Fed. Cl. 129, 136 n.11 (2006) (quotations omitted).

159. United States Court of Federal Claims Amendments to Rules of the Court of Federal Claims (RCFC), 2008 US Order 64 (C.O. 64) (originally effective Oct. 1, 1982, as revised and reissued May 1, 2002, and as amended through Nov. 3, 2008).

160. *See* RCFC app. C.

161. RCFC app. C, ¶ II. The contents of the pre-filing notice should comply with RCFC app. C, ¶ II(3), and be transmitted in accordance with RCFC app. C, ¶ II(2).

162. RCFC app. C, ¶ II.

163. RCFC app. C, ¶ IV.

164. RCFC app. C, ¶¶ IV(8)(c), V.

165. RCFC App. C, ¶¶ IV(8)(d), VI. Protective orders are the "principal vehicle relied upon by the court to ensure protection of sensitive information." Motions for protective orders must meet the requirements of RCFC 10, and are issued at the court's discretion. Once a protective order is issued, individuals who seek access to protected information—with the exception of the Court, the procuring agency, and the DOJ—must file an appropriate application to be admitted to the protective order. If admitted to the protective order, an individual becomes subject to the terms of the order. RCFC app. C, ¶ VI. Forms for protective orders, and applications for admission to the order, are available in the RCFC Appendix of Forms. *See* RCFC, app. Forms, Forms 8, 9 & 10.

166. RCFC app. C, ¶ III(4). The procedures for filing documents under seal is detailed in RCFC 5.4(d) and RCFC app. C, ¶¶ III(5)–(7).

167. *See* Heritage of Am., LLC v. United States, 77 Fed. Cl. 66, 72–73 (2007); see also Griffy's Landscape Maint. LLC v. United States, 46 Fed. Cl. 257 (2000) (protest sustained by COFC, despite fact protest had already been deemed untimely when filed with GAO); Wackenhut Servs., Inc. v. United States, 85 Fed. Cl. 273, 289 (2008) (doctrine of laches cannot be used to bar a protest in the COFC, unless the defendant demonstrates that the delay in bringing the protest is unreasonable and that it caused prejudice to the defendant).

168. *See* Blue & Gold Fleet, L.P. v. United States, 492 F.3d 1308 (Fed. Cl. 2007) (citing Transatl. Lines LLC v. United States, 68 Fed. Cl. 48, 52, 57 (2005) (considering "delay in procurement process" in balance of hardships prong of injunctive relief); Wit Assocs., Inc. v. United States, 62 Fed. Cl. 657, 662 n.5 (2004) ("[I]n some cases, serious delay in raising a claim may impact the equities in determining whether an injunction should issue or lead to the imposition of laches."); CW Gov't Travel, Inc. v. United States, 61 Fed. Cl. 559, 568–69 (2004) (considering delay as part of laches analysis); Software Testing Solutions, Inc. v. United States, 58 Fed. Cl. 533, 535–36 (2003) (stating that delay may be "considered in the multi-factored analysis of whether injunctive relief is warranted" or in "the application of equitable doctrines such as laches"); Miss. Dep't of Rehab. Servs. v. United States, 58 Fed. Cl. 371, 372–73 (2003) (same)).

169. Blue & Gold Fleet, L.P. v. United States, 492 F.3d 1308, 1313 (Fed. Cir. 2007). The "close of the bidding process" has been interpreted as the closing date for receipt of proposals. *See* Allied Materials & Equip. Co., Inc. v. United States, 81 Fed. Cl. 448, 459 (2008).

170. *See Allied Materials*, 81 Fed. Cl. at 459–60.

171. RCFC 24(a)(2).

172. Northrop Grumman Info. Tech., Inc. v. United States, 74 Fed. Cl. 407, 412 (2006).

173. DESKBOOK FOR PRACTITIONERS, *supra* note 13, at 28 (citing RAMCOR Servs. Group, Inc. v. United States, 185 F.3d 1286, 1287 (Fed. Cir. 1999)); Nortel Gov't Solutions, Inc. v. United States, 84 Fed. Cl. 243, 247 (2008) ("This court therefore has jurisdiction to review an agency's override decision issued under 31 U.S.C. § 3553(d)(3)(C).").

174. *See, e.g.,* Reilly's Wholesale Produce v. United States, 73 Fed. Cl. 705, 708 (2006).

175. DESKBOOK FOR PRACTITIONERS, *supra* note 13, at 28–29.

176. *Reilly's,* 73 Fed. Cl. at 709.

177. *Id.* (quoting Citizens to Preserve Overton Park, Inc. v. Volpe, 401 U.S. 402, 416 (1971)).

178. Analytical & Research Tech., Inc. v. United States, 39 Fed. Cl. 34, 41 (1997) (citing Cubic Applications, Inc. v. United States, 37 Fed. Cl. 339, 341 (1997)); Charles H. Tompkins Co. v. United States, 43 Fed. Cl. 716, 719 (1999); S.K.J. & Assocs., Inc. v. United States, 67 Fed. Cl. 218, 224 (2005); The Ravens Group, Inc. v. United States, 78 Fed. Cl. 390, 403 (2007).

179. *S.K.J. & Assocs.,* 67 Fed. Cl. at 224 (citing Honeywell, Inc. v. United States, 870 F.2d 644, 647 (Fed. Cir. 1989); Charles H. Tompkins Co. v. United States, 43 Fed. Cl. 716, 719 (1999)).

180. *Tompkins Co.,* 43 Fed. Cl. at 719 (quoting *Cubic Applications,* 37 Fed. Cl. at 342).

181. *Analytical & Research Tech.,* 39 Fed. Cl. at 41 n.7 (citing *Honeywell,* 870 F.2d 644).

182. Advanced Data Concepts, Inc. v. United States, 43 Fed. Cl. 410, 416 (1999) (citing Camp v. Pitts, 411 U.S. 138, 142 (1973)); E-Management Consultants, Inc. v. United States, 84 Fed. Cl. 1, 11 (2008) ("the focal point for judicial review should be the administrative record already in existence" (quoting *Camp,* 411 U.S. at 142)).

183. *E-Management,* 84 Fed. Cl. at 11 (citing Esch v. Yeutter, 876 F.2d 976, 991 (D.C. Cir. 1989)).

184. *Id.* (citing Impresa Construzioni Geom. Domenico Garufi v. United States, 238 F.3d 1324, 1338 (Fed. Cir. 2001) (supplementation of record appropriate where "required for meaningful judicial review"); Portfolio Disposition Mgmt. Group, LLC v. United States, 64 Fed. Cl. 1, 12 (2005) ("We may allow supplementation of the administrative record in limited circumstances where the record is insufficient for the court to render a decision.").

185. *Impresa Construzioni,* 238 F.3d at 1339, *as reiterated in E-Management,* 84 Fed. Cl. at 11.

186. *E-Management,* 84 Fed. Cl. at 11 (quoting Al Ghanim Combined Group Co. Gen. Trad. & Cont. W.L.L. v. United States, 56 Fed. Cl. 502, 508 (2003)).

187. *Id.* (citing *Esch,* 876 F.2d at 991).

188. Centech Group, Inc. v. United States, 554 F.3d 1029, 1037 (Fed. Cir. 2009).

189. *Id.; see also* Precision Images, LLC v. United States, 79 Fed. Cl. 598, 614 (2007) ("courts have recognized that contracting officers are 'entitled to exercise discretion upon a broad range of issues confronting them' in the procurement process." (citations omitted)).

190. *Precision Images,* 79 Fed. Cl. at 614 (quoting Keeton Corrs., Inc. v. United States, 59 Fed. Cl. 753, 755 (2004); Motor Vehicle Mfrs. Ass'n v. State Farm Mut. Auto. Ins. Co., 463 U.S. 29, 43 (1983)); *see also* Advanced Data, 216 F.3d at 1058 (standard of review is "highly deferential" and requires the reviewing court to sustain agency action "evincing rational reasoning and consideration of relevant factors" (citing Bowman Transp., Inc. v. Ark.-Best Freight Sys., Inc., 419 U.S. 281, 283 (1974))).

191. *Precision Images,* 79 Fed. Cl. at 614–15 (citations omitted); *see also* Honeywell, Inc. v. United States, 870 F.2d 644, 648 (Fed. Cir. 1989) (Where the court finds

a reasonable basis for an agency's action, it "stay[s] its hand even though it might, as an original proposition, have reached a different conclusion as to the proper administration and application of the procurement regulations." (quoting M. Steinthal & Co. v. Seamans, 455 F.2d 1289, 1301 (D.C. Cir. 1971))).

192. *Precision Images*, 79 Fed. Cl. at 615 (quoting AmerisourceBergen Drug Corp. v. United States, 60 Fed. Cl. 30, 35 (2004)). In the case of a "best value" contract, the contracting agency is afforded even greater discretion. *Id.* (citing Galen Med. Assocs., Inc. v. United States, 369 F.3d 1324, 1330 (Fed. Cir. 2004)). And, where "a negotiated procurement is involved and at issue is a performance evaluation, the greatest deference possible is given to the agency—what [the COFC] has called a 'triple whammy of deference.'" *Id.* (quoting Gulf Group Inc. v. United States, 61 Fed. Cl. 338, 351 (2004); Overstreet Elec. Co. v. United States, 59 Fed. Cl. 99, 117 (2003)).

193. *Centech*, 554 F.3d at 1037; *Precision Images*, 79 Fed. Cl. at 615.

194. *Precision Images*, 79 Fed. Cl. at 615 (citing Bannum, Inc. v. United States, 404 F.3d 1346, 1351 (Fed. Cir. 2005)); see also Int'l Outsourcing Servs., L.L.C. v. United States, 69 Fed. Cl. 40, 46 (2005) ("[T]o prevail in a protest the protester must show not only a significant error in the procurement process, but also that the error prejudiced it." (quoting Data Gen. Corp. v. Johnson, 78 F.3d 1556, 1562 (Fed. Cir. 1996))).

195. *Precision Images*, 79 Fed. Cl. 598, 615–16 (2007) (citations omitted).

196. 28 U.S.C. § 1491(b)(2).

197. DESKBOOK FOR PRACTITIONERS, *supra* note 13, at 28 n.46.

198. RCFC app. C, ¶ V.9; DESKBOOK FOR PRACTITIONERS, *supra* note 13, at 27.

199. Labatt Food Serv., Inc. v. United States, 84 Fed. Cl. 50, 64 (2008) (citing Textron, Inc. v. United States, 74 Fed. Cl. 277, 287 (2006); *Bannum*, 60 Fed. Cl. at 723–24); GraphicData, LLC v. United States, 37 Fed. Cl. 771, 779 (1997).

200. *Labatt*, 84 Fed. Cl. at 64 (quoting U.S. Ass'n of Imps. of Textiles and Apparel v. U.S. Dep't of Commerce, 413 F.3d 1344, 1346 (Fed. Cir. 2005)).

201. Centech Group, Inc. v. United States, 554 F.3d 1029, 1037 (Fed. Cir. 2009) (citing PGBA, LLC v. United States, 389 F.3d 1219, 1228–29 (Fed. Cir. 2004)); *see also* Somerset Pharms., Inc. v. Dudas, 500 F.3d 1344, 1346 (Fed. Cir. 2007) ("To establish entitlement to a preliminary injunction a movant must establish a reasonable likelihood of success on the merits.").

202. CNA Corp. v. United States, 83 Fed. Cl. 1, 5 (2008) (quoting E.W. Bliss Co. v. United States, 77 F.3d 445, 447 (Fed. Cir. 1996)); *see also* Cal. Indus. Facilities Res., Inc. v. United States, 80 Fed. Cl. 633, 640 (2008) (for an award of bid preparation and proposal costs, there must be finding of unreasonable action by the procuring agency).

203. *See* 28 U.S.C. § 1491(b)(2); *see, e.g.*, Ala. Aircraft Indus., Inc.–Birmingham v. United States, 85 Fed. Cl. 558, 563 (2009) ("Precedents conclusively established that a disappointed offeror did not have a right to recover *lost profits* from the government in a bid protest."); Rotech Healthcare Inc. v. United States, 71 Fed. Cl. 393, 430 (2006) (unsuccessful bidder could not recover lost profits since contract under which bidder would have made such profits never came into existence (citing Keco Indus., Inc. v. United States, 428 F.2d 1233, 1240 (Ct. Cl. 1970))); La Strada Inn, Inc. v. United States, 12 Cl. Ct. 110, 115 (1987) (lowest bidder had no right to recover

lost profits bidder expected if it had been awarded contract; "the law grants damages to the disappointed bidder based on his reliance interest, not his expectation interest.").

204. *See* 28 U.S.C. § 2412 (Equal Access to Justice Act, providing for attorney's fees and costs for successful protesters whose net worth and/or workforce do not exceed statutory limits); Dubinsky v. United States, 44 Fed. Cl. 360, 364 (1999).

205. *Centech*, 554 F.3d at 1037.

206. *Id.* (quoting Bannum, Inc. v. United States, 404 F.3d 1346, 1351 (Fed. Cir. 2005)).

207. *Id.* (citing 28 U.S.C. § 1491(b)(4); Impresa Construzioni Geom. Domenico Garufi v. United States, 238 F.3d 1324, 1332 (Fed. Cir. 2001)).

208. *Centech*, 554 F.3d at 1037 (quoting PGBA, LLC v. United States, 389 F.3d 1219, 1223–24 (Fed. Cir. 2004)).

209. Blue & Gold Fleet, L.P. v. United States, 492 F.3d 1308, 1312 (Fed. Cir. 2007) ("Where the Court of Federal Claims makes factual findings from the administrative record in the first instance, however, 'this court reviews such findings for clear error,' 'like any finding in a bench trial.'" (quoting *Bannum*, 404 F.3d at 1357)).

CHAPTER 7

Provisions and Issues Unique to Government Contracts

DAVID F. INNIS
AARON P. SILBERMAN

Many aspects of construction contracting in general and federal construction contracting in particular are unique, or at least different from other types of contracting. These aspects are discussed throughout the other chapters in this book. Unlike those chapters, this chapter focuses on provisions and issues that are unique to federal contracting generally, regardless of whether the government is procuring construction, supplies, services, or some combination thereof. The issues in this category to be dealt with here include the *Christian* doctrine, the authority to bind the government, prohibitions on improper business practices (including mandatory disclosures), conflicts of interest, the impact of the Freedom of Information Act, and suspension and debarment from government contracting.

I. *The* Christian *Doctrine*

One of the distinctive characteristics of all contracting with the federal government is the degree to which it is regulated. Unlike commercial contracts, for which few contract clauses are required to be included by law, government contracts are subject to many statutes and regulations. Most notable are the Federal Acquisition Regulation (FAR)[1] and agency procurement regulations,[2] mandating that certain requirements be included in those contracts and often prescribing specific contract language containing those requirements.

What happens when the law requires a clause to be included in a government contract but the parties omit it? In the commercial world, where the parties are free, for the most part, to include or not include whatever they want in their contracts, the general rule is that an omitted provision will not be read into a contract (subject to limited exceptions, such as mutual mistake). In the government contracting world, the answer is different. In a nutshell, if a clause is omitted from a contract between the federal government and a contractor, it will be read into that contract if (a) the clause is required by law and (b) the requirement is based on fundamental procurement policy. This is the essence of the *Christian* doctrine.

The *Christian* doctrine is named after the Court of Claims decision in *G.L. Christian & Assocs. v. United States.*[3] In that case, the contractor entered into a fixed-price construction contract with the Army Corps of Engineers. The contracting officer (CO) failed to include a termination for convenience provision, even though that provision was required by the then-applicable regulations (the Defense Acquisition Regulations, or DAR, which were the predecessor to the current FAR). As a result, when the government terminated the contract, the contractor sought to recover anticipated profits. Those profits would have been recoverable under common law but were not recoverable under the required, but omitted, clause. The court held that since the omitted termination clause was required under the DAR, it therefore had the force and effect of law. As a result, the court incorporated the clause as a matter of law into the contract. The rationale for the court's decision was that, since the regulation was published in the Federal Register, contractors were on notice of its requirements and would be charged with knowing that COs had no authority to ignore a regulation having the force and effect of law.

For some time after the *Christian* decision, it was believed that all contract provisions required by federal acquisition regulations would be read into government contracts under the *Christian* doctrine. However, in *Chamberlain Manufacturing Corp.*,[4] the Armed Services Board of Contract Appeals (ASBCA) limited the scope of the *Christian* doctrine by holding that only those mandatory provisions that constitute fundamental procurement policy would be incorporated by law into government contracts.[5]

The more limited rule introduced in *Chamberlain* has been adopted by the Federal Circuit and remains the law today. In *General Engineering & Machine Works v. O'Keefe*, the Federal Circuit held that "the Christian doctrine does not permit the incorporation of every required clause . . . It applies to mandatory clauses which express a significant or deeply ingrained strand of public procurement policy."[6] Other decisions have recognized that merely because a required clause may address substantive rights does not necessarily mean that it meets this standard.[7]

Several courts and boards have rejected attempts by the government to read omitted clauses into contracts where those clauses were permitted but not required. The *Christian* doctrine never applies to an optional clause because the contractor would not be on notice of any regulatory requirement that such a clause be included in its contract.[8] In *Michael M. Grinberg*, for example, the Department of Transportation Board held that a termination for convenience clause would not be read into a services contract under the

Christian doctrine because the clause was not then mandatory in such contracts.[9] Likewise, where the CO has discretion regarding which among several prescribed clauses to include in a contract, a court or board will not read the selected clause out and another clause into the contract.[10]

Courts and boards have continued to recognize that termination for convenience clauses, where required by applicable regulations (as they are for most government contracts, including most construction contracts[11]), express a fundamental procurement policy and so will be read into those contracts under the *Christian* doctrine.[12] Similarly, the doctrine has been applied where the parties included the wrong termination for convenience clause in their contract.[13]

In addition to termination for convenience clauses, the following are among the clauses that have been held to represent fundamental procurement policy, such that they will be read into government contracts under the *Christian* doctrine: the Changes Clause (FAR 52.243),[14] the Payments Clause (FAR 52.232-1),[15] the Disputes Clause (FAR 52.233-1),[16] the Protest After Award Clause (FAR 52.233-3),[17] the Assignment of Claims Clause (FAR 52.232-23),[18] the Termination for Default Clause (FAR 52.249-8 through -10),[19] the Limitation of Price and Contractor Obligations Clause (FAR 52.217-1),[20] small business set-aside clause (FAR 52.219-14),[21] and clauses implementing the Buy American Act (FAR 52.225-1 through 12),[22] the Truth In Negotiations Act (FAR 52.215-12),[23] and the Service Contract Act (FAR 52.222-43).[24]

Courts and boards are divided over whether government-furnished property clauses represent fundamental procurement policy and must be read into the contract under the *Christian* doctrine. In *Chamberlain*, the ASBCA found that such clauses will not be read in, and several other decisions agree.[25] But the ASBCA has issued three decisions after *Chamberlain* holding that those clauses will be read into government contracts.[26]

Among the clauses that courts and boards found do *not* represent fundamental procurement policy, so are *not* subject to the *Christian* doctrine, are the following: the Variation in Estimated Quantities Clause (FAR 52.212.11),[27] the Cancellation Under Multiyear Contracts Clause,[28] and other mandatory provisions under agency-specific regulations.[29]

As a practical matter, the *Christian* doctrine means that contractors will be bound by most clauses that the FAR requires for the contract they enter into with the government, regardless of whether the clauses are in fact included in the contract documents. Contractors would be wise to determine what those clauses are and what they require before entering into the contract. When the contractor reviews a solicitation that does not contain a required clause, it should seek clarification from the CO. If a properly authorized deviation has been granted (allowing the CO not to include the clause), then the contractor should request documentation of the deviation authorization. Absent such documentation, the contractor should assume that the clause is incorporated into the contract and bid and plan accordingly.

II. *The Authority of Officials to Bind the Government*

Another unique aspect of contracting with the government is the issue of authority. This issue frequently arises when a government representative directs a change in the work or in inspection and testing requirements.[30]

Whether a contractor is dealing with a private, commercial owner or the government, construction projects by necessity involve many individuals with many different obligations and responsibilities. The owner is generally not a single natural person but, rather, is a business entity that operates though representatives, such as officers and employees, or agents, i.e., third parties such as architects or construction managers.

In private contracting, an owner representative may bind the owner if he has actual, implied, or apparent authority. In government contracting, however, the government may only be bound through actions or omissions by representatives with actual or implied authority; it will not be bound based on apparent authority.

A. Actual Authority

Heads of government agencies or their designees are required by statute to appoint COs with actual authority to bind their agencies.[31] Agency heads typically delegate this responsibility.

The CO is the person who speaks for the United States in its contracting capacity and is specifically delegated authority to bind the government contractually. The CO has the authority "to enter into, administer, or terminate contracts and make related determinations and findings."[32] While this authority is extremely broad,[33] it is not limitless.[34]

The CO's binding authority, often called the CO's "warrant," is documented on a Standard Form 1402, Certificate of Appointment. Any limitations on the CO's authority to bind his agency must be stated in that form.[35] For example, a CO's warrant may be limited to a specified maximum dollar amount, in which case he would not be authorized to bind the government to any contracts over that amount. Contractors are on notice of any limit on the CO's warrant. Information regarding any limits on a CO's authority must be made available to the contractor upon request.[36]

The CO also may delegate some or all of his contract administration authority to subordinates, often referred to as contracting officer's representatives (CORs).[37] This delegation should be in writing. The CO is required to advise the contractor of the scope of the delegation "as necessary."[38] The authority of a COR may also be limited under agency-specific provisions that specifically disavow the authority of a COR[39] and/or under provisions in the parties' contract.[40]

Whether the contractor receives direction from the CO or a COR, it should make sure that direction is within that official's authority. It can do so by requesting and reviewing the CO's warrant or the COR's written delegation. A contractor who follows an official's direction without confirming his actual authority does so at its risk.[41]

B. Ratification

The government may be bound through an action by a government representative, even if it was unauthorized initially, if that action is later ratified.[42] Under the FAR, a government commitment is only subject to ratification if the sole reason it was not binding initially was "because the Government representative who made it lacked the authority to enter into that agreement on behalf of the Government."[43] In addition, only the head of the contracting activity, or a higher level official if one is designated by the agency, may ratify an unauthorized commitment.[44]

C. Implied Authority

The requisite authority to obligate the government may be implied from a course of government conduct or from an assignment of managerial duties to a government official, if such authority is normally an integral part of the delegated responsibility. The Court of Claims found implied authority in *Max Drill v. United States*. In that case, a government technical representative was sent by the CO to guide contractors making onsite inspections. Finding an exception to the general rule requiring actual authority, the Court of Claims found the government bound under the doctrine of implied authority: "When an official of the contracting agency is not the contracting officer, but has been sent by the contracting officer for the express purpose of giving guidance in connection with the contract, the contractor is justified in relying on his representations."[45]

In other situations, unauthorized government actions have been found to bind the government based on implied authority. For example, in *Clevite Ordinance Division of Clevite Corp.*, the ASBCA held that the official was, in fact, an authorized representative of the CO where a government official was in general superintendence of all contract activities of a particular project but was not an officially appointed, authorized representative.[46] The government has also been held liable where the CO authorized technical personnel to guide or instruct contractors with specification problems, even though their guidance was erroneous.[47] In other instances, boards have recognized implied authority to direct changes where the government representative who directed the change derived his authority to make decisions from the CO.[48] However, when the government's course of conduct changes, casual dealings and instructions, although previously honored, cannot be relied on by the contractor, especially when the CO cautions the contractor to wait for a formal agreement.[49]

D. Apparent Authority

The doctrine of apparent authority is not applicable to the government. This means that a contractor may not rely on the appearance that a government official has the authority to make changes or give directions. That "apparent" authority, in the absence of actual authority provided by regulation or by a written delegation, or those limited circumstances in which implied authority may be found, is without any binding effect.[50]

The Court of Federal Claims explained the difference between implied and apparent authority in *Stevens Van Lines v. United States*.[51] In that case, the court awarded damages to contractors that had relied on government employees' statements that the government would pay for work until a new procurement became effective. Implied authority is a form of actual authority that arises from government action and intent, as opposed to apparent authority, which is based on contractor reliance regardless of the owner's intent. The court held that, because the government employees' statements were made frequently, were "appropriate and essential" to the employees' duties, and were within the known duties they performed, the employees had implied authority, and the government was bound by their statements.[52]

III. *Improper Business Practices*

Like other federal contracts, government construction contracts are subject to the prohibitions on improper business practices set out in the FAR. The prohibitions collected in FAR Part 3 can be divided into two broad categories: (1) improper pricing practices and (2) improper interactions with federal employees. The full impact of those prohibitions cannot be understood, however, without first considering the transparent environment in which government contracting takes place. Recent additions to the FAR have made government contractors subject to mandatory disclosure obligations that increase the importance of carefully addressing these prohibitions when establishing procedures for bid and proposal preparation, and in managing awarded contracts.

A. Transparency of Federal Construction Contracting Environment

Government construction contractors need to assume that they are operating in an environment fully transparent to both the government and the public. The FAR warns federal employees that they should behave this way: "While many Federal laws and regulations place restrictions on the actions of Government personnel, their official conduct must, in addition, be such that they would have no reluctance to make a full public disclosure of their actions."[53] With the recently promulgated mandatory disclosure rules, combined with the long-standing impact of the Freedom of Information Act (discussed in

Section V below), government construction contractors should also conduct themselves with the expectation of full public disclosure of their actions.

1. Mandatory Disclosure and Internal Control Systems

Under FAR provisions that went into effect in December 2008, essentially every federal government contractor is required to disclose to the appropriate agency Office of Inspector General (OIG) all "credible evidence" related to a government contract of (1) violations of criminal law involving fraud, conflict of interest, or bribery; (2) violations of the civil False Claims Act (FCA);[54] or (3) significant overpayments. This disclosure obligation is not only mandatory; it also applies retroactively to violations or overpayments occurring before these disclosure provisions went into effect, so long as they relate to any contract that has not been closed out for at least three years. And to ensure that contractors do not try to avoid making disclosures by ignoring potential problems, the FAR now requires most contractors to establish internal control systems that actively seek disclosable evidence to be passed on to agency OIGs.[55] Because these mandatory disclosure requirements and internal control systems are new, there is little detailed guidance as to how to satisfy these broad obligations.

The government has three separate tools to enforce mandatory disclosure. First, the contract clause at FAR 52.203-13 is included in all prime contracts worth more than $5 million and lasting more than 120 days.[56] This clause must also be "flowed down" to all subcontracts over $5 million and with a duration of more than 120 days.[57] The clause requires the contractor and significant subcontractors to disclose to the agency's OIG any criminal violations or civil FCA violations that relate to that contract and any subcontracts.

Second, the same FAR clause also imposes specific obligations on the contractor and significant subcontractors to establish and maintain an internal control system, unless the contract is with a small business or is for commercial items. As a baseline, contractors must have a written code of business ethics and conduct.[58] All contractors should have an employee business ethics and compliance training program and an internal control system that

(1) Are suitable to the size of the company and extent of its involvement in government contracting;
(2) Facilitate timely discovery and disclosure of improper conduct in connection with government contracts; and
(3) Ensure corrective measures are promptly instituted and carried out.[59]

If a contract includes the clause at FAR 52.203-13, the contractor takes on the additional obligation to "(i) Exercise due diligence to prevent and detect criminal conduct; and (ii) Otherwise promote an organizational culture that encourages ethical conduct and a commitment to compliance with the law."[60] These internal control obligations require the contractor to implement the

Justice Department's corporate compliance standards set out in the United States Sentencing Guidelines. To meet those standards, the internal control system must actively look for potential violations of criminal law and the civil FCA. Violations that are uncovered that relate to *any* of the contractor's federal contracts or subcontracts must then be disclosed to the relevant agency OIG.

Third, separate FAR provisions make the failure to disclose any violations and overpayments on any federal contract or subcontract grounds for suspension or debarment.[61] There is no dollar threshold for the debarment and suspension provisions, and they apply to all contractors and subcontractors on any government contracts that have not been closed out for more than three years. The definition of "contractor" for purposes of debarment and suspension applies to any party that submits an offer or is awarded a contract with the government, and any subcontractors to such a party, at any tier.[62] It also applies to anyone who "conducts business, or reasonably may be expected to conduct business, with the government or as an agent or representative of another contractor."[63]

Taken in combination, these FAR provisions essentially establish a mandatory disclosure system applicable to all companies that do business with the government. The disclosure requirements apply to all types of companies, including small businesses, commercial contractors, and companies that do all of their work outside the United States.

a. Mandatory Disclosure Requirements
(1) What Is to Be Disclosed

Under the mandatory disclosure requirements, contractors must disclose evidence of (1) certain types of misconduct; (2) that is credible; (3) that is known to principals of the contractor; (4) that is related to any federal contract awarded to the contractor, or subcontracts thereunder; (5) so long as the relevant contract is either current or was completed within the past three years.

Contractors must disclose three types of misconduct: (1) violations of criminal law involving fraud, conflict of interest, and bribery; (2) violations of the civil FCA; and (3) significant overpayments. Disclosures are required for "a violation of Federal criminal law involving fraud, conflict of interest, bribery, or gratuity violations found in Title 18 of the United States Code."[64] Violations of any part of the civil FCA are also subject to mandatory disclosure.[65] The disclosure obligation extends beyond violations by contractor employees to include violations by a contractor's agents and subcontractors. The requirement for disclosure of "significant overpayments" is found in the debarment and suspension rules. Payment clauses require contractors to notify the government if the contractor becomes aware of an overpayment, regardless of its significance.[66]

The disclosure requirements contemplate that contractors will make disclosures even if they are not certain that a violation has occurred. The requirement to disclose "credible evidence of a violation" of a criminal law or the civil FCA indicates that the government is unlikely to agree that a disclosure was not required if a contractor bases its failure to disclose on a legal theory that is subject to challenge by the Justice Department. A contractor must disclose more than violations of a law; it must disclose credible evidence of violations of a law.

The FAR mandatory disclosure provisions only require the disclosure of violations and significant overpayments that are known to a principal of the contractor. A "principal" is a relatively high level of management. Principals are defined as "officers; directors; owners; partners; and, persons having primary management or supervisory responsibilities within a business entity (e.g., general manager; plant manager; head of a subsidiary, division, or business segment, and similar positions)."[67]

A contractor is required to disclose violations and overpayments related to any government contract it performs. The mandatory disclosures required by the detailed internal control system and the new debarment and suspension rules cover all such contracts, not just the contracts containing the new FAR clause.[68] A contractor is also required to disclose credible evidence of known violations and overpayments by subcontractors under its government contracts.[69] However, the prime contractor is not required to review or approve subcontractor internal control systems.[70]

A major subcontractor on a government construction contract takes on its own independent mandatory disclosure requirement, even without having any direct contract with the government. The clause at FAR 52.203-13 must be flowed down to all subcontracts that meet the same threshold as the prime contract: a value in excess of $5 million and performance period of more than 120 days.[71] When subcontractors are required to make disclosures, they are required to do so directly to the agency OIG and not through the prime contractor.[72]

The FAR requires disclosure of earlier misconduct so long as it is related to a contract that either is still active or was closed out within the last three years.[73] This continuing obligation to disclose should not be confused with a statute of limitations. The obligation to disclose is determined not by the date of the misconduct but by the status of the contract to which the misconduct relates. Delays in the processing of final payments on completed contracts will lengthen the required disclosure period because the three-year cutoff of disclosure obligations does not begin to run until final payment is made.

(2) When Disclosures Should Be Made

The mandatory disclosure rules require "timely" disclosure of credible evidence of violations, but they do not set any specific boundaries on what would be considered a timely disclosure.[74] To avoid questions about the timeliness of a disclosure and have enough time to complete a thorough investigation if necessary before disclosure to the government, contractors should promptly initiate investigations of allegations of misconduct.

(3) To Whom Disclosures Must Be Made

Contractors should make all required disclosures to the relevant agency OIG, with a copy to the CO responsible for the contract.[75]

b. Internal Control System Requirements

The FAR internal control system requirement imposed by FAR 52.203-13 consists of both a training program and a detailed internal control system. A contractor must establish both within 90 days after receiving its first contract including this FAR clause, unless it can convince the CO that it needs a longer period of time.[76]

The contract clause calls the mandated training program "an ongoing business ethics awareness and compliance program."[77] The FAR provision only provides guidelines for the training program:

> This program shall include reasonable steps to communicate periodically and in a practical manner the Contractor's standards and procedures and other aspects of the Contractor's business ethics awareness and compliance program and internal control system, by conducting effective training programs and otherwise disseminating information appropriate to an individual's respective roles and responsibilities.[78]

No specific content is required. The training must be provided to the contractor's principals and employees and, as appropriate, to the contractor's agents and subcontractors.[79] But there is no guidance as to when providing training to agents or subcontractors would be appropriate.

The minimum requirements for the internal control system set out in the FAR are modeled on those in section 8B2.1 of the United States Sentencing Guidelines on the sentencing of organizations.[80] As a result, sources interpreting section 8B2.1, including application notes and case law, may be helpful in interpreting these FAR requirements.

The internal control system requires the following elements: (1) mandatory disclosure; (2) periodic reviews of company practices, procedures, policies, and internal controls; (3) a reporting mechanism that offers anonymity; (4) disciplinary action for improper conduct; (5) assignment of responsibility at a sufficiently high level of the organization; (6) devotion of adequate resources to ensure effectiveness of the ethics awareness and compliance program; (7) reasonable efforts to exclude individuals who have been engaged in illegal or unethical conduct from the contractor's senior management; and (8) periodic assessments of risk of criminal conduct.[81]

These internal control system requirements impose an active obligation on contractors to try to uncover violations of criminal law and the civil FCA that would require disclosure, including the requirement for monitoring and auditing to detect criminal conduct[82] and establishing an internal anonymous reporting mechanism.[83]

2. Whistle-Blower Protections for Contractor Employees

Government contractors also are required to protect their employees from reprisals for reporting these same types of violations to government officials. Employee whistle-blowers are protected under all government contracts for reporting substantial violations of contracting law to responsible government officials.[84] Even greater whistle-blower protections have been introduced to protect whistle-blowers working on federal contracts funded by the stimulus funds appropriated under the 2009 American Recovery and Reinvestment Act.[85]

a. Whistle-Blower Protections Applicable to All Federal Contracts

Federal statutes protect all contractor employees against retaliation by their employers in reprisal for disclosing information relating to a "substantial violation of law related to a contract" to a responsible government official:[86]

> Government contractors shall not discharge, demote or otherwise discriminate against an employee as a reprisal for disclosing information to a Member of Congress, or an authorized official of an agency or of the Department of Justice, relating to a substantial violation of law related to a contract (including the competition for or negotiation of a contract).[87]

Note that not all disclosures to federal employees are protected, only disclosures made to:

- Members of Congress
- "Authorized officials of an agency," defined as "an officer or employee responsible for contracting, program management, audit, inspection, investigation, or enforcement of any law or regulation relating to Government procurement or the subject matter of the contract"[88]
- "Authorized officials of the Department of Justice," defined as "any person responsible for the investigation, enforcement, or prosecution of any law or regulation"[89]

If an employee believes that he has been discriminated against in reprisal for disclosing a substantial violation of contracting law, there is a procedure established by the FAR for seeking relief. The aggrieved employee first files a detailed complaint with the OIG of the agency that awarded the contract.[90] The OIG determines whether the complaint merits further investigation and, if so, prepares a written report to the agency head.[91] The contractor is given a chance to respond after the report is issued.[92]

If the agency head concludes that there has been an improper reprisal against a contractor employee, the FAR sets out remedies the agency may impose. They include:

- abatement of the reprisal
- reinstatement of the employee to his position before the reprisal, with full compensation
- contractor payment of the employees' costs and expenses, including attorney's fees.[93]

b. Whistle-Blower Protections Applicable to Contracts Funded Through the American Recovery and Reinvestment Act of 2009

For contracts funded by the 2009 stimulus bill, even greater protections are afforded to whistle-blowers. The types of disclosures that are protected are much broader than information relating to "a substantial violation of law related to a contract." They include information that the employee reasonably believes is evidence of:

- gross mismanagement of the contract or subcontract related to stimulus funds,
- gross waste of stimulus funds,
- a substantial and specific danger to public health or safety related to the implementation or use of stimulus funds,
- an abuse of authority related to the implementation or use of stimulus funds, or
- a violation of law, rule, or regulation related to an agency contract (including the competition for or negotiation of a contract) awarded or issued relating to stimulus funds.[94]

Disclosures to a much broader range of persons are also protected, including:

- The Recovery Accountability and Transparency Board
- an Inspector General
- the Comptroller General
- a member of Congress
- a state or federal regulatory or law enforcement agency
- a person with supervisory authority over the employee
- other person working for the employer who has the authority to investigate, discover, or terminate misconduct
- a court or grand jury
- the head of a federal agency.[95]

The most significant additions to the list of recipients are the contractor's own employees, the employee's supervisor, and another employee with authority to investigate misconduct. Thus, an employee can now seek relief for reprisals based on internal disclosures alleging mismanagement, waste, or abuse of authority, when those disclosures are related to a contract involving stimulus funds.

Employees of contractors working on stimulus-funded contracts are also given greater procedural rights. In addition to the rights provided to whistle-blowers on all federal contracts, a whistle-blower disclosing informa-tion related to a contract involving stimulus funds is also given:

- access to the investigation file of the Inspector General[96]
- a very favorable burden of proof for establishing reprisal[97]
- the right to a jury trial in federal district court if the agency head does not rule in the employee's favor.[98]

Contractors receiving stimulus funds also are required to post a notice of employee rights and remedies for whistle-blowers.[99]

B. Improper Pricing Practices
1. Bid Rigging

The government prohibits all efforts to eliminate competition or restrain trade. The FAR specifically lists a number of anticompetitive practices as antitrust violations that must be reported to the Attorney General: collusive bidding, follow-the-leader pricing, rotated low bids, collusive price estimating systems, and sharing of business.[100] FAR 3.103-2(b)(2) requires rejection of offers sus-pected of being collusive.

The government requires contractors on fixed price solicitations to sign a certification disavowing bid rigging. For almost all fixed-price contracts, the FAR requires inclusion of clause 52.203-2—Certificate of Independent Price Determination. Each offeror is required to certify the following per FAR 52.203-2(a):

(1) The prices in this offer have been arrived at independently, with-out, for the purpose of restricting competition, any consultation, communication, or agreement with any other offeror or competi-tor relating to—
(i) Those prices;
(ii) The intention to submit an offer; or
(iii) The methods or factors used to calculate the prices offered.
(2) The prices in this offer have not been and will not be knowingly disclosed by the offeror, directly or indirectly, to any other offeror or competitor before bid opening (in the case of a sealed bid solici-tation) or contract award (in the case of a negotiated solicitation) unless otherwise required by law; and
(3) No attempt has been made or will be made by the offeror to induce any other concern to submit or not to submit an offer for the pur-pose of restricting competition.

The certification must be signed by the company official responsible for deter-mining the offered price, or another person formally authorized to certify on that official's behalf.[101]

An offeror may properly sign this certification so long as there is no inten-tion or purpose to restrict competition. Although the certification prohibits disclosure of bid prices, it does not restrict innocent or inadvertent disclosures regarding other aspects of an offeror's proposal, as long as the disclosures are not anti-competitive. FAR 3.103-2(a)(1) provides authorization for specific types of disclosures related to prices.

Contractors also are not permitted to enter into agreements restricting their subcontractors' ability to provide their goods or services directly to the government, and may not unreasonably preclude subcontractors from making direct sales to the government.[102] This restriction is implemented through the clause at FAR 52.203-6—Restrictions on Subcontractor Sales to the Government, which prohibits entering into

> any agreement with an actual or prospective subcontractor, nor otherwise act in any manner, which has or may have the effect of restricting sales by such subcontractors directly to the Government of any item or process (including computer software) made or furnished by the subcontractor under this contract or under any follow-on production contract."[103]

These prohibitions do not preclude contractors from asserting rights that are otherwise authorized by law or regulation.[104] This clause also has to be flowed down to subcontractors.[105]

2. Subcontractor Kickbacks

The Anti-Kickback Act of 1986[106] prohibits providing or accepting kickbacks, or attempting to do so. It also prohibits including the amount of any kickback either in the price of any prime contract with the government or in any price of a subcontract charged to a prime contractor or a higher tier subcontractor.[107] Kickbacks are defined broadly to include

> any money, fee, commission, credit, gift, gratuity, thing of value, or compensation of any kind which is provided, directly or indirectly, to any prime contractor, prime contractor employee, subcontractor, or subcontractor employee for the purpose of improperly obtaining or rewarding favorable treatment in connection with a prime contract or in connection with a subcontract relating to a prime contract.[108]

In *Morse Diesel International, Inc. v. United States*,[109] the Court of Federal Claims held that the Anti-Kickback Act covers any kind of benefits that might be provided to a prime contractor in return for participation on a government contract. According to the court, "Congress intended the language 'favorable treatment' be construed broadly to reach all conduct analogous to commercial bribery . . . even if it did 'not directly [impact] the federal treasury.'"[110] In that case a bond broker's sharing of commissions with a prime contractor was determined to be a kickback under the Act.[111] In the same opinion, the court also emphasized that the Anti-Kickback Act creates "a presumption that any kickback was included in the price of an affected federal contract or subcontract and therefore increased costs to the Government."[112]

Knowing and willful violations of the Anti-Kickback Act are subject to criminal liability.[113] Knowing violations can lead to civil penalties.[114] The Act also authorizes the CO to withhold the amount of a kickback and to direct a prime contractor to withhold the amount of a kickback from a subcontractor.[115]

The Act creates an obligation not only to avoid kickbacks but also to prevent, detect, and report them. A contractor is required to have and follow "reasonable procedures designed to prevent and detect violations of the Act in its own operations and direct business relationships."[116] Specific examples of reasonable procedures are listed:

- Company ethics rules prohibiting kickbacks by employees, agents, or subcontractors
- Education programs for new employees and subcontractors, explaining policies about kickbacks, related company procedures, and the consequences of detection
- Procurement procedures to minimize the opportunity for kickbacks
- Audit procedures designed to detect kickbacks
- Periodic surveys of subcontractors to elicit information about kickbacks
- Procedures to report kickbacks to law enforcement officials
- Annual declarations by employees of gifts or gratuities received from subcontractors
- Annual employee declarations that they have violated no company ethics rules
- Personnel practices that document unethical or illegal behavior and make such information available to prospective employers.

If the contractor has reasonable grounds to believe the Anti-Kickback Act has been violated, it is obliged to "promptly report in writing the possible violation . . . to the inspector general of the contracting agency, the head of the contracting agency if the agency does not have an inspector general, or the Department of Justice."[117]

The Anti-Kickback Act is enforced through the clause at FAR 52.203-7, Anti-Kickback Procedures. The Anti-Kickback clause must also be flowed down to all subcontracts worth more than $100,000.[118]

3. Buying-In

The FAR discourages but does not prohibit "buying-in." Buying-in is defined as submitting an offer below anticipated costs, expecting to make up the losses by increasing the amount of the contract after award or receiving follow-on contracts at artificially high prices.[119] The FAR directs the government to "minimize the opportunity for buying-in by seeking a price commitment covering as much of the entire program concerned as possible."[120]

4. *Contingent Fees*

Federal statutes restrict the type of contingent fee arrangements that may be used for soliciting or obtaining federal contracts, based on the fear that contingent fee arrangements encourage the exercise of improper influence.[121] But the focus on improper influence in the statute and regulations implementing it gives this restriction little real effect.

The prohibition on certain types of contingent fee arrangements is implemented through a warranty against contingent fees that is required of every contractor.[122] But the "Covenant Against Contingent Fees" actually permits contingent fee arrangements, so long as the arrangements are with "bona fide employees" or "bona fide agencies." A bona fide employee is

> [a] person, employed by a contractor and subject to the contractor's supervision and control as to time, place, and manner of performance, who neither exerts nor proposes to exert improper influence to solicit or obtain Government contracts nor holds out as being able to obtain any Government contract or contracts through improper influence.[123]

A bona fide agency is

> an established commercial or selling agency, maintained by a contractor for the purpose of securing business, that neither exerts nor proposes to exert improper influence to solicit or obtain Government contracts nor holds itself out as being able to obtain any Government contract or contracts through improper influence.[124]

"Improper influence" is defined as

> any influence that induces or tends to induce a Government employee or officer to give consideration or to act regarding a Government contract on any basis other than the merits of the matter.[125]

Provided the members of a contractor's sales force do not hold themselves out as being able to obtain federal contracts through improper influence, these definitions permit the Covenant Against Contingent Fees to be easily satisfied. If an actual breach of the covenant does occur, however, the government may annul the contract or deduct the cost of the contingent fee from the contract price.[126]

C. Interacting with Federal Employees

1. *General Restrictions on Federal Employee Interaction with Contractors*

Federal employees operate under broad restrictions against receiving anything of value from contractors or potential contractors. These restrictions are described generally in the FAR:

As a rule, no Government employee may solicit or accept, directly or indirectly, any gratuity, gift, favor, entertainment, loan, or anything of monetary value from anyone who (a) has or is seeking to obtain Government business with the employee's agency, (b) conducts activities that are regulated by the employee's agency, or (c) has interests that may be substantially affected by the performance or nonperformance of the employee's official duties.[127]

Agencies have authority to establish exceptions.[128] Nevertheless, federal construction contractors should take care in their marketing efforts to avoid putting federal employees in jeopardy by offering them goods or services that provide the employee with any personal benefit.

2. Prohibitions on Obtaining Competitors' Bid or Proposal Information and the Government's Source Selection Information

The Procurement Integrity Act[129] prohibits contractor employees from obtaining the following types of procurement information before award of the relevant contract: (1) contractor bid or proposal information; and (2) source selection information.[130] The Act specifically covers federal construction contracts.[131]

The FAR provides detailed definitions of both types of protected information. "Contractor bid or proposal information" is defined as the following types of information that (a) have been submitted to the government in connection with a procurement, and (b) have not yet already been made public:

- Cost or pricing data
- Indirect costs and direct labor rates
- Proprietary information about manufacturing processes, operations, or techniques marked by the contractor in accordance with applicable law or regulation
- Information marked by the contractor as "contractor bid or proposal information" in accordance with applicable law or regulation
- Information in a proposal marked for protection from public disclosure in accordance with FAR 52.215-1(e), Instructions to Offerors: Restriction on Disclosure and Use of Data[132]

"Source selection information" includes the following types of nonpublic information that have been prepared for use by the government to evaluate a bid or proposal:

- Bid prices submitted in response to an invitation for bids
- Proposed costs or prices submitted in response to a solicitation
- Source selection plans
- Technical evaluation plans
- Technical evaluations of proposals
- Cost or price evaluations of proposals

- Competitive range determinations that identify proposals that have a reasonable chance of being selected for award of a contract
- Rankings of bids, proposals, or competitors
- Reports and evaluations of source selection panels, boards, or advisory councils
- Other information marked as "Source Selection Information"—See FAR 2.101 and 3.104.[133]

The potential penalties for receipt of either type of protected information are severe. If a contractor obtains bid or proposal information or source selection information in violation of these rules, the government may disqualify the contractor if the contract has not yet been awarded.[134] If the contract has been awarded to the contractor, the government may recapture the contractor's profits,[135] or it may void or rescind the contract and recover all funds spent on the contract.[136] FAR Subpart 3.7 contains detailed policies and procedures for voiding or rescinding contracts. Finally, the government can also seek criminal penalties for Procurement Integrity Act violations.[137]

There are a few exceptions when information that would otherwise be protected under the Procurement Integrity Act may be received or disclosed by a federal contractor. A contractor may disclose its own bid or proposal information.[138] Information related to a canceled procurement that the government does not intend to resume may be disclosed, provided there are no other protections against disclosure.[139] The government also may use technical data consistent with its rights to do so.[140] Information must still be released in response to a proper request from Congress, any of its committees, a federal agency, the Comptroller General, or an Inspector General of any agency.[141] But these circumstances rarely permit release of either bid or proposal information or source selection information before an award has been announced.

3. Limitations on Recovering Costs of Influencing Federal Transactions

The Byrd Amendment[142] prohibits contractors from using appropriated government funds to pay the costs of the contractor's effort to influence federal officials in connection with the award of a federal contract, grant, loan, or cooperative agreement.[143] Under the Byrd Amendment, no federal funds may be expended by the recipient of a federal contract to pay any person for influencing or attempting to influence a federal officer in connection with the awarding of a federal contract.[144] FAR 3.801 defines "influencing or attempting to influence" as "making, with the intent to influence, any communication to or appearance before" a federal officer or employee in connection with a federal contract.

Contractors are also required to certify and disclose any payments out of appropriated funds made to influence such awards.[145] The certification is found at FAR 52.203-11, and the relevant contract clause is found at FAR 52.203-12. A contractor must also obtain such certifications from all subcontractors on subcontracts worth more than $100,000. At the same time, the contractor must disclose all lobbying contacts on behalf of the contractor by registered lobbyists using OMB Standard Form LLL, Disclosure of Lobbying Activities.[146]

Violation of the Byrd Amendment or failure to meet the certification and disclosure requirements could lead to the imposition of civil penalties of not less than $10,000 per violation.[147]

Despite the broad-sounding prohibition on contractor recovery of lobbying costs, significant regulatory exceptions to the Byrd Amendment greatly narrow the types of expenditures that are actually prohibited and must be disclosed. The most significant exception is that the prohibition does not apply to payments to influence award of contracts that are made out of profits or fees received on federal contracts.[148] The contractor can prove that the payments did not use appropriated funds by simply showing that the lobbying expenditures were less than the contractor's profits on federal contracts or its income from non-federal contracts.[149] The Byrd Amendment also contains an exception that allows recovery of the costs of a company employee's salary, so long as that employee's attempts to influence are "not directly related" to award of a specific contract.[150] This exception covers general agency and legislative liaison efforts by the contractor's own employees,[151] as well as the provision of professional and technical services directly in preparation of a bid or proposal.[152] Specific examples of activities for which costs may be recovered are listed in FAR 3.803(a) and 52.203-12(c).

4. Bribery and Gratuities

Offering a bribe or gratuity to a federal official is prohibited by 18 U.S.C. § 201 and 10 U.S.C. § 2207.[153] Even the acceptance of gifts is usually prohibited under 5 U.S.C. § 7353 and 5 C.F.R. § 2635.

FAR 3.202 also requires inclusion of a special contract clause in all federal construction contracts addressing gratuities[154] that permits the government to terminate a contract if it concludes that the contractor offered a gratuity to a government employee with the intent to obtain favorable treatment. It also allows the government to seek exemplary damages beyond breach of contract damages in Department of Defense contracts.[155]

IV. Conflicts of Interest

The government is very sensitive to the possibility of conflict of interest in the procurement process. "The general rule is to avoid strictly any conflict of interest or even the appearance of a conflict of interest in Government-contractor relationships."[156] The concern regarding conflicts of interest extends to two areas: (1) government employees' personal conflicts of interest; and (2) organizational conflicts of interest that give one contractor unfair advantages over others.

A. Government Employees' Personal Conflicts of Interest

1. Prohibitions Against Award to Contractors Associated with Government Employees

The government nearly always prohibits award of a contract to a government employee or an organization owned or controlled by a government employee. The FAR establishes as a matter of policy that the government will not award contracts to government employees or companies they substantially own or control.[157] There is an exception for special government employees[158] performing services as experts, advisors, or consultants, or as members of advisory committees, unless some other conflict of interest exists.[159] But otherwise, the head of an agency or head of a contracting activity may only grant an exception "if there is a most compelling reason to do so, such as when the Government's needs cannot reasonably be otherwise met."[160]

2. Revolving Door Restrictions

The government also protects against potential government employee personal conflicts of interest by restricting the ability of a federal contractor to gain special advantages in the procurement process by hiring government employees involved in that procurement process. Former government employees are prohibited from representing the contractor before the government in relation to any contract or matter on which they, while with the government, supervised or directly participated personally and substantially (though not necessarily determinatively) in the specifications, statement of work, solicitation, bid evaluation, contract or price negotiation, or award.[161]

Government construction contractors need to appreciate that contacting a government official involved in a procurement regarding possible employment for that official could lead that official to be disqualified from working on the procurement.[162] This is true even if the contact is made through an agent or some other intermediary.[163]

There is also a one-year ban on a former government official accepting any compensation from a contractor awarded a contract, when that official played a major role in the award or administration of that contract.[164] The ban does not apply, however, if the former government employee is working for a different division or affiliate of the contractor that does not produce similar products or services as the entity that received the relevant contract.[165]

B. Organizational Conflicts of Interest

Because of increasing use of contracts that require contractors to exercise judgment in a way that affects ongoing and future acquisitions, organizational conflicts of interest (OCIs) have become a significant concern and subject of frequent bid protest challenges. OCIs are defined by the FAR as situations where because of other activities or relationships with other persons, a person is unable or potentially unable to render impartial assistance or advice to the Government, or the person's objectivity in performing the contract work is or might be otherwise impaired, or a person has an unfair competitive advantage.[166]

FAR Subpart 9.5 sets out rules for handling OCIs. These rules are based on two underlying principles: "(a) Preventing the existence of conflicting roles that might bias a contractor's judgment; and (b) Preventing unfair competitive advantage."[167]

Construction procurements must be analyzed for potential OCIs. The FAR recognizes that OCIs are more likely to occur in contracts involving

(1) Management support services;
(2) Consultant or other professional services;
(3) Contractor performance of or assistance in technical evaluations; or
(4) Systems engineering and technical direction work performed by a contractor that does not have overall contractual responsibility for development or production.[168]

The government uses these types of contracts in connection with construction projects, but typically in specialized contracting arrangements that have built-in mitigations against potential OCIs. Potential OCIs are dealt with systematically in two types of specialized contracting arrangements: design-build contracts[169] and architect-engineer contracts.[170] Nevertheless, the general protections against OCIs in the FAR do apply to construction contracts, and are not preempted by either of these special construction contracting vehicles.[171] Construction contractors should be aware of the potential restrictions that could be placed on them to avoid OCIs in both construction management and design contracts.

1. CO Responsibilities to Avoid, Neutralize, or Mitigate Potential OCIs
COs are responsible for recognizing OCIs early in the procurement planning process and for modifying solicitations and contracts in order to minimize their impact. COs are required to (1) identify and evaluate potential organizational conflicts of interest as early in the acquisition process as possible; and (2) avoid, neutralize, or mitigate significant potential conflicts before contract award.[172]

The FAR does not encourage COs to disqualify offerors based on potential OCIs. Only if the CO determines that a conflict of interest "cannot be avoided or mitigated" may the CO cancel award to the apparent successful offeror. But before making that determination, the CO must "notify the contractor, provide the reasons therefor, and allow the contractor a reasonable opportunity to respond."[173] As a result, if a contractor is aware of any activity that might be construed as creating an OCI in a particular procurement, it would benefit from preparing a mitigation plan designed to allow it to continue to participate.

2. Rules Governing Specific Categories of OCIs

The FAR directly addresses four types of OCIs,[174] and provides specific examples to help contractors understand how particular situations might be handled.[175] The case law groups these four types of OCIs into three categories: (1) biased ground rules; (2) unequal access to information; and (3) impaired objectivity.[176] Each category is addressed below.

The FAR makes clear that these categories are not the only types of OCIs requiring intervention by a CO. Because conflicts may arise in other factual situations, COs are directed to examine each situation individually and to exercise "common sense, good judgment, and sound discretion" in assessing whether a significant potential conflict exists and in developing an appropriate way to resolve it.[177]

a. Biased Ground Rules

Biased ground rule cases involve contractors that set the ground rules for a subsequent procurement by providing the specifications or statement of work. The primary concern here is that the firm could skew the competition, whether intentionally or not, in favor of itself. There may also be a concern that the firm, by virtue of its special knowledge of the agency's future requirements, would have an unfair advantage in the competition for those requirements.

This category includes two of the OCI types described in the FAR, Providing Systems Engineering and Technical Direction[178] and Preparing Specifications or Work Statements.[179] A contractor that provides systems engineering or technical direction for a system without having overall contractual responsibility for the development and integration of the system cannot "(1) [b]e awarded a contract to supply the system or any of its major components; or (2) [b]e a subcontractor or consultant to a supplier of the system or any of its major components."[180]

For contractors that prepare specifications or work statements to be used in a competitive procurement, the FAR treats nondevelopmental work differently from developmental work. Generally, for nondevelopmental work, if the contractor prepares the complete specification, it will not be permitted to receive the contract utilizing that specification.[181] Likewise, a contractor not involved in development or design that prepares a work statement to be used in a competitive acquisition for a system or services generally will not be permitted to supply the system or services.[182] But for developmental work, the FAR states categorically that the competitive advantage derived from performing the development work is not unfair, because of the advantages to the government of selecting the developer to do the subsequent production.[183]

b. Impaired Objectivity

Impaired objectivity cases occur where a contractor is required to evaluate itself or a competitor, and the concern is that the firm's ability to render impartial advice to the government could appear to be undermined by its relationship with the entity whose work product is being evaluated. A contractor providing evaluation services will not be permitted to receive a contract for evaluation of its own offers for products or services, or those of a competitor, unless there are "proper safeguards to ensure objectivity to protect the Government's interests."[184] A firewall between the evaluating team and the rest of the contractor organization is not sufficient to overcome impaired objectivity, because a firewall will not eliminate the incentive to benefit the employer of the evaluators.[185]

c. Unequal Access to Information

The concern in unequal access to information cases is the risk of the firm gaining a competitive advantage. A contractor that obtains access to other companies' proprietary information in the course of providing services to the government is required to agree with those companies to protect that information from unauthorized use or disclosure.[186] The government is also suspicious that marketing consultants may provide proprietary and source selection information to a contractor that will give that contractor an unfair competitive advantage. To counter this possibility, contractors are directed to "make inquiries of marketing consultants to ensure that the marketing consultant has provided no unfair competitive advantage."[187]

3. Waiver

An agency head has the authority to waive application of any of the specific rules in FAR Subpart 9.5. Any of the general rules or procedures may be waived after a determination that applying such a rule or procedure would not be in the government's interest in a particular situation.[188] The determination that the government's interest outweighs the need to apply the FAR OCI provisions must have a reasonable basis.[189]

4. Solicitation Provision and Contract Clause

Unlike the norm under the FAR, when an OCI is anticipated there is no mandated provision to insert into a solicitation or contract. Instead, the CO is directed to craft provisions appropriate to the particular OCI. For the solicitation, the CO is required to include a provision that states (1) the potential conflict as seen by the CO; (2) the proposed restraint upon future contractor activities; and (3) whether or not the terms of any proposed OCI restriction are subject to negotiation.[190] The CO may permit offerors to provide additional information in response to the solicitation, and negotiate the final contract clause based on that information.[191] The contract clause will impose restraints upon future contractor activities, but "the restraint imposed by a clause shall be limited to a fixed term of reasonable duration, sufficient to avoid the circumstance of unfair competitive advantage or potential bias."[192] Moreover, "[i]n every case, the restriction shall specify termination by a specific date or upon the occurrence of an identifiable event."[193]

5. Legal Challenges to OCI Determinations

A CO's OCI determination may be challenged in a bid protest before the agency, the Government Accountability Office (GAO), or the COFC. The GAO and COFC have developed extensive case law in this area, with separate analyses for the three primary categories of OCIs. A useful summation of the case law can be found at Daniel I. Gordon, *Organizational Conflicts of Interest: A Growing Integrity Challenge*, 35 *Public Contract Law Journal* 25, 34–41 (2005).

V. The Freedom of Information Act

Another distinctive characteristic of contracting with the government is the ability of the public, including competitors, to obtain much of the information and documentation that bidders and contractors provide to the government. To varying degrees, the public has a right to access contractors' bids, contracts, and communications with the government, as well as documents regarding contract performance, including contract deliverables. The public's right to this information arises under the Freedom of Information Act (FOIA).[194]

As a general matter, FOIA requires the government "to make information available to the public."[195] FOIA contains nine exemptions to its presumption in favor of disclosure.[196] Among these are exemptions for commercial or financial information that is privileged or confidential (exemption 4) and for internal government documents protected by a common law or statutory privilege (exemption 5).[197] The party seeking to withhold requested information has the burden of proving that an exemption applies, and FOIA exemptions "are to be narrowly construed by the courts."[198] The identity of the requesting party is irrelevant to whether an exemption applies, even if the requestor is a direct competitor of the person who provided the information to the government.[199]

Commercial or financial information is exempt from disclosure under FOIA exemption 4 if disclosure is likely either "to impair the Government's ability to obtain necessary information in the future" or "to cause substantial harm to the competitive position of the person from whom the information was obtained."[200] If the information is reasonably available from other sources, the information will not be considered confidential, and the exemption will not apply.[201] Exemption 4 generally applies to proprietary information in contractor bids and contracts, such as technical solutions, financial information, and line-item or unit prices, though only upon an appropriate showing.[202] On the other hand, information will not be protected under this exemption where the bidder knew or should have known it would be incorporated into a public document.[203] This exemption also has been held not to prevent disclosure of the overall contract price.[204]

Exemption 5 is often invoked to avoid disclosure of government documents subject to the deliberative process privilege (referred to by some courts as the "executive privilege").[205] For the privilege, and thus the exemption, to apply, the documents must be both predecisional and deliberative.[206] This protection applies to government source selection documents, such as bid evaluations and scoring sheets.[207] Exemption 5 also prevents contractors and prospective bidders from obtaining the government's construction cost estimates both during bidding and during contract performance.[208]

Documents concerning contract performance generally will be subject to disclosure under FOIA, unless the government or the contractor can show that an exemption applies, e.g., that they reflect confidential business information (and thus are exempt under exemption 4) or are internal government documents subject to the deliberative process or other privilege (and thus are exempt under exemption 5).[209]

VI. Suspension and Debarment

Suspension and debarment are not unique to federal government contracting. Many state and local governments have procedures to bar contractors from doing business with them, and certainly private owners have been known to "blacklist" contractors for various reasons. What sets apart suspension and debarment at the federal level is the detailed rules and regulations governing the procedure to effect it.

Suspension is a temporary disqualification of the contractor while a debarment investigation and proceeding is pending.[210] Debarment is exclusion of the contractor from government contracting for a reasonable, specified period upon a finding that debarment is merited.[211]

The law of suspension and debarment has multiple sources. Depending on the circumstances, contractors can be suspended or debarred under a variety of statutes, under the FAR, or both.

A. Statutory Debarments

Some statutes (for example, the Service Contract Act,[212] Drug-Free Workplace Act of 1988,[213] and the Immigration and Nationality Act[214]) require or allow agency officials to debar contractors who have engaged in prohibited conduct. Statutory debarments are government-wide. Some of these statutes mandate debarment and prescribe the duration of the debarment.[215] The purpose of a statutory debarment is to punish the contractor for violating the statute.

B. Administrative Debarments

The FAR also authorizes debarment and suspension of contractors. Administrative debarment is not a punishment. Rather, it is discretionary and can be imposed only for the government's protection, as a matter of protecting the public interest.[216] Agencies can use administrative agreements instead of debarment and can continue current contracts of debarred contractors. The seriousness of a debarment's cause determines its length, which generally cannot exceed three years, and agency heads can waive debarment for compelling reasons.

C. Debarment Procedures and Standards

FAR Part 9 deals with contractor qualifications, and Subpart 9.4 contains the procedures for suspension and debarment.[217] Agencies seeking debarment or suspension must give written notice to the contractor of the reasons for the proposed action.[218] The contractor then has 30 days in which to make its case.[219] Generally, suspension pending debarment is not to exceed 18 months.[220] If a suspension is made permanent, it is converted to debarment, which generally can last up to three years.[221]

While *prior* conduct is generally what triggers a debarment proceeding, the standard for debarment is *present* responsibility. This is an important distinction. Prior misconduct, no matter how severe, is not a proper basis for debarment unless the government shows that the contractor is not presently a responsible contractor.[222]

Over time, the government has expanded its definition of responsibility to include more than just what a company does on its government contracts. In order to be "responsible," a contractor must have, among other things, a satisfactory performance record; a satisfactory record of integrity and business ethics; the necessary accounting and operational controls; and production control procedures, property control systems, and quality assurance measures.[223]

FAR 9.407-2 and FAR 9.406-2 list the causes for suspension and debarment, respectively.[224] Each list divides into two categories: (1) convictions or civil judgments and (2) other grounds proven upon a preponderance of evidence. In the first category, the following convictions or civil judgments are grounds for both suspension and debarment:

- Commission of fraud or a criminal offense in connection with a public contract or subcontract;
- Violation of antitrust statutes relating to the submission of offers;
- Commission of embezzlement, theft, forgery, and related crimes; and
- Intentionally affixing a label bearing a "Made in America" inscription to a product sold in or shipped to the United States, when the product was not made in the United States.[225]

Grounds for suspension or debarment upon sufficient proof by the government include the above violations where they have not resulted in a conviction or judgment. Also included are the following:

- Commission of an unfair trade practice;
- Violation of the terms of a government contract or subcontract so serious as to justify debarment;
- Commission of any other offense indicating a lack of business integrity or business honesty that seriously and directly affects the present responsibility of a government contractor or subcontractor;
- Delinquent federal taxes in an amount that exceeds $3,000;
- Knowing failure by a principal of a contractor to timely disclose to the government credible evidence of a violation of federal criminal law involving fraud, conflict of interest, bribery, or gratuity violations in connection with a government contract; a violation of the civil False Claims Act; or a significant overpayment on the contract (see Section III.A.1, above); or
- Any other cause of so serious or compelling a nature that it affects the present responsibility of the contractor or subcontractor.[226]

In addition, a contractor may be ineligible for award of a government contract due to its affiliation with a debarred contractor or with a contractor proposed for debarment.[227]

D. Effect of Suspension and Debarment

Suspension and debarments render contractors ineligible from receiving any new federal government contracts. COs are not allowed to solicit offers from, award contracts to, or consent to subcontracts with suspended or debarred contractors.[228] They may avoid this prohibition only if they have "compelling reasons" to do so.[229] Likewise, a contractor may not use a debarred or suspended subcontractor unless it can demonstrate both that there are "compelling reasons" to do so and that the contractor has adequate systems and procedures in place to ensure that the government's interests are fully protected.[230]

Suspension and debarment do not require the termination of existing contracts unless a proper determination to do so is made under the FAR. FAR 9.405-1 makes termination of existing contracts a matter of agency discretion. Circumstances in which terminating a contract based on a suspension or debarment may be appropriate include those where the underlying conduct both involved the contract being terminated and renders the contract void ab initio; where the conduct constitutes a default under the terms of the contract; or where termination would otherwise serve the interests of the government.

The government maintains an Excluded Parties List System (EPLS) to identify contractors that are suspended or debarred.[231] Agencies enter data directly into EPLS concerning those contractors, and are responsible for updating that information as appropriate. FAR 9.105-2 requires the CO to make a present responsibility determination regarding the proposed contractor before making a contract award to that contractor. Checking the EPLS is a step in that responsibility determination.[232]

Finally, as an additional check within the federal procurement system to ensure that the government does not contract with suspended or debarred contractors, contractors are required to certify that they are not suspended or debarred when they submit offers to the government. Since January 1, 2005, the government has required that contractors make this certification through the Online Representations and Certifications Application (ORCA) database.[233]

Notes

1. 48 C.F.R. § 1.000, *et seq.*

2. *E.g.,* Defense Federal Acquisition Regulation Supplement (DFARS), 48 C.F.R. ch. 2.

3. 312 F.2d 418 (Ct. Cl. 1963).

4. ASBCA No. 18103, 74-1 BCA ¶ 10,368.

5. *Id.*

6. 991 F.2d 775, 779 (Fed. Cir. 1993). *Accord* S.J. Amoroso Constr. Co. v. United States, 12 F.3d 1072 (Fed. Cir. 1993); Ryco Constr., Inc, v. United States, 55 Fed. Cl. 184, 199 (2002); United States v. Schlesinger, 88 F. Supp. 2d 431, 443 (D. Md. 2000); Lambrecht & Sons, Inc., ASBCA No. 49515, 97-2 BCA ¶ 29,105; F2M, Inc., ASBCA No. 49719, 97-2 BCA ¶ 28,982.

7. *E.g.,* Computing Application Software Tech., ASBCA No. 47554, 96-1 BCA ¶ 28,204 (NASA clause—Liability for Government Property Furnished for Repair or Other Services—does not express significant strand of procurement policy).

8. *E.g.,* Michael M. Grinberg, DOTBCA No. 1543, 87-1 BCA ¶ 19,573; Amfac Resorts, LLC v. U.S. Dep't of Interior, 282 F.3d 818, 2002 WL 312831 *10 (D.C. Cir. Mar. 1, 2002).

9. *Grinberg,* 87-1 BCA ¶ 19,573; *see also* Johnson v. United States, 15 Cl. Ct. 169 (1988) (convenience termination clause not inserted into a real estate lease by operation of law since the regulations did not explicitly state that such a termination clause was applicable to real property leases); Montana Ref. Co., ASBCA No. 44250, 94-2 BCA ¶ 26,656 (reading standard termination for convenience recovery provisions into a contract not required where the absence of those provisions was an "authorized deviation" from the required clause).

10. *E.g.,* Arrow, Inc., ASBCA No. 41330, et al., 94-1 BCA ¶ 26,353 (CO had made the prescribed determination under FAR to justify use of short-form clause instead of long-form clause).

11. FAR 52.249-2, Alternate 1 "Termination for Convenience of the Government (Fixed-Price)," FAR 52.249-3, "Termination for Convenience of the Government (Dismantling, Demolition, or Removal of Improvements)," and FAR 52.249-6, Alternate 1 "Termination (Cost-Reimbursement)." For further discussion of terminations for convenience, see Chapter 13.

12. *E.g.,* C & J Assocs., VABCA 3892, et al., 95-2 BCA ¶ 27,834 (termination for convenience clause was required by regulation to be included in a purchase order, so it had to be read into contract).

13. Carrier Corp., GSBCA 8516, 90-1 BCA ¶ 22,409; DWS, Inc., ASBCA 29742, 29865, 90-2 BCA ¶ 22,696 (long-form termination clause read in and short-form deleted, since the short-form clause was improperly included); Guard-All of Am., ASBCA 22167, 80-2 BCA ¶ 14,462 (same); *but see* ITT Commc'ns Servs., Inc., GSBCA 9072, 91-3 BCA ¶ 24,337 (contractor challenging government's right to terminate equitably estopped from attacking as invalid inclusion of short-form termination clause in its contract despite finding that CO failed to make required determination for use of short-form clause at award).

14. GAI Consultants, Inc., ENGBCA No. 6030, 95-2 BCA ¶ 27,620.

15. Gen. Eng'g & Mach. Works, ASBCA No. 38788, 92-3 BCA ¶ 25,055.

16. Fireman's Fund Ins. Co., ASBCA No. 38284, 91-1 BCA ¶ 23,439; Poindexter, d/b/a Modern Prop. Serv., HUDBCA No. 77-6, 78-1 BCA ¶ 12,904 (*dicta*).

17. Labat-Anderson, Inc. v. United States, 42 Fed. Cl. 806, 857 (1999).

18. Rodgers Constr., Inc., IBCA No. 2777, 92-1 BCA ¶ 24,503.

19. OFEGRO, HUDBCA No. 88-3410-C7, 91-3 BCA ¶ 24,206; H&R Machinists Co., ASBCA No. 38440, 91-1 BCA ¶ 23,373.

20. Tech. & Mgmt. Servs. Corp., ASBCA No. 39999, 93-2 BCA ¶ 25,681 (inclusion of the cancellation ceiling clause in a fixed price multiyear contract is mandated by FAR so will be read into the contract).

21. Unit Data Serv. Corp. v. Dep't of Veterans Affairs, GSBCA No. 10775-P-R, 93-3 BCA ¶ 25,964.

22. S.J. Amoroso Constr. Co. v. United States, 12 F.3d 1072, 1075–76 (Fed. Cir. 1993).

23. Univ. of Cal., VABCA 4661, 97-1 BCA ¶ 28,652.

24. Telesec Library Servs., ASBCA No. 42968, 92-1 BCA ¶ 24,650; Ace Servs., Inc. v. Gen. Servs. Admin., GSBCA No. 11331, 92 2 BCA ¶ 24,943; Miller's Moving Co., ASBCA No. 43114, 92-1 BCA ¶ 24,707.

25. Chamberlain Mfg. Corp., ASBCA No. 18103, 74-1 BCA ¶ 10,368; Am. Bank Note, AGBCA No. 2004-146-1, 05-1 BCA ¶ 32,867; Computing Application Software Tech., Inc., ASBCA No. 47554, 96-1 BCA ¶ 28,204.

26. Dayron Corp., ASBCA No. 24919, 84-1 BCA ¶ 17,213; Hart's Food Serv., Inc., d/b/a Delta Food Serv., ASBCA No. 30756, 89-2 BCA ¶ 21,789; Rehab. Servs. of N. Cal., ASBCA No. 47085, 96-2 BCA ¶ 28,324.

27. Lambrecht & Sons, 97-2 BCA ¶ 29,105.

28. Applied Devices Corp., ASBCA No. 18384, 77-1 BCA ¶ 12,347.

29. *E.g.*, Ward Meat Co., ASBCA No. 20847, 77-1 BCA ¶ 12,249 (amendment to published military specification not read into contract under *Christian* doctrine).

30. For further discussion of changes and inspection and acceptance, see Chapters 8 and 10.

31. 41 U.S.C. § 414(4); *see also* FAR 1.603-1.

32. FAR 1.602-1(a).

33. Arthur Venneri Co. v. United States, 180 Ct. Cl. 920, 924, 381 F.2d 748, 750 (1967).

34. *E.g.*, Gen. Elec. Co., ASBCA 11990, 67-1 BCA ¶ 6377.

35. FAR 1.603-3(a).

36. FAR 1.602-1(a).

37. FAR 42.202(a), (c).

38. FAR 42.202(b).

39. DFARS 252.201-7000; Winter v. Cath-dr/Balti Joint Venture, 497 F.3d 1339, 1344, 1346–48 (Fed. Cir. 2007); Silverman v. United States, 679 F.2d 865, 870 (Ct. Cl. 1982); Telenor Satellite Servs., Inc. v. United States, 71 Fed. Cl. 114 (2006); States Roofing Corp., ASBCA Nos. 55500, 55503, 08-2 BCA ¶ 33,970 (Dec. 9, 2008); Romac, Inc., ASBCA No. 41150, 91-2 BCA ¶ 23,918; Parking Co. of Am., Inc., GSBCA No. 7656, 87-2 BCA ¶ 19,823.

40. S&M Mgmt., Inc. v. United States, 82 Fed. Cl. 240 (2008).

41. *Winter*, 497 F.3d at 1346–48; Richard & Assocs. v. United States, 177 Ct. Cl. 1037, 1051 (1966); Niko Contracting Co. v. United States, 39 Fed. Cl. 795, 800 (1997); Henry Burge, ASBCA No. 2431, 89-3 BCA ¶ 21,910; States Roofing, 08-2 BCA ¶ 33,970.

42. FAR 1.602-3. Mgmt. Tech., Inc., DOTCAB 73-28, 73-29, 76-1 BCA ¶ 11,791; Henry Burge, 89-3 BCA ¶ 21,910; Corners & Edges, Inc., ASBCA No. 55767, 09-1 BCA ¶ 34,019; Dan Rice Constr. Co. v. United States, 36 Fed. Cl. 1 (1996).

43. FAR 1.602-3(a).

44. FAR 1.602-3(b)(2).

45. Max Drill v. United States, 192 Ct. Cl. 608, 625 (1970); *see also* Ctr. Mfg. Co. v. United States, 183 Ct. Cl. 115, 392 F.2d 299, 236 (1968).

46. ASBCA No. 5859, 1962 BCA ¶ 3330.

47. *Ctr. Mfg.*, 183 Ct. Cl. 115; J.R. Chesire Janitorial, ENG BCA No. 5487, 91-3 BCA ¶ 24,351.

48. Gricoski Detective Agency, GSBCA No. 8901, 90-3 BCA ¶ 23,131; Gonzales Custom Painting, Inc., ASBCA No. 39529, 90-3 BCA ¶ 22,950.

49. *See* Stevens Van Lines v. United States, 80 Fed. Cl. 276 (2008).

50. Edwards v. United States, 22 Cl. Ct. 441 (1991); ECC Int'l Corp. v. United States, 43 Fed. Cl. 359 (1999); LaCoste Builders, Inc., ASBCA No. 30085, 88-1 BCA ¶ 20,360.

51. 80 Fed. Cl. 276 (2008).

52. *Id.*

53. FAR 3.101-1.

54. 31 U.S.C. §§ 3729–3733. For further discussion of FCA issues, see Chapter 24.

55. *See* 73 Fed. Reg. 67,064.

56. FAR 3.1004(a).

57. FAR 52.203-13(d).

58. FAR 3.1002(b).

59. *Id.*

60. FAR 52.203-13(b)(2).

61. FAR 9.403. Debarment is discussed further in section VI.

62. *Id.*

63. *Id.*

64. FAR 3.1003(a)(2); 9.406-2(b)(1)(vi)(A); 9.407-2(a)(8)(i); 52.203-13(b)(3)(i)(A); 52.203-13(c)(2)(ii)(F).

65. FAR 3.1003(a)(2); 9.406-2(b)(1)(vi)(B); 9.407-2(a)(8)(ii); 52.203-13(b)(3)(i)(B); 52.203-13(c)(2)(ii)(F).

66. *E.g.*, FAR 52.212-4, 52.232-25 to -27.
67. FAR 52.209-5(a)(2).
68. FAR 3.1003(a)(2); 52.203-12(c)(2)(ii)(F).
69. *Id.*
70. 73 Fed. Reg. 67,064, 67,084.
71. FAR 52.203-13(d)(1).
72. FAR 52.203-13(d)(2).
73. FAR 3.1003(a)(2); 52.203-13(c)(2)(ii)(F)(3).
74. *E.g.*, FAR 3.1003(a)(2); 52.203-13(b)(3)(i).
75. *See* FAR 52.203-13(b)(3)(i), -13(c)(2)(ii)(F).
76. FAR 52.203-13(c).
77. FAR 52.203-13(c)(1).
78. *Id.*
79. FAR 52.203-13(c)(1)(ii).
80. U.S. Sentencing Guidelines Manual § 8B2.1 (2008).
81. FAR 52.203-13(c)(2)(ii).
82. FAR 52.203-13(c)(2)(ii)(C)(1).
83. FAR 52.203-13(c)(2)(ii)(D).
84. FAR 3.903.
85. FAR 3.907.
86. 10 U.S.C. § 2409; 41 U.S.C. § 265.
87. FAR 3.903.
88. FAR 3.901.
89. *Id.*
90. FAR 3.904.
91. FAR 3.905(a)–(c).
92. FAR 3.905(d).
93. FAR 3.906(a).
94. FAR 3.907-1.
95. FAR 3.907-2.
96. FAR 3.907-5.
97. FAR 3.907-6(a).
98. FAR 3.907-6(c).
99. FAR 3.907-7; 52.203-15.
100. FAR 3.301, 3.303(c); *see* FAR 3.303(a); 41 U.S.C. § 253b(i); 10 U.S.C. § 2305(b)(9).
101. FAR 52.203-2(b).
102. 10 U.S.C. § 2402; 41 U.S.C. § 253g.
103. FAR 52.203-6(a).
104. FAR 3.501-1; 52.203-6(b).
105. FAR 52.203-6(c).
106. 41 U.S.C. §§ 51–58.
107. FAR 3.502-2(a); 52.203-7(b).
108. FAR 3.502-1; 52.203-7(a).
109. 66 Fed. Cl. 788 (2005).
110. *Id.* at 800, quoting United States v. Purdy, 144 F.3d 241, 244 (2d Cir. 1998).
111. *Id.* at 801.

112. *Id.* at 800.
113. FAR 3.502-2(b).
114. FAR 3.502-2(c).
115. FAR 3.502-2(d); 52.203-7(c)(4).
116. FAR 3.502-2(i)(1); 52.203-7(c)(1).
117. FAR 52.203-7(c)(2); *see* FAR 3.502-2(g).
118. FAR 52.203-7(c)(5).
119. FAR 3.501-1.
120. FAR 3.501-2(b).
121. 10 U.S.C. § 2306(b); 41 U.S.C. § 254(a); *see* FAR 3.402.
122. *See* FAR 52.203-5.
123. FAR 3.401.
124. *Id.*
125. *Id.*
126. FAR 3.402.
127. FAR 3.101-2.
128. FAR 3.101-3(a)(1).
129. 41 U.S.C. § 423.
130. FAR 3.104-3(b).
131. FAR 3.104-1.
132. *Id.*
133. FAR 2.101.
134. FAR 3.104-7(d)(1).
135. FAR 3.104-7(d)(2)(i); FAR 52.203-10.
136. FAR 3.104-7(d)(2)(ii); FAR 52.203-8.
137. FAR 3.204-8(b).
138. FAR 3.104-4(e)(1).
139. FAR 3.104-4(e)(2).
140. FAR 3.104-4(e)(4).
141. FAR 3.104-4(f).
142. 31 U.S.C. § 1352.
143. FAR 3.802(a).
144. 31 U.S.C. § 1352(a)(1), (a)(2)(A).
145. FAR 3.802(b).
146. FAR 52.203-11(d).
147. FAR 52.203-11(e).
148. FAR 3.802(a); 52.203-12(b).
149. *See* FAR 3.802(a)(2); 52.203-12(b)(2).
150. 31 U.S.C. § 1352(d)(1)(A).
151. FAR 3.803(a)(1).
152. FAR 3.803(a)(2).
153. *See* FAR 3.104-2(b)(1).
154. *See* FAR 52.203-3.
155. *Id.; see also* FAR 3.204(c).
156. FAR 3.101-1.
157. FAR 3.601(a).
158. As defined in 18 U.S.C. § 202.

159. FAR 3.601(b).

160. FAR 3.602.

161. 18 U.S.C. § 207; 5 C.F.R. pts. 2637 & 2641; *see also* FAR 3.104-2(b)(3).

162. *See* FAR 3.104-3(c).

163. *See* FAR 3.104-5(a).

164. FAR 3.104-3(d).

165. FAR 3.104-3(d)(3).

166. FAR 2.101.

167. FAR 9.505.

168. FAR 9.502(b).

169. 10 U.S.C. § 2305a; 41 U.S.C. § 253m; FAR subpt. 36.3.

170. 40 U.S.C. §§ 1101–1104; FAR subpt. 36.6.

171. SSR Eng'rs, Inc., B-282244, June 18, 1999, 99-2 CPD ¶ 27 ("the fact that FAR part 36 does not specifically address conflict of interest provisions in the context of construction and architect-engineer services contracts does not somehow render the organizational conflict of interest provisions of FAR subpart 9.5 inapplicable to such contracts.")

172. FAR 9.504(a).

173. FAR 9.504(e).

174. FAR 9.505.

175. *See* FAR 9.508.

176. *E.g.,* Aetna Gov't Health Plans, Inc.; Found. Health Fed. Servs., Inc., B-254397 et al., July 27, 1995, 95-2 CPD ¶ 129; Vantage Assocs., Inc. v. United States, 59 Fed. Cl. 1, 10 (2003); see also Daniel I. Gordon, *Organizational Conflicts of Interest: A Growing Integrity Challenge*, 35 Pub. Cont. L.J. 25, 32 (2005).

177. FAR 9.505.

178. FAR 9.505-1.

179. FAR 9.505-2.

180. FAR 9.505-1(a).

181. FAR 9.505-2(a)(1), (2). But note that this prohibition will not be applied to contractors that (1) furnish their specifications at government request, or (2) act as industry representatives in helping the government agencies develop the specifications.

182. FAR 9.505-2(b)(1). But the prohibition does not apply if the contractor: (1) is the sole source; (2) has participated in the development and design work; or (3) was one of multiple contractors involved in preparing the work statement.

183. FAR 9.505-2(a)(3); 9.505-3(b)(3).

184. FAR 9.505-3.

185. Nortel Gov't Solutions, Inc., B-299522.5, B-299522.6, Dec. 30, 2008, 2009 CPD ¶ 10 at 7.

186. FAR 9.505-4(b).

187. FAR 9.505-4(c).

188. FAR 9.503.

189. Jones/Hill Joint Venture, B-286194.4 et al., Dec. 5, 2001, 2001 CPD ¶ 194 at 13 n.13.

190. FAR 9.507-1.

191. FAR 9.506(d).

192. FAR 9.507-2(b).

193. *Id.*

194. 5 U.S.C. § 552.

195. 5 U.S.C. § 552(a).

196. 5 U.S.C. § 552(b)(1)–(9).

197. 5 U.S.C. § 552(b)(4), (5).

198. Frazee v. U.S. Forest Serv., 97 F.3d 367, 370–71 (9th Cir. 1996) (quoting GC Micro Corp. v. Defense Logistics Agency, 33 F.3d 1109, 1112–13 (9th Cir. 1994)); *see also* Coastal States Gas Corp. v. Dep't of Energy, 617 F.2d 854, 862 (D.C. Cir. 1980).

199. U.S. Dep't of Justice v. Reporters Comm., 489 U.S. 749, 771–72 (1989).

200. *Frazee,* 97 F.3d at 371 (quoting Nat'l Parks & Conservation Ass'n v. Morton, 498 F.2d 765, 770 (D.C. Cir. 1974); *GC Micro,* 33 F.3d at 1112.

201. *Id.*

202. Canadian Commercial Corp. v. Dep't of the Air Force, 514 F.3d 37, 40–42 (D.C. Cir. 2008) (detailed pricing data, such as line-item pricing, exempt from disclosure); McDonnell Douglas Corp. v. Dep't of the Air Force, 375 F.3d 1182 (D.C. Cir. 2004) (same); *Frazee,* 97 F.3d at 369, 372 (proposal exempt where it contained proposed operating plan and financial information and solicitation informed bidders that their proposals would be confidential). *But see* R&W Flammann GmbH v. United States, 339 F.3d 1320 (Fed. Cir. 2003); (government's disclosure of unit prices upheld); Pacific Architects and Eng'rs, Inc. v. U.S. Dep't of State, 906 F.2d 1345, 1347–48 (9th Cir. 1990) (same); Acumenics Research & Tech. v. Dep't of Justice, 843 F.2d 800, 807–08 (4th Cir. 1988) (same).

203. *Frazee,* 97 F.3d at 372 (final plan not exempt after it was accepted and attached to a special use permit, a public document).

204. *Canadian Commercial,* 514 F.3d 37.

205. Coastal States Gas Corp. v. Dep't of Energy, 617 F.2d 854, 862 (D.C. Cir. 1980).

206. *Id.* at 866.

207. SMS Data Prods. Group, Inc. v. U.S. Dep't of the Air Force, No. 88-481, 1989 WL 201031, at *1–2 (D.D.C. Mar. 31, 1989) (technical scores and technical rankings of competing contract bidders predecisional and deliberative); *see also* Prof'l Review Org., Inc. v. Dep't of Health and Human Servs., 607 F. Supp. 423, 427 (D.D.C. 1985).

208. Quarles v. Dep't of the Navy, 893 F.2d 390, 392–93 (D.C. Cir. 1990); Hack v. Dep't of Energy, 538 F. Supp. 1098, 1100 (D.D.C. 1982); Taylor Woodrow Int'l v. United States, No. 88-429, 1989 WL 1095561, at *3 (W.D. Wash. Apr. 6, 1989) (disclosure would permit requester to take "unfair commercial advantage" of agency).

209. *But see* Legal & Safety Employer Research, Inc. v. U.S. Dep't of the Army, No. CIV. S-00-1748, 2001 WL 34098652, at *6 (E.D. Cal. May 4, 2001) (contractor performance evaluations, which were required to be considered in future government contract award determinations, were not "the type of policy decision contemplated by Exemption 5").

210. FAR 9.407.

211. FAR 9.406.

212. 41 U.S.C. § 354(a).

213. 41 U.S.C. ch. 10.

214. 8 U.S.C. §§ 1101–1503.

215. *E.g.*, 41 U.S.C. § 354(a).

216. FAR 9.402(a), (b).

217. FAR Part 9.4.

218. FAR 406-3(c).

219. FAR 406-3(c)(4).

220. FAR 9.407-4(b).

221. FAR 9.406-4(a)(1).

222. Lion Raisins, Inc. v. United States, 51 Fed. Cl. 238 (2001) (debarment based on conduct before contractor corrected its business practices and agency affirmed revised practices met its present responsibility standards was arbitrary and capricious).

223. FAR 9.104-1.

224. FAR 9.407-2, 9.406-2.

225. *Id.*

226. *Id.*

227. Aardvark Keith Moving, Inc., B-290565, 2002 CPD ¶ 134, 44 GC ¶ 353.

228. FAR 9.402(a), 9.405(a).

229. FAR 9.405(a).

230. FAR 9.405(b), 9.405-2(a), (b).

231. FAR 9.404. The list is available at https://www.epls.gov.

232. FAR 9.105-1(c)(1); *see also* FAR 9.405(d) (requires the CO to check the EPLS once after opening of bids or receipts of proposals and then again immediately before award).

233. FAR 4.1202, 52.204-8.

CHAPTER 8

Changes

GEOFFREY T. KEATING

I. Introduction

The Changes clause may have the most significant impact on contract performance of any of the standard federal government construction contract provisions. The Changes clause not only provides the basis and authority for added work, specifications changes, and new technical requirements, but also may be used to reduce or delete contract requirements. It empowers the contracting officer (CO) unilaterally to change contract requirements and obligates the contractor to accept these changes and continue work. Change orders may be express or implied, formal or informal. A change order may be a specific written directive to change technical requirements. It also may evolve from disagreements over the meaning of specifications or drawings, defective specifications, impossibility of performance, acceleration of work, or failure of the government to disclose information.

The Changes clause benefits both the government and the contractor. The clause provides the government with a significant and essential degree of operating flexibility. By empowering the CO unilaterally to change the contract specifications without the need for a new, separate contract, the Changes clause provides the government with a means to rapidly and effectively respond to the ever-changing requirements and problems inherent in a construction project.

A contractor is obligated to proceed with the changed work but, at the same time, is entitled to an "equitable adjustment" to the contract price and completion date, if a change order adds to the construction costs or justifies an extension to the completion date. Not to be overlooked is that change orders provide the contractor with new work and additional profit, without competition.

The Changes clause also is made applicable by other government construction contract clauses. The best example is the Government Property clause, which provides that the government may supply identified equipment or materials to the contractor for use and installation in the work. If the government fails to timely deliver the material or delivers defective material, the contractor is entitled to an equitable adjustment. The clause expressly provides that the price and time adjustment will be made "in accordance with the procedures of the Changes clause."[1]

The Changes clause, together with the Disputes clause,[2] also provides the contractor with an administrative procedure to resolve disputes over whether a change has occurred, the scope of changed work, and the reasonable equitable adjustment due for the change. For example, where a change order has been issued and the contractor is dissatisfied with the CO's suggested price for the additional work, the contractor, under the Changes and Disputes clauses, has the right to obtain a final decision from the CO on the issue and the right to appeal that decision to an agency administrative board of contract appeals (BCA) or the U.S. Court of Federal Claims (COFC).[3]

Use of a Changes clause in federal construction contracts has been widespread since at least World War II. The origin of the clause dates back to the Civil War era and the construction of U.S. warships. The clause was initially included in these contracts so the government could receive the benefit of advancements in technology, such as the shifts from wooden to ironclad ships and from sail to steam power.

No longer optional, the clause is mandatory in all government firm fixed-price construction contracts.[4]

The early introduction of the Changes clause into government construction contracts was unique. As a general proposition, in all contracts, not only construction, the parties agree to terms before starting the project. Absent a Changes clause, if either party unilaterally changes or alters its obligation without the consent of the other, that party would be in breach of contract. Changes to original contract terms can be made only by mutual agreement. In contrast, a construction contract that includes a typical Changes clause provides that the government can unilaterally direct changes to the work, within broad boundaries, without the consent of the contractor. The federal Changes clause specifically grants this authority to the CO. While the CO and contractor typically agree on desired contract changes and the associated equitable adjustment, the government's unilateral right to direct changes to the contractor's work is the hallmark of the federal Changes clause, and this right has been incorporated into nearly every other public and private contract Changes clause.

The specific terms of the Changes clause have evolved over the years to become more refined and explicit. The current standard construction Changes clause in government contracts was last revised in 1987. The revisions were minor. Before 1987, the most significant revisions to the Changes clause in a generation were made in 1967, as a result of a General Services Administration (GSA) study. The GSA study incorporated suggestions of both federal agencies and the construction industry. The primary revisions in the 1967 clause addressed the following areas:

- Clarification of the authority of the CO to make changes
- Express recognition of the doctrine of "constructive changes"
- Elimination of the "Rice" doctrine that had precluded consideration of the effect of a change upon other contract work not specifically addressed by the change
- Revision of notice requirements imposed on contractors, especially for constructive changes

These revisions produced some noticeable differences between the standard construction Changes clause and the standard supply contract Changes clause.[5] The construction clause expressly entitles the contractor to equitable adjustments or constructive change orders, while the standard supply contract remains silent on these subjects. In addition, the 1967 amendments broadened the scope of expressly permitted changes to include, without limitation, changes in specifications, method or manner of performance, government-furnished facilities or materials, and acceleration of performance. In contrast, the supply contract, then and still, provides for only three types of changes: to specifications, method of shipment, and place of delivery.

Despite long and widespread usage, the meaning and coverage of the construction Changes clause is not without controversy. Of all the cases before the various BCAs, nearly one-third deal with this clause. Contractors, therefore, must be well acquainted with its meaning, its scope, its application, and the procedures by which contract price and time adjustments are obtained.

In many respects, the federal construction Change clause is similar in purpose and application to many other such clauses used in both public and private construction settings. However, the Federal Acquisition Regulation (FAR) and the federal tribunals have created an extensive body of federal law on change orders, so much so that state courts from time to time have looked to federal case interpretations for guidance.

Any discussions of the federal Changes clause normally would address defective specifications, acceleration of work, and the equitable adjustment and related time extensions. Inasmuch as this book includes separate chapters on these important subjects, only brief mention of them is made here.[6]

II. The Clause

The federal Changes clause has five constituent sections that control the government's right to make changes and the contractor's obligation to proceed and to receive reasonable compensation for changes. First, the clause invests the CO with the authority to make any change in the work that is "within the general scope of the contract." Second, the clause identifies four types of change orders: changes (1) in specifications, (2) in the method or manner of performance, (3) in government-furnished facilities or materials, and (4) in acceleration of the work; the clause does not exclude other possible changes. Third, the clause recognizes constructive changes, thereby giving recognition to changes resulting from informal, oral "orders," directives, or other actions of the CO. Fourth, the clause provides the contractor with the right to an "equitable adjustment" in the contract price and time. Fifth, the clause makes the right to an equitable adjustment contingent upon a contractor's giving timely notice to the government that it will seek a price or time adjustment for the change.

The text of the federal Changes clause is set out below:

(a) The Contracting Officer may, at any time, without notice to the sureties, if any, by written order designated or indicated to be a change order, make changes in the work within the general scope of the contract, including changes—

(1) In the specifications (including drawings and designs);
(2) In the method or manner of performance of the work;
(3) In the government-furnished facilities, equipment, materials, services, or site; or
(4) Directing acceleration in the performance of the work.

(b) Any other written order or oral order (which, as used in this paragraph (b), includes direction, instruction, interpretation, or determination) from the Contracting Officer that causes a change shall be treated as a change order under this clause; *provided*, that the Contractor gives the Contracting Officer written notice stating (1) the date, circumstances, and source of the order and (2) that the Contractor regards the order as a change order.

(c) Except as provided in this clause, no order, statement, or conduct of the Contracting Officer shall be treated as a change under this clause or entitle the Contractor to an equitable adjustment.

(d) If any change under this clause causes an increase or decrease in the Contractor's cost of, or the time required for, the performance of any part of the work under this contract, whether or not changed by any such order, the Contracting Officer shall make an equitable adjustment and modify the contract in writing. However, except for an adjustment based on defective specifications, no adjustment for any change under paragraph (b) of this clause shall be made for any costs incurred more than 20 days before the Contractor gives written notice as required. In the case of defective specifications for which the government is responsible, the equitable adjustment shall include any increased cost reasonably incurred by the Contractor in attempting to comply with the defective specifications.

(e) The Contractor must assert its right to an adjustment under this clause within 30 days after (1) receipt of a written change order under paragraph (a) of this clause or (2) the furnishing of a written notice under paragraph (b) of this clause, by submitting to the Contracting Officer a written statement describing the general nature and amount of proposal, unless this period is extended by the government. The statement of proposal for adjustment may be included in the notice under paragraph (b) above.

(f) No proposal by the Contractor for an equitable adjustment shall be allowed if asserted after final payment under this contract.[7]

The meaning and significance of the various phrases and sections of the Changes clause are not as clear as may appear from casual reading. The vast number of disputes involving change order claims has produced substantial litigation. The resulting large number of BCA and court decisions have created a substantial and substantive body of interpretative law, which has extended the plain meaning of the Changes clause.

III. *Issuing the Change Order*

A. The Contracting Officer

Because the Changes clause opens with the statement that "the Contracting Officer may . . . make changes in the work," the CO is the central figure in the government's change order process. The CO is the only government official mentioned in the clause as having the authority to issue and execute contract changes.[8]

The Changes clause and the FAR provide the CO with the general power to administer and modify government contracts. Since this power is limited, questions frequently are raised concerning the actual extent of such authority.[9] Consistent with the FAR, the COFC has recognized the CO's broad authority in the contracting process:

> In recognition of the important role the contracting officer plays in the operation of government contracts, the courts have liberally interpreted this "authority to enter into and administer contracts." . . . "The [contracting] officer in a sense is a party to the contract, not only representing but speaking for the impersonality of the Government.". . .
>
> Consequently, the statutory authority of the contracting officer has been construed to include the power to settle claims arising under the contract. . . . He also has the power to modify the contract. . . . "In general, an officer authorized to make a contract for the United States has the implied authority thereafter to modify the provisions of that contract particularly where it is clearly in the interest of the United States to do so."[10]

There are limits to such authority, however, as expressed by the Armed Services Board of Contract Appeals (ASBCA) in a 1967 decision:

> The mere fact that the contract provides that the contracting officer may take an action which binds the Government does not mean that the contracting officer may act without regard to the limitation of his

authority or without regard to restrictions imposed by regulations
or law. For example, the changes articles often provide that a change
within the scope of the contract may be ordered by the contracting
officer and, if ordered, the contractor is obligated to comply. The standard form of contract does not say that the contracting officer may
not order the change unless he has authority to do so and funds are
available. This is, of course, primarily a matter between the contracting officer and the Government but the fact that some limitations on
contracting officers' authority exist is well known.

The decision to order or not to order a change is a discretionary
administrative decision not entailing any fact-finding and not com-
ing within the Disputes clause, and in issuing or deciding not to issue
the change order, the contracting officer must act in accordance with
the requirements of applicable law and regulations and the specific
instructions of his higher authority.[11]

The above-stated position reflects that a CO's authority to issue changes
is not without limitations. It is confined by various federal regulations and
specifically is limited by restrictions stated in the designation document. For
example, if a CO agrees to pay for work later determined to be clearly within
the scope of the existing contract, the commitment is not binding. If there is
consideration for the payment, however, it may be binding.[12] Nevertheless, a
contracting officer's bad business decision will not be vacated on the basis that
"authority" was lacking.[13]

B. Unauthorized Actions of Government Representatives

The determination of who has what authority is extremely important because
the government is not legally bound by unauthorized acts or acts of unauthorized agents.

The burden of knowledge is accentuated by the fact that the doctrine of
apparent authority is not applicable to the government. Accordingly, if a CO
has authority only to issue change orders not exceeding $25,000 in value, but
in fact directs a change for additional work with a value of $50,000, then the
contractor may not get paid in full even if the changed work is completed.

One contractor was fortunate to receive a more-generous interpretation
when the CO negotiated an equitable adjustment for additional work, and
the government later refused to pay on the ground that, according to internal operating instructions, the CO lacked such authority. The ASBCA stated:
"An internal instruction does not have the effect of a statute or a regulation
published in the Federal Register and is not binding on appellant absent his
knowledge of its existence."[14]

Nevertheless, the government often refuses to pay, and properly so, for
work ordered by one of its representatives who did not have the requisite
authority.[15] This can produce harsh results.

Good contract administration, therefore, requires that every contractor know who the CO is, the limitations of the CO's authority, and whether any authority has been delegated to others. Doing so avoids the risk of completing change order work directed by an unauthorized person, only to find you have worked "voluntarily" and incurred the costs without any right to reimbursement. This type of situation can easily occur when government inspectors or technical representatives offer suggestions or order changes in the method of work. The best approach is to request and receive a written directive from an authorized official before proceeding with any change in the work. Additionally, the contractor should request evidence of the official's authority if such authority has not previously been communicated.

C. Actions by Representatives of the Contracting Officer and Other Government Officials

The practice of issuing change orders by authorized representatives of the CO acting within the limits of their delegated authority[16] burdens both contractors and the government. For the contractor, it means that the CO is not the only government official from whom it must take direction. For the government, it means that the CO is not the only official for whose actions it assumes responsibility. An added complication is that, in some instances, the actions of unauthorized government officials will obligate the government as if the official in fact had proper authority. A 2007 Federal Circuit case provides additional warnings where conflicting information is given. Overruling the ASBCA, the Court held that two clauses exclusively reserving change order authority to the CO prevailed over the CO's designation of the resident officer in charge of construction (the ROICC) as contract administrator. This outcome nullified ROICC approvals of compensable change orders.[17] Unauthorized actions may be binding on the government based on (1) a course of government conduct, or (2) ratification of the unauthorized directive by an official who has proper authority.[18]

Ratification has been implied from circumstances where the CO (1) knew that a change had been directed by an unauthorized official, and (2) failed to countermand the change directive.[19] Noting that actual knowledge is not always necessary, in *U.S. Federal Engineering and Manufacturing, Inc.*, the ASBCA stated:

> The fact that the contracting officer did not have actual knowledge of the additions to be made to the device does not insulate the Government from the consequences that actual knowledge would impose. His various representatives are his eyes and ears (if not his voice) and their knowledge is treated for all intents and purposes as his.[20]

The ASBCA concluded that the CO had constructive knowledge of the added work and awarded the contractor an equitable adjustment. Similarly, the Court of Claims has imputed knowledge of a contractor's difficulty in performance to the CO on the basis that the government's representatives were assumed to have, or at least should have, informed the CO of these difficulties as was required under the terms of their delegated authority.[21]

As a general rule, the BCAs have not based their decisions on the concept of ratification as frequently as has the COFC and its predecessors. The boards instead have permitted recovery by way of the broader "constructive change" approach. That is, if the government participated or acquiesced in a decision to perform extra or modified work, the government may be found to have issued a change order.[22] This concept of constructive change is discussed in greater detail below. Regardless of the rationale applied, the Changes clause has acquired meaning beyond its clear terms through years of litigation. However, the contractor can avoid most controversies by finding out the express authority of the government representatives with whom it deals.

IV. Scope of Changes

A. The Timing and Form of Change Orders

Although the Changes clause provides that the CO may change the work "at any time, by written order," the time frame in which changes may be ordered is not defined in the clause or elsewhere in the contract. Generally, it is understood that a change order may be issued any time between the execution of the contract and final payment. Given this broad time span, the timing of change orders only occasionally raises controversy. The Court of Claims has referenced a contractor who asserted that the government acted improperly by issuing 33 change orders after the work was 84 percent completed. Rejecting a specific time limit, the court held:

> We know of no contract requirement or any obligation imposed by law which restricts the government's power to order changes within a specified period. . . . [t]he contracting officer can order changes as long as the finished product was substantially the same as originally contracted for. The fact that performance was extended is immaterial, since the contractor is not precluded from recovering overhead in its cost for the added work.[23]

Final payment acts as a positive bar to contractor equitable adjustment claims and, likewise, acts as a bar to further government change orders.[24] Change orders of a minor nature sometimes are issued after substantial completion and during punch list work. Contractors often do not object. However, when a contractor has moved off-site and no work force or equipment remains, a contractor likely would have just cause for refusing significant additional work, even if final payment had not been made. Occasionally, a contractor is directed to perform a change under the guise of warranty work. If the work is extra work, and not a warranty repair, additional payment may be sought under the Changes clause.

More significant than the time of issuance is the Changes clause requirement that a change is to be issued "by written order." Historically, the requirement for written changes was absolutely essential. As far back as 1913, the U.S. Supreme Court, in *Plumley v. United States*,[25] held that a contractor's lack of a written change order prevented the contractor from receiving payment for the work, even though the work was found to be extra and of considerable expense. When a change order is formally issued in writing there should be no dispute relating to this requirement of the Changes clause. Nonetheless, contractors do receive oral directives or change orders without any written direction or written confirmation. In some instances the CO may refuse to issue a written directive and, at the same time, insist that the order be followed. The obvious merit of many requests for change orders in such situations gradually eroded the harsh decisions born of strict adherence to the requirement for written orders.

Thus, contrary to the *Plumley* decision, the federal courts and BCAs in recent years rarely have rejected a contractor's equitable adjustment claim solely because the contractor lacked a written directive. For example, to avoid the harsh consequences of *Plumley*, the Court of Claims recognized that an enforceable oral contract was created so as to permit a contractor to recover its additional costs:

> What we have said leads us to this conclusion: when the contracting officer orally directed plaintiff to use the steel plant bricks, and promised it that an adjustment would be made when the work was completed and the fair amount determined, and when plaintiff by doing the work and using the required material, a contract to pay plaintiff the reasonable cost of the work resulted.[26]

The approach most frequently used to avoid the necessity for written change orders is the doctrine of constructive change. Although the concepts of constructive change are discussed in more detail later in this chapter, under this approach the absence of a formal written change order or other express formalities may be overlooked where circumstances fairly compel the conclusion that an oral directive, or erroneous interpretation of contract terms, was imposed that caused extra work and costs. Hence, abandoning strict adherence to the formal requirement of a written order, the courts in the early cases began to "regard as done that which should have been done."[27] For example, when presented with the issue of constructive change in *Len Co. & Associates v. United States*, the Court of Claims discussed the background and succinctly explained the doctrine as follows:

> Where [the standard clause] (or a comparable one) is included, we, as well as the Armed Services Board of Contract Appeals, have held

that, if a contracting officer compels the contractor to perform work not required under the terms of the contract, his order to perform, albeit oral, constitutes an authorized but unilateral change in the work called for by the contract and entitles the contractor to an equitable adjustment in accordance with the "Changes" provision. The court has considered it to "be idle for contractor to demand a written order from the contracting officer for an extra when the contracting officer was insisting that the work required was not additional . . ." and, therefore, has often dispensed, on these occasions, with the formality of issuing a written change order under the standard clause.[28]

Although the courts and boards for years have granted contractors an equitable adjustment in the absence of a written directive, the existence of an oral directive or some compelling circumstances is nevertheless necessary. Absent both written and oral directives, the contractor may be considered to have performed extra work as a volunteer and thus an equitable adjustment will be denied.[29]

B. The Intended Scope of Work

The extent to which the CO may order changes in the work is limited by the Changes clause provision that changes must be "within the general scope of the contract." The clause does not define "general scope." The four specified categories of allowable changes are not limiting, but case law is instructive. A good working definition of "within the general scope of the contract" was given many years ago in *Freund v. United States*,[30] and this definition is still often cited. The Supreme Court stated in *Freund* that "within the general scope" means that work that "should be regarded as having been fairly and reasonably within the contemplation of the parties when the contract was entered into."[31] The U.S. Court of Claims has defined the term to mean that the CO can order changes provided the work was essentially the same as that for which the government originally contracted.

The basic standard, as the Court has put it, is whether the modified job "was essentially the same work as the parties bargained for when the contract was awarded. . . . Our opinions have cautioned that the problem is a matter of degree varying from one contract to another and can be resolved only by considering the totality of the change and this requires recourse to its magnitude as well as its quality. . . . There is no exact formula."[32]

These general standards, while consistent over the years, offer no single test. What is meant by "within the general scope" varies from case to case. "There is no exact formula. . . . Each case must be analyzed on its own facts and in light of its own circumstances, giving just consideration to the magnitude and quality of the changes ordered and their cumulative effect upon the project as a whole."[33]

In federal construction contracting the prohibition on issuing changes outside the general scope of the contract prevents significantly different work from being undertaken without open competition. It also prevents an agency from avoiding new solicitations for new projects. The vast majority of change orders, large and small, are within the general scope of the contract, and the case law over the years reflects a consistent application of this limitation on the CO's authority.[34] Influential factors for evaluating whether a change is within the general scope are the original work requirements, the nature of work as changed, comparative costs, the time consumed, the level of expertise and effort required to comply with the change, and occasionally, the contractor's capabilities.

Because the Changes clause places no numerical limit on authorized changes, large numbers of changes alone are not the measure of what is "within the general scope," nor are they an abuse of authority. As the U.S. Court of Claims stated in *Air-A-Plane Corp. v. United States,* where an estimated 1,000 changes were ordered, "[W]e have repeated that [the] number of changes is not, in and of itself, the test."[35]

Nor is the amount of work required by any one change usually a significant factor. The measure of "general scope" generally is not quantitative either in the number of changes or the amount of work. Rather, qualitative factors are most relevant to analyzing the issue. Longer and more costly work, even wasteful work, is within the general scope so long as the nature of the work is not changed.[36]

C. Changes Outside the General Scope of the Contract

Government-directed changes that are beyond the general scope of the contract generally are referred to as "cardinal changes." A cardinal change therefore exceeds the authority of the CO to order, and insisting on implementation constitutes a government breach of contract. As the Court of Claims stated in *Keco Industries, Inc. v. United States:*

> Since the right to make changes was reserved to the defendant (the Government), it cannot be liable for breach of contract for its exercise of that right, unless the changes were cardinal changes and exceeded the discretion which the Changes article vested in the contracting officer.[37]

Examples of a cardinal change are as follows: a change order more than doubling the amount of earth fill required for a levee embankment, where the contractor was a small company without the equipment or resources to comply with the 150-percent increase in work;[38] an order requiring a foundation to be placed on piles when such a requirement was not previously specified;[39] and an order changing the destination of a moving company's shipments such that 673 of the shipments were outside of the contractually prescribed zones.[40] Again, the determination of whether a given change order is within the general scope of the contract is generally qualitative in nature. The nature and effect of the change are the relevant factors to consider, and not necessarily the quantity of work, number of changes, or cumulative effect of the changes ordered.[41]

Where defective specifications are the root of the problem, the corrective change orders issued usually are considered within the general scope of the contract. However, where defective specifications result in extraordinary corrective work, the Court of Claims has made an exception. In *Edward R. Marden Corp. v. United States*, the contractor alleged that the government specifications were defective and caused the collapse of the completed aircraft hangar. The contractor was directed to rebuild the entire building. Although the rebuilt hangar was the same basic structure, the court found that it was proper to examine "the entire undertaking . . . rather than the product" and held the corrective change orders to be beyond the general scope of the contract because "where drastic consequences follow from defective specifications, we have held that the change was not within the contract."[42]

Whenever a change order is issued that significantly affects the nature of the undertaking, the contractor should consider whether the change is a cardinal change. The issuance and acceptance of change orders does have a bearing on this decision. *Amertex Enterprises, Ltd. v. United States*[43] points out that executing a series of bilateral change orders for corrective work can undermine the contractor's cardinal change claim. Amertex received a government supply contract for the production of 2,400,000 chemical warfare protective suits. The government specifications contained numerous errors and there were unreasonable overinspections and tolerances. Ambiguities in the specifications also plagued the contract. These problems resulted in a constant stream of changes, delays, waivers, and clarifications. Over the course of the work, the government issued 42 modifications and 8 amendments to the contract encompassing a multitude of changes. A new delivery schedule also was approved by the government and agreed to by the contractor. Still the contractor failed to perform and was defaulted.

In addition to challenging the government's grounds for default, the contractor asserted that the specification's defects were so extensive as to constitute a cardinal change to the contract. The U.S. Court of Federal Claims recognized that the unreasonable acts by the government hampered mass production and arguably "fundamentally changed the nature of this contract." Nonetheless, the court denied the cardinal change claim because the contractor's position was "fatally undercut by the bilateral modifications made to the delivery schedule." In one modification, the contractor agreed to produce and deliver the products as modified, under the extended schedule, including all the changes that had been made up to that point. The lesson is clear: A cardinal change that requires a contractor to perform materially different duties from those bargained for in the contract may be lost by agreeing to the cardinal change. Amertex agreed to the changes, accepted payment for the added costs, and thus agreed that the changes were within the coverage of the Changes clause.

To direct a cardinal change and insist on its performance is a breach of contract by the government. The damages may include a contractor's anticipated profits, which are not allowable in an equitable adjustment under the Changes clause. If the contractor believes that a change is outside the general scope of the contract, it has the option to abandon performance or proceed with the work under protest, and later institute a suit for damages. This is not a casual decision. The contractor who abandons performance runs a high risk that the court may find that decision to have been incorrect, and instead find the contractor in breach of contract for refusing to perform a legitimate change and for abandoning the contract.[44] Abandoning performance is an extremely serious step and should not be taken without a careful legal analysis and due regard for the consequences. The safer approach is to proceed with the work under protest and pursue a claim.[45]

V. Equitable Adjustment

An important aspect of the Changes clause is determining the equitable adjustment. As the term is used in federal construction contracts, an equitable adjustment has both cost and time elements. The contractor is entitled to an equitable adjustment whenever a change order increases or decreases the cost of, or the time required for, performance.

A. The Equitable Adjustment in Price

The Changes clause includes neither a definition of "equitable adjustment" nor any criteria for its calculation. It only provides:

> (d) If any change under this clause causes an increase or decrease in the Contractor's cost of, or the time required for, the performance of any part of the work under this contract, whether or not changed by any such order, the contracting officer shall make an equitable adjustment and modify the contract in writing.

The Court of Claims and various boards have supplied a working definition of the term. In the landmark case *Bruce Construction Co. v. United States*,[46] the Court of Claims stated that the basic goal of an equitable adjustment is to "keep [a] contractor whole when the government modifies the contract." In other words, the objective is to leave the parties in the same cost and profit position as if there had been no change, preserving to each the advantages or disadvantages of their original bargain. The Court of Claims three years later stated that the proper measurement of the cost element is the "actual, reasonable cost" ordinarily incurred by reason of the change.[47] This means that in determining the monetary equitable adjustment in advance of doing the work the best estimate of the actual cost to be incurred must be used. The relevant costs include the cost of labor, materials, equipment, bonding, and overhead, to which profit is then added.

The policy and goal of the government is to fix the equitable adjustment in advance of performing the change, based on estimates of added costs and time. As a practical matter, however, the equitable adjustment is not determined in many instances until after the changed work is completed, either because the government and the contractor cannot agree on the adjustment or the change involved is a constructive change. In these situations, the costs typically already have been incurred when equitable adjustment negotiations are under way. The problem then becomes one of segregating the costs that were incurred as the result of the change order.

Often when performing change order work, a contractor not only incurs the added cost of performing the changed work but also sustains added costs in performing related work on the same system in or nearby work areas. Such added costs in performing "unchanged" work stemming from a change order also are properly part of the equitable adjustment.

B. The Equitable Adjustment in Time

Each construction contract typically establishes a time frame within which the work must be completed.[48] The deadline for completion may be enforced by the government's right to terminate the contract for default and/or by an assessment of liquidated damages for late completion. Often, when a change order adds significant new work to the contract, an extension of the completion date of the contract may be appropriate. Under the Changes clause additional time for performance is part of the equitable adjustment. Without such an equitable adjustment to the completion date, the contractor would be under the threat of liquidated damages and thereby would be pressured to accelerate the work to achieve the unadjusted completion date. The equitable adjustment in time makes such costs unnecessary and allows the contractor to proceed at the originally scheduled pace of performance.

Construction delays frequently result from the issuance of change orders. These delays also may occur prior to issuance of a change order while waiting for a major change to be issued. Delays incurred prior to issuance of a change are addressed by the Suspension of Work clause.[49] Delays or additional performance time, as well as time-related costs incurred as the result of the change order, are properly part of the equitable adjustment under the Changes clause.[50]

With respect to any equitable adjustment, the contractor has a duty to mitigate or reduce expense and delays where possible. As the ASBCA has stated:

> A contractor has the duty to minimize its costs in the execution of a change order in the same manner, as he must mitigate his damages

after breach. Normally, he would be required to transfer or discharge idle men and find uses for his equipment pending the time that work can commence.[51]

C. Processing the Formal Change Order

The change order process may be initiated by either the contractor or the government. Once the need is acknowledged, the government and the contractor typically confer and exchange ideas and technical information concerning the proposed change. These discussions define the scope of the change, help to coordinate the changed work, and generally result in an agreed adjustment in the contract price and time.

Federal guidelines and policy for contract modifications and change orders are set forth in the FAR.[52] Actions taken under the Changes clause generally are formalized by one of two types of change orders: bilateral agreements and unilateral change orders.[53]

Bilateral change orders are issued by the CO when both the government and the contractor have agreed to the equitable adjustment. This formal modification, signed by the government and the contractor, binds the contractor regardless of whether the change is within or outside the scope of the original contract. Government policy as expressed in the FAR favors the issuance of bilateral change orders over unilateral orders and the pricing of changes before issuing and performing them:

> Contract modifications, including changes that could be issued unilaterally, shall be priced before their execution if this can be done without adversely affecting the interest of the Government. If a significant cost increase could result from a contract modification and time does not permit negotiation of a price, at least a maximum price shall be negotiated unless impractical.[54]

Despite the government's preference for bilateral change orders, there are several potential drawbacks to bilateral changes from the contractor's perspective. The total effect of the change on other, unchanged contract work may not be apparent prior to performance of the change, and unforeseen problems may arise. The risk of the contractor underestimating the change order cost and time is ever present. The contractor's signing of a mutually agreed change order will cut off any further recovery unless an express reservation of rights is made.[55] At the same time, a contingency in the change order to cover unforeseen costs or delays usually is unacceptable to the government.

Although government policy favors the issuance of bilateral change orders, the government retains the right to unilaterally direct a change to the work where change order negotiations break down or time does not allow them. Unilateral changes are signed only by the CO, not by the contractor.[56]

Unilateral change orders frequently are issued where all the necessary details of the change can only be determined as the modified work progresses. Even then the unilateral order should describe the change in as much detail as is known at the time of issuance and establish an equitable adjustment in the contract price and time based on the government's estimate. A maximum price should be negotiated if possible. Although the contractor is not required to, and seldom does, agree to the government's cost and time estimate, it must acknowledge receipt of, though not agreement with, the order. Promptly after receipt of a unilateral change the contractor should return it unsigned and notify the CO, in writing, that it will seek a different equitable adjustment. The changed work must then proceed.

VI. Constructive Changes

Simply stated, any conduct of the CO or his authorized representative that requires the contractor to change, alter, or modify the original contract work may be a constructive change order. These actions or conduct may be written or oral, affirmative acts or omissions by the government, the cumulative effect of suggestions from the CO to the contractor, or a government course of conduct. A constructive change order differs from a formal change order in that the constructive change order does not meet the formality requirements of paragraph (a) of the Changes clause. A constructive change order is *not* clearly designated as a "change order," often is not in writing, and frequently does not explicitly direct a change in performance.[57]

Constructive changes have caused contractors and the government untold problems over the years. Contractors may have performed additional work and incurred additional costs and delays in circumstances where, lacking a written order, the government direction failed to trigger a notice and a proposal for an equitable adjustment. Such additional work can be difficult to identify because of its informal nature, such as in the case of field level directives that cause added work. Many consider identification of these constructive change events to be at the heart of the problem. If oral orders are only later identified as the cause of added work, contractors are put in the position of pursuing claims without having complied with important notice requirements. COs, too, may be faced with requests for additional costs and time based on past events that may not have been called to their attention, and over which they did not exercise any technical and administrative control at the time of occurrence.

Subsection (b) of the Changes clause addresses such informal changes.[58] The phrase "constructive change order" apparently was first used in the early 1960s,[59] and since that time, the phrase and the concept have gained widespread acceptance. Early on, however, the boards had difficulty in accepting the constructive change concept, particularly because there existed no written directive or formal authorization by the CO. Nonetheless the concept was helpful as a means to provide a remedy within the contract without resort to concepts of breach of contract.[60] As a result of the increasing and widespread recognition of the constructive change order doctrine, the revised 1967 federal Changes clause expressly recognized in subparagraph (b) that changes could occur and be compensated without the trappings of written and formal authorization by the CO:

> (b) Any other written order or oral order (which, as used in this paragraph (b), includes direction, instruction, interpretation, or determination) from the Contracting Officer that causes a change shall be treated as a change order under this clause; *provided,* that the Contractor gives the Contracting Officer written notice stating (1) the date, circumstances, and source of the order and (2) that the Contractor regards the order as a change order.[61]

Recognition of the constructive change order is a result of the increasingly complex and highly competitive nature of government construction contracting. For the contractor, the difference between profit and loss may depend on recognizing and recovering the additional expenses of constructive change events. For agencies, controlling constructive changes improves the ability to control costs and meet budgets.

A. Elements of a Constructive Change Order

Like formal change orders, constructive changes have the basic elements of a change. While a clearly written change order so designated is not necessary, a constructive change order nevertheless requires all the elements otherwise present in a formal change order. The Court of Claims stated:

> But each of the other elements of the standard "Changes" or "Extras" clause has been present—the Contracting officer has the contractual authority unilaterally to alter the contractor's duties under the agreement, the contractor's performance requirements are enlarged and the additional work is not volunteered but results from a direction of the Government's officer. *See* Spector, "An Analysis of the Standard 'Change' Clause", 25 Fed.B.J. 177, 179 (1965).[62]

In other words, the government must either issue a directive or act in such a way as to require the contractor to increase or alter its contract work. A contractor must establish four factors in order to be entitled to an equitable adjustment: (1) the minimum performance required under the contract; (2) the government's "ordered" work exceeded the required contract minimum; (3) the government required the additional work (i.e., the contractor was not a volunteer); and (4) the additional work caused increased cost or time.

The government may also use the doctrine of constructive change for its benefit, typically in the case of deductive changes.[63] The nearly infinite variety of events giving rise to constructive changes requires that contractors be constantly on the alert for oral instructions by government representatives; conflicts between specifications and drawings; and any directives to change sequences of work, to substitute materials, to test or redo work already installed, to change dimensions, to provide more support, or to change the methods of work. In summary, contractors must identify any act or failure to act that imposes more work, costs, or time than the original contract required, because any such requirement may constitute a constructive change order. COs also must be aware of the importance of their own and their staff's informal or indirect actions in order to minimize the circumstances that may lead to changed work and increased payment obligations for the government.

B. Types of Constructive Changes

1. Interpretation of Specifications and Drawings

Interpretation of contract requirements frequently creates a context in which constructive change orders arise. Two basic issues generally are involved: (1) the contract specifications contain an element of ambiguity pertaining to the scope or method of work to be done; and (2) the government and the contractor disagree over the correct interpretation of the specifications. Contract ambiguities are of two types: patent and latent. A patent ambiguity is one that is obvious on its face. A latent ambiguity is subtle or obscure in that the language employed may seem clear at first, yet more than one reasonable interpretation is possible.

If an ambiguity, discrepancy, or inconsistency is obvious on the face of the drawings or specifications, it is a patent ambiguity, and the contractor has a duty to seek clarification from the CO before submitting a bid. Failure to do so will preclude any claim based on the patent ambiguity. The Court of Claims described this process years ago:

> If the bidder fails to resort to the remedy preferred by the Government, a patent and glaring discrepancy (like that which existed here) should be taken against him in interpreting the contract. We do not mean to rule that, under such contract provisions, the contractor must at his peril remove any possible ambiguity prior to bidding; what we do hold is that, when he is presented with an obvious discrepancy of significance, he must consult the Government's representatives if he intends to bridge the crevasse in his favor.[64]

If a contractor performs work according to its own interpretation of a patent ambiguity without seeking clarification from the CO, a request for equitable adjustment may be denied.[65]

Every minor inconsistency, however, need not be brought to the government's attention, and the Court of Claims has so stated:

Although the potential Contractor may have some duty to inquire about a major patent discrepancy, or obvious omission, or a drastic conflict in provisions . . . he is not normally required (absent a clear warning in the contract) to seek clarification of any and all ambiguities, doubts or possible differences in the interpretation. The government, as the author, has to shoulder the major task of seeing that within the zone of reasonableness the words of agreement communicate the proper notions as well as the main risk of a failure to carry that responsibility.[66]

The contractor's duty to inquire also is tempered by the amount of time available for bid preparation, the sheer magnitude of the details to be reviewed in the government specifications, and the ease with which the government could have stated the requirements more precisely and clearly.[67]

Where an ambiguity is not patent but latent, the contractor has the right to perform the work according to its own reasonable interpretation. If the CO insists on a different interpretation, a constructive change may exist even if the CO believed his interpretation was proper or merely was stating that the specifications have a particular meaning.

When an interpretation conflict arises, the issue is whether a reasonable bidder would have interpreted the specifications as the contractor did. More than one reasonable interpretation may exist, but the issue is not which party's interpretation is the best. Rather, it is whether the contractor's interpretation is reasonable, because ambiguous language is the responsibility of the government as the drafter of the drawings and specifications. The Court of Claims has stated:

> To prevail . . . it is not essential that plaintiff demonstrate his position to be the only justifiable or reasonable one. A specification susceptible to more than one interpretation, each interpretation found to be consistent with the contract's language and the parties' objectively ascertainable intentions, becomes convincing proof of an ambiguity; the burden of that ambiguity falls solely upon the party who drew the specifications.[68]

The objective issue is "what plaintiff would have understood as a reasonable construction contractor," not what the drafter of the contract terms subjectively intended. This concept was aptly described by the Corps of Engineers Board of Contract Appeals: "The Government cannot insist in the midst of performance that the contractor should comply with what it subjectively intended to say rather than what it specified."[69]

Finally, the contractor is not expected to be an expert. Even if a certified professional engineer or geologist might interpret the specifications differently than a contractor, such an expert interpretation is not controlling. So long as the contractor was "reasonable" in its interpretation, based on experience and general knowledge of the industry, specialized expertise is not necessary.[70]

In summary, if the CO directs the contractor to perform the work in accordance with the government's interpretation that is different from and imposes more work than the contractor's reasonable interpretation of ambiguous specifications, the direction may be a constructive change. The contractor will be entitled to an equitable adjustment if either its interpretation is reasonable or the government's interpretation is erroneous.

2. Inconsistencies Between Drawings and Specifications

Particularly important to contractors faced with discrepancies between drawings and specifications is the application of the standard FAR Order of Precedence clause. That clause provides that:

> (a) ... Anything mentioned in the specifications and not shown on the drawings, or shown on the drawings and not mentioned in the specifications, shall be of like effect as if shown or mentioned in both. In case of difference between drawings and specifications, the specifications shall govern.[71]

In *Hensel Phelps v. United States*,[72] the contractor bid a government specification that called for 18 inches of nonexpansive fill under the concrete floor slabs of a jet engine blade facility. A note on the drawings called for 36 inches of nonexpansive fill. The contractor used the lesser amount in its bid.

After award, when the contractor pointed out the discrepancy, the CO directed the contractor to place 36 inches of fill as shown on the drawings. Hensel Phelps complied and submitted a claim for equitable adjustment, pointing out that the Order of Precedence clause required that the specifications be followed, which eliminated the apparent conflict with the drawings. The government asserted that the Order of Precedence clause was inapplicable because the contractor was aware of the obvious discrepancy between specifications and drawings prior to bidding and should have sought clarification. The court rejected the government's position and allowed the equitable adjustment, reasoning that:

> an order of precedence clause may be relied on to resolve a discrepancy between the specifications and the drawings even though the discrepancy is known to the contractor prior to bid or is patent. If the contractor is required to perform work in addition to that called for by application of the order of precedence clause, he may seek an *equitable* adjustment in the price of the contract for such work and, in that event, *equitable principles* would apply to overreaching or profiteering.[73]

In a similar case, *McGhee Construction*,[74] the contractor was faced with a classic specifications and drawing conflict, but it disregarded the Order of Precedence clause and instead based its bid on the drawings. The specifications described removal of 13,000 square feet of asbestos sealant, whereas the drawings indicated only 4,300 square feet of sealant to be removed. The contractor's choice was correct, but when the government learned of the discrepancy, it issued a change order reducing the quantity stated in the specifications and asserting a claim for an appropriate credit.

The government contended that, since the Order of Precedence clause required the contractor to rely on the specifications, the government was due an equitable adjustment for the reduction in work quantity. The contractor argued that the Order of Precedence clause did not require it to bid deliberately on the basis of an erroneous quantity. In rejecting the government's position, the ASBCA noted that since the government already had obtained the benefit of the lower price and the bid was based on the correct quantity, equitable principles would not allow the government to benefit again.

Thus, whether the contractor relies on the Order of Precedence clause or disregards the Order of Precedence clause and relies instead on the drawings assumed to be correct, the contractor must be prepared to show that it has not reaped a windfall.

The *Hensel Phelps* and *McGhee* decisions both fit within the larger framework of a "common sense" contract interpretation rule. Simply put, contractors should not rely on strained or hypertechnical contract interpretations when bidding.[75]

The key to the board's holding in *McGhee* is that the contractor made the correct assessment of objectively verifiable facts when faced with a clear government error. In contrast, another contractor[76] misread a clear requirement for painting indicated on the drawings and consequently bid too little for that work. After award, the government deleted the painting and demanded a credit. The board specifically rejected the contractor's assertion that the government should not be permitted to take a deductive equitable adjustment for deleted work that the contractor had erroneously omitted from its bid price.

3. Defective Specifications

Defective specifications provide a fertile context for constructive changes. Defective specifications are those that contain errors, omissions, inadequate work descriptions, misleading provisions, or inconsistencies, or, in more severe circumstances, render the work impossible or commercially impractical to perform. As a general rule, the party that prepares the plans and specifications (the government) warrants their adequacy to produce the intended results. When the contractor incurs additional costs in attempting to comply with defective government specifications, it is entitled to an equitable adjustment under the Changes clause.

The concept of implied warranty stems from the Supreme Court decision in *United States v. Spearin,* which held that "a contractor who is bound to build according to plans and specifications prepared by the owner, will not be responsible for the consequences of defects in the plans or specifications."[77]

Hence, if the government-furnished specifications are defective, the contractor is entitled to an equitable adjustment for costs incurred both in attempting to comply with the defective specifications and in performing corrective work.[78] Defective plans and specifications often are considered as constructive changes. Chapter 18 discusses this important subject in greater detail.

4. Inspection and Testing

Although inspection and quality assurance are primarily a contractor responsibility, the government has the final say on whether the work meets the contract requirements. CO's representatives at the job site are frequently authorized to reject work where materials or workmanship are not in compliance with contract requirements. Where representatives lack proper authority, the government will not pay for changed or corrective work.[79] Terms and provisions relating to contractor rework, repair, or replacement are prescribed in the Inspection of Construction clause.[80]

Testing methods, test results, measurements, finishes, and appearance are all areas of potential disagreement and thus potential constructive changes. Government inspection actions that may result in constructive changes to the work include excessive inspection[81] and improper or incompetent rejection.[82] In such instances, the central issue is whether the contractor was required unnecessarily to perform repair or rework, or change its work method as a result of an inappropriate inspection process.

Similarly, testing requirements may result in constructive changes where the government directs a change in the testing method or frequency.[83] The contractor may be entitled to an equitable adjustment for the cost of conducting added tests, regardless of whether the new testing requirement reveals that contract specifications are not being met.[84] Likewise, an alteration in the testing standard or an interpretation of test results that requires the contractor to meet "a more stringent testing procedure or standard for demonstrating compliance than is set forth in contract" also will constitute a constructive change.[85]

Testing requirements go hand in hand with the quality or performance standard the government specifies. The Court of Claims described this relationship as an equation in *Southwest Welding & Manufacturing Co. v. United States,* a case involving the quality of pipe welding:

> These were typically design (and not performance) specifications. They specified, in detail, the type of steel to be employed, and this steel was inspected and approved in accordance with the specifications.

Welding procedures, the method of obtaining approval thereof, and the qualifications of welders, were set forth in great detail. . . . The testing and inspection requirements for the welding were set forth in great detail and were unusually stringent. They established, in effect, the performance standard to which the contractor was obliged to adhere. The test specified to determine compliance with contract requirements, and the contract requirements, constitute an equation.[86]

In 1998, the U.S. Court of Federal Claims picked up this theme in *M.A. Mortenson v. United States*:

At issue in *Southwest Welding* was whether the government imposed a "more severe inspection procedure or a more severe standard for acceptable welding than was set forth in the original contract." The court insisted that when the government examines if the "original specifications were sufficient to insure the integrity of the welding" (cite omitted) by undertaking a more stringent application and interpretation of contract specifications the conclusion would be "inescapable that the Government is questioning the sufficiency of its own specifications."[87]

Where the work conforms to the specifications, the CO's rejection of that work constitutes a constructive change.[88] The result is that the contractor may recover the cost and additional time for the rework or additional work resulting from the improper rejection. Although the government may inspect the contractor's work at any time, the government may not unreasonably interfere with the contractor's performance. Unreasonable government interference via the inspection process may also constitute a constructive change.[89]

5. Limitations on or Change to Work Method or Manner

Government construction contracts usually do not stipulate a method of performance. Rather, a contractor may perform in any reasonable manner so long as the completed work meets the specification requirements and standards. A constructive change can therefore arise where the CO insists on a departure from a contractor's planned method of performance.[90] As the court explained in *Consolidated Diesel Electric Co. v. United States*:

We find that defendant (the government) . . . unlawfully stopped and thereafter continued to stop plaintiff from substituting engines manufactured by Allis-Chalmers . . . for engines manufactured by Caterpillar, . . . even though plaintiff had the right to substitute during the Step 2 period Allis-Chalmers manufactured engines . . . under the contract. Defendant's action constituted a constructive change to the contract.[91]

In such instances, a contractor may recover the difference between the cost of the government-directed method and the cost of the contractor-preferred method or manner of performance. Other examples of constructive changes arising from limitations on or changes to the work method include the following: the CO required a two-step method of performance when a one-step method was the standard practice and would have sufficed;[92] the CO directed shipment by truck, although the contract allowed shipment by rail;[93] and the CO required removal of wet earth and replacement with sand as the only approved method for meeting compaction density requirements.[94]

6. Constructive Acceleration

The Changes clause expressly provides in subsection (a)(iv) that the CO may direct "acceleration in the performance of the work." In other words, the government may order the contractor to work at a pace faster than that required to meet the original contract completion date. In a variety of circumstances, however, acceleration is imposed less directly by the government, but recovery of resulting acceleration costs has been justified under the constructive change doctrine as "constructive acceleration." This important subject is discussed in another chapter in more detail.[95]

7. Breach of Implied Duties

Implied terms or obligations are unwritten duties owed by each party to a contract to the other party. They exist because of the express agreement, and they underlie the intentions and actions of the contracting parties.[96] Particularly relevant to the constructive change doctrine are the implied duties of each party to cooperate, to act in good faith and deal fairly, and not to interfere with the other parties' performance. These obligations are present in many everyday project activities, such as shop drawing approval, providing site access, timing of inspections, and so on. Failure to meet these obligations on the part of the government may constitute a constructive change.[97]

A contractor whose work was unjustifiably stopped by a CO's representative based on noncompliance with the contract, when the true but undisclosed reason for the work stoppage was a safety concern, constituted a breach of the government's duty to cooperate. Recovery was grounded on the constructive change doctrine.[98] Breaches of implied duties are the basis for constructive changes in several recent cases. Consider the following.

In *R.W. Jones Construction Inc.*, the Bureau of Reclamation (BOR) awarded a contract to R.W. Jones (Jones) for the construction of a fish passageway structure at a dam and canal owned and operated by a local water and power company (WPC). When Jones received the notice to proceed and attempted to enter the work site, the WPC gatekeeper denied Jones access, and both the BOR and the WPC refused to provide Jones with a key to the site. The BOR erroneously and repeatedly interpreted the contract to require that Jones coordinate its work directly with the WPC, thereby cutting off communications between Jones and the BOR. The BOR then declined to communicate with the WPC when Jones sought the BOR's assistance in construction of the dam abutment protection system. The board held that the BOR acted unreasonably in its administration of the contract and that it breached its implied duty to cooperate with Jones. The government's violation of its implied duty to cooperate constituted a constructive change entitling the contractor to an equitable adjustment under the Changes clause.[99]

In *Odebrecht Contractors of California, Inc.*, the Corps of Engineers awarded a contract to construct a dam where the specifications stated that the contractor could use existing wells depicted on the drawings as a source of water for the dam construction, provided that the contractor first rehabilitated the wells. Prior to award, the Corps learned that local water authorities would not agree to this use of the wells. The Corps took a series of actions to prevent the contractor from using the wells, and it withheld notice of any restrictions on using the wells until after the contractor had rehabilitated the wells. The Corps then asserted that the problem was the contractor's responsibility under the Permits and Responsibilities clause and devised a strategy to restrict permits for use of the wells. The board held that these actions were a constructive change, based on its finding that "the Corps failed miserably in its duties of good faith and fair dealing in this contract before and after award. It completely failed to meet the duties of cooperation and noninterference it owed to [the contractor]."[100]

8. Cumulative Impact of Many Changes

When contractors receive large numbers of closely related changes, it can become difficult to determine where work on a single change begins and ends. For the same reasons, it may become difficult to separate changed work from the original contract work. Identifying the extent of new work and related costs for each individual change becomes a very difficult and imprecise task. For this reason, contractors sometimes submit claims premised on the cumulative impact of many changes.

Several boards have addressed this issue by differentiating between the "direct impact" arising from each change and "cumulative impact" arising from many changes.[101] The impact of many changes is typically addressed in terms of delays, disruption, and losses of efficiency. The Claims Court has described this concept as follows:

> Such costs [of cumulative impact], as the board defined them, addressed the inefficiencies and disruptions associated with changes that, when viewed cumulatively (that is, retrospectively), were so large in number and/or magnitude as to give rise to a separately compensable impact claim. The term "ripple effect" has also been used to describe such impact costs.[102]

The BCAs and the COFC have had some difficulty grappling with the application of this concept. Some decisions condition recovery on proof that the numerous changes taken together exceeded the general scope of the contracts and thus constituted a cardinal change.[103] These cases seem to avoid the Changes clause as the basis for recovery.

Other decisions have found that the cumulative impact of numerous changes is an independent constructive change separate from those on which it is based.[104] Using the Changes clause as the basis for a cumulative impact claim is both logical and consistent with the purpose and terms of that clause. The cardinal change concept may apply in some factual situations, but is not the exclusive standard for recovery. A review and analysis of cumulative impact cases has shown the courts and boards to be inconsistent about what it takes to prove these claims.[105] The cumulative impact theory is clearly recognized in the law of federal contracts, although it has a relatively low rate of success and has been criticized as being "somewhat amorphous."[106] The multiplicity of changes that provides the premise for such claims presents an inherent challenge in that it often precludes satisfactory proof of causation.[107] Large numbers of requests for equitable adjustment and changes alone are not enough to establish government liability.[108]

Another important and related issue involves whether the changes underlying such a cumulative impact claim were closed out by releases signed on each change or were the subject of an express reservation of right. The accord and satisfaction defense is alive and well in cumulative impact claims as shown in the 2009 Federal Circuit's decision in *Bell BCI Co.*[109]

VII. *Notice Requirements*

The Changes clause requires a contractor to notify the CO of any additional costs incurred as the result of any change order, whether formal or constructive. Notice must be timely, in writing, sent to the proper individual, and must state the nature of the claim. Failure to comply with these notice requirements may be fatal to recovery. In addition, the applicable notice requirements are somewhat different for express changes as compared to constructive changes. The Changes clause notice requirements state:

> (b) Any other written or an oral order . . . shall be treated as a change order under this clause, *provided, that the Contractor gives the Contracting Officer written notice* stating (1) the date, circumstances, and source of the order and (2) that the Contractor regards the order as a change order.
>
>   . . . .
>
> (d) If any change under this clause causes an increase or decrease in the Contractor's cost of, or the time required for, the performance of any part of the work under this contract, whether or not changed by

any such order, the Contracting Officer shall make an equitable adjustment and modify the contract in writing. However, except for an adjustment based on defective specifications, *no adjustment for any change under paragraph (b) of this clause shall be made for any costs incurred more than 20 days before the Contractor gives written notice* as required. In the case of defective specifications for which the Government is responsible, the equitable adjustment shall include any increased cost reasonably incurred by the Contractor in attempting to comply with the defective specifications.

(e) *The Contractor must assert its right to an adjustment* under this clause *within 30 days after (1) receipt of a written change order under paragraph (a)* of this clause or (2) the furnishing of a written order under paragraph (b) of this clause, by submitting to the Contracting Officer a written statement describing the general nature and amount of proposal, unless this period is extended by the Government. The statement . . . may be included in the notice under paragraph (b) above.

(f) No proposal by the Contractor for an equitable adjustment shall be allowed if asserted after final payment under the contract. (Emphasis added)[110]

There are actually two different notice requirements in the Changes clause. First, if a directive is a formal change in writing and signed by the CO under paragraph (a) of the Changes clause, then the notice requirement contained in paragraph (e) must be followed. That is, the contractor must submit a written notice to the CO within 30 days of receipt of the change order if it intends to assert a claim for costs incurred as the result of the change. That statement must include the general nature and monetary amount of the claim. This 30-day notice requirement may be extended by the CO, although it may not be extended beyond final payment.

Second, if the directive is a constructive change provided for in paragraph (b), which includes any written or oral directive from the CO not in the form of a formal change, both of the notice requirements are applicable. Per paragraph (b), the contractor must provide written notice to the CO that a constructive change has been directed, providing the relevant date, circumstances, and source of the directive. No absolute time requirement is stated in terms of days. However, the notice should be prompt because under paragraph (d) any expense attributed to the constructive change that is incurred more than 20 days prior to the notice is not recoverable. As such, the right to claim additional costs is not extinguished or waived by late notification, but the amount recoverable may be limited by the 20-day window. Time extensions are not expressly subject to the 20-day notice limit, although the associated delay costs may be curtailed by the 20-day limit. Additionally, the notice requirement in paragraph (e) also applies to constructive changes, so within 30 days of a constructive change order the contractor must also provide a written statement setting forth the general nature and monetary extent of the claim.

As a general rule, the above notice requirements must be observed in order to preserve the right to an equitable adjustment. However, certain exceptions are well recognized where the failure to strictly adhere to the notice requirements will not be fatal to the equitable adjustment claim. As stated in *Korshoj Construction Co.*,[111] the exceptions are as follows: (1) the government, in fact, knew of the circumstances that form the basis of the claim; (2) the CO actually considered the claim on its merits without raising the lack of notice; or (3) the lack of notice was not prejudicial to the government. As between the two notice provisions in the Changes clause, the 20-day notice requirement for constructive change order costs is somewhat more strictly applied than the 30-day notice of paragraph (e), especially if the contractor offers no excusable justification for its failure to give notice and the government had no actual knowledge of the problem.[112]

A. Government Knowledge of the Basis of Claim

Notice requirements may be avoided where the CO or his authorized representative in fact knew of the circumstances that form the basis of the alleged change.[113] Actual knowledge of the change can be shown from government records,[114] the presence of government personnel on the job site where the alleged difficulties took place, affirmative acts of government personnel in cooperating with the contractor to alleviate the causes of the difficulty,[115] conferences between the government and the contractor where the causes of the difficulty were discussed,[116] or from the nature of the delay where the causes impute knowledge to the government.[117] Whether actual knowledge will in fact be imputed to government personnel is a function of the degree of expertise required to recognize the cause and the degree of the participation of government representatives.[118]

B. Claim Considered on the Merits

Notice requirements may be waived if a CO considered the associated equitable adjustment claim on its merits without objecting to the contractor's noncompliance with the notice provisions.[119] The rationale for this rule was set forth in an early Court of Claims decision:

> The Contracting Officer received and considered the claim on its merits as did the appeals board, without any point being made as to lack of protest, timeliness of the assertion of the claim, or lack of a written order. Under the express contract provisions, as well as the circumstances involved in the processing of the claim thereunder, this defense is, therefore, not well taken.[120]

C. Failure to Notify Not Prejudicial

A failure to provide timely notice will not adversely affect the right to obtain an equitable adjustment if the government was not prejudiced.[121] The principal purpose of the notice requirement is protection against changed work, added costs, and project delays that are adverse to the government's interest. If there is little or no actual harm to the government or the government knows of or participated in the event, then there is no point in strictly applying the notice requirement.[122]

The burden of proof is upon the government to establish that it was prejudiced by lack of notice. In *Eggers & Higgins*, the board defined prejudice as "much more than an inability to appraise the records of work already performed." Instead, prejudice "occurs when one of the parties loses a valuable right or suffers a detrimental change in position as a result of the unjustified act, delay or omission of another."[123] Therefore, prejudice typically occurs where the lack of notice either prevents the CO from reducing or avoiding possible extra expense to the government or where the passage of time limits the government's options.[124]

D. The Contractor's Duty to Perform Changed Work

As a practical matter, the contractor can neither challenge the government's right to order changes within the general scope of the contract nor refuse to perform the extra ordered.[125] Two standard contract clauses mandate this obligation to continue work. First, the Default clause provides:

> (a) If the Contractor refuses or fails to prosecute the work or any separable part, with the diligence that will ensure its completion within the time specified . . . , the Government may, by written notice to the Contractor, terminate the right to proceed with the work.[126]

Second, the Disputes clause expressly requires that work proceed pending final resolution of a claim: "(i) The Contractor shall proceed diligently with performance of this contract, pending final resolution of any request for relief, claim . . . , and comply with any decision of the Contracting Officer."[127]

Additionally, the FAR itself provides that "[t]he contractor must continue performance of the contract as changed."[128] This duty to continue performance in the face of a change order substantially limits the contractor's bargaining power to negotiate equitable adjustments during the performance of the contract. Nonetheless, refusing to proceed with the work pending resolution of the dispute is not an option.[129] This rule was discussed in *United States v. Stoeckert*, in which the court stated:

> At the outset, it must be observed that the termination of plaintiff's contract for default was proper if for no other reason than that he

refused to follow the contracting officer's directive to remove and replace the tile floor. Instead, he elected to abandon the work and refused to proceed with the removal and replacement of the floor except at Government expense and by a method of installation specified by the Government. However, such an election is simply not open to a Government contractor.[130]

Where advance agreement cannot be reached on the amount of an equitable adjustment before the changed work is under way, the government usually issues a unilateral change order. A unilateral order, if challenged by the contractor, transfers the issues of the appropriate added costs or time extension to the disputes process involving a final decision by the CO and appeal to a BCA or the COFC for resolution under the Disputes clause. Only if the ordered change is outside the general scope of the contract does the contractor have the right to stop work and file suit against the government for damages. As a practical matter, this "right" is more theoretical than real. Abandonment is a dangerous course of action, especially considering the very broad reach of the Changes clause. In any event, refusing changed work directives should be considered only as a last resort, and even then only after careful legal analysis and assessment.

VIII. Value Engineering

Value engineering (VE) is a method used by the government to obtain the best value for an item or service under a government contract. The contractor is presented with an incentive or mandate to submit either a Value Engineering Proposal (VEP) or a Value Engineering Change Proposal (VECP). Through VE a contractor proposes changes that will provide savings or increase the value of the contract for the government. The contractor, in turn, receives a benefit by sharing in the government's savings.

VE became a federal mandate for all executive agencies in February 1996 when Congress passed the "National Defense Authorization Act."[131]

SEC.36. VALUE ENGINEERING.
(a) In General.—Each executive agency shall establish and maintain cost-effective value engineering procedures and processes.
(b) Definition.—As used in this section, the term "value engineering" means an analysis of the function of a program, project, system, product, item of equipment, building, facility, service, or supply of an executive agency, performed by qualified agency or contractor personnel, directed at improving performance, reliability, quality, safety, and life cycle costs.
(c) Regulations.—Each Secretary shall prescribe regulations to carry out this section. Such regulations shall be prescribed in consultation with the Director of the Office of Management and Budget.

Additional laws and regulations relevant to VE in government construction contracts can be found in various government documents.[132] For example, the May 21, 1993 OMB A-131 Circular ". . . requires Federal Departments and Agencies to use value engineering (VE) as a management tool, where appropriate, to reduce program and acquisition costs." The Circular states that agencies "shall use value engineering as a management tool, where appropriate, to ensure realistic budgets, identify and remove nonessential capital and operating costs, and improve and maintain optimum quality of program acquisition functions." Federal agencies are required to report annually on the status of their VE programs if they have budget or procurement obligations exceeding $10 million. *Id.*

FAR Part 48 provides the details regarding the government's use and administration of VE contract techniques. Notably, FAR 48.103(c) states that decisions regarding the acceptance or rejection of a VECP and the determination of collateral costs or savings are unilateral decisions made solely at the discretion of the government.

The FAR provides a VE clause for federal construction contracts, and requires its use for most such contracts. Specifically, FAR 48.202, Clause for Construction Contracts, provides:

> The contracting officer shall insert the clause at 52.248-3, Value-Engineering—Construction, in construction solicitations and contracts when the contract amount is estimated to be $100,000 or more, unless an incentive contract is contemplated. The contracting officer may include the clause in contracts of lesser value if the contracting officer sees a potential for significant savings. The contracting officer shall not include the clause in incentive-type construction contracts. If the head of the contracting activity determines that the cost of computing and tracking collateral savings for a contract will exceed the benefits to be derived, the contracting officer shall use the clause with its Alternate I.

The contractor should be encouraged to develop, prepare, and submit VECPs voluntarily.[133] Where a VE clause is included in the prime contract, the contractor is required to include a VE clause in any of its own subcontracts that exceed $50,000.[134] The contractor can restrict the government's right to use any part of its VECP, or supporting data, if the contractor marks the drawings, or affected parts of the VECP, with FAR-specified wording.[135] Once the VECP is accepted, the government has unlimited rights to the VECP, excepting technical data.[136]

Construction contractors are allowed to share with the government in instant contract savings or collateral savings.[137] Instant contract savings are considered a form of acquisition savings that are generated from the contract being performed.[138] FAR 52.248-3(f) provides the formula to determine the rates and payment for instant contract savings. The contractor has to prove that cost savings occurred.[139]

Savings to an agency's costs of operation, maintenance, logistics support, or government-furnished property are considered collateral savings.[140] The CO is the "sole determiner" of collateral savings amounts.[141] A contractor is not entitled to collateral savings if the collateral savings clause is not included in the contract.[142]

FAR 52.248-3(d) provides the technical instructions regarding how a contractor must submit a VECP. However, case law indicates that lack of technical conformity may not be dispositive if other factors, such as the intent of the parties and treatment of a proposal by the parties, show a VE understanding between the government and the contractor.[143]

A contractor has the right to withdraw its VECP any time before it is accepted by the government.[144] However, once a proposal is withdrawn, the government has the right to use the proposal on the instant contract, or any other contracts, without sharing the cost savings with the contractor.[145] Therefore, a contractor may wish to preserve his rights by not withdrawing the proposal.[146]

Notes

1. FAR 52.245-2(h).

2. FAR 52.233-1.

3. 41 U.S.C. § 605 (for a decision by the CO); 41 U.S.C. §§ 606, 607, 609 (for appeals to a BCA or the COFC).

4. Trowbridge vom Baur, *The Origin of the Changes Clause in Naval Procurement*, 8 Pub. Cont. L.J. 175 (1976); Trowbridge vom Baur, *The Breakdown of the Changes and Disputes Process*, 9 Pub. Cont. L.J. 143 (1977); *see also* GAI Consultants. Inc., ENG BCA No. 6030, 95-2 BCA ¶ 27,620 (stating that, as a matter of law, the appropriate Changes clause would be incorporated into the contract under the *Christian* doctrine).

5. FAR 52.243-4; FAR 52.243-1.

6. See chapters 15, 18, and 19.

7. FAR 52.243-4.

8. FAR 43.102(a), (b). For further discussion of issues regarding the authority of COs and their representatives to bind the government, see Chapter 7.

9. FAR 1.602-1.

10. Arthur Venneri Co. v. United States, 180 Ct. Cl. 920, 924, 381 F.2d 748, 750 (1967) (citations omitted).

11. Gen. Elec. Co., ASBCA No. 11 990, 67-1 BCA ¶ 6377.

12. Elias Pamfilis Painting Co., ASBCA No. 30839, 87-3 BCA ¶ 20,189.

13. Sterling Millwrights, Inc. v. United States, 25 Cl. Ct. 49 (1992).

14. Electrospace Corp., ASBCA No. 14520, 72-1 BCA ¶ 9455; Tex. Instruments, Inc. v. United States, 922 F.2d 810 (Fed. Cir. 1990).

15. Henry Burge, ASBCA No. 2431, 89-3 BCA ¶ 21,910.

16. FAR 43.202; FAR 42.202.

17. Winter v. Cath-dr/Balti Joint Venture, 497 F.3d 1339 (Fed. Cir. 2007); States Roofing Corp., ASBCA No. 55500, 09-1 BCA ¶ 34,036, 2008 WL 5392082.

18. See Chapter 7.

19. FAR 1.602-3 (c); Lox Equip. Co., ASBCA No. 8985, 1964 BCA ¶ 4463.

20. U.S. Fed. Eng'g & Mfg., Inc., ASBCA No. 19909, 75-2 BCA ¶ 11,578.

21. Gresham & Co. v. United States, 200 Cl. Ct. 97 (1972).

22. Jordan & Nobles Constr. Co., GSBCA No. 8349, 91-1 BCA ¶ 23,659.

23. J.D. Hedin Constr. Co. v. United States, 171 Ct. Cl. 70, 106, 347 F.2d 235, 258 (1965).

24. Electro-Tech Corp., ASBCA No. 42495, 93-2 BCA ¶ 25,750; Design & Prod., Inc. v. United States, 18 Cl. Ct. 168 (1989).

25. Plumley v. United States, 226 U.S. 545 (1913).

26. W.H. Armstrong Co. v. United States, 98 Ct. Cl. 519, 530 (1943); *but see* Globe Indem. Co. v. United States, 102 Ct. Cl. 21 (1944) (court was bound by *Plumley* decision).

27. U.S. Fed. Eng'g, 75-2 BCA ¶ 11,578; Lincoln Constr. Co., IBCA No. 438-5-64, 65-2 BCA ¶ 5234; Lillard's, ASBCA No. 6630, 61-1 BCA ¶ 3053.

28. Len Co. & Assocs. v. United States, 385 F.2d 438, 443 (Ct. Cl. 1967).

29. Woodcraft Corp. v. United States, 146 Ct. Cl. 101, 173 F. Supp. 613 (1959); Dittmore-Freimuth Corp. v. United States, 182 Ct. Cl. 507 (1968); Blake Constr. Co., VACAB No. 1725, 83-1 BCA ¶ 16,431.

30. Freund v. United States, 260 U.S. 60, 63 (1922).

31. *Id.*

32. Air-A-Plane Corp. v. United States, 408 F.2d 1030, 1033, 187 Ct. Cl. 269, 276 (1969) (internal quotations omitted); Thermocor, Inc. v. United States, 35 Fed. Cl. 480 (1996); J.D. Hedin Constr. Co. v. United States, 171 Ct. Cl. 70, 106, 347 F.2d 235, 258 (1965).

33. Wunderlich Contracting Co. v. United States, 351 F.2d 956, 966 (Ct. Cl. 1965); Becho, Inc. v. United States, 46 Fed. Cl. 595 (2000).

34. Universal Constr. & Brick Pointing Co. v. United States, 19 Cl. Ct. 785, 792 (1990).

35. *Air-A-Plane*, 408 F.2d 1030; PCL Constr. Servs. v. United States, 47 Fed. Cl. 745 (2000); *Wunderlich*, 351 F.2d 956; S.J. Groves & Sons Co. v. United States, 228 Ct. Cl. 598, 661 F.2d 170 (1981); Coley Props. Corp., PSBCA No. 291, 75-2 BCA ¶ 11,514.

36. Melville Energy Sys., Inc. v. United States, 33 Fed. Cl. 616 (1995).

37. Keco Indus., Inc. v. United States, 364 F.2d 838 (Ct. Cl. 1966); Gen. Dynamics Corp. v. United States, 218 Ct. Cl. 40, 585 F.2d 457 (1978); AT&T Commc'ns, Inc. v. Wiltel, Inc., 1 F.3d 1205 (Fed. Cir. 1993).

38. Saddler v. United States, 152 Ct. Cl. 557 (1961).

39. Stapleton Constr. Co. v. United States, 92 Ct. Cl. 551 (1940).

40. Embassy Moving & Storage Co. v. United States, 424 F.2d 602 (Ct. Cl. 1970).

41. PCL Constr. Servs., Inc. v. United States, 96 Fed. Appx. 672, 2004 WL 842984 (Fed. Cir. 2004) (unpublished opinion), 46 GC ¶ 235; Aragona Constr. Co. v. United States, 165 Ct. Cl. 382 (1964).

42. Edward R. Marden Corp. v. United States, 442 F.2d 364, 369 (Ct. Cl. 1971); In re Boston Shipyard Corp., 886 F.2d 451, 456 (1st Cir. 1989).

43. Amertex Enters., Ltd. v. United States, 108 F.3d 1392 (Fed. Cir. 1997) (non-precedential opinion).

44. *See* Schmid v. United States, 351 F.2d 651, 655 (Ct. Cl. 1965).

45. United States v. Callahan-Walker Constr. Co., 317 U.S. 56 (1942); Air-A-Plane Corp. v. United States, 408 F.2d 1030, 187 Ct. Cl. 269 (1969).

46. Bruce Constr. Co. v. United States, 324 F.2d 516 (Ct. Cl. 1963).

47. Keco Indus., Inc. v. United States, 364 F.2d 838 (Ct. Cl. 1966).

48. For further discussion of schedule and delay issues, see Chapter 19.

49. FAR 52.242-14; Berrios Constr. Co., VABCA No. 3152, 92-2 BCA ¶ 24,828; Robert McMullan & Son, Inc., ASBCA No. 19023, 76-1 BCA ¶ 11,728.

50. Sw. Marine, Inc, ASBCA No. 34058, 91-1 BCA ¶ 23,323; Paul Hardeman, Inc. v. United States, 406 F.2d 1357 (Ct. Cl. 1969).

51. Hardeman-Monier-Hutcherson, ASBCA No. 11785, 67-1 BCA ¶ 6210.

52. FAR subpt. 43.2—Change Orders.

53. FAR 43.103.

54. FAR 43.102(b).

55. Ed Zueblin v. United States, 44 Fed. Cl. 228, 234 (1999).

56. FAR 43.103(b).

57. L. Spector, *An Analysis of the Standard "Changes" Clause*, 25 FED. BAR. J. 177 (1965); F. Trowbridge vom Baur, *Constructive Change Orders/Edition II*, Briefing Papers 73-5 (Fed. Pubs. 1973).

58. FAR 52.243-4; Lehigh Chem. Co., ASBCA No. 8427, 1963 BCA ¶ 3749.

59. *See* note 66.

60. Johnson & Son Erectors, ASBCA No. 24564, 81-1 BCA ¶ 15,082, *aff'd*, 231 Ct. Cl. 753 (1982); Bruce-Anderson Co., ASBCA No. 35791, 89-2 BCA ¶ 21,871; JEM Dev. Corp., DOTBCA No. 1961, 88-3 BCA ¶ 21,022.

61. FAR 52.243.4.

62. Len Co. & Assocs. v. United States, 385 F.2d 438, 443 (Ct. Cl. 1967).

63. Dawson Constr. Co., VABCA No. 3558, 94-1 BCA ¶ 26,362; Worldwide Parts, Inc., ASBCA No. 38896, 91-2 BCA ¶ 23,717; Porshia Alexander of Am., GSBCA No. 9604, 91-1 BCA ¶ 23,657.

64. Beacon Constr. Co. v. United States, 161 Ct. Cl. 1, 7, 314 F.2d 501, 504 (1963).

65. SOG of Ark. v. United States, 546 F.2d 367 (Ct. Cl. 1976); B.D. Click Co., ASBCA No. 18647, 74-2 BCA ¶ 10,714; Co-op Constr. Co., ASBCA No. 18663, 74-2 BCA ¶ 10,917; Tom Shaw, Inc., DOTBCA No. 2109, 90-2 BCA ¶ 22,861.

66. WPC Enters., Inc. v. United States, 323 F.2d 874 (Ct. Cl. 1963).

67. Bridge Constr. Corp, DOTCAB No. 70-30, 71-1 BCA ¶ 8857; D&L Constr. Co. v. United States, 406 F.2d 990 (Ct. Cl. 1967); Dawson Constr. Co., GSBCA No. 3820, 75-1 BCA ¶ 11,339; Gorn Corp. v. United States, 424 F.2d 588 (Ct. Cl. 1970).

68. George Bennet v. United States, 371 F.2d 859, 861 (Ct. Cl. 1967); Big Chief Drilling Co. v. United States, 25 Ct. Cl. 1276 (1992).

69. A.A. Conte & Son, Inc., ENG BCA Nos. 6104, 6227, 96-2 BCA ¶ 28,581; Hoffman Constr. Co., VABCA No. 3676, 93-3 BCA ¶ 26,110; Corbetta Constr. Co. v. United States, 461 F.2d 1330, 1336 (Ct. Cl. 1972); *see also* Piracci Constr. Co. GSBCA No. 3477, 74-2 BCA ¶ 10,800.

70. Salem Eng'g & Constr. Corp. v. United States, 2 Ct. Cl. 803 (1983).

71. FAR 52.236-21.

72. Hensel Phelps Constr. Co. v. United States, 886 F.2d 1296 (Fed. Cir. 1989).

73. *Id.* at 1299–1300.

74. McGhee Constr., Inc., ASBCA No. 45175, 93-3 BCA ¶ 26,154.

75. Shemya Constructors, ASBCA No. 45251, 94-1 BCA ¶ 26,346.

76. Bruce Anderson Co., ASBCA Nos. 29412, 32247, 89-2 BCA ¶ 21,872.

77. United States v. Spearin, 248 U.S. 132, 135 (1918).

78. A.A. Conte & Son, Inc., ENG BCA Nos. 6104, 6227, 96-2 BCA ¶ 28,581; Sterling Millwrights, Inc. v. United States, 26 Ct. Cl. 49 (1992); Ziebarth & Alper, ASBCA No. 25040, 82-1 BCA ¶ 15,777; Hol-Gar Mfg. Corp. v. United States, 360 F.2d 634 (Ct. Cl. 1966); F.J. Stokes Corp., ASBCA No. 6532, 1963 BCA ¶ 3944; Husman Bros., Inc., AGBCA No. 216, 77-2 BCA ¶ 12,563.

79. S&M Mgmt. Inc. v. United States, 82 Fed. Cl. 240 (2008).

80. FAR 52.246-12. For additional discussion of government inspection and acceptance, see Chapter 10.

81. Ramsdell Constr. Co., ASBCA No. 87-118-1, 92-2 BCA ¶ 24,770.

82. Harvey C. Jones, Inc., IBCA No. 2070, 90-2 BCA ¶ 22,762; Batteast Const. Co., ASBCA No. 25841, 91-3 BCA ¶ 24,352.

83. *See* Orion Elec. Corp., ASBCA No. 18495, 75-1 BCA ¶ 11,193; Electro Plastic Fabrics, Inc., ASBCA No. 14762, 71-2 BCA ¶ 8996.

84. Centric/Jones Constructors, IBCA No. 3139, 94-1 BCA ¶ 26,404; J.W. Bateson Co., GSBCA No. 2434, 68-2 BCA ¶ 7333.

85. United Techs. Corp. v. United States, 27 Fed. Cl. 397 (1992); Sw. Welding & Mfg. Co. v. United States, 188 Ct. Cl. 925, 951–53, 413 F.2d 1167, 1183–84 (1969); *see also* SMS Data Prods. Group, Inc. v. United States, 17 Cl. Ct. 1, 10 (1998) ("imposing a changed acceptance test" upon plaintiff constituted contractual interference).

86. 188 Ct. Cl. at 952, 413 F.2d at 1183.

87. 40 Fed. Cl. 389 (1998).

88. *See* Fox Valley Eng'g, Inc. v. United States, 151 Ct. Cl. 228 (1960); Kollsman Instrument Corp., ASBCA No. 14849, 74-1 BCA ¶ 10,837; Algernon Blair, Inc., GSBCA No. 2700, 71-2 BCA ¶ 9156.

89. *See* Heritage Co., VABCA No. ¶ 3004, 91-1 BCA ¶ 23,482; Adams v. United States, 358 F.2d 986 (Ct. Cl. 1966); G.A. Karnavas Painting Co., NASA BCA No. 28,1963 BCA ¶ 3633; GW Galloway Co., ASBCA No. 16975, 73-2 BCA ¶ 10,270.

90. Metric Constr. Co. v. United States, 81 Fed. Cl. 804 (2008); A.A. Conte & Son, Inc.; ENG BCA Nos. 6104, 6227, 96-2 BCA ¶ 28,581; Frank Briscoe Co., GSBCA No. 3455, 73-1 BCA ¶ 10,008; C.H. Leavell & Co., GSBCA No. 3031, 70-2 BCA ¶ 8529.

91. 533 F.2d 556, 209 Ct. Cl. 521 (1976).

92. Leavell, 70-2 BCA ¶ 8529; *see also* DRC Corp. v. Dep't of Commerce, GSBCA No. 14,919, 00-1 BCA ¶ 30,649.

93. Goslin-Birmingham, Inc., ENG BCA No. 2800, 67-2 BCA ¶ 6402.

94. Wilco Constr. Co., ASBCA No. 13163, 69-2 BCA ¶ 7857.

95. See Chapter 19.

96. RESTATEMENT (SECOND) OF CONTRACTS § 205; 8 ARTHUR L. CORBIN, CORBIN ON CONTRACTS § 32.1 (1962).

97. Blinderman Constr. Co. v. United States, 695 F.2d 552, 558 (Fed. Cir. 1982).

98. A.A. Conte & Son, Inc.; ENG BCA Nos. 6104, 6227, 96-2 BCA ¶ 28,581; Chris Berg, Inc. & Assocs., ASBCA No. 3466, 58-1 BCA ¶ 1792.

99. IBCA No. 3656-96, 99-1 BCA ¶ 30,268.

100. ENG BCA No. 6327, 2000 BCA ¶ 30,299.

101. Pittman Constr. Co., GSBCA Nos. 4897, 4923, 81-1 BCA ¶ 14,847, *recon. denied,* 81-1 BCA ¶ 15,111, *aff'd,* 2 Ct. Cl. 211 (1983).

102. Pittman Constr. Co. v. United States, 2 Ct. Cl. 211, 216 (1983).

103. F.H. McGraw & Co. v. United States, 130 F. Supp. 394 (Ct. Cl. 1955); Dyson & Co., ASBCA No. 21673, 78-2 BCA ¶ 13,482, *recon. denied,* 79-1 BCA ¶ 13,661.

104. Coates Indus. Piping, Inc., VABCA No. 5412, 99-2 BCA ¶ 30,479; Sw. Marine Inc., DOTBCA No. 1663, 94-3 BCA ¶ 27,102.

105. Geoffrey T. Keating & Thomas F. Burke, *Cumulative Impact Claims: Can They Still Succeed?,* CONST. LAW., Apr. 2000, at 30.

106. Centex Bateson Constr. Co., VABCA Nos. 4612, 5162, 5165, 99-1 BCA ¶ 30,153.

107. *Id.*

108. Hensel Phelps Constr., ASBCA No. 49270, 99-2 BCA ¶ 30,531 at 150, 795.

109. 2009 U.S. App. LEXIS 13625 (June 25, 2009); J.T. Constr. Co., ASBCA No. 54352, 06-1 BCA ¶ 33,182.

110. FAR 52.243.4.

111. Korshoj Constr. Co., IBCA No. 321, 1963 BCA ¶ 3848.

112. Southland Constr. Co., VABCA No. 2217, 89-1 BCA ¶ 21,548; Harris Sys. Int'l Inc., ASBCA No. 33280, 88-2 BCA ¶ 20,641; Power Regulator Co., GSBCA No. 4668, 80-2 BCA ¶ 14,463.

113. Niko Contracting Co., IBCA No. 2368, 91-1 BCA ¶ 23,321.

114. Korshoj Constr. Co., IBCA No. 321, 1963 BCA ¶ 3848.

115. Cent. Mech. Constr., ASBCA No. 29431, 85-2 BCA ¶ 18,061; Xplo Corp., DOTBCA No. 1244, 86-2 BCA 18,869.

116. J.D. Abrams, ENG BCA No. 4332, 89-1 BCA ¶ 21,379.

117. Inet Power, NASA BCA No. 566-23, 68-1 BCA ¶ 7070.

118. *Id.*

119. Power Regulator Co., GSBCA No. 4668, 80-2 BCA ¶ 14,463; Fox Valley Eng'g, Inc. v. United States, 151 Ct. Cl. 228 (1960); Robertson-Henry Co., 61-2 BCA ¶ 3156 (1961).

120. *Fox Valley Eng'g,* 151 Ct. Cl. at 228, 237–38.

121. SIPCO Servs. & Marine Inc. v. United States, 41 Fed. Cl. 196 (1998); Monarch Lumber, IBCA No. 217, 60-2 BCA ¶ 2674; *see also* Precision Tool & Eng'g Corp., ASBCA No. 14148, 71-1 BCA ¶ 87,38l; E.W. Bliss Co., ASBCA No. ¶ 9584, 65-1 BCA ¶ 4610; Arvel Elecs., ASBCA No. 18990, 76-1 BCA ¶ 11,067.

122. Turnagain Paint & Const. Co., ASBCA No. 17884, 74-2 BCA ¶ 10,872; Hartford Acc. and Indem. Co., IBCA No. 1139, 77-2 BCA ¶ 12,604; Santa Fe, Inc., VABCA No. 1983, 84-3 BCA ¶ 17,538.

123. Eggers & Higgins, VACAB No. 537, 66-1 BCA ¶ 5525, at 25,876; Atl. Constr. Co., ASBCA No. 22647, 79-1 BCA ¶ 13,612.

124. Imbus Roofing Co., GSBCA No. 10,430, 91-2 BCA ¶ 23,820; Toland & Sons, IBCA No. 2716, 91-3 BCA ¶ 2368.

125. FAR 43.201(b).

126. FAR 52.249-10. For further discussion of terminations for default, see Chapter 14.

127. FAR 52.233-1(i).

128. FAR 43.201(b).

129. Twigg Corp., NASA BCA No. 62-0192, 93-1 BCA ¶ 25,318.

130. Stoeckert v. United States, 391 F.2d 639, 645 (Ct. Cl. 1968); *see also* Am. Dredging Co., ENG BCA No. 2920, et al., 72-1 BCA ¶ 9316, *aff'd*, 207 Ct. Cl. 1010 (1975); D.H. Dave & Gerben Constr. Co., ASBCA No. 6257, 1962 BCA ¶ 3493.

131. National Defense Authorization Act of 1996, Pub. L. No. 104-106, § 4306, Value Engineering for Federal Agencies, 110 Stat. 186, 665–666; Office of Federal Procurement Policy Act (41 U.S.C. §§ 401, 432) as amended by § 4203, Section 36—Value Engineering.

132. Office of Federal Procurement Policy Act (41 U.S.C. § 432); OFFICE OF MGMT. & BUDGET, OMB Circular A-131, SUBJECT: VALUE ENGINEERING (1993), *available at* http://www.whitehouse.gov/omb/circulars_a131; FAR Part 48; FAR 52.248-3.

133. FAR 52.248-3(a).

134. FAR 52.248-3(h).

135. FAR 52.248-3(i).

136. *Id.*

137. FAR 52.248-3(f); FAR 52.248-3(g).

138. FAR 48.001.

139. Adam Elec. Co., GSBCA No. 6988, 85-2 BCA ¶ 17,997; Lockheed Aircraft Serv. Co., ASBCA No. 16269, 74-1 BCA ¶ 10,601.

140. FAR 48.001.

141. FAR 52.248-3(g).

142. Lawson & Co., ASBCA No. 15266, 73-2 BCA ¶ 10,062, 15 GC ¶ 335; Johnson Constr. Co., ENG BCA No. 3279, 73-2 BCA ¶ 10,295, 16 GC ¶ 44.

143. Covington Indus., Inc., ASBCA No. 12426, 68-2 BCA ¶ 7286, 11 GC ¶ 24; Syro Steel Co., ASBCA No. 12530, 69-2 BCA ¶ 8046, 12 GC ¶ 85.

144. FAR 48.103(b).

145. *See* § 9:26. *See generally*, RALPH C. NASH, JR. & STEVEN W. FELDMAN, GOVERNMENT CONTRACT CHANGES, 304–305 (3rd ed. 2007).

146. *Id.*

CHAPTER 9

Differing Site Conditions

DONALD G. GAVIN
DANIEL J. DONOHUE
BRIAN P. WAAGNER

I. Introduction

A significant risk in lump-sum construction contracting is the risk that unanticipated hidden site conditions will cost the contractor substantial additional cost and time to complete the contract. Contractors try to assess and cover this risk prior to bid by reviewing contract data, inspecting the site, and including contingencies in their bids to protect against unforeseen problems.

Construction contracts with the federal government contain the Differing Site Conditions (DSC) clause—formerly known as the Changed Conditions clause. This clause is the result of a federal policy to discourage bidders from including such contingencies.[1] It provides for an equitable adjustment in the contract price and time for performance if the contractor encounters either (1) subsurface or latent physical conditions at the site that differ materially from those indicated in the contract (called a Type I differing site condition); or (2) unknown physical conditions at the site, of an unusual nature, that differ materially from those ordinarily encountered and generally recognized as inhering in work of the character provided for in the contract (called a Type II differing site condition).[2] According to a 1970 Court of Claims decision, this policy saves money for the government over the long term because it requires the government to pay for unknown or unusual hidden conditions only when they are actually encountered.[3]

Federal agencies increasingly have included in their construction contracts general and specific exculpatory clauses, warnings, site inspection provisions, and other limitations seeking to limit their exposure for differing site conditions. Federal construction contracts also frequently include or refer to subsurface studies, prior contracts, as-built data, surveys, borings, soil reports, historical data, and other physical data that may or may not disclose hidden conditions to bidders. Whether this approach avoids or minimizes the risk of differing site conditions is a policy question. Alternatively, government-funded studies have concluded that sharing with prospective contractors a thorough site investigation report by local geotechnical engineers as part of the contractual documents may be a more effective means of curtailing the risk and cost to the government.[4] When the contractor encounters difficulties, disputes may arise about whether the contractor is entitled to relief under the DSC clause or, instead, whether such exculpatory clauses or physical data reasonably indicated that the contractor should have expected to encounter the very conditions actually encountered. This is an area of increasing concern for contractors and the government, and is the subject of much of the current differing site conditions litigation.

This chapter discusses the common law concerning risks of unknown site conditions and how those risks are shifted by the DSC clause. It also reviews the Federal Acquisition Regulation (FAR) provisions that mandate the DSC clause and some common factual and legal issues that affect claims under the DSC clause and case law interpreting it.

II. Common Law Background

To understand the risk allocation in the DSC clause, it is helpful to review the common law applicable to a fixed-price construction contract in the absence of such a provision. First, in the absence of a DSC clause, a contractor generally would be obligated to complete the work and overcome any unexpected subsurface or hidden conditions without any increase in the contract price or time. Second, in the absence of a DSC clause, the government generally would have no obligation to disclose subsurface information to the contractor unless the government had "superior knowledge," unavailable to bidders, of hidden conditions that will materially increase the price of the work. Third, in the absence of a DSC clause, the government generally would be liable for inaccuracies in information about the site only if the contractor proved fraud, i.e., both that the government knew not only that the information was inaccurate but also that the contractor would rely on it and that the government failed to tell the contractor in order to obtain a low price.

A. The Common Law Rule

Under common law, the general rule applicable to lump-sum federal construction contracts is that the contractor takes the risk of unexpected problems. The Supreme Court summarized this general rule in the landmark decision *Spearin v. United States*.[5] The Court ruled that, when "one agrees to do, for a fixed sum, a thing possible to be performed, he will not be excused or become entitled to additional compensation, because unforeseen difficulties are encountered."[6] In a fixed-price construction contract, the contractor assumes all performance risks unless the risk specifically is shifted away from the contractor by a provision of the contract.

As a result, unexpected difficulties caused by unknown and unusual subsurface or hidden conditions are generally not compensable absent a DSC clause.

B. Superior Knowledge

Some construction contractors believe that the government must obtain and disclose information about subsurface or hidden site conditions and that the contractor is entitled to additional compensation if the government fails to do so. That is not correct.

In fact, the rule is the opposite: The government generally is not obligated to obtain or disclose subsurface or hidden conditions unless it actually has "superior knowledge"—knowledge that is important to successful contract performance that is in the possession of the government and unavailable to the contractor.[7] In order to invoke the "superior knowledge" doctrine, a contractor must produce specific evidence of the following:

1. That it undertook to perform without vital information of a fact that affects performance, costs, or direction;
2. That the government was aware the contractor had no knowledge and had no reason to obtain such information;
3. That any contract specifications supplied misled the contractor, or did not put it on notice to inquire; and
4. That the government failed to provide the relevant information.[8]

The contractor thus has a relatively difficult burden to prove entitlement to relief for the government's failure to disclose "superior knowledge." As such, the "superior knowledge" doctrine provides relief only in limited circumstances.

C. Fraud and Misrepresentation

In the absence of a DSC clause, contractors have to prove that the government's positive representations about site conditions were the result of knowing or negligent misrepresentation in order to recover for additional costs of unanticipated conditions. In 1914, the Supreme Court held that the government would be liable to a dam construction contractor for additional costs incurred when the government misrepresented the site conditions by making positive representations about subsurface conditions that were incorrect.[9] While this misrepresentation theory provides some avenue for relief, it places contractors in the difficult position of having to accuse their customer of misrepresentation in order to obtain relief for the costs of unanticipated conditions.

International Technology Corp. v. Winter[10] is a recent example of a claim for breach of contract alleging that the government misrepresented subsurface conditions in a contract that lacked a DSC clause. In that case, ITC had a contract for treatment of pesticide-contaminated soil at the Naval Communication

Station in Stockton, California. ITC claimed additional costs of treating soil that had unexpectedly high concentrations of clay. On appeal to the Federal Circuit Court of Appeals, ITC claimed that the government breached the contract because the government misrepresented the subsurface conditions to the contractor by understating concentrations of clay in the soil report given to bidders.

The Federal Circuit held that a misstatement of site conditions can support a breach of contract claim, but that the elements for such a breach claim were the same as the elements for a Type-1 Differing Site Conditions claim, discussed in greater detail later in this chapter. The contractor must prove that: a reasonable contractor would interpret the contract as making a representation of site condition; the actual site conditions were not reasonably foreseeable by the contractor; the contractor in fact actually relied upon the misrepresentation; and the conditions actually encountered differed materially from those represented by the government.

The Federal Circuit denied the claim and held that ITC failed to prove the first two elements—that the government represented the site conditions or that the actual conditions were not reasonably foreseeable. Citing *Renda Marine, Inc. v. United States*,[11] the court held that whether a contract represents site conditions is a question of contract interpretation, which is a question of law for the court to decide. The court held as a matter of law that the soil report in the solicitation, which included data in a table showing soil with less than 10 percent concentrations of clay, did not represent the soil conditions but "were intended to identify the range of soil types that were actually treated in the pilot study [that produced the soil report]."[12]

The court also noted that the soil report mentioned wide variations of clay content in different locations and that expert testimony showed that a site visit would have alerted a reasonable contractor to the likelihood of wide variations in clay content in different locations at the site. The court held that, given these indications of variations in clay content, the contractor could not prove that, even if the soil report misrepresented soil conditions, the contractor's reliance upon the soil report was reasonable.[13]

D. Consequences of the Common Law Rule

By making it so difficult for contractors to recover for unforeseen site conditions, the common law rule has led contractors to include contingencies in their bids to cover potential costs of overcoming unforeseen conditions. When the low bid includes such contingency amounts, the government pays the contingency whether or not unusual conditions are encountered on the project. Thus, this practice cost the government additional money over the long term.

This practice of including contingency amounts in bids hurt contractors as well. Such contingencies inflated bid prices and made bids uncompetitive. In order to be low bidder, a contractor had to eliminate any such contingency amounts for unforeseen conditions. Thus, the low bidder may have had to accept the risk of unforeseen conditions and hope for the best. If unforeseen conditions caused additional costs, the contractor would suffer the loss.

III. *The Differing Site Conditions Clause and Its Requirements*

Unlike the common law, which places risks of unforeseen site conditions on the contractor, the DSC clause shifts some of these risks to the government. The DSC clause is intended to discourage bidders from inflating their bids with contingencies to cover potential unknown risks of subsurface or other hidden conditions.[14]

A. Origin of the Differing Site Conditions Clause

The current Differing Site Conditions clause first emerged as a clause entitled Changed Conditions. The government first adopted a standard Changed Conditions clause in 1926, when it promulgated for government-wide use Standard Form 23A, containing standard terms and conditions for federal construction contracts.[15] The FAR now mandates that all fixed price construction contracts include the DSC clause (FAR 36.502).

B. The Differing Site Conditions Clause

The text of the DSC clause, prescribed in FAR 52.236-2, reads:

DIFFERING SITE CONDITIONS (APR 1984)

(a) The Contractor shall promptly, and before the conditions are disturbed, give a written notice to the Contracting Officer of—

(1) Subsurface or latent physical conditions at the site which differ materially from those indicated in this contract; or

(2) Unknown physical conditions at the site, of an unusual nature, which differ materially from those ordinarily encountered and generally recognized as inhering in work of the character provided for in the contract.

(b) The Contracting Officer shall investigate the site conditions promptly after receiving the notice. If the conditions do materially so differ and cause an increase or decrease in the Contractor's cost of, or the time required for, performing any part of the work under this contract, whether or not changed as a result of the conditions, an equitable adjustment shall be made under this clause and the contract modified in writing accordingly.

(c) No request by the Contractor for an equitable adjustment to the contract under this clause shall be allowed, unless the Contractor has given the written notice required; *provided*, that the time prescribed in paragraph (a) of this clause for giving written notice may be extended by the Contracting Officer.

(d) No request by the Contractor for an equitable adjustment to the contract for differing site conditions shall be allowed if made after final payment under this contract.

Paragraph (a) of the clause defines the two types of differing site conditions covered by the clause. A Type I differing site condition is an unexpected condition that differs materially from conditions indicated in the contract. A Type II differing site condition is an unexpected condition that differs materially from conditions normally to be expected in similar work.

Paragraph (a) requires the contractor to give written notice to the contracting officer (CO) when a differing site condition is encountered and states that such notice must be given before the condition is disturbed. As discussed below, a failure to give such written notice may not necessarily bar a contractor from obtaining relief under the clause.[16]

Paragraph (b) of the clause requires the CO to investigate the site condition promptly after receiving notice. If the condition is a differing site condition covered by the clause, the CO is required to grant an equitable adjustment in the contract price and/or time and to modify the contract accordingly. Paragraph (b) specifies that the equitable adjustment is to cover the effect of the condition on a part of the work, whether or not the work is changed by the condition or the government was prejudiced by the lack of formal notice.

Paragraph (c) of the clause states that no equitable adjustment in the contract shall be allowed unless the contractor has given the required written notice and that the CO may extend the time for the giving of such notice. The clause has been interpreted, however, so that a lack of written notice will not preclude the contractor from obtaining an equitable adjustment if the contractor proves that the government had actual knowledge of the condition.[17]

Paragraph (d) states that no equitable adjustment shall be allowed after final payment under the contract.

C. The Site Investigation Clause

The DSC clause is not the only mandatory contract language that addresses site conditions. The FAR also mandates that each fixed-price construction contract include a standard clause requiring the contractor to inspect the site.[18] The text of the standard Site Investigation clause is at FAR 52.236-3 and provides as follows:

SITE INVESTIGATION AND CONDITIONS AFFECTING THE WORK (Apr 1984)

(a) The Contractor acknowledges that it has taken steps reasonably necessary to ascertain the nature and location of the work, and that it has investigated and satisfied itself as to the general and local conditions which can affect the work or its cost, including but not limited to

(1) conditions bearing upon transportation, disposal, handling, and storage of materials;

(2) the availability of labor, water, electric power, and roads;

(3) uncertainties of weather, river stages, tides, or similar physical conditions at the site;

(4) the conformation and conditions of the ground; and

(5) the character of equipment and facilities needed preliminary to and during work performance.

The Contractor also acknowledges that it has satisfied itself as to the character, quality, and quantity of surface and subsurface materials or obstacles to be encountered insofar as this information is reasonably ascertainable from an inspection of the site, including all exploratory work done by the Government, as well as from the drawings and specifications made a part of this contract. Any failure of the Contractor to take the actions described and acknowledged in this paragraph will not relieve the Contractor from responsibility for estimating properly the difficulty and cost of successfully performing the work, or for proceeding to successfully perform the work without additional expense to the Government.

(b) The Government assumes no responsibility for any conclusions or interpretations made by the Contractor based on the information made available by the Government. Nor does the Government assume responsibility for any understanding reached or representation made concerning conditions which can affect the work by any of its officers or agents before the execution of this contract, unless that understanding or representation is expressly stated in this contract.[19]

Paragraph (a) of the Site Investigation clause is essentially a certification that the contractor has conducted a visual inspection of the site and the local working environment and that the contractor is familiar with the subsurface investigation conducted by the government. It is intended to avoid claims by contractors who bid in ignorance of information that is reasonably available to them.

Paragraph (b) of the Site Investigation clause seeks to disclaim government responsibility for a contractor's interpolations, guesses, or unstated understandings as to the likely subsurface conditions. It also disclaims responsibility for any oral representations made by government representatives.[20] These limitations are consistent with the DSC clause and the general scope of the government's liability for unanticipated subsurface conditions. While a contractor will be entitled to an equitable adjustment under the DSC clause if the conditions encountered on the site differ materially from those in the contract documents or from those ordinarily encountered in the particular type of work undertaken, subjective expectations and understandings will not support a DSC claim.

D. The Physical Data Clause

Another important clause that must be considered as part of any discussion of differing site conditions is the Physical Data clause. FAR 36.504 requires a Physical Data clause whenever the government includes site information in a fixed-price construction contract. FAR 36.504 provides:

> The contracting officer shall insert the clause at 52.236-4, Physical Data, in solicitations and contracts when a fixed-price construction contract is contemplated and physical data (*e.g.*, test borings, hydrographic data, weather conditions data) will be furnished or made available to offerors.

The Physical Data clause at FAR 52.236-4 specifies the type of data that is available for inspection and states that the government will not be responsible for the contractor's interpretation of it. It provides:

PHYSICAL DATA (APR 1984)

Data and information furnished or referred to below is for the Contractor's information. The Government shall not be responsible for any interpretation of or conclusion drawn from the data or information by the Contractor.

(a) The indications of physical conditions on the drawings and in the specifications are the result of site investigations by _____ [*insert a description of investigational methods used, such as surveys, auger borings, core borings, test pits, probings, test tunnels*].

(b) Weather conditions _____ [*insert a summary of weather records and warnings*].

(c) Transportation facilities _____ [*insert a summary of transportation facilities providing access from the site, including information about their availability and limitations*].

(d) _____ [*insert other pertinent information*].

When it is included in a construction contract, the Physical Data clause advises contractors as to the availability of additional information that may be relevant to their analysis of the expected subsurface conditions. The first paragraph states that subsurface data "identified below" is for the contractor's information. It then repeats the general disclaimer of liability for a contractor's interpretation of that data.

The lettered paragraphs in the boilerplate Physical Data clause specify the general type of information that would normally be furnished to the contractor. Paragraph (a) gives the government an opportunity to describe the geotechnical investigation that may have been conducted or the source of the subsurface data furnished as part of the contract documents. Paragraph (b) would contain weather-related data. Paragraph (c) would contain information about transportation issues that might affect the construction, such as geological or physical limitations imposed by the location of the work. And Paragraph (d) is the catchall provision that would include information not easily placed in a more specific category.

In recent years, the Physical Data clause has taken a prominent role in government efforts to avoid liability for differing site conditions. One illustrative case involved a contractor's claim for the additional costs of dewatering a Corps of Engineers construction site.[21] The contractor's bid was based exclusively on its pre-bid site visit and on the boring logs that were furnished in the contract documents. In defending the claim, the government argued that a set of gradation curves available for inspection during the pre-bid period would have revealed the dewatering problem. The government argued that the claim failed because the contractor had failed to inspect the gradation curves. In support of its argument, the government relied on language in the Physical Data clause, which stated that "soil test results" were available for inspection. The court rejected the claim because the contractor had failed to inspect the gradation curves, even though that term did not appear anywhere in the contract documents. The court reasoned that language in the Physical Data clause indicating that "soil test results" were "available for inspection" invoked the contractor's duty to inspect all available subsurface information, whether it is furnished as part of the contract documents or not.

IV. Practical Issues in Differing Site Conditions Cases

The DSC clause shifts the risk of unforeseen conditions to the government. "The purpose of the changed conditions clause is thus to take at least some of the gamble on subsurface conditions out of bidding."[22] DSC clause interpretations by the Court of Claims, the Court of Federal Claims, and the Federal Agency Boards of Contract Appeals have developed the general principles summarized below.

A. Notice

When a differing site condition claim is made, notice is the first issue that confronts the contractor, the government, and their counsel. The question is whether the contractor timely notified the government of a differing site condition and, if not, whether that affects the contractor's rights under the clause.

By its terms, the DSC clause requires the contractor to notify the CO in writing of a differing site condition "promptly, and before the conditions are disturbed."[23] Of course, it is best for the contractor literally to comply with the text of the clause and to give written notice of the differing site condition immediately after it is discovered. It is also best to leave the condition undisturbed until the CO responds and directs the contractor how the government wants the contractor to proceed. This scenario is not always possible in real life.

The contractor may not realize right away that site conditions differ materially from those indicated in the contract or in the description of work to be performed. The contractor may spend a long time and a lot of money continuing the work, trying various methods and equipment to try to overcome the site conditions. The contractor may not determine that it encountered a differing site condition until some or all of the work has been performed. The contractor may then be in a position of trying to assert a differing site conditions claim without being able to prove literal compliance with the notice requirement. Even if the contractor provides written notice, the CO may not respond to it in a timely fashion or may not investigate the claim. Sometimes the CO is far removed from the day-to-day activities on the job site. Government personnel at the job site may believe that no differing site condition exists and thus may tell the contractor to get on with the work. The contractor then may be in a quandary. Should the contractor disturb the conditions and proceed with the work as directed by job site personnel, even though the CO has not responded to the written notice? Or should the contractor continue to await direction from the CO and ignore the directives of government job site personnel to continue working?

The DSC clause requirement for written notice raises a series of potential government defenses:

1. The contractor gave no notice at all.
2. The contractor's notice was not "prompt."
3. The contractor's notice was not furnished "before the conditions [were] disturbed."
4. The notice was given orally, not in writing.
5. The notice was given to a third-party engineer, site superintendent, or project manager, rather than to the CO.

As a practical matter, technical arguments like these normally are rejected unless the lack of notice results in actual prejudice to the government.[24] The purpose of the notice requirement is to ensure that the government has an appropriate opportunity to change the design, select alternative construction techniques, or take other measures to mitigate the expense of completing the work in adverse subsurface conditions.[25] As long as this purpose is accomplished—whether through oral notice by the contractor or observation by government employees—the government's notice arguments usually will be unsuccessful.

A contractor should give written notice to the CO of a differing site condition as soon as possible. However, if the contractor has not done so, the Contracting Officer then must determine whether the government has been prejudiced by the lack of written notice. If government job-site personnel were aware of the condition and its effect on the contractor's work, the lack of written notice should not be fatal to a DSC claim.

B. The Condition Must Predate the Contract

The DSC clause applies only to conditions that existed at the time of contract formation. In 1942, the Court of Claims held that the clause does not provide for an equitable adjustment for events that take place during contract performance.[26] In a subsequent decision, the Court of Claims likewise rejected a differing site conditions claim even though a hurricane and flood took place during execution of the contract. The court found that, at the time of contract formation, the bed of the river was as stated in the contract.[27] Similarly, in *John McShain Inc. v. United States*,[28] the Court of Claims held that a water main break after the formation of the contract was not a differing site condition. To be covered by the DSC clause, the condition must exist when the contract is formed.

C. The Condition May Be Natural or Man-Made

1. Weather and Acts of God

The DSC clause does not cover weather or acts of God that take place during performance of the contract. "[C]onditions resulting from unusually severe weather generally do not constitute changed conditions pursuant to which a contractor is entitled to an equitable adjustment under the DSC Clause."[29] Although some other contract provisions may address the consequences of weather on the time and cost of performing the work, the DSC clause does not shift the risk of severe weather to the government.[30]

For example, in *Overland Elec. Co.*,[31] the Armed Services Board of Contract Appeals (ASBCA) held that a differing site condition does not exist where unusually harsh weather froze the subsurface soil. Similarly, in *E.W. Jackson Contracting Co.*, the ASBCA held that surplus water at the site resulting from a hurricane was not a differing site condition.[32]

Weather conditions do not normally constitute differing site conditions because weather occurs during contract performance.[33] Severe weather alone does not establish a differing site condition.[34] Hurricanes, excessive rainfall, frozen ground, and heavy snow are all severe weather conditions for which differing site conditions claims have been rejected.[35] Such conditions may, however, be grounds for a time extension under the excusable delay provisions in the Default Termination clause.

2. Artificial Conditions

The DSC clause covers artificial conditions that exist at the time the contract is formed. A Type I differing site condition exists, for example, when the contract indicates the locations of utility lines and the contractor encounters them in different locations.[36] Unknown utility lines also have been held to be Type II differing site conditions when they were unknown and unusual and increased the costs and time of performance.[37] Likewise, "[a]sphalt, abandoned foundations, concrete, rebar, and debris" have been found to be Type II conditions when the work was done in an area where artificial materials were not reasonably expected.[38]

3. Interaction of Site Conditions and Weather

A differing site condition may occur when weather during contract performance combines with an existing condition to create a condition that was unforeseeable and affects the contractor's work during contract execution. For example, a differing site condition was found when a road disintegrated due to the unanticipated effect of spring thaw on the known soils.[39] The contractor was entitled to an equitable adjustment because the severe weather prevented him from achieving the specified soil compaction. Similarly, a differing site condition exists when unusually heavy rains flood a site because of previously unknown inadequacies in the drainage system.[40] In another case, the court found a differing site condition when unusual conditions resulted from the combination of unanticipated water on known clay.[41]

D. The Condition Must Be "at the Site"

By its terms, the DSC clause covers only conditions "at the site." However, the clause has been interpreted to cover areas that are off-site, such as borrow pits and quarries, if the contract requires the contractor to use them. The clause does not cover such off-site areas that merely are made available to the contractor but are not required to be used. In *L.G. Everist, Inc. v. United States*,[42] the Court of Claims denied relief under the DSC clause because an off-site quarry "was not designated or even mentioned by the contract." Thus, it was not "at the site" for purposes of the clause. However, when the off-site area "is necessarily so bound up with the contractor's performance [that the government] should be responsible for the conditions," the area will be covered by the DSC clause. A differing site condition also may be found when the contract documents indicate that borrow pits or quarries are approved sources of material.[43]

A borrow area will be considered "at the site" and covered by the DSC clause if either (1) the contract requires the contractor to use the borrow area,[44] or (2) the contract designates it as an "approved source" of borrow.[45] Where the contract merely gives the contractor the option of using a borrow area and does not affirmatively state that the resulting borrow will be approved, the borrow area is not considered "at the site." For example, a differing site conditions claim was denied where the contract designated several potential borrow areas but stated that the contractor was required to determine whether the borrow material was suitable.[46]

The term "at the site" has particular significance when the contract imposes "pay limits" or a "pay template" for excavation or for dredging. In *Weeks Dredging & Contracting, Inc. v. United States*,[47] the Claims Court interpreted "at the site" to mean only the area within a pay template for dredging. The contractor alleged that during dredging, it encountered unanticipated gravel and clay not indicated in the contract documents. The contractor admitted that it dredged inside and outside the pay template and failed to prove that it encountered the material within the pay template rather than outside it. The court denied the claim as a result of that failure of proof.

E. Interpretation of "Differing Materially"

The contractor must establish that the conditions at the site "differ materially" from the conditions indicated in the contract—a Type I condition—or from conditions that normally would be expected in similar work—a Type II condition.

To prove a material difference, the contractor must at least show that the anticipated conditions were more favorable than those actually encountered.[48] Some decisions measure the materiality of the difference by the amount of additional work required.[49] Other decisions state that a change in design necessary to overcome an unanticipated condition also proves a "material difference."[50] The determination of materiality is to be made on the basis of objective criteria—the contrast between actual conditions and those that a reasonable contractor would have expected.[51] It may not be judged subjectively. For example, the Corps of Engineers Board of Contract Appeals rejected a Type II differing site condition claim based not on objective criteria but, rather, on the fact that the "hard" materials encountered on the project had never before been encountered by a dredging contractor with 100 years experience.[52]

Quantitative comparisons of the actual conditions with the expected baseline can be used to prove that a condition differs materially. A differing site conditions claim must describe what the conditions would be if they were as indicated in the contract or under normal conditions.[53] For instance, in *Northwest Painting Services, Inc.,*[54] the contractor could not prove that the site differed materially because he was unable to prove that depressions and chipping caused by sandblasting were deeper than the amount to be reasonably expected.

Some court decisions allow the contractor to prove that a condition differs materially by showing that a different performance method had to be used than originally anticipated.[55] Conditions may be materially different if the contractor demonstrates prolonged exertion to overcome the situation.[56] For example, if the contractor can prove that there are additional costs as a result of the unforeseen site condition, the contractor can receive an equitable adjustment.[57] The determination of materiality depends upon all the facts of each specific case.

V. Proving a Type I DSC Claim

As stated in the DSC clause, a Type I differing site condition exists when a contractor encounters "[s]ubsurface or latent physical conditions at the site which differ materially from those indicated in this contract."[58] Despite this simple formulation in the text of the clause itself, proof of a Type I claim requires a contractor to meet a four-part test:

1. The conditions indicated in the contract differ materially from those actually encountered during performance.
2. The conditions actually encountered were reasonably unforeseeable based on all information available to the contractor at the time of bidding.
3. The contractor reasonably relied upon its interpretation of the contract and contract-related documents.
4. The contractor was damaged as a result of the material variation between expected and encountered conditions.[59]

Proof of a differing site condition is not a question of fault or negligence, but one of risk allocation. The contractor does not have to prove that the site investigation was inadequate or that the contract documents were negligently or unreasonably prepared.[60]

A. Requirement for "Indications" in "Contract Documents"

The baseline against which actual conditions are compared in a Type I claim is the affirmative description of the expected subsurface conditions that appears in the contract documents. Specifications, contract drawings, bidding documents, and the materials in these items are considered contract documents, as are all documents specifically identified as such. "Contracts and documents are to be read as a whole, so as to give meaning to all provisions."[61]

The contract documents typically include the results of any subsurface investigation conducted by the government before awarding the construction contract. This would include, for example, boring logs and test pit results, geotechnical reports, and in some cases the results of soil tests. Information furnished as part of the contract documents may include information derived from earlier construction activities in the area or affirmative statements as to the historical uses of the property. A contractor that encounters unexploded ordnance in an area previously used as an Army live-ammunition testing range, for example, would not likely prevail on a Type I claim.

The question of what constitutes a contract document for purposes of the DSC clause has been a frequent subject of dispute. Contractors typically argue that the government should furnish all available subsurface data as part of the contract documents so that all bidders may compete on an equal playing field. But some agencies have adopted a strategy of limiting the information furnished to contractors in an effort to reduce their exposure to Type I claims.

Several significant court decisions have adopted a broad definition of contract documents. The court in *Randa/Madison Joint Venture III v. Dahlberg*[62] held that bidders have a general duty to inspect not only information in the contract itself but also any subsurface data identified in the contract documents.

When the contract identifies the availability of additional subsurface data that may be relevant to anticipated subsurface conditions, contractors bear the responsibility for reviewing that information, even if it is not itself a contract document. The claim in *Randa/Madison* was rejected, at least in part, because the contractor had failed to inspect a set of gradation curves that described the permeability of the on-site soils in greater detail than did the boring logs alone. The contract did not use the term "gradation curves," but the contractor was expected to review them because the contract documents alerted it to the availability of "soil test results."

In *Comtrol, Inc. v. United States*,[63] the contractor asserted that quicksand encountered during excavation constituted a Type I differing site condition. In support, the contractor pointed to language in the contract documents as to the conditions of the soil and the likely construction methods. The contractor also asserted that the geotechnical report supported the conclusion that quicksand was not expected. The court flatly rejected the claim because the contractor had never reviewed the geotechnical report. The court reasoned that the geotechnical report was a contract document because it had been identified in the solicitation. The contractor's failure to review the report barred its claim.

In *Renda Marine, Inc. v. United States*,[64] the Federal Circuit affirmed denial of a Type I claim because the contractor failed to prove that the contract documents represented the soil conditions and because the contractor did not prove that it could not have reasonably foreseen the conditions it encountered in performing the work. The case involved a contract to dredge the Houston-Galveston navigation channel. The contractor claimed additional costs for dredging "stiff" clays rather than the soft clay the contractor argued was represented in soil borings included in the contract documents. The court noted that the soil borings showed soft clay at elevations much lower than the elevations the contract required the contractor to dredge and that some logs showed stiff clay at the elevations at which Renda encountered it. The court held that the soil borings included in the contract documents could not reasonably be interpreted as representing that the contractor could expect to hit soft clay in the dredging area and that, as a result, the contractor could not prove that the stiff clays were not reasonably foreseeable.

The Federal Circuit's decisions hold that interpretation of the contract documents is a question of law for the judge to decide. But certain board decisions appear to treat the matter sometimes as a question of law and sometimes as a question of fact. In *Billington Contracting, Inc.*,[65] for example, the ASBCA granted the government's motion for summary judgment on the contractor's differing site conditions claim. The board held the claim was barred as a matter of law because the contractor failed to review records of previous dredging contracts, which were available for review in government offices.

In *American Renovation & Construction Co.*,[66] on the other hand, the ASBCA concluded that there were factual issues concerning the effect on the contractor's claim of contract provisions other than the DSC clause. The contract was for the design and construction of 60 units of family housing at Mountain

Home Air Force Base in Idaho. The contract included the DSC clause, and the solicitation included geotechnical and test pit information showing no rock in the first two feet below grade. Paragraph 5.2.1 of the contract stated that rock encountered above that depth would be treated as an extra under the Changes clause.[67] The contract was silent as to payment for rock encountered below that depth. The contractor argued that rock below two feet should also be treated as an extra, and it claimed that during the pre-proposal site visit the government representative told bidders that all rock encountered would be treated as an "extra." The government argued that the contract did not represent conditions below two feet, thus precluding the contractor's Type-1 DSC claim based on those conditions. It also argued that the contract clearly did not provide any compensation for rock below two feet. Alternatively, the government argued that the contract was patently ambiguous about such rock and that the contractor should have inquired but failed to inquire about the issue, thus precluding any claim for additional compensation. The board denied the government's motion, stating:

> Under FAR 1.4, contracting officers are prohibited from including a provision in a solicitation or contract which is inconsistent with a mandatory FAR clause without obtaining a deviation. The record does not reflect whether the government obtained a deviation in connection with paragraph 5.2.1 or include any information regarding the development, intended scope or prior usage, if any, of the clause. On this record, we cannot determine whether paragraph 5.2.1 impermissibly abrogates the [DSC] clause. (citations omitted)[68]

B. Types of Contract Indications

A Type I differing site condition is established by comparing the actual conditions encountered during construction with the anticipated conditions identified in the contract documents. This may be easily accomplished if the contract documents clearly state the expected subsurface conditions or include a geotechnical report or a set of boring logs for the site. Often the contractual indications are not so clear. The question then arises whether the contract made any affirmative representation at all or whether an affirmative representation may be implied.

1. Express Indications

As stated in the DSC clause, a Type I differing site condition exists if the actual conditions encountered during construction differ materially from those indicated in the contract documents. The simplest differing site conditions situations thus involve cases where the contractual representations as to the expected conditions are clearly and affirmatively stated. A contractor that encounters subsurface rock on-site, for example, encounters a Type I differing site condition if the contract expressly states that the subsurface materials are limited to sand and gravel.[69] In that case, the rock encountered during construction differs materially from the sand and gravel identified in the contract documents. This type of affirmative statement is commonly referred to as an "express indication" as to the expected subsurface conditions.

Another type of express indication arises when the contract documents affirmatively state that particular conditions will *not be* encountered. In these cases, a Type I differing site condition exists when the contractor encounters them. A contractor that encounters artesian water pressure encounters a Type I differing site condition, for example, when the geotechnical report referenced in the contract documents affirmatively states that there will be no artesian water at the site.[70]

Equivocal language as to the expected subsurface conditions may not be sufficient to support a Type 1 claim. In *Kato Corp. v. Roche*,[71] the contract documents provided that "[it] is not anticipated that contaminated soil will be encountered." The court held that this language identified only the government's expectations. It was not an affirmative representation that there would be no contaminated soil.

2. Implied Indications

Type I differing site conditions claims also may arise from implied indications in the contract documents. The relevant legal inquiry in such a case is whether a reasonable bidder was justified in relying on "reasonably plain or positive" indications of the anticipated subsurface conditions, even if those indications are not expressly stated. A "reasonably plain" indication means that the contract documents provide sufficient grounds to justify the bidder's expectations.[72] "[A]ll that is required is that there be enough of an indication on the face of the contract documents for a bidder reasonably not to expect subsurface or latent physical conditions at the site differing materially from those indicated in this contract."[73]

Since implied indications are judged from the perspective of a reasonable bidder, industry practice is a key factor in determining whether particular contract documents contain an implied indication as to the anticipated subsurface conditions. In *Titan Atlantic Construction Corp.*, for example, the board held that, when the contract documents indicate that soil will have a "California Bearing Ratio" of 10, it is implied that the soil would be easy to compact.[74] The term "California Bearing Ratio" has a specific meaning to the industry, and the board concluded that the use of the term allowed the contractor to use a simple logical process to determine what subsurface conditions to expect. The board sustained the Type I differing site conditions claim because the actual site conditions differed materially from those that were impliedly indicated in the contract documents.

Implied indications may also arise from statements as to potential uses of the on-site materials or from specified construction techniques. An example of this type of implied indication appears in *Sierra-Pacific Builders*,[75] which involved a contract that specified a particular blasting method for tightly joined rock. The board held that the contractor could reasonably assume that the use of this blasting method would leave a smooth rock surface after blasting. The board sustained the contractor's Type I claim when the resulting surface was not smooth.

An implied contract indication similarly exists if the omission of any reference to a particular subsurface condition would lead a reasonable bidder to believe that it will not be encountered. In *Rottau Electric Co.*,[76] for example, the contractor properly assumed that no subsurface concrete structures would be encountered because they were not identified in the contract documents. But one piece of information in a contract can provide only so much of an "indication." In *Stuyvesant Dredging Co. v. United States*,[77] the contractor argued that the contract indications as to the average density of the materials constituted an implied indication as to the dredging difficulty. The court rejected the argument. Among other things, the contractor had ignored other relevant information about the gravel content, which also has an effect on dredging difficulty and productivity.

The absence of subsurface data will not always support the conclusion that subsurface conditions are favorable. In *Servidone Construction Corp. v. United States*,[78] for instance, the contract documents made no representations as to the soil strength. The court held that this omission did not reasonably constitute any implied guarantee as to the expected strength of the soil.

C. Foreseeability

To prove a Type I differing site condition, a contractor must show that the actual subsurface conditions were reasonably unforeseeable. This inquiry is an objective one. Essentially, it asks whether a reasonable and prudent contractor would have anticipated the actual subsurface conditions encountered at the site.[79] As part of the analysis, the contractor is charged with knowledge of everything in the contract documents and all additional subsurface data identified in the contract documents, along with information that could have been gained during a visual inspection of the site.[80] No differing site condition exists if a reasonable bidder would have foreseen the actual conditions encountered.[81]

The application of the foreseeability test is illustrated by the decision in *Stuyvesant Dredging Co. v. United States*.[82] In this case, which involved a dredging contractor, the contract documents included an express requirement that bidders examine records of previous dredging projects. The court concluded that this express requirement made it unreasonable to rely solely on the contract indications: "Such an express instruction to bidders obligated them to perform the necessary investigation to decide for themselves the character of the materials to be removed." Had the contractor conducted the investigation mandated by the contract documents, it would have foreseen the difficult dredging conditions encountered on the site.

Observations made during a pre-bid site visit may also defeat a contractor's foreseeability argument. In *Weeks Dredging & Contracting Inc. v. United States*,[83] the contractor had observed "extensive commercial gravel operations" near the site where the contract required dredging. Because the gravel content of the material at the site was critical to anticipated production rates, the court held that the contractor's failure to inquire further into the commercial gravel operations was unreasonable. In the court's words, a contractor cannot rely solely on boring logs when other information "would put a reasonable and prudent contractor . . . on notice that there may be subsurface conditions different than those indicated in the contract boring logs." Having failed to make the additional inquiry that a reasonable contractor would have made, the contractor could not argue that the subsurface conditions were unforeseeable.

D. Reasonable Reliance

A contractor can establish a Type I differing site condition only if it can prove that it relied on the contractual indications as to the anticipated subsurface conditions and that this reliance was reasonable.[84] This is a two-part requirement. First, the contractor must, in fact, have reviewed and used the contractual subsurface data in the preparation of its bid. Second, its reliance on the information must have been reasonable. In other words, the subsurface data must be the type of information that a reasonable and prudent bidder would use to prepare a bid.

The reasonableness of a contractor's reliance is judged from the perspective of a hypothetical prudent bidder.[85] A contractor that is new to a construction discipline or a geographical area is expected to be at least as educated as others in the industry. The contractor is not required to know all that an expert (such as a geologist or a geotechnical engineer) would know,[86] nor is it required to retain an expert or to consult expert treatises in preparing a bid.[87]

The contractor generally has no obligation to investigate or to verify the accuracy of contract indications or to undertake an independent investigation of the site conditions.[88] On the other hand, the contractor must review all subsurface data included in the solicitation, as well as other relevant information cited in the solicitation. In *Comtrol, Inc. v. United States*,[89] for example, the contractor argued that quicksand was a differing site condition because the contract documents indicated only dense and cohesive clays and weathered rock. The court denied the claim because the contractor had failed to review the soils report that was made available for inspection at the offices of the government's architect. Without having reviewed the soils report, the court held that the contractor could not establish the reasonable reliance element of its differing site conditions claim.

Beyond this general duty of inquiry, contractors must investigate obvious errors in the contract documents,[90] such as conflicts between drawings or between boring logs and the geotechnical report. The contractor will not be able to recover under the DSC clause if there is a patent ambiguity that negates reasonable reliance on the soil descriptions included in the contract.[91] A contractor's reliance upon contract indications may be unreasonable if "relatively simple inquiries might have revealed contrary conditions."[92] Type I differing site conditions claims have been denied when the contractor could have learned of actual conditions merely by asking for clarification of information in the contract documents.[93]

The general rule is that a contractor may reasonably rely upon positive statements in bidding documents despite general cautionary language in other parts of the documents.[94] The government has nevertheless taken steps in recent years to disclaim responsibility for differing site conditions through disclaimer language, and some court decisions can be read to support this effort. In *Millgard v. McKee/Mays*,[95] the court enforced language disclaiming all government liability for Type I differing site conditions. The decisions in *Randa/Madison Joint Venture III v. Dahlberg*[96] and *Comtrol* also give a broad reading to the contractor's duty to inquire beyond the contract documents. These developments mark a departure from the risk allocation model originally contemplated by the DSC clause. While there are sound reasons for the results in these particular cases, it is not clear that they are consistent with the intended purpose of the DSC clause.[97] In particular, they limit the scope of the contractor's right to rely on subsurface data furnished as part of the contract.

In *Kilgallon Construction Co., Inc.*,[98] the ASBCA granted a Type I claim for "cement treated base" underlying asphalt paving but not disclosed in the contract documents. The board found that the contract documents characterized the soil to be excavated as "hard crust," which is materially different from the "cement treated base" encountered, that a site visit would not reveal the cement treated base under existing asphalt, that in its bid the contractor relied upon the contract description of the materials and that the contractor was damaged as a result.

VI. Proving a Type II DSC Claim

Though similar in many respects to Type I claims, Type II claims are different because they extend beyond the contract documents. To prevail on a Type II claim, the contractor must show not only that the conditions were not indicated in the contract but also that they were more generally unknown, unusual, and unforeseeable.[99] One court articulated the contractor's burden of proof on a Type II claim in three parts:

> The contractor must show three elements to prove a Type II claim. First, [the contractor] must show that it did not know about the physical condition. Second, [the contractor] must show that it could not have anticipated the condition from inspection or general experience. Third, [the contractor] must show that the condition varied from the norm in similar contracting work.[100]

A Type II claim thus compares the actual conditions encountered on the site with the subsurface conditions normally expected in the type of work undertaken.[101] Unanticipated subsurface conditions do not alone present a basis for a Type II differing site condition claim. The conditions must also be unusual.[102]

Generally, a Type II claim is more difficult to prove than a Type I claim.[103] A contractor asserting a Type II claim is "confronted with a relatively heavy burden of proof."[104] Proof of an unusual subsurface condition will often depend on expert testimony. The contractor will have to present evidence about the usual and ordinary conditions expected in the particular type of work. This evidence forms the baseline against which the actual conditions are measured. If the actual conditions are materially different from the normal conditions encountered in that type of work and they are sufficiently unusual, an equitable adjustment is in order.

A. Requirement That the Site Condition Be Unknown

The first requirement for a Type II claim is that the condition be unknown to the contractor at the time of bid. This is a fact-based inquiry to determine whether the contractor had actual knowledge of the subsurface conditions that were encountered. This inquiry typically involves the testimony of the individuals involved in reviewing the bid documents, visiting the site, and preparing the bid. It may also call for a review of the work papers developed by the contractor in connection with the bid and a review of the contractor's previous experience in the area or on the particular site. A contractor that has completed several similar projects in a small geographical area will have difficulty arguing that it did not have actual knowledge of the likely subsurface conditions.

Contractors should not try to preserve a Type II claim by shielding themselves from available subsurface information, as this technique is rarely effective. Contractors have an affirmative duty to review all of the information in the contract documents, all of the relevant information that is "available for inspection," and all of the information that would be available from a reasonable investigation of the site. If the contract documents describe the condition or if they refer to information available for the contractor's inspection, the condition is not unknown.[105]

B. Requirement That the Site Condition Be Unforeseeable

The second element of a Type II claim is that the actual conditions encountered during construction were unforeseeable. This determination goes beyond the contractor's actual knowledge and asks whether a reasonable and prudent bidder would have foreseen the subsurface conditions that were encountered. Certainly, a condition is not unforeseeable if a reasonable site inspection would have revealed it.[106] However, the contractor is not required to "antici-pate the worst."[107] Where a reasonable site inspection would not reveal the conditions, this is one factor in determining them to be unforeseeable.[108]

However, the general expertise of experienced contractors in the type of work at issue and the general knowledge of local contractors is also relevant to the question of foreseeability. A condition that would have been expected by an experienced contractor or by a member of the local community is foresee-able, even if it would not have been revealed by a site inspection.[109]

C. Requirement That the Condition Be Unusual

The final element of proof for a Type II claim is that the unanticipated sub-surface conditions must be unusual. Although quantitative differences and unusual soil characteristics may constitute a valid Type II differing site condi-tion,[110] the question of whether conditions are "unusual" is typically treated as a qualitative question. A marginally lower production rate or marginally less favorable soil compaction characteristics, for example, are not normally considered unusual and would not normally form the basis for a Type II dif-fering site condition.[111]

The requirement that Type II differing site conditions be unusual is best illustrated by examples. Unusual subsurface conditions existed where a con-tractor hired to clean air ducts encountered "beer cans and jars of jam to gunpowder, live ammunition" and other obstructions.[112] Because typical con-struction techniques use bricks only on exterior surfaces, one board found an interior brick wall to be unusual.[113] But under other circumstances—for example, the time period in which the building was constructed or the type of construction—an interior brick wall might be considered ordinary and expected.[114] In another case, "Asphalt, abandoned foundations, concrete, rebar, and debris" were Type II conditions when the work was done in an area where artificial materials were not reasonably expected.[115]

Notes

1. Foster Constr. C.A. v. United States, 435 F.2d 873, 887 (Ct. Cl. 1970).

2. The current clause is entitled "Differing Site Conditions (APR 1984)," FAR 52.236-2.

3. *Foster Constr.*, 435 F.2d at 887.

4. The Lev Zetlin Engineers Report to the General Services Administration in the late 1970s reached this conclusion after addressing in detail a series of large construction projects, all with sizable differing site conditions claims.

5. 248 U.S. 132 (1918).

6. *Id.*

7. Helene Curtis Indus. v. United States, 312 F.2d 774, 778 (Ct. Cl. 1963); *see also* Eshelman & Sanford, *The Superior Knowledge Doctrine: An Update*, 22 Pub. Cont. L.J. 477 (1993).

8. GAF Corp. v. United States, 932 F.2d 947, 949 (Fed. Cir. 1991), *cert. denied*, 502 U.S. 1071 (1992); Lopez v. A.C. & S., Inc., 858 F.2d 712, 717 (Fed. Cir. 1988); Am. Ship Bldg. Co. v. United States, 654 F.2d 75, 79 (Ct. Cl. 1981).

9. Hollerbach v. United States, 233 U.S. 165 (1914); *see also* Christie v. United States, 237 U.S. 234 (1915); United States v. Atl. Dredging Co., 253 U.S. 1, 10–11 (1920).

10. 523 F.3d 1341 (Fed. Cir. 2008).

11. 509 F.3d 1372 (Fed. Cir. 2008).

12. *Id.*

13. *Id.*

14. H.B. Mac, Inc. v. United States, 153 F.3d 1338 (Fed. Cir. 1998).

15. *See generally* 4 BRUNER & O'CONNOR ON CONSTRUCTION LAW, § 14:1 (2002). The Changed Conditions clause was cited in the following early cases: Joseph Meltzer, Inc. v. United States, 77 F. Supp. 1018 (Ct. Cl. 1948) (1933 contract); Cassidy & Gallagher v. United States, 95 Ct. Cl. 504 (1942) (1934 contract); Hirsch v. United States, 94 Ct. Cl. 602 (1941) (1928 airport construction contract).

16. See Section III.A. below.

17. See Section III.A. below.

18. FAR 36.503

19. FAR 52.236-3.

20. *See* N. Slope Technical Ltd. v. United States, 14 Cl. Ct. 242 (1988) (government warnings at pre-bid meeting not sufficient to notify contractor of adverse subsurface conditions).

21. Randa/Madison Joint Venture III v. Dahlberg, 239 F.3d 1264 (Fed. Cir. 2001).

22. Foster Constr. C.A. v. United States, 435 F.2d 873, 887 (Ct. Cl. 1970).

23. FAR 52.236-2.

24. Power Contracting & Eng'g Corp. v. Gen. Servs. Admin., GSBCA No. 12741, 96-1 BCA ¶ 28,125 (six-month delay prejudicial because it prevented verification and correction actions); T. Brown Constructors, Inc., DOTBCA No. 1986, 95-2 BCA ¶ 27,870 (prejudice resulting from lack of opportunity to verify site conditions encountered).

25. Whiting-Turner/A.L. Johnson v. Gen. Servs. Admin., GSBCA No. 15401, 02-1 BCA ¶ 31,708.

26. Arundel Corp. v. United States (*Arundel I*), 96 Ct. Cl. 77 (1942).

27. Arundel Corp. v. United States (*Arundel*), 103 Ct. Cl. 688, *cert. denied*, 326 U.S. 752 (1945).

28. 375 F.2d 829 (Ct. Cl. 1967).

29. Turnkey Enters., Inc. v. United States, 597 F.2d 750 (Ct. Cl. 1979).

30. Hardeman-Monier-Hutcherson, A Joint Venture, ASBCA No. 12392, 68-2 BCA ¶ 7220.

31. ASBCA No. 9096, 1964 BCA ¶ 4359.

32. ASBCA No. 1606, (1954) 1650 WL 194; *see also Arundel II*, 103 Ct. Cl. 688.

33. *See* F.E. Booker Co., ASBCA No. 15767, 71-2 BCA ¶ 9025.

34. *Id.*

35. *See* Concrete Constr. Corp., IBCA No. 432-3-64, 65-1 BCA ¶ 4520.

36. Jay & Saw Constr., Inc., ASBCA No. 9812, 65-1 BCA ¶ 4858, Dale Constr. Co. v. United States, 168 Ct. Cl. 692 (1964).

37. VEC, Inc., ASBCA No. 35988, 90-3 BCA ¶ 23,204.

38. Parker Excavating, Inc., ASBCA No. 54637, 06-1 BCA ¶ 33,217.

39. *See* D.H. Dave & Gerben Contracting Co., ASBCA No. 6257, 1962 BCA ¶ 3493.

40. Phillips Constr. Co. v. United States, 394 F.2d 834 (Ct. Cl. 1968).

41. Paccon, Inc., ASBCA No. 7643, 1962 BCA ¶ 3546.

42. 231 Ct. Cl. 1013 (1982), *cert. denied*, 461 U.S. 957 (1983).

43. *See, e.g.*, Morrison-Knudsen Co. v. United States, 397 F.2d 826 (Ct. Cl. 1949); Tobin Quarries, Inc. v. United States, 84 F. Supp. 1021 (Ct. Cl. 1949).

44. Differing site conditions claims involving required sources of borrow were sustained in Stock & Grove, Inc. v. United States, 493 F.2d 629 (Ct. Cl. 1974), and Barrett Rds. Corp., ASBCA No. 1297, 51-1 BCA ¶ 1256.

45. A differing site conditions claim involving approved sources of borrow was sustained in R.A. Heinz Constr. Co., ENG BCA No. 3380, 74-1 BCA ¶ 10,562.

46. Sims Paving Corp., DOTCAB No. 1822, 90-3 BCA ¶ 22,942.

47. 13 Cl. Ct. 193, *aff'd*, 861 F.2d 728 (Fed. Cir. 1983).

48. Pac. Alaska Contractors, Inc. v. United States, 193 Ct. Cl. 850 (1971).

49. McCormick Constr. Co. v. United States, 18 Cl. Ct. 259, *aff'd*, 907 F.2d 159 (Fed. Cir. 1990) ("Evidence of a material difference most commonly illustrates that a larger amount of work was performed . . . or that an alternative method of workmanship must be implemented").

50. Foster Constr. C.A. & Williams Bros. Co. v. United States, 193 Ct. Cl. 587 (1970) (changes in bridge design proved the materiality of the difference in conditions); *see also* Currie, Abernathy & Chambers, *Changed Conditions*, CONSTRUCTION BRIEFINGS, Dec. 1984.

51. *Pac. Alaska Contractors*, 193 Ct. Cl. 850.

52. Great Lakes Dredge & Dock Co., ENG BCA No. 5606, 91-1 BCA ¶ 23,613.

53. *See* Guy F. Atkinson Constr. Co., ENG BCA No. 4693, 87-3 BCA ¶ 19,971; Nw. Painting Serv., Inc., ASBCA No. 27854, 84-2 BCA ¶ 17,474.

54. ASBCA No. 27854, 84-2 BCA ¶ 17,474.

55. *See, e.g.*, Bick-COM Corp., VACAB 1320, 80-1 BCA ¶ 14,285; Wall St. Roofing, VABCA No. 1373, 81-2 BCA ¶ 15,417.

56. Dunbar & Sullivan Dredging Co., ENG BCA No. 3165, 73-2 BCA ¶ 10,285.

57. *See* N. Slope Technical Ltd. v. United States, 14 Cl. Ct. 242 (1988); Dawco Constr. Inc., ASBCA No. 31990, 88-2 BCA ¶ 20,606.

58. FAR 52.236-2(a)(1).

59. Comtrol, Inc. v. United States, 294 F.3d 1357 (Fed. Cir. 2002).

60. Titan Atl. Constr. Co., ASBCA No. 23588, 82-2 BCA ¶ 15,808, at 78,317.

61. Allied Tech. Group, Inc. v. United States, 39 Fed. Cl. 125, 144 (1997); B.D. Click Co. v. United States, 614 F.2d 748, 753 (Ct. Cl. 1980).

62. 239 F.3d 1264 (Fed. Cir. 2001).

63. 294 F.3d 1357.

64. 509 F.3d 1372 (Fed. Cir. 2008).

65. ASBCA Nos. 54147 & 54149, 05-1 BCA ¶ 32,900.

66. Am. Renovation & Constr. Co., ASBCA No. 54526, 06-1 BCA ¶ 33,156.

67. FAR 52.243-4.

68. *Am. Renovation*, 06-1 BCA ¶ 33,156.

69. Dunbar & Sullivan Dredging Co., ENG BCA No. 3165, 73-2 BCA ¶ 10,285.

70. Ilbau Constr., Inc., ENG BCA No. 5465, 92-1 BCA ¶ 24,476.

71. No. 02-1315, 2003 WL 475671 (Fed. Cir. Feb. 21, 2003).

72. P.J. Maffei Bldg. Wrecking Corp. v. United States, 732 F.2d 913, 916 (Fed. Cir. 1984).

73. Foster Constr. C.A. v. United States, 435 F.2d 873, 875 (Ct. Cl. 1970) (internal quotations omitted).

74. ASBCA No. 23588, 82-2 BCA ¶ 15,808.

75. AGBCA No. 78-161, 80-2 BCA ¶ 14,609.

76. ASBCA No. 20283, 76-2 BCA ¶ 12,001.

77. 834 F.2d 1576 (Fed. Cir. 1987).

78. 19 Cl. Ct. 346 (1990), *aff'd*, 931 F.2d 860 (Fed. Cir. 1991).

79. H.B. Mac, Inc. v. United States, 153 F.3d 1338 (Fed. Cir. 1998); United States v. Stuyvesant Dredging Co., 834 F.2d 1576.

80. Comtrol, Inc. v. United States, 294 F.3d 1357 (Fed. Cir. 2002); Randa/Madison Joint Venture III v. Dahlberg, 239 F.3d 1264, 1274 (Fed. Cir. 2001); Youngdale & Sons Constr. Co. v. United States, 27 Fed. Cl. 516 (1993); Mojave Enters. v. United States, 3 Cl. Ct. 353, 356–57 (1983).

81. *See, e.g., Randa/Madison*, 239 F.3d 1264.

82. 834 F.2d 1576.

83. 13 Cl. Ct. at 239.

84. *Comtrol*, 294 F.3d 1357; Dravo Corp., ENG BCA No. 3901, 80-2 BCA ¶ 14,757; Perini Corp., ENG BCA No. 4635, 86-1 BCA ¶ 18,524.

85. H.B. Mac, Inc. v. United States, 153 F.3d 1338 (Fed. Cir. 1998).

86. Pleasant Excavating Co. v. United States, 229 Ct. Cl. 654, 656 (1981).

87. N. Slope Technical, Ltd. v. United States, 14 Cl. Ct. 242 (contractor not required to hire an expert); Stock & Grove, Inc. v. United States, 493 F.2d 629 (Ct. Cl. 1974).

88. Hollerbach v. United States, 233 U.S. 165 (1914).

89. 294 F.3d 1357.

90. Beacon Constr. Co. v. United States, 314 F.2d 501 (1963); *see also* Allied Tech. Group, Inc., 39 Fed. Cl. 125, 139 (1997).

91. *See* Consultores Professionales de Ingenieria, S.A., ENG BCA No. PCC-78, 94-2 BCA ¶ 26,652.

92. Foster Constr. C.A. v. United States, 435 F.2d 873, 888 (Ct. Cl. 1970).

93. Leal v. United States, 276 F.2d 378, 384 (Ct. Cl. 1960).

94. *See, e.g.,* Carl W. Linder Co., ENG BCA No. 3526, 78-1 BCA ¶ 13,114 (citing Hollerbach v. United States, 233 U.S. 165 (1914)). *See also* N. Slope Technical Ltd. v. United States, 14 Cl. Ct. 242 (1988).

95. 49 F.3d 1070 (5th Cir. 1995).

96. 239 F.3d 1264.

97. Foster Constr. C.A. v. United States, 435 F.2d 873, 887 (Ct. Cl. 1970).

98. ASBCA No. 52583, 03-2 BCA ¶ 32,380.

99. FAR 52.236-2.

100. Latham Co. v. United States, 20 Cl. Ct. 122, 128 (1990). *See also* Fuel Tank Maint., Inc., ASBCA No. 54402, 08-02 BCA ¶ 33,888.

101. Spruce Constr., Inc., ASBCA No. 30679, 86-3 BCA ¶ 19,106.

102. Youngdale & Sons Constr. Co. v. United States, 27 Fed. Cl. 516 (1993); Kos Kam, ASBCA No. 34037, 88-3 BCA ¶ 21,000 (unknown waste lines not a basis for differing site conditions because they were not unusual for the type of work).

103. CCI Contractors, Inc., AGBCA No. 84-314-1, 91-3 BCA ¶ 24,225, *aff'd*, 979 F.2d 216 (Fed. Cir. 1992).

104. Charles T. Parker Constr. Co. v. United States, 433 F.2d 771 (Ct. Cl. 1970).

105. *Youngdale & Sons*, 27 Fed. Cl. 516; Randa/Madison Joint Venture III v. Dahlberg, 239 F.3d 1264 (Fed. Cir. 2001).

106. *See* Huntington Constr., Inc., ASBCA No. 3526, 89-3 BCA ¶ 22,150; C&L Constr. Co., ASBCA No. 22993, 81-1 BCA ¶ 14,943, *aff'd*, 81-2 BCA ¶ 15,373 (contractor should have been aware of the loosely packed soil due to the boring samples); D.J. Barclay & Co., ASBCA No. 28909, 88-2 BCA ¶ 20,741 (contractor did not perform an adhesion test, thus coatings deterioration was not unforeseeable).

107. Redman Servs., Inc., ASBCA No. 8853, 1963 BCA ¶ 3897; Kinetic Builders, Inc., ASBCA No. 32627, 88-2 BCA ¶ 20,657.

108. *See* Yamas Constr. Co., ASBCA No. 27366, 86-3 BCA ¶ 19,090 (floor under a carpet was uneven and contained excessive glue that could not have been discovered because of the carpet).

109. *See* Servidone Constr. Corp. v. United States, 19 Cl. Ct. 346 (contractor did not ask about local conditions, where this inquiry would make the contractor aware of the soil problems); CCI Contractors, 91-3 BCA ¶ 24,225 (contractor did not look for information from contractors in the area).

110. R.J. Crowley, Inc., GSBCA No. 11080 (9251)-REIN, 92-1 BCA ¶ 24,499; Paccon, Inc., ASBCA No. 7643, 1962 BCA ¶ 3546 (noting that clay had higher plasticity than expected).

111. *See, e.g.*, Jack Walser v. United States, 23 Cl. Ct. 591 (1991) (increase in beaver-generated debris not unusual).

112. Cmty. Power Suction Furnace Cleaning Co., ASBCA No. 13803, 69-2 BCA ¶ 7963.

113. Hercules Constr. Co., VABCA No. 2508, 88-2 BCA ¶ 20,527.

114. *See, e.g.*, J.J. Barnes Constr. Co., ASBCA No. 27876, 85-3 BCA ¶ 18,503.

115. Parker Excavating, Inc., ASBCA No. 54637, 06-1 BCA ¶ 33,217.

CHAPTER 10

Inspection, Acceptance, and Warranties

STEVEN L. REED
THOMAS J. KELLEHER, JR.

I. Introduction

Unlike the manufacture of goods, the construction of almost any facility, including a federal government facility, is unique or has unique characteristics or components. Many projects have one-of-a-kind architectural or engineering features that reflect the particular use and/or location of the facility. Even if a standard design is being employed, the geographic location will require site adaptation; thus, geophysical considerations, foundation design, and other constructability matters must be taken into account during design and construction. Moreover, the workforce, subcontractors, materials suppliers, and equipment sources will change over time and by geographic location.

Much of the construction work in any sizable project is performed by both major component subcontractors, such as a foundation system subcontractor, and specialty subcontractors, such as a security system subcontractor. Those subcontractors, in turn, will sometimes have lower-tier subcontractors to perform, for example, installation of pre-cast concrete or specialized electrical systems. Separate suppliers and vendors provide materials and equipment to the general (prime) contractor and subcontractors. Materials provided by others are incorporated into the work, and equipment provided by or leased from others is used to accomplish the work. However, the prime contractor remains generally responsible for the quality and timeliness of performance by its own forces and by that of subcontractors, suppliers, and vendors at every tier.

All of these factors contribute to the complexity of the construction process and introduce the potential for quality control issues and risks. Effective inspection policies, programs, and practices are crucial, even on projects that are otherwise very well managed and controlled. The inspection, acceptance, and warranty provisions of any government contract seek to address, manage, and allocate responsibilities and risks related to such issues.

A significant objective for the participants in every construction project is timely completion, within budget, and in accordance with the project's plans

273

and specifications. Contractors seek to make a profit, to obtain a favorable performance evaluation, and to sustain a positive working relationship with the government agency that owns and/or operates the project, as well as with any government construction agency that oversees design and construction efforts (e.g., the U.S. Navy's Naval Facilities Engineering Command). Given the requirement for performance evaluations and their prominent consideration in the context of best-value procurements, a contractor's ability to deliver the required level of quality, on schedule, has become increasingly critical to the ongoing viability of a government construction contractor. During performance of the work, quality control problems can disrupt the best planned schedule, can adversely impact a properly estimated budget, and can frustrate a workforce that only wants to do the job once.

Long after occupancy of a project for its intended use, quality control issues may arise in the form of warranty or latent defect claims. Consequently, every participant on a federal construction project should appreciate that inspection by the contractor of its work and inspection (or surveillance of contractor inspections) and acceptance by the federal government are integral parts of contract performance. Inspection and acceptance should be viewed as critical milestones for both the contractor and the government in their mutual desire for overall quality management of construction work and acceptability of the project.

Understanding the parties' respective rights and obligations begins with the contractual requirements for inspection, acceptance, and warranty. It is not possible to perform appropriate inspections and to determine whether the work conforms to the contract requirements without a clear understanding of the plans and specifications. Similarly, a failure to appreciate the total scope of the contract's quality control responsibilities and requirements can result in a substantial gap by the contractor in its budget for general conditions and in performing the required tests and inspections.

Not unlike the project's plans and specifications, the project's quality control requirements are often tailored or adapted to the unique requirements of the specific project. However, as with many other federal government contract clauses, the starting point is the standard clause entitled "Inspection of Construction (AUG 1996)," found in the Federal Acquisition Regulation (FAR) at FAR 52.246-12. This provision is key to understanding the parties' fundamental rights and obligations under the contract. The clause provides the basic contractual framework on numerous key topics, including:

- Contractor quality control inspection system responsibilities;
- Government surveillance of contractor inspections;
- Time, place, and extent of government inspections, if any;
- Contractual implications of government inspection;
- Authority of government inspectors, aka quality assurance representatives, project engineers, and authorized representatives of the contracting officer (CO);

- Contractor responsibilities for support of government inspections and inspectors;
- Treatment of nonconforming work; and
- Timing and consequences of acceptance.

Inspections are for the government's benefit, not the contractor's. The primary responsibility for inspection (quality control) and compliance, as well as the attendant risks of failure and noncompliance, are assigned to the contractor. As a result, the contractor typically conducts those tests and inspections. On some occasions, however, additional inspections are performed by government representatives or jointly by the contractor and the government.

Regardless of which party conducts a test or inspection, the purpose is to ensure compliance with the requirements of the contract. At the proposal stage, the determination of the appropriate cost of the work overall can be significantly influenced by an evaluation of the technical requirements, including pertinent inspection, acceptance, and warranty requirements. Such requirements can include post-acceptance warranty obligations as well as ongoing system training, operation, and performance requirements.

As with the prime contract work, it is essential to understand quality control requirements and processes as they apply to subcontracted work. Subcontracts must include comprehensive scopes of work that take into account inspection, acceptance, and warranty obligations. Purchase orders for materials or rented equipment must provide for and address such matters to the extent applicable. While the prime contractor is responsible to the government agency, any allocation of risk or recourse by a prime contractor to a subcontractor, supplier, or vendor should be clearly specified in written agreements between the prime and those entities.

During the inspection, acceptance, and warranty process, questions may develop regarding the authority of the government's inspectors, the nature of the proposed inspection or test, and the scope of any corrective action. After completion of the project, issues may arise regarding acceptance of portions of the work, warranty responsibilities, and post-acceptance responsibility for alleged defects or performance deficiencies. Subcontracts and purchase orders should make clear whether the subcontractor or vendor remains responsible to the prime contractor in such instances.

II. Inspection Clauses

A. Inspection of Construction

In federal government construction contracts, standard inspection clauses provide the framework for the inspection and acceptance process. The primary provision setting forth the contracting parties' inspection obligations and rights, the Inspection of Construction clause, FAR 52.246-12, must be included pursuant to FAR 46.312 in all fixed-price construction contracts for which the contract amount is expected to exceed the simplified acquisition threshold. The Inspection of Construction clause may be included in solicitations for other government construction contracts at the CO's discretion. FAR 52.246-12 provides as follows:

INSPECTION OF CONSTRUCTION (AUG 1996)

(a) Definition. "Work" includes, but is not limited to, materials, workmanship, and manufacture and fabrication of components.

(b) The Contractor shall maintain an adequate inspection system and perform such inspections as will ensure that the work performed under the contract conforms to contract requirements. The Contractor shall maintain complete inspection records and make them available to the government. All work shall be conducted under the general direction of the Contracting Officer and is subject to government inspection and test at all places and at all reasonable times before acceptance to ensure strict compliance with the terms of the contract.

(c) government inspections and tests are for the sole benefit of the government and do not—

    (1) Relieve the Contractor of responsibility for providing adequate quality control measures;

    (2) Relieve the Contractor of responsibility for damage to or loss of the materials before acceptance;

    (3) Constitute or imply acceptance; or

    (4) Affect the continuing rights of the government after acceptance of the completed work under paragraph (i) below.

(d) The presence or absence of a government inspector does not relieve the Contractor from any contract requirement, nor is the inspector authorized to change any term or condition of the specification without the Contracting Officer's written authorization.

(e) The Contractor shall promptly furnish, at no increase in contract price, all facilities, labor, and material reasonably needed for performing such safe and convenient inspections and tests as may be required by the Contracting Officer. The government may charge to the Contractor any additional cost of inspection or test when work is not ready at the time specified by the Contractor for inspection or test, or when prior rejection makes re-inspection or retest necessary. The government shall perform all inspections and tests in a manner that will not unnecessarily delay the work. Special, full size, and performance tests shall be performed as described in the contract.

(f) The Contractor shall, without charge, replace or correct work found by the government not to conform to contract requirements, unless in the public interest the government consents to accept the work with an appropriate adjustment in contract price. The Contractor shall promptly segregate and remove rejected material from the premises.

(g) If the Contractor does not promptly replace or correct rejected work, the government may (1) by contract or otherwise, replace or correct the work and charge the cost to the Contractor or (2) terminate for default the Contractor's right to proceed.

(h) If, before acceptance of the entire work, the government decides to examine already completed work by removing it or tearing it out, the Contractor, on request, shall promptly furnish all necessary facilities, labor, and material. If the work is found to be defective or nonconforming in any material respect due to the fault of the Contractor or its subcontractors, the Contractor shall defray the expenses of the examination and of satisfactory reconstruction. However, if the work is found to meet contract requirements, the Contracting Officer shall make an equitable adjustment for the additional services involved in the examination and reconstruction, including, if completion of the work was thereby delayed, an extension of time.

(i) Unless otherwise specified in the contract, the government shall accept, as promptly as practicable after completion and inspection, all work required by the contract or that portion of the work the Contracting Officer determines can be accepted separately. Acceptance shall be final and conclusive except for latent defects, fraud, gross mistakes amounting to fraud, or the government's rights under any warranty or guarantee.

While the above standard provision is the basic FAR contract clause defining inspection-related rights and responsibilities, government construction contracts often contain additional clauses generally addressing so-called quality control system and contractor quality control (CQC) requirements.

B. CQC Contract Specifications

Any evaluation of the parties' respective rights and obligations under a federal construction contract requires a detailed examination of its CQC provisions. In many federal government contracts, such provisions consist of extensive, multipage specifications, usually contained in Division 1 of the project's technical specifications. Such provisions may be tailored to a specific project; however, many are generic. Like the standard Inspection of Construction clause, CQC provisions also affect the parties' obligations as well as the contractor's cost for performance of work related to inspections, acceptance, and warranty.

CQC provisions provide additional definition to the obligations set forth in the Inspection of Construction clause that the contractor "maintain an adequate system" to "ensure" that the work conforms to the requirements of the contract.[1] CQC clauses may include requirements for one or more of the following:

- Use of government information technology systems such as the U.S. Army Corps of Engineers Resident Management System (RMS) for Windows;
- CQC plan submittals describing the contractor's CQC programs and procedures;
- Submittals of results of required tests and inspections;
- Government forms and formats for daily documentation of availability at the site of labor, equipment, and materials; the conduct of tests and inspections; and progress on the overall job and work features;
- Notifications of any noncompliance or test/inspection failure;
- CQC management staffing requirements and qualifications;
- CQC meetings and minutes of such meetings;
- CQC phased inspections and controls during performance of each segment of the work;
- CQC test procedures and parameters, test facilities, and/or testing sources such as laboratories; and
- CQC completion inspections, punch list preparation, and final inspection.

In addition, each separate section of the technical specifications usually requires:

- Specific inspections, tests, and acceptance criteria for the work described by that particular specification section;
- Government operation or observation of contractor operation, prior to acceptance, of equipment and systems incorporated into the project;
- Special and/or extended warranty periods or guarantees of performance within certain specified parameters; and
- CQC compliance certifications.

Unless the contractor has evaluated these requirements and factored the cost of performance of the CQC program into its budget, it may find its oversight to be extremely costly. Moreover, the failure to provide the government with a CQC plan and program that complies with the detailed specifications may justify a CO's decision to preclude the contractor from proceeding with performance.[2]

Before submitting a bid or proposal for a government construction contract, the entire contract, including every division of the specifications, should be reviewed using a checklist that addresses topics such as the following:

- Must the CQC plan be approved by the government before the government is obliged to issue the notice to proceed overall or as to any specific feature of the work?
- Is there a requirement for a separate CQC staff distinct from the contractor's project management staff?
- How many and what type of individuals must be on the CQC staff?

- What level of experience or education must each member of the CQC staff possess?
- May the members of the CQC staff have any duties in addition to their CQC functions?
- Must the members of the CQC staff be employees of the general contractor, or may they be independent contractors or employees of a subcontractor or supplier, or independent consultants or manufacturers' representatives?
- Must each member of the CQC staff be present on-site at all times when work at the site is under way, only when work pertinent to that CQC person is ongoing, or at least once each working day or for a specified time period? (Some contracts require a "full time" CQC representative while others state that the CQC staff must be present when "work is ongoing." The latter can be troubling if "work" is later interpreted to mean "any work," rather than work that falls within the technical purview of that member of the CQC staff.)
- Are there requirements for approval inspections by other entities or persons outside of the government, such as a government consulting laboratory, a federal agency that will use the facility after acceptance (e.g., medical professionals), or local cooperating entity (e.g., a local utility responsible for maintenance after acceptance by the government)? If so, is there a specified time within which such inspections must be performed or within an undefined "reasonable" time? Must all work (or the pertinent work to be tested) cease during any such inspections?

The answers to these and similar questions can directly and materially affect a contractor's (or a subcontractor's) budget for the project, pricing considerations, and negotiation of subcontracts and purchase orders. If the provisions are not clear, a prudent contractor should seek clarification as early as possible during the proposal preparation period before submitting a proposal.

Provided the scope of all applicable CQC requirements is covered in the contractor's price or cost recovery line items, a qualified CQC staff should be able to reduce costs during performance by early identification of potential defects in or deviations from the contract's requirements. Deviations include either performance to a stricter standard than required by the contract (which may constitute a compensable change if required by an authorized government official[3]) or a failure by the contractor (or a subcontractor or supplier) to meet the contract's quality standard. For example, a knowledgeable CQC representative may be able to identify, at an early stage, possible misinterpretations of the contract requirements by a government inspector. This information, combined with a thorough explanation of the contract's requirements, may enable the contractor to provide an early written notification and explanation to the government that potentially can avoid the cost and difficulties associated with disputed claims for extra work.[4]

C. Other Applicable Provisions

Inspection, acceptance, and warranty provisions may be found throughout the contract, including within notes on contract drawings, in technical specifications, and/or in standard publications that are incorporated into or referenced in the contract. An evaluation of the contract should not be limited to provisions specifically designated as inspection, acceptance, or warranty clauses. Instead, the entire contract should be read and analyzed. An attempt must be made to reconcile all provisions that address inspection, acceptance, and warranty matters related to certain work. If reconciliation is not possible by a reasonable reading of the contract in its entirety, a request for information must be submitted to the government.

When soliciting terms from a prospective subcontractor or supplier, the prime contractor should provide the entire contract for analysis. If the prime contractor only provides those portions thought to be pertinent, and provisions that actually are necessary for a full understanding of the scope of inspection, acceptance, and warranty requirements are omitted, then the prime contractor will be responsible to its subcontractor or supplier for any additional costs and time caused by those omissions.

Any requirement for a project operations manual must also be tailored to reflect all pertinent provisions. Obviously, the CQC plan must be equally comprehensive.

Even though the organization of government construction contract documents is relatively uniform, contractors should be careful and comprehensive in developing scopes of work for prospective subcontractors, suppliers, and purchase order vendors. Although the FAR does not require all contract provisions to be flowed down, prime contractors should flow down all applicable requirements in their subcontracts and purchase orders, to the extent practicable, as should subcontractors at every tier. For example, drafting a purchase order for materials around only the technical specifications may result in an unanticipated purchasing or cost and pricing gap when the special conditions or contract drawing notes contain specific requirements affecting the inspection, acceptance, or warranty applicable to the materials to be covered by the purchase order.

Other standard FAR clauses can significantly affect the inspection and acceptance process. These include:

- Materials and Workmanship, FAR 52.236-5. This clause requires the contractor to provide materials and equipment that are "new and of the most suitable grade for the purpose intended."
- Use and Possession Prior to Completion, FAR 52.236-11. This clause permits the government to use portions of the work without such use or possession constituting acceptance.

Finally, as mentioned briefly above, the inspections, acceptance, and warranty process may result in the contractor (or its subcontractors or vendors) concluding that the government's action or lack of timely action increased the cost of performance or delayed the work. Under those circumstances, the contractor may be entitled to an adjustment in the contract price and/or time. This necessarily involves the consideration of either the standard Changes clause found at FAR 52.243-4 or the standard Suspension of Work clause found at FAR 52.242-14. Each of the provisions is addressed in other chapters of this book.[5]

III. Inspection Rights and Duties

A. Government's Right—But Not Duty

The Inspection of Construction clause, FAR 52.246-12(b), provides that government inspection and testing may occur "at all places and at all reasonable times before acceptance to ensure strict compliance with the terms of the contract." However, subparagraph (c) of the clause states that government inspections and tests are for the "sole benefit" of the government and do not relieve the contractor from its contract obligations. Under the Inspection of Construction clause, the government has no obligation or duty to inspect the work for noncompliance with the plans and specifications; however, if an obvious defect is accepted by the government, such acceptance will be conclusive, subject to warranty provisions.[6]

Inspection is a right reserved to, but not required of, the government. It does not become a duty unless the government assumes that responsibility under a different contract provision (or has knowledge of a defect but does not reject it within a reasonable time).[7] The fact that the government approves a submittal[8] or fails to reject the contractor's work does not necessarily establish that the work meets the requirements of the contract.[9] An interim inspection may not preclude the government from subsequently rejecting the work. As a general principle, the Inspection of Construction clause places the basic obligation for the conformance with the plans and specifications on the contractor.

As previously noted, the Inspection of Construction clause also obligates the contractor to maintain an "adequate inspection system" to ensure that the work conforms to the requirements of the contract. This obligation can present a significant expense, but it can be even more costly if the system is deficient. For example, the contractor's failure to maintain an adequate inspection system may preclude recovery from the government for errors in the specifications that otherwise would have been detected by the contractor if it had an "adequate inspection system."[10]

B. Government Inspections: Manner, When, and Where

The standard Inspection of Construction clause provides that the government may perform its inspections "at all places and at all reasonable times before acceptance." Although this language gives the government substantial latitude in performing its inspections of the work and equipment or materials being incorporated in the project, the government's inspection activities must be reasonable.[11] What constitutes a reasonable inspection involves consideration of time, nature, and expense associated with the government's actions.[12] The government inspector may not inspect in a manner that is extraordinarily intrusive or burdensome to the contractor or its lower-tier subcontractors and suppliers.[13] The government does not have the right to create extra work by requiring tests that are expensive, arbitrary, and not recognized by the industry. Ultimately, the "reasonableness" standard involves interpretation of the contract, and its application is highly case-specific.

A contract may specify a particular location for government inspection. However, under the Inspection of Construction clause, the government retains the right to conduct inspections at a reasonable place other than the one specified in the contract. In addition, the government's inspection at a place other than the one specified in the contract does not preclude later government inspections at the specified locations.[14]

C. Government Punch Lists

On most construction projects, including federal projects, the preparation of punch lists is customary. These may be prepared at particular phases of the work, for example, before ceilings are installed or walls are closed up, or at the time of a more comprehensive pre-final or final inspection. In either situation, the form, content, and clarity of the punch list can be a source of frustration and controversy. Inconsistent, conflicting, or multiple punch lists may be a source of claims. Punch list items are typically minor deficiencies that do not preclude beneficial use and occupancy of a facility. However, more substantial deficiencies can appear on such lists.

The government bears the burden of proving noncompliance with the contract's requirements.[15] If the government can prove noncompliance, the contractor is responsible not only to remedy the noncompliance but also for any delay or extra costs attendant to resolving the noncompliance. As explained elsewhere in this book, when the government asserts that there is a noncompliance in the work and directs the contractor to correct it, the contractor is generally obligated to proceed with performance of work (and claim later) even if the government is ultimately incorrect.[16]

There is no FAR provision that obligates the government to provide a punch list or a list of deficiencies to a contractor. The contract's specifications or custom and practice may result in the preparation of a punch list. Generally, the government's failure either to provide a punch list or to include an item on the punch list will not, in and of itself, excuse noncompliance with the plans and specifications.[17] However, if there is a question about the requirements of the contract, or there is an ambiguity in the plans or specifications, the omission of a later disputed item from a government-generated punch list may be important evidence of the reasonable interpretation of the contract.

Such omission would tend to show that the government did not consider the omitted item to be noncompliant at a time when no dispute was evident.

D. Government Inspectors: Conduct and Authority

Generally, government inspectors are not personally liable for extra costs or delay resulting from their actions or inactions while performing tests and inspections. The contractor's recourse in those circumstances is to seek reimbursement from the government for any costs incurred as a result of the acts and omissions of the government's inspectors.[18] The contractor generally may accomplish this by establishing that the inspector changed the requirements of the contract by its actions and seeking compensation under the Changes clause. If the inspector's conduct delays the work, any request for time or money due to the delay would be submitted pursuant to either the Changes clause or the Suspension of Work clause, depending upon the circumstances and nature of the delay. While each of these provisions is addressed elsewhere in this book,[19] it is important that the contractor and its counsel are aware of the notification and documentation requirements of these and other clauses in order to preserve the contractor's ability to obtain an equitable adjustment of the contract's price and/or time.

Since government inspectors have a primary role of ongoing contact with contractor personnel and in the interpretation of the contract's requirements, it is not unusual for a contractor to assert that a government inspector changed the requirements of the contract. This change may be a relaxation of the specifications. More likely, it is the imposition of a more stringent requirement. Subparagraph (d) of the Inspection of Construction clause specifically addresses these situations by stating that an inspector does not have authority to change the contract's requirements (plans or specifications) without the "written authorization" of the CO.[20] Generally, this provision has been sufficient to shield the government from liability for an unauthorized change directed by an inspector.[21] Recovery based on an action or omission of an otherwise unauthorized government employee may be obtained by the contractor only if the CO subsequently ratifies that action or omission. Further, there is a renewed emphasis by the government on disavowing "unauthorized" directives by government personnel lacking express authority under the contract to bind the government. It is crucial to be aware of agency-specific provisions (such as DFARS 252.201-7000) that specifically disavow the authority of a "contracting officer's representative."[22]

Given this basic principle, it is important that a contractor maintain a documentation system that satisfies the notice requirements of the Changes clause and other contract clauses, adequately recording and giving notice of any directives provided by the government's inspector and the effect of those directives on the cost and time for performance. A good CQC documentation system should provide this information.

There are two interrelated issues here. The first is whether a directive from the government has been issued by a person with requisite authority to bind the government. It is imperative that the CO be notified in writing that a directive from an inspector, for example, is considered by the contractor to cause extra work or delay. Such notice allows the CO the opportunity to ratify, to correct, or to rescind the directive.

The second issue is whether the CO agrees that directed work is compensable as an extra. The CO may interpret the matter as required by the contract without modification. If, after giving written notice to the CO of the inspector's instruction, a contractor is directed to proceed with the requirement by the CO, the contractor should proceed under protest if it is not clear that the CO agrees that the directed work is extra.[23] If not already stated by the contractor in the initial written notice to the CO, addressing authority issues, the contractor should notify the CO that the work is considered extra and compensable.[24]

Simply accepting that a government inspector has the authority to order changes can be a costly error for a contractor. Failure by the contractor to notify the CO in writing of its disagreement with the inspector's position may result in a determination that there was no ratification by the CO,[25] the contractor was a volunteer,[26] or it agreed with the inspector's interpretation of the contract's requirements under the principle of concurrent interpretation.[27] The converse is also true. For example, a government inspector's contemporaneous failure to object to a contractor's method of testing may be evidence of the parties' mutual interpretation of the contract.[28] While not obvious from the detailed requirements of the typical CQC specification, the documentation developed in that process may be the best evidence of this mutual contemporaneous interpretation of the contract.

IV. Who Pays the Cost?

A. Initial Inspections

The government typically pays the costs of its own inspections and tests. However, the Inspection of Construction clause obligates the contractor to furnish "all facilities, labor, and material" reasonably needed by the government to perform its inspections and tests. The key phrase is "reasonably needed." While the contractor is expected to absorb these costs as part of the performance of work, it may be possible to recover any increase in these costs caused by the government's inspection. For example, the government has the contractual right to conduct a test at a location other than the one specified in the contract. However, depending on the language in a contract that specifies particular conditions of inspection, if a change in location increases the contractor's performance costs the contractor may recover those costs.[29] If the contractor can prove that the facilities, labor, or material required by the government were unreasonable or would amount to economic waste, the government must bear the expense of conducting such inspections or tests.[30] A government test may be deemed to be economically wasteful if it is arbitrary, unusually expensive, or not recognized by the industry as valid.

1. Government Reinspection

If the work is not ready at the time specified for inspection, or if the contractor fails to provide the government with the facilities, labor, or material reasonably necessary for inspection, the contractor may be required to pay for retesting. If a proper rejection of defective work necessitates further testing and inspections, then the government must notify the contractor of the rejection and provide it an opportunity to correct or replace the work before conducting a reinspection.[31]

2. Government Destructive Tests of Completed Work

The Inspection of Construction clause also allows the government to perform destructive inspections. Consistent with that clause, the government is entitled to tear out completed work to inspect it for contract compliance. If the work does not conform to the contract, the contractor must pay the costs and absorb any delay caused by both the destructive inspection and the reconstruction. If, on the other hand, the work is satisfactory, the contractor is entitled to an equitable adjustment for (1) any additional costs associated with the destructive inspection and the reconstruction; and (2) a time extension for any delay caused.[32]

3. Inspection Obligations Imposed on the Contractor

Although the government has broad rights to conduct its own inspections, a more typical practice is for the government to exercise its right to inspect by (1) conducting surveillance of the contractor's required inspection system; and (2) approving tests performed by the contractor. When the government takes this approach, the government inspector's main responsibility is to interpret contract specifications and make evaluations as to whether the contractor's inspection system and test procedures are adequate to assure quality control and compliance with the plans and specifications.

The contractor's inspection obligations may be defined in the CQC provision in the specification by means of specific tests or, if no specific test is called out, by some reasonably standard test that the contractor must perform.[33] In that context, if the government changes the test or inspection procedure, the contractor may be entitled to an equitable adjustment.[34] Whether that action is a compensable change under the Changes clause may depend on the answer(s) to the following questions:

1. Did the use of a different inspection procedure or test only facilitate the discovery of deviations from the contract's requirements or did it effectively change (increase) the standard of performance (i.e., did it "raise the bar")?

2. Did the implementation of the new test procedure increase the con-
    tractor's cost for inspection or the time for performance of the test
    from that required by the tests specified in the contract?

If the new inspection test or procedure alters the requirements of the contract, the consequences of complying with the new test may entitle the contractor to recover time and/or money. In contrast, if the contractor elects to
"voluntarily" perform tests beyond those detailed in the contract, the government is not obligated to reimburse those costs or extend the performance
time.[35] Therefore, if the government's inspector "suggests" a particular test
that differs from those required by the contract, it is important that the contractor obtain written documentation of that suggestion and provide notice to
the CO *before* following the inspector's "suggestions."

The government has a duty not to hinder the performance of the contractor's work. This basic duty means that the government may not *unreasonably*
interfere with the contractor's inspection system by preventing a contractor
from performing tests that would determine if the work complied with the
requirements of the contract.[36]

4. CQC Staffing Costs

As set forth above, most government contracts include detailed contractor
CQC specifications. If the contract specifies a CQC person or a CQC organization, the government has the right to require the contractor to retain the CQC
person or to establish and maintain the CQC organization without additional
compensation.[37] The specified CQC person or organization may vary depending on the nature of the particular work. Thus, it is possible to have requirements for separate CQC personnel to inspect line and grade, compaction,
rebar placement, concrete, welds, mechanical, plumbing, electrical systems, or
any combination of specified work features.

B. Inspection and Constructive Changes

Inspection of construction under a government contract creates a variety
of situations in which a contractor may incur additional costs that were not
anticipated in the initial project estimate. Extra work may be created by the
government inspector's application of a heightened standard of performance
or the government's inconsistent application of inspection standards. Erroneous rejection of work may also cause the contractor to incur costs due to
unnecessary correction and reinspection. If a government inspection alters
the method of performance or requires performance to a different standard
that results in extra work, the contractor may have a valid compensable
change request. The contractor's ability to obtain an equitable adjustment for
inspection-related changes depends on not only satisfying any applicable
notice obligations (including notice to the CO for the purposes of proper
authority to direct a change) but also demonstrating that the required work
was, in fact, a change to the requirements of the plans and specifications.

1. Compliance with Notice Requirements

Contractors should not ignore the basic principle that government inspectors do not have the authority to order changes to the work.[38] Even if the government's inspector knows or acknowledges that his conduct or direction has effected a change to the contract, the government is not bound by the inspector's act alone. The Inspection of Construction clause seeks to confirm this lack of authority by specifically providing that a change to the contract may be made only by obtaining written authorization from the CO.

This limitation on the inspector's authority is not an absolute bar to recovery for inspection-related changes because relief may still be available under the Changes clause for the performance of a "constructive change." Compliance with the notice requirements specifically related to constructive changes is addressed in another chapter of this book,[39] although we cannot overemphasize the notice requirements, described above, related to authority of government personnel and to CO ratification. A contractor's failure to notify the CO that the contractor regards certain inspection-related work as a change may extinguish any possibility of recovering for inspection-related costs.[40] If the contractor fails to provide timely notice to the CO, the contractor will have to prove, or at least satisfy an augmented burden of persuasion, that the "lack of notice" did not "prejudice" the government[41] or it will have to prove that the CO knew or should have known the facts giving rise to the claim.[42] Again, the failure contemporaneously to signify disagreement with the inspector's interpretation may be viewed as evidence that the contractor agreed with the interpretation or acted as a volunteer. Finally, if the contract contains a specially drafted strict-notice provision, even those traditional exceptions to the requirement for the written notice may not apply.[43]

The Changes clause applies to any written or oral order from the CO that causes extra costs or time for overall performance. Since the government inspector is acting as the CO's agent for the limited purposes of inspection, the orders of the government's inspector may trigger application of the Changes clause and its notice requirements. To avoid notice defenses, a contractor should provide any notice in writing and should state (1) the date, circumstances, and source of the order; and (2) that the order is considered to be as a change to the contract. If there is an opportunity for the CO to rescind the inspector's directive, it is appropriate to set forth the time frame for a response from the CO and to advise the CO that the contractor will proceed as directed by the inspector *unless* advised to the contrary by the CO. Once the CO is on notice that the government inspector has ordered extra work, the failure to take any action to reverse that direction is often considered to be a ratification of the inspector's actions.[44]

While preferable for the purposes of notice, demonstrating that the CO had actual knowledge of the inspector's actions is not always required. Actual notice to a representative of a CO may be imputed to the CO since the representative may be characterized as the "eyes and ears" of the CO for many purposes.[45] However, there is no absolute rule that notice to an inspector will satisfy the notice requirements of the Change clause under all circumstances.[46] The better course of action for the contractor is to provide written and expeditious notice directly to the CO.

Notice to the CO requires no particular form. Notice should be factual and sufficiently detailed to explain the circumstances and the basis for the conclusion that the directive is a change to the contract. Contractors should be counseled to avoid being argumentative or derogatory in describing the people involved. A contractor should deal with *facts* and avoid dealing with *feelings* or *personalities*. The primary functions of the notice are:

- To fully inform the CO of the events;
- To make the change order directive one that was, in effect, issued by the CO; and
- To give the CO an opportunity to exercise remedial action by modifying or reversing the inspector's directive.

2. Government Rejection of Work: Basis and Timing

In rejecting work, the government generally is expected to provide timely notice of the rejection and the rationale for its action.[47] Inadequate information regarding the basis for a rejection that creates extra work for the contractor may be the basis for a compensable change.[48] However, even if a rejection notice provides an incorrect reason or no reason, such failure by the government may be compensable only if the contractor has been misled by the erroneous information or was not otherwise on notice of the defects.[49]

Rejection of work by the government may be considered improper for a variety of reasons. The most obvious situation involves the rejection of conforming work. Historically, the government has the burden to prove that "rejected" work was not in compliance with the contract.[50] However, under the current approach to government inspection, i.e., surveillance by the government of the contractor's CQC system, this burden may be shifted to the contractor. Consequently, proper and comprehensive documentation of the CQC system will tend to establish that the work conformed to the contract's requirements and that the government's later rejection for an alleged nonconformance was not justified.[51]

Rejection of work is erroneous where it was a result of the government's failure to apply the appropriate testing standard[52] or was based solely on destructive tests when reasonable and continuous surveillance during performance by the government inspector would have revealed the nonconformity.[53] A rejection also may be deemed improper if the contractor proves that the CO had *actual knowledge* that work being performed did not comply with the contract but allowed the contractor to continue with the work without taking any action.[54] This argument is one of detrimental reliance and prejudice to the contractor. In reality, it may be difficult to establish the facts that would support a claim by a contractor based on a theory of "actual knowledge" of the deficiency, particularly given the language of subparagraphs (c) and (d) of the Inspection of Construction clause. Not only is the CO rarely present at the site of the work, but government inspection is solely for the benefit of the government. Further, the presence or absence of a government inspector does not relieve the contractor from compliance with contract requirements.

Passage of time may provide a basis to establish that a rejection was improper. Paragraph (i) of the Inspection of Construction clause requires the government to "accept, as promptly as practicable after completion and inspection, all work required by the contract." Notice of rejection must be furnished by the government within a reasonable period of time, or the work may be deemed to have been accepted.[55] The key to prevailing is to establish what constitutes a "reasonable" time frame for action by the government. This is fact-intensive unless the contract stipulates a time frame for action by the government. However, once that period of time is defined, the government has the *duty* to reject work within that time upon learning of the nonconformity. An unreasonable delay in rejection may constitute either constructive acceptance of the work or a constructive change to the contract.[56]

3. Government Imposition of Different Performance Criteria

It is the contractor's responsibility to comply with the contract requirements.[57] The government *may* inspect or test the work under the Inspection of Construction clause to ensure strict compliance with the plans and specifications.[58] If performance criteria are not clearly defined by the contract, then performance standards are subject to interpretation. If interpretation by the government results in the imposition or use of increased (i.e., stricter) standards of performance that effectively change the contract and create extra work, then the contractor may be entitled to recover its costs.[59] The key is reasonable interpretation of the contract. In addition, satisfaction of notice obligations and documentation of any extra costs or time caused by the change are essential.

The government is not required to follow industry practice or trade custom (although the meaning of terms may be informed by trade practice and custom). As long as a requirement is specifically incorporated or set forth in the plans or specifications, the government is free to impose a strict standard of acceptance.[60] If the contract is unambiguous, trade custom is not controlling. While trade meaning, usage, and custom may explain or define contract language, such evidence may not be used to vary or contradict contract language.[61] The government may require more stringent performance standards than are customarily required as long as such standards are unambiguously set out in the contract.[62]

If the contract does not contain a statement of the acceptance criteria, the standard of workmanship customarily followed within the industry controls.[63] If the performance criteria set forth in the contract are not clear, the interpretation relied upon by the government is generally measured against industry practice to determine whether application of such criteria constitutes a constructive change.[64] If the criteria followed by the government are found to constitute a change and if the contractor can establish that its work would have been accepted based on the application of customary trade practice, then the contractor may recover for its increased expense associated with any rework.[65] An example of this type of constructive change is insistence on tolerances that are *not* required by the contract and are more strict than those customarily used in the trade. Such insistence by the government entitles the contractor to recover its extra costs.[66]

The government inspector's imposition of an impractical performance standard may be a compensable change. However, to prove the impracticability, the contractor must demonstrate that compliant materials installed in a workmanlike manner while following sound construction and engineering practices repeatedly produced a test failure. In other words, the contractor must show that the government's plans and specifications were defective.[67]

4. *Government's Use of Varied Inspection Procedures*

The government may exercise discretion in selecting its inspection and testing methods. Often, the particular tests are specifically set forth in the contract. If such procedures are described clearly and unambiguously, the contractor cannot successfully assert a claim on the theory that the specified tests are too strict and overly burdensome.[68] Compliance with any contractually specified method of testing is the government's right and the contractor's obligation. If the contractor believes that a particular test is unreasonable, the appropriate time to raise the question is during the proposal or bid phase.

The government may change the test method from that specified in the contract. The contractor must be notified by the government that the new test will be applied.[69] The new test must not increase the standard of performance or the cost of performance without allowing additional compensation to the contractor. If the test is changed by the government, then the government has the burden of proving that the results achieved under the newly chosen test procedures are comparable to those results that would have been achieved under the contractually specified procedure.[70] If the government is unable to sustain its burden and the new test increases the time or cost of performance, this action constitutes a change to the contract and a basis for an equitable adjustment to the contract price and/or time.[71]

If the contract fails to specify a test, the test used by the government inspector must be "accurate and reasonably calculated to determine compliance with the specification."[72] There is a presumption that government tests are conducted correctly. Tests that conform to generally accepted industry practices are considered reasonable. For a contractor to recover for claimed extra costs, it must show that the test methods employed by the government were unreasonable and did not comply with trade custom.[73]

5. *"Overinspection" by the Government*

It is difficult to make a successful "overinspection" claim under a government contract. Government inspections generally can be thorough and very strict without exceeding contract requirements. For example, a government increase in the number of inspectors at the job site does not, in and of itself, constitute overinspection.[74] However, if the contractor demonstrates that it experienced multiple inspections performed by different individuals and that these inspectors applied different acceptance standards, it may be entitled to an equitable adjustment for extra costs and/or time.[75] Similarly, the government's use of stricter or different inspection guidelines for second and third inspections after a previous rejection may entitle the contractor to an equitable adjustment if the use of such guidelines increases the level of performance above that required in the contract.

Varying inspection standards from one test to another may be considered a constructive change unless the first test was for detection of nonconformity and the second was used to determine the extent of the nonconformities.[76] If the different tests produce varying results, the failure to follow customary industry test practice to reconcile the test discrepancies may negate the government's rejection of the work.[77]

The government has the right to inspect the contractor's work at any reasonable time or place during contract performance. While not explicitly set forth in the Inspection of Construction clause, the government may not exercise a test or inspection right at a time or place that *unreasonably* interferes with, disrupts, or delays performance of the work. If tests or inspections are unreasonable as to time or place and thereby cause extra costs or time for performance, the contractor is entitled to an equitable adjustment in the contract terms.[78]

Claims for extra compensation for inspection-related delays, disruption, or interference are fact-specific and usually require detailed supporting documentation in order to be successful. This includes documentation of the actions or inactions by the government's inspectors, the effect of that conduct on the contractor's work, and the effect on the cost and/or time for performance. General allegations of overinspection or nonspecific general opinions of adverse impact on the cost of work or time for performance are rarely sufficient.[79]

Examples of conduct that have formed the basis for successful overinspection claims include the following:

- Government inspector, who was also the resident engineer, gave detailed orders directly to contractor's employees.[80]
- Government's representatives required compliance with overstrict interpretation of specifications coupled with an unrealistic striving for perfection.[81]
- An unsuccessful overinspection claim involved the use of harsh and vulgar language by government inspectors, without additional evidence of substantive misconduct or interference by the inspectors.[82]

As a general rule, because the government is entitled to receive performance in full compliance with the contract's plans and specifications, "disparate treatment" claims are not usually successful. Thus, the fact that the government may not have enforced a similar contract requirement on a different contract with a different contractor is rarely a basis to argue that the same requirement is not enforceable on a subsequent contract.[83] However, the principle of strict performance has limits. While the government has the right to reject nonconforming work, total replacement of that work may not be justified if an alternative fix would have been satisfactory and less costly to the contractor.[84]

C. Government Inspection Delays

Government inspection conduct may delay performance either by action or inaction. Claims for equitable adjustment for money and/or time on account of inspection-related extra work are most often brought under the Changes clause. However, claims for pure delay may also be brought under the Suspension of Work clause. In that context, it may be appropriate to seek a time extension and monetary compensation for a work stoppage caused by the government's inspection activity or lack of timely action. The principles regarding notice and documentation discussed above with respect to inspection-related changes are equally applicable to delay claims. Under the right set of documented circumstances, a situation that warrants compensation under the Changes clause may also entitle the contractor to recovery under the Suspension of Work clause if there is an unreasonable work stoppage preceding the government's direction to perform the inspection-related change.[85] More comprehensive treatment of matters related to both clauses and recovery for delays are addressed in other chapters of this book.[86]

As with any claim for time or money, compliance with applicable notice requirements may be critical. The contractor should promptly notify the CO when a government inspection activity or inactivity delays the work. Although proving the CO's on-site representatives had actual knowledge of the delay or work suspension may suffice to show notice to the government, it is always better to notify the CO directly and in writing.

Examples of conduct resulting in claims for delay compensation or extensions of time include the following:

- Government's failure to establish inspection standards within a reasonable time[87]
- Extremely limited availability of the government's inspector[88]
- Post-award imposition by the government of a requirement for advance notice of the need for inspection where no such requirement was contained in the contract[89]
- Government inspectors' requiring two to four hours of advance notice for inspections where no notice was required by the contract[90]

As with inspection-related changes, it is important to demonstrate that the government's actions were unreasonable and caused a delay or disruption in performance. Demonstrating "unreasonable delay" is an express element of recovery under the Suspension of Work clause.

There is no substitute for complete and factual documentation. The contractor's CQC staff can be an excellent resource to document both the time lost and the effect on the work. Complete and factual contemporaneous records are invaluable in any claim. Conversely, gaps in the contractor's records may give the government an effective counterargument that no problems or delays were experienced during the relevant time periods.

V. Nonconforming Work: Government Rights

The Inspection of Construction clause provides a basic statement of the parties' respective rights and responsibilities if any portion of the work is rejected.

A. Government Notice of Rejection

If the work does not comply with the contract requirements, the contractor is entitled to receive timely notice of the rejection that sets forth the reasons therefor.[91] Generally, the government bears the burden of proving that the work does not conform to the contract's requirements.[92] If the rejection is untimely and prejudices the contractor, the government may lose its rights with respect to requiring corrective action.[93] This is a fact-specific inquiry.

B. Contractor Opportunity to Cure Deficiencies or Defects

Once the government has provided notice of rejection, the contractor generally is entitled to have a reasonable period of time to correct the deficiency prior to reinspection by the government, or the issuance of a deductive change by the government to account for performance of remedial work by government workers or by another contractor on behalf of the government.[94] If the contractor is not given an opportunity to correct nonconforming work, it is important that the contractor provide written notice of its objection to the CO. In that context, the contractor should formally offer to correct the deficiencies and propose a realistic and reasonable proposal for making the needed corrections.[95]

C. Economic Waste Issues

The Inspection of Construction clause provides that the government is entitled to require prompt replacement of rejected work. This right of strict compliance may be limited by the doctrine of "economic waste." Therefore, even though the government is normally entitled to require replacement of nonconforming work, that right is qualified or limited if (1) the cost of corrections is economically wasteful; and (2) the nonconforming work is otherwise adequate for its intended purpose. In those instances, the government's remedy is a downward price adjustment.[96]

D. Credits of Deductive Changes

When ordering replacement of defective work would be economically wasteful, the government may make a downward price adjustment. As an alternative, the Inspection of Construction clause permits the government to reduce the contract price in conjunction with accepting nonconforming work, with or without directing performance of corrective work. In that context, the government's right to monetary compensation is not limited to a credit for the reduced value; rather, the government is generally entitled to recover all reasonably foreseeable costs that are the proven consequences of the defect.[97]

VI. Government Acceptance

Under the Inspection of Construction clause, government acceptance of the work is "final and conclusive except for latent defects, fraud, gross mistake amounting to fraud, or the government's right under any warranty or guarantee."[98] Acceptance of the work is also significant to the allocation of risks under the Permits and Responsibilities clause.[99] As previously noted, that clause provides that the contractor is responsible for all materials delivered and work performed until the entire work, or a completed unit of work, has been accepted. Therefore, government acceptance is a critical matter on every project. The central questions related to acceptance are as follows:

A. Who Has Authority to Accept the Work?

The Inspection of Construction clause does not identify the government representative with the authority to accept the work. This information may be specified in other contract provisions. The government is not bound when an unauthorized government representative accepts the work.[100] However, it is possible—for example, when a government inspector orders unauthorized changes—for the government to ratify an unauthorized acceptance. Acceptance by an unauthorized government employee may be final and conclusive if the CO knew or should have known of the situation and failed to act in a way inconsistent with the acceptance.[101] Again, basic factual notice to the CO may be sufficient to bind the government. Such notice should be timely, written, directed to the CO, and specific as to acceptance.

B. When and Where Does Acceptance Occur?

Paragraph (i) of the Inspection of Construction clause provides, "Unless otherwise specified in the contract, the government shall accept, as promptly as practicable after completion and inspection, all work required by the contract or that portion of the work the Contracting Officer determines can be accepted separately." For a construction contract, the obvious and usual place for acceptance is the site of the work; however, the contract can specify a separate or specific location for final inspection and/or acceptance. In any event, final inspection and acceptance may occur at any reasonable location.

Usually, the government is required to accept a facility when it is substantially complete (i.e., when no or only minor items of work remain and the facility is capable of serving its intended purpose).[102] If the project comprises distinct phases, it is fairly common for the government to accept the project in phases, in which case government acceptance of each phase is final as to that phase.[103] In dealing with phased completion and acceptance, contractors should be conscious of the terms of any express warranty in the contract. It is not unusual to find that the specifications in the contract expressly state that the warranty on equipment runs from the date of final acceptance of the *entire project* rather than an earlier phase. If equipment was installed during the earlier phase and was placed in operation, that event may trigger a manufacturer's standard warranty, creating a contingent warranty gap liability. The contractor should analyze potential warranty gaps and secure extended warranties during the project bidding phase and not at the end of the job or when an item of installed equipment malfunctions or fails.

C. How Does Acceptance Occur?

There is no requirement that acceptance be made in any specific manner. Some contracts spell out the procedures to be followed for final acceptance. Normally, government acceptance is acknowledged by the execution of a written document.

While formal acceptance may be customary, it is not always mandated in the contract. Acceptance may be implied by various government acts or simply by the passage of time. The standard for determining whether an acceptance was timely made is whether it occurred within a reasonable period of time, considering all relevant circumstances.[104]

Final payment by the government for the work may also imply or signify acceptance of the work. Absent any known defects or nonconformities in the work or warranty rights, the government cannot compel a contractor to examine or to correct completed work once the government has made final payment. Progress payments, on the other hand, do not constitute acceptance by the government of the completed work.[105]

D. What Are the Consequences of Acceptance?

Basically, government acceptance is final and conclusive as to clearly observable or detectable defects in the work. These types of obvious defects are generally labeled "patent defects."[106] Once the government has accepted the work, it usually cannot recover, even under a typical warranty clause, for patent defects that could have been detected by a reasonable pre-acceptance inspection.[107] Generally, the government is bound by acceptance of the work as to patent defects regardless of whether the government had knowledge of such defects. Similarly, even if the contract requires the contractor to maintain an inspection system to ensure compliance with the specifications and plans, the government's rights under a warranty provision may be negated if the government failed to discover an "obvious" error during construction.[108]

VII. Exceptions to the Finality of Acceptance

A. Latent Defects

A latent defect is generally a defect in existence at the time of final acceptance by the government, but one that the government could not have ascertained from reasonable examination. The government's failure to make an examination or test that would have revealed a defect does not make a defect "latent."[109] If the contract specifies that the contractor has the responsibility for all inspection and testing, the government may look to the contractor to detect nonconformities. Thus, a defect, even if discoverable through a government inspection, may be considered latent if the contractor is responsible for testing and does not discover that deficiency.[110]

Once the government discovers a latent defect, it has the duty to act promptly in exercising its rights to require correction. For example, an unexplained lengthy delay between the government's discovery of the latent defect and the government's decision to order performance of any corrective work may operate to waive the government's rights against the contractor.[111] "Lengthy delay" should be understood in the context of the relevant statute of limitations under federal contracts. Pursuant to 41 U.S.C. § 605(a) and FAR 33.206, a claim by either party must be submitted to the CO within six years of the date the claim accrues.

B. Fraud and Gross Mistakes Amounting to Fraud

If the government's acceptance is induced by fraud, the Inspection of Construction clause states that the government may revoke its acceptance.[112] In order to revoke final acceptance based on fraud, the government has the burden of proving (1) that acceptance was induced by government reliance on (2) a misrepresentation of fact, actual or implied, or the concealment of a material fact; (3) made with the knowledge of its falsity or in reckless or wanton disregard of the facts; (4) with intent to mislead the government into relying on the misrepresentation; and, (5) that the government suffered injury thereby.[113] Performance that is nonconforming, without proof of misrepresentation, intent to defraud, and reliance, does not constitute fraud.[114]

A "gross mistake amounting to fraud" is a major mistake "so serious or uncalled for it was not to be reasonably expected, or justifiable, in the case of a responsible contractor"[115] or a mistake that cannot be reconciled in good faith.[116] The government may revoke final acceptance for a gross mistake amounting to fraud.[117] Not all misrepresentations constitute gross mistakes amounting to fraud.[118] To revoke final acceptance based on a gross mistake amounting to fraud requires that the government prove the same elements as fraud listed above, except that the requirement to prove intent to deceive or mislead is not required.[119] Accordingly, all the facts and circumstances surrounding the contractor's performance will be considered in determining whether there was a degree of recklessness or a lack of good faith in making any mistaken representations to the government. The mere fact that the work was defective does not mean that the contractor has made a gross mistake amounting to fraud. Indeed, a post-acceptance showing that there is noncompliance with the requirements of the contract is not sufficient to overcome an otherwise proper acceptance unless one of the exceptions discussed above applies.[120]

C. Warranties

Contracts include warranty clauses to protect the government from certain types of defects for a period of time beyond final acceptance. Traditionally, warranty and guarantee clauses are strictly construed against the government. Therefore, the government is protected only from those defects specifically enumerated in the clauses. The current Warranty of Construction[121] clause at FAR 52.246-21 provides in part as follows:

WARRANTY OF CONSTRUCTION (April 1994)

(a) In addition to any other warranties in this contract, the Contractor warrants, except as provided in paragraph (i) of this clause, that work performed under this contract conforms to the contract requirements and is free of any defect in equipment, material, or design furnished, or workmanship performed by the Contractor or any subcontractor or supplier at any tier.

(b) This warranty shall continue for a period of 1 year from the date of final acceptance of the work. If the government takes possession of any part of the work before final acceptance, this warranty shall continue for a period of 1 year from the date the government takes possession.

(c) The Contractor shall remedy at the Contractor's expense any failure to conform, or any defect. In addition, the Contractor shall remedy at the Contractor's expense any damage to government-owned or controlled real or personal property, when that damage is a result of—

(1) The Contractor's failure to conform to contract requirements; or

(2) Any defect of equipment, material, workmanship, or design furnished.

(d) The Contractor shall restore any work damaged in fulfilling the terms and conditions of this clause. The Contractor's warranty with respect to work repaired or replaced will run for 1 year from the date of repair or replacement.

(e) The Contracting Officer shall notify the Contractor, in writing, within a reasonable time after the discovery of any failure, defect, or damage.

(f) If the Contractor fails to remedy any failure, defect, or damage within a reasonable time after receipt of notice, the government shall have the right to replace, repair, or otherwise remedy the failure, defect, or damage at the Contractor's expense.

(g) With respect to all warranties, express or implied, from subcontractors, manufacturers, or suppliers for work performed and materials furnished under this contract, the Contractor shall—

(1) Obtain all warranties that would be given in normal commercial practice;

(2) Require all warranties to be executed, in writing, for the benefit of the government, if directed by the Contracting Officer; and

(3) Enforce all warranties for the benefit of the government, if directed by the Contracting Officer.

(h) In the event the Contractor's warranty under paragraph (b) of this clause has expired, the government may bring suit at its expense to enforce a subcontractor's, manufacturer's, or supplier's warranty.

(i) Unless a defect is caused by the negligence of the Contractor or subcontractor or supplier at any tier, the Contractor shall not be liable for the repair of any defects of material or design furnished by the government nor for the repair of any damage that results from any defect in government-furnished material or design.

(j) This warranty shall not limit the government's rights under the Inspection and Acceptance clause of this contract with respect to latent defects, gross mistakes, or fraud.

While the scope of this warranty provision is very broad, other warranties are often found in the specifications, as indicated in the opening phrase of the standard provision at subparagraph (a). As previously stated, it is essential on projects with phased completion and acceptance to determine the commencement date on all of the warranties, especially those on equipment. In addition to time, these special warranties may address system or equipment performance.

In the event the government detects a nonconformity or defect covered by a warranty or guarantee clause, the contractor remains responsible for correcting that condition, despite final acceptance under the Inspection of Construction clause. In addition, a typical warranty clause may obligate the contractor to remedy, at its own expense, any damage to government property caused by the defect or nonconformity and the repair of any defective work performed by the contractor. To avoid warranty gaps in terms of the scope of those obligations, the contractor should compare the scope of the warranty running to the government to the scope of the warranty provided by a vendor.

By their very nature, government warranty claims usually arise after final acceptance, and warranty provisions survive final acceptance. The government has the burden to prove liability, causation, and injury or damages.[122] The facts the government must prove depend on the specifications governing the allegedly defective equipment, materials, or design furnished or the workmanship performed and the alleged defect.[123]

VIII. False Claims Exposure

A. The Government's Arsenal

In connection with inspection and acceptance (as well as any and all other performance and documentation issues related to federal government construction contracts), government contractors need to be conscious of the several federal antifraud statutes available to the government in its campaign against fraud, waste, and abuse. These include, but are not limited to, the Contract Disputes Act, the Civil False Claims Act, the Criminal False Claims Act, False Statements Act, and the Forfeiture of Claims Act. Another chapter of this book gives more detailed treatment of the subject of criminal and civil liability under the False Claims Acts.[124] It is essential that every contractor be aware of these statutes when considering questions related to conformance of the work to the contract's requirements, submitting a routine request for payment, or preparing a request for equitable adjustment or claim for additional time or money due to an inspection-related issue.

B. Payment Certification of Conformity

Any contractor certification associated with the performance of a government contract has a potential for a false claims exposure. The Payments Under Fixed-Price Construction Contracts clause[125] contains a seemingly benign certification that must accompany every application for payment. This certification of conformance reads, in relevant part, as follows:

> I hereby certify, to the best of my knowledge and belief, that—
> (1) The amounts requested are only for performance in accordance with the specifications, terms, and conditions of the contract;
>
> . . . .

In conjunction with relevant CQC provisions of the typical federal government construction contract, this could be construed to mean that 100 percent of the work performed has been inspected and is fully compliant with all contract requirements. The bottom line is that no certification should be considered as merely routine or simply a form that may be signed without consideration of the potentially serious consequences if it is false.

IX. Conclusion

Completion of federal government construction on time and within budget will require close attention to the contract's requirements for inspection by the contractor. The construction work may also be inspected by the government. Proper work that conforms to the contract's technical requirements and careful inspection and follow-up can avoid rework, reinspection, adverse issues at the time of final inspection and acceptance by the government, and later warranty or latent defect claims by the government.

Government construction contracts typically include extensive, comprehensive, and detailed inspection and acceptance provisions. Inspections are almost always conducted at the contractor's expense. Careful attention must be paid to such provisions. The costs of required inspection procedures and personnel must be included in the price offered prior to award.

To the extent that inspection by government personnel causes extra work or delays completion of the overall project, the contractor must assure that the government employee has authority to order extra work. Written notice to an authorized representative of the government is essential when the contractor believes that extra work has been required as a result of a government inspection.

Acceptance by an authorized government representative usually occurs at the conclusion of all work unless otherwise specified in the contract. Such acceptance is final and conclusive except for latent defects, fraud, or gross mistakes amounting to fraud. In addition, warranty provisions in many government contracts protect the government from certain defects after final acceptance.

Notes

1. FAR 52.246-12(b).
2. Avedon Corp. v. United States, 15 Cl. Ct. 648, 653–58 (1988); Elter S.A., ASBCA No. 52451, 01-1 BCA ¶ 31,373; *cf.* Tidewaters Contractors, Inc. v. Dep't of Transp., CBCA No. 50-R, 07-2 BCA ¶ 33,618 (no contractual requirement that QC plan be approved before notice to proceed would be issued).
3. For further discussions of changes and authorized government officials, see Chapters 7 and 8.
4. For further discussion of claims, see Chapter 15.
5. See Chapters 8 and 19.
6. Kaminer Constr. Corp. v. United States, 488 F.2d 980 (Ct. Cl. 1973); Teller Envtl. Sys., Inc., ASBCA No. 25550, 85-2 BCA ¶ 18,025; concerning acceptance of patent defects, *see* Bromley Contracting Co., DOTCAB No. 78-1, 81-2 BCA ¶ 15,191, Conrad Weihnacht Constr., Inc., ASBCA No. 20767, 76-2 BCA ¶ 11,963; however, concerning recovery by the government under warranty provisions, *see* Kordick & Son, Inc. v. United States, 12 Cl. Ct. 662, 668 (1987).

7. Cone Bros. Contracting Co., ASBCA No. 16078, 72-1 BCA ¶ 9444; William F. Klingensmith, Inc., GSBCA No. 5451, 83-1 BCA ¶ 16,201.

8. Elter S.A., ASBCA No. 52451, 01-1 BCA ¶ 31,373. Federal government contracts often include provisions stating that approval of submittals does not relieve the contractor of strict compliance with contract requirements. Martin Paving Co., ASBCA No. 48279, 97-2 BCA ¶ 29,085.

9. Ryan Co., ASBCA No. 53385, 03-1 BCA ¶ 32,077.

10. Pac. W. Constr., Inc., DOTBCA No. 1084, 82-2 BCA ¶ 16,045; Coastal Structures, Inc., DOTBCA 1693, 88-3 BCA ¶ 20,943; Donald C. Hubbs, Inc., DOTBCA No. 2012, et al., 90-1 BCA ¶ 22,379.

11. E.W. Eldridge, Inc., ENG BCA No. 5269, 89-3 BCA ¶ 21,899; Robert L. Rich, DOTBCA No. 1026, 82-2 BCA ¶ 15,900; Contract Maint., Inc., ASBCA No. 19603, 75-1 BCA ¶ 11,097; *see* Grumman Aerospace Corp., ASBCA No. 50090, 01-1 BCA ¶ 31,316 (aircraft manufacturing contract); J.D. Pirrotta Co., ASBCA No. 37939, 94-2 BCA ¶ 26,726 (reasonableness of government inspections measured against number of contractor deficiencies found).

12. Murdock Constr. Co., IBCA No. 1050-12-74, 77-2 BCA ¶ 12,728.

13. Centric/Jones Constructors, IBCA No. 3139, 94-1 BCA ¶ 26,404.

14. Red Circle Corp. v. United States, 398 F.2d 836 (Ct. Cl. 1968); *but see* Allomatic Indus., Inc., ASBCA No. 30301, 87-1 BCA ¶ 19,380 (a reasonable inspection procedure having been established by the government, the government was obliged to give notice to the contractor if that procedure was no longer to be followed).

15. Sw. Welding & Mfg. Co. v. United States, 413 F.2d 1167 (Ct. Cl. 1969); Mitchell Enters., Inc., ASBCA Nos. 53569, et al., 06-1 BCA ¶ 33,277.

16. See Chapters 8 and 15.

17. Ryan Co., ASBCA No. 53385, 03-1 BCA ¶ 32,077; J.D. Steele, Inc., GSBCA No. 1416, 65-2 BCA ¶ 5154.

18. Ove Gustavsson Contracting Co. v. Floete, 299 F.2d 655 (2d Cir. 1962); *cert. denied*, 374 U.S. 827 (1963).

19. See Chapters 8 and 19.

20. For further discussion of authority of the CO and other government representatives to bind the government, see Chapter 7.

21. Franklin Pavkov Constr. Co., HUDBCA Nos. 93-C-C13, 93-C-C14, 94-3 BCA ¶ 27,078; Carothers Constr. Co., ASBCA No. 41268, 93-2 BCA ¶ 25,628.

22. Winter v. Cath-dr/Balti Joint Venture, 497 F.3d 1339, 1344, 1346–48 (Fed. Cir. 2007); Silverman v. United States, 679 F.2d 865, 870 (Ct. Cl. 1982); Telenor Satellite Servs., Inc. v. United States, 71 Fed. Cl. 114 (2006); States Roofing Corp., ASBCA No. 55500, 2008 WL 5392082 (Dec. 9, 2008); Romac, Inc., ASBCA No. 41150, 91-2 BCA ¶ 23,918; Parking Co. of Am., Inc., GSBCA No. 7656, 87-2 BCA ¶ 19,823.

23. Triax Co., ASBCA No. 31231, 89-1 BCA ¶ 21,485; Zablocki & Assocs., VABCA No. 2438, 88-2 BCA ¶ 20,675.

24. For further discussion of requests for equitable adjustments, see Chapter 15.

25. *Winter*, 497 F.3d at 1347; *States Roofing* 08-2 BCA ¶ 33,970.

26. Standard Coating Serv., Inc., ASBCA Nos. 48611, 49201, 00-1 BCA ¶ 30,725.

27. Crown Coat Front Co. v. United States, 292 F.2d 290 (Ct. Cl. 1961); Northrop Grumman Corp., ASBCA No. 56399, 05-2 BCA ¶ 32,992; Fentress Bradburn Architects, Ltd. v. GSA, GSBCA No. 15898, 02-2 BCA ¶ 32,011.

28. Maxwell Dynamometer v. United States, 386 F.2d 855 (Ct. Cl. 1967); Transtechnology Corp., Space Ordnance Div. v. United States, 22 Cl. Ct. 349, 361 (1990).

29. Wash. Technological Assocs., Inc., ASBCA No. 10048, 65-2 BCA ¶ 4892; Gordon H. Ball, Inc., ASBCA No. 8316, 1963 BCA ¶ 3925.

30. Blair v. United States, 99 Ct. Cl. 71 (1942); Corbetta Constr. Co., ASBCA No. 5045, 60-1 BCA ¶ 2613.

31. *See, e.g.*, Okland Constr. Co., GSBCA No. 3557, 72-2 BCA ¶ 9675.

32. FAR 52.246-12(h); *see* Capital Westward, Inc., AECBCA No. 89-5-71, 71-2 BCA ¶ 9153 (contract for supply and installation of nuclear waste storage vessels).

33. EM Sys., Inc., ASBCA No. 51782, 01-2 BCA ¶ 31,586; Gen. Time Corp., ASBCA No. 22306, 80-1 BCA ¶ 14,393.

34. Ace Constructors, Inc. v. United States, 70 Fed. Cl. 253, 287–88 (2006); H.E. Johnson Co., ASBCA No. 48248, 97-1 BCA ¶ 28,921 (unreasonable not to allow work to proceed following test completion).

35. Sancolmar Indus., Inc., ASBCA No. 16193, 74-1 BCA ¶ 10,426.

36. Alonso & Carus Iron Works, Inc., ASBCA Nos. 38312, 40334, 90-3 BCA ¶ 23,148.

37. *See, e.g.*, Glover Contracting Co., ASBCA No. 24973, 84-1 BCA ¶ 16,994.

38. Max Drill, Inc. v. United States, 427 F.2d 1233, 1249 (Ct. Cl. 1976); L.B. Samford v. United States, 410 F.2d 782, 788 (Ct. Cl. 1969); FAR 52.246-12(d); *see* Winter v. Cath-dr/Balti Joint Venture, 497 F.3d 1339 (Fed. Cir. 2007); States Roofing Corp., ASBCA No. 55500, 2008 WL 5392082 (Dec. 9, 2008) (lack of authority further affected by agency (DoD) regulations).

39. See Chapter 8.

40. Calfon Constr. Inc. v. United States, 18 Cl. Ct. 426, 438–39 (1989), *aff'd*, 923 F.2d 872 (Fed. Cir. 1990).

41. *Id.*; R.P. Richards Constr., Inc., DOTCAB Nos. 4019, 4032, 4048, 01-2 BCA ¶ 31,594.

42. G.M. Shupe Inc. v. United States, 5 Cl. Ct. 662, 727 (1984); Scientific Coating Co., VABCA No. 2400, 87-3 BCA ¶ 20,161.

43. Universal Dev. Corp. v. GSA, GSBCA Nos. 12138 (11520)-REIN, 12139 (11529)-REIN, 93-3 BCA ¶ 26,100; *but see* Robichaud v. GSA, GSBCA No. 13975, 97-2 BCA ¶ 29,209.

44. Williams v. United States, 127 F. Supp. 617 (Ct. Cl.), *cert. denied*, 349 U.S. 938 (1955); Kumin Assocs., Inc., LBCA No. 94-BCA-3, 98-2 BCA ¶ 30,007; Parking Co. of Am., Inc., GSBCA No. 7654, 87-2 BCA ¶ 19,823; HFS, Inc., ASBCA Nos. 43750, et al., 92-3 BCA ¶ 25,198; *but see* Harbert/Lummus Agrifuels Projects v. United States, 142 F.3d 1429, 1433 (Fed. Cir. 1998) (no ratification of implied-in-fact contract formation based on silence).

45. KRW, Inc., DOTCAB No. 2572, 94-1 BCA ¶ 26,435; Selma Apparel Corp., ASBCA No. 30011, 88-3 BCA ¶ 20,928.

46. Canadian Commercial Corp., ASBCA No. 17187, 77-2 BCA ¶ 12,758.

47. Caddell Constr. Co., VABCA No. 5608, 03-2 BCA ¶ 32,257.

48. *Id.*

49. Standard Elecs. Corp., ASBCA No. 14753, 73-2 BCA ¶ 10,137; Manhattan Lighting Equip. Co., ASBCA No. 6533, 61-2 BCA ¶ 3140; *see* Teledyne McCormick-Selph, ASBCA No. 15664, 73-2 BCA ¶ 10,243 (supply contract).

50. Sw. Welding & Mfg. Co. v. United States, 413 F.2d 1167, 1176 n.7 (Ct. Cl. 1969); *see* Northrop Grumman Corp., ASBCA Nos. 53699, et al., 05-2 BCA ¶ 32,992, *aff'g* 04-2 BCA ¶ 32,804 (manufacturing contract).

51. *See* Miller Elevator Co. v. United States, 30 Fed. Cl. 662, 696–99 (waiver by government of strict compliance after long-standing acceptance of contractor work reports).

52. CEMS, Inc. v. United States, 59 Fed. Cl. 168, 201 (2003).

53. Ahern Painting Contractors, Inc., GSBCA No. 7912, et al., 90-1 BCA ¶ 22,291, citing Kaminer Constr. Corp. v. United States, 488 F.2d 980 (Ct. Cl. 1973).

54. Malone v. United States, 849 F.2d 1441, 1445–46 (Fed. Cir. 1988); Walsky Constr. Co., ASBCA No. 36940, 90-2 BCA ¶ 22,934.

55. FAR 46.407(g); *see* Cudahy Packing Co. v. United States, 75 F. Supp. 239 (Ct. Cl. 1948) (inspection and acceptance of food products based on "Uniform Sales Act"); White Tiger Graphics, Inc., VABCA No. 7208, 05-2 BCA ¶ 33,016 (supply contract).

56. Wash. Constr. Co., ENG BCA No. 5318, 89-3 BCA ¶ 22,077; William F. Klingensmith, Inc., GSBCA No. 5451, 83-1 BCA ¶ 16,201; *see* Max Bauer Meat Packer, Inc. v. United States, 458 F.2d 88 (Ct. Cl. 1972) (food products; applicability of Uniform Commercial Code); Snowbird Indus., Inc., ASBCA No. 31368, 88-2 BCA ¶ 20,618 (supply contract); Cone Bros. Contracting, ASBCA No. 16078, 72-1 BCA ¶ 9444 (government had testing responsibility and failed to publish test results promptly).

57. Amigo Bldg. Corp., ASBCA No. 54329, 05-2 BCA ¶ 33,047; FAR 52.246-12(b).

58. FAR 52.246-12(b).

59. WRB Corp. v. United States, 183 Ct. Cl. 409 (1968); Gonzales Custom Painting, Inc., ASBCA No. 39547, et al., 90-3 BCA ¶ 22,950.

60. Gholson, Byars & Holmes Constr. Co. v. United States, 351 F.2d 987, 999 (Ct. Cl. 1965); Tomahawk Constr. Co., ASBCA No. 41717, 93-3 BCA ¶ 26,219.

61. Sw. Welding & Mfg. Co. v. United States, 513 F.2d 639 (Ct. Cl. 1975); W.G. Cornell Co. v. United States, 376 F.2d 299, 311 (1967); Jimenez, Inc., VABCA No. 6611, et al., 02-2 BCA ¶ 32,019.

62. *WRB Corp.*, 183 Ct. Cl. 409; Rodan Commercial Contractors, Inc., ASBCA No. 34853, 88-2 BCA ¶ 20,579.

63. A&D Fire Prot., Inc., ASBCA Nos. 53103, 53838, 02-2 BCA ¶ 32,053.

64. Batteast Constr. Co., ASBCA No. 35841, 91-3 BCA ¶ 24,352.

65. Williams & Dunlap, Gen. Contractors, ASBCA No. 6145, 1963 BCA ¶ 3834; Warren Painting Co., ASBCA No. 6511, 61-2 BCA ¶ 3199; *see* Vi-Mil, Inc., ASBCA No. 25111, 82-2 BCA ¶ 15,840 (manufacturing contract).

66. *WRB Corp.*, 183 Ct. Cl. 409; J.J. Barnes Constr. Co., ASBCA No. 27876, 85-3 BCA ¶ 18,503.

67. Montgomery Ross Fisher, Inc., H.A. Lewis, Inc., J.V., PSBCA No. 3261, 93-2 BCA ¶ 25,834.

68. *See* Gen. Time Corp., ASBCA No. 22306, 80-1 BCA ¶ 14,393 (manufacturing contract).

69. Pinay Flooring Prods., Inc., GSBCA No. 9286, 91-2 BCA ¶ 23,682; Tester Corp., ASBCA No. 21312, 78-2 BCA ¶ 13,373, *recon. denied*, 79-1 BCA ¶ 13,725, *aff'd*, 227 Ct. Cl. 648 (1981); *see* Smith of Galeton Gloves, Inc., ASBCA No. 50580, 99-1 BCA ¶ 30,269 (supply contract).

70. Mega Constr., Inc., ASBCA No. 32127, 88-1 BCA ¶ 20,427.

71. *See* Emerson-Sack-Warner Corp. v. United States, 416 F.2d 1335, 1344–45 (Ct. Cl. 1969) (supply contract); Technical Ordnance, Inc., ASBCA No. 34748, 89-2 BCA ¶ 21,818 (supply contract).

72. Centex Bateson Constr. Co., VABCA No. 4802, 97-2 BCA ¶ 29,194.

73. Toombs & Co., ASBCA Nos. 37620, et al., 91-1 BCA ¶ 23,403; *see* Goal Chem. Sealants Corp., GSBCA Nos. 8627, 8628, 88-3 BCA ¶ 21,083 (supply contract); Cal. Reforestation, AGBCA No. 88-254-1, 91-3 BCA ¶ 24,306 (tree planting contract).

74. Shipco Gen., Inc., ASBCA Nos. 29206, 29942, 86-2 BCA ¶ 18,973.

75. WRB Corp. v. United States, 183 Ct. Cl. 409 (1968); Steele & Sons, Inc., ASBCA No. 49077, 00-1 BCA ¶ 30,387.

76. Donald C. Hubbs, Inc., DOTCAB Nos. 2012, et al., 90-1 BCA ¶ 22,379; *see* H&H Enters., Inc., ASBCA Nos. 27081, 26864, 27920, 86-2 BCA ¶ 18,794 (supply contract).

77. Praoil, S.r.L., ASBCA Nos. 41499, 44369, 94-2 BCA ¶ 26,840.

78. *WRB Corp.*, 183 Ct. Cl. 409.

79. *See* NMS Mgmt., Inc., ASBCA No. 53444, 03-2 BCA ¶ 32,340 ("anecdotal" examples of overinspection insufficient to prove that government imposed overly stringent inspection standards).

80. Roberts v. United States, 357 F.2d 938, 942–43 (Ct. Cl. 1966); Stanley W. Wasco, ASBCA No. 12288, 68-1 BCA ¶ 6986.

81. K&M Constr., ENG BCA Nos. 3121, et al., 72-1 BCA ¶ 9195.

82. Southland Constr. Co., VABCA Nos. 2217, 2543, 89-1 BCA ¶ 21,548.

83. Sw. Welding & Mfg. Co. v. United States, 513 F.2d 639 (Ct. Cl. 1975); *see* Int'l Verbatim Reporters, Inc. v. United States, 9 Cl. Ct. 710, 718–19 (1986) (nonconstruction services contract).

84. Granite Constr. Co. v. United States, 962 F.2d 998, 1005–07 (Fed. Cir. 1992), *cert. denied*, 506 U.S. 1048 (1993).

85. Len Co. & Assocs. v. United States, 385 F.2d 438, 450–51 (Ct. Cl. 1967); Canadian Commercial Corp., ASBCA No. 17187, 76-2 BCA ¶ 12,145.

86. See Chapters 8 and 19.

87. *See* Allomatic Indus., Inc., ASBCA No. 30301, 87-1 BCA ¶ 19,380 (supply contract).

88. Maint. Eng'rs, ASBCA No. 17474, 74-2 BCA ¶ 10,760.

89. Russell R. Gannon Co. v. United States, 417 F.2d 1356, 1358–59 (Ct. Cl. 1969).

90. G.W. Galloway Co., ASBCA Nos. 16656, 16975, 73-2 BCA ¶ 10,270.

91. Craig Enters., AGBCA Nos. 93-227-1, 92-183-1, 95-2 BCA ¶ 27,766; FAR 46.407(g).

92. Towne Realty, Inc. v. United States, 1 Cl. Ct. 264, 268 (1982).

93. *See* Exquisite Serv. Co., ASBCA No. 21058, 77-2 BCA ¶ 12,799 (non-construction services contract).

94. Lionsgate Corp., ENG BCA No. 5809, 92-2 BCA ¶ 24,983.

95. Southland Constr. Co., VABCA No. 2579, 89-2 BCA ¶ 21,704; *see* Frederick P. Warrick Co., ASBCA Nos. 9644, 9685, 65-2 BCA ¶ 5169.

96. Granite Constr. Co. v. United States, 962 F.2d 998, 1006–07 (Fed. Cir. 1992), *cert. denied*, 506 U.S. 1048 (1993).

97. Wickham Contracting Co., ASBCA No. 32392, 88-1 BCA ¶ 20,559.

98. FAR 52.246-12(i).

99. RNJ Interstate Corp. v. United States, 181 F.3d 1329, 1331–32 (1999); FAR 52.236-7.

100. Miller Elevator Co. v. United States, 30 Fed. Cl. 662, 693–95 (1994); Peter Bauwens Bauunternehmung GmbH & Co., ASBCA No. 44679, 98-1 BCA ¶ 29,551; *but see* Design & Prod., Inc. v. United States, 18 Cl. Ct. 168, 207–08 (1989) (other contract provision expressing authority in the contracting officer's technical representative to accept work).

101. *See* Allstate Leisure Prods., Inc., ASBCA No. 35614, 89-3 BCA ¶ 22,003 (manufacturing contract; CO ratification of government quality assurance representative's contract interpretation resulting in extra work).

102. Blinderman Constr. Co. v. United States, 39 Fed. Cl. 529, 571–74 (1997).

103. L.A. Barton & Co., ASBCA No. 17547, 73-2 BCA ¶ 10,249 (considering a three-month delay in inspecting paintwork on fuel tanks unreasonable and therefore constituting constructive acceptance of the work by the government).

104. John C. Kohler Co. v. United States, 498 F.2d 1360, 1366 (Ct. Cl. 1974).

105. Lobar, Inc., ASBCA No. 48699, 96-1 BCA ¶ 28,079; FAR 52.232-5(f).

106. Sentell Bros., Inc., DOTBCA No. 1824, 89-3 BCA ¶ 21,904, *recon. denied*, 89-3 BCA ¶ 22,219; FAR 52.246-12(i).

107. *See* Instruments for Indus., Inc. v. United States, 496 F.2d 1157, 1159–62 (2d Cir. 1974) (bankruptcy proceeding under a supply contract).

108. Kaminer Constr. Corp. v. United States, 488 F. 2d 980, 987 (Ct. Cl. 1973).

109. Dale Ingram, Inc., ASBCA No. 12152, 74-1 BCA ¶ 10,436.

110. *Kaminer*, 488 F. 2d at 987.

111. Perkin-Elmer Corp. v. United States, 47 Fed. Cl. 672, 675 (2000).

112. FAR 52.246-12(i).

113. Bender GmbH, ASBCA No. 52266, 04-1 BCA ¶ 32,474.

114. Henry Angelo & Co., ASBCA No. 30502, 87-1 BCA ¶ 19,619.

115. Catalytic Eng'g & Mfg. Corp., ASBCA No. 15257, 72-1 BCA ¶ 9342.

116. Warren Beaves d/b/a Commercial Marine Servs., DOTCAB No. 1160, 84-1 BCA ¶ 17,198.

117. FAR 52.246-12(i).

118. *Catalytic*, 72-1 BCA ¶ 9342, *recon. denied*, 72-2 BCA ¶ 2518; *see also* Peters Mach. Co., ASBCA No. 21857, 79-1 BCA ¶ 13,649.

119. Bender GmbH, ASBCA No. 52266, 04-1 BCA ¶ 32,474.

120. *Catalytic*, 72-1 BCA ¶ 9342.
121. FAR 52.246-21.
122. Land O'Frost, ASBCA Nos. 52012, 52241, 03-2 BCA ¶ 32,395.
123. FAR 52.246-21(a).
124. See Chapter 24.
125. FAR 52.232-5.

CHAPTER 11

Payment and Contract Funding

LORI ANN LANGE

I. Introduction

Receiving full and timely payment from the government for work performed under government contracts is often essential to continued contract performance and the financial integrity of the contractor. The Federal Acquisition Regulation (FAR) contains detailed clauses regarding the payment of contractors. This chapter addresses five issues relating to payment on government contracts: progress payments, late payment of contractor invoices, assignment of payments to financial institutions, contractual limitations on funding, and government setoff.

II. Progress Payments

Government contracts, especially construction contracts, require a large outlay of money by the contractor prior to the receipt of any payment from the government. If a contractor had to wait until completion of a project in order to receive any payment, it is unlikely that many contractors would be able to perform government construction contracts. In order to help the contractor finance performance, government construction contracts provide for the disbursement of progress payments.

There are two types of progress payments: payments based on costs incurred and payments based upon a percentage or stage of completion of the work. Progress payments based upon a percentage or stage of completion are used for fixed-price construction contracts,[1] and the applicable FAR clause is the Payments Under Fixed-Price Construction Contracts clause.[2]

Generally speaking, progress payments on construction contracts are made monthly and are based on estimates of the completed work that meets the contract standards.[3] The contractor must submit a request for progress payment that includes:

1. An itemization of the amounts requested, related to the various elements of work required by the contract covered by the payment requested;
2. A listing of the amount included for work performed by each subcontractor under the contract;
3. A listing of the total amount of each subcontract under the contract;
4. A listing of the amounts previously paid to each such subcontractor under the contract; and
5. Any additional supporting data in a form and detail required by the contracting officer.[4]

The contractor must certify each progress payment request.[5] The language of the certification is contained in the Payments Under Fixed-Price Construction Contracts clause,[6] which requires the contractor to certify that:

1. The amounts requested are only for work performed in accordance with specifications, terms, and conditions of the contract;
2. All payments due to subcontractors and suppliers from previous payments received under the contract have been made, and timely payments will be made from the proceeds of the payment covered by the certification in accordance with subcontract agreements and the requirements of 31 U.S.C. chapter 39;
3. The request does not include any amounts that the prime contractor intends to withhold or retain from a subcontractor or supplier in accordance with the terms and conditions of the subcontract; and
4. The certification is not to be construed as final acceptance of a subcontractor's performance.[7]

The contractor is permitted to delete the last item from the certification.[8] If the government agrees with the stage of completion in the contractor's progress payment request, it will pay the amount of the request. If the government disagrees with the stage of completion or believes that some aspect of the contractor's performance does not comply with the contract requirements, it will pay the amount it determines to be due the contractor.

The government may withhold a portion of the progress payment as retainage if the government determines that satisfactory progress has not been achieved.[9] Retainage acts as an incentive for the contractor to complete the contract and protects the government's interests against potential default.[10] The amount to be withheld is determined by the contracting officer on a case-by-case basis, but it cannot exceed 10 percent of the amount of the progress payment otherwise due the contractor.[11]

If the contractor discovers that a portion of the certified progress payment request includes payment for work that fails to conform to the contract requirements (that is, an unearned payment), the contractor must notify the contracting officer of the performance deficiency.[12] If the contractor has already received the unearned payment, it must pay the government interest on the unearned payment.[13] The interest is calculated at the rate of average bond equivalents of 91-day Treasury bills auctioned at the most recent auction.[14] Interest will continue to accrue until the contractor notifies the contracting officer that the deficiency has been corrected, or the contractor reduces the amount of any subsequent certified progress payment request by the unearned amount.[15]

Once the government makes a progress payment, it acquires title to the material and work covered by the progress payment. The Payments Under Fixed-Price Construction Contracts clause states that all material and work covered by a progress payment become the sole property of the government at the time of the payment.[16] In *G&R Service Co. v. Department of Agriculture,*[17] however, the Civilian Board of Contract Appeals held that the transfer of title provision in the Payments Under Fixed-Price Construction Contracts clause was intended to apply solely to material delivered on-site for incorporation in the work. If the on-site material is not ultimately incorporated into the work, title to the unincorporated material revests in the contractor after final acceptance of the work. As the Civilian Board of Contract Appeals explained:

> For fixed price construction contracts, progress payments *generally* are geared to completion of work. In other words, the contractor is provided periodic progress payments as construction progresses, based on estimates of the percentage of completion achieved, as measured against a schedule of values, a breakdown of the contract price into various work items. *See* Appeal File at 100. Progress payments for such contracts, however, are not restricted to work completion. As noted earlier, the "Payments Under Fixed Price Construction Contracts" clause contemplates additional consideration being given to materials delivered to the site (or to an offsite storage facility, with the approval of the contracting officer). At least for purposes of materials-related progress payments, the payments are based on costs incurred. Thus, even though the "Progress Payments" clause under FAR 52.232-16 may not be specifically incorporated into federal fixed price construction contracts, because that provision is mandatory whenever progress payments under any fixed price contract are to be cost-based, FAR 32.502-4(a)(1)(ii), the clause's treatment of title reversion upon contract completion should be applied to any progress payments that the Government may make on the basis of costs—such as the December 12, 2005, progress payment in this case. Accordingly, we refuse to read the "sole property of the Government" language of the present contract's "Payments Under Fixed Price Construction Contracts" clause as prescribing a permanent divestiture of title to any materials other than those that have become part of "discrete work items completed during the course of construction." *See Reddick & Sons of Gouverneur, Inc. v. United States*, 31 Fed. Cl. 558, 561 (1994). Payments for materials under the construction contract Payments clause is intended solely for

"material delivered on site *for incorporation in the work." See C. Lawrence Construction Co.*, ASBCA 45270, 93-3 BCA ¶ 26,129, at 129,886 (emphasis supplied). Thus, although the Payments clause language created "a security interest in favor of the Government in materials . . . covered by the progress payments" made to G&R, *Skip Kirchdorfer, Inc. v. United States*, 6 F.3d 1573, 1581 (Fed. Cir. 1993), once G&R completed the construction and the construction was inspected and accepted by the Forest Service, there no longer was a need to secure the prior progress payments, and, as with other contracts where progress payments have been completely liquidated by a final delivery of product, title to any unused materials in this case revested in the contractor, G&R. At that juncture, the terms and conditions of the Payments clause of the instant contract regarding Government property interests in materials covered by interim progress payments, other than materials incorporated into the completed construction, no longer were operative.[18]

It should be noted that the transfer of title does not relieve the contractor from the sole responsibility for all material and work upon which payments have been made and the restoration of any damaged work.[19] In other words, if the materials or work are damaged or destroyed prior to final acceptance, the contractor is obligated to repair or replace them. Nor does the transfer of title constitute a waiver of the government's right to require the contractor to fulfill all of the terms of the contract.[20] The contractor remains obligated to fully perform the contract in accordance with the contract terms and conditions.

In order to receive final payment after completion and acceptance of the work, the contractor must present a properly executed voucher to the contracting officer.[21] The contractor must also provide the contracting officer with a release of all claims against the government arising by virtue of the contract, other than claims, in stated amounts, that the contractor has specifically excepted from the operation of the release.[22] If the contractor has assigned its right to payment to a financing institution, the contracting officer may also require a release from the assignee.[23]

III. Prompt Payment

Enacted in 1982, the Prompt Payment Act[24] (PPA) was intended to ensure that the federal government pays its contractors in a timely manner. The PPA requires the government to pay proper invoices by a required due date. The required due date for the payment of invoices is generally 30 days, although certain construction contract invoices have a shorter required due date.[25] For example, progress payments based on the contracting officer's approval of the estimated amount and value of work or services performed are due within 14 days.[26] If the government fails to make timely payment, the contractor is entitled to interest.[27] The contractor need not specially request that the interest be paid.[28]

A. Proper Invoices

The requirement to pay interest is contingent upon the contractor's submission of a proper invoice.[29] To be considered proper, the Prompt Payment for Construction Contracts clause requires that an invoice include the following items:

1. Name and address of the contractor;
2. Invoice date and number;
3. Contract number or other authorization for supplies delivered or services performed (including order number and contract line item number);
4. Description of the work or services performed;
5. Delivery and payment terms (such as any discount for prompt payment terms);
6. Name and address of contractor official to whom payment is to be sent (this must be the same as that in the contract or in a proper notice of assignment);
7. Name (where practicable), title, phone number, and mailing address of the person to be notified in event of a defective invoice;
8. Taxpayer Identification Number (TIN) if required by the contract;
9. Electronic funds transfer banking information if required by the contract; and
10. Any other information or documentation required by the contract.[30]

Invoices for progress payments based upon the stage of completion must substantiate the amounts requested and be certified in accordance with the Payments Under Fixed-Price Construction Contracts clause, FAR 52.232-5.[31]

If the contractor's invoice is defective, the government is required to return the invoice and notify the contractor of the defect within seven days.[32] Interest will then run from the submission of the corrected invoice if the payment is late. If the government fails to inform the contractor that the invoice is defective, the agency must pay the invoice by the required due date or be liable for prompt payment interest.[33]

B. Disputed Invoices

PPA interest does not accrue when the government's failure to make timely payments is due to a dispute between the government and contractor as to an amount claimed or compliance with the contract requirements.[34] In order to create a bona fide dispute, the government need only have a good-faith basis to question the contractor's invoice.[35] The fact that the contractor may ultimately prevail on the merits does not mean that the government did not have a good-faith basis to withhold or delay the payment. When there is a disputed invoice, in order to recover interest on disputed amounts, the contractor must turn the unpaid invoice into a claim under the Contract Disputes Act (CDA).[36]

Conversely, a construction contractor must pay the government an amount equal to a PPA interest penalty if the contractor is paid for work that fails to conform to the contract requirements.[37] The interest penalty runs from the date the contractor received the payment to the date the contractor (1) notifies the agency that the performance deficiency has been corrected; or (2) reduces the amount of any subsequent certified application for payment to the agency by an amount equal to the unearned amount.[38]

C. Required Due Date Under Construction Contracts

As previously noted, the government must pay the contractor by the required due date or be subjected to an interest penalty. The Prompt Payment for Construction Contract clause[39] sets forth the required due dates for payments under construction contracts. The clause provides that the due date for making payment on progress payment invoices is 14 days after the designated billing office receives a proper payment request.[40] The due date for final payment is 30 days after receipt of a proper invoice or 20 days after government acceptance of the work, whichever is later.[41] Retainage, to the extent that it has not been released prior to final payment, is priced with the final payment.[42] If final payment is subject to contract settlement actions such as a release of claims, acceptance is deemed to have occurred on the effective date of the settlement.[43]

D. Payment Date

Payment is considered made on the day a check is dated or on the date of an electronic funds transfer.[44] When the payment due date falls on a Saturday, Sunday, or legal holiday, payment is not due until the following working day.[45]

E. Rate of Interest and Length of Payment

Interest on invoices is paid automatically by the procuring agency's payment office.[46] The interest rate is set by the Treasury Department under Section 12 of the CDA[47] and is compounded monthly.[48]

Interest begins to run the day after the required payment is due and continues to run until the date of payment.[49] After 30 days, the amount of the interest penalty is added to the principal amount of the debt, and the interest penalty thereafter accrues on the added amount.[50] Interest will cease to accrue after one year or after the contractor files a claim for unpaid interest under the CDA.[51] When the contractor files a CDA claim, the CDA's interest provisions apply.[52]

A contractor is entitled to an additional penalty if the government pays the invoice but does not pay the interest penalty within 10 days after the invoice amount is paid.[53] In order to receive the additional penalty, the contractor must make a written request for payment within 40 days of when the principal amount is paid, include a copy of the original invoice, and state when the payment of the principal amount was received.[54] The amount of the additional penalty is 100 percent of the original interest penalty or $5,000, whichever is less.[55]

F. Contract Financing Payments

Under the PPA and its implementing regulations, prompt payment interest does not apply to contract financing payments.[56] Contract financing payments are defined as disbursements of monies to a contractor prior to acceptance of the supplies or services.[57] Contract financing payments include advance payments, performance-based payments on non-construction contracts, commercial advance and interim payments, progress payments based on cost on non-construction contracts, and progress payments based on a percentage or stage of completion on non-construction contracts.[58] The FAR excludes progress payments made under construction and architect-engineer contracts from the definition of contract financing payments.[59] Thus, prompt payment interest applies to late progress payments made under the construction and architect-engineer contracts.

G. Prompt Payment to Subcontractors and Suppliers

The PPA and its implementing regulations contain provisions requiring construction contractors to pay their subcontractor and supplier invoices in a timely manner. Prime contractors must pay their subcontractors and suppliers for work performed satisfactorily within seven days of receipt of payment from the government.[60] If the prime contractor fails to pay the subcontractor/ supplier in a timely manner, the prime contractor must pay an interest penalty on the amount due.[61] Interest runs from the day after the required payment date to the date payment actually is made. It is computed at the rate set by the Treasury Department under Section 12 of the CDA.[62]

In addition to the obligation to pay interest to subcontractors and suppliers, a contractor may be liable to the government if it falsely certifies that it has timely paid its subcontractors. As described above, the standard certification contained in the Payments Under Fixed-Price Construction Contracts clause requires the contractor to certify that payments due to subcontractors and suppliers from previous payments received from the government under the contract have been made, and timely payments will be made from the proceeds of the payment covered by the certification in accordance with subcontract agreements and the PPA.[63]

The PPA also contains a flow-down requirement.[64] The prime contractor must require each first-tier subcontractor to include a flow-down clause mandating prompt payment of lower-tier subcontractors.[65] In turn, the first-tier subcontractors must include a flow-down clause in their subcontracts.

The requirement to pay subcontractors/suppliers promptly does not preclude the prime contractor from withholding retainage, even if the subcontractor's work complies fully with the subcontract, when retainage is permitted under the subcontract.[66] Nor does it prohibit the prime contractor from withholding payment for deficient work in accordance with the terms of the subcontract. The prime contractor, however, must give timely notice to the subcontractor of the withholding and the reason for the withholding, and provide a copy of the notice to the contracting officer.[67] The prime contractor must pay the subcontractor the amount being retained within seven days after the correction of the deficiency, unless the prime contractor must recover the funds from the government.[68] In such a case, the prime contractor must pay the subcontractor within seven days after receiving the funds from the government.[69]

Contractors that elect to withhold funds from their subcontractors and suppliers must not only comply with the FAR withholding requirements and procedures set forth above, but must also pay special attention to the details of their progress payment applications to the government. This is because the standard certification contained in the Payments Under Fixed-Price Construction Contracts clause requires the contractor to certify that the application does not include any amounts that the contractor intends to withhold or retain from a subcontractor.[70]

H. Government Overpayments

The Prompt Payment for Construction Contracts clause requires the contractor to reimburse the government for duplicate payments or overpayments.[71] If the contractor becomes aware of a duplicate contract financing or invoice payment, or learns that the government has otherwise overpaid on a contract financing or invoice payment, the contractor is required to:

1. Remit the overpayment amount to the payment office; and
2. Provide the payment office with a description of the overpayment including
   a. the circumstances of the overpayment (e.g., duplicate payment, erroneous payment, liquidation errors, date(s) of overpayment);
   b. the affected contract number and delivery order number, if applicable;
   c. the affected contract line item or subline item, if applicable; and
   d. the contractor point of contact.[72]

In addition, the contractor must provide a copy of the remittance and supporting documentation to the contracting officer.

If the contractor discovers that a portion of the certified progress payment request includes payment for work that fails to conform to the contract requirements (that is, an unearned payment), the contractor must notify the contracting officer of the performance deficiency.[73] If the contractor already has received the unearned payment, it must pay the government interest on the unearned payment.[74] The interest is calculated at the rate of average bond equivalents of 91-day Treasury bills auctioned at the most recent auction.[75] Interest will continue to accrue until the contractor notifies the contracting officer that the deficiency has been corrected, or the contractor reduces the amount of any subsequent certified progress payment request by the unearned amount.[76]

IV. *Assignment of Payments*

Government contractors often receive financing from banks and other financial institutions in order to fund performance of their contracts. As security for the loan, the contractor may be required to assign its right to payments from the government to the financial institution. In order for that assignment to be effective against the government, the assignment must comply with the Anti-Assignment Acts and their implementing regulations in FAR Subpart 32.8, Assignment of Claims.[77]

A. Validity of Assignments

Assignments to financial institutions are valid as long as the following conditions are met:

1. The aggregate contract payments exceed $1,000 or more;
2. The assignment is made to a bank, trust company, or other financing institution, including any federal lending agency;
3. The contract does not prohibit the assignment;
4. The assignment covers all unpaid amounts payable under the contract unless otherwise permitted in the contract;
5. The assignment is made only to one party unless otherwise permitted in the contract;
6. The assignment is not subject to further assignment unless otherwise permitted by the contract;
7. The assignee sends written notice of the assignment together with a true copy of the assignment instrument to the contracting officer or agency head, the surety on any bond applicable to the contract, and the disbursing officer designated in the contract.[78]

The contract may prohibit assignment only if the agency determines that it is in the best interest of the government.[79]

B. Requirement for a Valid Assignment

FAR 32.805(a) states the requirements for a valid assignment. Generally, an assignment by a corporation must be executed by an authorized representative of the corporation, attested by the secretary or the assistant secretary of the corporation, and impressed with the corporate seal or accompanied by a true copy of the resolution authorizing the representative to execute the assignment.

As previously noted, the assignee must provide notice of the assignment to the contracting officer or agency head, the surety on any bond applicable to the contract, and the disbursing officer designated in the contract.[80] The assignee must provide each party with an original and three copies of the notice of assignment, and one true copy of the instrument of assignment.[81] FAR 32.805(c) contains a suggested format for the notice of assignment. Upon receipt of the notice of assignment, the government will acknowledge its receipt and thereafter make payments to the assignee. If there is a valid assignment and the government pays the contractor instead of the assignee, the assignee may recover the amount of the improper payment from the government.[82]

C. The Government's Right to Set Off Assigned Payments

Payments made by the government to the assignee cannot be recovered on account of any liability of the contractor to the government.[83] In other words, the government cannot recover money paid to an assignee under a valid assignment when the contractor who assigned the payment owes money to the government.

If the contract contains a no-setoff commitment, the assignee has even greater rights. In such cases, the government may not set off or reduce the payment to the assignee as a result of any contractor liability arising independently of the contract (that is, debts of the contractor unrelated to the performance of the particular contract).[84] The government also may not set off or reduce payments for the following contractor liabilities arising from the assigned contract:

1. Renegotiation under any statute or contract clause;
2. Fines;
3. Penalties, exclusive of amounts that may be collected or withheld from the contractor under, or for failure to comply with, the terms of the contract;
4. Taxes or social security contributions; and
5. Withholdings or nonwithholdings of taxes or social security contributions.[85]

Even if the contract contains a no-setoff commitment, the government still may set off debts owed by the assignor contractor arising under the contract.[86] For example, the government may set off liquidated damages, reprocurement costs, and government claims. The government also may reduce payment if the assignee has not made a loan under the assignment, has not made a commitment to make a loan, or the amount due on the contract exceeds the amount of any loans made or expected to be made.[87]

D. Obligations of the Parties

The creation of an assignment does not alter the relationship between the government and the contractor nor does the assignee become a party to the contract.[88] The contractor is still required to perform the work in accordance with the contract requirements. The assignee is not in privity with the government as a result of the assignment and is not a contractor under the CDA.[89]

An assignee may sue the government in order to recover for work performed by the assignor contractor.[90] Any recovery is limited to the assignor's outstanding debt to the assignee.[91] Claims by assignees must be brought under the Assignment of Claims Act,[92] not the CDA.[93]

V. Contractual Limitations on Funding

In order to prevent federal agencies from obligating the United States to make payments for goods and services received by the agencies when the agencies have insufficient funds to pay for those goods and services, Congress enacted various statutes curtailing the agencies' authority to make or authorize an expenditure of funds. Two of these statutes are collectively known as the Anti-Deficiency Act.[94] According to the Government Accountability Office (GAO), the purpose of the Anti-Deficiency Act is to prevent government officials from making payments or committing the United States to make payments at some future time for goods or services, unless there is an available appropriation to cover the cost in full.[95] The Anti-Deficiency Act prohibits:

1. Making or authorizing expenditure from, or creating or authorizing an obligation under, any appropriation or fund in excess of the amount available in the appropriation or fund unless authorized by law;[96]
2. Involving the government in any obligation to pay money before funds have been appropriated for that purpose, unless otherwise allowed by law;[97]
3. Accepting voluntary services for the United States, or employing personal services not authorized by law, except in cases of emergency involving the safety of human life or the protection of property;[98]
4. Making obligations or expenditures in excess of an apportionment or reapportionment, or in excess of the amount permitted by agency regulations.[99]

Government contract payments generally are to be charged to the fiscal year appropriation current at the time the legal obligation arose, i.e., the fiscal year in which the agency had the bona fide need for the good or service and in which the parties entered into a valid contract or agreement.[100] This may not be the fiscal year in which payment was earned. As a general rule, construction contracts represent the genuine needs of the year in which the contract was made.[101]

Once an appropriation is exhausted, the agency may not make any further payments unless it has other general funding available.[102] Nor may the agency "borrow" from an appropriation for the next fiscal year.[103]

A. Limitation/Availability of Funds Clauses

Many construction contracts are fully funded even though the work will be performed over more than one fiscal year. If the agency does not have sufficient appropriations to fully fund the contract, however, it may incrementally fund the contract. When a contract is incrementally funded, the contract will contain a clause advising the contractor of the extent of the available funding. The clause will limit the government's liability for payment up to the amount funded. Such clauses are lawful and enforceable.[104] Examples of such clauses include the Availability of Funds clause,[105] the Availability of Funds for the Next Fiscal Year clause,[106] the Department of Defense's Limitation of Government's Obligation clause,[107] and the Corps of Engineers' Incremental Funding clause.[108]

Under these types of clauses, the contractor is only required to perform work up to the point at which the total amount payable, including reimbursement in the event of a termination for convenience, approximates the total amount currently funded. The contractor must give the contracting officer written notice (usually 90 to 120 days) before the point at which the funds (or a specified percentage of the funds) will be exhausted. The contractor is entitled to stop work when the funds are exhausted. When additional funds are obligated, the contractor must resume work. Depending on the specific clause used, the contractor may or may not be entitled to an equitable adjustment for any additional costs resulting from the cessation of work, including delay damages.[109] The contractor generally will be entitled to a time extension. If the government cannot provide additional funding, the contract will be terminated for convenience.

B. Corps of Engineers' Continuing Contract Clause

The Corps of Engineers (COE) has the unique authority to award continuing contracts for civil works construction. Under a continuing contract, the COE obligates the full price of the contract in advance of appropriations where the project is specifically authorized by Congress.[110] As a result, the COE can award a multiyear construction contract when it only has appropriations to cover the first year's work. The credit of the United States is pledged to the ultimate payment of contract earnings, regardless of whether there is a shortage of funds at any given time.[111]

Continuing contracts include the Continuing Contracts clause,[112] which authorizes the contractor to continue work even though appropriations are not currently available. When funds become available, the COE will pay the contractor's costs plus interest. If the government fails to reserve sufficient additional funds to cover payments otherwise due 60 days after the beginning of the fiscal year following the exhaustion of funds, the contractor may elect to treat its right to proceed with the work as having been terminated for convenience.[113] If the amount due as a result of the termination is not paid within one year, the contractor may sue the government for breach of contract.[114]

As a result of the COE's perceived overuse of continuing contracts, Congress restricted the types of projects and work for which the COE was authorized to use continuing contracts in the Energy and Water Development Appropriations Act of 2006.[115] This Act restricted the use of continuing contracts to projects funded from the COE's operation and maintenance account or the operation and maintenance subaccount of the Mississippi River and tributaries account. It also prohibited the COE from entering into a continuing contract or modifying an existing contract to commit an amount for a project in excess of the amount appropriated for such project. In order to use a continuing contract, the COE must receive approval from COE headquarters and the Assistant Secretary of the Army (Civil Works) or his delegate. Continuing contracts may be used only if other acquisition methods cannot meet the COE's needs.

VI. Government Setoff and Debt Collection

Government contractors may become indebted to the government for a variety of reasons. The government has various self-help remedies that it can use to collect money owed to it. These remedies include setoff and withholding. Generally, setoff occurs when the government holds money due a contractor on one contract and applies those funds to debt arising under a different contract with the same contractor. Withholding, on the other hand, occurs when the government holds money due the contractor under a contract for debts arising under that contract. Often, however, the terms are used interchangeably.

A. Common Law and Statutory Right of Setoff

The government has a common law right of setoff unless there is a statutory or contractual provision barring the right.[116] The Debt Collection Act[117] did not alter or eliminate the government's common law set-off right.[118] Rather, the act merely made additional debt recovery procedures available to the government.

In order to effectuate a common law setoff, the government must take the following three steps:

1. Decide to effectuate a setoff;
2. Take some action to accomplish the setoff; and
3. Record the setoff.[119]

In *Johnson v. All-State Construction, Inc.*,[120] the Court of Appeals for the Federal Circuit held that the government satisfied these three steps when it withheld a progress payment from the construction contractor and advised the contractor that it would not make the progress payment because the amount to be retained for liquidated damages exceeded the amount of the invoice.

In addition to the common law right of setoff, there is a federal statute that allows the Treasury Department to offset money owed the government against judgments that a contractor has against the government, such as judgments by the boards of contract appeals and the Court of Federal Claims.[121] Under the statute, the Treasury Department may withhold paying part of a judgment against the United States that is equal to a debt the contractor owes the government.[122] If the contractor does not agree to the setoff or disputes the debt, the Treasury Department may also withhold any additional amount to cover the legal costs of bringing a civil action for the debt.[123]

B. FAR Debt Collection Procedures

FAR Subpart 32.6 contains the procedures used by the government to collect contract debts. Contract debts include:

1. Billing and price reductions resulting from contract terms for price redetermination or for determination of prices under incentive type contracts;
2. Price or cost reductions for defective cost or pricing data;
3. Financing payments determined to be in excess of the contract limitations;
4. Increases to financing payment liquidation rates;
5. Overpayments disclosed by quarterly statements required under price redetermination or incentive contracts;
6. Price adjustments resulting from Cost Accounting Standards (CAS) noncompliance or changes in cost accounting practice;
7. Reinspection costs for nonconforming supplies or services;
8. Duplicate or erroneous payments;
9. Damages or excess costs related to defaults in performance;
10. Breach of contract obligations concerning progress payments, performance-based payments, advance payments, commercial item financing, or government-furnished property;
11. Government expense of correcting defects;
12. Overpayments related to errors in quantity or billing or deficiencies in quality;
13. Delinquency in contractor payments due under agreements deferral or postponement of collections; and
14. Reimbursement of protest costs under FAR 33.102(b)(3) and FAR 33.104(h)(8).[124]

The first step in government debt collection is a government determination that a debt is due and the amount owed.[125] The determination of the amount due must be based on the merits of the case and must be consistent with the contract terms.[126]

Once the contracting officer determines that a debt is due and the amount of the debt, the contracting officer will issue a demand for payment.[127] The contracting officer will issue the demand even if:

1. The debt is or will be the subject of a bilateral modification;
2. The contractor is otherwise obligated to pay the money under the contract; or
3. The contractor has agreed to repay the debt.[128]

The demand for payment will include:

1. A description of the debt and the amount due;
2. A distribution of the principal amount of the debt by line(s) of accounting;
3. The basis for and amount of any accrued interest or penalty;
4. Notice that any amounts not paid within 30 days will bear interest, unless an applicable contract clause contains a different payment and interest period;
5. A statement advising the contractor to contact the contracting officer if the contractor believes the debt is invalid or the amount is incorrect, and if the contractor agrees, to remit a check payable to the agency's payment office;
6. Notice that the payment office many initiate procedures to offset the debt against any payments otherwise due the contractor;
7. Notice that the debt may be subject to administrative charges; and
8. Notice that the contractor may submit a request for installment payments or deferment of collection if immediate payment is not practicable or if the amount is disputed.[129]

The contracting officer should not issue a demand for payment if the contracting officer only becomes aware of the debt when the contractor provides a lump sum payment, submits a credit invoice, or notifies the contracting officer that the payment office overpaid on an invoice payment.[130] The contracting officer will subsequently issue a demand for payment after at least 30 days unless the contractor has liquidated the debt, the contractor requested an installment payment agreement, or the payment office issued a demand for payment.[131]

The contracting officer will issue a final decision on the debt if:

1. The contracting officer and the contractor are unable to reach agreement on the existence or amount of the debt in a timely manner;
2. The contractor fails to liquidate a debt previously demanded by the contracting officer within the timeline specified in the demand for payment; or
3. The contractor requests a deferment of collection.[132]

The contractor is generally required to pay the debt by making a lump-sum cash payment or by issuing a credit against existing unpaid bills. If the contractor does not make full payment within 30 days of the due date, request installment payments, or request deferment of collection, the payment office may begin collecting by withholding payment of the contractor's invoices.[133] The contracting officer need not issue a second final decision before the government can exercise its rights to withholding and setoff.[134] Any debt that is delinquent by more than 180 days will be transferred to the Department of the Treasury for collection.[135]

The contractor may request that the government defer collection of the debt or may propose payment through installment payments.[136] If the contractor has not appealed the debt or filed an action under the Disputes clause, the government may agree to installment payments or deferment if the contractor is unable to pay in full at once, or the contractor's operations under national defense contracts would be seriously impaired.[137] Even when the contractor has disputed the debt under the Disputes clause, the contractor still may request deferment of collection.[138]

Any installment or deferment agreement must include a specific schedule or plan for payment, as well as permit the government to make periodic financial reviews of the contractor and require earlier payments if the government considers that the contractor's ability to pay has improved.[139]

Requests for deferment must be submitted in writing to the contracting officer.[140] If the contractor has appealed the debt, the information supporting the request may be limited to an explanation of the contractor's financial condition.[141] If the contractor has not disputed the debt, its deferment request must contain the following information:

1. The contractor's financial condition;
2. The contract backlog;
3. The contractor's projected cash receipts and requirements;
4. The feasibility of immediate payment of the debt; and
5. The probable effect on the contractor's operations of immediate payment in full.[142]

The government will review the contractor's deferment request in order to determine if deferment is appropriate. When the contractor has disputed the debt, deferment may be granted to avoid possible overcollection.[143] Deferment pending an appeal also may be granted to small business contractors and financially weak contractors.[144] If the contractor has not disputed the debt, deferment may be granted if requiring immediate payment in full would seriously impair the contractor's operations under national defense contracts.[145]

If the government agrees to a deferment, the parties will enter into a deferment agreement.[146] If a contractor appeal of the debt determination is pending, the deferment agreement will include a requirement that the contractor diligently prosecute the appeal and pay the debt in full when the appeal is decided or the parties reach agreement on the debt amount.[147] The agreement may also provide for the contractor's right to make early payments without prejudice, for refund of overpayments, and for crediting of interest.[148]

The government will generally apply interest charges to contract debts that are unpaid after 30 days from the issuance of the demand or payment.[149] The interest rate is the CDA rate established by the Secretary of the Treasury.[150]

Notes

1. *See* G&R Servs. Co. v. Dep't of Agric., CBCA No. 121, 2007 WL 994561 (Mar. 14, 2007).

2. FAR 52.232-5.

3. FAR 52.232-5(b).

4. FAR 52.232-5(b)(1).

5. A contractor who submits a false or fraudulent progress payment invoice may be subject to False Claims Act liability. Morse Diesel Int'l Inc. v. United States, 74 Fed. Cl. 601 (2007).

6. FAR 52.232-5(c).

7. *Id.*

8. *Id.*

9. FAR 32.103; FAR 52.232-5(e).

10. Fireman's Fund Ins. Co. v. United States, 909 F.2d 495 (Fed. Cir. 1990); Nat'l Sur. Corp. v. United States, 118 F.3d 1542 (Fed. Cir. 1997).

11. FAR 52.232-5(e).

12. FAR 52.232-5(d)(1).

13. FAR 52.232-5(d)(2).

14. FAR 52.232-5(j)(1).

15. FAR 52.232-5(d)(2).

16. FAR 52.232-5(f).

17. CBCA No. 121, 2007 WL 994561 (Mar. 14, 2007).

18. *Id.*

19. FAR 52.232-5(f)(1).

20. FAR 52.232-5(f)(2).

21. FAR 52.232-5(h)(2).

22. FAR 52.232-5(h)(3).

23. *Id.*

24. 31 U.S.C. §§ 3901–3907. Implementing regulations are in FAR Subpart 32.9 and OMB Circular A-125, "Prompt Payment," 5 C.F.R. § 1315.

25. FAR 32.904(b).

26. FAR 32.904(d)(1)(i); FAR 52.232-27(a)(1)(A).

27. 31 U.S.C. § 3902; FAR 32.097.

28. Sarang Corp. v. United States, 76 Fed. Cl. 560 (2007) (PPA interest may be awarded even if the contractor does not specifically request it from the contracting officer.).

29. Gavosto Assocs., Inc., PSBCA Nos. 4058 et al., 01-1 BCA ¶ 31,389 (PPA interest commences only upon submission of a proper invoice) (citing Young

Enters. of Ga., Inc. v. Gen. Servs. Admin., GSBCA Nos. 14437, 14603, 00-2 BCA ¶ 31,148).

30. FAR 52.232-27(a)(2).

31. FAR 52.232-27(a)(2)(viii).

32. FAR 32.905(b)(3); FAR 52.232-27(a)(2).

33. *Young Enters.*, 00-2 BCA ¶ 31,148.

34. 31 U.S.C. § 3907(c); FAR 32.907(d)(1); FAR 52.232-27(a)(4)(ii); Gutz v. United States, 45 Fed. Cl. 291 (1999); L & A Jackson Enters. v. United States, 38 Fed. Cl. 22 (1997), *aff'd,* Jackson v. United States, 135 F.3d 776 (Fed. Cir. 1998); Cargo Carriers, Inc. v. United States, 34 Fed. Cl. 634 (1995), *aff'd,* 135 F.3d 775 (Fed. Cir. 1998); Wilner v. United States, 23 Cl. Ct. 241 (1991); Active Fire Sprinkler Corp. v. Gen. Servs. Admin., GSBCA No. 15318, 01-2 BCA ¶ 31,521.

35. Dick Pac./GHEMM, JV, ASBCA No. 55829, 08-2 BCA ¶ 33,937; Ross & McDonald Contracting, GmbH, ASBCA Nos. 38154 et al., 94-1 BCA ¶ 26,316.

36. FAR 52.232-27(a)(4)(ii); Sprint Comms. Co. v. Gen. Servs. Admin., GSBCA No. 15139, 01-2 BCA ¶ 31,464 (submission of a CDA claim for PPA interest is a jurisdictional prerequisite to consideration of any PPA interest award); Allstate Prods. Co., ASBCA No. 52014, 00-1 BCA ¶ 30,783 (in order to have CDA jurisdiction, the contractor must submit an independent claim for PPA interest); Int'l Bus. Inv., Inc., ASBCA No. 38639, 91-2 BCA ¶ 23,899.

37. 31 U.S.C. § 3905(a); FAR 52.232-5(d)(2).

38. *Id.*

39. FAR 52.232-27.

40. FAR 52.232-27(a)(1)(i)(A).

41. FAR 52.232-27(a)(1)(ii)(A).

42. FAR 52.232-5(e).

43. FAR 52.232-27(a)(1)(ii)(A)(2).

44. FAR 52.232-27.

45. FAR 32.906(a)(3); FAR 52.232-27(a)(3).

46. FAR 32.907(a); FAR 52.232-27(a)(3).

47. 41 U.S.C. §§ 601–613.

48. 31 U.S.C. § 3902(a).

49. *Id.* § 3902(b).

50. *Id.* § 3902(e).

51. *Id.* § 3907(b)(1); 5 C.F.R. § 1315.10(a)(5).

52. 31 U.S.C. § 3907(b)(2); Sol-Mart Janitorial Servs., Inc., ASBCA No. 32873, 87-3 BCA ¶ 20,120.

53. FAR 52.232-27(a)(6).

54. 5 C.F.R. § 1315.11(a).

55. *Id.* § 1315.11(b).

56. FAR 32.901(b).

57. FAR 32.001.

58. *Id.*

59. *Id.*

60. 31 U.S.C. § 3905(b)(1); FAR 52.232-27(c)(1).

61. 31 U.S.C. § 3905(b)(2); FAR 52.232-27(c)(2).

62. 31 U.S.C. § 3905(b)(2); FAR 52.232-27(c)(2).

63. FAR 52.232-5(c).

64. 31 U.S.C. § 3905(c); FAR 52.232-27(c)(3).

65. 31 U.S.C. § 3905(c); FAR 52.232-27(c)(3).

66. 31 U.S.C. § 3905(d); FAR 52.232-27(d)(1).

67. 31 U.S.C. § 3905(d)(2)–(3); FAR 52.232-27(d)(2)–(3).

68. 31 U.S.C. § 3905(e)(4); FAR 52.232-27(e)(4).

69. 31 U.S.C. § 3905(e)(4); FAR 52.232-27(e)(4).

70. FAR 52.232-5(c).

71. FAR 52.232-27(*l*).

72. *Id.*

73. FAR 52.232-5(d)(1).

74. FAR 52.232-5(d)(2).

75. FAR 52.232-5(j)(1).

76. FAR 52.232-5(d)(2).

77. The Assignment of Claims Act, 31 U.S.C. § 3727, and the Assignment of Contract Act, 41 U.S.C. § 15, are collectively known as the Anti-Assignment Acts. These Acts govern assignment of government contracts and payments under those contracts. The intent of the Anti-Assignment Acts is to allow the government to deal exclusively with the original claimant and to eliminate the risk of multiple payments or liability for payments. *See* Nelson Constr. Co. v. United States, 79 Fed. Cl. 81 (2007).

78. 31 U.S.C. § 3737; 41 U.S.C. § 15; FAR 32.802.

79. FAR 32.803(b).

80. FAR 32.805(b).

81. *Id.*

82. Bank of Am. v. United States, 23 F.3d 380 (Fed. Cir. 1994) (erroneous payment made by the government is no bar to rightful assignee); D&H Distrib. Co. v. United States, 102 F.3d 542 (Fed. Cir. 1996); Tuftco Corp. v. United States, 614 F.2d 740 (Ct. Cl. 1980); Kawa v. United States, 77 Fed. Cl. 294 (2007); Banco Bilbao Vizcaya-P.R. v. United States, 48 Fed. Cl. 29 (2000); Norwest Bank Ariz. v. United States, 37 Fed. Cl. 605 (1997).

83. FAR 32.804(a).

84. FAR 32.804(b)(1).

85. FAR 32.804(b)(2).

86. First Nat'l City Bank v. United States, 548 F.2d 928 (Ct. Cl. 1977).

87. FAR 32.804(c).

88. Produce Factors Corp. v. United States, 467 F.2d 1343 (1972); Thomas Funding Corp. v. United States, 15 Cl. Ct. 495 (1988).

89. First Commercial Funding, L.L.C., ENG BCA No. 6447, 00-1 BCA ¶ 30,769 (citing Banco Disa, S.A., ASBCA No. 49167, 96-2 BCA ¶ 28,278); Ft. Carson Nat'l Bank, ASBCA No. 38789, 89-3 BCA ¶ 22,192; Tolson Oil Co., ASBCA No. 28327, 84-3 BCA ¶ 17,576.

90. Merchs. Nat'l Bank v. United States, 689 F.2d 181 (Ct. Cl. 1982); Tuftco Corp. v. United States, 614 F.2d 740 (Ct. Cl. 1980); *First Nat'l City Bank*, 548 F.2d 928; *Thomas Funding*, 15 Cl. Ct. 495; Fla. Nat'l Bank v. United States, 5 Cl. Ct. 396 (1984).

91. Am. Nat'l Bank & Trust Co. v. United States, 22 Cl. Ct. 7 (1990).

92. 31 U.S.C. § 3737; 41 U.S.C. § 15.

93. Innovative Tech. Sys., Inc. v. Dep't of the Treasury, GSBCA No. 13474-TD, 97-1 BCA ¶ 28,971 (citing *Thomas Funding*, 15 Cl. Ct. 495).

94. 31 U.S.C. §§ 1341–1344, 1511–1517.

95. Gov't Accountability Office, 2 Principles of Federal Appropriations Law 6–37 (3d ed. 2006).

96. 31 U.S.C. § 1341(a)(1)(A).

97. *Id.* § 1341(a)(1)(B).

98. *Id.* § 1342.

99. *Id.* § 1517(a).

100. B-208730, 83-1 CPD ¶ 75 (Jan. 6, 1983).

101. 60 Comp. Gen. 219 (1981).

102. Gov't Accountability Office, *supra* note 95, at 6–41.

103. B-236667, 1990 WL 277766 (Jan. 26, 1990).

104. *See* PCL Constr. Servs., Inc. v. United States, 41 Fed. Cl. 242 (1998), *aff'd*, 96 Fed. Appx. 672 (Fed. Cir. 2004).

105. FAR 52.232-18. This clause is to be used only for operation and maintenance and continuing services contracts that are necessary for normal operations and for which Congress previously had consistently appropriated funds unless specific statutory authority exists permitting applicability to other requirements. FAR 32.703-2(a).

106. FAR 52.232-19. While the implementing regulation implies that this clause might be restricted to indefinite-quantity or requirements contracts, the Court of Federal Claims has held that the clause may be used beyond these contracting situations. *PCL Constr. Servs.*, 41 Fed. Cl. 242.

107. DFARS 252.232-7007.

108. EFARS 52.232-5004.

109. For example, the Limitation of Government's Obligation clause provides that an equitable adjustment will be made if the contractor incurs additional costs solely by reason of the government's failure to allot additional funds by the specified date. DFARS 252.232-7007(e). The COE Incremental Funding clause, on the other hand, states that any suspension, delay, or interruption of the work arising from exhaustion or anticipated exhaustion of funds shall not constitute a breach of contract, and shall not entitle the contractor to any price adjustment under a Suspension of Work or similar clause or in any other manner under the contract. EFARS 52.232-5003(f).

110. The River and Harbor Act of 1922, 33 U.S.C. § 621, created an exemption from the Anti-Deficiency Act for certain specifically authorized projects on canals, rivers, and harbors. The scope of this exemption was expanded under the Water Resources Development Act of 1999. The Water Resources Development Act broadened the type of projects and work covered by the continuing contracts authority. It authorized the COE to award continuing contracts if sufficient funding was not available to complete a project funded from the COE's (1) construction; (2) operation and maintenance; or (3) flood control, Mississippi River, and tributaries appropriations accounts. The Energy and Water Development Appropriations Act of 2004 further expanded the COE's authority to use continuing contracts to include contracts funded from its investigations appropriations account.

111. J.A. Jones Constr. Co., ENG BCA No. 4977, 86-2 BCA ¶ 18,806.

112. EFARS 52.232-5001.

113. EFARS 52.232-5001(i).

114. EFARS 52.232-5001(c).

115. Pub. L. No. 109-103 (codified as amended at 33 U.S.C. § 2331).

116. Munsey Trust Co. v. United States, 332 U.S. 234 (1947); Johnson v. All-State Constr., Inc., 329 F.3d 848 (Fed. Cir. 2003); Applied Cos. v. United States, 37 Fed. Cl. 749 (1997), *aff'd*, 144 F.3d 1470 (Fed. Cir. 1998); D.L. Kaufan, Inc., PSBCA Nos. 4159 et al., 00-1 BCA ¶ 30,846; Massapequa Partners Ltd. P'ship, MPL Group, Inc., PSBCA No. 3817, 97-2 BCA ¶ 29,058.

117. 31 U.S.C. §§ 3701 *et seq.* The Debt Collection Act allows the government to collect delinquent debts owed by persons, including government contractors, through administrative offset after the government has tried to collect the money due. The Act also permits the government to recover interest on contractor debts and government claims. DKW Constr. v. Gen. Servs. Admin., CBCA No. 438, 2007 WL 4661156 (Dec. 20, 2007); Advanced Injection Molding, Inc., GSBCA No. 16504-R, 05-2 BCA ¶ 33,097.

118. Cecile Indus., Inc. v. Cheney, 995 F.2d 1052 (Fed. Cir. 1993).

119. Citizens Bank of Md. v. Strumpf, 516 U.S. 16 (1995).

120. 329 F.3d 848 (Fed. Cir. 2003).

121. 31 U.S.C. § 3728.

122. *Id.* § 3728(a).

123. *Id.* § 3728(b).

124. FAR 32.601(b).

125. FAR 32.603.

126. FAR 32.603(b).

127. FAR 32.604(a)(1).

128. FAR 32.604(a)(2).

129. FAR 32.604(b).

130. FAR 32.604(c).

131. FAR 32.604(d).

132. FAR 32.605(a).

133. FAR 32.606(a).

134. Applied Cos. v. United States, 37 Fed. Cl. 749 (1997), *aff'd*, 144 F.3d 1470 (Fed. Cir. 1998); Kearfott Guidance & Navigation Corp., ASBCA No. 49263, 99-2 BCA ¶ 30,518.

135. FAR 32.606(b).

136. FAR 32.607.

137. FAR 32.607(b).

138. FAR 32.607-2.

139. FAR 32.607(b)(2).

140. FAR 32.607-2(a).

141. FAR 32.607-2(a)(1).

142. FAR 32.607-2(a)(3).

143. FAR 32.607-2(d).

144. FAR 32.607-2(e).

145. FAR 32.607(b).

146. FAR 32.607-2(g).
147. FAR 32.607-2(h).
148. FAR 32.607-2(i).
149. 31 U.S.C. § 3717; FAR 32.608-1.
150. FAR 32.604(b)(4)(ii).

CHAPTER 12

Socioeconomic Issues in Government Contracting

DENISE E. FARRIS[1]

I. Introduction

Government contracting practitioners must be familiar with a very wide variety of special contracting programs created by Congress or Executive Order. This chapter provides a brief overview of the most significant programs, including: (a) Small Business Contracting Program; (b) Disadvantaged and Minority/Woman Business Enterprise Program; (c) 8(a) Business Development Program; (d) Equity in Contracting for Women Act of 2000; (e) Service-disabled veteran business programs; (f) Historically Underutilized Business Zone (HUBZone) contracting program; (g) domestic supply requirements under the Buy American Act and Trade Agreements Act; (h) the Davis-Bacon Act; and (i) the McNamara-O'Hara Service Contract Act. While each program could justify a chapter in itself, the intent here is to provide a general overview and sufficient references to permit further in-depth research.[2]

II. The Constitutionality of Affirmative Action Programs

Affirmative action programs in government contracting, based on race and gender classifications and the remediation authority of Congress under the 5th and 14th Amendments of the U.S. Constitution, include the Minority, Women, Disadvantaged Business Enterprise programs, 8(a) Business Development (BD) Program, and the recently enacted Equity in Contracting for Women Act of 2000. Where these programs center on race and gender classifications, thus excluding certain other classes, they have been aggressively litigated at the local, state, and federal level. The following briefly addresses the constitutional issues impacting the programs and the major issues subject to legal challenge.

A. Legal Parameters of Race- and Gender-Based Programs

1. *Constitutional History*

Broadly defined as a "formal effort to provide increased opportunities for women and ethnic minorities to overcome past patterns of discrimination,"[3] affirmative action dates back to the passage of Title VII, Civil Rights Act of 1964, which prohibited discrimination by private employers on the basis of race, color, religion, sex, or national origin.[4] Executive Orders issued through the 1960s, including Executive Order 11246,[5] implemented the advent of affirmative action plans for recruitment, employment, and promotion. The early 1970s ushered in regulations requiring written affirmative action plans, goals, and timetables by those entities wishing to do business with government agencies.[6] New programs, including Section 8 of the Small Business Act, permitted "set-asides" of specific government contracts, meaning restricted competition or sole source awards to small businesses owned and controlled by socially and economically disadvantaged individuals.[7] From 1960 through the mid-1980s, race and gender preferences and set-asides were the law of the land, with ever-aggressive goals losing touch with the limited availability of qualified minority and women-owned businesses capable of performing the required work. This situation resulted in a number of cases specifically challenging the use of these race- and gender-based programs.[8]

a. *City of Richmond v. J. A. Croson*

The constitutional validity of a local and state affirmative action program was addressed by the U.S. Supreme Court in the 1989 decision of *City of Richmond v. J.A. Croson Construction Co.*,[9] where a 5-to-4 majority settled on "strict scrutiny" as the proper review standard. The decision invalidated not only Richmond's 30 percent set-aside for minority-owned businesses but also a spate of local and state affirmative action programs nationally.[10] *Croson* suggested that "race-conscious" remedies properly could be legislated in response to proven past discrimination by the affected state and local governmental entities but that "racial balancing" untailored to "specific" and "identified" evidence of minority exclusion was impermissible. The *Croson* decision implied that, for federal programs, the inherent authority of Congress suggested a constitutional review standard more tolerant of racial line-drawing (i.e., "intermediate scrutiny"). This conclusion was reinforced when a year later, in *Metro Broadcasting, Inc. v. FCC*,[11] the Court utilized intermediate scrutiny to uphold certain preferences for minorities in broadcast licensing proceedings.[12]

b. *Adarand Constructors v. Pena*

This two-tiered approach to equal protection analysis was short-lived. In *Adarand Constructors, Inc. v. Pena*,[13] the Court applied "strict scrutiny" review to a federal transportation program that gave financial incentives to prime contractors who subcontracted to firms owned by "socially and economically disadvantaged individuals." Although the Court refrained from addressing the merits of the particular program, remanding for further proceedings, it determined that all racial classifications by government at any level must be both justified by a "compelling governmental interest" and "narrowly tailored" to that end. Two subsequent trips of *Adarand* to the Supreme Court ultimately resulted in the holding that racial preferences in local, state, and federal law are remedies of "last resort," subject to strict scrutiny review. To pass constitutional muster, such programs require not only a compelling interest supported by reliable evidence of past discrimination but also a remedial program "narrowly tailored."[14]

c. Proving "the Compelling Government Interest"

Under *Croson* and *Adarand*, an affirmative action program must serve a "compelling governmental interest."[15] Remedying a governmental entity's own past discrimination can constitute a "compelling government interest."[16] However, governments must have a "strong basis in evidence for its conclusion that remedial action [is] necessary."[17]

The *Croson* decision required programs to be supported by reliable statistical studies proving availability, capability, and past discrimination against minorities and women in the relevant locale. Justice O'Connor suggested: "Where gross statistical disparities can be shown, they alone in a proper case may constitute prima facie proof of a pattern or practice of discrimination under Title VII."[18] By implication, this same standard would apply to set-aside plans. However, in cases such as *Coral Construction Co. v. King County*,[19] lower courts have preferred to rely on a combination of statistical and testimonial or anecdotal evidence. In cases in which testimonial evidence is given, that evidence must be reliable and relate to specific instances of discrimination, rather than the kinds of generalized allegations of discrimination rejected in *Croson*.[20]

Challenges to existing programs over the past 20 years have relied heavily on the lack of reliable statistical studies to refute proof of a "compelling government interest."[21] These challenges demonstrate that courts will not accept conclusory or anecdotal allegations of discrimination. Rather, courts require proof of the compelling interest via detailed and statistically sound disparity impact studies that are probative and regionally relevant, and that analyze discriminatory factors such as company size versus nondiscriminatory factors (such as race or gender of owners).[22] The cases additionally suggest that the statistical studies must be current, which in turn implies an affirmative duty upon governmental sponsors of affirmative action programs to update the studies on a regular basis.[23] This ongoing duty to update studies has defeated some programs. Courts have concluded that, where years of goals and detailed record keeping show achievement of affirmative action goals, this "proves" that past discrimination has been "remedied."[24]

d. Proving a Program Is "Narrowly Tailored"

In addition to providing proof of "compelling interest" (i.e., evidence of past discrimination), the proponent of affirmative action must prove that the program is "narrowly tailored" to address and remedy that discrimination. This analysis looks at whether the program, on its face or as applied, is overbroad or underinclusive, as well as whether it contains race-neutral factors. *Croson*, as followed by *Adarand*, states that a "narrowly tailored" analysis must examine whether the program is:

1. More than a mere promotion of racial balancing;
2. Based on the number of qualified minorities in the area capable of performing the scope of work identified in the set-aside plan;
3. Not overinclusive by presuming discrimination against certain minorities;
4. Complete with race-neutral alternatives to set-aside programs; and
5. Not based upon numerical quotas.[25]

To avoid strict scrutiny review, many programs followed Justice Scalia's suggestion in *Croson* that the programs "adopt a preference for small businesses, or even for new businesses—which would make it easier for those previously excluded by discrimination to enter the field. Such programs may well have a racially disproportionate impact, but they are not based on race."[26] Because they are entirely race-neutral, such programs are not subject to strict scrutiny and are thus more easily sustained.[27]

Other race-neutral features of a procurement may include, but are not limited to, advertising contracting opportunities in publications and media targeted to minorities, women, and small businesses; providing written notice to small companies in sufficient time to allow them to bid; educating small businesses on how to do business with the governmental entity; encouraging joint ventures between majority/minority and/or small business firms; assisting small contractors in obtaining bonds, lines of credit, and insurance; segmenting larger contracts into smaller jobs capable of performance by small contractors; and utilizing the resources of business development organizations to assist small contractors to grow their businesses.[28]

e. Recent Developments

While *Croson* and *Adarand* remain the law of the land on affirmative action, a notable 2008 Federal Circuit decision, *Rothe Development Corp. v. Dept. of Defense*, is touted as potentially fatal to many affirmative action programs, if ultimately upheld after remand and anticipated subsequent appeals.[29] In *Rothe*, the Federal Circuit invalidated Section 1207 of the National Defense Authorization Act of 1987,[30] which established a 5 percent participation goal for Disadvantaged Business Enterprises (DBEs) in Department of Defense contracts; incorporated Section 8(a)'s presumption of racial disadvantage of certain designated minority groups; and authorized the Department of Defense to apply a price evaluation adjustment of 10 percent to attain the 5 percent DBE goal.[31]

Key holdings in *Rothe* include the following: (a) the district court's standard of review was too deferential, concluding that proper inquiry was whether a "strong basis in evidence" supported Congress's conclusion that discrimination existed; (b) that a "mere listing" of evidence before Congress when it enacted the statute was insufficient; (c) that detailed statistical information regarding existence of discrimination in the 1992 reauthorization was necessary to find reauthorized Section 1207 constitutional; and (d) that the government must also produce evidence of pre-enactment discrimination.[32] While the full impact of *Rothe* is not yet known, the decision potentially invalidates many elements of various small disadvantaged business programs existing under the Small Business Act that are based on a socioeconomic designation. Within months of *Rothe*, the Small Business Administration (SBA) and the Justice Department used arguments and concerns about constitutionality in *Rothe* to explain delays in implementing the Equity in Contracting for Women Act of 2000.[33] Agencies such as NASA, which also had employed the 10 percent price credit, have rolled back that practice as a result of the ongoing lawsuit. It is expected that *Rothe* eventually will be heard by the Supreme Court.

III. *Other Socioeconomic Contracting Programs*

A. Small Business Contracting Program

1. Purpose

As noted in a recent report by the SBA, the federal government has been examining the extent to which small businesses receive their "fair share" of government procurement dollars for more than half a century.[34]

2. Program Summary

This concern takes a number of legislative forms, from the establishment of the SBA in 1953 to the Business Opportunity Development Reform Act of 1988, which established a goal that at least 20 percent of overall direct federal procurement contract dollars would be awarded to small businesses.[35] This race- and gender-neutral goal was raised to 23 percent in 1997 as part of the Small Business Reauthorization Act of 1997,[36] and includes a variety of underlying programs that aim to remedy evidence of past discrimination against, and create contracting opportunities for, minorities and women who also qualify as "small businesses" under SBA size standards.[37] Utilization of minority- and women-owned businesses at the federal level permits capture as both "small business utilization" and credit under the applicable federal minority and women business enterprise contracting programs.

3. Certification Criteria

For purposes of federal certification as a "small business," a contractor must be independently owned and operated, organized for profit, and not dominant in its field.[38] It must also be "small," which is defined by industry category under the North American Industry Classification System (NAICS), published by the SBA. Depending on the industry, size standard eligibility is based on the average number of employees for the preceding 12 months or on sales volume averaged over a three-year period. In construction, there are several SBA size standards that are frequently encountered:

1. Manufacturing (suppliers): Maximum number of employees may range from 500 to 1,500, depending on the type of product manufactured;
2. Wholesaling (suppliers): Maximum number of employees may range from 100 to 500, depending on the particular product being provided;
3. Services (construction managers, design professionals, consultants): Annual receipts may not exceed $2.5 to $21.5 million, depending on the particular service being provided;
4. Retailing (suppliers): Annual receipts may not exceed $5 to $21 million, depending on the particular product being provided;
5. General and heavy construction (contractors): General construction annual receipts may not exceed $13.5 to $17 million, depending on the type of construction; and
6. Special trade construction (contractors): Annual receipts may not exceed $7 million.[39]

4. Administration

Certification as a "small business" is done through registration using the federal Central Contractor Registration (CCR),[40] on a "self certifying" basis using the government's Online Representations and Certifications Application (ORCA) process.[41] These self-certifications are subject to administrative challenge.[42]

B. Disadvantaged Business Enterprise Program

1. Purpose

The present-day Disadvantaged Business Enterprise (DBE) program grants preferential treatment in the award of government contracts to "socially and economically disadvantaged" small businesses. In 1983, Congress enacted the first DBE statutory provision, which required the Department of Transportation (DOT) to ensure that at least 10 percent of the funds authorized for federal highway and transit assistance programs be expended with DBEs. In 1987, Congress reauthorized and amended the DBE program by adding women to the groups presumed to be disadvantaged. Since 1987, DOT has established a single federal DBE goal to encompass firms owned by both women and minority group members.[43] The DBE goal varies on a state-by-state basis.

2. Program Summary

Under the DOT's program, DBE regulations require recipients of DOT financial assistance (namely, state and local transportation agencies) to establish goals for the participation of disadvantaged entrepreneurs and to certify the eligibility of DBE firms to participate in their DOT-assisted contracts within that particular agency.[44] Each DOT-assisted state and local transportation agency is required to establish narrowly tailored DBE goals.[45] These DOT-assisted agencies then evaluate their DOT-assisted contracts throughout the year and establish contract-specific DBE subcontracting goals where needed to ensure nondiscrimination in federally assisted procurements. The level of DBE subcontracting may vary from the approved DBE goal. However, the annual amount of contract/subcontract awards to DBEs across all contracts for that agency should be consistent with the overall goal.

3. Certification Criteria

In order for small disadvantaged firms, including those owned by minorities and women, to participate in DOT-assisted contracts awarded by state and local transportation agencies under the DBE program, they must apply for and receive certification as a DBE. Current DBE program criteria can be found under both SBA and DOT regulations, as those two agencies operate under a memorandum of understanding that streamlines certification by employing reciprocal certification standards.[46] An applicant business must demonstrate the following:[47]

1. The individual qualifying as "socially and economically disadvantaged" must be the 51 percent owner and controller of the applicant.[48] Discussion of "social disadvantage" is addressed below.
2. The individual must be a U.S. citizen.[49]
3. Owners of the applicant who are women and not members of one of the designated groups presumed to be socially disadvantaged under 13 C.F.R. § 124.103(b) must provide personal statements relating to their individual social disadvantaged status as required by 13 C.F.R. § 124.103(c)(2)(ii). This element is further discussed below.
4. Disadvantaged owners must have a net worth less than $750,000, must lack adequate access to credit and capital, and may not have made any asset transfers within the past two years contrary to the requirements of 13 C.F.R. § 124.104(c).
5. The applicant must have been in business in its primary industry classification for at least two full years or satisfy the requirements for a waiver under 13 C.F.R. § 124.107(b).
6. The applicant and its principals must demonstrate good character under 13 C.F.R. § 124.108(a).
7. The applicant, its disadvantaged owners, or the disadvantaged owners' family members must not own more than 20 percent equity own-

ership interest in another 8(a) disadvantaged firm, as required by 13 C.F.R. § 124.105(g).

8. An applicant with non-disadvantaged owners must not own more than 10 percent of another 8(a) participant in the developmental stage of program participation or 20 percent of another 8(a) participant in the transitional stage of program participation, as required by 13 C.F.R. § 124.105(h).

9. With respect to airport concessions, the applicant must meet the SBA size standard corresponding to its primary SIC code. Note: The SBA may certify a firm or an individual claiming disadvantaged status for participation in the 8(a) program only one time.

An applicant can establish "social disadvantage" if he or she is a member of a group of persons that DOT considers "disadvantaged." Groups presumptively considered "disadvantaged" include women, African-Americans, Hispanic Americans, Native Americans, Asian-Pacific Americans, Subcontinent Asian-Pacific Americans, or other minorities found by the SBA to be disadvantaged. Persons who are not members of one of the above groups and own and control their business may be eligible if they establish their "social" and "economic" disadvantage via submission of a personal statement as identified above. The DOT notes, for example, that people with disabilities have disproportionately low incomes and high rates of unemployment, and that many may be socially and economically disadvantaged. A determination of whether an individual with a disability meets DBE eligibility criteria is made on a case-by-case basis.[50] Any presumption of "social and economic disadvantage" is rebuttable.[51]

To determine "business size," a firm (including its affiliates) must be a small business as defined by SBA standards, adjusted by DOT regulations. It must not have annual gross receipts over $20,410,000 in the previous three fiscal years ($47,780,000 for airport concessionaires, with some exceptions). Under SAFETEA-LU,[52] this threshold may be adjusted annually for inflation by the DOT Secretary. Be warned that under DOT regulations, the definition of "small" may deviate from those standards concurrently employed by the SBA. For DOT projects, the DOT definitions of "small" must be followed.

Finally, the DBE program establishes limits on personal net worth (PNW). Only disadvantaged persons having a PNW of less than $750,000 can be considered as a potential qualified DBE. Items excluded from the PNW calculation include an individual's ownership interest in the applicant firm and equity in his or her primary residence. Additional exclusions are available for owners of airport concessionaires.[53]

4. Administration

Three major DOT operating administrations have primary responsibility for implementing the DBE program: the Federal Highway Administration, the Federal Aviation Administration (FAA), and the Federal Transit Administration. State and local transportation agencies carry out the DOT DBE program under the rules and guidelines in the Code of Federal Regulations.[54] The FAA also maintains a separate DBE program for concessions in airports.[55] DBE certification typically is handled by the state department of transportation in the state where the business entity is located or does business.[56] Consulting with a local transportation official regarding both PNW criteria and applicable DBE goals is advisable, as these sometimes vary depending on the agency, program, and certifying entity implementing DOT-funded projects.

Appeals from the DBE certification process require review per the applicable state department of transportation administrative procedures. Appeals are also processed through the federal DOT.[57]

C. 8(a) Business Development Program

1. Purpose

As noted above, the current 8(a) program originated in conjunction with the Disadvantaged Business Enterprise Program under Section 8(a) of the Small Business Act of 1958. The decision to convert 8(a) into a minority business development program occurred in 1978 with the passage of Public Law 95-507, which broadened the range of assistance that the government in general, and SBA in particular, would provide to minority businesses.

2. Program Summary

Originally, the program only applied to traditional minorities and excluded nonminority or female applicants. The program now permits application by nonminority male and female applicants who can demonstrate they are "socially and economically disadvantaged."

The 8(a) legislation authorizes the SBA to enter into various construction, supply, and service contracts with other federal departments and agencies, under which the SBA acts as the prime contractor. SBA then "subcontracts" the performance of these contracts, on either a "sole source" or "restricted competition" basis, to small business concerns owned and controlled by "socially and economically disadvantaged" individuals or various Native American tribes or organizations.[58] Accordingly, 8(a) status can be extremely lucrative to qualifying firms.

3. Certification Criteria

To qualify for 8(a) status, the applicant firm must:

1. Qualify as a "small business" under applicable size standards;
2. Have been in business for at least two full years or satisfy the requirements for a waiver;[59]
3. Be unconditionally owned and controlled by one or more socially and economically disadvantaged individuals who are of good character and citizens of the United States; and
4. Demonstrate potential for success.

Definitions of "social and economic disadvantage" essentially track the same definitions used in the DBE program, except that the applicant's personal net worth may not exceed $250,000 excluding the interest in the business and equity of the applicant's principals in their homes.

"Ownership and control" criteria are outlined in the applicable regulations, which provide a lengthy detailed description of both the criteria and the factors governing their interpretation. Criteria examined for an applicant's "potential of success" include:

1. Technical and managerial experience of the applicant firm's managers;
2. Operating history;
3. Ability to access credit and capital;
4. Financial capacity;
5. Record of performance; and
6. Whether the applicant firm or individuals employed by the firm hold required licenses, if any.[60]

4. Administration

The application can be printed online and is submitted to the SBA regional office that services the applicant's domicile state.[61] In addition, each federal agency has an Office of Small and Disadvantaged Business Utilization (OSDBU), as required by the 1978 amendment of the Small Business Act of 1953.[62] OSDBU staff provide technical assistance and information to small and disadvantaged businesses seeking contracting opportunities. Such staff include Small and Disadvantaged Business Utilization Specialists and Small Business Technical Advisors. These advisors have the specific purpose of ensuring that small businesses—including Small Disadvantaged Businesses (SDBs), Disadvantaged Business Enterprises (DBEs), Minority-owned Business Enterprises (MBEs), Veteran-Owned Small Businesses (VOSBs) and Service-Disabled Veteran-Owned Small Businesses (SDVOSBs), or Women-Owned Business Enterprises (WBEs)—actively participate in contracts let by the agency and subcontracts awarded by the agency's prime contractors. These advisors do not, however, control any contract awards.[63] An SBA 8(a) certification denial can be appealed through procedures set forth in 13 C.F.R. § 124.206.

D. Equity in Contracting for Women Act of 2000

1. Purpose

The Equity in Contracting for Women Act (ECWA) was enacted in 2000 to ensure that small businesses owned by women receive a full and fair opportunity to compete and participate in federal contracts.[64] Enacted specifically to address the inability to achieve a 5 percent WBE goal set in 1988, ECWA creates a specific set-aside, similar to the 8(a) program, that authorizes governmental agencies to implement restricted or sole source competition in certain industries.

ECWA was placed on hold in 2001 pending the Bush administration's request for additional disparity studies to justify the Act's implementation. In 2004, the U.S. Women's Chamber of Commerce filed suit, receiving an injunctive order requiring that the SBA immediately implement the Act.[65] A proposed rule issued by the SBA in 2008 would have limited the effect of the ECWA, but was opposed and tabled due to an unprecedented backlash from women's business organizations and owners.[66] The Obama Administration has instructed the SBA to return to the drafting table to implement the program as legislatively intended.

2. *Program Summary*

ECWA permits a contracting officer (CO) to restrict competition to certified women-owned small businesses (WOSBs) if: (a) the CO has a reasonable expectation that two or more WOSBs will submit offers for the contract; (b) the contract is for procurement of goods or services with respect to an industry identified by the SBA as having WOSB underutilization in the past (which includes construction); and (c) the anticipated award does not exceed $5 million in the case of manufacturing or $3 million for all other types of contracts.[67]

3. *Certification Criteria*

The WOSB must certify that it is at least 51 percent owned and controlled by women, and that the business is "small" according to the NAICS size standards published by the SBA.[68]

4. *Administration*

Until the SBA completes a final rule implementing the program, administrative details are uncertain. Registration as a WOSB would be handled through the CCR,[69] on a "self certifying" basis handled through the government's ORCA process.[70] The self-certification would be subject to administrative challenge as set forth in 13 C.F.R. § 121.1001 through .1003.

E. Service-Disabled Veteran Business Programs

1. *Purpose*

The Service-Disabled Veteran-Owned Small Business (SDVOSB) Procurement Program was established under the auspices of the Veterans Benefit Act of 2003,[71] specifically to benefit small business concerns owned and controlled by service-disabled veterans. The Veterans' Entrepreneurship and Small Business Development Act of 1999 (Public Law No. 106-50) and the SDVOSB Procurement Program (Subpart 19.14 of the Federal Acquisition Regulation (FAR)) authorize COs to implement procedures designed to attain or exceed the mandated 3 percent goal for prime contracts and subcontracts awarded to SDVOSBs.

2. Program Summary

To provide contracting opportunities to SDVOSBs, government agencies may implement set-aside procedures limiting competition to SDVOSBs, or in certain instances utilize sole-source awards.[72] The Act requires consideration of set-asides (i.e., restricted competition) before utilizing sole-source awards.[73] To justify the set-aside, the CO must have a reasonable expectation that offers will be received from two or more SDVOSBs and that the award will be made at a fair market price.[74] If the CO receives only one acceptable offer from an SDVOSB in response to the set-aside, the CO is directed to make an award to that concern.[75] If the CO receives no acceptable offers from SDVOSBs, the SDVOSB set-aside must be withdrawn and the requirement, if still valid, set aside for other small business concerns, as appropriate.[76] SDVOSB awards or non-awards may be appealed by the SBA.[77]

The government may make a sole-source award to an SDVOSB under FAR 19.1406, provided:

1. Only one SDVOSB can satisfy the requirement;
2. The anticipated award price of the contract (including options) will not exceed—
   a. $5.5 million for a requirement within the NAICS codes for manufacturing; or
   b. $3 million for a requirement within any other NAICS code;
3. The SDVOSB has been determined to be a responsible contractor with respect to performance; and
4. Award can be made at a fair and reasonable price.[78]

The program does not apply to requirements that can be satisfied through award to:

1. Federal Prison Industries, Inc.;[79]
2. Javits-Wagner-O'Day (JWOD) Act participating nonprofit agencies for the blind or severely disabled;[80]
3. Orders under indefinite delivery contracts;[81]
4. Orders against the Federal Supply Schedule (FSS);[82] or
5. Requirements currently being performed by an 8(a) participant or requirements that SBA has accepted for performance under the authority of the 8(a) program, unless SBA has consented to release the requirements from the 8(a) program.[83]

3. Certification Criteria

To qualify as an SDVOSB, the applicant must be "small" under the NAICS code assigned to the procurement,[84] and be owned by a service-disabled veteran (SDV) meeting the following criteria:

1. Have a service-connected disability that has been determined by the Department of Veterans Affairs or Department of Defense;

2. Unconditionally own 51 percent of the applicant;
3. Control the management and daily operations of the applicant; and
4. Hold the highest officer position of the applicant.[85]

Under certain circumstances the wife of a deceased SDV is permitted to continue the operation of the business for a limited time period. A prime or subcontractor SDVOSB is permitted to subcontract a portion of its contract, subject to the following limitations:

1. For supply contracts, the SDVOSB must perform at least 50 percent of the cost of manufacturing supplies;
2. For general construction, it must use SDVOSB personnel for at least 15 percent of contract performance; and
3. For special trade construction, it must use SDVOSB personnel for at least 25 percent of contract performance.[86]

4. Administration

Protests of SDVOSB status determinations may be raised on the basis of either size or eligibility.[87] Protests may be submitted by the SBA contracting officer or SBA regarding procurement award or non-award under an SDVOSB status. Protests are submitted to the SBA Associate Administrator for Government Contracting. Protests may also be submitted by any concern that submits an offer by submitting the protest to the CO for the agency letting the contract.[88]

F. Historically Underutilized Business Zone (HUBZone)

1. Purpose

The Historically Underutilized Business Zone (HUBZone) Act of 1997 created the HUBZone Program (sometimes referred to as the "HUBZone Empowerment Contracting Program").[89] This program was enacted as part of the Small Business Reauthorization Act of 1997 and falls under the auspices of the SBA. The program encourages economic development in historically underutilized business zones—"HUBZones"—through preferences in contracting. The HUBZone Program applies to all federal agencies that employ one or more COs.[90]

2. Program Summary

To justify a HUBZone set-aside for acquisitions exceeding the simplified acquisition threshold of $25,000, there must be a reasonable expectation that offers will be received from two or more HUBZone small businesses and that the award will be made at a fair market price.[91] Sole source awards to HUBZone entities may be made only where:

1. Only one HUBZone small business concern can satisfy the requirement;
2. The anticipated price of the contract, including options, will not exceed:

    a. $5.5 million for a requirement within the NAICS codes for manufacturing; or

    b. $3.5 million for a requirement within any other NAICS code;

3. The requirement is not currently being performed by a non-HUBZone small business concern;

4. The acquisition is greater than the simplified acquisition threshold;[92]

5. The HUBZone small business concern has been determined to be a responsible contractor with respect to performance; and

6. Award can be made at a fair and reasonable price.[93]

The SBA retains the right to appeal the CO's decision not to utilize a HUBZone sole source award.[94]

In addition, where a HUBZone entity participates in full and open competition, the agency CO is allowed to utilize a price evaluation preference for the HUBZone entity's bid.[95] This preference is not allowed:

1. In acquisitions expected to be less than or equal to the simplified acquisition threshold;

2. Where price is not a selection factor so that a price evaluation preference would not be considered (e.g., architect-engineer acquisitions);

3. Where all fair and reasonable offers are accepted (e.g., the award of multiple award schedule contracts).[96]

Also, HUBZone preferences do not apply to:

1. Requirements that can be satisfied through award to Federal Prison Industries, Inc.;[97] or JWOD Act participating nonprofit agencies for the blind or severely disabled;[98]

2. Orders under indefinite delivery contracts;[99]

3. Orders against Federal Supply Schedules;[100]

4. Requirements currently being performed by an 8(a) participant or requirements SBA has accepted for performance under the authority of the 8(a) program, unless SBA has consented to release the requirements from the 8(a) program; or

5. Requirements for commissary or exchange resale items.[101]

In setting the price evaluation preference, the CO

> shall give offers from HUBZone small business concerns a price evaluation preference by adding a factor of 10 percent to all offers, except—
>
> (1) Offers from HUBZone small business concerns that have not waived the evaluation preference; or
>
> (2) Otherwise successful offers from small business concerns.[102]

The factor of 10 percent is to be applied on a line-item basis or to any group of items on which the award may be made. Other evaluation factors, such as transportation costs or rent-free use of government property, are added to establish the base offer before adding the 10 percent preference.[103] In addition, a business that is both a HUBZone small business and a small disadvantaged business receives the benefit of both the HUBZone small business price evaluation preference and the small disadvantaged business price evaluation adjustment.[104] Each applicable price evaluation preference or adjustment is to be calculated independently from the offeror's base offer. Each of these individual preference and adjustment amounts is then added to the base offer to arrive at the total evaluated price for that offer. Federal solicitations typically will contain notices to indicate whether they include HUBZone preferences.[105]

3. Certification Criteria

To qualify as a HUBZone small business concern, the entity must:

1. Be unconditionally 51 percent owned and controlled by persons who are either: (a) Native American, (b) U.S. citizens, or (c) a small agricultural cooperative;
2. Qualify as a small business under the size standard corresponding to its primary industry classification as defined in 13 C.F.R. Part 121 (SBA Small Business Size Standards);
3. Have its principal office located in a HUBZone area;[106]
4. Have at least 35 percent of its employees residing in a HUBZone;
5. Agree that it will "attempt to maintain" having 35 percent of its employees reside in a HUBZone during the performance of any HUBZone contract it receives;[107] and
6. Agree to comply with certain contract performance requirements in connection with contracts awarded to it as a qualified HUBZone Small Business Concern.[108]

4. Administration

Certification is done under the auspices of the SBA and can be accomplished online. An applicant not granted HUBZone status may reapply after one year if the basis for nonqualification has been corrected.[109] After the SBA has verified a HUBZone entity's certification, the HUBZone entity will be added to a list of qualified HUBZone Small Businesses. Firms on the list are eligible for HUBZone program preferences without regard to the place of performance. Joint ventures may also be considered HUBZone entities if they meet all the criteria in 13 C.F.R. § 126.616. Protests of HUBZone awards may be made by either the SBA, the agency CO, or any other interested party according to the procedures set forth under 13 C.F.R. § 126.800.

G. Buy American Act and Trade Agreements Act

1. Purpose

In 1933, in response to the Great Depression, Congress enacted the original Buy American Act (BAA).[110] The BAA provides that, except to the extent that it is inconsistent with the public interest or is cost-prohibitive, only articles, materials, and supplies mined, produced, or manufactured in the United States can be used for federal government projects, and all contractors for government construction projects in the United States must use only domestic materials.[111]

The intent of the BAA was to protect the U.S. industrial base and strengthen the nation's economy. The BAA does not apply to professional or personal services, only products.[112] Other Buy America requirements are found in other statutory or regulatory provisions, such as the Surface Transportation Assistance Act of 1982.[113]

2. *Program Summary*

FAR Subparts 25.1 and 25.2 implement the Buy American Act, and FAR Subpart 25.4 addresses various trade agreements. The FAR defines a domestic end product for the BAA using a two-part test: (1) the article must be manufactured in the United States, and (2) the cost of domestic components (i.e., components mined, produced, or manufactured in the United States) must exceed 50 percent of the cost of all components.[114] Therefore, a product may include as much as 49 percent of foreign component cost and still be considered a domestic purchase.[115] A foreign end product may be purchased if the CO determines that the price of the lowest domestic offer is unreasonable or if another exception applies.[116]

"End products" means those articles, materials, and supplies to be acquired for public use under the contract.[117] "Components" means those articles, materials, and supplies incorporated directly into the end products.[118] Labor is not a component of the end product within the definition set forth at FAR 25.003.[119]

The Trade Agreements Act (TAA) exempts acquisitions under certain international trade agreements from the restrictions of the BAA.[120] In acquisitions subject to this exception, end products and construction materials from certain countries receive nondiscriminatory treatment in evaluation, with the same treatment as domestic offers. Generally, the dollar value of the acquisition determines which of the trade agreements applies. Exceptions to the applicability of the trade agreements are described in Subpart 25.4.[121]

Under the TAA, the test to determine country of origin is "substantial transformation" (i.e., transforming an article into a new and different article of commerce, with a name, character, or use distinct from the original article). For the reporting requirement at FAR Section 25.004, the only criterion is whether the place of manufacture of an end product is within or outside of the United States, without regard to the origin of the components except as modified by the TAA, discussed below.

The FAR lists exceptions to the BAA. The government may purchase a foreign end product if one of the following conditions exists:[122]

- The products acquired are for use outside of the United States.
- It would be in the public's interest to do so.[123]

- The product is not reasonably available in sufficient commercial quantities in the domestic market.
- The cost of the domestic product is unreasonable.
- The product is for resale.
- The products are commercial information technology items.
- The products are eligible products acquired under the TAA.

The reasonableness of cost is determined by adding certain evaluation factors to the foreign offeror's proposed price before performing the price evaluation.[124] If the price of the domestic offer exceeds the price of the foreign offer after addition of the evaluation factors, then the agency may purchase the foreign end product.

3. Trade Agreements Act Impact on Buy American Act

As noted above, provisions of the BAA may be expressly waived by the TAA. Enacted in 1979, the TAA governs trade agreements negotiated between the United States and other countries under the Trade Act of 1974.[125] The stated purposes of the TAA are:

(1) To approve and implement the trade agreements negotiated under the Trade Act of 1974 [19 U.S.C. 2101 et seq.];
(2) To foster the growth and maintenance of an open world trading system;
(3) To expand opportunities for the commerce of the United States in international trade; and
(4) To improve the rules of international trade and to provide for the enforcement of such rules, and for other purposes.[126]

Per FAR 52.225-12, the TAA waives the BAA for acquisitions above a certain dollar threshold,[127] and allows use of eligible products in those acquisitions that are obtained from countries that have signed an international trade agreement with the United States. Thus, in many ways the TAA supersedes the BAA, because the TAA enables waiver of the BAA as provided in FAR Subpart 25.4.

FAR Subpart 25.4[128] designates products as "TAA compliant" if made in the United States or a "Designated Country." Designated Countries include:

1. Those with a free trade agreement with the United States (e.g., Canada, Mexico, Australia, and Singapore);
2. Countries that participate in the World Trade Organization Government Procurement Agreement (WTO GPA), including Japan and many countries in Europe;
3. Least developed countries (e.g., Afghanistan, Bangladesh, Laos, Ethiopia, and many others); and
4. Caribbean Basin countries (Aruba, Costa Rica, Haiti, and others).

Notably, the People's Republic of China is not a Designated Country.[129]

4. *The American Recovery and Reinvestment Act*

Recent economic recovery legislation, the American Recovery and Reinvestment Act of 2009 (ARRA), provides for over $460 billion in federal spending, and states that none of the funds appropriated by the ARRA may be used for the construction, alteration, maintenance, or repair of a public building or public work unless all of the iron, steel, and manufactured goods used in the project are produced in the United States.[130]

The ARRA's Buy America provision, however, is required to be "applied in a manner consistent with our United States obligations under international agreements." As implemented in FAR Subpart 25.6, all ARRA-funded construction contracts over $7,433,000 accordingly may utilize materials from the TAA list of Designated Countries, other than the Caribbean Basin countries.[131]

The ARRA also allows an agency to waive the application of the Buy America provision upon a finding that (1) the provision would be inconsistent with public interest; (2) iron, steel, and the relevant manufactured goods are not produced in the United States in sufficient and reasonably available quantities and of a satisfactory quality; or (3) inclusion of iron, steel, and manufactured goods produced in the United States will increase the cost of the overall project by more than 25 percent.[132]

H. Davis-Bacon Act

1. *Purpose*

The Davis-Bacon Act of 1931 establishes a minimum wage requirement for workers on federally subsidized public works construction projects. This minimum wage is referred to as "prevailing wage," in that it is determined based upon wage levels that prevail, or are paid in a particular locality for similar work. The rationale is that since the U.S. government is the largest purchaser of construction services, its contracting impacts wage levels. The Davis-Bacon Act was modeled on the prevailing wage law enacted in the state of Kansas in 1891.[133] The Kansas Act recognized that state projects had the potential to impact wages, and gave local contractors fair opportunity to compete in the bidding process for, and local workers fair opportunities to work on, state construction work. When such work was awarded to local contractors, the benefits of imposing a prevailing wage were confirmed by statistics revealing increased expenditures by the workers and contractors within that region, thus benefiting the local economies.

2. *Program Summary*

The Davis-Bacon Act requires that all government construction contracts, and most contracts for federally assisted construction over $2,000, include provisions for paying workers on-site no less than the locally prevailing wages and benefits for similar projects.[134] In addition to the Davis-Bacon Act itself, Congress has added Davis-Bacon prevailing wage provisions to approximately 60 other laws—the "related Acts"—under which federal agencies enable construction projects through grants, loans, loan guarantees, and insurance. Examples of related acts are the Federal-Aid Highway Act,[135] the Housing and Community Development Act of 1974,[136] and the Federal Water Pollution Control Act.[137] Generally, the same Davis-Bacon prevailing wage requirements are applied to projects receiving federal assistance under the related acts.

Originally, the government relied upon wages paid to the majority of local workers to determine the prevailing wage. If a majority wage did not exist, a 30 percent rule was applied, meaning if 30 percent of workers in the area are paid the same rate, that rate becomes the prevailing wage. Today, wages are determined by surveys conducted by the Department of Labor (DOL) that indicate the wages and number of workers employed during a peak week. The peak week is that in which the highest number of workers were employed at the same classification. If a majority wage does not exist, then a weighted average may be used. Wage determinations utilize a survey of a specified geographical area within a set calendar period, generally one year, from which is produced a general wage determination, or for a specific project, a project wage determination.[138] In addition, Davis-Bacon rates are established for each trade classification. The classifications were previously found in the *Dictionary of Occupational Titles*, which has been replaced by an online system known as O*NET, reflecting the same information.[139]

"Little Davis-Bacon Acts," which are state prevailing wage laws, have been enacted by the some 33 states, including the District of Columbia, requiring the payment of prevailing wages for state and local government construction projects.[140]

3. Administration

The DOL has oversight responsibilities to ensure coordination of administration and consistency of enforcement of the Davis-Bacon and related acts, and DOL regulations accordingly establish standards and procedures for the administration of the Davis-Bacon labor standards provisions.[141] The contracting agencies have day-to-day responsibility to administer and enforce the Davis-Bacon provisions in their own covered contracts and contracts for which they provide federal assistance. This includes ensuring proper Davis-Bacon wage determination(s) are applied to all such construction contracts.[142] Instructions as to wage determination issues can be found on various DOL Web sites.[143] Program administration for Little Davis-Bacon Acts is handled by the state where the work is located. Generally, local or county administrative offices are responsible for compliance with state law.

I. McNamara-O'Hara Service Contract Act

1. Purpose

The McNamara-O'Hara Service Contract Act (SCA) of 1965[144] is the services contract equivalent of the Davis-Bacon Act and is intended to ensure that government contractors compensate their service workers fairly.[145] The SCA

"establishes standards for minimum compensation and safety and health protection of employees performing work for contractors and subcontractors on
service contracts entered into with the Federal Government . . . ."[146] The SCA
accordingly prevents the government from "subsidizing" substandard levels
of compensation by awarding contracts to those who were able to bid low by
paying less.[147]

2. Program Summary

The SCA extends to "[e]very contract . . . entered into by the United States . . .
the principal purpose of which is to furnish services in the United States
through the use of service employees . . . ."[148] Administration and enforcement of the SCA is handled by the Wage and Hour Division of DOL.[149]

A "service employee" is any person engaged in the performance of a
service contract, other than any person employed in a bona fide executive,
administrative, or professional capacity, as those terms are defined in 29 C.F.R.
Part 541. The term "service employee" includes all such persons regardless of
any contractual relationship that may be alleged to exist between a contractor
or subcontractor and such persons.[150] As such, the SCA extends to qualifying
persons categorized as independent contractors instead of employees.

A "service contract" is any government contract, the principal purpose of
which is to furnish services in the United States through the use of service
employees, except as exempted under section 7 of the Act,[151] or any subcontract at any tier thereunder.[152] The SCA does not apply if the principal purpose of the contract is to provide something other than the services of service
employees.[153]

Contractors and subcontractors performing services on covered contracts
in excess of $2,500 must pay service employees in various classes no less than
the monetary wage rates and provide the fringe benefits found to be prevailing in the locality, or the rates (including prospective increases) contained in a
predecessor contractor's collective bargaining agreement.[154]

Safety and health standards also apply to covered services contracts.[155]
No part of the services may be performed in buildings or surroundings or
under working conditions, provided by or under the control or supervision of
the contractor or subcontractor, that are unsanitary or hazardous or dangerous to the health or safety of the covered service employees.[156] The safety and
health provisions of the SCA are administered by the Occupational Safety
and Health Administration.[157]

The wage rates and fringe benefits required are usually specified in the
contract, but in no case may employees doing work necessary for the performance of the contract be paid less than the applicable minimum wage established in section 6(a)(1) of the Fair Labor Standards Act.[158]

The DOL maintains a Web site that provides appropriate SCA wage determinations for each official contract action.[159] Every employer performing work
covered by the SCA is required to post a notice of the compensation required

in a prominent and accessible location at the work site, where it may be seen by all employees working on the contract.[160] On the date a service employee commences work on a covered contract, the contractor or subcontractor must provide the employee with a notice of the compensation required by the SCA.

A covered contractor is required to insert in all subcontracts the labor standards clauses specified by the regulations.[161] Prime contractors are liable for violations of the Act committed by their covered subcontractors.[162]

The SCA regulations point out that "the Act does not define, or limit, the types of 'services' which may be contracted for."[163] The list of covered services in the regulations is accordingly illustrative, not exhaustive.[164] The SCA does not apply to construction contracts covered by the Davis-Bacon Act,[165] among other exceptions.[166] There is no specific exception for construction management, design, engineering, or consulting services. However, to the extent contracts for such services are to be performed primarily or exclusively by administrative, executive, or professional employees (such as architects or engineers), with the use of service employees only being a minor factor in performance of the contract, the SCA will not apply to that contract. If there is, instead, a "significant or substantial" use of employees within the SCA definition of "service employee," the contract is covered by the SCA.[167]

3. Administration

The DOL (not the contracting agency) has the authority and responsibility for administering and interpreting the SCA, including making determinations of coverage.[168] The Supreme Court has recognized that such interpretations "provide a practical guide to employers and employees as to how the office representing the public interest in its enforcement will seek to apply it" and "constitute a body of experience and informed judgment to which courts and litigants may properly resort for guidance."[169] DOL interpretations are generally afforded deference by the courts.[170]

As with the Davis-Bacon Act, each contractor and subcontractor performing SCA-covered work is required to maintain records for each employee performing work on the covered contract for three years from completion of the work.[171]

IV. Conclusion

This chapter makes clear that there are many socioeconomic programs that have a significant impact on government construction contracts. It is important to be aware, however, that many of these programs are in a constant state of flux, depending on administrative policy, which party controls Congress, and court decisions. It is accordingly important to check on the current state of a program before making important decisions, by updating case law research and conducting an Internet search regarding the latest program developments, typically identified in agency postings or the news media.

Notes

1. The author acknowledges the contributions of attorney Michelle Illig, Esq., of the Illig Law Firm, Overland Park, Kansas, in the preparation of this chapter.

2. A comprehensive and detailed analysis of all federal small business contracting programs can be found in FAR Part 19, accessible online at http://www.acquisition.gov/far/current/html/FARTOCP19.html.

3. 1 ACADEMIC AMERICAN ENCYCLOPEDIA 132 (1995); 7 *id.* at 223–24.

4. Pub. L. No. 88-352, 78 Stat. 241, 253 (1964). Title VII is codified at 42 U.S.C. §§ 2000e–2000e-17.

5. Exec. Order No. 11,246, 3 C.F.R. 339 [1964–1965], *reprinted in* 42 U.S.C.A. § 2000e app. at 28–31 [1982], *amended by* Exec. Order No. 11,375, 3 C.F.R. 684 [1966–1970], *superseded by* Exec. Order No. 11,478, 3 C.F.R. 803 [1966–1970], *reprinted in* 42 U.S.C.A. § 2000e app. at 31–33 [1982]. Reassignment of Civil Rights Functions, 1 WEEKLY COMP. PRES. DOC. 305 (Sept. 27, 1965).

6. ANDORRA BRUNO, CONG. RESEARCH SERV. REPORT 98-992: AFFIRMATIVE ACTION IN EMPLOYMENT: BACKGROUND AND CURRENT DEBATE (1998).

7. 15 U.S.C. § 631.

8. Affirmative action is typically defined as those programs that "attempt to equalize the opportunity for women and racial minorities by explicitly taking into account their defining characteristics—sex or race—which have been the basis for discrimination." T. Mullen, *Affirmative Action,* in THE LEGAL RELEVANCE OF GENDER, 244–66 (S. McLean & N. Burrows eds., 1988). Thus, "affirmative action" is expressly based upon those race- and gender-specific classifications that are "inherently suspect" under the concept of a gender-neutral and color-blind Constitution. Regents of the Univ. of Cal. v. Bakke, 438 U.S. 265, 318–20 (1978); U.S. CONST. amend. V; U.S. CONST. amend. XIV. Early cases developing the constitutional review standards of these programs include *Bakke* (one of the first cases identifying the appropriate level of review for equal protection challenges to non-federal affirmative action programs); Fullilove v. Klutznick, 448 U.S. 448 (1980) (developing an "intermediate" or middle level of scrutiny to review the constitutionality of a federal minority set-aside program included in The Public Works and Employment Act of 1977); Wygant v. Jackson Bd. of Educ., 476 U.S. 267 (1986) (applying "strict scrutiny" to invalidate a local board of education's minority hiring preference set out in its collective bargaining agreement).

9. City of Richmond v. J.A. Croson Constr. Co., 488 U.S. 469 (1989).

10. For a summary of state *Croson* challenges of affirmative action programs from 1989 through 2004, *see* Farris Law Firm, L.L.C., Socio-Economic Issues in Government Contracting: The ABC's of Affirmative Action Compliance, app. A, http://www.farrislawfirm.com/Default.aspx?PageID=56.

11. Metro Broad., Inc. v. FCC, 497 U.S. 547 (1990).

12. The *Metro Broadcasting* program approved by Congress was found not to remedy past discrimination but as promoting the "important" governmental interest in achieving "broadcast diversity" under an intermediate scrutiny standard of review. 497 U.S. 547 (1990).

13. Adarand Constructors, Inc. v. Pena, 515 U.S. 200 (1995).

14. CHARLES DALE, CONG. RESEARCH SERV. REPORT RL30470: AFFIRMATIVE ACTION REVISITED: A LEGAL HISTORY AND PROSPECTUS 6–8 (2000).

15. *Adarand*, 115 S. Ct. at 2113; City of Richmond v. J. A. Croson Constr. Co., 488 U.S. 469, 492 (1989).

16. *Croson*, 488 U.S. at 492.

17. *Id*. at 500 (quoting Wygant v. Jackson Bd. of Educ., 476 U.S. 267, 277 (1986)). In *Wygant*, the Court stated that a local or state public employer need not "convinc[e] the court of its liability for prior unlawful discrimination; nor does it mean that the court must make an actual finding of prior discrimination based on [the Government's] proof before the [the Government's] affirmative action plan will be upheld." 476 U.S. 267, 292–93 (O'Connor, J., concurring). In *Croson*, the Court stated that courts should follow a flexible approach in determining whether there exists a "firm basis" for determining that affirmative action is warranted. 488 U.S. at 501 (quoting Hazelwood Sch. Dist. v. United States, 433 U.S. 299, 307–08 (1977)). The Government's burden of establishing a compelling governmental interest is thus satisfied by establishing reliable statistical proof of past discrimination against specifically identified minority or gender-based groups. *Id*.

18. *Croson*, 488 U.S. at 501 (quoting *Hazelwood Sch. Dist.*, 433 U.S. at 307–08).

19. Coral Constr. Co. v. King County, 941 F.2d 910, 929 (9th Cir. 1991), *cert. denied*, 112 S.Ct. 875 (1992).

20. *Id*. at 919.

21. *See* Associated Gen. Contractors of Am. v. City of Columbus, 936 F. Supp. 1363 (S.D. Ohio 1996) (challenging the validity of the city's statistical study), *vacated at* 172 F.3d 411 (6th Cir. 1999).

22. *See* City of Richmond v. J. A. Croson Constr. Co., 488 U.S. 469 (1989); Monterey Mech. Co. v. Wilson, 125 F.3d 702 (9th Cir. 1997); L. Tarango Trucking v. County of Contra Costa, 181 F. Supp. 2d 1017 (N.D. Cal. 2001); Concrete Works of Colo., Inc. v. City and County of Denver, 86 F. Supp. 2d 1042 (D. Colo. 2000); Cortez III Serv. Corp. v. Nat'l Aeronautics & Space Admin., 950 F. Supp. 357 (D.D.C. 1996); Eng'g Contractors Assoc. of S. Fla., Inc. v. Metro. Dade County, 122 F.3d 895 (11th Cir. 1997); Rothe Dev. Corp. v. Dept. of Defense, 545 F.3d 1023 (Fed. Cir. 2008), *rev'd in part and remanded*, 606 F. Supp. 2d 648 (2009).

23. *See* cases cited *supra* note 22.

24. *See* cases cited *supra* note 22.

25. 488 U.S. at 507–08.

26. *Id*. at 526 (Scalia, J., concurring); *see* State of Kansas Exec. Order 08-08 (implementing Small Business, Minority and Women Business Program) (June 2008); *see also* Small Business Set-Aside Program, State of Illinois, Frequently Asked Questions, at http://www.sell2.illinois.gov/FAQ_SBSP.cfm.

27. *Croson*, 488 U.S. at 507–08.

28. *See generally* CITY OF KANSAS CITY, MISSOURI DISPARITY STUDY, at E15, E40 (1994).

29. Rothe Dev. Corp. v. Dept. of Defense, 545 F.3d 1023 (Fed. Cir. 2008), *rev'd in part and remanded*, 606 F. Supp. 2d 648 (2009).

30. National Defense Authorization Act for Fiscal Year 1987, Pub. L. No. 99-661, 100 Stat. 3816, 3973, § 1207 (1986).

31. *Id.*

32. The *Rothe* case may also be reviewed on the Internet at http://caselaw
.lp.findlaw.com/data2/circs/fed/081017p.pdf.

33. See *supra* section D, "Equity in Contracting for Women Act of 2000" discussion. Additional references to Department of Justice testimony can be viewed
at http://www.youtube.com/watch?v=69W5Df2T1FU.

34. ELAINE REARDON, NANCY NICOSIA & NANCY Y. MOORE, KAUFFMAN-RAND
INSTITUTE FOR ENTREPRENEURSHIP PUBLIC POLICY, THE UTILIZATION OF WOMEN-
OWNED SMALL BUSINESSES IN FEDERAL CONTRACTING 1 (2007).

35. Pub. L. No. 100-656, § 502 (1988).

36. Pub. L. No. 105-135, § 603 (1997).

37. 13 C.F.R. § 121; *see also* the SBA's TABLE OF SMALL BUSINESS SIZE STANDARDS
MATCHED TO NORTH AMERICAN INDUSTRY CLASSIFICATION SYSTEM CODES, at
http://www.sba.gov/idc/groups/public/documents/sba_homepage/serv_sstd_
tablepdf.pdf.

38. 13 C.F.R. § 121.

39. For specific size standards per industry, see TABLE OF SMALL BUSINESS
SIZE STANDARDS, *supra* note 37.

40. For CCR registration, see http://www.ccr.gov/.

41. For ORCA registration, see https://orca.bpn.gov/.

42. 13 C.F.R. §§ 121.1001–.1003.

43. 49 C.F.R. §§ 23, 26. *See also* 49 C.F.R. § 23, app. A; 49 C.F.R. § 26, app. A.

44. 49 C.F.R. §§ 23, 26.

45. *Id.*

46. U.S. Dept. of Transp., Office of Small & Disadvantaged Bus. Utilization,
Memorandum of Understanding SBA/DOT, at http://osdbu.dot.gov/dbeprogram/
memofunder.cfm.

47. 13 C.F.R. pt. 124, and DOT regulations, 49 C.F.R. pts. 23, 26.

48. 13 C.F.R. pt. 124, and DOT regulations, 49 C.F.R. pts. 23, 26.

49. 13 C.F.R. § 124/101.

50. 49 C.F.R. pt. 26, app. E.

51. Rules, General Services Administration, National Aeronautics and Space
Administration, Department of Defense, 62 Fed. Reg. 25,786 (1997). Note 1
explained:

> FASA and 10 U.S.C. § 2323 (which, in language similar to that in FASA,
> permits the Department of Defense, NASA, and the Coast Guard to use
> less than full and open competition in order to aid SDBs) incorporate
> by explicit reference the definition of social and economic disadvan-
> tage contained in Section 8(d) of the Small Business Act. Pursuant to
> Section 8(d), members of designated groups are presumed to be both
> socially and economically disadvantaged; those presumptions are
> rebuttable.

52. 23 C.F.R. ch. 1, Safe, Accountable, Flexible, Efficient Transportation Equity
Act: A Legacy for Users (2008). This Act can be reviewed with commentary at
http://www.fhwa.dot.gov/safetealu/legis.htm.

53. 49 C.F.R. pt. 23.

54. 49 C.F.R. pt. 26.

55. 49 C.F.R. pt. 23.

56. A listing of the certification sites can be found at http://www.osdbu.dot .gov/DBEProgram/StateDOTDBESites.cfm.

57. *See* OSDBU DOT assistance at http://www.dotcr.ost.dot.gov/asp/dbe .asp. Other benefits provided under the DBE program are educational in nature. State Departments of Highways and Transportation (SDH&T) receive supportive services funds from the Federal Highway Administration to help increase DBE participation in federal-aid highway contracts. The SDH&T may decide to use the funds in-house to provide the supportive services or hire consultants. Supportive services, whether done by the state agency or by consultants, assist the DBE firms to compete in winning contracts. The services include research and development, training and on site-technical assistance, business management assistance, estimating assistance, and assistance in obtaining necessary financing and bonding. *Id.*

58. 15 U.S.C. § 637(a).

59. Waivers may be granted where the applicant can show: (a) substantial business management experience; (b) technical experience to carry out its business plan with a substantial likelihood of success; (c) adequate capital to sustain its operations and carry out its business plan; (d) a record of successful performance on contracts from governmental or nongovernmental sources in its primary industry category; and/or (e) possession of, or the ability to timely obtain, the personnel, facilities, equipment, and any other requirements needed to perform on contracts if it is admitted to the 8(a) program.

60. 13 C.F.R. §§ 124.105, .106.

61. The SBA 8(a) application may be found at http://www.sba.gov/idc/ groups/public/documents/sba_homepage/form_tr1010.pdf.

62. Pub. L. No. 95-507, 15 U.S.C. § 644(k).

63. *See* http://www.va.gov/OSDBU/.

64. Pub. L. No. 106-554 § 811 (2000).

65. The ruling can be viewed online at https://ecf.dcd.uscourts.gov/cgi-bin/ show_public_doc?2004cv1889-24.

66. Women-Owned Small Business Federal Contract Assistance Procedures, 72 Fed. Reg. 73,295 (Dec. 27, 2007).

67. Pub. L. No. 106-554 (2000); 15 U.S.C. §§ 637, 811.

68. For NAICS size codes by industry, see TABLE OF SMALL BUSINESS SIZE STANDARDS, *supra* note 37.

69. For CCR registration, *see* http://www.ccr.gov/.

70. *See* https://orca.bpn.gov/.

71. Pub. L. No. 108-183; 15 U.S.C. § 657(f).

72. FAR 19.1405.

73. FAR 19.1406.

74. *Id.*, subpt. 19.5.

75. *Id.*

76. *Id.*

77. FAR 19.402.

78. FAR 19.501(d), 6.302-5.

79. FAR 19.1404 subpt. 8.6.

80. *Id.*, subpt. 8.7.

81. *Id.*, subpt. 16.05.

82. *Id.*, subpt. 18.04.

83. FAR 19.1404.

84. 13 C.F.R. §§ 125.8–.10; FAR 19.1403(a)–(b).

85. See references, *supra* note 84.

86. 13 C.F.R. § 125.6(b).

87. 13 C.F.R. §§ 121, 125.

88. 13 C.F.R. § 125.25(d).

89. 15 U.S.C. § 631; FAR 19.1301.

90. FAR 19.3102.

91. FAR 19.1305.

92. *See* FAR pt. 13.

93. FAR 19.1306.

94. *Id.*

95. FAR 19.1307.

96. *Id.*

97. *Id.*, subpt. 8.06.

98. *Id.*, subpt. 8.07.

99. FAR subpts. 8.07, 16.05.

100. *Id.*, and subpts. 8.04, 8.07, 16.05.

101. FAR 19.1304.

102. FAR 19.1307(b).

103. *Id.*

104. *Id. See also* subpt. 19.11.

105. FAR 19.1308 requires the CO on a HUBZone eligible project to insert FAR clause 52.219-3, Notice of Total HUBZone Set-Aside, in solicitations and contracts for acquisitions that are set aside for HUBZone small business concerns under FAR 19.1305 or 19.1306. The CO is also required to insert the clause at FAR 52.219-4, Notice of Price Evaluation Preference for HUBZone Small Business Concerns, in solicitations and contracts for acquisitions conducted using full and open competition. The clause is not to be used in acquisitions that do not exceed the simplified acquisition threshold.

106. HUBZone areas may be located by entering the address at the following Web site: http://map.sba.gov/HUBzone/init.asp#address.

107. 13 C.F.R. § 126.103.

108. 13 C.F.R. § 126.700. HUBZone certification assistance may be found at http://www.certassist.net/HUBZone.html.

109. 13 C.F.R. §§ 126.309, 126.304(c).

110. 41 U.S.C. § 10a–d.

111. *Id.; see* http://buyamericacoalition.org/history.html.

112. 41 U.S.C. § 10a–d.

113. Pub. L. No. 97-424, § 165, 96 Stat. 2097 (1983), *implemented at* 23 C.F.R. § 635.410.

114. FAR 25.003, 25.101.

115. Although the BAA did not define the term "substantially all," Executive Order No. 10582, as amended by Executive Order Nos. 11051 and 12148, established that 50 percent of the component cost must consist of domestically produced items.

116. Subpt. 25.1.

117. FAR 25.003.

118. *Id.*

119. Consol. Tanneries, Ltd., BCA 166786 (June 24, 1969); *see also* Glazer Constr. Co. v. United States, 50 F. Supp. 2d 85, 98 (D. Mass. 1999); City Chem. LLC, B-296135.2, B- 296230.2, June 17, 2005, 2005 CPD ¶ 120.

120. 15 U.S.C. § 2502; FAR subpt. 25.4. *See* Buy American Act, 48 § 25.001; Trade Agreements Act, 48 C.F.R. § 25.004.

121. 15 U.S.C. § 2502; FAR subpt. 25.004.

122. *Id.*

123. FAR does not specifically define "in the public's interest," except to say that this exception applies when an agency has an agreement with a foreign government that provides a blanket exception to the BAA. *See* FAR 25.103(a).

124. FAR subpt. 25.501–25.504.

125. Pub. L. No. 93-618, 88 Stat. 1978 (1975), codified at 19 U.S.C. ch. 12. This Act is intended to help industry in the United States become more competitive or phase workers into other industries or occupations. It created fast-track authority for the President to negotiate trade agreements that Congress can approve or disapprove but cannot amend or filibuster. The fast-track authority created under the Act extended to 1994 and was restored in 2002 by the Trade Act of 2002.

126. 19 U.S.C. § 2502 (Congressional statement of purposes).

127. FAR § 52.225-12.

128. FAR subpt. 25.4.

129. A complete list of the Designated Countries can be found at FAR 25.003 and 48 C.F.R. § 52.225-11 (2009).

130. The American Recovery and Reinvestment Act of 2009, 123 Stat. 115 (2009), *available at* http://frwebgate.access.gpo.gov/cgi-bin/getdoc.cgi?dbname=111_cong_public_laws&docid=f:publ005.pdf.

131. FAR 25.603(c), 25.4.

132. American Recovery and Reinvestment Act of 2009, 123 Stat. 115 (2009).

133. KAN. STAT. ch. 114 (1891).

134. 29 C.F.R. § 1-7.

135. Federal-Aid Highway Act of 1956, popularly known as the National Interstate and Defense Highways Act (Pub. L. No. 84-627, 70 Stat. 374 (1956)).

136. The Housing and Community Development Act of 1974, among other provisions, authorizes "Entitlement Communities Grants" to be awarded by the Department of Housing and Urban Development.

137. 33 U.S.C. §§ 1251–1376; 33, Ch. 26 (as amended through Pub. L. 110–288, July 29, 2008).

138. Wage determinations can be found at http://www.wdol.gov/.

139. O*NET classifications can be found at http://online.onetcenter.org/.

140. For a reference chart of related state statutes, see http://www.lsc.state.oh.us/membersonly/126prevailingwagelaws.pdf.

141. Detailed compliance summaries can be found at the DOL's Web site at http://www.dol.gov/compliance/laws/comp-dbra.htm.

142. 29 C.F.R. §§ 1.5, 1.6(b).

143. Davis-Bacon prevailing wage determination instructions and guidelines can be found at http://www.gpo.gov/davisbacon/referencemat.html.

144. 41 U.S.C. § 351 (2008). Further Department of Labor guidelines may be found at http://www.dol.gov/compliance/laws/comp-sca.htm.

145. *Id.*

146. 29 C.F.R. § 4.102.

147. Saavedra v. Donovan, 700 F.2d 496 (9th Cir.), *cert. denied*, 464 U.S. 892, 104 S. Ct. 236 (1983). *See also* S. REP. No. 89-798, *reprinted in* 1965 U.S.C.C.A.N. 3737.

148. 41 U.S.C. § 351(a) (2008).

149. http://www.dol.gov/esa/WHD.

150. Subpt. 22.10—Service Contract Act of 1965, as amended.

151. 41 U.S.C. § 356. *See also* 41 U.S.C. § 356; 48 C.F.R § 22.1003-3, -4.

152. *See* 48 C.F.R. § 22.1003-5 and 29 CFR § 4.130 for a partial list of services covered by the Act.

153. JEROME S. GABIG, HUNTSVILLE NAT'L CONTRACT MGMT ASS'N, THE SERVICE CONTRACT ACT (2006)—A TRAP FOR UNSUSPECTING CONTRACTORS (2006), http://www.ncmahsv.org/pdfs/scanotes.pdf.

154. 41 U.S.C. § 351.

155. *Id.*

156. *Id.*

157. http://www.dol.gov/.

158. 41 U.S.C. § 351.

159. http://www.wdol.gov/.

160. http://www.dol.gov/esa/whd/regs/compliance/posters/sca.htm.

161. 29 C.F.R. pt. 4.

162. *Id.*

163. 29 C.F.R.§ 4.111(b).

164. *Id.* at § 4.130.

165. http://www.dol.gov/esa/whd/contracts/sca.htm.

166. *Id.*

167. 29 C.F.R. § 4.113(a)(2)–(3).

168. Woodside Vill. v. Sec'y of Labor, 611 F. 2d 312 (9th Cir. 1980).

169. Skidmore v. Swift & Co., 323 U.S. 134 (1944).

170. Griggs v. Duke Power Co., 401 U.S. 424, 433–34 (1971); Udall v. Tallman, 380 U.S. 1 (1965).

171. 29 C.F.R. pt. 4.6.

CHAPTER 13

Termination for Convenience

JOSEPH D. WEST
CHRISTYNE K. BRENNAN

I. Introduction

Contrary to standard commercial contracts, the government, as a party to a contract, may unilaterally terminate a contract for convenience or partially eliminate work from a contract using a partial termination for convenience. In the event of such a termination, the contractor has certain duties to close out the terminated work, as well as rights to recover certain associated costs. In this regard, the Federal Acquisition Regulation (FAR) sets forth specific duties and rights of both the government and contractor upon a termination for convenience. These are generally summarized below.

II. The Right to Terminate for Convenience

A. *Corliss* and the Absolute Right to Terminate

Over 130 years ago in *United States v. Corliss Steam Engine Co.*,[1] the Supreme Court acknowledged the government's right to terminate a contract when completion of the contract is no longer in the government's best interest, notwithstanding the absence of a termination for convenience clause in the contract or specific statutory authority to do so.

The legal basis for the *Corliss* decision was that the unpredictability of war justified termination of useless military procurements. This view was extended to peacetime military contracts and to civilian procurement in the 1950s, and in 1967 to all situations in which the expectations of the parties were substantially altered after contract execution.

In 1974, the Court of Claims in *Colonial Metals Co. v. United States*[2] extended *Corliss* even further by allowing a termination for convenience where there were no altered circumstances. In that case, the government had terminated the contract before performance for a reason that was known to it prior to contract award. Subsequent decisions interpreted *Colonial Metals* to provide

357

contracting officers (COs) with virtually unlimited discretion to terminate contracts for convenience. As a practical matter, a termination for convenience was viewed as an absolute right of the government. The only potential bases to overturn a convenience termination were government bad faith or abuse of discretion, and contractors challenging such terminations were held to a heightened standard of proof and were very rarely successful.[3]

B. Torncello's Limit to Convenience Terminations

In 1982, the Court of Claims departed from *Corliss* and *Colonial Metals* and articulated a seemingly broad limitation on the government's right to terminate a contract for convenience. In *Torncello v. United States*,[4] the Navy awarded an exclusive requirements contract for janitorial and other services, including pest control, to Soledad Industries. Soledad's bid was unbalanced, bidding $500 per call on one of the pest control items. The Navy accepted the entire contract knowing that the same services could be obtained at a lower price from an in-house Navy organization. The Navy then proceeded to procure pest control services from the in-house organization, even though Soledad had offered to reduce its price for this aspect of the contract. The contract was subsequently terminated for default. The Navy claimed, and the Armed Services Board of Contract Appeals (ASBCA) agreed, that "constructive convenience termination" was available as a valid excuse for procuring the pest control services from the cheaper source.[5] Soledad's bankruptcy trustee, Mr. Torncello, appealed the ASBCA's decision to the Court of Claims.

No clear decision emerged from the trustee's appeal. Three judges formed a plurality opinion, and each of the other three members of the en banc panel wrote concurring opinions. The plurality expressly overruled *Colonial Metals*, holding that the government had no right to invoke a termination for convenience merely to escape its obligation under a requirements contract to purchase goods or services.[6] The plurality stated that the government may not terminate a contract for convenience unless a "change in circumstances" occurred between the time of contract award and contract termination.[7] Without a limit on the government's right to terminate, the plurality reasoned, the contract would become "illusory and therefore unenforceable."[8]

Some cases interpreted the *Torncello* decision broadly, requiring a "change in circumstances" as a prerequisite to a valid termination for the convenience of the government. For instance, in *Municipal Leasing Corp. v. United States*,[9] the Air Force considered whether to repair its own computers or to lease new computers and, ultimately, decided to lease new computers. After entering into a sole-source contract with Municipal to lease Intecolor computer terminals, however, the Air Force decided instead to repair its own computers and to terminate Municipal's contract for convenience. The Claims Court held that the government could not terminate for convenience to utilize a different product the government had previously considered without "some kind of change from the circumstances of the bargain or in the expectations of the parties."[10]

Similarly, in *Maxima Corp. v. United States*,[11] the Federal Circuit agreed with a broad interpretation of the "change in circumstances" standard enunciated in *Torncello* by ruling that the government could not retroactively terminate a contract for convenience in order to limit its contractual liability.[12] In *Maxima*, the Environmental Protection Agency (EPA) contracted with Maxima for word processing services. The contract guaranteed Maxima a minimum amount of work and set a minimum sum to be paid by the EPA. The EPA failed to request the minimum amount of work, even though Maxima was fully capable of providing that work.[13] Over a year after completion of the contract, in order to recoup some of the minimum sum paid to Maxima, the EPA constructively terminated the contract for convenience due to its failure to request the minimum amount of work over the term of the contract. The Federal Circuit ruled that the government could not invoke a constructive termination for convenience in order to escape its contractual obligation to pay a guaranteed minimum sum on a contract that had already been performed.[14]

C. *Krygoski* and the Impact of the Competition in Contracting Act on Convenience Terminations

In *Krygoski Construction Co. v. United States*,[15] the Federal Circuit explicitly limited the reach of the *Torncello* decision, "render[ing] the plurality's dicta in *Torncello* inapplicable to the present regime of contract administration."[16] In *Krygoski*, the Army Corps of Engineers awarded Krygoski Construction a contract to demolish an abandoned Air Force missile site and to remove an estimated amount of asbestos located on the site. After contract award, but before contract performance, additional asbestos was discovered that, according to the contractor's estimate, would raise the asbestos removal costs from approximately 10 percent to almost 50 percent of the overall contract price. Considering the additional asbestos removal requirement to represent a change in the scope of the contract, the CO terminated the contract for convenience in order to solicit new bids reflecting the revised specifications. The Court of Federal Claims agreed with Krygoski's claim that the Corps had breached its contract. Relying on *Torncello*, the court found no "change in circumstances" to justify the termination for convenience and consequently awarded Krygoski $1.46 million in damages.

On appeal, the Federal Circuit reversed, holding *Torncello* inapplicable to the present case because "*Torncello* applies only when the Government enters a contract with no intention of fulfilling its promises."[17] The Federal Circuit predicated this narrow reading of *Torncello* on the 1984 enactment of the Competition in Contracting Act (CICA).[18] The Federal Circuit viewed CICA as allaying the *Torncello* plurality's concerns regarding the government's ability to shop for lower prices after contract award.[19] "CICA permits a lenient convenience termination standard," the Federal Circuit noted, in order to foster its objective of compelling full and open competition throughout the procurement process.[20] In light of the statutory enactment and policy goals of CICA, the Federal Circuit concluded that it "will avoid a finding of abused discretion when the facts support a reasonable inference that the CO terminated for convenience in furtherance of statutory requirements for full and open competition."[21]

Since *Krygoski*, the courts and boards have held contractors to a restrictive standard when challenging convenience terminations. In order to get such a termination overturned, a contractor must generally prove, with clear and convincing evidence, that the government acted in bad faith with the intent to injure the contractor.[22]

III. *The Parties' Rights and Duties upon Termination for Convenience*

A. Duties of the Government

The termination for convenience clauses at FAR 52.249-2, Alternate 1 "Termination for Convenience of the Government (Fixed-Price)," FAR 52.249-3 "Termination for Convenience of the Government (Dismantling, Demolition, or Removal of Improvements)," and FAR 52.249-6, Alternate 1 "Termination (Cost-Reimbursement)" authorize COs to terminate construction contracts for convenience and to enter into termination settlement agreements with contractors. After the CO issues a notice of termination, the termination contracting officer (TCO) is responsible for negotiating any settlement with the contractor, including a no-cost settlement, if appropriate.

1. *Termination by Written Notice*

The regulations provide that the CO "shall terminate contracts for convenience . . . only by a written notice to the contractor."[23] The notice must specify (1) that the contract is being terminated for the convenience of the government; (2) the effective date of termination; (3) whether the contract is being completely or partially terminated; (4) any special instructions (for example, concerning the disposition of inventory and special tooling); and (5) the steps the contractor should take to minimize the impact on personnel if the termination will result in a significant reduction in the contractor's work force.[24] FAR 49.601 suggests that the CO send this notice to the contractor by telegraph or letter.[25] However, in *Maibens, Inc.*,[26] a termination for convenience was upheld even though the contractor denied receiving written notice because the contractor admitted receiving the CO's oral notice during which the contractor was referred to the contract page and clause.

The CO may amend the termination notice to correct nonsubstantive mistakes, add supplemental data or instructions, or rescind the notice.[27] The CO may also reinstate a terminated contract upon written consent of the contractor, if circumstances clearly indicate a requirement for the terminated items and reinstatement is advantageous to the government.[28] This situation may arise when the contractor, after receiving a complete termination notice, advises the CO that the project is near completion and it would be less costly to complete than to terminate.

2. Termination Through Operation of Law

In addition to written notices, convenience terminations may arise by operation of law. Examples of convenience terminations through operation of law include the termination of improperly awarded contracts, the judicial conversions of improper terminations for default, and constructive terminations based on the actions of the parties.

a. Improperly Awarded Contracts

If the illegality in the contract award is "plain" or "palpable,"[29] then the award may be "canceled without liability to the government except to the extent recovery may be had on the basis of quantum meruit."[30] If the contractor did not contribute to the mistake resulting in improper award and was not on direct notice before award that the procedures being followed were incorrect, then the contract must be terminated for the convenience of the government.[31]

b. Conversion of a Termination for Default

If the government terminates a contract for default and it is later determined that the contractor was not in default, the rights of the parties are determined as though the termination was one for the convenience of the government.[32] If the contract does not expressly contain a termination for convenience clause, the clause will be incorporated into the contract pursuant to the *Christian* doctrine.[33]

c. Constructive Termination

It is possible to terminate for convenience in the absence of a formal notice. The constructive termination doctrine is used in situations where contracts have been erroneously canceled or not properly entered into by the government. Typically, this situation arises when the government cancels awards that it believes are illegal, but that prove to be valid.[34]

The Boards of Contract Appeals (BCAs) have held constructive terminations to be effective by analogizing them to constructive changes. In constructive change situations, the CO instructs a contractor to perform work, and the contractor complies with the instruction (perhaps under protest) even though the change was not in writing. In the constructive termination for convenience situation, the services, which were required and should have been performed by the contractor, were performed by someone else. In these circumstances, the absence of a notice of termination is not fatal to the government's request for a convenience settlement.[35]

Additionally, constructive terminations have been used to terminate contract changes outside the scope of the contract, referred to as "cardinal changes." In *American Air Filter Co.*,[36] the Comptroller General held that where

a cardinal change is believed to be necessary, the contract should be terminated for convenience and resolicited.

3. Settlement of the Terminated Contract

Negotiating the settlement of terminated contracts falls on the TCO. Auditors
and the TCO must promptly schedule and complete audit reviews and negotiations, giving particular attention to the need for timely action on all settlements, especially those involving small businesses.

The TCO must take three primary actions in settling a contract terminated
for convenience: (1) direct the action required of the contractor; (2) examine
the contractor's settlement proposal; and (3) negotiate the settlement with the
contractor and enter into a settlement agreement.[37] If a complete settlement is
not reached, the TCO must settle those elements on which agreement exists.[38]

B. Duties of the Contractor

After receipt of the notice of termination, the contractor must comply with the
notice and the applicable termination for convenience clause of the contract,
except as otherwise directed by the TCO. Generally, the termination for convenience clause requires the contractor to stop work immediately or advise
the TCO of any special circumstances precluding the stoppage of work, terminate all subcontracts, perform any continued portion of the contract (in the
case of partial terminations), protect and preserve government property in the
contractor's possession, settle outstanding liabilities, submit a settlement proposal, and dispose of termination inventory, as directed or authorized by the
TCO.[39]

1. Cessation of Work

Work must be terminated as of the effective date of the termination notice.[40]
However, in the case of a partial termination, the contractor remains obligated
to complete the unterminated portion of the contract.[41]

2. Termination of Subcontracts
a. Prime's Obligation to Terminate Subcontractors

FAR 49.104(b) requires that the contractor terminate all subcontracts related to
the terminated portion of the contract. From this also stems the contractor's
duty to settle all subcontractor claims arising from the termination.

b. Subcontract Termination Absent a Termination for Convenience Clause

It is important for the contractor to realize that termination for convenience
clauses are not mandatory flow-down provisions. As a result, contractors
must negotiate to include them in their subcontracts. Absent a termination for
convenience clause, the prime may be liable under state law for breach of the
subcontract if the government terminates the prime's contract for convenience.
Settlements between the prime and the subcontractor require government

approval, and "in no event . . . will the Government pay the prime contractor any amount for loss of anticipatory profits or consequential damages resulting from the termination of any subcontract."[42]

When a termination for convenience clause is included in the subcontract, the principal question concerns the breadth of the prime contractor's rights to terminate the subcontract. Where a subcontract termination for convenience clause is coextensive with that of the prime contract, the prime's right to terminate will be dictated by the contours of the *Torncello/Krygoski* line of cases as discussed in Section I.

Subcontractors may attempt to structure the termination for convenience clause in the subcontract to limit the prime's right to terminate to those situations where the prime's contract has been terminated for convenience.

For a subcontractor, the termination provisions of its contract with the prime are of paramount importance because subcontractors lack privity with the government. Thus, subcontractors cannot bring legal action directly against the government in the event the prime contract is terminated for convenience.

3. Preparation of Termination Inventory

No later than 120 days from the effective date of termination, the contractor generally is required to submit complete termination inventory schedules to the CO.[43] Termination inventory includes any property purchased, supplied, manufactured, furnished, or otherwise acquired for the performance of a contract subsequently terminated and properly allocable to the terminated portion of the contract.[44] Termination inventory may include raw materials, work in progress, and completed supplies that do not conform to the contract specifications, provided the amount or percentage of nonconforming inventory is reasonable.[45]

As a practical matter, the contractor should immediately segregate the inventory allocable to the terminated contract. It is best to physically segregate the inventory, not simply allocate inventory through the use of control documents. It is important that all inventory is accounted for and not accidentally overlooked or used on another project. This protects both the government and the contractor by assuring complete compensation for work performed, while avoiding double payments for inventory the contractor uses elsewhere.

4. Disposal of Termination Inventory

The contractor is required to follow the TCO's direction as to disposal of termination inventory.[46] However, after expiration of the plant clearance period, which is the period beginning on the effective date of termination and ending 90 days (or such longer period as may be agreed to) after receipt by the CO of acceptable inventory schedules for each property classification, the contractor may request that the government remove the inventory not previously disposed of or enter into a storage agreement. Within 15 days of such request, the government is required either to accept title to the inventory not previously disposed of or to enter into an agreement with the contractor for storage of that inventory.

5. Submittal of Settlement Proposals

One of the stated objectives of FAR Part 49 termination procedures is a negotiated agreement between the contractor and the government as to the costs and profit due the contractor. If an agreement cannot be reached, the TCO may resort to a unilateral determination of the amount due.[47] However, before issuing such a determination, the TCO must give the contractor at least 15 days' notice to submit written evidence substantiating the amount previously proposed.[48] This unilateral determination is subject to appeal by the contractor under the Contract Disputes Act (CDA).

The contractor must take the first step toward reaching a negotiated settlement by preparing and submitting a settlement proposal to the TCO that sets forth the contractor's termination claim for costs and profits. Settlement proposals must be certified and submitted to the TCO within one year from the effective date of termination, unless this time period is extended in writing by the CO prior to the end of the one-year period.[49] If the convenience termination is a conversion from a default termination, the one-year period begins to run as of the date of conversion.[50] Typically, subcontract termination clauses require that settlement proposals be submitted in significantly less than one year.

There are two primary bases used for the preparation of settlement proposals: (1) the inventory basis method and (2) the total cost method. While the inventory basis is the preferred method for preparing settlement proposals for most kinds of government contracts,[51] that method is generally impractical, and so not used, on convenience terminations of construction contracts. Rather, for those terminations, the parties will usually agree that the total cost method is appropriate.

FAR 49.206-2(b) provides examples of when a total cost settlement is appropriate, including where a construction contract is completely terminated.[52] Total cost settlement is generally appropriate if the inventory method would unduly delay settlement or is impracticable.[53] However, the TCO must approve this method prior to its use. In the total cost method, all of the incurred costs are itemized, and allowable costs up to the bid price are paid to the contractor. It is improper to calculate the settlement by deducting the value of the unfinished work from the contract price.[54] This method is disfavored because it rewards contractors for inefficiency. When the total cost method is used in a partial termination, the settlement proposal may only be submitted after the unterminated portion of the contract is completed.[55]

In some circumstances, a no-cost settlement may be the best alternative. If a contractor has not incurred costs pertaining to the terminated portion of the contract or agrees to waive its costs and no costs are due the government under the contract (e.g., as a compromise where the government threatens and the contractor disputes a termination for default), the parties may execute a no-cost settlement agreement.[56] However, where the contractor stands ready and able to perform, the government may not force a no-cost settlement.[57]

IV. Monetary Recovery

When a contract is terminated for convenience, the contractor has a general right to costs and profit on the pre-termination work.[58] The articulated goal in the settlement of fixed-price contracts is "fair compensation."[59] As a result, business judgment, not strict accounting principles, guides the determination.[60]

A. Negotiated Settlements

The contractor is free to negotiate with the TCO for any reasonable amount up to the contract price. The termination settlement should include reasonable profit on the work completed, as reduced by the payments already made and adjusted for any losses.

The agreement may be limited to a total amount to be paid without agreeing on or segregating particular elements of costs or profit.[61] Costs and expenses may be estimated and resolved through negotiation. Cost and accounting data may provide guides, but are not dispositive.[62]

B. Settlements by Contracting Officer (Unilateral) Determination

The TCO must make a unilateral determination of the amount to which the contractor is entitled if a negotiated settlement cannot be reached. This amount will include:

1. The cost of all contract work performed before the effective date of termination, not to exceed the total contract price, reduced by the amount of payments otherwise made, the contract price of work not terminated, and the fair value of items in the termination inventory that are undeliverable to the government or a potential buyer because of loss, destruction, or damage;

2. The cost of settling and paying subcontractor claims arising out of the termination of subcontractor work, exclusive of amounts paid or payable for supplies, materials, or services furnished before termination, that are payable under the cost of contract work as discussed in item (1) above;

3. A fair and reasonable profit, as determined by the TCO pursuant to FAR 49.202, unless the contractor would have sustained a loss on the entire contract if completed, in which case the settlement will be reduced to reflect the rate of loss;

4. Reasonable costs of preservation and protection of the termination inventory; and

5. Any other reasonable costs incurred because of the termination, as well as expenses incurred in providing data required for the TCO's determination of the amount due.[63]

1. Contractor Costs

Allowable costs under the termination for convenience clauses are determined by the cost principles of FAR Part 31. In general, those costs that have arisen from the contract and have not yet been reimbursed are compensable if they are reasonable, allocable, and not specifically designated as unallowable when incurred.[64] FAR 31.205-42 specifically addresses termination costs and serves as a supplement and modification to other cost provisions. This FAR provision establishes a framework for determining the allowability of these costs if termination occurs.

a. Termination Inventory

Inventory costs are generally payable to the contractor subject to the general and specific FAR Part 31 limitations on cost. The TCO directs the disposition of inventory under the standard termination for convenience clause. The TCO may require the contractor to transfer title and deliver to the government or sell any material included in the inventory.[65] The government may accept completed items for payment in the usual manner by the TCO.

Costs of termination inventory remaining after disposal in accordance with the TCO's directions are recoverable. Compensable costs include reasonable handling and transportation costs as well as reasonable cancellation and restocking charges.[66]

Before the transfer of title to the government, the risk of loss is on the contractor. Any part of the termination inventory that is lost, stolen, destroyed, or otherwise undeliverable will be deducted from the contractor's settlement, except for normal spoilage.[67] Even after the government takes title to the inventory, the contractor remains responsible for items in its possession.[68]

b. Common Items

The cost of items *reasonably usable* on the contractor's other work [is] not allowable unless the contractor submits evidence that the items could not be retained at cost without sustaining a loss."[69] These items are referred to as "common items." FAR 2.101 defines a "common item" as "material that is common to the applicable government contract and the contractor's other work." "Material" is defined in FAR 45.101 as:

property that may be consumed or expended during the performance of a contract, component parts of a higher assembly, or items that lose their individual identity through incorporation into an end-item.

Material does not include equipment, special tooling, and special test equipment.

Because of this broad definition, the deduction for common items may be significant.

Contractors may be able to prove an item is not "reasonably usable" by proving their orders are insufficient to merit retention of the material or by showing they have no other work despite efforts to obtain work.[70] If the contractor can show that retaining the items would cause a loss, then the burden shifts to the government to show that the contractor's projected business will provide a use for this material.[71]

c. Costs Continuing after Termination

FAR 31.205-42(b) allows costs incurred after termination to be claimed if they are unavoidable through reasonable diligence, and then only for a reasonable period of time based on the particular circumstances. Any continuing costs the contractor incurs after the effective date of termination are unallowable to the extent they are due to contractor's negligent or willful failure to discontinue the costs.

Costs such as salaries for employees awaiting reassignment and travel expenses for employees returning from remote locations have been allowed.[72]

Severance pay is an allowable cost if it is required by law, existing employment agreements, established policy, or the circumstances of the particular employment.[73] However, severance pay may be disallowed if it is discretionary and not based on a standard formula.[74]

Other costs, such as those incurred in taking inventory and preparing materials for storage or transportation, are also allowable. These costs may be considered to be either "settlement expenses" or "continuing costs." If these costs are treated as settlement expenses, then the contractor will receive only limited overhead costs on those expenses. If, on the other hand, the costs are treated as "continuing costs," the normal overhead rate applies. At least one case has accepted inventory costs as a direct cost subject to normal overhead, even though the cost was characterized as a "settlement expense."[75]

d. Costs under a Cost-Sharing Contract

Government contracts occasionally call for cost sharing between the government and the contractor. This type of arrangement is especially likely where the contractor may gain other valuable consideration from the completion of the contract, such as intellectual property rights. In *Jacobs Engineering Group, Inc. v. United States*, the Federal Circuit held that where such a cost-sharing contract was terminated for convenience, the contractor was entitled to recover all allowable costs, rather than only the percentage of the costs that the government would have been obligated to pay had performance been completed.[76]

2. Initial Costs

In the initial stages of a project, the contractor often incurs higher unit costs due to necessary organizational and preparatory activities such as mobilization as well as the inefficiency attributable to inexperience. If termination occurs, these initial costs, although expended, are not amortized as they would have been over the course of the entire contract. Therefore, FAR 31.205-42(c) allows the contractor to claim nonrecurring costs attributable to the contractor's initial preparation and performance, exclusive of special machinery and equipment.[77]

The definition of "preparatory costs" is flexible, and the determination of reasonable initial costs will depend on the equities of the individual cases.[78]

3. Loss of Useful Value of Special Tooling, Special Machinery, and Equipment

Loss of useful value of "special tooling" and "special machinery and equipment" is generally allowable per FAR 31.205-42(d). "Special tooling" is defined as:

> jigs, dies, fixtures, molds, patterns, taps, gauges, and all components of these items including foundations and similar improvements necessary for installing special test equipment, and which are of such a specialized nature that without substantial modification or alteration their use is limited to the development or production of particular supplies or parts thereof or to the performance of particular services.[79]

"Special test equipment" specifically excludes items that fall into the "special tooling" category but includes "single or multipurpose integrated test units engineered, designed, fabricated, or modified to accomplish special purpose testing in performing a contract."[80] Neither special tooling nor special test equipment includes material, buildings, or nonseverable structures.

The government may require transfer of title or otherwise require the contractor to relinquish its interest in some manner to the special tooling or test equipment to allow those costs to be claimed. The costs claimed are the difference between the underappreciated book value and the salvage value or fair market value.[81]

4. Rental Cost under Unexpired Leases

Rental costs are allowable if (1) the contractor can show that the lease was reasonably necessary for contract performance under the "ordinarily prudent" business operator standard; (2) the amount of the rental claimed is commensurate with the reasonable use of the property; and (3) the contractor has made reasonable efforts to discontinue the costs.[82] This last requirement, however, does not oblige the contractor to dispose of the lease at the first opportunity where it reasonably believes a greater recovery may be possible if it waits to do so.[83]

The cost of alterations and reasonable restorations required by the lease may be allowed when the alterations are necessary to perform the contract.[84]

5. Settlement Expenses

Settlement expenses are those costs that would not have been incurred by the contractor had the contract been completed. The FAR provides that these expenses include inventory and disposal of contract items, settling claims against the subcontractors, claims preparation, and "[i]ndirect costs related to salary and wages incurred as settlement expenses in [the above-mentioned items]."[85] With regard to indirect costs, the FAR limits recovery to "payroll taxes, fringe benefits, occupancy costs, and immediate supervision costs."[86]

Recovery of settlement expenses has been frequently litigated. Most of the disputes have centered on the cost of claims preparation or settlement proposal and presentation expenses. Where reasonably incurred, legal, accounting, clerical, and other costs are recoverable.[87] Claim preparation costs are recoverable whether incurred by in-house personnel or by outside contractors.[88]

When a claim reaches the stage where a contractor brings a suit against the government (under the CDA, an "appeal" of CO's final decision), however, the costs become unallowable.[89] The cost principles allow only those costs involved in preparation and presentation of the settlement proposal to the TCO.

Costs associated with an appeal contesting a termination for default and a conversion to a termination for convenience are not allowable. This includes both consultant and in-house expenses.[90] Similarly, the costs associated with litigating a TCO's unilateral determination regarding termination settlement are not recoverable. The dividing line between allowable and unallowable costs is the line between negotiation and litigation.[91] Costs for materials prepared for negotiation are allowable, even if those materials are subsequently used in litigation.[92]

Contingent fee agreements based on recovery of costs against the government are not allowed,[93] but fees paid on retainer for preparation of termination claims (but not in prosecuting the claim against the government) are allowable.

In *Kalvar Corp. v. United States*, the court found legal fees to be recoverable when the preexisting retainer was equivalent to the reasonable value of services.[94] As such, legal and accounting fees are reimbursable in full even if portions of the claim are found to be unallowable, unless the termination claim is patently invalid.[95] Similarly, the contractor may not recover costs for "continuing to beat a dead horse" when there is no possibility of agreement with the TCO on a settlement proposal.[96] The touchstone is the reasonableness of the costs claimed.

6. Subcontractor Claims

FAR 31.205-42(h) provides for the recovery of subcontractor claims and related indirect expenses but only to the extent that the prime contractor actually has liability to the subcontractors.[97]

Subcontractor claims negotiated by the prime contractor and approved by the government are proper costs to be included in the prime contractor's settlement proposal. If the subcontractor claim is pending trial, but settlement is reached before trial begins, allowable costs will be based on both practical and legal considerations such as the likelihood of success, the forum, anticipated expenses, and uncertainties regarding legal rulings.[98]

The prime contractor must show the reasonableness of the settlement. The "competence and good faith with which the settlement negotiations were conducted and the adequacy of the information on which the settlement" is based are the basic factors in evaluating the reasonableness.[99]

If the subcontractor has received a judgment or arbitration award, then these costs are allowable based on the underlying facts as well as satisfaction of the conditions in FAR 49.108-5.[100] In addition, reasonable legal fees incurred in defending suit by subcontractors are allowable.[101]

7. Interest

Interest accruing on money borrowed for performance of the contract is unallowable.[102]

Although unallowable under other circumstances, interest is allowed on costs resulting from a termination for convenience, starting from the date of certification of the claim.[103]

Section 12 of the CDA allows interest to be paid on all claims from the date the claim is submitted until it is paid. By definition, a submission to the TCO must be properly certified to be considered a claim.[104] Although termination settlement proposals must be certified,[105] the required certification is different from, and inadequate to suffice as, the certification required for a CDA claim.[106] Therefore, interest may not begin to accrue until a properly certified settlement proposal is submitted to the TCO as a certified claim.[107]

8. Costs If Contract Is Terminated Prior to Performance

In *Nicon, Inc. v. United States*, the contractor entered into a contract with the U.S. Army to repair an on-base dormitory.[108] The government delayed Nicon from beginning performance on the contract for over 100 days and then terminated the contract for convenience.[109] The Federal Circuit held that Nicon could recover the unabsorbed overhead costs incurred during the government-caused delay as long as it could offer a reasonable method for allocating the overhead costs.[110]

C. Profit and Loss

1. Profit

Profit recovery is determined in accordance with FAR 49.202, which allows the application of a "fair and reasonable" profit factor to costs of preparations made and work done by the contractor for the terminated portion of the contract. Any reasonable method may be used to arrive at a profit factor, but FAR 49.202 identifies the following factors for consideration when negotiating profit:

(1) Extent and difficulty of the work done by the contractor as compared with the total work required by the contract (engineering estimates of the percentage of completion ordinarily should not be required but if available should be considered);

(2) Engineering work, production scheduling, planning, technical study and supervision, and other necessary services;

(3) Efficiency of the contractor, with particular regard to—
  (i) Attainment of quantity and quality production;
  (ii) Reduction of costs;
  (iii) Economic use of materials, facilities, and manpower; and
  (iv) Disposition of termination inventory;

(4) Amount and source of capital and extent of risk assumed;

(5) Inventive and developmental contributions, and cooperation with the government and other contractors in supplying technical assistance;

(6) Character of the business, including the source and nature of materials and the complexity of manufacturing techniques;

(7) The rate of profit that the contractor would have earned had the contract been completed;

(8) The rate of profit both parties contemplated at the time the contract was negotiated; and

(9) Character and difficulty of subcontracting, including selection, placement, and management of subcontracts, and effort in negotiating settlements of terminated subcontracts.

2. Adjustment for Loss Contracts

Where a contractor would have sustained a loss had the contract been completed, the termination settlement amount must be adjusted to reflect that fact.[111] In other words, not only will the contractor be denied profit, but the contractor will not be able to recover all of its incurred costs.

The procedure and factors for application of the loss ratio (or formula) are specified in FAR 49.203. Generally, the ratio reduces the contractor's recovery, exclusive of its termination settlement costs, by a percentage equal to the percentage loss it would have suffered on the entire contract had it been completed. Consideration is given to expected production efficiencies and to other factors affecting the cost to complete.[112]

a. Settlement on a Total Cost Basis

When the settlement is on a total cost basis, the contractor will not be paid more than the negotiated or determined amount of its settlement expenses, plus the remainder of the total settlement amount reduced by the ratio of the total contract price to the remainder plus the estimate of the contractor's cost to complete the contract.[113] The loss formula for a total cost settlement can be reduced to this equation:

Contractor's Recovery = $SE + IC [CP / (IC + ETC)]$
where: SE = Settlement Expenses
         IC = Costs Incurred to Termination
         CP = Contract Price
         ETC = Estimated Cost to Completion

To provide an example of application of the loss ratio to a settlement under the total cost basis, assume that the contract price is $1 million. After the contractor has incurred costs of $700,000, the contract is terminated. At this point, the estimated cost to complete the contract is $600,000. Assume that there are $100,000 of reasonable settlement expenses associated with the termination. Had the contractor actually completed the contract, it would have lost $300,000. Its recovery, again using the total cost basis, is computed as follows:

$$\$100,000 + \$700,000 [\$1,000,000 / (\$700,000 + \$600,000)]$$

or

$$\$100,000 + \$539,000$$

or

$$\$639,000$$

This means that even after being reimbursed for $100,000 in settlement expenses, the contractor loses $161,000 on the contract. That is, the contractor's $639,000 recovery is $161,000 less than his incurred costs of $600,000, plus his settlement expenses of $100,000.

The $161,000 loss, however, would place him in the same position (losses relative to incurred costs) as he would have been in had he actually completed the contract. The loss-ratio concept is based on the principle that the contractor should not be in a better position as a result of the government's termination for convenience than had the contract been fully completed.[114] The contractor may avoid the application of the loss ratio by proving that any loss that would have occurred was attributable to the government. For example, contractors have successfully argued against use of the loss ratio based on constructive changes, defective specifications, and government-caused delay.[115]

b. Inventory-Basis Settlement

If the settlement is on an inventory basis, the loss adjustment does not apply to settlement expenses, and the contract price for acceptable products is excluded before the loss adjustment is made.[116] This can be expressed by the formula as follows:

Contractor's Recovery = $SE + VCEI + [IC - VCEI] [CP/(IC + ETC)]$
where: SE = Settlement Expenses
         IC = Costs Incurred to Termination
         CP = Contract Price
         ETC = Estimated Cost to Completion
         $VCEI$ = Value of Completed End Items

V. *Partial Terminations for Convenience*

Under the termination for convenience clauses, the CO may terminate either the entire contract or a part of the contracted work.[117] Contractors can recover costs to date for the terminated work, a "fair and reasonable" profit on those costs, and settlement expenses. Contractors may also be entitled to equitable adjustments on a showing that the unit price of the remaining work rises when a portion is eliminated. Claims for equitable adjustments of the price of the continued portion of the contract must be made within 90 days from the effective date of termination, unless a written extension is granted by the CO.[118] Expenses allocable to the terminated portion of the contract may be recoverable only in a final termination settlement.

Just as with a complete termination, a partial termination for convenience should not operate to leave the contractor in a better position than if the work had not been canceled. It should not increase the profit margin or reverse a loss position.[119] For this reason, a loss-adjustment factor may also be applied to partial termination.

A. Partial Termination for Convenience versus Deletion of Work by Change Order

The government unilaterally may eliminate work from a contract using either a partial termination for convenience or a deductive change under the Changes clause. There are no clear guidelines for determining which procedure is more appropriate. However, the choice of procedure often will have a significant impact on the amount of compensation allowed the contractor.

1. *Criteria for Determining Which Clause Is Appropriate*
a. "Within the Scope"
Under the "within the scope" test, as long as the proposed deletion meets all of the requirements of the Changes clause, the deletion should be considered a deductive change.[120]

b. Major versus Minor Plan Variations
The test enunciated in *J.W. Bateson v. United States*[121] states that the proper criterion for differentiating between a deductive change and a partial termination for convenience is whether the modification has a "major" or "minor" impact

on the overall contract. Minor impacts are generally treated as deductive changes and major impacts as partial terminations.[122]

c. Continuing Needs Test

In *Skidmore, Owings & Merrill*,[123] the ASBCA formulated a third possible test. This test bases the distinction between partial termination and deductive change on whether the government need for the item continued. If the government no longer had a need for the item, then a partial termination for convenience is the board's preferred method. If the need continued at a reduced level, then a contract change is in order. This distinction is not often followed.

2. *Differences in Measure of Recovery*

One of the most significant reasons for distinguishing between a deductive change and a partial termination for convenience is that there will usually be different monetary recoveries under the two approaches. Generally, contracts partially terminated for the government's convenience are repriced on the basis of work performed and to be performed. Administrative costs, attorneys' fees, and other expenses incurred in preparing cost estimates and negotiating a settlement are allowed when the government has terminated a contract for its own convenience. These costs may not be allowed when the deletion is made under the Changes clause. Deductive change orders are handled by reducing the existing contract price by the expected savings as well as the estimated profit from the deleted work.[124]

a. Partial Termination

FAR 52.249-2(l) provides for an equitable adjustment in case of termination. In an equitable adjustment situation, contractors may receive reasonable profit on the work they have done but cannot recover anticipatory profits or consequential damages.[125] Equitable adjustments are intended to leave the contractor in proportionately the same position it would have been in had the parties originally contracted for the lesser amount of work. Thus, equitable adjustments do not operate to increase the profit margin or to reverse a loss position.[126]

b. Recovery for Deduction of Work under the Changes Clause

If the government deletes work under the Changes clause, the original contract price is reduced by the cost of the deleted work and the profit reasonably attributable to the deleted work. Deleting the cost of the work and the profits from the work leaves the contractor in the same position it would have been in if the deleted work had never been part of the contract.

3. *Summary of Advantages*
a. Anticipated Profits

A contractor may be able to retain a greater portion of its anticipated profits in a deductive change situation. Under the Changes clause, only profit attributable to the deleted work is lost. In the partial termination for convenience situation, the contractor receives a "reasonable" profit on the work actually performed.

b. Anticipated Loss

In a loss situation, if there is a deductive change, the contract price is reduced by the cost of the deducted items and the corresponding anticipated profit, leaving the contractor to suffer the whole loss and possibly aggravating the situation.[127] In the partial termination situation, the loss ratio applies, reducing the amount of the anticipated loss by the amount of reduction in the work called for by the contract.[128] In the loss situation, the contractor bears only part of the loss.

c. Settlement Costs

Costs associated with the preparation and negotiation of settlement terms, including direct labor, administrative costs, and attorney's fees, are recoverable under a partial termination for convenience.[129] Under the Changes clause, these costs can be recovered only indirectly as part of the increased overhead due to the change.[130]

4. Summary

In general, if the contractor anticipates a large profit, then the deductive change is preferable because it allows the preservation of maximum profit. In a small profit or loss situation, the partial termination is preferable because the contractor will recover settlement and other special costs not recoverable under the Changes clause, as well as reduce projected losses proportionately.

VI. Conclusion

Upon receipt of a notice of termination for convenience, the contractor should consult the TCO and FAR regarding its rights and duties and consider retaining counsel to assist in preparing and negotiating a termination settlement proposal. Indeed, as discussed previously, the FAR sets forth very specific obligations of the contractor upon a termination for convenience, and such requirements must be timely performed and the settlement proposal must be timely submitted in order to preserve the contractor's right to recover its costs.

Notes

1. United States v. Corliss Steam Engine Co., 91 U.S. 321 (1876).
2. Colonial Metals Co. v. United States, 204 Ct. Cl. 320 (1974).
3. Kalvar Corp., Inc. v. United States, 543 F.2d 1298 (Ct. Cl. 1976); John Reiner & Co. v. United States, 325 F.2d 438 (Ct. Cl. 1963).
4. Torncello v. United States, 681 F.2d 756 (Ct. Cl. 1982).
5. Soledad Enters., ASBCA No. 20376, 77-2 BCA ¶ 12,552; *see also Torncello,* 681 F.2d at 759–60.
6. *Torncello,* 681 F.2d at 772.

7. *Id.*

8. *Id.* at 761.

9. Mun. Leasing Corp. v. United States, 7 Cl. Ct. 43 (1984).

10. *Id.* at 47 (citing *Mun. Leasing*, 1 Cl. Ct. at 774–75; *Torncello*, 681 F.2d 756).

11. Maxima Corp. v. United States, 847 F.2d 1549 (Fed. Cir. 1988); *see also* Ace-Fed. Reporters, Inc. v. Barram, 226 F.3d 1329, 1333–34 (Fed. Cir. 2002); *cf.* White v. Delta Constr. Int'l, Inc., 285 F.3d 1040, 1044 (Fed. Cir. 2002) (noting that *Maxima* does not establish the difference between the guaranteed minimum and the actual amount paid as the measure of damages).

12. *Id.* at 1553–54.

13. *Id.* at 1551.

14. *Id.* at 1557.

15. Krygoski Constr. Co. v. United States, 94 F.3d 1537 (Fed. Cir. 1996).

16. *Id.* at 1542.

17. *Id.* at 1545 (citing Salsbury Indus. v. United States, 905 F.2d 1518, 1521 (Fed. Cir. 1990)).

18. *See* 10 U.S.C. §§ 2304–2305.

19. *Krygoski*, 94 F.3d at 1542.

20. *Id.* at 1543.

21. *Id.* at 1554.

22. *See generally* Schweiger Constr. Co. v. United States, 49 Fed. Cl. 188 (2001); Travel Centre v. Gen. Servs. Admin., GSBCA No. 14057, 98-1 BCA ¶ 29,422, 98-1 BCA ¶ 29,536, *recon. denied*, 98-1 BCA ¶ 29,541, 98-2 BCA ¶ 29,849, *rev'd on other grounds*, 236 F.3d 1316 (Fed. Cir. 2001); *but see* Applied Cos., Inc., ASBCA No. 50749, 01-1 BCA ¶ 31,325.

23. FAR 49.102; *see also* FAR 49.601.

24. FAR 49.102.

25. FAR 49.601.

26. Maibens, Inc., ASBCA No. 25915, 82-1 BCA ¶ 15,668, *recon. denied*, 82-1 BCA ¶ 15,796.

27. FAR 49.102(c).

28. FAR 49.102(d).

29. John Reiner & Co. v. United States, 325 F.2d 438, 440 (Ct. Cl. 1963); Warren Bros. Rds. Co. v. United States, 355 F.2d 612, 615 (Ct. Cl. 1965).

30. United States v. Amdahl Corp., 786 F.2d 387, 395 (Fed. Cir. 1987); *cf.* Trauma Serv. Group v. United States, 33 Fed. Cl. 426 (1995) (quantum meruit recovery is not applicable where the status of the agreement as a contract has not been shown).

31. *Amdahl*, 786 F.2d at 395 (citing *John Reiner & Co.*, 325 F.2d at 440); *accord*, Memorex Corp., B-213430.2, Oct. 23, 1984, 84-2 CPD ¶ 446.

32. FAR 52.249-8(g).

33. G.L. Christian & Assocs. v. United States, 312 F.2d 418 (Ct. Cl. 1963).

34. Trilon Educ. Corp. v. United States, 578 F.2d 1356 (Ct. Cl. 1978).

35. Shader Contractors, Inc., ASBCA No. 3957 et al., 58-1 BCA ¶ 1579.

36. Am. Air Filter Co., B-188408, Feb. 16, 1978, 78-1 CPD ¶ 136.

37. FAR 49.105(a).

38. *Id.*

39. FAR 49.104.

40. FAR 49.104(a).

41. FAR 49.104(d).

42. FAR 49.108-3; *but see* FAR 49.108-5.

43. *See, e.g.,* FAR 52.249-2(b)(9).

44. FAR 2.101.

45. Best Lumber Sales, ASBCA No. 16737, 72-2 BCA ¶ 9661 (15 percent of termination inventory was defective but was accepted for full cost recovery).

46. FAR 49.104(i).

47. FAR 49.105(a)(4).

48. FAR 49.109-7(b).

49. *See, e.g.,* FAR 52.249-2(e).

50. Space Dynamics Corp., ASBCA No. 25106, 81-2 BCA ¶ 15,205.

51. FAR 49.206-2(a)(1).

52. FAR 49.602-2(b)(4).

53. Hi-Shear Tech. Corp. v. United States, 356 F.3d 1372, 1383 (Fed. Cir. 2004); Parsons of Cal., ASBCA No. 20867, 82-1 BCA ¶ 15,659.

54. Technology, Inc., DCAB No. NBS-1-78, 79-1 BCA ¶ 13,752.

55. FAR 49.206-2(b)(3).

56. FAR 49.109-4.

57. Eugene M. Keane, AGBCA 770150, 78-1 BCA ¶ 12,975.

58. Blue Ridge Leasing Co., ENG BCA No. 4666, 82-1 BCA ¶ 15,734.

59. FAR 49.201(a); Am. Elec., Inc., ASBCA No. 16635, 76-2 BCA ¶ 12,151, *aff'd in part and modified in part,* 77-2 BCA ¶ 12,792.

60. RHC Constr. Co., ASBCA No. 2083, 88-3 BCA ¶ 20,991 (CO use of strict accounting approach by demanding exact documentation and refusing to negotiate violated intent and purpose of regulations that require use of business judgment approach); Scope Elec., Inc., ASBCA No. 20359, 77-1 BCA ¶ 12,404 (costs of unsuccessful work, if reasonable, are recoverable).

61. FAR 49.201(b).

62. FAR 49.201(c).

63. *See, e.g.,* FAR 52.249-2(g).

64. FAR subpt. 31.2; Racquette River Constr., Inc., ASBCA No. 26486, 82-1 BCA ¶ 15,769; Pac. Architects & Eng'rs, Inc., ASBCA No. 21043, 76-2 BCA ¶ 11,953.

65. *See, e.g.,* FAR 52.249-2(b)(8)–(9).

66. Essex Electro Eng'rs Inc., DOTCAB No. 1025 et al., 81-1 BCA ¶ 14,838, *aff'd,* Essex Electro Eng'rs, Inc. v. United States, 701 F.2d 998 (Fed. Cir. 1983).

67. FAR 49.204(b).

68. *See, e.g.,* FAR 52.249-2(b)(8).

69. FAR 31.205-42(a) (emphasis added).

70. Southland Mfg. Corp., ASBCA No. 16830, 75-1 BCA ¶ 10,994.

71. Fiesta Leasing & Sales Inc., ASBCA No. 29311, 87-1 BCA ¶ 19,622; *Essex Electro Eng'rs,* 81-1 BCA ¶ 14,838.

72. Hugo Auchter GmbH, ASBCA No. 39642, 91-1 BCA ¶ 23,645; Sys. Dev. Corp., ASBCA No. 16947, 73-1 BCA ¶ 9788; DCAA CONTRACT AUDIT MANUAL (DCAM) § 6-304.7 (1995).

73. FAR 31.205-6(g).

74. Engineered Sys., Inc., ASBCA No. 18241, 74-1 BCA ¶ 10,492.

75. Condec Corp., ASBCA No. 14232, 73-1 BCA ¶ 9808; *see also* Amplitronics, Inc., ASBCA No. 20545, 76-1 BCA ¶ 11,760.

76. Jacobs Eng'g Group, Inc. v. United States, 434 F.3d 1378 (Fed. Cir. 2006).

77. *See* Fil-Coil Co., ASBCA No. 23137, 79-1 BCA ¶ 13,618, *aff'd on recon.*, 79-1 BCA ¶ 13,683; *see also* Robt. M. Tobin, HUDBCA No. 79-388-C20 et al., 84-3 BCA ¶ 17,651 (costs for recruitment of nonlocal personnel qualified as nonrecurring preparatory costs in preparing to perform contract).

78. *Condec*, 73-1 BCA ¶ 9808.

79. FAR 2.101; *see also* Dairy Sales Corp. v. United States, 593 F.2d 1002 (Ct. Cl. 1979); Hugo Auchter GmbH, ASBCA No. 39642, 91-1 BCA ¶ 23,645.

80. FAR 2.101.

81. Southland Mfg. Corp., ASBCA No. 16830, 75-1 BCA ¶ 10,994 (useful value was determined as the difference between the net book value and the salvage or sale value).

82. FAR 31.205-42(e); *see also* Q.V.S. Inc., ASBCA No. 7513, 1963 BCA ¶ 3699; *see also Southland Mfg.* 75-1 BCA ¶ 10,994.

83. *See Southland Mfg.*, 75-1 BCA ¶ 10,994.

84. FAR 31.205-42(f).

85. FAR 31.205-42(g).

86. *Id.*

87. *But see* Robt. M. Tobin, HUDBCA No. 79-388-C20 et al., 84-3 BCA ¶ 17,651 (settlement expenses unallowable where contractor failed to submit evidence sufficient to assess their reasonableness).

88. Chesterfield Assoc., DOTCAB No. 1028, 80-2 BCA ¶ 14,580; *see* L.K. Ferguson, AGBCA No. 79-122-4, 81-1 BCA ¶ 14,915; *see also* Paul E. McCollum, Sr., ASBCA No. 23269, 81-2 BCA ¶ 15,311 (bills or receipts showing payment of legal fees necessary for entitlement).

89. FAR 31.205-33(b).

90. Q.V.S. Inc., ASBCA No. 7513, 1963 BCA ¶ 3699; Frigitemp Corp., VABCA No. 646, 68-1 BCA ¶ 6766.

91. Acme Process Equip. Co. v. United States, 347 F.2d 509 (Ct. Cl. 1965).

92. E.A. Cowen Constr. Inc., ASBCA No. 10669, 66-2 BCA ¶ 6060.

93. Hugo Auchter GmbH, ASBCA No. 39642, 91-1 BCA ¶ 23,645; Manuel M. Liodas, ASBCA No. 12828, 71-2 BCA ¶ 9015.

94. Kalvar Corp. v. United States, 218 Ct. Cl. 433 (1978).

95. Engineered Sys., Inc., ASBCA No. 18421, 74-1 BCA ¶ 10,492.

96. Henry Spen & Co., ASBCA No. 20766, 77-2 BCA ¶ 12,784.

97. Atl., Gulf & Pac. Co., ASBCA No. 13533, 72-2 BCA ¶ 9415.

98. The Boeing Co., ASBCA No. 10524, 67-1 BCA ¶ 6350; Lockheed Ga. Co., ASBCA No. 8652, 1964 BCA ¶ 4325.

99. Bos'n Towing & Salvage Co., ASBCA No. 41357, 92-2 BCA ¶ 24,864 (holding that allowing costs for which the prime was not responsible was not reasonable and therefore such costs would be excluded).

100. TransWorld Airlines, NASA No. 472-2, 75-1 BCA ¶ 11,146; Dade Bros. Inc., ASBCA No. 4315, 58-1 BCA ¶ 1741, *aff'd*, Dade Bros. Inc. v. United States, 163 Ct. Cl. 485 (1963).

101. R-D Mounts, Inc., ASBCA No. 17422 et al., 75-1 BCA ¶ 11,077, *aff'd on recon.*, 75-1 BCA ¶ 11,237.

102. FAR 31.205-20; Southland Mfg. Corp., ASBCA No. 16830, 75-1 BCA ¶ 10,994; Breed Corp., ASBCA No. 15163, 87-3 BCA ¶ 19,999.

103. Essex Electro Eng'rs, Inc., DOTCAB No. 1025 et al., 81-1 BCA ¶ 14,838, *aff'd*, Essex Electro Eng'rs, Inc. v. United States, 702 F.2d 998 (Fed. Cir. 1983).

104. *See, e.g.,* Lehman v. United States, 673 F.2d 352 (Ct. Cl. 1982); *see also Essex Electro Eng'rs*, 702 F.2d at 1003.

105. FAR 53.301-1435, FAR 53.301-1436, FAR 53.301-1437, FAR 53.301-1438.

106. FAR 33.207(c); 41 U.S.C. § 605(c)(1).

107. Kahn Commc'ns, Inc., ASBCA No. 35768, 88-2 BCA ¶ 20,706.

108. Nicon, Inc. v. United States, 331 F.3d 878 (Fed. Cir. 2003).

109. *Id.* at 881–82.

110. *Id.* at 888.

111. FAR 49.203.

112. FAR 49.203(a).

113. FAR 49.203(c).

114. *See* Power Generators, Inc., ASBCA No. 7607, 1962 BCA ¶ 3358; Caskel Forge, Inc., ASBCA No. 7638, 1962 BCA ¶ 3318; *see also* FAR 52.249-2(e).

115. Allied Specialties Co., ASBCA No. 10335, 67-2 BCA ¶ 6657 (constructive change); Scope Elec., Inc., ASBCA No. 20359, 77-1 BCA ¶ 12,404 (defective specifications); M.E. Brown, ASBCA No. 40043, 91-1 BCA ¶ 23,293 (government-caused delay).

116. FAR 49.203(b).

117. *See, e.g.,* FAR 52.249-2(a).

118. *See, e.g.,* FAR 52.249-2(l).

119. Fairchild Stratos Corp., ASBCA No. 9169, 68-1 BCA ¶ 7053.

120. Nolan Bros., Inc., ASBCA No. 4378, 58-2 BCA ¶ 1910; Gen. Contracting & Constr. Co., Inc. v. United States, 84 Ct. Cl. 570 (1937).

121. J.W. Bateson v. United States, 308 F.2d 510, 513 (5th Cir. 1962).

122. *See* Fred A. Arnold, ASBCA No. 7761, 1962 BCA ¶ 3508; Am. Constr. & Energy, Inc., ASBCA No. 34934, 88-1 BCA ¶ 20,361 (a 12 percent reduction in total work to be performed was a minor modification); Capital Elec. Co., GSBCA No. GS-04B-16555, 81-2 BCA ¶ 15,281 (reduction in quantity involving major specification variation was properly treated as partial termination despite contractor request under the Changes clause).

123. Skidmore, Owings & Merrill, ASBCA No. 5115, 60-1 BCA ¶ 2570.

124. Bruce Constr. Corp. v. United States, 324 F.2d 516 (Ct. Cl. 1963).

125. Ted J. Grimsrud & Claude Corp., ASBCA No. 7971, 1962 BCA ¶ 3562.

126. Fairchild Stratos Corp., ASBCA No. 9169, 68-1 BCA ¶ 7053.

127. S.N. Nielson Co. v. United States, 141 Ct. Cl. 793 (1958).

128. Power Generators, Inc., ASBCA No. 7607, 1962 BCA ¶ 3358; *Skidmore, Owings & Merrill*, 60-1 BCA ¶ 2570.

129. *Power Generators*, 1962 BCA ¶ 3358.

130. *Id.*

CHAPTER 14

Termination for Default

LARRY D. HARRIS

I. *The Standard*

A. Historical Background

The government has the right to terminate a construction contract for default when a contractor fails to perform in accordance with the contract's material or significant terms. This right is based on the Default clause that the Federal Acquisition Regulation (FAR) mandates be included in virtually every federal government construction contract.[1] Even where the contracting officer (CO) omits the required Termination for Default clause from a contract, it will be read in and enforced by the boards and courts under the *Christian* doctrine.[2]

The government's power to terminate a contract for default has a long history, tracing its lineage at least as far back as the Civil War.[3] By the end of the 19th century, default clauses appeared in most federal government contracts.[4]

Early versions of the Default clause were intended to ensure the government of the termination-for-default remedy. The government's need for this remedy arose from the urgent demand for construction services and supplies in wartime, and the failure of common law remedies to correspond to and take account of the government's needs.

Unlike today, the early Default clauses varied substantively from one contract to another. In 1864, one version provided as follows:

> In case of failure on the part of the party of the second part to deliver the articles within the time and in the manner specified in this agreement, the party of the first part is authorized to make good the deficiency by purchase in the open market at the expense of the said party of the second part.[5]

The author would like to thank Nick Hoogstraten, who assisted with the updating of this chapter.

The remedy stated in this early version, reprocurement at the contractor's expense, eventually became standard in all Default clauses. This version noticeably fails to address termination of the contract. Nonetheless, the Court of Claims held that termination was appropriate because time had been "of the essence" in the contract.[6]

Another version of a Default clause appearing in an early government contract stated:

> [I]f, in any event, the contractor shall delay, or be unable to proceed with the work in accordance with its terms, the engineer officer in charge shall have full right and authority to take away the contract, and employ others to complete the work, deducting the expenses from any money that may be due and owing him, and the contractor will be responsible for any damages caused to others by his delay or noncompliance.[7]

This clause expressly empowered the government to terminate the contract if the contractor failed to perform or performed deficiently.

By the end of the 19th century, Default clauses typically granted the government four basic rights, which have become standard:

- To terminate the contract if the contractor failed to perform in accordance with the material terms of the contract;
- To terminate the contract if the contractor's lack of diligence threatened contract performance;
- To reprocure the contract work following termination; and
- To proceed against the contractor for both the excess costs of reprocurement and other damages.

The early 20th century witnessed much experimentation with the terms of Default clauses, and saw the introduction of clauses excusing the contractor for delays beyond its control.[8] Uniformity of government contracts in general, and the Default clause specifically, began with the introduction of standard form contracts after World War I. Two new remedies appeared in the standard form Default clause:

- In contracts involving construction, the government could take possession of, and utilize in completing the work, any "materials, appliances, and plant" that existed at the site and were necessary for work completion; and
- In contracts involving construction and supply, the concept of excusable delay was expanded to include delays caused by government acts within its sovereign capacity (that is, acts that affect the public as a whole), strikes, freight embargoes, and other similar occurrences.

The modern era of government contracting arrived with the passage of the Armed Services Procurement Act of 1947, ch. 65, 62 Stat. 21 (codified as amended in scattered sections of 10 U.S.C. (1982)), implemented via the Armed Services Procurement Regulation (ASPR), renamed in 1982 as the Defense Acquisition Regulation (DAR), and subsequently superseded by the Federal Acquisition Regulation (FAR) in 1984.

B. Comparison to Commercial Contracts

The failure to materially perform a government contract is a default, with the government's right to terminate as expressed in the Termination for Default clause. This right generally compares with material breach of contract disputes in commercial contracts. However, most standard private construction contracts contain termination for default (or for cause) clauses. While comparable to the FAR clauses in government contracts, commercial provisions typically contain some notable differences. For example, the American Institute of Architects (AIA) family of construction documents all contain a termination for default clause that differs in important respects from the FAR clause.[9]

The AIA clauses spell out, in detail that is analogous to FAR clauses, the grounds for termination for default and the process for giving notice of termination. However, unlike the remedy provided to a successful contractor challenging a default termination under a federal contract, the AIA documents do not provide for conversion of a wrongful termination for default to a termination for convenience. The contractor's remedy under commercial contracts, such as the AIA documents, is to recover breach of contract damages under common law principles, which could include the recovery of anticipated profit on the entire project and, in some instances, reliance damages and litigation costs.

Additionally, the AIA documents, as well as other similar commercial construction contracts, provide the contractor with the ability to terminate or suspend the contract based upon default by the owner.[10] In other words, the default provisions are mutual. In contrast, federal government contract clauses do not allow the contractor to terminate the contract based upon a default by the government. Federal common law provides only a limited exception to this rule in the cardinal change doctrine (discussed *infra*); which, where applicable, will allow a contractor to cease performance based upon the government's breach.

Like the FAR clauses for Termination for Default of federal government construction contracts, the AIA documents do not expressly require the construction owner to give the contractor a cure notice or cure period prior to terminating the contract for default. In contrast, the FAR clauses for fixed-price supply and service contracts, and fixed-price research and development contracts, require the government to issue a cure notice prior to a default termination.[11]

As a practical matter, most government agencies will issue a cure notice for a construction contract termination and copy the performance bond surety before formally defaulting the contractor. The AIA default clauses require the construction owner to give notice of default within 10 days of its occurrence, which some interpret as a cure notice. However, unlike the FAR clauses that require cure notices, the AIA clause does not provide that the owner must forbear its right to terminate the contract for default pending a responsive plan to cure the defaults identified in the cure notice. Both public and private construction practitioners suggest that it is a best practice to issue a cure notice before terminating a construction contract for default, given the significant consequences of a wrongful default termination and the protracted litigation that usually follows from the owner exercising what the courts and boards have characterized as a "drastic remedy."[12]

C. Nature of Default Termination

A default termination is a serious government action that subjects the contractor to a so-called forfeiture.[13] The Court of Claims (now the Court of Federal Claims) referred to the government's right of default as a "forfeiture clause," which cut off the contractor's rights under the contract and subjected the contractor to liability for monetary damages. The court noted that forfeiture is not favored in the law and that "parties who seek to assert a forfeiture are generally held to the very letter of their authority."[14]

While forfeiture does not necessarily cut off all of the contractor's rights, because default involves very serious consequences for a contractor, the courts and boards have followed the principle that a default termination is a drastic sanction that requires the government to strictly account for its actions.[15]

D. Current Clauses

The FAR provides Default (or Termination) clauses for all types of contracts. Specifically, the FAR contains Default clauses for the following fixed-price contracts utilized in connection with construction:

- fixed-price architect-engineer, FAR 52.249-7;
- fixed-price construction contracts, FAR 52.249-10; and
- dismantling and demolition contracts, FAR 52.249-10 with Alternate I.

The FAR also contains a Termination clause for cost-reimbursement construction contracts, FAR 52.249-6 with Alternate I. The Termination clause utilized in cost-reimbursement contracts covers both default and convenience termination of the contract by the government. Excusable delays under cost-reimbursement contracts mirror those provided under fixed-price contracts.[16] Excusable delay results in a conversion of the default termination into a termination for convenience.[17]

The FAR clauses are the government's contractual basis for Default Termination. When used in conjunction with the Disputes clause (FAR 52.233-1) and the Contract Disputes Act of 1978,[18] they provide a government contractor with a process for challenging a wrongful termination and a remedy (conversion of a wrongful termination for default to a termination for convenience) that allows the successful appellant to recover its costs of performance and profit on the work performed prior to termination.[19]

E. Workings of the Standard Clause

1. Grounds for Termination

The standard fixed-price construction Termination for Default clause (FAR 52.249-10) provides that the government may terminate the contract for default, in whole or in part, if the contractor refuses or fails to prosecute any or all of the work with sufficient diligence to ensure timely completion or if it fails to complete the work within the specified time. Boards and courts have also upheld default terminations of government contracts based on a contractor's anticipatory repudiation of the contract,[20] failure to comply with other provisions of the contract,[21] and failure to maintain acceptable standards of skill and workmanship.[22]

When the termination is based on a failure to prosecute the work diligently, the boards and courts require the government to establish that the CO had a reasonable belief that there was no reasonable possibility of timely completion.[23] For example, in *Ralph Rosedale*,[24] the board found default termination to be proper where the contractor had completed only 17 percent of the contract work after 69 percent of the contract performance time had elapsed. In a termination for failure to make progress, the CO will be default terminating before the completion date of the contract. The government thus bears some risk. However, the CO's assessment of the contractor's ability to complete the work need not be "correct" in an objective sense provided the termination decision was reasonable.[25] If the contractor "repudiates" performance, the government can also terminate for default. A contractor can repudiate a contract by:

- Refusing to perform;[26]
- Engaging in conduct indicating an intent not to perform;[27]
- Failing to provide adequate assurances that performance will proceed;[28] or
- Failing to continue performance during a dispute.[29]

In *Precision Cable Manufacturing Co.*,[30] the government terminated the contract for default when the contractor informed the government that it would not continue performance without an increase in the contract price. While the board noted that the contractor could no longer employ its low-price supplier, it nonetheless concluded that the contractor's only proper remedy was to perform the contract and later file a claim for the additional costs.

A contractor's failure to perform its obligations under the contract also justifies termination for default. For example, in *Kelso v. Kirk Bros. Mechanical*

Contractors, Inc.,[31] the violation of federal labor reporting requirements under both the contract and the regulations implementing the Davis-Bacon and Anti-Kickback Acts justified termination of the contract for default.[32]

2. Alternative Grounds for Termination

While "[i]t is improper to base the decision to terminate for default on materially erroneous information or analysis,"[33] the courts and boards have upheld default terminations as valid if a ground for termination existed when the termination decision was made, even if that ground was not known to the CO at the time the contract was terminated. The right of the government to "reach back" to justify the default termination has been well established. For example, terminations have been sustained by the Court of Federal Claims (COFC) on the basis of a contractor's fraud conviction involving a change order to the contract, although at the time of the termination the CO was unaware of the contractor's fraudulent act.[34] Similarly the government's termination has been upheld where the contractor violated other statutes, such as the Davis-Bacon Act,[35] even though the violations were found after the issuance of the CO's final decision effecting the termination for default.[36]

On the other hand, it is possible for the government to waive the right to assert fraud as a post hoc alternative ground for termination if it was aware of the fraud but allowed the contractor to proceed notwithstanding. In *Aptus Co. v. United States*,[37] after terminating the contractor for failure to make progress, the government asserted that the contractor had made fraudulent representations prior to performance. The COFC held that the government had waived the right to assert fraud because it was aware of the misrepresentations at the time, but nevertheless allowed the contractor to work for 10 months before terminating, including agreeing to a contract modification to extend the performance period.

3. Government's Right to Terminate Is Discretionary

The Termination for Default clause expressly provides that the "government may terminate" a contract for default upon certain circumstances. Thus, the government is not required to terminate on a finding of default. The permissive language gives rise to a right on the part of the government to exercise discretion in making a decision to terminate.[38] "Before exercising its discretion to terminate under the Default clause, the government should consider all the relevant circumstances."[39]

In terminating a contract for default, the CO must be careful not to abuse this discretion. For instance, in *L&H Construction Co.*,[40] the contractor was to replace a pipeline system at McGuire Air Force Base in New Jersey. The chief of operational contracting provided a memorandum with materially erroneous information regarding the contractor's performance to the termination contracting officer (TCO). Based upon this memorandum, the contract was terminated for default. The board stated that termination based on materially erroneous information cannot be a reasonable exercise of discretion because it would reward the ignorance of COs and deception by their subordinates.[41]

The FAR lists seven specific factors to be considered in arriving at a decision to terminate for default. These factors are generally the kinds of issues that a commercial contractor would consider if faced with a delinquent supplier. Failure to consider these factors is probative evidence that the decision was made unreasonably, and without consideration of the relevant circumstances leading to the termination.[42] Although the CO must use his discretion when applying the FAR factors to determine if a termination for default is appropriate, a mistake in the evaluation of one factor may not be enough to overturn an otherwise reasonable termination for default.[43] The FAR factors are:

- The terms of the contract and applicable laws and regulations;
- The specific failure of the contractor and the excuses, if any, made by the contractor for such failure;
- The availability of the supplies and services from other sources;
- The urgency of the need for the supplies or services and the period of time that would be required to obtain the supplies or services from other sources as compared with the time in which delivery could be obtained from the delinquent contractor;
- The degree of essentiality of the contractor in the government procurement program and the effect of a termination for default upon the contractor's capability as a supplier under other contracts;
- The effect of a termination for default on the ability of the contractor to liquidate guaranteed loans, progress payments, or advance payments; and
- Any other pertinent facts and circumstances.

4. Options to Termination

The Default clause merely states that the government "may" default terminate upon the occurrence of the stated events. There are three alternatives in lieu of a termination for default that the CO may consider when "in the best interest of the government":[44]

- Permit the contractor, its surety, or the guarantor, to continue performance of the contract under a revised delivery schedule;
- Permit the contractor to continue performance of the contract by means of a subcontract or other business arrangement with an acceptable third party, provided the rights of the government are adequately preserved; or
- If the requirement for the supplies and services specified in the contract no longer exists, and the contractor is not liable to the government for damages, execute a no-cost termination settlement agreement.[45]

The most common alternative to default is to allow the contractor to continue performance under a revised contract schedule. Note, however, that in order to implement this alternative, the CO must receive adequate consideration from the contractor for the contract modification.

F. Conversion into Convenience Termination

Excusable delay or improper termination for default by the government entitles the contractor to conversion of the default termination into a termination for convenience. The convenience recovery by the contractor will likely be much more favorable than the results of a termination for default.[46] The FAR Default clauses expressly provide for the conversion of default terminations into terminations for convenience.[47] This remedy is unique to government contracts and has no common law parallel. However, a request to change a default termination into a termination for convenience must be submitted to the CO, and cannot be first requested before a board of contract appeals.[48]

1. Basis of the Remedy

The conversion of a default termination into a termination for convenience may result from a variety of circumstances, including (1) contractor impossibility of performance; (2) government fault; or (3) defects in performance so minor as to preclude termination for default. For example, in *Insul-Glass, Inc.,*[49] the board converted a default termination into a termination for convenience where the government did not show that the contractor's alleged failure to submit acceptable drawings was a valid basis for default.

2. Measure of the Recovery

The FAR sets forth the measure of recovery available under a termination for convenience. This amount is greater than that which is recoverable under a default termination. For example, the FAR expressly provides for recovery of:

- The costs of performance up to the effective date of termination;
- The costs of performance to the effective date of termination;
- Some "continuing costs," generally defined as unavoidable costs incurred after termination; and
- Settlement expenses.[50]

It should be noted, however, that the contractor may not recover anticipated profits in a convenience termination.

3. Time for Filing Termination for Convenience Settlement Proposal

Pursuant to the Termination for Convenience clause, FAR 52.249-2(e), a contractor has one year from the effective date of a termination for convenience to submit a termination settlement proposal to the CO. In *Ryste & Ricas, Inc.,*[51] the ASBCA held that the one year runs from the date the contractor received the board's decision converting the default to a termination for convenience.

II. *Failure to Timely Perform*

A. The Government Right

1. Standard Clause

Paragraph (a) of the Default clause for fixed-price construction contracts provides: "If the Contractor refuses or fails to prosecute the work . . . with the diligence that will insure its completion within the time specified in this contract including any extension, or fails to complete the work within this time, the government may, by written notice to the Contractor, terminate the right to proceed with the work . . . that has been delayed." The FAR provides alternate but similar language for contracts involving the dismantling, demolition, or removal of improvements.

2. Contractor Defenses

The contractor has three defenses if the government terminates for failure to complete the contract on time: (1) the contractor may claim that, despite its failure to deliver on time, it substantially performed the contract, and that this was all that was required; (2) causes of excusable delay were present; or (3) the government waived the completion date.

3. Contracting Officer Discretion

Despite the existence of a default, the CO may allow a contractor to continue performance for a variety of reasons:

- Reasonable doubt as to whether a court or board would sustain the termination;
- Unknown excusable delays justifying the failure to perform or deliver;
- Concern that the government waived the delivery date;
- Doubt as to whether damages can be collected from the defaulted contractor;
- Need for the contractor on another contract, and the belief that termination on one contract may bankrupt the contractor;
- Concern about the capability of a reprocurement contractor; or
- Desire to avoid a lawsuit.

The contractor must distinguish between the CO's discretionary power to terminate and the right to terminate. If the contract is terminated, any factors that would have justified the CO to continue the contract for practical reasons will not be considered by a court or a board. These factors are irrelevant.

4. "Show Cause" Notice

The Default clause for supply and service contracts[52] provides that where the contractor fails to "make progress" or fails to "perform any of the other provisions" of the contract, the government must give the contractor a written notice that, if it does not cure the failure within 10 days, the government may terminate the contract. A "show cause" notice is not expressly required, however, in the case of default termination of a construction contract. Nevertheless, the CO may issue a show cause letter before default terminating a construction contractor. The issuance of the letter, however, is not required.[53] Typically, a show cause letter in a construction context advises the contractor of the grounds for default and requests that it inform the government of any excuses for the default.

A poorly worded show cause notice by the CO may be construed as an inducement to continue performance. This, in turn, may open the door for the contractor to argue in litigation that the government waived the completion date.[54]

B. Construction Contracts: Substantial Completion

As a general rule in construction contracting, both public and private, "substantial completion" occurs when the contract work is completed to the extent that the building or facility may be used or occupied for its intended purpose.[55] Under paragraph (a) of the Default clause for fixed-price construction contracts, the government's right to terminate the contract is limited by the doctrine of substantial completion. Generally, substantial completion limits or precludes the government's ability to terminate the contract for default or assess liquidated damages. Courts and boards developed this concept to reduce the inevitable and inequitable consequences to contractors running afoul of the strict compliance doctrine.[56] These consequences included economic waste and severe forfeiture.

Although the contractor remains obligated to complete the remaining punch list work, the government cannot terminate the entire contract for default. Rather, the government's remedy for a failure to complete the punch list work is either reduction of the contract price or a partial termination for default limited to the uncompleted work.[57] If the work is substantially complete when the default notice is given, termination for failure to complete the entire contract will be deemed wrongful.

1. Determining Substantial Completion

The burden of proving substantial completion is on the contractor.[58] In determining whether substantial completion was achieved, the court or board must determine the quantity of work remaining to be done and the extent to which the project was capable of serving its intended purpose at the time of termination.[59]

The first factor that must be proven to establish a construction contract's substantial completion is a high percentage of completion. In *Mitchell Engineering & Construction Co.*,[60] the board held that liquidated damages would not be assessed when the project was 97 percent complete because only minor punch list items remained to be corrected. In *Spruill Realty/Construction Co.*,[61] liquidated damages could not be assessed beyond the point at which only cleanup and punch list items remained to be performed, because the board found that these items did not preclude beneficial occupancy. Moreover, 98 percent of the project was complete. While no bright line may be drawn, if completion falls below 80 to 85 percent, the contractor will likely have to present extensive proof of usability. Even if some benefit may be obtained from the uncompleted project by the government, a court or board will not find substantial completion where a large percentage of work remains.[62]

The second factor is the availability for use. Courts and boards require that the government have the opportunity for so-called beneficial occupancy before they will find substantial completion.[63] Beneficial occupancy occurs "when the government occupies or uses a facility before its final acceptance for the purpose for which it was intended, provided it was satisfactorily completed to that extent."[64] Moreover, the Federal Circuit has recently emphasized that substantial completion goes beyond mere beneficial use, and requires contractors to ensure that their performance provides the government with the benefits bargained for under the contract.[65]

2. Duty to Complete

Although the government may be precluded from default terminating a construction contract as a result of substantial completion, the government retains the right to demand full completion of the project and the correction of any and all defects.[66] The government's remedies for the contractor's failure to complete punch list items include (1) termination of the uncompleted portion of the contract for default and assessment of completion costs against the contractor; and (2) reducing the contract price through an equitable adjustment.

Note that the government must still receive the reasonably anticipated benefits of the contract. In *Triple M Contractors, Inc.*,[67] the board rejected the contractor's assertion of substantial performance defense on this basis. The contract required the contractor to replace a drainage gutter, but defects in the gutter initially installed resulted in a five-year loss of useful life. Thus, the government did not receive the benefit of its bargain.

III. Failure to Make Progress

The government may exercise its right to terminate a contract for failure to make satisfactory progress only if (1) the contractor's lack of progress or inability to obtain financing or material threatens successful and timely completion; and (2) the contractor fails to address the deficiencies stated in a cure notice or fails to provide sufficient assurances of timely completion to the CO.

A. Standard Clause

The Default clause for fixed-price construction contracts[68] grants the government the right to terminate if the contractor fails to "[m]ake progress, so as to endanger performance of the contract." The right to terminate for failure to make progress is independent of the right to terminate for failure to timely deliver.

B. Standards for Termination

To sustain a termination for lack of progress, the government must prove facts that establish the lack of progress and show that the CO, relying on those facts and after giving notice to the contractor, reasonably believed that no reasonable likelihood existed that the contractor could perform in accordance with the schedule.

1. Traditional versus Modern View

Early cases required the government to prove that a contractor was incapable of timely completing the contract work before a termination for failure to make progress would be sustained.[69] Some boards, however, began to recognize that the government could terminate for failure to progress even if it could not show that timely performance was physically impossible at the time of termination. For example, in *General Products Co.*,[70] the board sustained a default termination despite the fact that, in theory, the contractor could have subcontracted for plastic bowls in time to meet the delivery date. The board noted that the threat to performance came from two factors: (1) the contractor itself could not timely complete manufacture of the bowls by the delivery date, even operating at 100 percent capacity, and (2) the specifications rendered manufacture and timely delivery of the bowls by a supplier highly improbable.

The more modern view is for the courts and boards to test whether the CO's belief that timely performance is highly improbable is reasonably based upon facts establishing: (a) inadequate progress; and (b) the contractor's failure to cure noted deficiencies and to provide reasonable assurances of timely completion. Thus, the following lines of inquiry appear critical to cases of termination for failure to progress:

- Was the contractor ready, capable, and willing to perform the work in a timely manner?
- Did the contractor make known to the CO its ability and willingness to perform such that concerns over timely performance were dispelled?

The leading case applying the modern interpretation of the Default clause's failure to progress provision established the following test:

[T]he default clause in this contract did not require a finding that completion within the contract's time limitations was impossible. Rather, under this contract default termination was appropriate if a demonstrated lack of diligence indicated that the government could not be assured of timely completion.[71]

This test guides the court in evaluating whether the CO was justified in believing that the contractor was unlikely to complete performance in a timely manner. For example, in *Dave's Aluminum Siding, Inc.*,[72] the contractor had completed painting and aluminum siding on only 20 of 249 houses through five months of the eight-and-one-half-month performance period. Much of the completed work contained deficiencies that the contractor left unremedied after receiving cure and show-cause notices. The board noted that the "government need only show that the contractor's demonstrated lack of diligence reasonably indicated that the government could not be assured of timely completion." This lack of diligence, combined with the CO's reasonable belief that timely completion of the work was unlikely, rendered the termination proper.[73]

In *Hanson & Sons*,[74] the government properly terminated a tree-thinning contract for failure to progress because no work was performed during the time allowed for performance.[75] In *Engineering Technology Consultants, S.A.*,[76] the government properly terminated the contractor where the contractor failed to obtain the required insurance. The board treated this omission as a failure to make progress because the contractor could not satisfactorily proceed without the insurance.

The government, however, may not terminate a contract solely because timely completion is less than absolutely certain.[77] The CO must have a reasonable belief, under the totality of circumstances and in light of the contractor's track record and responses to the cure notice, that no reasonable likelihood existed that the contractor could perform the contract within the contract schedule. Furthermore, the termination for failure to make progress must be supported by tangible, direct evidence of the impairment of timely completion.[78]

In *NECCO, Inc.*,[79] the contractor sought in preconstruction meetings to have the notice to proceed delayed until the spring of 2004, despite award of the contract in August 2003. The contractor then failed to submit an appropriate schedule and to commence work in September 2003. In response to a cure notice, the contractor did not provide adequate assurance that the work would be performed in the required time.

The board held that there was no agreement to delay the start date or extend the completion date, and that the contractor bore the burden of demonstrating that its failure to make progress was excusable. The board held that the decision to terminate for default was based upon a reasoned consideration of the totality of the circumstances and full consideration of the factors set forth in FAR 49.203-3(f).

2. Construction versus Supply Contracts

The legal standards controlling the proof of default termination based upon a failure to make progress appears to be the same for both supply and construction contracts. Compare the similar language of both clauses: Supply contracts permit termination for failure to "[m]ake progress, so as to endanger

performance"; construction contracts permit termination for failure to "prosecute the work . . . with the diligence that will ensure completion within the
time specified." Both clauses permit termination for intentional and uninten-
tional failures, and both clauses relieve the contractor of responsibility for
excusable delay.

Because the performance of supply and construction contracts differs, the
facts necessary to support a default termination for failure to make progress
also vary. Proof in construction contract progress failure terminations is usually generated by applying a percentage-of-completion test.

C. Proving Failures to Make Progress

Some of the factors and methods the government employs to determine progress failure are discussed below, as are the defense of waiver of the due date,
the effect of negotiations, and the parties' burden of proof.

1. Percentage of Completion

Comparing the amount of work completed with the amount of time remaining under the contract constitutes the predominant method for determining
whether timely completion is threatened by lack of progress. The following are examples of default terminations based on percentage-of-completion
analysis.

- 58 percent of work completed when at least 80 percent was scheduled
  to be completed[80]
- Only 53 percent of highway construction work completed within 90
  percent of elapsed contract time[81]
- Only 8.7 percent of tree-planting work completed in 53.3 percent of
  the performance period[82]
- Only 42 percent of tree-planting work completed in 66 percent of the
  performance period[83]

Caution should be used before relying too heavily upon a percentage-
of-completion analysis, however. First, the reviewing tribunal will have to
choose between the inevitably conflicting percentages of completion offered
by the government and the contractor. Also, intervening factors such as better working conditions or increases in productivity may assist a contractor in
completing the remaining work in less time. The percentages generated by the
completion analysis will usually fail to reflect these possibilities.[84]

Proof that the portion of work remaining is less than or equivalent to
the percentage of time needed for completion may result in the reversal of a
default termination.

2. Failure to Meet Progress Milestones

Failure to meet progress benchmarks strongly indicates that timely completion is in jeopardy. The government's right to terminate on this basis depends

upon the CO's reasonable belief that final delivery is unlikely under the totality of circumstances. The untimely submission of data items (such as drawings and test plans), the failure to submit a performance schedule, and the failure to skillfully manage the work to ensure timely completion all may indicate a lack of progress.[85] Although the right to terminate is not based upon the failure to meet interim submission requirements, the government may raise these failures as evidence of a continuing lack of progress.[86]

Of course, the nearer the scheduled completion date, the more significant the failure to meet a project milestone can be. In *Star Painting & Contracting Co.*,[87] the government terminated for default a cleaning and repairing contract. Six days before the scheduled contract completion date, and 114 days into the contract, the contractor had yet to make the required submittals on 17 items. The board upheld the termination for failure to progress.

Contractors should be aware that the government's failure to take action upon a missed milestone or interim submission does not equate to a waiver of the government's rights.

3. Subcontractor and Supplier Problems

The government may raise a contractor's difficulties in securing subcontracts and in resolving problems with subcontractors and suppliers in order to prove a failure to progress. In *Nichols Dynamics, Inc.*,[88] the board upheld the default termination of a contract for a jet engine test cell. The contract provided for a specific subcontractor for the supply of certain instruments. The CO default terminated the contract when the contractor could not reach an agreement with the required subcontractor. In another case, the board sustained the termination for default of a contract for the supply of sweaters when a yarn supplier refused to extend the contractor credit for the yarn necessary to manufacture 120,000 sweaters.[89]

A contractor's inability to acquire the requisite materials at a price that will allow it to achieve a profit on the contract will not excuse a failure to progress.[90] In a similar vein, the contractor's delayed performance due to the unavailability of the cheapest of several acceptable materials provides sufficient grounds for terminating the contract for failure to progress.[91]

4. Management and Labor Problems

The contractor's management and labor problems may support a determination by the CO that lack of progress jeopardizes contract performance. In *C.C. Galbraith & Son, Inc.*,[92] the government cited numerous deficiencies in terminating the contract, including failures to employ competent engineers, to respond to cure and deficiency notices, and to execute subcontracts. In upholding the termination, the board noted the company's mismanagement and its high turnover among executives. Implicitly, the board found that the unexcused inadequacy of management hindered contract progress.

The loss of employees or the failure to hire essential personnel may be cause for termination if it results in hindered progress or if timely performance is thereby rendered improbable.[93] Additionally, the inability to pay workers and suppliers inevitably causes delays and shortages that prevent progress and provides the basis for terminating for lack of progress.

5. Waiver

It almost goes without saying that a termination for failure to make progress is keyed to a determination that the contractor will miss the future completion date. The contractor, however, may be able to show that this date is not binding, if it can prove that through the government's action or inaction the CO waived the completion date. The central question regarding waiver of the completion date is whether the government's action or silence constituted a mere forbearance, or was instead a legal waiver of the schedule.

For example, in *Jack Spires & Sons Electrical Co.*,[94] the contract required repair of three motors in six months, but did not require a specific delivery date for the motors. While fixing the first motor, the contractor found existing irreversible damage and advised the government of the discovery and the need for further work. The contractor continued to work on that motor, delivering it to the government three and one-half months into the contract. Six weeks later, the government notified the contractor of faults in the first motor and complained that the contractor had delivered only one motor, while the contract required all three to be completed in less than two more months. The government then terminated for default. The board found that the government's delay in testing the motor and its failure to respond to the contractor's notification of existing irreversible damage in the fan and motor indicated that time had not been of the essence. The contractual completion date was thus waived. Because the government bore the burden of setting a new delivery date, and never did so, no new deadline existed.

Most allegations of waiver fail. In *Missile Systems, Inc.*,[95] the board upheld a default termination based largely on the lack of detrimental reliance by the contractor on the alleged waiver of the delivery schedule. Although the government knew for more than two months that the contractor could not meet the delivery date, the board refused to find a waiver because for the entire two-month period the contractor failed to proceed with performance. The board described the contractor's nonperformance as an attempt to forestall termination, rather than reasonable and detrimental reliance on the government's waiver of the delivery schedule.

6. Government's Burden of Proof

When the government terminates a contract for failure to make progress, it must prove with clear and convincing evidence that (1) the contractor was failing to proceed; and (2) timely completion was highly improbable. The preceding sections discuss the facts typically invoked by the government to prove lack of progress, that is, inability to commit subcontractors, failure to submit plans for approval, absence of adequate workforce, and failure to respond adequately to a cure notice. The government's burden of proving endangerment of performance is a heavy one. As the court explained in *J.D. Hedin Construction Co. v. United States*,[96] "a default termination is a drastic sanction . . . which should be imposed (or sustained) only for good grounds and on solid evidence."

Once the government has met its burden, the contractor must show that the lack of progress was excusable or did not endanger performance. The critical factor for the contractor is that the contractor must show that termination was unjustified as of the date of its issuance. The courts and boards will ask whether at the time of termination, the CO justifiably believed that the contractor would not complete on time. Later events that would have rendered timely performance possible will not be considered by a board or court unless the CO should have reasonably anticipated them or actually knew of them. For this reason, the contractor's response to the cure notice is highly significant. The board or court will assess the situation from the vantage point of the CO at the time of termination.

D. Cure Notice

FAR 49.402-3(e) provides that, when practicable, a show cause letter "should" be sent to the contractor whenever termination appears justifiable. However, termination for failure to timely complete may be made immediately upon expiration of the performance period.

Although not expressly required by the construction contract Default clause, the government will generally send a notice to the contractor prior to terminating for failure to progress as a precautionary measure.[97]

Where the contractor's failure to make progress also constitutes an abandonment of the work, or facts suggest that the issuance of a cure notice would be a "futile act," the government may terminate a contract for default without issuing a cure notice.[98]

IV. Contractor's Duty to Proceed

Unique to government contracts is the contractor's duty to proceed with contract performance pending resolution of a contract dispute. The obligation exists in virtually every standard form Disputes clause. For instance, FAR 52.233-1 states as follows in paragraph (i):

> The Contractor shall proceed diligently with performance of this contract, pending final resolution of any request for relief, claim, appeal, or action arising under the contract, and comply with any decision of the Contracting Officer.

This particular duty to proceed is limited to disputes "arising under the contract." Thus, only if no remedy-granting clause exists in the contract for the government's actions may the contractor cease performance because of a dispute. Because most government actions are covered by a corresponding remedy-granting clause, few situations allow the contractor to suspend performance in the face of a dispute.

Alternate I to the Disputes clause further extends this duty to proceed with performance in the event of a dispute to include breach of contract disputes. This particular contractor obligation is to be used only when the agency determines that continued performance is necessary pending resolution of any claim relating to the contract.

A. Failure to Proceed versus Anticipatory Repudiation

Whether a contractor's failure to proceed with contract performance pending resolution of a dispute constitutes a repudiation was addressed in *Lon E. Nelson*.[99] In that case, the contract required the contractor to appraise certain parcels of property. The contractor delivered the appraisals nearly one year late. The CO requested that the contractor remedy certain deficiencies in the appraisals. When the contractor refused, the CO terminated the contract for default based upon the contractor's duty to continue performance pending resolution of the dispute.

Noting the contractor's obligation to proceed with performance pending resolution of disputes, the board upheld the termination. According to the board, "[The contractor's] refusal to comply and complete the report in accordance with the specifications constituted an abandonment of performance during the course of the dispute and an anticipatory breach of the contract, justifying the termination for default."[100]

The contractor must proceed even where the government incorrectly advises the contractor that no relief is possible for its claim.[101] Moreover, the contractor cannot stop work on one contract because of a pending dispute on another contract.[102]

B. Exceptions to Duty to Proceed

Three limited exceptions exist to the broad, general obligation to continue performance when a dispute arises. These exceptions are detailed below.

1. Material Breach

The contractor need not continue performance where the government's actions amount to a material breach of the contract. Where the government severely impacts the contractor's ability to perform by failing to perform itself, the contractor is relieved of the obligation to continue performance pending resolution of the dispute. For example, the government's failure to make progress payments after the lapse of a reasonable time period constitutes a material breach excusing the contractor's continued performance obligation. In *General Dynamics Corp.*[103] a radar and antennae production and installation contractor did not receive the contractually required cost reimbursement from the government on monthly invoices totaling $800,000. The board held that the contractor was justified in suspending its performance instead of continuing the work and pursuing its claim according to the Disputes clause.[104] But mere delay in payments must be distinguished from "a total failure to pay over months."[105] Only the latter constitutes a material breach that would suspend the contractor's duty to proceed pending a dispute. Where the government makes a cardinal change—a change that fundamentally alters the agreement of the parties—the contractor may also be discharged from its obligation to continue performance.[106]

Additionally, unreasonable and untimely inspections by the government may justify a contractor's refusal to proceed. In *Brand S. Roofing*,[107] the government delayed for three months to inform the contractor of defects in performance. The delay greatly increased the costs of remedying the defects and constituted a material breach justifying a refusal to proceed.

2. Lack of Guidance

The contractor is relieved of the duty to continue performance pending a dispute where the government fails to provide the guidance necessary to proceed.[108] The government has the obligation to provide clarification or direction where the contractor questions the meaning of the specifications. After a contractor makes a reasonable request for clarification, the government's refusal to provide assistance justifies the contractor's ceasing performance. The contractor, though, must be certain that ambiguities exist and must make some attempt to remedy the problem.[109]

3. Impossible or "Impractical" to Proceed

Where performance becomes impossible or impractical because of government acts, the contractor may be excused from a failure to perform. One example may be the government's furnishing of defective specifications.[110] In *D.E.W., Inc.*,[111] the board found that the government supplied the contractor with a defective design for bearing bolts. The board further found that the defective design rendered impossible performance of the hangar construction contract in accordance with the government's specifications. The default termination was thus held improper.

Impracticality of performance under the circumstances—instead of actual impossibility—may also constitute a reasonable basis to refuse to proceed with performance.[112] In *Ned C. Hardy*,[113] the government denied permission for workers to camp at the project site in a national forest. This denial rendered performance impractical because unattended equipment was subject to vandalism and the contractor reasonably suspected that a single night watchman would be in personal danger from environmental activists.[114] In these circumstances the contractor was entitled to abandon performance, but only after the government had been informed of the circumstances and given an opportunity to correct the problem.[115] In claiming practical impossibility, the contractor must meet both an objective and subjective test to show that neither it nor anyone else could perform in accordance with the contract's specifications.[116]

V. Excusable Delay

The contractor may be able to avoid the consequences of default if its performance deficiency resulted from an excusable delay. Some of the causes of excusable delay are: fires, epidemics, floods, strikes, unusually severe weather, acts of God or of the public enemy, quarantine restrictions, deficiencies attributable to a subcontractor of any tier that is beyond the control of both subcontractor and contractor, acts of the government in either its sovereign or contractual capacity, freight embargoes, and impossibility of performing the contract specifications.[117] Financial inability to perform is generally not a ground for excusable delay.[118] As this topic is also covered in Chapter 19 regarding delay issues, however, the discussion here is limited to the default context.

A. Foreseeability

In order to limit contractor excuses under the excusable delay provision, the government has developed the rule that an excusable cause of delay must not have been foreseeable. Foreseeability is generally thought of as having knowledge, or having reason to know, prior to bidding.[119] Often, the dispositive factor in deciding issues of foreseeability is whether an existing or a supervening event caused the delay. The former will be held to be within the contractor's implied knowledge, and thus is not considered unforeseeable.

In *Brooks-Callaway Co. v. United States*,[120] the contractor encountered high water and suffered delays as it constructed levees on the Mississippi River. Some of the delays emanated from unforeseeable events, and some from conditions the contractor should have anticipated. The Default clause stated that the contract could not be terminated "because of any delays . . . due to unforeseeable causes beyond the control and without the fault or negligence of the contractor, including, but not restricted to . . . floods." The contractor asserted that because the clause explicitly enumerated floods as an excusable delay, foreseeability was not relevant. The Supreme Court disagreed, holding that the foreseeability requirement still applied as an enumerated ground for excusable delay.

B. Subcontractor Delay

Normal and foreseeable subcontractor difficulties cannot be relied upon as grounds for excusable delay. In *Walsh Bros. v. United States*,[121] the contractor failed to timely install equipment as a result of the subcontractor's difficulty in obtaining parts and labor, but the scarcity of parts that created the delay was held not to be unforeseeable.

The definition of "subcontractor" includes a "subcontractor of any tier."[122] If a subcontractor were to fail to deliver as a result of a dispute between it and the prime contractor, the subsequent delay will not be excusable. The delay will be considered to be the fault of the parties.[123]

C. Proving Excusable Delay

1. Elements

The contractor must prove three elements to successfully assert excusable delay. Those elements are: (1) the occurrence of an event that was unforeseeable, beyond its control, and without its fault or negligence; (2) a delay in performance caused by the event that prevented timely completion of the contract; and (3) a time extension of a specific length was justified by the delay.

A delay is not excusable if, at the time of contracting, the contractor could have foreseen the event that actually delayed performance. In *Marine Transport Lines, Inc.*,[124] heavy rainfall and fog delayed blasting and painting of a cargo ship's hull. The board found that the conditions encountered were normal for the particular time of year at that location. The foreseeability requirement is a critical factor in evaluating whether an occurrence was beyond the control or without the fault or negligence of the contractor. The impact, as well as the event itself, must be unforeseeable.[125]

Boards have found the following to be unforeseeable: high and persistent levels of vandalism; unusually severe weather; unusually long delays in obtaining parts and equipment; and the loss of essential and unique personnel. Boards have found the following to be foreseeable: unavailability of components of which the contractor was aware; severe weather that nevertheless fell within expectable levels for the location at issue; and holiday work stoppages.

In *J.D. Hedin Construction Co. v. United States*,[126] the court held that the CO improperly terminated a contract for default where a delay was caused by a nationwide shortage of cement. The court found that the shortage was an unforeseeable event, stating that a contractor does not need to have "prophetic insight and take extraordinary preventative action which it is simply not reasonable to ask of the normal contractor."

Some cases indicate that foreseeability requires knowledge on the contractor's part beyond the mere possibility of an event's occurrence. For instance, in *Marine Transport Lines, Inc.*,[127] where several workers walked off the job, thereby causing delays, the board found that the workers were members of a trade well-known for walking off jobs and that it should have been foreseeable to the contractor when they did so. The board, nevertheless, held that the contractor's inability to foresee exactly when the workers would abandon the job constituted an excusable delay.

Courts and boards expect the contractor to know or have reason to know of facts that are within the scope of its business operations relating to the contract. For instance, in *Diversified Marine Technologies, Inc.*,[128] a ship repair contractor encountered adverse weather that delayed contract completion. The contract allowed the contractor to select the location of performance. Because the contractor knew of the weather trends in the area of its chosen facility, the board required it to take those weather expectations into account in making its pre-bid determination as to the work schedule. The claim of excusable delay was accordingly denied.

The requirement that the cause of delay be beyond the contractor's control is generally interpreted to mean that both the event and the consequences must be beyond the contractor's control.[129] In *Fox Construction, Inc. v. General Services Administration*,[130] the contractor was held responsible for freeze damage to air-conditioning condenser units, although the contractor was technically correct in alleging a breach of the implied warranty of the specifications. The board held that the contractor failed to show that the damage was directly attributable to the design defect, finding instead that the freeze damage was due to the contractor's failure to anticipate the need to perform pressure testing with a liquid other than water or air.

Where the contractor's or subcontractor's acts or omissions cause the delay, the board or court will deny a claim of excusable delay. For example, the board denied a claim of excusable delay based upon the connection of a 110-volt power source to a 24-volt fire alarm because the contractor's negligence was the cause of the delay.[131]

In *KARPAK Data & Design*,[132] the contractor claimed that government-imposed work excused the delay. The contract required the contractor to inform the government that it would not honor the government's excessive orders. The board held that because the delay resulted from the contractor's failure to notify the government of the contractor's inability to perform all the work, the delay was not without the contractor's fault or negligence and was, therefore, not excusable.

2. *Causation and Burden of Proof*

The contractor must produce evidence showing how and to what extent the unforeseen event delayed performance. The contractor must establish excusable delay by a preponderance of the evidence.[133] However, the contractor may be excused from meeting this burden if the government controls the evidence necessary for the presentation of the contractor's defense.[134]

After the contractor meets its burden, the government must then rebut the contractor's case. Most frequently, the government attempts to do so by presenting evidence that the delay was foreseeable, that it resulted from the contractor's fault or negligence, or that the contractor assumed the risk of the delay.[135] The government may also assert affirmative defenses of its own, including the contractor's failure to notify the government at the onset of an excusable delay.[136]

D. Enumerated Delays

The Default clause for fixed-price construction contracts[137] incorporates a list of excusable causes of delay as follows:

1. Acts of God or a Public Enemy

An act of God is generally defined as "some inevitable accident which cannot be prevented by human care, skill, or foresight, but results from natural causes."[138] Some "acts of God" appear elsewhere in the enumerated causes. Other natural disasters, however, do not appear in the list and may be categorized as an act of God. For example, in *Nogler Tree Farm*,[139] the board found that the government should have granted a time extension to the contractor because of the eruption of Mount St. Helens.

The act of a public enemy—such as war, armed conflict, or hostile act of a foreign government—that causes delay may constitute an excusable delay. For instance, in *Gibson Manufacturing Corp.*,[140] the board held that delays resulting from the scarcity of materials arising after the outbreak of the Korean War were excusable.

2. Acts of the Government

"Acts of the government in either its sovereign or contractual capacity" constitute the second enumerated ground for excusable delay. The government acts in its sovereign capacity when it purports to act in the public arena and does not direct its actions at the contractor.[141] The following are examples of sovereign acts:

- Wage regulations or price controls;[142]
- The operation of the federal allocation or priority system for copper, steel, and other materials;[143]
- The imposition of security restrictions in a hostile area;[144] and
- The raising of the minimum wage by a larger percentage than could have been anticipated and the unanticipated cancellation of a loan.[145]

The government acts in its contractual capacity where the object of the government's action is a single contractor.[146] The contractor must prove that the government's act was wrongful. The improper acts of the government during the contractor's performance of one contract that cause default in another contract may also be grounds for excusable delay.

3. Fires and Floods

Fires constitute the third listed cause of excusable delay.[147] The contractor must present evidence that the fire caused its failure to perform.[148] Where the contractor's negligence causes the fire, the claim for excusable delay may be denied.[149]

The fourth enumerated cause of excusable delay is floods. In *Molony & Rubien Construction Co.*,[150] the contractor's performance was delayed by flooding in the basement of an air-traffic control tower where the contractor was installing an elevator. The board found that rainfall had entered a conduit system deficiently installed by a different contractor, causing damage to equipment stored in the basement. The board held that the event was an excusable delay.

4. Epidemics and Quarantine Restrictions

The fifth enumerated cause of excusable delay is epidemics. The contractor must prove that an epidemic in the area of performance affected its work.[151] In *Ace Electronics Associates, Inc.,*[152] the board denied a claim of excusable delay when an influenza epidemic caused 30- to 40-percent absenteeism among the workforce. The contractor failed to present evidence as to when the epidemic occurred, its duration, or its effect on the production testing program. The contractor also failed to present evidence that it attempted to perform despite the epidemic. Similarly, in *Tommy Nobis Center, Inc.,*[153] the contractor failed to show that a flu epidemic that struck its locale was of sufficient duration or that it adversely affected production. The board thus denied the claim of excusable delay.

The sixth enumerated cause of excusable delay is quarantine restrictions. No case has been found discussing this excusable event, although the analysis would most likely parallel that of epidemics and freight embargoes.

5. Strikes and Freight Embargoes

The seventh enumerated cause of excusable delay is strikes. The Court of Claims has stated that a strike must "substantially impair" performance of the contract for it to constitute an excusable delay.[154] The extent of the excusable delay resulting from a strike may not necessarily correspond to the actual strike length. The reassignment of managers and supervisors away from a strike-afflicted contract may work delays on the contract even after the dispute is settled. Moreover, the contractor must still prove that the strike delayed performance.[155]

Other labor disputes that boards have found constituted strikes within the meaning of term as utilized in the Default clause are:

- Work stoppages by the contractor's labor force;[156]
- Sympathy strikes by the contractor's employees who refuse to cross a picket line set up in protest of another contractor's labor practices;[157]
- A refusal to work by a subcontractor's laborers during "informational" picketing by a union at the job site;[158] and
- The refusal of employees of a predecessor contractor for guard services to work for a new contractor after being trained.[159]

Strikes caused by the contractor's unfair labor practices will not be recognized as excusable delay. These strikes are generally not considered to be "beyond the control and without the fault or negligence of the contractor."[160] Shortages of material caused by a strike may also provide the basis for excusable delay. In these circumstances, the material must actually be unavailable to support an excusable delay. An increase in price or an inability to procure less expensive material does not suffice to excuse nonperformance.[161]

The eighth enumerated cause of excusable delay is freight embargoes. A shortage of material caused by an embargo may be sufficient grounds for excusable delay. In *Automated Extruding & Packaging Co.*,[162] the board excused a contractor's default because of unforeseen, unusual, and severe shortages caused by the 1973 oil embargo.

6. Unusually Severe Weather

The ninth enumerated cause of excusable delay is unusually severe weather. The contractor will not be granted relief if the weather that caused the delay was not unusual for the location at the particular time of year, was foreseeable, or could have been reasonably anticipated.[163] The following are examples in which a board has granted relief for unusually severe weather:

- An extension of time should have been granted where weather unusual for fall caused brush to freeze in ice and then thaw during the course of the contractor's tree-clearing obligations.[164]
- The contractor's default on a contract to wash aircraft was excused where weather caused the cleaner to freeze.[165]

In *Sealtite Corp.*,[166] however, the board denied the contractor's claim of excusable delay resulting from unusually severe weather where a heavy snowstorm in late November hit Denver, Colorado.

7. Delays of Subcontractors or Suppliers

The Default clause strictly limits the situations in which subcontractor or supplier delay may serve as a basis for excusable delay. First, the subcontractor's delay must be beyond the control of and without the fault or negligence of both the subcontractor and the contractor, as well as unforeseeable.[167] Second, the ability to assert this ground is limited by whether the supplies or services subcontracted for were obtainable from other sources in time to permit the contractor to meet the performance schedule. If they were, then the contractor may not assert subcontractor delay. The fact that the services were available only at a higher price than originally anticipated does not serve to undermine the rule.[168] In like manner, a claim for excusable delay will be denied where the contractor fails to take reasonable steps to locate alternate sources. These rules do not apply where the supplier was the sole source of the material.[169] However, the requirement remains that the contractor show that the cause of the delay was beyond its and the sole-source supplier's control and not due to either's fault or negligence.

8. Acts of Another Contractor

In construction contracts, "acts of another contractor in the performance of a contract with the government" is an enumerated cause of excusable delay. The clause thus recognizes that the delay of one contractor may carry over and affect the ability of another contractor to perform according to schedule. For instance, in *Modern Home Manufacturing Corp.*,[170] the site preparation contractor failed to complete its work at the time that the construction contractor was scheduled to begin its work. The CO extended the time for completion upon recognition of the excusable delay caused by the first contractor.

E. Unenumerated Excusable Delays

The enumerated excusable delays in the Default clause are not exhaustive.[171] Attempts to rely on grounds outside the enumerated causes, however, are generally not successful. Nonetheless, unenumerated causes may provide the basis for excusable delay in the right case. For example, in *Xplo Corp.*,[172] the board granted a time extension for delay caused by the illegal arrest of several contractor employees at the project site by city police officers.

1. Financial Difficulties

Because the contractor is required to have sufficient financial ability to perform the contract, claims of excusable delay based on financial difficulties often fail. In *Centennial Leasing Corp.*,[173] for example, a lender's failure to provide supplemental financing to the contractor to purchase vehicles necessary to fill contract delivery orders did not excuse the default. The board held that it is the responsibility of the contractor to secure financing, regardless of whether earlier credit had been withdrawn by the lender.[174] However, if the contractor can prove that improper government actions constituted the primary or controlling cause of the default, and rendered the contractor financially incapable of performing, the default will be excused. The following are examples of such government actions:

- The wrongful denial of progress payments due the contractor;
- The improper denial of a contractor's valid claim of extra work;
- The wrongful cancellation of a Small Business Administration loan; and
- Misapplication of the criteria for inspection that caused financial collapse.

In *El Greco Painting Co.*,[175] the contractor argued that the government's refusal to compensate it for extra work caused financial dislocation and therefore should excuse the default. The board, however, held that the contractor failed to prove a causal link between the extra work and the failed performance. The board instead found that the contractor's initial undercapitalization of the project, and the diversion of money to a different project, caused the failure to perform.

2. Labor and Other Performance Difficulties

The contractor has the responsibility to hire and retain qualified employees to perform the contract.[176] Thus, a claim of excusable delay based on the loss of key personnel will generally fail. The same rationale applies to the illness of company officers, key personnel, or the owner. For instance, in *M&T Construction Co.,*[177] the board held that a heart attack suffered by one of the subcontractor's personnel did not excuse nonperformance. The obvious exception is where the government enters into a contract with a sole proprietor.

Labor shortages also fail as bases for excusable delay because of the contractor's responsibility to adequately staff a contract, unless the government causes the labor shortage.[178] In one case, however, where a competitor hired away the contractor's highly skilled operators, and the contractor diligently searched for new employees, the board concluded that an excusable delay occurred.[179]

The contractor has the responsibility to obtain the material, facilities, and equipment necessary for performance.[180] Therefore, shortages of material and machinery breakdown or loss do not constitute grounds for excusable delay unless they arise from causes beyond the control of and without the fault or negligence of the contractor or its subcontractors.

F. Notice Requirements

The Default clause requires the contractor to give the government notice of any alleged excusable delays.[181] Courts and boards have held, however, that imperfect or delayed notification will not automatically bar the assertion of excusable delay.[182] Failure to give notice will not usually defeat the contractor's claim, unless the government can prove prejudice. The government is not prejudiced if it is actually aware of the conditions under which the contractor is laboring.[183]

VI. Specific Types of Government-Caused Delays

The government's responsibilities under the contract create another category of excusable delay. The government must not interfere with or impede the contractor's performance of its work. Moreover, the government must not fail to grant the requisite approvals in a timely manner and must otherwise satisfy its contractual obligations. Delays may be excusable if they result from the government's violation of these duties. Accordingly, a delay may be excusable if it results from (1) delay in required government approval; (2) improper government inspection; or (3) breach of the government's duty to cooperate.

Under the Default clause, if the government's contractual acts cause delay, the default is excused.

A. Delayed Government Approval of Drawings or Samples

The government's failure to grant timely approval at any stage of the contract may result in an excusable delay. The situation arises where, for example, the contract requires government approval of the contemplated subcontractors.[184]

This defense is not available where the government grants a needed extension after untimely approval and the contractor still fails to perform.

The burden rests with the contractor to prove that its delay resulted from the lag in government approval and that delay in performance resulted in default.[185] The contractor must show that the delay in approval was unreasonable and thus prevented timely performance, or that the late approval increased the time required to complete the work.[186]

Government contracts often require approval of drawings, plans, or samples to be used in the course of performance. Failure to respond to submissions of these items within the specified time will often entitle the contractor to an extension of time.[187] If no specific time is mentioned, the government must respond within a "reasonable time."[188] The contractor must prove that it was injured as a result of the delay in approval.[189] Even where the contractor is late in submitting drawings, the government's failure to approve them within the period of review may constitute a breach.[190] The contractor must, however, notify the government of the delay in the submission.

B. Delayed Contract Award by the Government

Where the government delays awarding the contract beyond the date specified in an invitation for bids, the contractor may assert a claim of excusable delay. However, the contractor must prove that the delay in award caused the contractor's late performance.[191] A delay in award will not result in a time extension if the delay did not affect the contractor's ability to perform. The delay may, on the other hand, be the basis for an equitable adjustment.[192]

In *Ordnance Parts & Engineering Co.*,[193] the government delayed award of the contract for nearly one year during which time material prices increased. The contractor, however, had extended its acceptance period, and the government properly acted within that extended period. Because the contractor had the opportunity to retract its offer instead of extending it, the board refused to find excusable delay resulting from a delay in contract award. Similarly, in *Alpine Aggregate Associates*,[194] the government delayed award for eight months. The CO ultimately default terminated the contract for untimely performance. The contractor claimed that the CO acted unfairly by not waiving the completion date after the contractor granted the government extensions of the bid-acceptance period. The board sustained the default termination, ruled that the CO did not abuse his discretion, and found no evidence that the delay in award caused the performance problems.

Even in situations where the government delayed award beyond the start-work date, boards have been unwilling to find an unreasonable delay.[195] However, some contracts contain a Time of Delivery clause that automatically extends the contractor's completion schedule by the number of days after the award date that the contract was, in fact, awarded. For instance, in *Turnco Machine Co.*,[196] the board overturned a default termination because the contract had been awarded 98 days after the date anticipated in the Time of

Delivery clause, and the CO had default terminated the contract before the extended delivery date.

C. Government Delay in Payment

Government delay in making progress payments to the contractor is a frequent cause of excusable delay.[197] The contractor typically has two courses of action when faced with unreasonably late or stopped payments. It may be entitled to cease performance altogether, or it may have a defense against a subsequent assessment of liquidated damages or a default termination.

As always, the contractor must prove that the difficulties it encountered are attributable to the government. Moreover, the contractor must at least establish that its financial problems were primarily caused by the government's nonpayment.[198] In *Environmental Devices, Inc.*,[199] the contractor claimed that the government delayed progress payments and ceased progress payments after paying approximately 10 percent of the contract price. The contractor asserted that these actions amounted to a breach of contract. However, the board found that the government's actions were not the "primary or controlling" cause of the contractor's default. Poor management was the principal reason for the contractor's failure to perform.

Preexisting financial distress, and even problems that exacerbate those difficulties, are not grounds for excusable delay. In *Sterling Millwrights, Inc. v. United States*,[200] the government improperly withheld earned progress payments. The court found that the CO's intent in such withholding was to gain leverage for negotiation. Because no law, regulation, or contract provision created the authority to withhold payments for this purpose, the court found that the withholding dried up the contractor's cash flow and forced it to suspend work. Default termination in these circumstances was improper.

The contractor has the duty to proceed pending a dispute. Nevertheless, if a contractor ceases performances as a result of the government's improper refusal to make progress payments, a default termination may be converted to a termination for convenience.[201] In *Jones Plumbing & Heating, Inc.*,[202] the board established the following factors in considering whether the government's failure to pay constituted a breach of the contract:

- The amount of money involved;
- The duration of the nonpayment;
- The payment procedure agreed to by the parties; and
- Whether the withholding of money impeded the contractor's ability to perform.

The contractor must also show that the government's refusal to make timely payments was unjustified.[203] The contractor must additionally prove that the failure to pay caused an inability to perform. If these aspects are shown, then even a temporary improper suspension of payments can result in excusable delay.

In addition, the failure to provide an adequate price adjustment may excuse the contractor's nonperformance if the failure causes financial deterioration and an inability to perform.[204] These situations are similar to those that excuse nonperformance because of the government's refusal to make progress payments. A contractor does not need to prove that the withholding resulted in the outright inability to perform, or that it would have been able to perform but for the government's default.

D. Delayed Government Approval of Subcontractors

Often, the government must approve the use of subcontractors or materials before the contractor may proceed. If the government improperly delays or withholds such approval, this may be an excusable delay.[205] The usual remedy is a conversion of the default termination into one for the convenience of the government.[206]

E. Government Issuance of Constructive Changes

When the CO orders or requires, in writing or verbally, the contractor to perform work different from that described by the contract and does not issue a formal change order, a constructive change has occurred. If the order somehow drives the contractor to default, the contractor may (1) attempt to prove that the CO did issue a constructive change; and (2) assert the constructive change as the basis for excusable delay. If it is successful, the default termination may be converted into a termination for convenience.

The contractor must prove two basic elements in order to prove a constructive change: (1) a change in the contract work (that is, that the work done or to be done exceeds the minimum required); and (2) a government order or requirement to perform the change. Constructive changes may be grounds for excusable delay in a default termination context. For instance, in *Ned C. Hardy*,[207] the CO terminated the contract for default after the contractor, claiming that performance was impossible without a particular permit, refused to continue performing without one. The board found that the government's refusal to grant the permit constituted a constructive change. The board then held that the change excused the contractor's failure to perform and converted the default termination into one for convenience.[208] The constructive change doctrine is discussed in more detail in Chapter 13.

F. Government Overinspection Actions

The government has the duty not to interfere with the contractor's performance. The government, therefore, must not exercise its right to inspect contract work in an unreasonable manner. If it does so, and a delay results for which the contractor is default terminated, the contractor may assert excusable delay.

1. Standard and Special Clauses

Under the standard Inspection clause, the government has the broad right to inspect the contractor's work during performance at all reasonable times and places to ensure conformity with the specifications. The standard Inspection clause requires the government to accept or reject contract work as promptly as practicable. If the government fails to reject work within a reasonable time after inspection, the delay may be construed as an acceptance, thus barring a subsequent default termination. For example, in *Tranco Industries, Inc.*,[209] the board overturned a default termination where the government unreasonably delayed inspection of the contractor's work for three months and failed to reject within a reasonable time. The board held that the government constructively accepted the work before final acceptance. The government's failure to perform in accordance with established inspection standards or within the time specified in the contract may excuse the delay and render any default termination improper.[210]

2. Failure to Inspect or Object to Defective Work

Despite the general rule that the government has no duty to inspect, in certain circumstances the failure to inspect may excuse a default termination. In *Curtis L. Holt*,[211] the contract required government inspection of the work within five days of receipt. The government had the duty to give notice of deficiencies and provide an opportunity to correct them. The government, however, delayed the inspections until after the allotted time and the contractor was neither paid nor informed of deficiencies. The board held that the contractor's abandonment of the work was therefore excusable.

The government's promise to make available a government inspector, and its subsequent failure to provide that inspector upon completion of the work, may result in a breach of contract. However, the government has no duty to inform the contractor of defects observed during inspection but before submission for acceptance.[212] In *Holt Roofing Co.*,[213] the contractor claimed that government inspectors knew that it was installing a product incorrectly, and that their silence ratified the contractor's actions. The board found that even if the inspectors had kept silent, the contractor could not claim excusable delay. The Inspection clause provided that inspections were for the sole benefit of the government and did not relieve the contractor of the duty to comply with the contract. Thus, the government had no obligation to accept work that it had conditionally approved or about which it had remained silent after an inspector observed defects.[214] If, however, the government inspector rejects a contractor's work and knows at the time how to correct the failures, the omission to inform the contractor may violate the government's duty to share superior knowledge.

3. Inspection Standards and Methods

The government has the duty to refrain from conducting inspections and tests that unduly delay or interfere with the contractor's performance.[215] A default termination after improper inspection will usually be converted into a termination for convenience. However, the inspection must be more than an inconvenience or irritant to the contractor.

Additionally, the government may not terminate a contract for default where faulty testing resulted in a determination that the end items did not conform to specifications. In *Communications Ltd.*,[216] the government default terminated a contract for antennae multicouplers. The board found that faulty testing had been the basis for the government's conclusion that the products did not conform to the specifications. Thus, the government wrongfully terminated the contract for default.

A presumption exists, however, that the government's tests were proper.[217] The burden of showing that the method was flawed and produced incorrect results is on the contractor.[218]

The government may not subject the contractor's work to unduly rigid or strict inspection standards. The following have been found to constitute breaches of the government's duty not to hinder performance:

- Overly close supervision of work;
- Too many inspection visits;
- Excessive numbers of inspectors; and
- Multiple inspections that produce inconsistent results.

These actions and conditions may then be the basis for relief under the Default clause.

The government is liable for overinspection where the contractor proves that (1) the government's inspection practices were unjustified and unreasonable; or (2) the government retained no contractual right to inspect at the particular level. The contractor must also show the inappropriate inspection's disruptive impact on performance. Where the government discovers substantial numbers of defects in the work, an increased level of inspection may be appropriate.

Although the government does have the right to establish strict inspection standards, absent a specification of the standards to be applied, the government may not require testing more stringent that its own prevailing standards and those accepted in the industry.[219]

G. Defective Specifications and Impossibility

Where defective contract specifications disrupt the contractor's performance and lead to a termination for default, then (1) the contractor is not liable for excess costs incurred by the government as a result of the default; and (2) the termination will usually be converted to one for government convenience. Defective specifications, or those that are impossible to perform, provide a special defense for a defaulted contractor. If the defects cannot be resolved and performance becomes impossible, the contractor may be entitled to stop performance.

Under the implied warranty of specifications, the government guarantees that the specifications, if followed, will result in construction that can be accomplished and that fulfills its intended purpose.[220] The government's duty also encompasses an obligation to ensure that ambiguities are not present in the specifications.[221] Again the discussion here is abbreviated as this topic is discussed more extensively in Chapter 18.

The implied warranty of specifications attaches to both design specifications and situations where performance specifications are mixed with design specifications. Examples where the government has been found responsible for defective specifications in construction cases include:

- River flow rates and drainage information provided to a fish hatchery construction contractor failed to reflect normal conditions and underestimated the flow rate at the worksite.[222]
- Design specifications for the size of a gasket needed for the construction of a subway tunnel were defective because the dimensions provided in the specifications were not proper to the job.[223]
- The contract-specified weather stripping failed to work.[224]
- A tack coat applied according to specification failed to properly cure.[225]
- The requirement that the contractor employ a particular type of tractor led to inefficiencies and decreased productivity.[226]

Additionally, the government's implied *Spearin* warranty of design specifications extends to each specified alternative method of performance. In *S&M Traylor Bros.*,[227] the specifications permitted two alternative methods to produce columnar beams. The method selected by the contractor proved impossible. The board held that the government breached the warranty because the warranty extended to both methods of performance.

Notification must be forthcoming within a "reasonable time" after the contractor knows or should know of the deficiency or impossibility. Thus, if a reasonably competent contractor would have been led by certain circumstances to believe that the contract specifications could not be met, the contractor will be held to have knowledge of the defect or impossibility. If the contractor waits too long to notify the CO of the deficiency or impossibility, it may effectively waive its claim against the government.[228]

VII. *Waiver of Completion Date*

A. Forbearance versus Waiver

Waiver of the contract completion date is a useful and important defense for defaulted contractors. However, the government does not waive the right to terminate by failing to terminate for default immediately after the completion date passes. The government may forbear from terminating for a reasonable time while retaining its right to do so.[229] This so-called forbearance period provides the government with the time necessary to assess the situation (that is, whether the delay was excusable) and consider its options.[230] During this period, the government may default terminate the contract at any time and without notice to the contractor.

The forbearance period lasts for only a reasonable time. The determination of such time must be made on a case-by-case basis depending on the nature of the contract. In addition, where the government acts in a manner inconsistent with forbearance by, for example, issuing change orders or setting new due dates, it may be held to have waived the right to terminate.[231]

The following factors appear critical in board determinations of what constitutes a reasonable time:

- The period of time between the accrual of the right to default terminate the contract and the attempt to exercise the right;
- Whether the government continued to encourage contractor performance; and
- Whether the contractor relied on the government's forbearance.

By way of example, the board held reasonable a four-month-long forbearance period where the government did not waive the contract's delivery schedule and did not encourage the contractor to perform.[232] In contrast, the board found that the government had waived its right to terminate for default where the contractor relied on the government's encouragement and continued to work for 180 days after the scheduled completion date.[233] In *Container Systems Corp.*,[234] the board held that, in the absence of demonstrable reliance by the contractor, the government had not waived the contract schedule by negotiating toward a new schedule.

Where the contractor ceases to perform after missing the completion date, it will generally not be able to successfully assert a waiver defense. In these circumstances, the boards will grant the government a longer forbearance period than if the contractor had continued performance.[235]

B. Time of Waiver after Completion Date

The government has three options to select from when the contractor misses the completion date:

- Terminate for default without issuing a warning;
- Forbear for a reasonable time; or
- Permit the contractor to continue performance, thereby waiving the right to terminate.

The most important case involving waiver after the completion date is *DeVito v. United States*.[236] Delivery of 1,000 units was due at the end of November 1960, and 2,000 more units were due each month afterwards. The contractor missed the first delivery date, made a partial delivery the next month, and then missed the following month's delivery date entirely. The government terminated for default in mid-January 1961. The court held that the government: (1) waived its right to terminate for default by waiting 48 days after the contractor failed to deliver before terminating; (2) accepted deliveries during that time; and (3) knew the contractor was trying to catch up on deliveries. Once the government reasonably appeared to have decided to permit continued performance, it surrendered its "alternate and inconsistent" right to terminate the contract for failure to deliver on time. The Court of Claims later interpreted the *DeVito* test as directing an examination of whether the contractor could reasonably infer that the government chose not to terminate.[237]

Construction contracts usually contain a Liquidated Damages clause that assesses a predetermined amount against the contractor for each day that completion is late. Because of this, the waiver doctrine only rarely appears in construction contract cases.[238] However, courts and boards do seem to be applying it more frequently. Often, the government's threatened or actual use of the right to assess liquidated damages is relevant in construction cases.

In *Overhead Electric Co.*,[239] the board found waiver where the government: (1) let the completion date pass without taking action; (2) failed to establish a new completion date; and (3) never informed the contractor of its intent to assess liquidated damages. The board implied that the government's failure to assess liquidated damages evidenced a determination that time is no longer of the essence.[240]

In contrast, in *Indemnity Insurance Co. v. United States*,[241] the board did not find waiver where the government waited 84 days to terminate the contract and the CO continued to administer the contract. The government, the board found, had informed the contractor that it would assess liquidated damages.

When the government communicates its intent to assess damages, the contractor will have a more difficult time trying to prove that time was no longer considered of the essence.[242] Additionally, if the government continues to make progress payments after the completion date, the contractor will face an uphill battle in trying to prove detrimental reliance. Some hope of proving detrimental reliance exists if the contractor can show that it incurred costs that the government did not reimburse through progress payments.

Martin J. Simko Construction, Inc. v. United States[243] is one case where the construction contractor successfully asserted waiver. The government waited 13 months after the revised completion date to terminate for default. Although the contractor had breached the contract by performing deficiently, the government had permitted the contractor to perform after the completion date and directed the contractor to perform in accordance with the contract requirements. The Claims Court found that the contractor relied on the government's actions, and held that the government waived the completion date.

VIII. Excess Costs of Reprocurement

The Default clause grants the government the right to complete the work by contract or otherwise, and then recover from the contractor any increased costs associated with reprocurement. The amount of recovery is determined by measuring the cost of the reprocured work against the original contract price. The repurchase, however, must be accomplished as soon as practicable and at a reasonable price in order to limit the contractor's liability.

The government's right to employ this remedy is "in addition to any other rights and remedies provided by law" or under the contract.[244] Therefore, if the government does not seek reprocurement costs for whatever reason, it may still seek to recover actual damages resulting from the contractor's default.[245] However, the government's task in proving actual damages in a common law breach of contract claim will be much more difficult than in assessing the costs of reprocurement. All procedures having otherwise been followed, the government need only show that it incurred reasonable excess costs in the reprocurement to recover them from the contractor.

A. The Requirement

The CO must stay within the parameters of the original procurement when reprocuring the defaulted contract work. FAR 49.402-6, "Repurchase Against Contractor's Account," establishes the basic process the CO should follow in pursuing a repurchase.

The government also has the duty to mitigate damages—that is, to limit the amount of excess costs as much as reasonably possible. In that light, the Federal Circuit set forth the following requirements the government must prove before it can recover excess costs of reprocurement:

- The reprocured supplies are the same as or similar to those of the original contract;
- The government actually incurred excess costs; and
- The government acted reasonably to minimize the excess costs resulting from the contractor's default.[246]

The first requirement involves a simple comparison of the items. The second requirement involves determining the amount spent in the reprocurement. The third requirement is based on whether the government acted within a reasonable time of the default, employed the best and most efficient methods to complete the repurchase, obtained a reasonable price, and mitigated the losses.[247]

1. Similarity

As noted above, a reviewing tribunal will compare the original contract work scope with that bought in the reprocurement. Although the construction contract Default clause does not expressly limit the reprocurement, the government may not make material changes in the desired performance if it takes over the work as permitted in the contract.[248]

Generally, a reprocurement is similar if the reprocured work, services, or supplies are "similar in physical and mechanical characteristics as well as functional purpose."[249] Courts and boards will direct their focus to whether substantial and material differences exist between the reprocured and original work. Minor differences will not defeat the assessment of damages, but substantial and material differences may.

For instance, in *California Bridge & Construction Co. v. United States*,[250] the government default terminated a construction contract and subsequently reprocured the remaining work. The court compared the original work with the reprocured work and held that the work performed according to the subsequent contract "was so substantially" equivalent to that of the first contract as to permit recovery of excess costs.

Although the terms and conditions of the subsequent contract should be similar to those of the defaulted contract, they do not need to be identical. The reprocurement contract may alter the specifications to correct mistakes and to incorporate updates.[251] An analysis of the bid prices received by the government for the reprocured contract provides a useful tool in determining whether the original and subsequent contracts are similar. Thus, the presence of terms and conditions in the repurchase contract that do not affect the cost of performance will usually not lead to a holding that the contracts are dissimilar.

In *Alfred T. Scevers, Jr.*[252] the government reprocured a tree-thinning contract. The board disallowed recovery of excess reprocurement costs because the government materially altered the terms of the contract by dividing the remaining acreage into two tracts, extending the time of performance, and awarding two separate contracts.

2. Relaxed Specifications

Where the government relaxes the specifications in the reprocurement contract, and the defaulted contractor proves that it could have performed given those relaxed requirements, the government will probably not be able to assess excess costs.[253] For example, in *F.R. Schultz Construction Co.*,[254] the government granted the reprocured contractor additional time within which to perform the contract, beyond what it had given the original contractor. The board held that it was reasonable to conclude that the original contractor could have performed the same amount of work as the reprocured contractor, given the extra time. The board therefore refused to let stand an assessment of excess costs.

3. Government Mitigation of Damages

The government must attempt to minimize the defaulted contractor's liability for the excess costs of reprocurement. The board in *Solar Labs, Inc.*[255] enumerated at least five considerations that factor into the determination of whether the government has adequately mitigated its damages:

- Whether the government conducted give-and-take negotiations;
- Whether the government obtained a cost breakdown;
- Whether the government undertook cost or price analyses;

- ■ "Whether the government obtained a satisfactory explanation for any substantial increase over the reprocurement contractor's bid on the original contract"; and,
- ■ Whether "vigorous" arm's-length bargaining occurred, even where no price reductions resulted.

The defaulted contractor frequently challenges the assessment of excess costs on the ground that the time the government took to reprocure or the method of reprocurement resulted in an unreasonably high price. The relevant time is measured from the date of default termination to the date of award of the reprocurement contract.[256] In *Arctic Corner*, the board found that a six-and-one-half-month delay in reprocurement did not provide grounds for overturning an assessment of excess reprocurement costs.

In *Clifford La Tousle*,[257] the contractor successfully contended that the government's initial decision to allow continued performance after default, subject to assessment of liquidated damages, breached the government's duty to mitigate damages. The delayed termination forced the contractor to pay disproportionately high damages and allowed the damages to run for an unreasonable period of time.

4. Excess Costs Incurred

The government must actually incur excess costs in order to assess them against the defaulted contractor. These amounts cannot be determined accurately until the reprocurement contract is completed and paid for. Thus, until such costs are incurred and paid for by the government, the defaulted contractor is not liable for them.

5. Liability for Cost of Second Reprocurement Contract

Where the government terminates for default a fixed-price contract and completes the work by itself or through another contractor, the government may recover from the defaulted contractor the difference in cost between the reprocurement contract and the defaulted contract.[258] Where the government terminates the reprocurement contract, either for default or for convenience, the original contractor will generally not be liable for the excess costs of the second reprocurement contract. The boards in such instances appear to ask whether the need for a second reprocurement contract was due to the original contractor's fault. In other words, boards will examine whether the second reprocurement was a foreseeable, direct, natural, or proximate result of the original contractor's breach.[259]

For instance, in *Logan Electric Corp.*,[260] the government terminated for default the original contractor and terminated the reprocurement contractor for convenience. The government reprocured the contract a second time. The board held that the government could not recover the difference between the original contract and the second repurchase contract because the termination of the second repurchase contract was a voluntary action unrelated to the original contractor's actions. The original contractor remained liable, however, for the difference in price between its contract and the first reprocurement contract.

In *Interstate Forestry, Inc.*,[261] the government default terminated the original contractor. The first reprocurement contractor abandoned the project, but the CO did not default terminate the contract. The government entered into a second reprocurement contract and attempted to assess those excess costs against the original contractor. The board found that the government failed to mitigate its damages when it did not terminate the first reprocurement contractor for default and then assess to that contractor the difference between its contract price and the second reprocurement. The board held that "[t]he need for a second reprocurement contract at a higher price was not due to [the original contractor's] default."

The government also retains other rights and remedies not listed in the Default clause. One such remedy is the right to recover liquidated damages under a Liquidated Damages clause. Courts and boards often turn to the Uniform Commercial Code (UCC) and apply these common law principles to government contracts. UCC-provided remedies include recovery of the price paid and recovery of damages for nondelivery.[262]

B. Calculating Excess Costs

1. Measure of Excess Costs

The usual recovery for the government when it seeks the excess costs of reprocurement is the difference in price between the defaulted contract and the reprocurement contract.[263] A defaulted contractor may be entitled to a price adjustment for changes made or necessary to be made in the contract work before the contract's termination for default. The effect is to increase the original contract price, thereby reducing the excess costs of reprocurement. In *Big Star Testing*,[264] the government demanded that the original contractor continue performing despite the impossibility of recertifying necessary equipment. The contractor instead defaulted. The default had the effect of saving the government money in the form of costs that would have resulted from the contractor's fruitless attempts to comply. The board held that the additional expense that would be incurred by the contractor prevented an assessment of excess costs, because there would be no difference in price between the defaulted contract as adjusted and the price of the reprocurement contract.[265]

The government must also adjust the assessment of excess costs where it directs changes in the work under the reprocurement contract. In *Steelship Corp.*,[266] the defaulted contractor claimed that a ship reprocurement involved a far more expensive design than the original contract contemplated. The board held that it would be unfair to essentially replace the original contract's Yugo with a Rolls-Royce and charge the defaulted contractor with the difference. The board reduced the assessed excess costs to take into account the improvements in the reprocurement contract's specifications.

2. *Other Recoverable Costs*

The government's right to recover the excess costs of reprocurement extends beyond the price difference between the reprocurement contract and the original contract. The government's recovery may include transportation costs, temporary rental allowances, and interest from the date of payment under the reprocurement contract. However, costs incurred in connection with advertising, awarding, or supervising the contract and other related administrative expenses are not generally part of the excess costs.

C. Government Completion of Work

The boards sometimes allow the government to recover the excess costs of reprocurement when the government completes the work using government personnel instead of private contractors. Indeed, the construction contract Default clause specifically allows the government to "take over the work and complete it by contract or otherwise."[267] Where an urgent need exists for speedy completion of the work, and no alternative appears, the government can usually justify its decision to use government resources to complete the defaulted work.[268]

In *Brent L. Sellick*,[269] the CO terminated the contract when a demolition contractor delayed removing debris after destroying a building. The CO employed government personnel to clear the debris, claiming that it constituted a safety hazard. The board denied the government excess costs and found that the CO waited four months to terminate the contract for default. The board also found that the debris was more an eyesore than a hazard. Under these circumstances, the board ruled that the CO could have resolicited the completion work using competitive procedures and thus obtained a lower price. The board noted that "[t]he government . . . failed to establish that, under all of the relevant circumstances, use of public works forces on a noncompetitive basis was the only reasonable alternative."

D. Challenging Assessment of Excess Costs—*Fulford* Doctrine

The contractor's right to appeal the assessment of excess costs is distinct from its right to appeal a default termination. Thus, if the contractor fails to timely appeal a default termination, or simply decides not to, it may still appeal the subsequent assessment of the excess costs of reprocurement.

In *Fulford Manufacturing Co.*[270] the board held that a contractor that timely appealed the CO's separate decision assessing excess reprocurement costs could contest both the assessment of excess costs and the underlying default termination, despite the fact that the contractor failed to timely appeal the default termination. Similarly, other boards have recognized that the contractor may wait to challenge a default termination until after the government assesses excess reprocurement costs.

The *Fulford* doctrine has been interpreted to permit a contractor to challenge the default termination on procedural or other grounds in an appeal regarding the assessment of excess reprocurement costs.[271] However, the contractor may not similarly defer appeal of other claims the government may have against the contractor.

The *Fulford* doctrine is recognized by most boards and by the Court of Federal Claims (COFC).[272]

IX. Other Remedies

The government has the right not only to assess excess costs, but also to recover in accordance with "any other rights and remedies provided by law or under [the] contract."[273] The following are some remedies seen in construction contract cases:

- Confiscation of the contractor's project-related inventory;
- Common law damages for breach of contract; and
- Deferred payment agreements (applicable in some situations where the contractor cannot immediately pay the damages demanded by the government in full).

The government also may find the contractor nonresponsible based on a default termination, and render it ineligible to receive future contracts.

A. Damages for Breach of Contract

The CO must take action to recover any ascertainable damages that the government suffers due to a contractor's default.[274] Thus, the government may pursue a breach of contract action against the contractor. The damages for breach of contract in the case of default include those available following contractor delay.[275]

In a common law breach of contract action, the government may recover from the contractor all damages that are "foreseeable, direct, natural and proximate results of the breach."[276] Damages that are only remotely tied to the breach, or that are highly attenuated, are not recoverable. In *Gibson Forestry*,[277] the Forest Service default terminated a contract for planting tree seedlings after the contractor abandoned performance. The Forest Service sought the costs for 77,000 lost seedlings that could not be planted within their three-week life expectancy. The board held that the government failed to show that the contractor had knowledge or constructive knowledge of the three-week life span of the seedlings. Thus, the costs were not a foreseeable result of the breach.

B. Deferred Payment Agreements

The FAR in some instances permits a contractor who cannot immediately pay damages to the government to defer payment of the debt through a deferred-payment agreement. The following limitations apply to such an agreement:

- If the contractor did not file an appeal, the government may enter into the agreement only if the contractor is unable to pay in full at once, or if the contractor's operations under defense contracts would be seriously impaired as a consequence of the debt.[278]
- If the contractor did file an appeal, the government may enter into the agreement if the arrangement might help avert the possible overcollection of debts.[279]
- If the contractor is a small business or financially weak, the government may grant deferment where "a reasonable balance of the need for government security against loss and undue hardship on the contractor" exists.[280]

The contractor must submit documentation of financial condition, contract backlog, cash flow, the inability to pay short-term debts, and the effect that immediate and full payment will have on current operations.[281]

C. Cost-Reimbursement Contracts

The government's remedies under a defaulted cost-reimbursement contract are more limited than those available under a fixed-price contract. In particular, the government may not recover excess costs of reprocurement; the defaulted contractor retains all payments received for incurred costs.

The contractor must submit a termination settlement proposal to the CO within one year of the termination. If the CO and the contractor cannot agree on a settlement, the CO determines the sum owed to the contractor, including the following:

- All costs, not previously paid, that are reimbursable under the contract;
- All costs of settling terminated subcontracts chargeable to the terminated contract; and
- That portion of the total fee payable under the contract that is the proportion of the deliveries made and accepted by the government to the total number required by the contract.

D. Nonresponsibility Determinations

Once the government terminates a contractor for default, the government may find the contractor nonresponsible and thereby render it ineligible for award of future contracts. FAR 9.104-3(c) provides:

A prospective contractor that is or recently has been seriously deficient in contract performance shall be presumed to be nonresponsible, unless the contracting officer determines that the circumstances were properly beyond the contractor's control or that the contractor has taken appropriate corrective action.

The Government Accountability Office (GAO) has held that the government may find a contractor nonresponsible based upon prior default terminations, even if the contractor performed well on other contracts.[282]

X. *Liquidated Damages*

The contractor's failure to complete in accordance with the contract gives the government the right to terminate for default or to permit continued performance and recover delay damages. In construction contracts, these are generally liquidated damages. Liquidated Damages clauses appear frequently in government construction contracts.

The Liquidated Damages–Construction clause[283] contains a reasonableness requirement: the government may assess liquidated damages until such "reasonable time" as may be needed for final completion. Moreover, the government may continue to assess liquidated damages until the contract is completed or accepted, even if those damages exceed the contract price.[284] In practical terms, however, such an assessment would probably constitute a violation of the reasonableness requirement.

Contractors attacking a Liquidated Damages clause in construction contracts usually claim that the use of the clause was unreasonable because the government knew that it would suffer no damages if delay occurred. The contractor must show that the government had information that indicated that no damages could be expected.[285]

A. Assessment of Liquidated Damages

Where delinquent performance results in a default termination of the contractor, the government may assess liquidated damages from the original contract's date of completion until completion of the reprocured contract.[286] The government's duty to reprocure the contract in a reasonable period of time remains.[287] Also, the government has a general duty to mitigate damages. Furthermore, liquidated damages may not be assessed upon reprocurement unless (1) the work is actually completed by the reprocurement contractor; and (2) the reprocured work is similar to the work that was called for under the defaulted contract.

The defaulted contractor cannot be assessed liquidated damages for a reprocurement contractor's delay.[288]

B. Enforceability

A breach of a contract containing a Liquidated Damages clause raises two questions: (1) was the amount of liquidated damages stipulated by the contract a reasonable forecast of the damages that would result from breach; and (2) was the harm resulting from breach difficult to estimate?

A reviewing tribunal will assess the reasonableness of the forecast by viewing the circumstances at the time the contract was made, not the time of breach.[289] For instance, in *Schouten Construction Co.*,[290] the panel invalidated a Liquidated Damages clause because the government knew at the time the contract was awarded that no damages would result from late completion of part of a tower's construction. Likewise, in *Sunflower Landscaping & Garden Center*,[291] the board denied the assessment of liquidated damages where the government terminated the contract for default for failure to make progress and 130 days remained before the completion date. Immediately after the termination, the government reprocured the contract with a 50-day performance period.

The liquidated damages may exceed actual damages as a result of this standard of enforceability. In *Connell Rice & Sugar Co.*[292], the board assessed $289,549 in liquidated damages despite the fact that the government's expenses were less than one-tenth that amount. Similarly, in *Lane Co.*,[293] the government's actual damages were also lower than the liquidated damages. The board sustained the assessment of liquidated damages because the stipulated amount was reasonable at the time the contract was formed.

The stipulated amount must be a reasonable estimate of the consequences of the breach. If it is otherwise, a board may consider it punitive and therefore invalid. The amount must be reasonable when analyzed in view of the harm the government believed would result from breach.[294]

Although boards have split on the issue in the past, the current trend seems to be that the government may not recover liquidated damages if it sustained no actual damages.[295] Although a disparity between actual loss and liquidated damages does not by itself render a liquidated damages provision invalid, it may, nevertheless, attract greater attention and scrutiny by a board or court as to the reasonableness of the stipulated amount.[296]

Because a presumption exists that a Liquidated Damages clause is reasonable, the contractor bears the burden of proving that the provision or the sum stipulated is unreasonable.[297] The government, of course, bears the initial burden of proving that the contractor caused the delay.[298]

In the absence of a ceiling placed on the total amount of liquidated damages recoverable, the amount assessed may quickly become quite high. If the sum is too great, a board may decide that it is punitive rather than compensatory, and invalidate the provision.[299] The government will therefore often place a cap on the amount of damages it may assess. Note that the cap is not required and that the liquidated damages may properly exceed actual loss.

C. Contractor Defenses

1. Penalty to Compel Performance

If the government uses liquidated damages for a purpose other than compensation for losses that are difficult to ascertain, the clause may be deemed punitive. Thus, where the government employs the clause in an effort to compel contractor performance, the board or court may invalidate the clause.[300]

Although the concept of liquidated damages inherently spurs a contractor to perform, the provision may not be enforced where it is included primarily to encourage performance. Thus, where the government's concern with prompt performance supersedes its interest in compensation, the liquidated damages provision may be deemed a penalty.[301]

2. Substantial Completion

Substantial completion of a contract occurs when a significant portion of the work is completed and the project is available for its intended use.[302] In such circumstances, the assessment of liquidated damages is unreasonable. For example, in *R.J. Crowley, Inc.*,[303] the board denied the assessment of liquidated damages where the government had beneficial use of the building in question, despite the fact that a fire alarm system had not yet been installed.

As earlier noted, no preestablished percentage governs when a contract is substantially complete. That determination hinges on whether a project is fit to be employed for its intended purpose.

XI. Contractor Appeals Challenging Default Terminations

When the government terminates a contract for default, the contractor has a right to challenge, or appeal, that termination under the contract's Disputes clause (FAR 52.233-1) and the Contract Disputes Act of 1978.[304] A contractor may appeal a termination for default in either the COFC or the appropriate board of contract appeals.[305]

In a contractor appeal, the government bears "the burden of proving, based on sound evidence and analysis," that the termination was justified.[306] If the government satisfies its burden of proof, then the contractor must prove that its default was excusable in order to overturn the termination.[307] In *Airport Industrial Park, Inc. v. United States*,[308] the COFC upheld a termination for default because the contractor failed to maintain solvent performance and payment bonds required by the contract.

Significantly, the court affirmed that, once the government has established the lack of adequate bonding, which is a material breach, the contractor has the burden of proving that the default was excusable under the terms of the contract.

The impact of the burden of proof was demonstrated in *Kostmayer Construction*,[309] a case decided by the ASBCA in May 2008. In that case, the Board determined that the termination of a construction contract was improper. The Board stated:

> We consider that the government has failed to sustain its burden of proving that the termination was justified. The decision to terminate here was unreasonable and an abuse of the contracting officer's discretion because it was based on a materially inaccurate, misleading

analysis by the contracting officer of the percentage of contract completion and a flawed assessment of appellant's capabilities to complete the work in the more than seven months remaining for performance. The government unreasonably underestimated appellant's ability to timely complete the project.

Notes

1. FAR 49.501.
2. OFEGRO, HUDBCA No. 88-3410-C7, 91-3 BCA ¶ 24,206; H&R Machinists Co., ASBCA No. 38440, 91-1 BCA ¶ 23,373. For further discussion of the *Christian* doctrine, see Chapter 7.
3. Jones v. United States, 11 Ct. Cl. 733 (1875), *aff'd*, 96 U.S. 24 (1877).
4. King v. United States, 37 Ct. Cl. 428 (1902).
5. *Jones*, 11 Ct. Cl. 733.
6. *Id.*
7. Quinn v. United States, 99 U.S. 30 (1879).
8. United States v. McMullen, 222 U.S. 460 (1912).
9. *See* AIA A201, § 14.2.
10. *See* AIA A207, § 14.1.
11. *See* FAR 52.249-8 Default (Fixed Price Supply and Service); FAR 52.249-9 Default (Fixed-Price Research & Development).
12. Lisbon Contractors, Inc. v. United States, 828 F.2d 759, 765 (Fed. Cir. 1987); Appeal of Kostmayer Constr., LLC, ASBCA No. 55053, 2008-2 BCA ¶ 33,869; FFR-Bauelemente & Bausanierung GmbH, ASBCA No. 52152, 2007-2 BCA ¶ 33,627.
13. DeVito v. United States, 188 Ct. Cl. 979, 990 (1960).
14. King v. United States, 37 Ct. Cl. 428, 436 (1902).
15. H.N. Bailey & Assocs. v. United States, 448 F.2d 387 (1971); K & M Constr., ENG BCA No. 2998 et al., 73-2 BCA ¶ 10,034.
16. FAR 52.249-14(a); FAR 49.505(d).
17. FAR 52.249-6(b).
18. 41 U.S.C. §§ 601 et seq.
19. *See* FAR 52.249-2, Termination for Convenience for the Government (Fixed-Price) (May 2004). For further discussion of terminations for convenience, see Chapter 13.
20. United States v. DeKonty Corp., 992 F.2d 826 (Fed. Cir. 1991); Twigg Corp., NASA BCA No. 67-0192, 93-1 BCA ¶ 25,318 (refusal to replace concrete).
21. Mega Constr. Co. v. United States, 29 Fed. Cl. 396 (1993); Santee Dock Builders, AGBCA No. 96-161-1, 99-1 BCA ¶ 30,190.
22. C.C. Galbraith & Son, Inc., ASBCA No. 10769, 67-2 BCA ¶ 6488.
23. Lisbon Contractors, Inc. v. United States, 828 F.2d 759, 765 (Fed. Cir. 1987); McDonnell Douglas Corp. v. United States, 76 Fed. Cl. 385 (2009); McDonnell Douglas Corp. v. United States, CAFC No. 2007-5111, 5131 (June 2, 2009); Kostmayer Constr., LLC, ASBCA No. 55053, 2008-2 BCA ¶ 33,869.
24. AGBCA No. 441, 77-1 BCA ¶ 12,344.

25. FFR-Bauelemente & Bausanierung GmbH, ASBCA No. 52152, 2007-2 BCA ¶ 33,627.

26. Swiss Prods., Inc., ASBCA No. 40031, 93-3 BCA ¶ 26,163.

27. Rex Conklin Reforestation, AGBCA No. 76-155, 78-1 BCA ¶ 13,070.

28. Twigg Corp., NASA BCA No. 62-0192, 93-1 BCA ¶ 25,318.

29. Boston Shipyard Corp. Military Sealift Command v. United States, 886 F.2d 451 (1st Cir. 1989).

30. ASBCA No. 39030, 90-2 BCA ¶ 22,833.

31. 16 F.3d 1173 (Fed. Cir. 1994).

32. Gittron Assocs., Inc., ASBCA No. 14561, 70-1 BCA ¶ 8316 (violation of Service Contract Act).

33. Kostmayer Constr., ASBCA No. 55053, 2008-2 BCA ¶ 33,869 (citing L&H Constr. Co., ASBCA No. 43844, 97-1 BCA ¶ 28,766 at 143,556).

34. Joseph Morton Co., Inc. v. United States, 3 Ct. Cl. 120 (1983).

35. 40 U.S.C. §§ 3141 et seq. For further discussion of the Davis Bacon Act, see Chapter 12.

36. Glenn Constr. Co. v. United States, 52 Fed. Cl. 513 (2002).

37. 61 Fed. Cl. 638 (2004).

38. *Kostmayer*, 2008-2 BCA ¶ 33,869; Ryan Co., ASBA No. 48151, 00-2 BCA ¶ 31,094 at 153,544, *aff'd on recon.*, 01-1 BCA ¶ 31,151; Walsky Constr. Co., ASBA No. 41541, 94-1 BCA ¶ 26,264 at 130,625, *aff'd on recon.*, 94-2 BCA ¶ 26,698 at 132,784.

39. *Kostmayer*, 2008-2 BCA ¶ 33,869; Ryan Co., ASBA No. 48151, 00-2 BCA ¶ 31,094 at 153,544, *aff'd on recon.*, 01-1 BCA ¶ 31,151; Walsky Constr. Co., ASBA No. 41541, 94-1 BCA ¶ 26,264 at 130,625, *aff'd on recon.*, 94-2 BCA ¶ 26,698 at 132,784.

40. ASBCA No. 43833, 97-1 BCA ¶ 28,766.

41. FAR 49.402-3(f).

42. DCX, Inc. v. Perry, 79 F.3d 132 (Fed. Cir. 1996); Darwin Constr. Co. v. United States, 811 F.2d 593, 598 (Fed. Cir. 1987).

43. Phoenix Petroleum Co., ASBCA No. 42763, 96-2 BCA ¶ 28,284.

44. FAR 49.402-4.

45. FAR 49.603 and 49.603-7 provided formats for this as a guide.

46. 2 NASH & CIBINIC, FEDERAL PROCUREMENT LAW 1143–62 (3d ed. 1980).

47. FAR 52.249-8(g); FAR 52.249-10; FAR 52.249-9(g).

48. J. C. Equip. Corp. v. England, 360 F.3d 1311 (Fed. Cir. 2004).

49. GSBCA No. 8223, 89-1 BCA ¶ 21,361.

50. FAR 52.249-2(d), (e), and (f).

51. Ryste & Ricas, Inc., ASBCA No. 54514, 06-1 B.C.A. ¶ 33,124, at 164,148; *aff'd* Ryste & Ricas, Inc. v. Sec'y of the Army, 477 F.3d 1337 (2007).

52. FAR 52.249-8.

53. Anchor/Darling Value Co., ASBCA No. 46019, 95-1 BCA ¶ 27,595. *But see* FAR 49.402-3(e) (specifying that a show cause letter or notice should be issued "if practicable").

54. Logan Elec. Corp., ASBCA No. 13054, 70-1 BCA ¶ 8083.

55. *See* Cent. Ohio Bldg. Co., Inc., PSBCA No. 2742, 92-1 BCA ¶ 24,399.

56. *See* H.L.C. & Assocs. Constr. Co. v. United States, 367 F.2d 586 (Ct. Cl. 1966); Cosmos Eng'g, Inc., ASBCA No. 19780, 77-2 BCA ¶ 12,713.

57. *See* Keith Crawford & Assocs., ASBCA No. 46893, 95-1 BCA ¶ 27,388.

58. Fidelity Constr. Co., DOT CAB No. 75-19, 75-19A, 72-2 BCA ¶ 12,831.

59. Blinderman Const. Co., Inc. v. United States, 39 Fed. Cl. 529, 573 (1997) (citing Elec. Enters., Inc., IBCA No. 972-9-72, 74-1 BCA ¶ 10,400).

60. ENG BCA No. 3785, 89-2 BCA ¶ 21,753.

61. ASBCA No. 30686, 85-3 BCA ¶ 18,421.

62. Societé de Constr. et de Distribution de Materiaux, ASBCA No. 31029, 87-1 BCA ¶ 19,468. *See* Cox & Palmer Constr. Corp., ASBCA No. 38739, 92-1 BCA ¶ 24,756 (contracting officer has discretion to exclude off-site materials in calculating performance completion).

63. J&A Pollin Constr. Co., GSBCA 2780, 70-2 BCA ¶ 8562; Preston-Brady Co., VACAB 1849, 86-2 BCA ¶ 18,860.

64. Rivera Constr. Co., ASBCA No. 29391 et al., 88-2 BCA ¶ 20,750.

65. MC&D Capital Corp. v. United States, 948 F.2d 1251 (Fed. Cir. 1991) (contractor's failure to satisfy certain contractual requirements, labeled "important requirements" by the court, was inconsistent with defense of substantial performance); Envtl. Data Consultants Inc., GSBCA Nos. 13244 et al., 96-2 BCA ¶ 28,614 (even where contract substantially completed, default termination justified where the contractor did not complete at least two significant punch list items).

66. Edward S. Good, Jr., ASBCA No. 10514, 66-1 BCA ¶ 5362.

67. ASBCA No. 42945, 94-3 BCA ¶ 27,003.

68. FAR 52.249-10(a).

69. United States v. O'Brien, 220 U.S. 321 (1911); Manhattan Lighting Equip. Co., ASBCA No. 5113, 60-1 BCA ¶ 2646.

70. Gen. Prods. Co., ASBCA No. 6522, 61-1 BCA ¶ 3003.

71. Disc. Co. v. United States, 554 F.2d 435 (Ct. Cl.), *cert. denied*, 431 U.S. 938 (1977).

72. ASBCA No. 29397, 86-1 BCA ¶ 18,623.

73. *E.g.*, Multi Roof Sys. Co., ASBCA No. 26464, 84-3 BCA ¶ 17,529 (focusing on the improbability that the contractor would timely complete the work, rather than on its incapability of doing so).

74. AGBCA No. 83-134-1, 87-2 BCA ¶ 19,698.

75. *E.g.*, Greater Am. Constr. Co., ASBCA No. 23028, 85-1 BCA ¶ 17,753.

76. ASBCA No. 43454, 94-1 BCA ¶ 26,586.

77. Lisbon Contractors, Inc. v. United States, 828 F.2d 759 (Fed. Cir. 1987).

78. *E.g.*, Surfside Builders, Inc., VABCA No. 1854, 85-2 BCA ¶ 18,024.

79. GSBCA 16354, 05-1 BCA ¶ 32,902.

80. Cox & Palmer Constr. Corp., ASBCA No. 38739 et al., 92-1 BCA ¶ 24,756.

81. Richey Constr. Co., IBCA No. 187, 60-1 BCA ¶ 2554.

82. Arthur L. Cruz, IBCA No. 2098, 87-3 BCA ¶ 20,142.

83. Arrowhead Starr Co., AGBCA No. 81-236-1, 83-1 BCA ¶ 16,320.

84. *E.g.*, Robert Hart, IBCA No. 659-8-67, 68-1 BCA ¶ 6984 (tree-planting contractor who fell behind due to difficult terrain showed it could have caught up in the remaining easier terrain).

85. *See* Disc. Co. v. United States, 554 F.2d 435 (Ct. Cl. 1977), *cert. denied*, 431 U.S. 938 (1977).

86. *E.g.*, Whittaker Corp., ASBCA No. 14191 et al., 79-1 BCA ¶ 13,805.

87. VABCA No. 1982, 85-3 BCA ¶ 18,393.

88. ASBCA No. 16400, 74-1 BCA ¶ 10,542.

89. ABC Knitwear Corp., ASBCA No. 22575, 81-1 BCA ¶ 14,826.

90. *E.g.*, Kaufman DeDell Printing, Inc., ASBCA No. 19268, 75-1 BCA ¶ 11,042.

91. Ala. Bridge & Iron Co., ASBCA No. 6124, 61-1 BCA ¶ 2970.

92. ASBCA No. 10769, 67-2 BCA ¶ 6488.

93. Cal-Pac. Foresters, AGBCA No. 250, 70-1 BCA ¶ 8088.

94. ENG BCA No. 5143, 87-3 BCA ¶ 20,069.

95. ASBCA No. 46079, 94-3 BCA ¶ 27,091.

96. 408 F.2d 424 (Ct. Cl. 1969).

97. *E.g.*, Olympic Painting Contractors, ASBCA No. 15773, 72-2 BCA ¶ 9549.

98. *See* Johnson & Gordon Sec., Inc. v. Gen. Servs. Admin., 857 F.2d 1435 (Fed. Cir. 1988) (contractor had abandoned its post and was unable and unwilling to continue to perform).

99. AGBCA No. 80-179-1, 86-3 BCA ¶ 19,077.

100. *E.g.*, Aero Prods. Co., ASBCA No. 44030, 93-2 BCA ¶ 25,868. *But see* Scott Aviation, ASBCA No. 40776, 91-3 BCA ¶ 24,123 (fact that contractor's engineers and managers continued to work in an attempt to resolve technical problems was inconsistent with repudiation).

101. Accu-Met Prods., Inc., ASBCA No. 19704, 75-1 BCA ¶ 11,123.

102. Kirk Caswan, AGBCA No. 76-192, 78-2 BCA ¶ 13,459.

103. DOTCAB No. 1232, 83-1 BCA ¶ 16,386.

104. *E.g.*, Drain-A-Way Sys., GSBCA No. 6473, 83-1 BCA ¶ 16,202.

105. Northern Helex Co. v. United States, 455 F.2d 546 (Ct. Cl. 1972).

106. Kakos Nursery, Inc., ASBCA No. 10989, 66-2 BCA ¶ 5733, *recon. denied*, 66-2 BCA ¶ 5909.

107. ASBCA No. 24688, 82-1 BCA ¶ 15,513.

108. Indus.-Denver Co., ASBCA No. 13735, 70-1 BCA ¶ 8118.

109. Fla. Sys. Corp., ASBCA No. 12443, 69-2 BCA ¶ 8028.

110. United States v. Spearin, 248 U.S. 132 (1918).

111. ASBCA No. 35876, 94-3 BCA ¶ 27,182.

112. Ned C. Hardy, AGBCA No. 74-111, 77-2 BCA ¶ 12,848.

113. *Id.*

114. *E.g.*, Hempstead Maint. Serv., Inc. GSBCA No. 3127, 71-1 BCA ¶ 8809.

115. Wise Instrumentation & Control, Inc., NASA BCA No. 1072-12, 75-2 BCA ¶ 11,478, *recon. denied*, 76-1 BCA ¶ 11,641; Suffolk Envtl. Magnetics, Inc., ASBCA No. 17593, 74-2 BCA ¶ 10,771.

116. Oak Adek, Inc. v. United States, 24 Cl. Ct. 502 (1991).

117. FAR 52.249-8(c), (d); FAR 52.249-10(b)(1); FAR 52.249-14(b).

118. *E.g.*, Epact Corp., GSBCA No. 3830, 73-2 BCA ¶ 10,329.

119. NASH & CIBINIC, ADMINISTRATION OF GOVERNMENT CONTRACTS 550 (3d ed. 1995).

120. 318 U.S. 120 (1943).

121. 69 F. Supp. 125, 107 Ct. Cl. 627 (1947).

122. FAR 52.249-8(d).

123. Fairfield Scientific Corp., ASBCA No. 21152, 78-1 BCA ¶ 12,869, *aff'd*, 611 F.2d 854 (1979).

124. ASBCA No. 28962, 86-3 BCA ¶ 19,164.

125. N. Va. Elec. Co., ASBCA No. 21446, 80-1 BCA ¶ 14,293.

126. 408 F.2d 424 (Ct. Cl. 1969).

127. ASBCA No. 28962, 86-3 BCA ¶ 19,164.

128. DOTBCA, No. 2455 et al., 93-2 BCA ¶ 25,720.

129. Clay Bernard Sys. Int'l, ASBCA No. 25382, 88-3 BCA ¶ 20,856.

130. GSBCA No. 11543, 93-3 BCA ¶ 26,193.

131. Dawson Constr. Co., VABCA No. 2322, 88-3 BCA ¶ 20,945.

132. IBCA No. 2944, 93-1 BCA ¶ 25,360.

133. Milcraft Mfg., Inc., ASBCA No. 19305, 74-2 BCA ¶ 10,840.

134. Meyer Labs., Inc., ASBCA No. 17061, 74-2 BCA ¶ 10,804.

135. Allied Paint Mfg. Co. v. United States, 470 F.2d 556 (Ct. Cl. 1972).

136. FAR 52.249-10(b)(2).

137. FAR 52.249-10(b).

138. B-169473, 1970 CPD ¶ 42.

139. AGBCA No. 81-104-1, 81-2 BCA ¶ 15,315.

140. ASBCA No. 1555 et al., 6 CCF ¶ 61,781 (1955).

141. Horowitz v. United States, 267 U.S. 458 (1925).

142. World Wide Meats, Inc., ASBCA No. 18891, 74-1 BCA ¶ 10,573.

143. Gyrotron Corp., ASBCA No. 16705, 73-1 BCA ¶ 9843.

144. Woo Lim Constr. Co., ASBCA No. 13887, 70-2 BCA ¶ 8451.

145. Southland Mfg. Corp., ASBCA No. 10519, 69-1 BCA ¶ 7714, *recon. denied*, 69-2 BCA ¶ 7968.

146. Sundswick Corp. v. United States, 75 F. Supp. 221 (Ct. Cl.), *cert. denied*, 334 U.S. 827 (1948).

147. Pat-Ric Corp., ASBCA No. 10581, 66-2 BCA ¶ 6026.

148. *E.g.*, Roflan Co., ASBCA No. 23141 et al., 80-1 BCA ¶ 14,342, *aff'd*, 17 Cl. Ct. 242 (1985); Hawk Mfg. Co., GSBCA No. 4025, 74-2 BCA ¶ 10,764.

149. *E.g.*, Simpson Transfer & Storage Co., ASBCA No. 24750, 82-2 BCA ¶ 15,949 (the contractor was negligent in that it stored packing crates containing combustibles too close to its warehouse).

150. DOTBCA No. 2486, 93-1 BCA ¶ 25,384, *recon. denied*, 94-2 BCA ¶ 26,727.

151. Crawford Dev. & Mfg. Co., ASBCA No. 17565, 74-2 BCA ¶ 10,660 (flu outbreak did not strike enough workers to cause the delay).

152. ASBCA No. 11496 et al., 67-2 BCA ¶ 6456.

153. GSBCA No. 89-88-TD, 89-3 BCA ¶ 22,112.

154. Int'l Elecs. Corp. v. United States, 646 F.2d 496 (Ct. Cl. 1981).

155. Santa Fe Eng'rs, Inc., PSBCA No. 902 et al., 84-2 BCA ¶ 17,377.

156. Bill's Janitor Serv., ASBCA No. 10345, 65-2 BCA ¶ 4916.

157. Montgomery Ross Fisher, Inc., ASBCA No. 16843 et al., 73-1 BCA ¶ 9799.

158. Andrews Constr. Co., GSBCA No. 4364, 75-2 BCA ¶ 11,598.

159. Carolina Sec. Patrol, Inc., GSBCA No. 5602, 81-1 BCA ¶ 15,040.

160. *See also* Diversacon Indus., Inc., ENG BCA 3284 et al., 76-1 BCA ¶ 11,875 (denying claim of excusable delay where the contractor's refusal to bargain collectively, despite the union's election victory, caused the strike).

161. Ala. Bridge & Iron Co., ASBCA No. 6124, 61-1 BCA ¶ 2970.

162. GSBCA No. 4036, 74-2 BCA ¶ 10,949.

163. Aulson Roofing, Inc., ASBCA No. 37677, 91-2 BCA ¶ 23,720.

164. L.K. Maint., DOTCAB No. 1560, 87-1 BCA ¶ 19,578.

165. Albert J. Jansen, ASBCA No. 6245 et al., 60-2 BCA ¶ 2793.

166. GSBCA No. 7458 et al., 88-3 BCA ¶ 2108.

167. Atlas Mfg. Co., ASBCA No. 15177, 71-2 BCA ¶ 9026.

168. C&M Mach. Prods., Inc., ASBCA No. 43348, 93-2 BCA ¶ 25,748; Smith Faison Military Sales Co., ASBCA No. 24229, 82-1 BCA ¶ 15,512.

169. Joseph J. Bonavire Co., GSBCA No. 4819, 78-1 BCA ¶ 12,877, *recon. denied,* 78-1 BCA ¶ 13,132.

170. ASBCA No. 6523, 66-1 BCA ¶ 5367.

171. Andrews Constr. Co., GSBCA No. 4364, 75-2 BCA ¶ 11,598.

172. DOTCAB No. 1242, 86-2 BCA ¶ 18,867.

173. GSBCA No. 12037, 94-1 BCA ¶ 26,398.

174. Swiss Prods., Inc., ASBCA No. 40031, 93-3 BCA ¶ 26,163.

175. ENG BCA No. 5693, 92-1 BCA ¶ 24,522.

176. KARPAK Data & Design, IBCA No. 2944 et al., 93-1 BCA ¶ 25,360; Yankee Telecomms. Lab., Inc., ASBCA No. 25240, 85-1 BCA ¶ 17,786; Lome Elecs., Inc., ASBCA No. 8642 et al., 1963 BCA ¶ 3833.

177. ASBCA No. 42750, 93-1 BCA ¶ 25,223.

178. Space Sys. Lab., Inc., ASBCA No. 12162, 68-1 BCA ¶ 6859.

179. Bannercraft Clothing Co., ASBCA No. 6247, 1963 BCA ¶ 3995.

180. Ace Elecs. Assocs., Inc., ASBCA No. 11496 et al., 67-2 BCA ¶ 6456.

181. FAR 52.249-10(b)(2).

182. Gulf & W. Indus., Inc. v. United States, 6 Cl. Ct. 742 (1984); *see* Hel-Stetton Constr. Co. v. United States, 496 F.2d 760 (Ct. Cl. 1972) (noting that notice provisions should "not be applied too technically and illiberally where the Government is aware of the operative facts").

183. *E.g.,* Phillips Constr. Co., IBCA No. 1295-8-79 et al., 81-2 BCA ¶ 15,256.

184. *E.g.,* Northeast Constr. Co., ASBCA No. 11109, 67-1 BCA ¶ 6282.

185. Mil-Craft Mfg., Inc., ASBCA No. 19305, 74-2 BCA ¶ 10,840.

186. *But see* Standard BlackBoard & Sch. Supply Co., GSBCA No. 7403 et al., 86-1 BCA ¶ 18,712 (contractor's late performance not excusable where contract required written approval by government and the government supplied only an oral approval).

187. Discovery Corp., ASBCA No. 36130, 89-1 BCA ¶ 21,189, *aff'd on recon.,* 89-1 BCA ¶ 21,403. *See* Levering & Garrigues v. United States, 73 Ct. Cl. 566 (1932).

188. Y.L. Malone & Assocs., Inc., VABCA No. 2335, 88-3 BCA ¶ 20,894.

189. KRW, Inc., DOTBCA No. 2572, 94-1 BCA ¶ 26,435.

190. L.B. Gallimore, Inc., GSBCA No. 3327, 72-1 BCA ¶ 9232.

191. *E.g.,* Hinkley & Powers v. United States, 49 Cl. Ct. 148 (1913).

192. *E.g.,* L.O. Brayton & Co., IBCA No. 641-5-67, 70-2 BCA ¶ 8510.

193. ASBCA No. 44327, 93-2 BCA ¶ 25,690.

194. AGBCA No. 83-141-3, 83-2 BCA ¶ 16,817.

195. *E.g.,* Bhd. Timber Co., AGBCA No. 83-153-1, 85-1 BCA ¶ 17,801.

196. ASBCA No. 33559, 88-2 BCA ¶ 20,551.

197. *E.g.*, Johnson v. United States, 618 F.2d 751 (Ct. Cl. 1980).

198. Bldg. Maint. Specialist, Inc., ASBCA No. 25453, 85-1 BCA ¶ 17,932.

199. ASBCA No. 37430, 93-3 BCA ¶ 26,138.

200. 26 Cl. Ct. 49 (1992).

201. Monarch Enters., Inc., VABCA No. 2239 et al., 86-3 BCA ¶ 19,281.

202. VABCA No. 1845 et al., 86-1 BCA ¶ 18,659.

203. *E.g.*, Coast Canvas Prods. II Co., ASBCA No. 27980, 83-2 BCA ¶ 16,804, *recon. denied*, 84-2 BCA ¶ 17,333.

204. Disc. Co., AGBCA No. 291, 74-1 BCA ¶ 10,511.

205. Ballenger Corp., DOTCAB 74-32 et al., 84-1 BCA ¶ 16,973, *recon. denied*, 84-2 BCA ¶ 17,277.

206. *E.g.*, Bristol Elec. Corp., ASBCA No. 24792 et al., 84-3 BCA ¶ 17,543.

207. ASBCA No. 74-11, 77-2 BCA ¶ 12,848.

208. *But see* Amertex Enters. Ltd. v. United States, No. 96-5070, 1997 WL 73789 (Fed. Cir. 1997) (contractor's cardinal change claim fatally undercut by the bilateral modification to the delivery schedule, because the contractor agreed to the changes and thus implicitly agreed that the changes were within the Changes clause of the contract).

209. ASBCA No. 26305 et al., 83-1 BCA ¶ 16,414, *recon. denied*, 83-2 BCA ¶ 16,679.

210. *E.g.*, Riverport Indus., Inc., ASBCA No. 28089 et al., 86-2 BCA ¶ 18,835, *aff'd on recon.*, 86-3 BCA ¶ 19,050.

211. HUDBCA No. 75-11, 76-2 BCA ¶ 11,999.

212. Gene Fuller, Inc., ASBCA No. 21682, 79-2 BCA ¶ 14,039.

213. GSBCA No. 8270, 91-1 BCA ¶ 23,361.

214. W.L. Spruill & Co., ASBCA No. 14390, 71-2 BCA ¶ 8930.

215. D.E.W., Inc., ASBCA No. 37232, 93-1 BCA ¶ 25,444.

216. ASBCA No. 23261 et al., 80-1 BCA ¶ 14,368.

217. *Id.*

218. Horn Waterproofing Corp., DOTCAB No. 73-24, 74-2 BCA ¶ 10,933.

219. Shirley Contracting Corp., ENG BCA No. 4650, 85-3 BCA ¶ 18,214.

220. United States v. Spearin, 248 U.S. 132 (1918).

221. United Pac. Ins. Co. v. United States, 204 Ct. Cl. 686 (1974).

222. PK Contractors, Inc. ENG BCA No. 4901 et al., 92-1 BCA ¶ 24,583.

223. Harrison W./Franki-Denys, Inc., ENG BCA No. 5523, 92-1 BCA ¶ 2458.

224. Parsons of Cal., ASBCA No. 20867, 82-1 BCA ¶ 15,659.

225. Valley Asphalt Corp., ASBCA No. 17595, 74-2 BCA ¶ 10,680.

226. Maitland Bros. Co., ASBCA No. 23849, 83-1 BCA ¶ 16,434.

227. ENG BCA No. 3852, 78-2 BCA ¶ 13,495.

228. *E.g.*, King Elecs. Co. v. United States, 341 F.2d 632 (Ct. Cl. 1965).

229. Pelliccia v. United States, 525 F.2d 1035 (Ct. Cl. 1975).

230. Eraklis Gaklidis, ASBCA No. 40110, 91-3 BCA ¶ 24,188.

231. Free-Flow Packaging Corp., GSBCA No. 3992 et al., 75-1 BCA ¶ 11,105, *recon. denied*, 75-1 BCA ¶ 11,332.

232. FXC Corp., ASBCA No. 33904, 91-2 BCA ¶ 23,928, *aff'd without op.*, 965 F.2d 1065 (Fed. Cir. 1992).

233. Jack L. Hartman & Co., AGBCA No. 84-126-1, 91-1 BCA ¶ 23,546.

234. ASBCA No. 40611, 94-1 BCA ¶ 26,354.

235. Mil-Craft Mfg., Inc., ASBCA No. 24966, 81-1 BCA ¶ 14,989; Scandia Mfg. Co., ASBCA No. 20888, 76-2 BCA ¶ 11,949.

236. 413 F.2d 1147 (Ct. Cl. 1969).

237. Pelliccia v. United States, 525 F.2d 1035 (Ct. Cl. 1975).

238. Nexus Constr. Co., ASBCA No. 31070, 91-3 BCA ¶ 24,303, *aff'd*, 92-1 BCA ¶ 24,578. *See* Olson Plumbing & Heating Co. v. United States, 602 F.2d 950 (Ct. Cl. 1979) ("when liquidated damages have been imposed by the [government], the [contractor] has a heavier burden of proving that the right to terminate for failure to deliver on time has been waived.").

239. ASBCA No. 25656, 85-2 BCA ¶ 18,026.

240. *See* Corway, Inc., ASBCA No. 20683, 77-1 BCA ¶ 12,357.

241. 14 Cl. Ct. 219 (1988).

242. Brent L. Sellick, ASBCA No. 21869, 78-2 BCA ¶ 13,510.

243. 11 Cl. Ct. 257 (1986), *vacated on other grounds*, 852 F.2d 540 (Fed. Cir. 1988).

244. *E.g.*, FAR 52.249-8(h).

245. Cascade Pac. Int'l. v. United States, 773 F.2d 287 (Fed. Cir. 1985).

246. *Id.*

247. The government's right to assess the excess costs of reprocurement is analogous to the U.C.C.'s concept of "cover." U.C.C. section 2-712(1) permits the buyer, after the seller's breach, to purchase in good faith and without unreasonable delay, "goods in substitution for those due from the seller."

248. *See* United States v. Cal. Bridge & Constr. Co., 245 U.S. 337 (1917).

249. Marmac Indus., Inc., ASBCA No. 12158, 72-1 BCA ¶ 9249.

250. 245 U.S. 337 (1917).

251. Skiatron Elecs. & Television Corp., ASBCA No. 9564, 65-2 BCA ¶ 5098.

252. IBCA No. 1358-5-80, 83-2 BCA ¶ 16,579.

253. Luis Martinez, AGBCA No. 86-148-1 et al., 87-3 BCA ¶ 20,219.

254. AGBCA No. 455, 79-2 BCA ¶ 13,890.

255. ASBCA No. 19957, 76-2 BCA ¶ 12,115.

256. Arctic Corner, Inc., ASBCA No. 38075, 94-1 BCA ¶ 26,317.

257. AGBCA No. 93-132-1, 94-1 BCA ¶ 26,509.

258. FAR 52.249-8(b).

259. Interstate Forestry, Inc., AGBCA No. 89-114-1, 91-1 BCA ¶ 23,660.

260. ASBCA No. 13054, 70-1 BCA ¶ 8083.

261. AGBCA No. 89-114-1, 91-1 BCA ¶ 23,660.

262. UCC §§ 2-711, 2-713.

263. *E.g.*, Trujillo Janitor & Carpet Shampoo Serv., DOTCAB 73-26, 74-1 BCA ¶ 10,367.

264. GSBCA No. 5793, 81-2 BCA ¶ 15,335, *recon. denied*, 82-1 BCA ¶ 15,635.

265. *E.g.*, Am. Dredging Co., ENG BCA No. 2920, 78-2 BCA ¶ 13,494 (assessment of excess reprocurement costs reduced to account for differing site conditions).

266. ENG BCA No. 3795 et al., 78-2 BCA ¶ 13,478.

267. FAR 52.249-10 (emphasis added).

268. *E.g.*, Guenther Sys., Inc., ASBCA No. 18343 et al., 77-1 BCA ¶ 12,501; Collins Elecs., Inc., ASBCA No. 16956, 72-2 BCA ¶ 9542.

269. ASBCA No. 21869, 78-2 BCA ¶ 13,510.

270. ASBCA No. 2143 et al., 6 CCF ¶ 61,815 (1955).

271. Fairfield Scientific Corp., ASBCA No. 21151, 78-1 BCA ¶ 13,082, *aff'd on recon.*, 78-2 BCA ¶ 13,429, *aff'd*, 655 F.2d 1062 (Ct. Cl. 1981).

272. Am. Telecom Corp. v. United States, 59 Fed. Cl. 467 (2004).

273. *E.g.*, FAR 52.249-8(h).

274. FAR 49.402-7(b).

275. Tester Corp. v. United States, 1 Cl. Ct. 370 (1982).

276. *Id.*

277. AGBCA No. 87-325-1, 91-2 BCA ¶ 23,874.

278. FAR 32.613(f).

279. FAR 32.613(d).

280. FAR 32.613(e).

281. FAR 32.613(c).

282. S.A.F.E. Export Corp., B-226111 et al., 87-1 CPD ¶ 400.

283. FAR 52.211-12.

284. Parker-Schram Co., IBCA No. 96, 59-1 BCA ¶ 2127.

285. Connell Rice & Sugar Co., AGBCA No. 85-483-1, 87-1 BCA ¶ 19,489, *rev'd on other grounds*, 837 F.2d 1068 (Fed. Cir. 1988).

286. D&S Roofing Co., ASBCA No. 28130 et al., 85-2 BCA ¶ 18,114.

287. Davidson Enters., IBCA No. 1835 et al., 88-1 BCA ¶ 20,267.

288. Rayco, Inc., ENG BCA No. 4792, 88-2 BCA ¶ 20,671, *aff'd*, 867 F.2d 615 (Fed. Cir. 1989); Mathews Co., AGBCA 459.76-2 BCA ¶ 12,164, *recon. denied*, 77-1 BCA ¶ 12,434.

289. Thermodyn Contractors, Inc. v. Gen. Servs. Admin., GSBCA No. 12510, 94-3 BCA ¶ 27,021; Sunflower Landscaping & Garden Ctr., ASBCA No. 87-342-1, 91-3 BCA ¶ 24,182.

290. FAACAP No. 65-20, 65-1 BCA ¶ 4803.

291. *Sunflower*, 91-3 BCA ¶ 24,182; *see also* Ford Constr. Co., AGBCA No. 241, 72-1 BCA ¶ 9275.

292. AGBCA No. 85-483-1, 87-1 BCA ¶ 19,489, *rev'd on other grounds*, 837 F.2d 1068 (Fed. Cir. 1988).

293. ASBCA No. 21691, 79-1 BCA ¶ 13,651.

294. Preston Brady Co., VABCA No. 1892 et al., 87-1 BCA ¶ 19,649, *clarified*, *recon. denied*, 87-2 BCA ¶ 19,925.

295. *E.g.*, Cavanaugh Co., GSBCA No. 7612, 86-2 BCA ¶ 18,878. *But see* Mix Constr., Inc., ASBCA No. 42884, 92-1 BCA ¶ 24,634; Cumbrum Constr. Co., ASBCA No. 28284, 84-1 BCA ¶ 17,178.

296. Connell Rice & Sugar Co., AGBCA No. 85-483-1, 87-1 BCA ¶ 19,489, *rev'd on other grounds*, 837 F.2d 1068 (Fed. Cir. 1988).

297. Truesdale Constr. Co., ASBCA No. 36645, 89-1 BCA ¶ 21,483.

298. N.Y. Shipyard Corp., DOTBCA No. 2070 et al., 91-1 BCA ¶ 23,365.

299. *See generally* Fred A. Arnold, ASBCA No. 21661 et al., 86-1 BCA ¶ 18,701.

300. Priebe & Sons v. United States, 332 U.S. 407 (1947).

301. *Id.*

302. Theon v. United States, 765 F.2d 1110 (Fed. Cir. 1985); Lindwall Constr. Co., ASBCA No. 23148, 79-1 BCA ¶ 13,822.

303. GSBCA No. 11080 (9521)-REIN, 92-1 BCA ¶ 24,499.

304. 41 U.S.C. §§ 601–613.

305. See chapters 15 and 16.

306. Kostmayer Constr., LLC, ASBCA No. 55053, 2008-2 BCA ¶ 33,869 (citing Lisbon Contractors, 828 F.2d at 765–66; J.D. Hedin Constr. Co. v. United States, 408 F.2d 424, 431 (Ct. Cl. 1969); Mich. Joint Sealing, Inc., ABSCA No. 41477, 93-3 BCA ¶ 26,011, *aff'd* 22 F.3d 1104 (Fed. Cir. 1994)).

307. DCX, Inc. v. Perry, 79 F.3d 132, 134 (Fed. Cir. 1996); *Kostmayer*, 2008-2 BCA ¶ 33,869.

308. Airport Indus. Park, Inc. v. United States, 59 Fed. Cl. 332 (2004).

309. *Kostmayer*, 2008-2 BCA ¶ 33,869.

CHAPTER 15

Equitable Adjustments and Claims

LAURENCE SCHOR
AARON P. SILBERMAN

I. Introduction

The government may cause changes to a contract either intentionally, i.e., by directing a change, or unintentionally, i.e., by action or inaction that has the effect of changing the contract (also known as a constructive change). Common examples of changes in the construction context include changed conditions, failure to identify differing site conditions, government-caused delays, suspensions of work and stop work orders, and acceleration. These changes, and the contract clauses in the Federal Acquisition Regulation (FAR) that concern them, are covered in detail elsewhere in this book.[1]

When the government causes a change, either by directive or constructively, the contractor has two means to obtain additional compensation and/or time: a request for equitable adjustment (REA) or a claim. These actions are not mutually exclusive. Frequently, a contractor will first seek an REA and, if that fails, will then submit a claim. That said, an REA is not a prerequisite to a claim, so contractors may, and sometimes do, submit a claim without having submitted an REA.

This chapter describes the law and regulation applicable to both the REA and claims processes under federal government contracts.

II. Requests for Equitable Adjustments

REAs may be made in several circumstances. In some cases, the government directs a change but does not recognize it as a change. In others, it directs a change, but the parties cannot reach an agreement on the cost and/or time impact of that change. In still others, the government makes constructive changes impacting the cost and/or duration of the work.

Mr. Silberman authored section II, "Requests for Equitable Adjustments."

When the government directs a change, the Changes clause in the contract will require the contractor to perform.[2] If that change increases the contractor's costs or the time it needs to perform, the change will entitle the contractor to additional compensation or time. If the parties agree to the price and/or time extension for that change, they will sign a bilateral modification to their contract. If the parties cannot reach agreement, however, then the government may issue a unilateral change order stating the amount the government is willing to pay and/or the time extension it is willing to give. In this circumstance, the contractor may then request an REA.

Where an REA arises from a directed change, it will be governed by the Changes clause, which provides in relevant part:

> (e) The Contractor must assert its right to an adjustment under this clause within 30 days after (1) receipt of a written change order under paragraph (a) of this clause . . . by submitting to the Contracting Officer a written statement describing the general nature and amount of proposal, unless this period is extended by the Government.[3]

The 30-day notice requirement often is not possible to meet, but, since the government is rarely prejudiced by lack of notice, the requirement has not been strictly enforced.[4]

The FAR contains no specified form requirements for an REA, although agency regulations and/or contracts may impose such requirements. A contractor typically will submit such a request in the form of a letter to the Contracting Officer (CO) with attached back-up documentation where appropriate. The typical REA will identify the change or changes on which it is based; explain how each change is outside the current contract scope of work; explain and, ideally, document the cost impacts of each change; and, if the contractor has given prior notice of the change, describe how and when that notice was provided.[5]

An REA is both a request and a negotiation. The contractor must not only submit its request but also persuade the CO to grant it. First, the contractor must show a *change*—i.e., a change to the contractor's obligations under the contract. Second, it must show *entitlement*—i.e., that the government caused or otherwise was responsible for the change under the parties' contract. Third, it must show *harm*—i.e., that the government change increased the cost or duration of the contractor's work. And, finally, it must show *quantum*—i.e., what those increased costs are and why they were reasonably and necessarily incurred. This is the difference between the reasonable cost of contract performance without the change and reasonable cost with the change.

Unlike claims, there is no FAR requirement that REAs be certified, although there is such a requirement in the Defense Federal Acquisition Regulation Supplement (DFARS) for all Department of Defense contracts exceeding the simplified acquisition threshold.[6] Contractors should note, however, that this does not lessen their responsibility to ensure that all material representations in their REAs are true. Knowing or reckless false statements made in such requests may subject a contractor to liability under the federal False Claims Act.[7]

There are three major issues to consider when deciding whether to submit an REA rather than a claim. One, there is no interest recovery with REAs while there is with claims. Two, there is no mandated timeframe within which the REA must be decided. Three, the costs to prepare the REA, including attorney and consultant costs, are recoverable under an REA and are much more difficult to recover under a claim.

III. Claims

A. The Contract Disputes Act of 1978

The Contract Disputes Act of 1978 (CDA)[8] is the statutory basis for claim submission and dispute resolution on claims against the federal government. It establishes the procedures and requirements for asserting and resolving claims subject to the Act. The CDA also governs the payment of interest on contractor claims, certification of contractor claims, and civil penalties for fraudulent claims.

The CDA applies to all express or implied contracts entered into by an executive agency of the federal government after March 1, 1979, for (a) procurement of property, other than real property in being; (b) procurement of services; (c) procurement of construction, alteration, repair, or maintenance of real property; or (d) disposal of personal property. The CDA requires that all contractor claims be submitted in writing to the CO for a decision within six years after accrual of the claim. The CDA extends the CO's authority to cover all matters "relating to the contract," which includes breach of contract claims.

Claims are governed by the Disputes clause. The FAR requires that this clause be included in all federal contracts and states that "[e]xcept as provided in the Act, all disputes arising under or relating to this contract shall be resolved under this clause."

B. The FAR Definition of a "Claim"

The CDA does not define the term "claim." However, that term is defined in FAR 52.233-1, which states:

> Claim, as used in this subpart, means a written demand or written assertion . . . seeking, as a matter of right, the payment of money in a sum certain, the adjustment or interpretation of contract terms, or other relief arising under or relating to the contract. A claim arising under a contract . . . is a claim that can be resolved under a contract clause that provides for the relief sought by the claimant. However, a written demand or written assertion by the contractor seeking the payment of

money exceeding $100,000 is not a claim under the Disputes Act until certified. A voucher, invoice or other routine request for payment that is not in dispute when submitted is not a claim under the Act. The submission may be converted to a claim under the Act by complying with the certification requirements of this clause, if it is disputed as to liability or amount or is not acted upon in a reasonable time.

This provision identifies the mandatory elements of a claim. It, however, does not instruct the contractor how to write a claim. Although there is no official claim format, the CO cannot consider a claim if the contractor's submission does not meet the minimum requirements for a claim set out in the FAR. Even if the CO considers and denies the contractor's purported claim, neither a board of contract appeals (BCA) nor the Court of Federal Claims (COFC) has jurisdiction to entertain an appeal if the claim does not meet the FAR requirements. In addition, interest does not begin to accrue until a valid claim has been submitted.

There are three main requisites of any claim: (1) a written demand to the CO asserting specific rights and requesting specific relief; (2) a statement of the amount sought in a sum certain; and (3) a request to the CO for a final decision. In addition, if the claim is in excess of $100,000, it must be certified.

1. Written Demand to the CO

Both the CDA and the FAR require the claim to be in writing. In *Contract Cleaning Maintenance v. United States*,[9] the Federal Circuit stated that the contractor must submit in writing to the CO "a clear unequivocal statement that gives the CO adequate notice of the basis and the amount of the claim."[10] This has been called the "common sense" analysis approach. However, this language has raised many issues as to the "adequacy" of a contractor's claim, which is judged on a case-by-case basis. To be adequate, the statement of the claim must be sufficient to inform the CO of what is being claimed and permit him to make a meaningful review of the claim. It must include basic factual allegations of what happened, the reasons why the contractor believes that it was not financially responsible for the additional work or cost, and documentation supporting those assertions, such as contract clauses, specification provisions, and drawing references. It also must be sufficiently detailed so that the negotiation process can begin. The contractor risks dismissal of its claim if adequate data and backup support are not submitted with its assertions.[11]

All claims by a contractor against the government must be submitted to the CO for decision. The "submit" language has been described as merely a requirement that once a claim is made, the parties must "commit" the claim to the CO and "yield" to his authority to make a final decision. Claims, however, do not have to be submitted directly to the CO. The Federal Circuit has held that if a contractor sends a claim to its primary contact with a request for a final decision and a reasonable expectation that its request will be honored, and the claim is delivered to the CO, the claim will be deemed submitted to the CO. The Armed Services Board of Contract Appeals (ASBCA) has concluded that the claim-submission requirement is satisfied if the claim is submitted in a manner reasonably calculated to ensure reception by the CO. This can be accomplished by giving the claim to a CO's representative with specific instructions that it be forwarded to the CO.

The government has argued, for jurisdictional purposes under the CDA, that the claim must be signed by the contractor itself, rather than by its lawyer. The ASBCA concluded that the lawyer may submit claims on behalf of the contractor.[12] This, however, may be inconsistent with decisions of the COFC, which have indicated that the CDA requires the claim to be signed by the contractor, not counsel.[13] To avoid problems, the claim should be signed by an authorized official directly employed by the contractor.

2. Sum Certain

A claim must seek a sum certain.[14] Therefore, a claim seeking monetary relief must include either the exact amount of the relief sought or contain sufficient information so that the amount can be determined by a simple mathematical calculation.[15] Claims seeking "in excess of" a certain amount have been deemed inadequate,[16] and claims that state an approximate amount run the risk of failing to state an amount in a sum certain. However, submitting a claim based on estimates is not a bar to the CO's consideration of the claim.[17] If there is supporting documentation that outlines the quantum, which can be determined by simple mathematical calculations, and the claim sets forth the basis of the estimate, the boards and COFC will entertain an appeal.[18] While most claims seek monetary relief, the FAR allows a claim that seeks "an adjustment or interpretation of contract terms or other relief arising under or relating to this contract."[19] However, a claim for damages cannot be disguised as a request for an interpretation of the contract.

3. Request for a Final Decision

The CDA requires that all claims be submitted to the CO for decision.[20] This has been interpreted to require that the contractor request a final decision from the CO. The request, however, does not have to be explicit.[21] In *Transamerica Insurance Corp. v. United States*,[22] the Federal Circuit stated that a request may be implicit, that a common-sense analysis should be made to determine the adequacy of a contractor's request for a final decision, that no "magic words" are necessary, and that the "intent of the claim governs."[23] If it is reasonably clear that the contractor wants a final decision, then it is sufficient to meet the FAR requirement. Accordingly, a request for a final decision does not have to be in any particular format as long as the basic requirements of the CDA are met. Nonetheless, a contractor should include an explicit request for a final decision in its claim unless there is a reason why it does not want one—for example, if the contractor is negotiating with the CO and would prefer to have its submission considered as something other than a formal claim (usually an REA, discussed above).

4. Matter in Dispute Requirement

Before 1980, the Disputes clause in federal contracts required that a matter be "in dispute" to qualify as a claim. In 1980, this clause was modified to provide that a request for payment that is not in dispute when submitted may be converted to a claim by the CO's failure to act or to dispute it in a timely manner. The clause was amended again in 1984 to provide that a request for payment applies only to "routine" requests.

The FAR is in accord and states that a voucher, invoice, or other routine request for payment that is not in dispute when submitted is not a claim. The submission may be converted to a claim, by written notice to the CO as provided in FAR 33.206(a), if it is disputed either as to liability or amount or is not acted upon in a reasonable time.[24] There has been considerable litigation regarding the "matter in dispute" language and whether claims submitted by contractors are sufficiently "in dispute" to be valid CDA claims.

Under the CDA, there is no preexisting dispute requirement as to either amount or liability when a contractor submits a "nonroutine" written demand seeking, as a matter of right, the payment of money in a sum certain. Routine requests for payment seeking, as a matter of right, the payment of money in a sum certain, must be in dispute when submitted to satisfy the definition of a claim. The critical distinction in identifying a claim is between "routine" and "nonroutine" submissions. A routine request for payment that initially does not qualify as a claim may be converted into a claim by complying with the submission and certification requirements of the CDA after the request for payment is disputed by the government.[25] It also may be converted to a claim by complying with the submittal and certification requirements of the CDA, if the CO does not act within a reasonable time.

The definition of reasonable time is made on a case-by-case basis. Case law, however, indicates that a CO's inaction for five and one-half months after submission of a routine invoice makes it a proper subject of a claim. Similarly, a six-month delay by a CO in responding to a simple REA was found to be unreasonable. One way to avoid this potential problem is for the contractor to state explicitly in its letter forwarding the payment request that the failure to pay or otherwise respond within 30 days will cause the contractor to proceed under claim rules and demand a final decision.

C. Interest on Claims

The CDA provides for interest on valid claims. The FAR specifies that the government will pay interest at the Treasury rate from the date on which the CO receives a valid claim or the date payment otherwise would be due, if that date is later, until the date of payment to the contractor. The Federal Circuit has held that interest begins to run upon receipt of the claim even if the costs have not been incurred.[26] Where a defective certification is submitted with the initial claim and is subsequently corrected, interest runs from the date of the initial submission.

D. Claim versus Request for an Equitable Adjustment

FAR 31.2005-47(1) makes unallowable costs incurred in prosecuting a claim against the government. However, in *Bill Strong Enterprises, Inc. v. Shannon*,[27] the Federal Circuit found that if a proposal is submitted to the government for negotiation as a proposal for a change or an REA, as opposed to a claim, then proposal preparation costs are recoverable as ordinary costs of contract administration. As a result, the contractor is left with a sometimes difficult choice in submitting its proposal to the government. If it submits the proposal as a claim, it starts the interest clock running and begins the process of getting its claim before the board or court. The contractor, however, will be precluded from collecting claim preparation costs. If it submits its proposal for negotiation and not as a claim, it may be able to collect its proposal preparation costs, which may be significant, but will be precluded from recovering interest until negotiations break down and it resubmits the proposal as a claim under the CDA.

E. Certification

1. The Requirement to Certify

For a claim in excess of $100,000, the CDA requires that the contractor certify that its claim is made in good faith, is accurate and complete, and reflects the amount the contractor believes it is entitled to recover. Past court and board decisions held that a proper certification was jurisdictional for claims seeking in excess of $100,000, and that defective certifications could not be corrected retroactively. As a result, interest did not begin to accrue, and the CO, board, and court did not have jurisdiction, until the claim was properly certified. In 1992, the Federal Court Administration Act (FCAA)[28] provided that a proper certification was no longer a jurisdictional requirement for a court or board to consider a claim and that defective certifications did not affect jurisdiction. This provision was retroactive and applied to all claims filed before, on, or after October 29, 1992. It did not apply to pending appeals to boards, or pending suits in the Claims Court (now the Court of Federal Claims).[29] The Act also relaxed the requirements regarding the person who could certify the claim, added language required in the certification, and provided the right to correct a defective certification.[30] The Act, however, did not eliminate the requirement for a certification for claims in excess of $100,000.

The FAR requires that a defective certification be corrected before entry of a final judgment by a court or a decision by a board. The correction of a defective certification relates back to the date of the original certification for purposes of interest calculations.

2. The $100,000 Threshold

The FAR states that the aggregate amount of the increases and decreases shall be used to determine if the claim exceeds the $100,000 threshold for certification. For example, the FAR would require certification of a claim for a change that added $40,000 in work and deleted $70,000 in work.

3. The Certification

FAR 33.207(c) requires that the certification contain the following language:

> I certify that the claim is made in good faith; that the supporting data are accurate and complete to the best of my knowledge and belief; that the amount requested accurately reflects the contract adjustment for which the contractor believes the Government is liable; and that I am duly authorized to certify the claim on behalf of the contractor.

The certification must contain the essence of the above statements, which means that it must substantially comply with the statutory certification language. Contractors are advised to use the exact language in FAR 33.207(c). Even though allowed, the correction of a defective certification may cause increased costs, attorney's fees, wasted time, and needless effort.

There are four basic elements of a claim certification. First, the contractor's claim must be made in good faith. This means that the contractor believes that its claim is well founded and is not intentionally inflated. It should be noted that while a contractor may modify its legal theories during litigation, it cannot raise new claims not presented to the CO. The contractor and its lawyer should remember that the CDA has provisions and penalties for fraudulent assertions and conduct in the claims process.[31]

Second, the certification requires that the supporting data be accurate and complete. This means that the contractor may not present documentation that it knows is incomplete or inaccurate. This does not require a contractor to make a full evidentiary presentation in its claim, but does require it to provide sufficient information to enable the CO to conduct a meaningful review.[32]

The third element is that the amount requested must accurately reflect the contract adjustment for which the contractor believes the government is liable. This prevents the contractor from knowingly submitting a false claim or one for which the government is not responsible. A contractor may nonetheless revise or modify the claim amount after it has been submitted to the CO. Generally, if a claim was properly certified, the amount may be increased without recertification. A claim that is less than the statutory threshold at the time of submission to the CO may not need to be certified if the increase to more than the statutory threshold is the result of actions of the government. However, no amendment increasing the amount will be permitted without recertification if information was available and should have been known to the contractor at the time it submitted its claim. The burden is on the contractor to prove that the additional information was not reasonably available at the time of the initial claim.

The fourth element is that the person executing the certification be authorized to certify the claim on behalf of the contractor.[33] Before the FCAA, which amended the CDA certification requirements, the issue of who could sign the certification was complex and strictly construed, causing dismissal of numerous appeals for improper certification. The CDA amendments substantially simplified the issue by providing that the person who certifies can be anyone familiar with the project who is authorized by the company to bind it legally with respect to the claim.

4. Defective Certification

The FAR defines a "defective certification" as "a certificate that alters or otherwise deviates from the statutory language in section 33.207(c) or that is not executed by a person duly authorized to bind the contractor with respect to the claim. Failure to certify shall not be deemed to be a defective certification." Congress intended to permit contractors to "cure technically defective certifications to avoid repetition of the entire administrative claims process and waste of judicial resources." Courts and boards, however, have not waived the certification requirement altogether: a contractor still must certify its claim or the court and boards will lack jurisdiction.

5. Contracting Officer Authority

The CO is arguably the most important player in the claims process. The CO deals directly with contractors who have disagreements with the government and with his staff. The CO must try to be objective in resolving the contractor's problems either through negotiation or settlement.[34] If the contractor and the CO are unable to resolve their differences, the contractor must formally present a claim to the CO and request a final decision. Thus, the CO also acts as a quasi-judicial official in rendering final decisions for the government under the Disputes clause of the contract. It is the CO's final decision that begins the appeal process and grants jurisdiction to either a board or the COFC. The contractor must know the authority of the person with whom it is dealing, since not every government representative can consider and resolve claims. Specifically, FAR 33.210 states:

> Except as provided in this section, COs are authorized, within specific limitations of their warrants, to decide or resolve all claims arising under or relating to a contract subject to the Act. . . . The authority to decide or resolve claims does not extend to (a) A claim or dispute for penalties or forfeitures prescribed by statute or regulation that another Federal agency is specifically authorized to administer, settle, or determine; or (b) The settlement, compromise, payment or adjustment of any claim involving fraud.

Thus, the FAR vests COs with broad settlement and decision-making powers, although it precludes them from dealing with claims involving fraud. In fact, FAR 33.209 specifically directs the CO to refer any claim that is suspected of being fraudulent to "the agency official responsible for investigating fraud."

F. Timeliness Requirements

The CDA requires that contractor or government claims against one another relating to a contract be submitted within six years after the claimant has actual knowledge of the basis for the claim.[35] The six-year statute of limitations, however, does not apply to claims by the government against a contractor for fraud.[36]

The Changes clause provides that no claim may be submitted after final payment is received. The other remedy-granting clauses in federal contracts contain similar provisions. The standard form for final payments under contracts permits contractors to list outstanding or even "to be submitted" claims, and contractors should complete the form and identify those claims in order to preserve them.

G. The CO's Final Decision

1. Timeliness of Final Decision

Under the CDA, the CO must issue a final decision on the contractor's claim within 60 calendar days of its receipt of the claim when it is less than $100,000. If the amount of the claim exceeds $100,000, then the CO must either issue a final decision or inform the contractor when a decision will be issued within 60 days of claim submission. In any event, the decision must be issued in a "reasonable time." If the CO fails to issue a final decision within the specified time limits, the contractor may consider the claim as "deemed denied" and file an appeal with the board or a suit in the COFC.[37] If an appeal is filed or a suit is commenced without a final decision, the board or court may suspend the proceedings for a time pending a final decision from the CO.[38] It should be noted that, if the contractor initially does not submit a valid claim, the fact that the CO subsequently issues a final decision does not confer jurisdiction on the court or board.

2. Form and Content of the Final Decision

The CDA provides that the final decision must state the reasons for the decision and inform the contractor of its rights. Specific findings of fact are not required, but, if they are made, they are not binding in any subsequent proceeding.[39] The FAR imposes certain obligations on the CO when a claim by or against a contractor cannot be satisfied or settled by mutual agreement.

Under the CDA, the authority to issue a final decision is granted solely to the CO, who is charged with exercising independent, unbiased judgment in reaching the final decision. While the CO can obtain advice or assistance from others, the CO must make a "personal and independent" judgment on the merits of the contractor's claim. If the contractor suspects that the CO has not made an independent decision but is reflecting the decision of a higher authority or others, then the contractor may challenge the decision and, if proven correct, have it voided. The FAR also requires that the final decision describe the contractor's rights with respect to the appeal process available to the contractor. The language "announcing" a final decision is "almost" mandatory. If it is not included in the CO's letter, the contractor may not assume that the letter is a final decision.

3. *Effect of Final Decision*

The CDA provides that the CO's decision on a claim is final and conclusive and not subject to review by any forum, tribunal, or government agency unless an appeal or suit is timely commenced as provided therein. Thus, if the contractor fails to file an appeal within the statutorily prescribed time periods, the CO's decision will be final, and the contractor will be without further recourse. Upon the issuance of a final decision with an amount determined as payable to the contractor, the amount should be paid without waiting for the contractor to appeal. Thus, in those cases in which a CO finds partial entitlement for the contractor, the contractor should invoice for the amount awarded while appealing to recover the remainder.

4. *Timeliness of Appeal of the CO's Final Decision*

Under the CDA, a contractor may elect to appeal the CO's final decision to the applicable board within 90 days of receipt of the final decision or file an action in the COFC within one year of receipt of the final decision. The limitations period for both forums begins on the same date, that is, the date the decision is physically delivered to the contractor's address.

Notes

1. See Chapters 8 ("Changes"), 9 ("Differing Site Conditions"), and 19 ("Delays, Suspension of Work, Acceleration, and Disruption").

2. FAR 52.243-4.

3. FAR 52.243-4(e).

4. For further discussion, see Chapter 8.

5. For further discussion of the notification of change requirement in the Changes clause, see Chapter 8.

6. DFARS 243.204-70 and 252.243-7002 (implementing 10 U.S.C. § 2410(a)).

7. 31 U.S.C. §§ 3729, et seq.; *see, e.g.,* United States ex rel. Wilkins v. N. Am. Constr. Corp., 101 F. Supp. 2d 500, 523–24 (S.D. Tex. 2000), *modified*, 173 F. Supp. 2d 601 (S.D. Tex. 2001) (holding that the government adequately stated an FCA claim against a contractor and its subcontractors where the government alleged that

they had falsely stated in their REA that the government was liable for costs related to differing site conditions). For further discussion of false claims issues, see Chapter 24.

8. 41 U.S.C. §§ 601–613.

9. 811 F.2d 586 (Fed. Cir. 1987).

10. RSH Constructors, Inc. v. United States, 14 Cl. Ct. 655 (1988); *distinguished by* Appeal of Bridgewater Constr. Corp., VABCA No. 2866, et seq., 90-2 BCA ¶ 22,764; *see also* Paragon Energy Corp. v. United States, 645 F.2d 966, 971, 227 Cl. Ct. 176, 186 (1981); *distinguished by* Parsons Transp. Group v. United States, 84 Fed. Cl. 779 (2008) ("A contractor cannot obtain CDA review of an agency's denial of a request for relief under Public Law No. 85-804. Here too, a claim based solely and directly on Public Law 85-804 is not before the Court. Rather, Parsons's claim is for a breach of contract provision that happened to have been based on authority conferred to the FRA under Public Law No. 85-804. The CDA covers Parsons's claims. 28 U.S.C. § 2501 does not apply. Defendant's Motion to Dismiss must be denied.")); Cubic Corp. v. United States, 20 Cl. Ct. 610 (1990) (Claims Court had a four-prong test to determine whether a claim is valid under the CDA).

11. H.L. Smith v. Dalton, 49 F.3d 1563 (Fed. Cir. 1995) (invoices, detailed cost breakdowns, and other supporting financial documentation need not accompany a CDA claim as a jurisdictional prerequisite); *see also* Transamerica Ins. Corp. v. United States, 973 F.2d 1572 (Fed. Cir. 1992); Kanag'Iq Constr., 51 Fed. Cl. 38 (2001).

12. Ebasco Envtl., ASBCA No. 44547, 93-3 BCA ¶ 26,220 (ASBCA stated that the government cited no authority for the proposition that the contractor itself must sign the claim. However, the ASBCA was careful to state that its holding did not address the question as to whether the contractor's lawyer could certify the claim on behalf of the contractor.).

13. Constr. Equip. Lease Co. v. United States, 26 Cl. Ct. 341 (1992).

14. FAR 52.233-1; *see also* Essex Electro Eng'rs, Inc. v. United States, 960 F.2d 1576, 158 (Fed. Cir. 1992); Van Elk, Ltd., ASBCA No. 45311, 93-3 BCA ¶ 25,995; *overruled on other grounds by* Reflectone, Inc. v. Dalton, 60 F.3d 1572 (Fed. Cir. 1995).

15. Hamza v. United States, 31 Fed. Cl. 315 (1994); United Tech. Corp., ASBCA Nos. 46880, et al., 96-1 BCA ¶ 28,226.

16. Corbett Tech. Co., ASBCA No. 47742, 95-1 BCA ¶ 27,587; *see also* Metric Constr. Co. v. United States, 14 Cl. Ct. 177 (1988); *distinguished by* J. Leonard Spodek, PSBCA No. 3964, 97-2 BCA ¶ 28,995 ("Respondent cites Corbett Technology Co., Inc., ASBCA No. 47742 for the proposition that Appellant's letter of November 16 does not satisfy the criteria of demanding a 'sum certain.' These cases are inapposite. In both of these cases the contractors claimed 'in excess of,' or 'in an amount exceeding' a specified sum, but without the further caveat that they would settle for a specific sum, as Appellant has done herein. Lacking a specified sum being claimed, the board and court dismissed the appeals."); AEC Corp., Inc., ASBCA No. 42920, 03-1 BCA ¶ 32,071.

17. *See, e.g.,* P.J. Dick Inc. v. Gen. Servs. Admin., GSBCA No. 11783, 94-3 BCA ¶ 27,172; Prod. Corp., DOTBCA No. 2424, 92-2 BCA ¶ 24,796.

18. *See* United Techs. Corp., ASBCA Nos. 47166, et. al., 96-1 BCA ¶ 28,226 (even though amount was stated "in approximate" amounts, court stated that contracting officer by simple mathematical calculations could determine sum

certain from the back-up documents filed with the claim); *but compare* Van Elk, Ltd., ASBCA No. 45311, 93-3 BCA ¶ 25,995 (board held amount in claim stated "in approximate" amount not a sum certain).

19. FAR 52.233-1; *see, e.g.,* Summit Contractors, AGBCA 81-136-1, 81-1 BCA ¶ 14,872 (Section 907 of the Federal Courts Administration Act of 1992, Pub. L. No. 102-572, amended 28 U.S.C. § 1491(a)(2), specifically grants jurisdiction to the COFC over nonmonetary claims "concerning claims of termination of a contract, rights in tangible or intangible property, compliance with cost accounting standards, and other nonmonetary disputes on which a decision of the contracting officer has been issued"); distinguished by John Hackler & Co., AGBCA No. 82-221-1, 85-2 BCA ¶ 17,957 ("This circumstance may be contrasted with cases previously decided by this Board where review of amended complaints did not show that claims had been made to a Contracting Officer.) [citation omitted] In *Summit, supra,* the appeal was also taken for protective purposes and dismissed as premature. While not apparent from the ruling, there in fact had been no claim made to the contracting officer. The letter dated June 17, 1980, alleged by the contracting officer to have been a claim, made no demand for payment or other relief, and was part of a long dialogue between the parties regarding disagreements over the administration of a timber sale.").

20. 41 U.S.C. § 605(a).

21. Transamerica Ins. Corp. v. United States, 973 F.2d 1572 (Fed. Cir. 1992); *see also* All Star Metals, LLC, v. Dep't of Transp., CBCA No. 53, 09-1 BCA ¶ 34,039.

22. 973 F.2d 1572 (Fed. Cir. 1992).

23. *Id.*

24. FAR 52.233-1.

25. Reflectone, Inc. v. Dalton, 60 F.3d 1572 (Fed. Cir. 1995); *declined to extend by* Johnson v. Advanced Eng'g & Planning Corp., Inc., 292 F. Supp. 2d 846 (E.D. Va. 2003) ("*Reflectone* overruled *Dawco* and its progeny by doing away with *Dawco*'s pre-existing dispute rule. *Bill Strong,* relying on *Dawco,* held that the claim at issue there was not a CDA claim because there was no pre-existing dispute when it was submitted to the CO. *Reflectone* reversed this holding, but it did not disturb *Bill Strong*'s further discussion of what constitutes legal, accounting, or consulting costs incurred in connection with the prosecution of a CDA claim against the government.").

26. Caldera v. J.S. Alberici Constr. Co., 153 F.3d 1381 (Fed. Cir. 1998); Servidone Constr. Corp. v. United States, 931 F.2d 860 (Fed. Cir. 1991; *distinguished by* Raytheon Co. v. White, 305 F.3d 1354 (Fed. Cir. 2002) ("*Raytheon* asserted they should be awarded prospective costs. *Raytheon* is correct that interest may not be denied merely because costs later found due had not been incurred at the time the claim was filed. In both *Servidone* and *J.S. Alberici,* however, the contractors completed their contract and thus actually incurred the costs upon which interest was later awarded. The Court never held that Section 611 permits interest to accrue on costs that, because of the termination of the contract, were never actually incurred by the contractor.").

27. 49 F.3d 1541 (Fed. Cir. 1998).

28. Federal Court Administration Act of 1992, Pub. L. No. 102-572, 106 Stat. 4506.

29. *See* Arnold M. Diamond, Inc. v. Dalton, 25 F.3d 1006 (Fed. Cir. 1994).

30. FAR 33.207.

31. 41 U.S.C. § 604; FAR 33.209. For further discussion, see Chapter 24.

32. Lee Ann Wyskiver, PSBCA No. 3621, 94-3 BCA ¶ 27,118.

33. 41 U.S.C. § 605(c)(1); FAR 33.207(c), (e).

34. 41 U.S.C. § 605(a); *see also* FAR 33.204.

35. 41 U.S.C. § 605(a).

36. *Id.*

37. 41 U.S.C. § 605(c)(5); *see* H. L. Smith, Inc. v. Dalton, 49 F.3d 1563 (Fed. Cir. 1995); Cincinnati Elecs., 32 Fed. Cl. 1496; Wyskiver, PSBCA No. 3621, 94-3 BCA ¶ 27,118; The Maxima Corp., EBCA No. C-9208139, 93-1 BCA ¶ 25,545; Cont'l Mar. of San Diego, ASBCA No. 37820, 89-2 BCA ¶ 21,694.

38. 41 U.S.C. § 605(a).

39. FAR 33.211(a).

CHAPTER 16

Litigating with the Federal Government

ADRIAN L. BASTIANELLI, III
LORI ANN LANGE

I. Introduction

The federal government awards more construction contracts than any other single owner in the country. The government has elected to include remedy-granting clauses, such as the Changes clause and the Differing Site Conditions clause, in its contracts, and it has established fair disputes procedures for resolving claims arising under its contracts. As a result, the large number of claims filed with the federal government represents a practice of law in and of itself. This chapter will summarize the procedures and pitfalls faced when litigating a contract claim with the federal government.

II. Forums in Which to Pursue the Claim[1]

The Contract Disputes Act of 1978 (CDA) provides the contractor with the option to pursue its claim in either the Court of Federal Claims or the appropriate agency board of contract appeals.[2] The government has no input into the contractor's forum selection.

A. The Court of Federal Claims

The Court of Federal Claims (COFC) is an Article I court created by the U.S. Federal Courts Improvement Act of 1982.[3] Initially named the U.S. Claims Court, it was renamed the Court of Federal Claims in 1992.[4] The court's jurisdiction encompasses suits for money against the United States, whether founded upon the Constitution, an act of Congress, an executive order, a regulation of an executive agency, a patent, or an express or implied-in-fact contract

---

Section XI, "Attorneys' Fees and Expenses," was adapted by the authors from Chapter 12 of the first edition of this book, which was authored by Sheila C. Stark.

with the United States.[5] In addition, the court has jurisdiction to hear pre-award and post-award bid protests. The COFC is located in Washington, D.C.

Significantly, the government may pursue claims of fraud against the contractor under the federal False Claims Act (FCA) in the COFC by way of a counterclaim in an action originally initiated by the contractor.[6] These cases can come as quite a surprise to the contractor, as the government frequently learns of the alleged violation only while conducting discovery relating to the contractor's CDA claim. It is important to note that an FCA counterclaim is not required to relate to the contractor's claim; rather, the FCA counterclaim need only relate to the underlying contract. As with the FCA, the COFC is also authorized to hear government claims under the Forfeiture Statute.[7] The Forfeiture Statute requires the forfeiture of all claims arising under a contract tainted by fraud against the government. (*See* Chapter 24 for detailed discussions of the FCA and the Forfeiture Statute.) This is an important distinction between the COFC and the boards of contract appeals. The boards are not authorized to hear claims of fraud under either the FCA or the Forfeiture Statute.

B. Boards of Contract Appeals

There are two boards of contract appeals that hear contract disputes arising out of contracts awarded by the executive agencies: the Armed Services Board of Contract Appeals (ASBCA) and the Civilian Board of Contract Appeals (CBCA).[8] Both boards are headquartered in the Washington, D.C. metropolitan area. The boards' jurisdiction is limited to appeals from contracting officers' final decisions.[9]

The ASBCA has jurisdiction to decide any appeal from a decision of a contracting officer (CO) of the Department of Defense, the Department of the Army (including the Corps of Engineers), the Department of the Navy, the Department of the Air Force, or the National Aeronautics and Space Administration, relative to a contract made by that department or agency.[10] By agreement, the ASBCA also has jurisdiction to hear appeals from the Washington Metropolitan Area Transportation Authority (WMATA).

The CBCA, which was established effective January 6, 2007, has jurisdiction to decide any appeal from a decision of a CO of any executive agency (other than those within the ASBCA's jurisdiction, plus the U.S. Postal Service, the Postal Regulatory Commission, and the Tennessee Valley Authority), relative to a contract made by that agency.[11] Prior to the CBCA's establishment, contract disputes involving the civilian executive agencies were decided by separate boards of contract appeals for the General Services Administration and the Departments of Agriculture, Energy, Housing and Urban Development, Interior, Labor, Transportation, and Veterans Affairs.

The U.S. Postal Service, a non-executive agency, has its own board of contract appeals—the Postal Service Board of Contract Appeals (PSBCA). This chapter does not address practice before the PSBCA.[12]

III. Implied Contracts

The Tucker Act and the CDA give the COFC the right to hear suits on implied contracts as well as express contracts.[13] The court's jurisdiction over implied contracts is limited to implied-in-fact agreements and not those implied by law.[14] In other words, the court does not have jurisdiction over quantum meruit claims for unjust enrichment.[15]

Under the CDA, the boards' jurisdiction includes some implied agreements,[16] and is concurrent with that of the COFC.[17] Thus, like the COFC, the boards do not have jurisdiction over implied-in-law contracts.[18]

IV. Six-Year Limitation on Appeals

In 1994, the CDA was amended to add a provision stating that each claim by a contractor against the government relating to a contract shall be submitted within six years after the accrual of the claim.[19] A claim accrues on the date when all events that fix the alleged liability and permit assertion of the claim were known or should have been known.[20] While some injury must have occurred, monetary damages need not have been incurred in order to fix liability. Similarly, with the exception of claims for fraud, the government must submit its claims to the contractor within six years from accrual.[21] The six-year limitation period applies to contracts awarded on or after October 1, 1995.[22]

The ASBCA and CBCA have held that this six-year limitation period is not a statute of limitations but rather is a time limit that is a prerequisite to the board's jurisdiction.[23] Under Section 605 of the CDA, the submission of a proper claim to the CO and a CO's final decision (or deemed denial) are prerequisites to an appeal either to the boards of contract appeals or to the COFC.[24] Without the submission of a proper claim, the CO does not have the authority to issue a final decision and hence there is no final decision to form the basis for an appeal. Section 605(a) of the CDA is the key provision in determining whether there is a proper claim. Since the six-year limitation is contained in Section 605 instead of the CDA sections establishing filing periods at the boards of contract appeals and the COFC, the boards have concluded that the six-year limitation is a jurisdictional requirement as opposed to a statute of limitations.[25]

V. Time to Appeal

The time in which the contractor has to appeal a CO's unfavorable final decision depends on the forum selected. If the contractor opts to go to the board, it must file its notice of appeal within 90 calendar days after receipt of the CO's final decision.[26] An appeal is filed when it is placed in the U.S. mail.[27] If the contractor uses a method of delivery other than the U.S. Postal Service, the notice of appeal is filed when the board receives it.[28] If the ninetieth day falls on a weekend or federal holiday, the deadline rolls over to the next business day.[29]

To appeal a final decision to the COFC, the contractor must file suit with the court. The allowable time for bringing suit in the COFC is far longer than the time for appeal of a decision to a board. The court action must be commenced by filing a complaint within 12 months of receipt of the CO's final decision.[30] The date of the court's receipt of the complaint generally controls, as compared to the date of mailing.[31] The time requirements, however, can be met if the complaint is sent by certified or registered mail in sufficient time to be received before the due date, and the contractor exercised no control over the complaint after mailing.[32]

Filing a timely notice of appeal with the board or suit in the COFC is a statutory prerequisite to jurisdiction. Thus, neither the board nor the court can waive the contractor's failure to file on time.[33] The final decision, however, must contain language apprising the contractor of its appeal rights and, if it does not contain the language and the contractor detrimentally relies on the defective notice, the final decision is valid and the time period in which to file an appeal does not begin to run.[34]

VI. Binding Election of Forum and Fragmentation of Claims

Once the contractor elects to proceed before either the board or the COFC and files an appeal or suit, its election is final. The contractor subsequently cannot change forums,[35] nor may it voluntarily dismiss the case and refile in the other forum.[36] The only exception is if the contractor's election was not informed and voluntary. If the contractor was not informed or was erroneously informed of its appeal rights, the contractor's election may not be binding.[37]

If the suit is dismissed because of lack of subject matter jurisdiction, the contractor did not make a binding election and therefore is permitted to subsequently file in the other forum.[38] There is no binding election when there was no valid claim initially.[39] Similarly, there is no binding election when the appeal to the board was untimely.[40]

A contractor may elect to pursue separate claims under the same contract in different forums.[41] The contractor, however, may not split a single claim based on a common set of facts into multiple claims in an attempt to avoid having to certify the claim under the CDA.[42] In order to determine if claims have been fragmented, the court or board will look to see if the same or related evidence controls the claims.[43]

If two or more claims by a contractor are filed in the board and the court, the COFC has statutory authority to order consolidation of the claims before either the board or the court for the convenience of the parties or witnesses, or in the interest of justice.[44] The contractor's preference does not control where the claims will be consolidated.[45] It should be noted that the court can transfer a case to the board even if the 90-day appeal period to the board has expired at the time the contractor filed suit in the court.[46]

VII. Prehearing Procedures

A. Rules of Procedure and Discovery

The COFC is not governed by the Federal Rules of Civil Procedure.[47] Rather, the court has adopted its own rules.[48] These rules, however, are patterned after the Federal Rules of Civil Procedure, and the court may look to decisions under the Federal Rules for guidance.[49] Thus, discovery in the COFC is very similar to discovery in federal district court. The parties can file requests for production of documents and interrogatories and can depose witnesses.

The ASBCA and CBCA also have their own rules of procedure.[50] Like the COFC, the boards are not bound by the Federal Rules of Civil Procedure.[51] Nevertheless, the boards also look to the Federal Rules and decisions thereunder for guidance.[52] The boards permit the use of requests for production of documents, interrogatories, depositions, and requests for admissions.[53] The board is authorized to issue subpoenas.[54] The subpoenas are enforceable through the federal district court at the location of the witness.[55] The boards encourage the parties to engage in voluntary discovery before seeking an order from the board.[56]

B. Appeal File

One unique feature of litigating at the board is the use of an appeal file, also known as a "Rule 4" file. After the contractor files its notice of appeal, the CO, within 30 days, must assemble and send to the board and the contractor all documents pertinent to the appeal, including the contract, the claim, and all relevant correspondence between the parties.[57] The contractor then has 30 days to supplement the appeal file with any documents it considers important.[58] Either party may object to the inclusion of such documents. The appeal file becomes part of the record and is used by the board in rendering its decision.[59]

C. Motions Practice

The COFC and the boards of contract appeals will entertain motions filed by either party, including motions to dismiss and motions for summary judgment. Like other forums, the COFC and the boards of contract appeals will grant summary judgment only when there is no genuine issue of material fact and the moving party is entitled to judgment as a matter of law.[60] All justifiable inferences and presumptions are made in favor of the non-moving party.[61]

There is a perception that the COFC is far more likely to decide a case on a motion for summary judgment based on a legal issue. Board decisions generally are more fact-oriented. If the case involves a legal issue and the facts are not in dispute, the contractor may prefer to proceed in the COFC, where the case likely would be decided by an early motion for summary judgment.

Contractors should be cautioned, however, that the Department of Justice (DOJ) has a reputation for attempting to dispose of cases on procedural issues or for alleged technical defects in the claims submission. These motions may result in the case being dismissed without a hearing on the merits. Even if the contractor defeats the motion, it will still incur the cost of filing briefs, preparing affidavits, and having oral argument (where required).

VIII. The Hearing

A. Accord to Be Given the Contracting Officer's Final Decision

The findings of fact and conclusions of law in the CO's final decision have no binding effect in a hearing at the board or court. The proceedings before the court or board are de novo.[62] In the past, decisions of the CO in favor of the contractor were treated as presumptively correct on appeal. The Federal Circuit, however, has reversed those decisions and held that both the government and the contractor start with a clean slate before the court or board.[63]

B. The Judges

There are 16 judges and 8 senior judges on the COFC. The judges are appointed by the President for 15-year terms. The judges generally do not have a government contracts or construction background. Further, the court's caseload is not predominantly government contract cases.

The boards are composed of a minimum of three administrative law judges. The ASBCA currently has 18 judges, while the CBCA currently has 16 judges. The judges must be full-time members of the board and may not have other, inconsistent duties.[64]

In contrast to the COFC, the board judges are selected because of their experience in government contracts. By statute, they must have a minimum of five years' experience in public contract law to be eligible for appointment.[65] The majority of board judges come from government agencies.

Unlike the court, the caseloads of the board judges are composed entirely of government contract cases. As a result, board judges generally bring a degree of expertise to the case that may not be equaled at the COFC. There is a perception, however, that some board judges may have a bias in favor of the government because of their past employment by the government. Fortunately, this perception generally does not appear to be supported by the boards' decisions.

The boards normally decide cases using three-judge panels. Only one judge hears the case, and he is assigned the task of initially drafting the decision. The judge who hears the case, however, is not required to prepare or participate in the decision.[66]

In selecting a forum, the contractor needs to consider whether it wants a judge with expertise in government contracts and whether it would be concerned if it were assigned a judge who previously worked for the agency against which it was claiming.

C. Senior Deciding Group/Full Board Consideration

The ASBCA has established a senior deciding group, consisting of the chairperson, vice chairs, and all division heads. The chairperson may refer an appeal to the senior deciding group if the appeal involves unusual difficulty, has significant precedential importance, or raises a serious dispute within the divisions of the ASBCA. Either of the parties also may request that an appeal or an issue in the appeal be referred to the senior deciding group. In addition, either party may file a motion for reconsideration with the senior deciding group after the decision has been issued. The ASBCA chairperson has the discretion to grant any such requests or motions.[67]

The CBCA has provisions for full board consideration of appeals.[68] A party may request full board consideration of an appeal.[69] Such requests are not favored and will be granted only when it is necessary to secure or maintain uniformity of board decisions or the matter to be referred is one of exceptional importance.[70] After a request is made, the board will poll the judges and, if a majority of the judges favors the request, the request will be granted.[71] Alternatively, a majority of the judges may initiate full board consideration of a matter at any time while the case is before the CBCA, no later than the last date on which any party may file a motion for reconsideration or relief from decision or order, or if such a motion is filed by a party, within 10 days after a panel has resolved it.[72]

D. Conduct of the Trial or Hearing

When conducting a trial on the contractor's claim, the COFC is bound by the Federal Rules of Evidence.[73] Indeed, the court conducts the trial in a manner similar to a trial in a federal district court.

The boards, on the other hand, are not bound by the Federal Rules of Evidence. However, they generally consider such rules in making evidentiary rulings.[74] Normally, the boards will receive most, if not all, evidence proffered at the hearing and give it the weight to which it is entitled. Thus, the boards have concluded that hearsay evidence is generally admissible.[75]

Similarly, the Administrative Procedure Act[76] does not apply to board proceedings.[77] The boards, however, do attempt to provide the litigant with procedural due process, including direct and cross-examination of witnesses.

In this respect, the CDA specifically gives the board the right to administer oaths and subpoena witnesses and documents.[78]

In making its forum selection, the contractor's lawyer should review the available evidence to determine whether the contractor or the government will rely on hearsay or other evidence that may be excluded in the COFC, but received by the board.

E. Bifurcation

The boards often bifurcate the issues of entitlement and quantum.[79] In other words, the board first will conduct a hearing and rule on the issue of entitlement alone. If the ruling upholds the claim, it will hold a second hearing on the issue of quantum (damages) if the parties are unable to reach a settlement. Bifurcation can benefit the contractor if the contractor can negotiate quantum with the government once entitlement has been decided. However, bifurcation may not be advantageous where quantum is hotly disputed and requires a second hearing, thereby unnecessarily prolonging the conclusion of the proceedings and ultimate resolution of the claim.

The COFC also has the authority to bifurcate the proceedings. The court, however, is far less likely to bifurcate its cases.

F. Hearing Location

The boards and the COFC are located in the metropolitan Washington, D.C., area. In most instances, the trials and hearings occur at the board's or court's headquarters. However, both the boards and courts will hold trials and hearings at other locations for the convenience of the parties and the witnesses.[80]

G. Trial Lawyers

In the COFC, the DOJ represents the government—not agency counsel. The DOJ usually assigns a career litigator who handles cases primarily before the COFC and the Court of Appeals for the Federal Circuit. The DOJ lawyers, however, generally litigate a wide variety of cases and may not necessarily be expert in government contract law. Typically, the DOJ lawyer's caseload is large, and if a settlement occurs, the settled case is quickly replaced by a new case. As a result, it is often difficult to get the DOJ lawyer to negotiate an early settlement.

Once suit is filed, the DOJ assumes control over the litigation. It has the authority to decide how the case will be litigated and whether the case should be settled.[81] Although the DOJ can proceed over the objection of the agency, this seldom occurs, and the agency usually has significant input into the proceedings.

The agencies have a wide variety of options for selecting trial counsel for litigation before the boards. Most agencies have lawyers dedicated to board litigation.[82] For example, the Judge Advocate General Corps represents the Army. The Navy has a contract litigation unit composed of civilian lawyers. Regardless of how the agency selects its counsel, the government trial lawyer works for the agency, and therefore does not have as much independence as a DOJ lawyer.

For settlement purposes, the introduction of a new lawyer not associated with the agency may provide a benefit. Thus, if the contractor believes that a new look at the case might result in settlement, the COFC offers a real advantage. On the other hand, the contractor may prefer to continue to deal with agency counsel if the contractor has established a good relationship with the agency and believes that discovery or the pressure of litigation is all that is needed to achieve a settlement.

H. Time to Receive a Decision

In the past, the general perception was that the contractor would receive a speedier trial and decision at the boards. This perception has changed, and many government contracts practitioners now believe that the time to trial and decision can be shorter in the COFC. However, neither forum is known for quick resolution of cases. The time it takes to receive a decision still very much depends on the individual judge and the nature of the case.

I. Accelerated and Simplified Procedures

Under the CDA, the boards are required to have an accelerated procedure for handling claims that have a dollar value of less than $100,000.[83] Under the accelerated procedure, the board's decision should be issued within 180 calendar days from the date the contractor elects the accelerated procedure, whenever possible.[84] That election is at the sole discretion of the contractor.[85]

The boards also are required to establish simplified procedures to handle small claims on an expedited basis. Small claims are claims with a dollar value of less than $50,000 for large businesses and $150,000 or less for small businesses.[86] The contractor has the sole discretion to opt for the small claims procedure.[87] Under the small claims procedure, the board's decision will be issued in 120 calendar days from the date of the election whenever possible.[88] However, that determination is final and conclusive and will be set aside only for fraud.[89]

The COFC has no similar procedures to adjudicate small claims. Therefore, the boards' procedures provide a more cost-effective way to handle small claims and may determine where the contractor proceeds.

IX. *Selection of Forum*

The selection of the forum in which to pursue a claim against the federal government frequently is influenced by the lawyer's previous experience.

Often that experience has more to do with the assigned judge than the forum. There are, however, real differences between the COFC and the boards, which should be considered in selecting the forum.

Counsel first should perform the legal research to determine whether the court or the board has been more favorable on the issues of law involved in the case. There can be significant differences in the applicable case law. Also, the lawyer should consider other factors when selecting the forum for a particular case:

- Has the 90 days for appeal from the CO's final decision to the boards of contract appeals expired?
- Is the judge's level of expertise in government contracts a significant issue? In other words, does the case depend on complex factual or legal issues?
- Can the client accept that the judge may have been a former agency counsel?
- Is the length of time to trial and a decision a concern?
- Can the contractor take advantage of the small claims procedure offered by the board?
- Is it preferable to have strict enforcement of the Federal Rules of Evidence or is it preferable to have more relaxed evidentiary rules?
- Is bifurcation of liability and quantum desirable?
- Can the case be disposed of through summary judgment or is a full trial on the merits required?
- What is the likelihood that the government will appeal from an unfavorable decision?
- Is there a concern that the DOJ will search through the contractor's records in order to identify evidence that may support a claim of fraud against the contractor under the False Claims Act or the Forfeiture Statute?

X. Getting Paid

Once a contractor prevails before either the COFC or a board, or it settles, the next step is to get paid. The government may make payments from agency funds already appropriated. To expedite payment when funds are not immediately available, Congress has created the Permanent Indefinite Judgment Fund (judgment fund). To obtain payment from the judgment fund, the following requirements must be met: (1) payment must not be otherwise provided for; (2) the judgment or compromise settlement must be final; (3) the Secretary of the Treasury must certify payment; and (4) the judgment must be one of the types listed in the Judgment Fund Statute.[90]

Payment under the judgment fund generally requires two submissions. One is from the government lawyer requesting payment, transmitting the applicable judgment or compromise agreement. The contractor then submits a statement to the effect that no further appeals will be taken. The agency must reimburse the judgment fund for the amounts paid to the contractor.[91]

As a general rule, the Treasury Department makes payments reasonably promptly once the judgment fund office has all of the necessary paperwork. Inasmuch as the government is responsible for initiating payment under the fund, contractors need to be diligent to make sure the necessary paperwork is promptly prepared and submitted.

XI. *Attorney's Fees and Expenses*

The Equal Access to Justice Act[92] (EAJA or Act) governs the recovery of attorney's fees and expenses arising out of contractual disputes between government contractors and the United States. The EAJA authorizes federal courts and the boards of contract appeals to award attorney's fees and other expenses to prevailing contractors who meet specified size standards when the government's position in the litigation was not substantially justified and there are no special circumstances that would make the award unjust. The Act is a statutory exception to the long-standing "American Rule" that parties to a lawsuit bear their own litigation expenses absent a contractual fee-shifting provision.[93]

Congress enacted the EAJA in 1980 as an attempt to mitigate the harsh impact of the American Rule on small businesses and non-wealthy individuals who successfully brought suit against the government.[94] The premise of the Act is that small businesses and individuals may be deterred from seeking a review of, or defending against, unreasonable government action because of the expense involved.[95] The EAJA, however, is not a mandatory fee-shifting statute.[96] The award of fees is not automatic but rather is within the discretion of the court or board.[97]

A. Requirements for Recovery

In order to be eligible to recover attorney's fees and expenses, the party must satisfy five criteria:

1. The applicant must meet the EAJA's net worth and size limits;
2. The applicant must be a "prevailing party" in a suit against the United States;
3. The government's position must not have been "substantially justified";
4. There are no "special circumstances" that would make an award unjust; and
5. The fee application must be submitted to the court or board within 30 days of final judgment in the action and be supported by an itemized statement from any attorney, agent, or expert witness representing or appearing on behalf of the applicant.[98]

Except for the substantial justification and special circumstances criteria, the applicant bears the burden of establishing that it meets these requirements.[99] The government has the burden to show that its actions were substantially justified or that special circumstances would make the award unjust.[100] The applicant only need allege that the government's position was not substantially justified and that no special circumstances exist.

An applicant's failure to initially allege in its application that the government's position was not substantially justified, however, is not fatal. In *Scarborough v. Principi*,[101] the U.S. Supreme Court held that a timely EAJA attorney fee application could be amended after the 30-day filing period had expired to cure the applicant's initial failure to allege that the government's position in the underlying litigation was not substantially justified. The Supreme Court based its holding on the fact that the no-substantial-justification allegation did not impose a burden of proof on the applicant and was nothing more than a "think twice" pleading requirement. The burden of establishing that the government's position was substantially justified is on the government.

Because the EAJA is viewed as a partial waiver of sovereign immunity, the eligibility requirements are strictly construed.[102] The courts and boards do not have the discretion to expand liability beyond the parameters of the EAJA in order to do equity.[103] While the EAJA is interpreted narrowly, this narrow construction cannot be used to avoid liability in all cases.[104] Thus, applicants have been permitted to amend timely applications to correct procedural defects in the application.[105]

B. Net Worth and Size Limits

The EAJA establishes precise eligibility standards for the recovery of fees and costs.[106] These standards are applied as of the time when the party filed the civil action or initiated the adversary adjudication.[107] An applicant who exceeds the net worth and size limits is ineligible for an EAJA award.[108]

An individual is eligible if his net worth does not exceed $2 million.[109] A business entity is eligible if it has a net worth not exceeding $7 million and no more than 500 employees.[110] The net worth and size of an applicant's affiliates will only be aggregated with the applicant when the underlying litigation substantially benefited another party, or if the applicant was not the real party in interest to the underlying litigation.[111] Net worth is calculated by subtracting total liabilities from total assets.[112]

The applicant bears the burden of proving that it meets the net worth and size limits.[113] An individual can prove his net worth through the submission of tax returns and property inventories. A business entity can establish its net worth through balance sheets or financial statements. Any information submitted must comport with generally accepted accounting principles.[114] Self-serving, nonprobative affidavits are insufficient to prove an applicant's net worth.[115]

C. Prevailing Party

In order to recover under the EAJA, the applicant must be a "prevailing party" in its litigation against the United States.[116] In order to be a prevailing party, an applicant must satisfy a three-part test:

1. The applicant must have accomplished at least some of the results sought to be achieved by the lawsuit;
2. The proceedings must involve an alteration in the legal relationship of the parties; and
3. The results must be effected by a necessary judicial imprimatur on the change.[117]

To meet this standard, the applicant must succeed on a significant issue in the litigation that achieves at least some of the benefits sought in the litigation.[118] So an applicant who achieves less than total victory still may qualify as a prevailing party. Indeed, an applicant who fails to obtain a judgment on the merits in its favor still may qualify as a prevailing party if the judgment reflects the applicant's success on a significant issue in controversy, such as a reduction in the amount of damages sought by the government.[119]

An applicant also may be deemed "prevailing" under the EAJA if it obtains a favorable settlement of its case.[120]

It is not sufficient that the applicant accomplish some of the results sought. The results must be caused by a judicial action. Enforceable judgments on the merits and court-ordered consent decrees (or their equivalents) reflect sufficient judicial action to support an EAJA award.[121] Preliminary injunctions, which merely preserve the status quo pending the litigation and do not materially alter the legal relationship of the parties, do not have sufficient judicial imprimatur to justify an EAJA award.[122] Similarly, a remand order generally lacks sufficient judicial imprimatur to justify an EAJA award unless the remand is to an administrative agency.[123] When the trial court remands the case to the agency, the remand grants relief on the merits sought by the applicant, and the trial court does not retain jurisdiction, then securing of the remand order is success on the merits.[124] If the trial court retains jurisdiction despite the remand order, the applicant is a prevailing party only if it succeeds before the agency.[125]

D. The Government's Position Was Not Substantially Justified

An EAJA applicant cannot prevail if the government's position was substantially justified. The government's position is substantially justified if it is justified in substance or in the main, i.e., justified to a degree that could satisfy a reasonable person.[126] The entirety of the government's conduct, including its conduct at the agency level, will be considered in determining whether the government's position was substantially justified.[127] As the COFC explained in *Metric Construction Co. v. United States*:[128]

The test for whether the government's position during the dispute was substantially justified is whether that position was reasonable. *Pierce v. Underwood*, 487 U.S. 552, 565, 108 S. Ct. 2541, 101 L. Ed.2d 490 (1988) (defining the EAJA term "substantially justified" to mean "justified to a degree that could satisfy a reasonable person"). In other words, the government's position must have had "a reasonable basis in law and fact" for the "substantial justification" defense to succeed. *Id.* at 566 n.2, 108 S. Ct. 2541. Both the agency action or inaction and the litigation conduct of the government are considered to constitute a singular position, to be reviewed as a whole. See [*Commissioner, INS v.*] *Jean*, 496 U.S. [154] at 159, 110 S. Ct. 2316 [(1990)] (stating that "the court need make only one finding about the justification of [the government's] position"); *Doty v. United States*, 71 F.3d 384, 386 (Fed. Cir. 1995) (stating that in EAJA disputes, "the term 'position of the United States' refers to the government's position throughout the dispute, including not only its litigating position but also the agency's administrative position") (citation omitted); *Chiu v. United States*, 948 F.2d 711, 715 (Fed. Cir. 1991) ("[T]rial courts are instructed to look at the entirety of the government's conduct and make a judgment call whether the government's overall position had a reasonable basis in both law and fact.").

The fact that the government lost the case does not mean that its position was not substantially justified.[129] Similarly, the government's willingness to settle or concede issues does not establish by itself that its litigation position was not substantially justified.[130] The Supreme Court has stated that substantial justification occurs somewhere between winning the case and being merely undeserving of sanctions for frivolousness.[131] The government's position is not substantially justified when explicit, unambiguous regulations directly contradict that position.[132]

The government bears the burden of proving that its position was substantially justified.[133] To meet its burden of proof, the government generally must make a detailed, factual presentation setting forth its conduct throughout the dispute process. If the government fails to proffer any evidence that its position was substantially justified, the court can either find for the applicant or focus on the actual merits of the government's position as shown by the record before the court.[134] In such instances, however, the court should consider the applicant's expectation that it will be awarded its attorney's fees and expenses unless the government successfully raises a defense of its position.[135]

E. No Special Circumstances

The final exception to EAJA coverage is special circumstances. An applicant is not entitled to recover its attorney's fees and costs when the government proves that special circumstances make an award unjust.[136] The EAJA does not establish any criteria to determine what special circumstances would make an award unjust.[137] Rather, the courts and boards have the discretion to deny an award for equitable considerations.[138]

Special circumstances include evidence that the applicant misused the dispute or appellate process. Mere discourteous or dilatory behavior on the part of an applicant, however, probably is insufficient to meet this exception. For example, in *Ideal Electronic Security Co.*,[139] the ASBCA rejected the government's argument that "special circumstances" existed. The evidence suggested that prior to the appeal, the contractor unreasonably challenged and delayed the government's audit of its proposal, and that the contractor's counsel was discourteous and insulting to the government's auditor. The board concluded that, even if true, these allegations should not bar the contractor from obtaining an award of fees and expenses incurred after the filing of the appeal.[140]

Under the EAJA, the court or board may reduce or deny an EAJA award if the prevailing party unduly and unreasonably protracted the litigation.[141] Thus, the ASBCA declined to award fees and expenses related to a contractor's discovery and subpoena requests that violated board rules and orders.[142]

F. Timely Application

In order to recover attorney's fees and expenses, an applicant must submit an application to the court or board.[143] The application must be submitted to the court or board within 30 days of final judgment in a civil action or final disposition in adversary adjudication before a board.[144]

For civil actions, final judgment is a judgment that is not appealable and includes an order of settlement.[145] The 30-day period begins to run on the date of the final judgment and not the date that the applicant receives a copy of that judgment.[146] When a court enters a judgment remanding the case to the agency and does not retain jurisdiction, the time in which to file an application begins with the remand order once it ceases to be appealable.[147] When a trial court remands a case but retains jurisdiction, the application clock begins to run when the trial court has entered its final disposition of the case after remand and that disposition is no longer appealable.[148]

In an unusual case, the Federal Circuit adopted a uniform rule for EAJA applications in the COFC that appeal rights from voluntary dismissals are presumed unless expressly disclaimed or specifically prohibited. In *Impresa Construzioni Geom. Domenico Garufi v. United States*,[149] the COFC sustained the applicant's bid protest but denied the applicant's claim for bid preparation and proposal costs. The applicant filed an appeal with the Federal Circuit, but subsequently filed a motion to withdraw the appeal. The unopposed motion was granted by the Federal Circuit, which issued a final decision. More than 30 days after the Federal Circuit issued its final judgment, the applicant filed its EAJA application with the COFC. The COFC held that the application was untimely and the applicant appealed. The Federal Circuit reversed. Holding that appeal rights from a voluntary dismissal are presumed, the Federal Circuit concluded that the period for filing the EAJA application started on the expiration of the period for filing a petition for certiorari to the Supreme Court from the final judgment of the Federal Circuit.

For adversary adjudications, final disposition occurs on the date the board's decision is no longer appealable to the Federal Circuit.[150] Appeals to the Federal Circuit must be filed within 120 days from the party's receipt of the board's decision.[151] When the case is settled, final disposition occurs when the applicant receives a copy of the board's order of dismissal.[152]

The failure to file a timely application bars recovery.[153] The filing of motion for enlargement of time in which to file an EAJA application does not toll running of this limitation period.[154] In some instances, a timely application may be amended to flesh out or correct information in the application.[155]

G. Recoverable Fees and Expenses

The EAJA provides that the prevailing party is entitled to recover its fees and expenses.[156] The types of recoverable fees and expenses are discussed below.

1. Attorney's Fees

Under the EAJA, the prevailing party is entitled to its reasonable attorney's fees calculated on the basis of the prevailing market rate for the kind and quality of services furnished.[157] The hourly rate for attorney's fees, however, is capped at $125 per hour.[158] This $125-per-hour cap can be exceeded if the court or board determines that an increase in the cost of living or a special factor, such as the limited availability of qualified attorneys for the proceedings involved, justifies a higher fee.[159]

The applicant bears the burden of establishing that special factors warrant an award in excess of $125 per hour.[160] The applicant must demonstrate that, due to the nature of the case—as opposed to the nature of the litigation strategy—the case can only be capably handled by a limited number of attorneys.[161] Expertise in government contracts litigation or bid protests is not the sort of distinctive knowledge or specialized skill that would warrant an award in excess of the $125-per-hour cap.[162]

The $125-per-hour cap may be adjusted upward to account for increases in the cost of living.[163] The applicant must allege that the cost of living has increased as measured by the Department of Labor's Consumer Price Index (CPI) and supply the court with the relevant CPI data.[164] The court or board has discretion to grant a cost of living adjustment (COLA) and such adjustments are the norm.[165]

The base date for calculating the COLA is March 1996—the effective date of the EAJA amendments establishing the $125-per-hour cap.[166] The end date is the date the legal services were finally rendered.[167] The COLA is calculated by multiplying the $125-per-hour rate by the current CPI and then dividing the product by the CPI for March 1996, the base date.[168] As the number of hours of legal fees often varies significantly from month to month, the COLA may be calculated separately for each month that services were provided.[169] When the hours are approximately even, however, the court may use a single COLA rate.[170]

The applicant is only entitled to recover attorney's fees incurred in the civil action or adversary adjudication. The prosecution of a claim before the CO is not a civil action or adversary adjudication. Thus, the applicant cannot recover attorney's fees and expenses incurred in preparing the claim or presenting the claim to the CO.[171] The applicant, however, may be able to recover attorney's fees incurred for legal and factual research prior to filing the complaint or notice of appeal.[172]

The applicant is entitled to recover attorney's fees incurred in preparing the EAJA application.[173]

The applicant has the burden of proving the amount of its attorney's fees. The applicant must provide an itemized statement from its attorney stating the actual time expended and the rate at which the fees were computed.[174] The attorney's hours must be reasonable. Thus, the reviewing court or board may reduce the total number of hours to a figure that the court or board deems reasonable for the dispute at hand.[175]

2. Paralegal Fees

Often attorneys will employ paralegals to assist them with the litigation. Such paralegal fees are recoverable. The Supreme Court has held that recovery is not limited to the attorney's cost for the paralegal services, but rather the prevailing market rate.[176] Similarly, applicants have recovered law clerk fees at market rates.[177]

3. Expert Witness Fees

An applicant may be entitled to recover the reasonable expense of expert witnesses and the reasonable cost of any study, analysis, engineering report, test, or project that the court or board determines was necessary for the preparation of the party's case.[178] Expert witness fees are limited to the highest rate of compensation for expert witnesses paid by the government.[179] Like attorney's fees, expert witness fees must be reasonable and adequately supported by documentation.[180]

4. Expenses

In addition to fees, an applicant also may be entitled to recover its expenses.[181] Common examples of recoverable expenses include:

- Travel
- Taxi fares
- Delivery services
- Telephone and fax charges
- Postage

- Copying
- Computerized research
- Models[182]

The prevailing party has the obligation to provide documentation supporting its costs.[183]

H. Partial Success

Often a case will involve several different issues and an applicant will not prevail on every claim before the court or board. Where the applicant has achieved only partial success, the court or board, in its discretion, may reduce the applicant's award and compute the appropriate recovery based on the applicant's degree of success. Often, the courts and boards will reduce the applicant's claimed amount by the attorney's fees incurred on the unsuccessful claim.[184] Such an apportionment, however, is not mandatory.[185]

XII. *Appellate Review of a Board or Court Decision*

If the government or the contractor believes the COFC's or board's decision is erroneous, it may appeal the unfavorable decision to the U.S. Court of Appeals for the Federal Circuit.[186] Before the government can appeal a board decision, the head of the agency must determine that an appeal should be taken and obtain the approval of the Attorney General.[187]

A. Time to Appeal

If the appeal is from the COFC, notice of appeal must be filed with the clerk of the COFC within 60 days after the judgment becomes final.[188] An appeal from an unfavorable board decision must be filed with the clerk of the Federal Circuit within 120 days from the receipt of the board decision.[189] The time to appeal a board decision, however, is extended by the filing of a timely motion of reconsideration.[190]

B. Standard of Review

The standard of review for findings of fact by the COFC is "clearly erroneous."[191] Conclusions of law are reviewed de novo. The Federal Circuit will review the granting of a motion for summary judgment on a de novo basis, and all factual inferences will be drawn in favor of the party who opposed the motion.[192]

The Federal Circuit applies the "substantial evidence" test to findings of fact in board decisions.[193] Substantial evidence is defined as relevant evidence that a reasonable mind could accept as adequate.[194] Conclusions of law are reviewed de novo.[195]

XIII. Binding Precedent

COFC judges are not bound by other COFC decisions[196] or board decisions.[197] As a result, decisions from different COFC judges can vary on the same legal issue.

Similarly, decisions of the COFC, the other boards, or other panels of the same board do not bind a board of contract appeals.[198] One board may give deference to another board's decisions, but is not bound by them. All ASBCA panels are bound by decisions of the senior deciding group.[199] Similarly, the CBCA is bound by decisions of the full panel.

XIV. Conclusion

Litigation with the federal government is a specialized area of practice with many perils and pitfalls for the unwary. The contractor's selection of the forum in which to litigate its claims can affect the ultimate result of the litigation and therefore must be considered prior to appealing a CO's final decision. By weighing the differences in practice outlined above, the contractor can make an informed decision.

Notes

1. For a discussion of federal forums for government contracts, *see* James F. Nagle & Bryan A. Kelley, *Exploring the Federal Forums for Government Contracts*, 2 JOURNAL OF THE AMERICAN COLLEGE OF CONSTRUCTION LAWYERS 189 (2008).

2. 41 U.S.C. §§ 601–613.

3. Pub. L. No. 97-164; Lockheed Martin Corp. v. United States, 50 Fed. Cl. 550 (2001).

4. Pub. L. No. 104-317, 10 Stat. 3847.

5. 28 U.S.C. § 1491; *Lockheed Martin*, 50 Fed. Cl. 550.

6. 31 U.S.C. § 3730.

7. 28 U.S.C. § 2514.

8. 41 U.S.C. § 607.

9. 41 U.S.C. § 607(d).

10. 41 U.S.C. § 607(d).

11. 41 U.S.C. § 607(d).

12. For information on the PSBCA's rules of practice, see 31 C.F.R. Part 955.

13. 28 U.S.C. §§ 1346(a), 28 U.S.C. § 1491, 41 U.S.C. § 602; Gould Inc. v. United States, 67 F.3d 925 (Fed. Cir. 1995); Johnson Controls World Servs., Inc. v. United States, 44 Fed. Cl. 334 (1999).

14. Hercules, Inc. v. United States, 516 U.S. 417 (1996); Merritt v. United States, 267 U.S. 338 (1925); Aero Union Corp. v. United States, 47 Fed. Cl. 677 (2000); Chaves v. United States, 15 Cl. Ct. 353 (1988).

15. AT&T v. United States, 124 F.3d 1471 (Fed. Cir. 1997), *rev'd en banc*, 177 F.3d 1368 (Fed. Cir. 1999); Cessna Aircraft Co. v. Dalton, 126 F.3d 1442 (Fed. Cir. 1997);

Enron Fed. Solutions, Inc. v. United States, 80 Fed. Cl. 382 (2008); Guardsman Elevator Co. v. United States, 50 Fed. Cl. 577 (2001).

16. 41 U.S.C. § 602(a).

17. 41 U.S.C. § 607(d); Means Co., AGBCA 95-182-1, 95-2 BCA ¶ 27,837; Farmers Grain Co. of Esmond, AGBCA 88-192-1, 92-3 BCA ¶ 25,072; Parking Co. of Am., GSBCA 7654, 87-2 BCA ¶ 19,823.

18. *See generally* Beyley Constr. Group Corp., ASBCA 55692, 08-2 BCA ¶ 33,999; Angel Menendez Envt'l Servs., Inc. v. Dep't of Veterans Affairs, CBCA 19,864, 2007 WL 4270845 (Nov. 28, 2007).

19. Pub. L. No. 103-355; 108 Stat. 3243, codified at 41 U.S.C. § 605(a).

20. FAR 33.201.

21. 41 U.S.C. § 605(a).

22. FAR 33.206.

23. Metlakatla Indian Cmty. v. Dep't of Health & Human Servs., CBCA 181-ISDA, et al., 2008 WL 3052446 (July 28, 2008), citing Greenlee Constr., Inc. v. Gen. Servs. Admin., CBCA 416, 07-1 BCA ¶ 33,514; Gray Personnel, Inc., ASBCA 54652, 06-2 BCA ¶ 33,378.

24. England v. Sherman R. Smoot Corp., 388 F.3d 844 (Fed. Cir. 2004).

25. *Metlakatla*, 2008 WL 3052446; *Gray Personnel*, 06-2 BCA ¶ 33,378.

26. 41 U.S.C. § 606; ASBCA R. 1(a); CBCA R. 2(b).

27. KAMP Sys., Inc., ASBCA 55317, 07-1 BCA 33,460; Butkin Precision Mfg. Corp., ASBCA 41961, 91-2 BCA ¶ 23,858; Micrographic Tech., Inc., ASBCA 25577, 81-2 BCA ¶ 15,357.

28. *KAMP Sys.*, 07-1 BCA ¶ 33,460; Innovative Refrigeration Concepts, ASBCA 48869; 96-1 BCA ¶ 28,231.

29. Lamb Enters., ASBCA 48314, 95-1 BCA ¶ 27,559; Images II, Inc., ASBCA 47943, 94-3 BCA ¶ 27,277; Interstate Constr., Inc., ASBCA 43261, 91-3 BCA ¶ 24,338; Vappi & Co., PSBCA 924, 81-1 BCA ¶ 15,080.

30. 41 U.S.C. § 609(a); W. Coast Gen. Corp. v. Dalton, 39 F.3d 312 (Fed. Cir. 1994); Ramah Navajo Sch. Bd., Inc. v. United States, 83 Fed. Cl. 786 (2008).

31. B.D. Click Co. v. United States, 1 Cl. Ct. 239 (1982).

32. COFC R. 3(b)(2)(C); Ross v. United States, 16 Cl. Ct. 378 (1989); B.D. Click Co. v. United States, 1 Cl. Ct. 239 (1982).

33. Cosmic Constr. Co. v. United States, 697 F.2d 1389 (Fed. Cir. 1982); Computer Prods. Int'l, Inc. v. United States, 26 Cl. Ct. 518 (1992); Kamp Sys., Inc., ASBCA 55317, 08-1 BCA ¶ 33,748; Carlson, ASBCA 48462, 95-2 BCA ¶ 27,880; Target Corp., ASBCA 42041, 91-2 BCA ¶ 23,806.

34. Decker & Co. v. West, 76 F.3d 1573 (Fed. Cir. 1996).

35. Bonneville Assocs., Ltd. v. Barram, 165 F.3d 1360, 1362 (Fed. Cir.), *cert. denied*, 528 U.S. 809 (1999); Frymire v. United States, 51 Fed. Cl. 450 (2002); Diamond Mfg. Co. v. United States, 3 Cl. Ct. 424 (1983); Santa Fe Eng'rs, Inc. v. United States, 677 F.2d 876 (Ct. Cl. 1982); Tuttle White Constr., Inc. v. United States, 656 F.2d 644 (Ct. Cl. 1981); Stewart Thomas Indus., Inc., ASBCA 38773, 90-1 BCA ¶ 22,481.

36. Aviation Transp. Props. v. United States, 11 Cl. Ct. 87 (1986); Prime Constr. Co. v. United States, 231, Ct. Cl. 782 (1982).

37. *Frymire*, 51 Fed. Cl. 450; Spodek v. United States, 51 Fed. Cl. 221 (2001).

38. *Bonneville Assocs.*, 165 F.3d 1360.

39. Skelley and Loy v. United States, 685 F.2d 414 (Ct. Cl. 1982).

40. Nat'l Neighbors, Inc. v. United States, 839 F.2d 1539 (Fed. Cir. 1988); Grinnell v. United States, 71 Fed. Cl. 202 (2006); Olsberg Excavating Co. v. United States, 3 Cl. Ct. 249 (1983).

41. Kanag'Iq Constr. Co. v. United States, 51 Fed. Cl. 38 (2001); Am. Nucleonics Corp., ASBCA 27894, 83-1 BCA ¶ 16,520.

42. Certification is discussed in Chapter 9. See Placeway Constr. Corp. v. United States, 920 F.2d 903 (Fed. Cir. 1990); *Kanag'Iq Constr.*, 51 Fed. Cl. 38.

43. Kinetic Builders, Inc. v. Peters, 226 F.3d 1307 (Fed. Cir. 2000).

44. 41 U.S.C. § 609(d); Precision Pine & Timber, Inc. v. United States, 45 Fed. Cl. 134 (1999); Blount, Inc. v. United States, 15 Cl. Ct. 146 (1988).

45. E.D.S. Fed. Corp. v. United States, 1 Cl. Ct. 212 (1983).

46. Glenn v. United States, 858 F.2d 1577 (Fed. Cir. 1988).

47. U.S. Court of Claims General Order No. 1 (Oct. 7, 1982); First Hartford Corp. Pension Plan & Trust v. United States, 42 Fed. Cl. 599 (1998).

48. The Rules of the U.S. Court of Federal Claims are available at http://www.cofc.uscourts.gov/rules-united-states-court-federal-claims.

49. *See generally* Abbey v. United States, 82 Fed. Cl. 722 (2008); CNA Corp. v. United States, 83 Fed. Cl. 1 (2008).

50. The Rules of the Armed Services Board of Contract Appeals are available at http://docs.law.gwu.edu/asbca/info/pdf/ASBCA%20RULES%202007.pdf. The CBCA's Rules of Procedure are available at http://www.cbca.gsa.gov/.

51. Omni Dev. Corp. v. Dep't of Agric., CBCA 609-C, 07-2 BCA ¶ 33,699; Grumman Aerospace Corp., ASBCA 46834, 48006, 98-1 BCA ¶ 29,591, n.1; Laka Tool & Stamping Co., ASBCA 21338, 84-2 BCA ¶ 17,326.

52. Beyley Constr. Group Corp., ASBCA 55692, 08-2 BCA ¶ 33,999; *Omni Dev.*, 07-2 BCA ¶ 33,699; Copy Data Sys., Inc., ASBCA 44058, 98-1 BCA ¶ 29,390; Elecs. & Space Corp., ASBCA 37352, 95-1 BCA ¶ 27,306.

53. ASBCA R. 14–15; CBCA R. 14–15.

54. 41 U.S.C. § 610; ASBCA R. 21; CBCA R. 16.

55. ASBCA R. 21(g) CBCA R. 16(h).

56. ASBCA R. 21(b) CBCA R. 16(a).

57. ASBCA R. 4(a); CBCA R. 4(a).

58. ASBCA R. 4(b); CBCA R. 4(d).

59. ASBCA R. 4(e); CBCA R. 4(g).

60. See generally, Water Reclaim Sys., Inc., ASBCA 55816, 08-2 BCA ¶ 34,000; Paris Bros., Inc. v. Dep't of Agric., CBCA 932, 08-2 BCA ¶ 33,991.

61. Nat'l Housing Group, Inc. v. Dep't of Housing & Urban Dev., CBCA 340, 341, 09-1 BCA ¶ 34,043; Comptech Corp., ASBCA No. 55526, 08-2 BCA ¶ 33,982.

62. 41 U.S.C. §§ 605(b), 609(a)(3); Wilner Constr. Co. v. United States, 24 F.3d 1397 (Fed. Cir. 1994); Hercules, Inc. v. United States, 49 Fed. Cl. 80 (2001); Safeco Credit v. United States, 44 Fed. Cl. 406 (1999); Reservation Ranch v. United States, 39 Fed. Cl. 696 (1997); Bay Shipbuilding Co. v. Dep't of Homeland Sec., CBCA 54, 07-2 BCA ¶ 33,678; Mass Constr. Group, Inc., ASBCA 55440, 06-2 BCA ¶ 33,439.

63. *Wilner Constr.*, 24 F.3d 1397.

64. 41 U.S.C. § 607(a).

65. 41 U.S.C. § 607(b)(1).

66. Tri-Cor Inc. v. United States, 458 F.2d 112 (Ct. Cl. 1972); Anthony P. Miller, Inc. v. United States, 161 Ct. Cl. 455 (1963); Salzburg Enters. of Cal., ASBCA 29509, 88-1 BCA ¶ 20,377; Massman Constr. Co., ENG BCA 4966, 88-1 BCA ¶ 20,262.

67. Empresa de Viacao Terceirense, ASBCA 49827, 01-1 BCA ¶ 31,243; M.A. Mortenson Co., ASBCA 40750, et al., 98-2 BCA ¶ 29,658; Bell-Boeing Joint Venture, ASBCA 39681, 94-1 BCA ¶ 26,383.

68. CBCA R. 28.

69. CBCA R. 28(a)(2).

70. CBCA R. 28(a)(1).

71. CBCA R. 28(a)(3).

72. CBCA R. 28(b).

73. 28 U.S.C. § 2503(b).

74. Libby Corp., ASBCA 40765, 42553, 96-1 BCA ¶ 28,255; Commercial Box & Lumber Co., ASBCA 47970, 96-1 BCA ¶ 28,026. ASBCA Rule 20 admits into the record evidence that the parties deem appropriate and that would be admissible under the Federal Rules of Evidence *or* admissible in the sound discretion of the presiding judge. CBCA Rule 10 provides that any relevant material evidence will be admitted into the record but that the board may exclude evidence to avoid unfair prejudice, confusion of the issues, undue delay, waste of time, or needless presentation of cumulative evidence. CBCA Rule 10 also states that the board will look to the Federal Rules of Evidence for guidance when making evidentiary rulings.

75. CBCA R. 10 (hearsay is admissible unless the board finds it unreliable or untrustworthy); C.F. Elecs., Inc., ASBCA 43212, 95-2 BCA ¶ 27,719; Maint. Eng'rs, ASBCA 23131, 83-1 BCA ¶ 16,411; Victoreen Instrument Co., ASBCA 14497, 72-2 BCA ¶ 9693.

76. 5 U.S.C. §§ 551–559.

77. McQuiston, ASBCA 24676, 83-2 BCA ¶ 16,602.

78. 41 U.S.C. § 610.

79. *See generally* CBCA R. 21(a)(4).

80. 28 U.S.C. § 173; ASBCA R. 17; CBCA R. 21.

81. 28 U.S.C. § 515.

82. There are some exceptions. For example, the Bureau of Prisons is represented by the Justice Department in litigation before the IBCA.

83. 41 U.S.C. § 607(f). See ASBCA R. 12; CBCA R. 53.

84. 41 U.S.C. § 607(f).

85. *Id.*

86. 41 U.S.C. § 608(a). See ASBCA R. 12; CBCA R. 52.

87. 41 U.S.C. § 608(a).

88. 41 U.S.C. § 608(c).

89. 41 U.S.C. § 608(d).

90. 31 U.S.C. § 1304(a).

91. 41 U.S.C. § 612(c).

92. 28 U.S.C. § 2412 (applicable to civil actions in federal courts) and 5 U.S.C. § 504 (applicable to agency adversary adjudications including board of contract appeals proceedings).

93. The U.S. Supreme Court consistently has given credence to the American Rule, which requires each party to bear its own attorney's fees unless a statute provides otherwise. See Hensley v. Eckerhart, 461 U.S. 424 (1983); Alyeska Pipeline Serv. Co. v. Wilderness Soc'y, 421 U.S. 240 (1975). There are two exceptions to the American Rule. Courts may award attorney's fees to prevailing litigants when the losing party acted in bad faith or when the prevailing litigant's action confers a benefit on a class of people. *See generally* Spencer v. NRLB, 712 F.2d 539 (D.C. Cir. 1983), *cert. denied*, 406 U.S. 936 (1984).

94. See Pub. L. No. 96-481, 28 U.S.C. § 2412 (Supp. IV 1980); see also H.R. Rep. No. 96-1418, at 9–10 (1980), *reprinted in* 1980 U.S.C.C.A.N. 4984, 4986–87. *See also* Cmty. Heating & Plumbing Co. v. Garrett, 2 F.3d 1143 (Fed. Cir. 1993); CEMS, Inc. v. United States, 65 Fed. Cl. 473 (2005); Filtration Dev. Co., LLC v. United States, 63 Fed. Cl. 612 (2005).

95. Gavette v. OPM, 808 F.2d 1456 (Fed. Cir. 1986).

96. *Id.*

97. Comm'r, Immigration & Naturalization Serv. v. Jean, 496 U.S. 154 (1990).

98. *See generally* Libas, Ltd. v. United States, 314 F.3d 1362 (Fed. Cir. 2003); ACE Constructors, Inc. v. United States, 81 Fed. Cl. 161 (2008).

99. *Ace Constructors*, 81 Fed. Cl. 161 (2008).

100. *See* White v. Nicholson, 412 F.3d 1314, 1315 (Fed. Cir. 2005); Hillensbeck v. United States, 74 Fed. Cl. 477, 479–80 (2006); Al Ghanim Combined Group Co. v. United States, 67 Fed. Cl. 494, 498 (2005).

101. 541 U.S. 401 (2004).

102. Ardestani v. INS, 502 U.S. 129 (1991); Fanning, Phillips & Molnar v. West, 160 F.3d 717 (Fed. Cir. 1998); Knowledge Connections, Inc. v. United States, 76 Fed. Cl. 612 (2007).

103. Levernier Constr., Inc. v. United States, 947 F.2d 497 (Fed. Cir. 1991).

104. Massie v. United States, 226 F.3d 1318, 1321 (Fed. Cir. 2000); Metric Constr. Co. v. United States, 83 Fed. Cl. 446 (2008).

105. Filtration Dev. Co., LLC v. United States, 63 Fed. Cl. 612 (2005), citing Bazalo v. West, 150 F.3d 1380, 1382 (Fed. Cir. 1998) (EAJA applicant may supplement its EAJA application to demonstrate eligibility); Cal. Marine Cleaning, Inc. v. United States, 43 Fed. Cl. 724 (1999) (applicant allowed to amend its EAJA application to satisfy the eligibility requirement as well as the "under oath" requirement).

106. 28 U.S.C. § 2412(d)(2)(B); 5 U.S.C. § 504(b)(1)(B)(ii).

107. 28 U.S.C. § 2412(d)(2)(B); 5 U.S.C. § 504(b)(1)(B)(ii).

108. Texas Instruments, Inc. v. United States, 991 F.2d 760 (Fed. Cir. 1993).

109. 28 U.S.C. § 2412(d)(2)(B); 5 U.S.C. § 504(b)(1)(B)(ii).

110. 28 U.S.C. § 2412(d)(2)(B); 5 U.S.C. § 504(b)(1)(B)(ii).

111. Lion Raisins, Inc. v. United States, 57 Fed. Cl. 505 (2003); Info. Sci. Corp. v. United States, 78 Fed. Cl. 673 (2007).

112. *See generally* Scherr Constr. Co. v. United States, 26 Cl. Ct. 248 (1992).

113. Asphalt Supply & Serv., Inc. v. United States, 75 Fed. Cl. 598 (2007).

114. Fields v. United States, 29 Fed. Cl. 376 (1993).

115. Al Ghanim Combined Group Co. Gen. Trad. & Cont. W.L.L. v. United States, 67 Fed. Cl. 494 (2005).

116. 5 U.S.C. § 504(a)(1); 28 U.S.C. § 2412(d)(1)(A).

117. Buckhannon Bd. & Care Home, Inc. v. W. Va. Dep't of Health & Human Resources, 532 U.S. 598 (2001); Brickwood Contractors, Inc. v. United States, 288 F.3d 1371 (Fed. Cir. 2002).

118. Davis v. Nicholson, 475 F.3d 1360 (Fed. Cir. 2007); JGB Enters., Inc. v. United States, 83 Fed. Cl. 20 (2008).

119. Precision Pine & Timber, Inc., 83 Fed. Cl. 554 (2008); Application Under Equal Access to Justice Act of JR & Assocs., ASBCA No. 41377, 92-3 BCA ¶ 25,121.

120. S.J. Thomas Co. v. United States, 47 Fed. Cl. 272 (2000).

121. *Buckhannon*, 532 U.S. 598; Rice Servs. Ltd. v. United States, 405 F.3d 1017 (Fed. Cir. 2005); Ryan v. United States, 75 Fed. Cl. 769 (2007).

122. Advanced Sys. Tech., Inc. v. United States, 74 Fed. Cl. 171 (2006).

123. *Rice Servs.*, 405 F.3d 1017.

124. *Id.*

125. *Id.*

126. Pierce v. Underwood, 487 U.S. 552 (1988).

127. Doty v. United States, 71 F.3d 384 (Fed. Cir. 1995); Manno v. United States, 48 Fed. Cl. 587 (2001); Application Under the Equal Access to Justice Act of Clauss Constr., ASBCA No. 51707, 05-1 BCA ¶ 32,809.

128. 83 Fed. Cl. 446, 449 (2008).

129. Scarborough v. Principi, 541 U.S. 401 (2004); Ace Constructors, Inc. v. United States, 81 Fed. Cl. 161 (2008).

130. Baldi Bros. Constructors v. United States, 52 Fed. Cl. 78 (2002).

131. *Pierce*, 487 U.S. 552. *See also* ACE Constructors, Inc. v. United States, 81 Fed. Cl. 161 (2008).

132. Geo-Seis Helicopters, Inc. v. United States, 79 Fed. Cl. 74 (2007).

133. *Scarborough*, 541 U.S. 401.

134. Metric Constr. Co. v. United States, 83 Fed. Cl. 446 (2008), citing Libas, Ltd. v. United States, 314 F.3d 1362 (Fed. Cir. 2003).

135. *Id.*

136. 5 U.S.C. § 504(a)(1), 28 U.S.C. § 2412(d)(1)(A).

137. Skip Kirchdorfer, Inc. v. United States, 35 Fed. Cl. 742 (1996).

138. H.R. Rep. No. 96-1418, at 11 (1980), *reprinted in* 1980 U.S.C.C.A.N. 4953, 4990.

139. ASBCA 49547, 99-1 BCA ¶ 30,228.

140. *Id.* at 149,545.

141. 28 U.S.C. § 2412(d)(2)(D); 5 U.S.C. § 504(a)(3).

142. Ideal Elec. Sec. Co., ASBCA 49547, 99-1 BCA ¶ 30,228.

143. 28 U.S.C. § 2412(d)(1); 5 U.S.C. § 504(a)(2).

144. 28 U.S.C. § 2412(d)(1)(B); 5 U.S.C. § 504(a)(2). *See also* Melkonyan v. Sullivan, 501 U.S. 89 (1991).

145. 28 U.S.C. § 2412(d)(2)(G).

146. Adam Sommerrock Holzbau, GmbH v. United States, 866 F.2d 427 (Fed. Cir. 1989).

147. Knowledge Connections, Inc. v. United States, 76 Fed. Cl. 612 (2007), citing Shalala v. Schaefer, 509 U.S. 292 (1993).

148. *Id.*

149. 531 F.3d 1367 (Fed. Cir. 2008).

150. ASBCA Equal Access to Justice Act Interim Procs. R. 6(b); CBCA R. 30; Tidewater Contractors, Inc. v. Dep't of Transp., CBCA No. 863-C, 2007 WL 3182519 (Oct. 16, 2007); SAWADI Corp., ASBCA 52973, 01-2 BCA ¶ 31,582.

151. 41 U.S.C. § 607(g).

152. Application Under Equal Access to Justice Act—Ideal Elec. Sec. Co., ASBCA 49547, 99-1 BCA ¶ 30,228, citing J.B. Eng'g Contractors, Inc., ASBCA No. 33390, 88-2 BCA ¶ 20,621.

153. *See* SAI Indus. Corp. v. United States, 421 F.3d 1344 (Fed. Cir. 2005) (EAJA application filed 31 days after the government's 60-day time for appeal expired was untimely).

154. J.H. Miles & Co. v. United States, 3 Cl. Ct. 10 (1983).

155. *See* Application Under the Equal Access to Justice Act of Oscar Narvaez Venegas, ASBCA No. 49291, 04-2 BCA ¶ 32,653 (2004) (deficiency in the applicant's net work statement could be fleshed out or corrected by amendment of its application).

156. 28 U.S.C. § 2412(b); 5 U.S.C. § 504(a)(1).

157. 28 U.S.C. § 2412(d)(2)(A); 5 U.S.C. § 504(b)(1).

158. 28 U.S.C. § 2412(d)(2)(A); 5 U.S.C. § 504(b)(1).

159. 28 U.S.C. § 2412(d)(2)(A); 5 U.S.C. § 504(b)(1).

160. Filtration Dev. Co., LLC v. United States, 63 Fed. Cl. 612 (2005).

161. *Id.*

162. JGB Enters., Inc. v. United States, 83 Fed. Cl. 20 (2008); ACE Constructors, Inc. v. United States, 81 Fed. Cl. 161 (2008); Chapman Law Firm Co. v. United States, 65 Fed. Cl. 422 (2005).

163. 28 U.S.C. § 2412(d)(2)(A)(ii); 5 U.S.C. § 504(b)(1).

164. Geo-Seis Helicopters, Inc. v. United States, 79 Fed. Cl. 74 (2007); Cal. Marine Cleaning, Inc. v. United States, 43 Fed. Cl. 724 (1999).

165. Oliveira v. United States, 827 F.2d 735 (Fed. Cir. 1987); CEMS, Inc. v. United States, 65 Fed. Cl. 473 (2005).

166. Phillips v. Gen. Servs. Admin., 924 F.2d 1577 (Fed. Cir. 1991).

167. Doty v. United States, 71 F.3d 384 (Fed. Cir. 1995).

168. *CEMS*, 65 Fed. Cl. 473 (2005).

169. *Id.*

170. *Id.*, citing Cal. Marine Cleaning, Inc. v. United States, 43 Fed. Cl. 724 (1999).

171. Levernier Constr., Inc. v. United States, 947 F.2d 497 (Fed. Cir. 1991).

172. Universal Fid. LP v. United States, 70 Fed. Cl. 310 (2006).

173. Schuenemeyer v. United States, 776 F.2d 329 (Fed. Cir. 1985).

174. 28 U.S.C. § 2412(d)(1)(B); 5 U.S.C. § 504(a)(2); Tidewater Contractors, Inc. v. Dep't of Transp., CBCA 982-C, 08-2 BCA ¶ 33,908 (2008).

175. St. Paul Fire & Marine Ins. Co. v. United States, 4 Cl. Ct. 762 (1984).

176. Richlin Sec. Serv. Co. v. Chertoff, ___ U.S. ___, 128 S. Ct. 2007 (2008).

177. JGB Enters., Inc. v. United States, 83 Fed. Cl. 20 (2008).

178. 28 U.S.C. § 2412(d)(2)(A); 5 U.S.C. § 504(b)(1)(A).

179. 28 U.S.C. § 2412(d)(2)(A); 5 U.S.C. § 504(b)(1)(A); ACE Constructors, Inc. v. United States, 81 Fed. Cl. 161 (2008).

180. Baldi Bros. Constructors v. United States, 52 Fed. Cl. 78 (2002); Esprit Corp. v. United States, 15 Cl. Ct. 491 (1988).

181. 28 U.S.C. § 2814(b); 5 U.S.C. § 504(b)(1).

182. *JGB Enters., Inc.*, 83 Fed. Cl. 20; Filtration Dev. Co., LLC v. United States, 63 Fed. Cl. 612 (2005); Design & Prod., Inc. v. United States, 20 Cl. Ct. 207 (1990).

183. *ACE Constructors*, 81 Fed. Cl. 161; Asphalt Supply & Serv., Inc. v. United States, 75 Fed. Cl. 598 (2007).

184. *See generally JGB Enters., Inc.*, 83 Fed. Cl. 20; Hubbard v. United States, 80 Fed. Cl. 282 (2008); Allen Ballew Gen. Contractor, Inc. v. Dep't of Veterans Affairs, CBCA3-C, 07-2 BCA ¶ 33,653 (2007).

185. Naekel v. Dep't of Transp., 884 F.2d 1378 (Fed. Cir. 1989); Loomis v. United States, 74 Fed. Cl. 350 (2006).

186. 41 U.S.C. § 1295(a)(3); 41 U.S.C. § 607(g)(1).

187. 41 U.S.C. § 607(g)(1).

188. 28 U.S.C. § 2107; Sofarelli Assocs. v. United States, 716 F.2d 1395 (Fed. Cir. 1983).

189. 41 U.S.C. § 607(g)(1).

190. Precision Piping, Inc. v. United States, 230 Ct. Cl. 741 (1982).

191. Hughes Commc'ns Galaxy, Inc. v. United States, 271 F.3d 1060 (Fed. Cir. 2001); Milmark Servs., Inc., 731 F.3d 855 (Fed. Cir. 1984).

192. Varilease Tech. Group, Inc. v. United States, 289 F.3d 795 (Fed. Cir. 2002); Cook v. United States, 86 F.3d 1095 (Fed. Cir. 1996); Winstar Corp. v. United States, 64 F.3d 1531 (Fed. Cir. 1995).

193. 41 U.S.C. § 609(b); Int'l Tech. Corp. v. Winter, 523 F.3d 1321 (Fed. Cir. 2008); Boeing N. Am., Inc. v. Roche, 283 F.3d 1320 (Fed. Cir. 2002).

194. Gen. Elec. Corp. v. United States, 727 F.2d 1567 (Fed. Cir. 1984).

195. Franklin Pavkov Constr. Co. v. Roche, 279 F.3d 989 (Fed. Cir. 2002); Motorola, Inc. v. West, 125 F.3d 1470 (Fed. Cir. 1997); Am. Elec. Labs., Inc. v. United States, 774 F.2d 1110 (Fed. Cir. 1985).

196. Pathman Constr. Co. v. United States, 817 F.2d 1573 (Fed. Cir. 1987); F. Alderete Gen. Contractors v. United States, 715 F.2d 1476 (Fed. Cir. 1983).

197. W. Coast Gen. Corp. v. United States, 19 Cl. Ct. 98 (1989); Universal Restoration v. United States, 16 Cl. Ct. 214 (1989); Structural Finishing, Inc. v. United States, 14 Cl. Ct. 447 (1988).

198. Hettich, GmbH, ASBCA 42602, 42604, 93-3 BCA ¶ 26,035; Cedar Lumber, Inc., AGBCA 85-214-1, 85-3 BCA ¶ 18,346, *rev'd on other grounds*, 779 F.2d 743 (Fed Cir. 1986).

199. Gaddell Constr. Co., ASBCA 49333, 99-1 BCA ¶ 30,702.

CHAPTER 17

Alternative Dispute Resolution

ADRIAN L. BASTIANELLI, III
LORI ANN LANGE

I. Introduction

The federal government has embraced the use of alternative dispute resolution (ADR) procedures to resolve claims through presidential proclamation, statutes, and regulations and through action at the working level. In 1998, the Interagency Alternative Dispute Resolution Working Group was established to facilitate and encourage the use of ADR.[1] While ADR with the government has some unique features and limitations, it has many similarities with ADR in the private sector.

The government always has attempted to resolve disputes without litigation through negotiation/settlement discussions. The government now also commonly uses mediation, mini-trials, hybrid forms of ADR, and partnering to resolve disputes. Although arbitration is permitted, it is not often used. These ADR procedures and the enabling legislation and regulations are discussed below.

II. Negotiation

A. Negotiating with the Government

The most common form of ADR with the government is negotiation, and most government construction disputes are resolved in this manner. Because of the unique nature of the government and its regulations, however, negotiating with the government is different from negotiating with a private party. Contractors doing business with the government must recognize and account for these differences.

Government negotiators generally are well schooled in the negotiating process.[2] Most of the government negotiators see their jobs as protecting the public interest. Government negotiators must justify in writing any negotiated settlement with the contractor. Seldom is a negotiation process or settlement based solely on business considerations. The contractor must convince the government that it is entitled not only to an equitable adjustment but also to payment of the amount demanded. However, if the contractor is entitled to compensation, the government generally has fewer budget constraints than private owners and seldom refuses to pay a fair amount due solely to its superior bargaining position.

There are various unique factors that play important roles in the negotiation process with the government. These include:

- While some government agencies consider the cost of litigation in the negotiation process, very few factor in the in-house costs of litigation.
- Generally, time is not of the essence with the government, even where a certified claim has been submitted and the government is obligated to pay interest on the claim under the Contract Disputes Act (CDA).[3]
- Many construction contracts contain requirements for the contractor to submit detailed certified cost or pricing data to support the claim.
- The contractor is required to certify that any claim in excess of $100,000 is made in good faith and that the supporting data are accurate and complete.[4]
- Contractors who submit false, inflated, or unsupported claims may be subject to the penalty provisions of the False Claims Act (FCA),[5] including fines and incarceration.

As a result of these factors, a contractor seeking a monetary recovery should submit an accurate, well-documented claim as early as possible. During negotiations, the contractor must convince the government of the correctness of the contractor's position and provide detailed support for its claimed costs. To reach a settlement, the contractor needs to submit adequate documentation to the government that justifies and supports its claim and to provide substantive responses to concerns expressed by the government in the negotiations. Emphasis on business considerations such as likelihood of success, future business, cost of litigation, substantial reductions in price (bargain-basement offer), and the other similar approaches used in private negotiations seldom advance the contractor's settlement prospects.

B. Timing of Negotiation

FAR 33.204 sets forth the government's policy of trying to resolve all contractual issues by mutual agreement at the contracting officer (CO) level. FAR 33.204 further provides that the government should make reasonable efforts to resolve disputes before a claim is submitted. This government policy reflects the conventional wisdom that it is best to resolve disputes by negotiation at the lowest level and earliest practical time whenever possible.

In most cases, the government follows its stated policy of attempting to resolve claims. As a result, it is in the best interest of the contractor to prepare a well-documented request for equitable adjustment (REA) for submission and negotiation at the field level.[6] Unlike private owners, the government generally does not delay settlement of valid REAs until the courthouse steps solely to gain leverage in the negotiations. As a result, the contractor's failure to fully articulate and support its claim at the earliest stage may unnecessarily delay its recovery on the claim.

III. Alternative Dispute Resolution Act

A. Authority to Use ADR

The federal government encourages the use of ADR to resolve disputes. FAR 33.204 specifically provides that agencies are encouraged to use ADR to the maximum extent practicable. So strong is the government's preference for ADR that, if the CO rejects a contractor's written request to use ADR, the CO is required to provide the contractor with a written explanation of why ADR is inappropriate for resolution of the dispute at issue.[7] Similarly, the contractor must provide the agency with its specific reasons for rejecting ADR when the agency requests it.[8]

The Administrative Dispute Resolution Act (ADRA)[9] provides the authority for the government to use ADR. While the Act does not apply to the courts of the United States, the Department of Justice (which represents the government in cases before the courts) has adopted regulations similar to the ADRA regarding the use of ADR in disputes in which it is involved.[10]

In addition, FAR 33.204 and 33.214 provide guidance on the use of ADR. Many agencies also have their own regulations or guidance on ADR.[11] The FAR lists the following essential elements of an ADR:

- Existence of an issue in controversy
- Voluntary election by both parties to use ADR
- Agreement on the procedures and terms to be used in lieu of formal litigation
- Participation in the process by officials of both parties who have the authority to settle the dispute[12]

The ADRA provides that an agency may use ADR to resolve issues in controversy if the parties agree to such proceeding. The agreement of both parties is the only absolute limitation on the use of ADR in the Act. The Act, however, does list instances when the agency may not want to use ADR, including:

- An authoritative decision is needed as future precedent.
- The dispute involves significant questions of government policy that require further development of the law.
- Maintaining established policies, without variation among individual decisions, is of special importance.
- The matter significantly affects persons who are not parties to the proceedings.

■ A full public record of the proceeding is important.
■ The agency has a significant need to maintain jurisdiction over the dispute.

B. Timing of ADR

The ADRA does not limit the time frame for use of ADR. ADR, therefore, can be used prior to submission of a claim or after the complaint is filed.

Obviously, the earlier a dispute is resolved, the lower the cost of the litigation. This should encourage the parties to hold the ADR proceeding at the earliest possible time. An ADR proceeding, however, can occur too early in the case and fail as a result. Ideally, both parties should have performed sufficient discovery and preparation for their cases, often including consultation with experts, so they both know and understand the strengths and weaknesses of their cases. A party that has not done its homework may have unreasonably high expectations and/or may be unable to mount a persuasive attack on the other party's case in order to reduce that party's expectations.

C. Confidentiality

Confidentiality in private ADR proceedings is the norm. Confidentiality in proceedings of any kind with the government, however, is troublesome because of the disclosure requirements of the Freedom of Information Act (FOIA).[13] The ADRA, however, contains provisions that offer protection against disclosure of certain communications in an ADR proceeding.[14]

The ADRA states that the ADR neutral shall not disclose or *through discovery or compulsory process be required to disclose* dispute resolution communications provided in confidence to the neutral unless:

■ The parties agree.
■ The communication already has been made public.
■ The communication is required by statute to be made public.
■ A court determines that such testimony or disclosure is necessary to prevent manifest injustice, help establish a violation of law, or prevent substantial harm to the public health or safety.[15]

By its own terms, the FOIA does not apply to matters that are exempt by statute from disclosure *if* the statute allows no discretion in whether to disclose the documents or establishes criteria for withholding documents.[16] The ADRA, however, specifically exempts communications between the neutral and a party from this section of the FOIA.[17] As a result, communications between the neutral and a party are not disclosable under the FOIA. The ADRA only prevents disclosure of communications between the neutral and one of the parties. It does not protect communications either between the neutral and more than one party or communications between the parties.

Under ADRA, the parties are prevented from disclosing dispute resolution communications with the same exceptions as the neutral.[18] The ADRA, however, adds the following exceptions to disclosure by a party:

- The communication was prepared by the party disclosing it.
- The communication is relevant to determining the existence or meaning of an agreement resulting from the proceeding or to enforce the agreement.
- Except for communications by the neutral, the communication was provided to or was available to all parties to the proceeding.

The last exception is the most important one. Communications, including documents, provided to the other party are not subject to the statutory confidentiality provisions. The parties may try to protect these communications through a confidentiality provision in the ADR agreement, but the agreement might not survive a FOIA challenge. The ADRA specifically allows the parties to revise the confidentiality requirements for disclosure by the neutral but does not address revisions relating to disclosure by the parties.[19] If the ADR agreement reduces the confidentiality of disclosures by the neutral, the parties lose the protection of the exemption from the FOIA exemption restrictions.[20]

If a demand for disclosure is made on the neutral through discovery or other legal process, the party who provided the information is obligated to agree in writing to defend the neutral within 15 days of receipt of notice, and, if the party does not so defend, the neutral is free to disclose the information.[21]

D. The Neutral

Selection of the neutral is probably the most important part of the ADR process. A competent neutral who is respected by all parties can often be the difference between success and failure in the ADR proceeding.

The ADRA provides authority for the government to enter into a contract with any person to provide the services of a neutral so long as the price for the neutral's services is fair and reasonable to the government.[22] The Act also requires the government to develop procedures to retain neutrals on an expedited basis.[23] As a result, the government does not have to procure the services of a neutral through competitive procurement procedures and has the ability to retain neutrals quickly and at the going rate.

The neutral may be an employee of the government, but he cannot have a conflict of interest with respect to the issues in controversy without disclosure of the conflict to the parties.[24] The neutral must be perceived as unbiased, fair, and ethical. The neutral should have training and experience in the type of ADR being used in the dispute. For example, even if a mediator has an excellent grasp of the subject matter in dispute and is perceived to be fair and reasonable, his lack of mediation skills, such as the ability to close or persuade, can doom the mediation to failure.

The ADR neutral also must have the appropriate personality traits. The neutral must have patience, the ability to persuade, the ability to gain the respect of the decision makers, and the ability to take charge and conduct the proceedings. A mediator also must have good negotiation skills. Finally, it generally is beneficial and sometimes necessary for the neutral to have subject matter expertise.

When selecting the neutral, the parties should take the time and make the effort to find the best available neutral for the dispute at issue and the type of ADR procedure selected by the parties.

IV. Types of Government ADR

The ADRA defines ADR as "any procedure that is used to resolve issues in controversy, including, but not limited to, conciliation, facilitation, mediation, factfinding, mini-trials, arbitration, and use of ombudsmen, or any combination thereof."[25] The Act contains no prohibition on the use of any form of ADR.[26]

A. Arbitration

The only form of ADR specifically addressed by the ADRA is arbitration.[27] The ADRA requires that both parties consent to the arbitration.[28] The parties may agree to submit limited issues to arbitration or arbitrate on the condition that the award be within a range of possible outcomes.[29] The arbitration agreement must be in writing and must specify the maximum award that may be issued by the arbitrator.[30] The requirement to specify a limitation on the award presents the parties with an interesting and unique negotiation before agreement is reached on arbitration that the parties in the private sector do not face.

The government employee agreeing to the arbitration must have authority to settle the matter in dispute or specifically be authorized by the agency to consent to arbitration.[31] Binding arbitration can be used only after the head of the agency, in consultation with the Attorney General and considering the factors in Section 572(b) of the ADRA, issues guidelines on the appropriate use of binding arbitration.[32]

Few agencies have specific authorization to enter into binding arbitration. One example is the Federal Aviation Administration (FAA). FAA regulations provide for the use of binding arbitration on a case-by-case basis.[33]

If arbitration is permitted, the agreement to arbitrate is enforceable under the Federal Arbitration Act.[34] The rules for the proceedings are set out in 5 U.S.C. § 579. The parties are entitled to present material evidence and to cross-examine witnesses appearing at the hearing.[35] The hearing is conducted in an expeditious and informal manner.[36] The arbitrator may exclude evidence that is irrelevant, immaterial, unduly repetitious, or privileged.[37]

The arbitrator is the interpreter of the statutes, regulations, legal precedents, and policy directives.[38] Ex parte communications, which address information relevant to the merits of the proceeding, are prohibited without the agreement of the other party.[39] The award must be made within 30 days of the close of the hearing or the filing of the briefs, unless otherwise agreed or provided by regulation.[40]

Unless the agency provides otherwise by rule, the award must include a brief, informal discussion of the factual and legal basis for the award; however, formal findings of fact and conclusions of law are not required.[41] The award becomes final 30 days after service on all parties.[42] The agency may extend the 30-day period for an additional 30 days by giving notice to the other party. A final award may be enforced under the Federal Arbitration Act.[43] The award may not serve as an estoppel in another proceeding, may not be used as precedent, and otherwise may not be considered in any other proceeding.[44] The award may be enforced and is subject to judicial review under the Act.[45]

In summary, by statute, the government may agree to arbitrate under certain circumstances. However, the reality is that the government seldom elects to use this form of ADR.

B. Mediation

Mediation is a nonbinding settlement process in which a third-party mediator attempts to facilitate a settlement between the parties. Typically, the parties submit short statements of their positions with supporting documents to the mediator. The mediator may discuss the matter with the parties or their counsel prior to the formal mediation.

Generally, the formal mediation commences with a joint session in which each party presents its position in any format it desires. After the joint session, the parties are separated, and the mediator meets with them individually in sessions referred to as caucuses. The mediator may bring the parties back together in a joint session, meet solely with the principals, meet solely with the lawyers, or proceed in any other fashion that he deems appropriate to achieve a settlement.

Mediators are generally divided into two categories: facilitative and evaluative. A facilitative mediator attempts to facilitate a settlement but does not provide evaluations of the parties' positions. Facilitative mediators do not necessarily need expertise in the subject matter of the dispute to be effective; rather, these mediators rely primarily on their mediation skills.

An evaluative mediator also attempts to facilitate a settlement but, in addition, is willing to provide the parties with some form of evaluation of their respective positions as part of the process. Evaluative mediators must have expertise in the subject matter in dispute to be effective. Such mediators must be careful when and how they provide their evaluations. If the mediator offers an evaluation before the parties have developed a trust in the mediator's opinions, the mediator can alienate a party, which can result in a failed mediation. Similarly, the manner in which the mediator expresses his evaluation can have a significant effect on how it is received and the outcome of the mediation.

One of the most important attributes of mediation is that it is nonbinding. While the mediator may provide an evaluation or suggest a settlement amount through a mediator's proposal, the parties are ultimately in control and make their own decision on whether to settle. Either party or the mediator can terminate the mediation at any time.

As in the private sector, the most common form of ADR used by the government is mediation, but there is a significant difference between mediation with the government and mediation between private litigants. In the private sector, the mediator often attempts to move away from the merits of the case to the business realities of litigation. While this has a place in mediating with the government, the government still must justify any settlement that results from the mediation. As a result, mediation with the government often concentrates more on the merits of the case than business considerations.

Because of this emphasis on the merits of the claim, an evaluative mediator is often better than a purely facilitative mediator for disputes involving the government. In many instances, in order to help justify the settlement internally, the mediator is asked by the government to issue an opinion setting forth the mediator's analysis of the claim and stating why the settlement is beneficial for the government.

C. Mini-trial

In the mini-trial, each party names a high-level executive, preferably a person who has not been involved in the dispute, to be a member of the panel. The government's representative is normally the CO. The parties also select a third-party neutral. The panel hears presentations by the parties of their respective positions. The presentations may be in any form. After completion of the presentations, the executives attempt to negotiate a settlement. If they cannot reach an agreement, the third-party neutral issues an opinion on the claims. The two representatives then try to reach a settlement.

The mini-trial was a favorite ADR method of the Army Corps of Engineers in the early development of government construction ADR. The mini-trial has certain unique features that can make it effective in government contract negotiations. The CO is the person with the ultimate responsibility to decide the claim before the start of litigation and often is not involved in the dispute prior to the mini-trial. In the mini-trial, the CO has an opportunity to hear an abbreviated version of the claim, as it will be presented at the board or in court, and to judge its merits. The negotiations then are commenced between executives at the highest level of the parties, that is, people who have not been involved in the heat of the battle, who have the ability to see the big picture, and who have the complete authority to settle the claim. The third-party neutral provides an independent opinion to help the parties in the event of a stalemate.

The mini-trial process is still in use but now often is combined with mediation to form a hybrid type of ADR. In this form of ADR, top-level executives are appointed to sit jointly with the mediator to hear the parties' positions. The mediator then mediates between the two executives, with the inclusion of other representatives of the parties at various points.

D. Dispute Resolution Board[46]

The Dispute Resolution Board (DRB) in its most common form is a panel of three neutrals selected prior to the start of the work to help resolve disputes. Each party selects a neutral who has expertise in the area of construction being performed. The other party must approve that neutral. The two neutrals then select a third neutral, who must be approved by both parties.

The board visits the job, generally on a quarterly basis, to observe the construction, discuss progress, and review in general potential future disputes. When claims arise, the parties submit brief written presentations and then hold an informal hearing where the parties present their position in any manner they want. Generally, within 30 days of the conclusion of the hearing, the DRB issues a nonbinding recommendation for resolving the dispute.

The DRB process has the advantage of having disputes decided in real time by experts who are approved by the parties and who have seen the construction as it was performed. In addition, the neutrals become a respected part of the construction team, which discourages the submission of frivolous claims and encourages the parties to resolve their disputes prior to bringing them to the DRB.

V. ADR at the Boards of Contract Appeals

The boards of contract appeals (BCAs) have adopted rules governing the use of ADR.[47] Requests for ADR must be made jointly. While the boards encourage the use of ADR, they will not require that a party consent to ADR.

A. The Armed Services Board of Contract Appeals (ASBCA)

Upon receipt of a joint request for ADR, the ASBCA board chairman selects a settlement judge or third-party neutral, normally another judge on the board. This neutral will not discuss the ADR proceedings with other members of the board and will be recused from the matter if it is not settled. Written materials submitted in the ADR proceedings are confidential and may not be admitted into evidence in a subsequent hearing. ADR proceedings are concluded within 120 days.

The ASBCA uses several basic forms of ADR described below:

- Settlement judge: The purpose of the settlement judge is to facilitate a settlement by a discussion of the strengths and weaknesses of each party's case. The settlement judge may meet individually or jointly with the parties to try to reach a settlement. The settlement judge process is essentially a form of evaluative mediation.
- Mini-trial: A third-party neutral judge is appointed to hear an abbreviated version of the case. After the presentation, the parties attempt to reach a settlement with the help of the third-party neutral. If they cannot reach a settlement, the third-party neutral issues a nonbinding decision.
- Summary trial with binding decision: In this form of ADR, the parties present their case in a summary fashion to a board judge, who issues a bench decision at the conclusion of the presentations. The parties must agree that the decision is final, binding, and conclusive on the parties. This procedure is useful in small cases where the cost of a full litigation can be greater than the possible award.
- Other agreed methods: The board specifically allows the parties and board to agree to any other method of ADR.

B. The Civilian Board of Contract Appeals (CBCA)

The CBCA will make its services available for ADR proceedings to help resolve issues in controversy and claims involving procurements, contracts (including interagency agreements), and grants.[48] Joint requests for ADR services for docketed appeals must be addressed to the board chairman, with a copy to the presiding judge.[49] ADR may be used concurrently with litigation or the presiding judge may suspend the litigation for a reasonable period of time while the parties attempt to resolve the appeal using ADR. Upon request, the CBCA also will make an ADR neutral available for an ADR proceeding even if a CO's decision has not been issued or is not contemplated.[50]

The parties may ask the board chairman to appoint a judge or judges to serve as the ADR neutral(s).[51] Alternatively, the parties may request the appointment of a particular judge, who may be the presiding judge if an appeal has been docketed. If the presiding judge is selected as the ADR neutral and the ADR involved ex parte contact, the presiding judge may retain the case for adjudication if ADR is unsuccessful, but only if the parties and judge all agree. If the ADR did not involve ex parte contact, the presiding judge, after considering the parties' views, may retain the case at his discretion.

The CBCA requires the parties to execute a written ADR agreement before ADR can occur.[52] Among other things, the agreement should set forth the identity of the ADR neutral, the role and authority of the neutral, the ADR techniques to be employed, the scope and extent of any discovery relating to ADR, the location and schedule for the ADR proceeding, and the extent to which dispute resolution communications in conjunction with the ADR proceeding are to be kept confidential. Written material prepared specifically for use in an ADR proceeding, oral presentations made at an ADR proceeding, and all discussions in connection with such proceedings are considered "dispute resolution communications" subject to the confidentiality requirements of 5 U.S.C. § 574.[53] Unless the parties specifically agree otherwise, confidential dispute resolution communications are inadmissible as evidence in any pending or future CBCA proceeding. This does not include evidence that is otherwise admissible.

The CBCA uses several basic forms of ADR described below:[54]

- Facilitative mediation: The parties usually will begin with a joint session in which they make informal presentations to one another and the ADR neutral. The ADR neutral, as a mediator, will aid the parties in settling their dispute by meeting with each party separately in confidential sessions and engaging in ex parte discussions with each of the parties for the purpose of facilitating the formulation and transmission of settlement offers.
- Evaluative mediation: The ADR neutral will discuss informally the strengths and weaknesses of the parties' respective positions in either joint sessions or confidential sessions.
- Mini-trial: The parties make abbreviated presentations to the ADR neutral who sits with the parties' designated principal representatives as a mini-trial panel to hear and evaluate evidence relating to an issue in controversy. The ADR neutral may meet with the principal representatives to attempt to mediate a settlement. The ADR neutral may issue a nonbinding advisory opinion or binding decision.

The CBCA also will consider the use of any ADR technique or combination of techniques proposed by the parties in their ADR agreement that is deemed to be fair, reasonable, and in the best interest of the parties, the CBCA, and the resolution of the issue in controversy.

C. Perceived Advantages and Disadvantages of ADR at the Boards

The use of the boards' ADR process has several advantages. The boards' ADR services are free to the parties. The board judge has instant credibility with the parties, especially the government. The judge knows and understands government contract law and will have a good feel for the likely outcome of the case if it goes forward. The summary trial is excellent for small claims.

The use of board judges, however, also has disadvantages. Board members may or may not be good mediators. The fact that the judges are on the board and have good judging skills does not mean that they have the necessary personality traits, skills, negotiation and mediation experience, time, and desire to be good mediators. While confidentiality is required, there is always the possibility that news of what happened in the unsuccessful ADR may leak to the trial judge who hears the appeal. Even if confidentiality is maintained, the client may still have the perception that the outcome of the ADR is known to the trial judge. Finally, payment to a mediator sometimes causes the parties to become vested in the settlement process and helps lead to the success of the process in some cases. As a result, the fact that the boards' ADR processes are free may not be a benefit in the long run.

VI. ADR at the Court of Federal Claims

The Court of Federal Claims (COFC) has ADR procedures very similar to those of the ASBCA.[55] Both parties must agree to use ADR. When the parties inform the presiding judge of their intentions and the presiding judge agrees, the presiding judge forwards the request for ADR to the clerk of the court, who appoints a settlement judge or other third-party neutral agreed to by the parties. The settlement judge or third-party neutral and the parties will develop ADR procedures appropriate to the case and a written statement outlining the terms of the settlement process, including an indication of assent to confidentiality by all parties.

ADR proceedings, including documents generated solely for the proceedings and communications within the scope of the proceedings, are confidential and will not be provided to a judge of the COFC who is not the settlement judge in the dispute. There is no transcript of any ADR proceeding. In addition, the parties agree not to subpoena or seek in any way the testimony of the settlement judge in any subsequent proceeding.

The types of processes contemplated by the COFC include the settlement judge and mini-trial procedures similar to those used by the ASBCA. The court also lists mediation and neutral evaluation as ADR procedures available to the parties. Like the boards, the court allows any other process or combination of processes agreed to by the parties and judge.

VII. ADR at the Government Accountability Office in Bid Protest Disputes

The Government Accountability Office (GAO) also has ADR procedures to resolve bid protest disputes either before or during the protest. GAO's bid protest regulations specially provide for the use of flexible alternative procedures to promptly and fairly resolve protests.[56] The most common form of ADR at GAO has been the early neutral evaluation or advisory opinion. In a somewhat unusual situation, the GAO lawyer who offers the advisory opinion may be the lawyer who decides the protest if it does not settle.

Notes

1. *See* www.adr.gov.
2. *See, e.g.,* U.S. ARMY CORPS OF ENGINEERS, NEGOTIATING CONSTRUCTION CONTRACT MODIFICATION (NCCM) GUIDE (1996).

3. 41 U.S.C. §§ 601–613.
4. FAR 33.207.
5. 31 U.S.C. §§ 3729–3732. For further discussion of the FCA, *see* Chapter 24.
6. REAs are discussed in Chapter 15.
7. FAR 33.214(b).
8. FAR 33.214(b).
9. 5 U.S.C. §§ 571–583.
10. 61 Fed. Reg. 36,895–913 (July 15, 1996).
11. *See* http://www.adr.gov/resources.htm.
12. 5 U.S.C. § 572(a).
13. 5 U.S.C. § 552. For further discussion of the FOIA, *see* Chapter 7.
14. 5 U.S.C. § 574.
15. 5 U.S.C. § 574(a).
16. 5 U.S.C. § 552(b)(3).
17. 5 U.S.C. § 574(j).
18. 5 U.S.C. § 574(b).
19. 5 U.S.C. § 574(d)(1).
20. 5 U.S.C. § 574(d)(2).
21. 5 U.S.C. § 574(e).
22. *Id.*
23. 5 U.S.C. § 573(c)(2).
24. 5 U.S.C. § 573(a).
25. 5 U.S.C. § 571(2).
26. For a discussion of some of the types of ADR used by the government, *see* Government Contract Law: The Deskbook for Procurement Professionals 525–28 (3rd ed. 2007).
27. 5 U.S.C. § 575.
28. 5 U.S.C. § 575(a).
29. 5 U.S.C. § 575(a)(1).
30. 5 U.S.C. § 575(a)(2).
31. 5 U.S.C. § 575(b). Issues of authority of government representatives are discussed in Chapter 7.
32. 5 U.S.C. § 575(c).
33. 14 C.F.R. § 17.33(f).
34. 5 U.S.C. § 576.
35. 5 U.S.C. § 579(c)(1).
36. 5 U.S.C. § 579(c)(3).
37. 5 U.S.C. § 579(c)(4).
38. 5 U.S.C. § 579(c)(5).
39. 5 U.S.C. § 579(d).
40. 5 U.S.C. § 579(e).
41. 5 U.S.C. § 580(a)(1).
42. 5 U.S.C. § 580(b).
43. *Id.*
44. 5 U.S.C. § 580(d).
45. 5 U.S.C. § 580(c); 5 U.S.C. § 581.
46. Also known as a Dispute Review Board.

47. *See* ASBCA's Notice Regarding Alternative Methods of Dispute Resolution, *available at* http://docs.law.gwu.edu/asbca/adr.htm; CBCA Rule 54, *available at* http://www.cbca.gsa.gov/.

48. CBCA Rule 54(a).

49. CBCA Rule 54(a)(2).

50. CBCA Rule 54(a)(1).

51. CBCA Rule 54(b)(1).

52. CBCA Rule 54(b)(2).

53. CBCA Rule 54(b)(3).

54. CBCA Rule 54(c).

55. *See* http://www.uscfc.uscourts.gov/sites/default/files/court_info/rules_071309_v8.pdf, Rules of the U.S. Court of Federal Claims, Appendix H.

56. 4 C.F.R. § 21.10(e).

CHAPTER 18

Defective Specifications— Impracticability/Impossibility of Performance

ROBERT K. COX

Construction plans and specifications are to inform bidders and, ultimately, the construction contractor what is to be built and the terms under which the project will be constructed. When the federal government provides the plans and specifications for construction and those plans or specifications prove to be defective, who bears the risk of any resulting added costs or delay? If the government's construction plans or specifications prove to be impossible to perform or are only achievable with an inordinate expenditure of resources, does the government or its construction contractor bear the risk of nonperformance? This chapter addresses the allocation of risk between the government and its contractors for defective specifications, commercial impracticability, and impossibility of performance in federal construction contracts.

I. The Government's Construction Contract Specifications

Part 36 of the Federal Acquisition Regulation (FAR), entitled "Construction and Architect-Engineer Contracts," prescribes policies and procedures for the government peculiar to contracting for construction and architect-engineer (A-E) services.[1] According to FAR 36.101, the government's construction and A-E contracts are subject to the other sections of FAR, but, in the event of a conflict, Part 36 takes precedence when the acquisition of construction or A-E services is involved.

A. Types of Specifications

FAR 36.102 defines the phrase "plans and specifications" to mean "drawings, specifications and other data for and preliminary to the construction."[2]

Within the plans and specifications for a federal construction contract there can be design specifications, performance specifications, purchase specifications, and a combination of two or all three types of specifications, sometimes referred to as "composite" specifications. It is not uncommon that a federal construction contract will contain more than one type of specification.[3] Likewise, within a specification, there may be both design and performance type features.[4] When federal construction contract specifications are alleged to be defective, the type of specification at issue, whether it be design, performance, purchase, or composite, is a critically important question to answer in determining the government's and its contractor's respective rights and obligations.[5]

1. Design Specifications

One type of specification used in federal construction contracts is a design specification. In essence, design specifications describe in precise detail the manner and method of the construction work to be performed and from which the contractor is not to deviate.[6] In *J.L. Simmons Co. v. United States*,[7] the then Court of Claims wrote of design specifications as follows:

> The specifications, which were prepared by the defendant, are a classic example of "design" specifications and not "performance" specifications. [Footnote omitted.] In other words, in these specifications, the defendant set forth in precise detail the materials to be employed and the manner in which the work was to be performed, and plaintiff was not privileged to deviate therefrom, but was required to follow them as one would a road map. In contrast, typical "performance" type specifications set forth an objective or standard to be achieved, and the successful bidder is expected to exercise his ingenuity in achieving that objective or standard of performance, selecting the means and assuming a corresponding responsibility for that selection.[8]

In many subsequent opinions, the courts and boards of contract appeals (BCAs) have used the characterization of "road map," from which the contractor has no discretion to deviate, when describing design specifications.[9] For example, in *Caddell Construction Co. v. United States*,[10] the Court of Federal Claims (COFC) concluded that the structural steel portion of the contract at issue was a design specification, writing:

> The court agrees with plaintiff that, at the very least, the structural steel portion of the contract was a design specification. Although the government did not dictate every aspect of the construction of the building and left certain key aspects of the construction, such as sequencing and scheduling, up to Caddell, the details and specifications for the structural steel were design specifications. Nine pages of

the contract are devoted to specifications for the structural steel with specific instructions on what type of bolts, washers, nuts, welds, finishes, and connections, among other things could be used for the construction. [Footnote omitted.] This was clearly a "road map" for the structural steel fabricator to follow.

In addition, the building itself was designed to meet specific earthquake proofing guidelines and the contractor had to strictly follow that design. The contract reads on its first page that the construction must be done "in strict accordance with specifications," making it clear that the contract was, at least in part, a design specification. [Footnote omitted.] The contractor could not deviate from the design of the structure because any variation could effect [sic] the load bearing ability of certain beams or the flexibility of joints and render the building seismically inadequate.[11]

In *Travelers Casualty & Surety of America v. United States*,[12] the COFC succinctly wrote: "Design specifications dictate the 'how' governing a contractor's tasks, in contrast to performance specifications, which concern the 'what' that is to be done."[13]

2. Performance Specifications

In comparison to design specifications, performance-type specifications set forth an objective, result, or standard to be achieved for the item or project.[14] The contractor is expected to exercise its ingenuity in achieving that objective or standard of performance, selecting the means and assuming a corresponding responsibility for that selection.[15] In performance-type specifications, design, measurements, and other specific details are typically neither stated nor considered important so long as the performance requirement is met. With performance specifications, the contractor has general discretion and election as to the means of performance, but the work is still subject to the government's right of final inspection and approval or rejection.[16]

For example, in *George Sollitt Construction Co. v. United States*,[17] the COFC addressed the terms of a contract for the renovation of a naval training center, including the specifications for revised chiller power in one of the training center buildings to be renovated. The court determined the chiller specification to be a performance-type specification, writing:

> "[T]ypical 'performance' type specifications set forth an objective or standard to be achieved, and the successful bidder is expected to exercise his ingenuity in achieving that objective or standard of performance, selecting the means and assuming a corresponding responsibility for that selection." *J.L. Simmons Co. v. United States*, 188 Ct. Cl. 684, 412 F.2d 1360, 1362 (1969).

Here, the chiller specification was a typical performance specification, where the government specified only the result, that of a particular cooling capacity, and [the contractor] was left the discretion to install an appropriate chiller.[18]

Another example of performance-type specifications was addressed in *Huber, Hunt & Nichols, Inc.*[19] In a General Services Board of Contract Appeals (GSBCA) case, the contract specifications called for the contractor to provide a legal elevator hoistway whose structure would withstand the forces and loads of elevator usage and also comply with American National Standards Institute (ANSI) code. The board concluded such specifications were standard industry, open, performance type specifications.

In another board decision, *Tri-State Consultants, Inc.*,[20] the contract was to repair a breach in a sand dune in Sedge Islands in Ocean County, New Jersey, using sand dredged from a designated borrow area. The Invitation for Bids (IFB) gave the bidders the alternatives of dredging with a hopper, clamshell, or hydraulic pipeline dredge and did not specify the means to transport and place fill material. During performance of the dredging, it was found that 10 times the displacement of the dredge originally anticipated would be required to operate in the tidal currents and meet the contract requirements for dredging depth and discharge of the dredged material. Among other claims, the contractor alleged the contract specifications to be defective as to the type and size of the dredge. The Armed Services Board of Contract Appeals (ASBCA) rejected the contractor's claim of defective specifications, writing:

> Appellant's proof of defective specification is unpersuasive. The contract gave Tri-State discretion to select the means of dredging, including the type and size of the dredge (findings 3–4). The contract requirements to dredge to -12' and to discharge a 150' embankment (finding 4) were performance-type specifications. We are not persuaded that Tri-State's or the Corps' interpretation of such requirements, for purposes of estimating, to be consistent with use of a 14" hydraulic pipeline dredge (findings 8, 13) changed those performance requirements to design requirements, as appellant contends. Nor do we agree that the CO's 6 December 1999 direction to Tri-State to comply with the express terms of the contract, "to obtain 'a dredge of sufficient size to withstand the strong tidal currents and large swells that are common in Barnegat inlet'" (finding 29), was a change order, as appellant asserted (finding 32).[21]

The design versus performance characterization has been applied to contracts as a whole and not just particular specifications. For example, in *P.R. Burke Corp. v. United States*,[22] the contractor was to repair and improve a sewage treatment plant. As part of its ruling, the Court of Appeals for the Federal Circuit considered whether the contract was a design or performance contract. In doing so, the court looked to the contract terms entitling the contractor to plan and schedule the manpower, materials, and methods of construction necessary to complete the project as specified, as well as the contract terms describing the work to include the furnishing of all labor, approved materials, and equipment required for the project and necessary to make the facilities fully operational. The court concluded that nothing in the contract's description dictated the manner in which the contractor was to perform. Since the contract merely identified what the contractor must have completed by the end of performance, the court found it was a performance contract.[23]

Similarly, in *PCL Construction Services, Inc. v. United States*,[24] the COFC concluded that the contract for the construction of a visitor center and parking structure at Hoover Dam for the Bureau of Reclamation (USBR) "was largely a performance specification . . . ."[25] Before citing some 22 examples from the contract, the court wrote: "In fact, PCL was contractually responsible for the design and/or engineering of significant portions and elements of the work. For example, the defendant provided a comprehensive list to the contractor of design and engineering obligations in the contract . . . ."[26] After citing examples of the contractor's design and engineering obligations, including design and engineering of precast concrete box girders and architectural wall panels, prefabricated steel stairs, metal floor and roof decks, handrails and railings, and various other elements of the visitor center and parking garage, the court wrote:

> It is apparent to the court from the contract requirements outlined above that the contract never contemplated that PCL's performance could be accomplished using only the contract documents.
>
> * * *
>
> In fact the contract expressly provided that the design package conveyed only the "design and engineering intent" for the project, and that the design drawings would be supplemented and detailed as necessary to construct the final product. Thus, the contract allocated a substantial amount of discretion and responsibility to PCL to participate in resolving design problems. The contract also stated performance goals that PCL was to meet, and did not tell PCL the methods or processes to use to achieve the specified end result. Indeed, it is evident that the drawings do not contain the level of detail necessary to actually construct the project in the field. It was up to PCL to provide the precise details of how the structures were to be built (including, but not limited to, the precise routing of electrical and mechanical systems, the number and locations of individual concrete pours, the sequence of construction activities, and details of all concrete reinforcement.)
>
> There were numerous decisions for PCL to make that required the exercise of discretion, based upon judgment and experience.
>
> * * *
>
> A cursory review of USBR's specifications and drawings demonstrates that USBR did not provide a "road map" in its IFB and contract documents telling PCL exactly how to perform the contract. Thus, the contract contained largely performance specifications. Consequently, notwithstanding PCL's frequent use of the terms "warranty"

and "build to design," and its expert's concentration on "standards of care," plaintiff did not substantiate its theory that the contract was a design specification.[27]

3. Purchase Specifications

Purchase specifications are specifications that designate a particular manufacturer's model, part number, or product.[28] Such specifications in federal government contracts may call for "brand name or equal" items, or in limited circumstances, items unique to one manufacturer.[29]

FAR 36.202 specifically allows for the use of "brand name or equal" specifications in federal government construction contracts but with the requirement that such specifications "clearly identify and describe the particular physical, functional, or other characteristics of the brand-name items which are considered essential to satisfying the requirement."[30]

FAR 36.202 also provides that construction specifications shall conform to the requirements in Part 11 of the FAR. In that regard, FAR 11.105 requires that agency requirements are not to be written around a particular brand-name product, or a feature of a product peculiar to one manufacturer, thereby precluding consideration of another manufacturer's product unless (a) the particular brand name product or feature is essential to the government's requirements, (b) other companies' similar products lack the particular feature and do not meet or cannot be modified to meet the agency's needs, or (c) there is written justification and approval for the brand name or peculiar item.[31]

There are also instances in which the government's specifications may designate use of a particular subcontractor or supplier.[32]

4. Composite Specifications

A composite specification is literally a composite of two or three of the specification types and, as such, may contain design characteristics, performance features, or purchase elements.[33]

In *Costello Industries, Inc.*,[34] the ASBCA addressed the composite characteristics of a joint sealant specification in a government contract for the preparation and sealing of cracks in asphalt pavement runways at an Air National Guard air base. The board wrote:

> We hold that Federal Specification SS-S-1401B is such a composite specification. Paragraph 3.1 has the characteristics of a performance specification because it merely sets forth the operational characteristics of the desired joint sealant without stating specific details concerning the material composition or formula for acceptable joint sealant, nor does it set forth the manufacturing process. Therefore, as stated in *Abramson & Sons, Inc.*, ASBCA Nos. 18848, 19205, 75-1 BCA ¶ 11,250, which dealt with the Federal Specification for jet fuel resistant joint sealant and which was in this respect similar to the Federal

Specification SS-S-1401B, the specification was essentially a performance specification. Nevertheless, the testing procedures set forth in the Federal Specification are very specific and detailed with respect to both the test procedures and performance standards for each of the specification performance requirements. In this respect, they are design specifications.[35]

In *Monitor Plastics Co.*,[36] the ASBCA found the specification at issue to be a composite of all three types of specifications:

This contract contains composite specifications since features of each of the three types appear. For example, Section 2.1 sets forth what was to be constructed but did not provide for any specific manufacturer's material, a performance specification. Section 2.1.1 provided for a brand name or equal for the core, a purchase description. Section 2.1.2 provided for a brand name or equal for the rubber and a vacuum processing of the rubber, a combination purchase description and design specification. Section 2.1.3 provided for a brand name or equal for the laminates, a purchase description. Section 2.2 provided for the size of the panels, a design specification. Section 2.3, the method of manufacture is both a design and performance specification. It leaves the method of manufacture with the exception of the requirement for vacuum pouring to the contractor. Sections 2.4.1 and 2.4.2, workmanship and appearance requirements respectively are performance specifications.[37]

As previously noted, it is not uncommon to find two or more of the different types of specifications within a government construction contract.

B. Distinguishing Among the Types of Specifications

In the world of federal government construction contracting, the liability that the government or the contractor takes on for the contract specifications or drawings varies greatly depending on the type of specifications at issue. Consequently, when there is a claim of defective specifications causing added costs or delay or preventing performance, one of the first questions a court or board is likely to consider is the type of specifications underlying the claim or defense.

For example, in *PCL Construction Services, Inc. v. United States (PCL Construction I)*,[38] the contractor for the construction of the visitor center and parking structure at the Hoover Dam alleged that the government had provided a "grossly inadequate" project design at the time of contract award in breach of the government's implied warranty of its specifications. When considering the contractor's allegation, the COFC in *PCL Construction I* wrote, ". . . whether the specifications in the instant actions were design or performance specifications is critically important for an understanding of the existence of any 'warranty' claims and the parties' respective rights and obligations."[39]

The determination whether a contract or a particular specification is a design or performance type is a mixed question of fact and law.[40] When undertaking that determination, the COFC in *Fru-Con Construction Corp. v. United States*[41] cautioned that "the distinction between design specifications and performance specifications is not absolute" and that courts should understand that "[i]t is the obligation imposed by the specification which determines the extent to which it is a 'performance' or 'design,' not the other way around."[42] The court further wrote: "Reluctance to adhere to the rigid constructs associated with each type of specification recognizes that contract language may not always fall squarely within the 'design' category or the 'performance' category and, moreover, that contracts may exhibit both design and performance characteristics."[43]

When distinguishing among the specification types, the courts and boards will direct their attention to the level of discretion inhering within a given specification. According to the COFC, "discretion serves as the touchstone for assessing the extent of implied warranty and attendant liability."[44] For example, in *PCL Construction Services, Inc. v. United States (PCL Construction II)*,[45] the court wrote:

> As the United States Court of Appeals for the Federal Circuit has noted, contractors typically are granted at least some discretion even when specifications are largely of the design variety, and the labels "design" and "performance," while helpful to some degree, are merely labels. It is the contract's provisions, and the amount of discretion that the contract affords the contractor that govern whether the contractor can recover for problems that occurred during performance. *See Blake Constr. Co. v. United States*, 987 F.2d at 746. The fact the specifications provided some details concerning how the work was to be performed does not convert what would otherwise be a performance specification into a design specification. *See, e.g., Penguin Indus., Inc. v. United States*, 209 Ct. Cl. 121, 123–25, 530 F.2d 934, 937 (1976) (noting that the specifications were "detailed"); *Aleutian Constructors v. United States*, 24 Cl.Ct. at 390; *Norwood Mfg., Inc. v. United States*, 21 Cl.Ct. at 308–09.
>
> Thus, where a specification does not tell a contractor how to perform a specific task, that part of the specifications can be a performance specification even if the rest of the specifications are design specifications. *See Penguin Indus., Inc. v. United States*, 209 Ct. Cl. at 123–25, 530 F.2d at 937.[46]

Similarly, in *Caddell Construction Co. v. United States*,[47] the court wrote:

> In order to determine whether a contract is a design specification, "[t]he relevant inquiry concerns the quality and quantity of the obligations that the specifications impose." *Travelers*, 74 Fed. Cl. at 89

(citing *Mega Constr. Co., Inc. v. United States*, 29 Fed. Cl. 396, 418 (1993)). "[D]etailed measurements, tolerances, materials, i.e., elaborate instructions on how to perform the contract" together may constitute a design specification. *Stuyvesant Dredging Co. v. United States*, 11 Cl.Ct. 853, 860 (1987). "Contracts may have both design and performance characteristics." *Blake Const. Co.*, 987 F.2d at 746.

* * *

The court agrees with plaintiff that, at the very least, the structural steel portion of the contract was a design specification. Although the government did not dictate every aspect of the construction of the building and left certain key aspects of the construction, such as sequencing and scheduling, up to Caddell, the details and specifications for the structural steel were design specifications. Nine pages of the contract are devoted to specifications for the structural steel with specific instructions on what type of bolts, washers, nuts, welds, finishes, and connections, among other things could be used for the construction. [Footnote omitted.] This was clearly a "road map" for the structural steel fabricator to follow.[48]

Distinguishing the type of specifications at issue is of great significance; it goes directly to the allocation of risk between the government and its contractors for defective specifications.

II. The Allocation of Risk between the Government and Its Contractors for Defective Construction Contract Specifications and Drawings

Depending on the type of construction contract specifications at issue, the government can be found to have (a) impliedly warranted its specifications, (b) provided no warranty obligation whatsoever for the specifications, or (c) limited responsibility for its specifications. Identifying the type of specifications at issue is not, however, the end of the analysis. It must be further assessed whether there are defenses available to the government arising from the contractor's knowledge, acts, or omissions that can relieve the government of what would otherwise be a liability for its defective specifications or drawings.

A. The Government's Implied Warranty of Its Design Specifications

1. The Spearin Doctrine

There is an implied warranty by the government for certain types of its specifications, sometimes referred to as the "*Spearin* doctrine," which provides that if the government furnishes specifications for the production or construction of an end product and proper application of those specifications does not result in a satisfactory end product, the contractor is not at risk for the unsatisfactory results.[49] The origin of this governmental warranty is the holding from the U.S. Supreme Court's decision in *United States v. Spearin*[50] that "if the contractor is bound to build according to plans and specifications prepared by the owner, the contractor will not be responsible for the consequence of defects in the plans and specifications."[51] Subsequent rulings have restated the warranty as follows:

> *Spearin* stands for the proposition that when the government includes detailed specifications in a contract, it impliedly warrants that (i) if the contractor follows those specifications, the resultant product will not be defective or unsafe, and (ii) if the resultant product proves defective or unsafe, the contractor will not be liable for the consequences.[52]

Similarly, other rulings have defined the warranty as follows: "Detailed design specifications contain an implied warranty that if they are followed, an acceptable result will be produced."[53]

As a result of this implied warranty, if the cause of faulty construction is a defective design specification, the government will be liable (barring a government defense) for the reasonable costs the contractor incurred performing to that defective design specification,[54] including costs attributable to any resulting period of delay.[55] The costs are recoverable as a constructive change.[56] Under the implied warranty, defective design specifications may also entitle a contractor to an equitable adjustment for the reparative work required to construct a satisfactory end product.[57] When pricing the costs recoverable for a termination for convenience, the contractor is entitled to recover its reasonable costs, including costs due to defective design specifications.[58] Be forewarned, however, the scope of the implied warranty does not encompass all types of specifications.[59] For example, the warranty can be disclaimed,[60] a contractor's knowledge or actions, actual or constructive, can vitiate the warranty,[61] and the warranty does not encompass all risks.[62]

Ironically, in providing its design services to the federal government, an A-E does not impliedly warrant the correctness or sufficiency of its design. At common law, generally, unless the A-E has contracted to another standard, it will be held to that degree of ordinary and reasonable care, skill, and diligence as would be expected from a typical, practicing member of the profession.[63] In federal government contracting, under FAR 36.609.2, the CO is to include the Responsibility of the Architect-Engineer Contractor clause in the government's fixed-price A-E contracts. This clause, at FAR 52.236-23, requires A-Es to make necessary corrections at no cost to the government (unless excused by the CO) when the designs, drawings, specifications, or other items or services provided contain any errors, deficiencies, or inadequacies.[64]

2. *Effect of Disclaimers and Other Clauses on the Government's Implied Warranty of Plans and Specifications*

As a general principle, courts and boards have not found the government's use of disclaimers or exculpatory clauses to be against public policy.[65] At the same time, however, the courts and boards are reluctant to enforce such clauses,[66] particularly when the clauses are broadly worded, stating the government does not guarantee the statements of fact contained in the specifications or drawings, or requiring the bidder to investigate the site and satisfy itself as to the conditions.[67] Of the courts and boards addressing disclaimer or exculpatory language, it appears that there are more rulings not upholding a governmental defense of disclaimer or exculpatory terms than there are rulings upholding such a defense.

For example, in *Bromley Contracting Co.*,[68] the Army Material Command contracted for the replacement of a roof at an arsenal facility. The government-furnished drawing described the ridge height of the roof monitors and provided scaled dimensions. The drawing also contained a note providing, "Contractor shall verify all dimensions and conditions prior to submission of bid."[69] The contractor visited the roof site prior to bidding as required, but it failed to notice the variance between the actual elevation of the roof monitors and the height shown on the contract drawing. The contractor claimed that the drawing error caused it to underestimate the roof area surface and cost an additional $43,370.64. The government contended that the drawing note required the contractor to verify the dimensions and roof conditions prior to bidding. The board rejected the government's defense, writing:

> In addition, the exculpatory provisions of the contract relied on by the Government are ineffective for the purpose intended. Governmental disclaimers of responsibility for the accuracy of specifications which it authors are viewed with disdain by the courts. The attitude of the Court of Claims is succinctly summarized in *Morrison-Knudsen Company, Inc. v. United States*, 184 Ct. Cl. 661, 686 (1968), 'But this Court has frequently held in comparable circumstances that broad provisions of this kind—stating that the Government does not guarantee the statements of fact contained in the specifications or drawings or requiring the bidder to investigate the site and satisfy himself of conditions, etc.—cannot be given their full literal reach and do not relieve the Government from liability.' See also *Hollerbach v. United States*, 233 U.S. 169, 172 (1914).
>
> The short of the matter is that the information contained in the drawing constituted positive representations upon which the appellant was justified in relying. This information was defective. The contractor is entitled to recover the additional cost of performance resulting from the defective drawing by way of a price adjustment under the Changes clause. See: *Pilcher, Livingston & Wallace, Inc.*, ASBCA No. 13391, 70-1 BCA ¶ 8331.[70]

In another case, *Enviroserve Inc.*,[71] the contract was for the reroofing of two structures for the Federal Prison Camp at Alderson, West Virginia. The specifications set forth an approximate square footage of roofing work and included provisions that the contractor was responsible for all field measurements and for undertaking a site investigation to satisfy itself of the conditions affecting the work and the character and quantity of all surface and subsurface materials or obstacles to be encountered.[72] The drawings proved to be inaccurate as to the actual area to be reroofed. While the Transportation Contract Appeals Board (DOTCAB) found the statements in the solicitation to mean the measurements were not precise, the board also found the estimated roof area to be "substantially erroneous," and beyond the variation one should reasonably anticipate.[73] As for the disclaimers in the solicitation and the opportunity for all bidders to take measurements at the pre-bid conference, the board determined that such clauses did not exonerate the government. The board wrote:

> The words of this Board in *TEC Construction, Inc.*, 94-2 BCA ¶ 26,924, 1994 DOT BCA LEXIS 16 (1994), are applicable to this issue:
>
> > Generally, a prospective contractor is chargeable with knowledge of conditions that could have been discovered upon a reasonable visual inspection, regardless of whether that firm has conducted a site investigation. *GIIS Corporation*, 85-1 BCA ¶ 17,810 (1984) at 89,003. However, that principle is not wholly applicable to situations in which the owner (the government) has in the plans and specifications made a positive statement about some condition and the contractor does not have actual knowledge to the contrary. A prospective contractor, in preparing a bid, is entitled to rely on the accuracy of the owner-issued specifications and drawings unless that bidder is aware of an error.
> >
> > "The rule is well established that where the Government makes positive statements in the specifications or drawings for the guidance of bidders, that [sic] a contractor has a right to rely on them regardless of contractual provisions requiring the contractor to make investigations. *Hollerbach v. United States*, 233 U.S. 165, 172; *United States v. Atlantic Dredging Co.*, 253 U.S. 1, 11; *Potashnick v. United States*, 123 Ct. Cl. 197, 218; *Binghamton Construction Co. v. United States*, 123 Ct. Cl. 804, 836–837." *Arcole Midwest Corporation v. United States*, 125 Ct. Cl. 818, 822 (1953).
>
> By having elected to include in the specifications the estimated roof area, the Bureau assumed the obligation of having that estimate reasonably correct, correct within a reasonable tolerance in the trade. In this, it failed.[74]

Similarly, in *Ball, Ball & Brosamer, Inc.*,[75] the contract was for the construction of an aqueduct, including a concrete canal lining. The contractor was to furnish specified aggregate material from approved sources to be processed and used in making the concrete. The designated source pit actually contained unusually dirty material and sand of suspect quality. The Interior

Board of Contract Appeals (IBCA) rejected the government's defense based on the contract's terms alerting bidders to investigate the site, plans, and scope, and difficulty of work. The board wrote: "As is well established, the Government impliedly warrants the correctness and adequacy of its design specifications. The warranty is not superseded by disclaimers such as those in this contract alerting bidders to investigate the site, plans, and scope, and difficulty of work."[76]

In a more recent case, *Edsall Construction Co.*,[77] the contract was for the construction of an Army Aviation Support Facility in Montana. At issue in the appeal was the design of the storage hangar tilt-up canopy doors and truss; in particular the three-pick-point design. The canopy doors drawings included the following note:

> CANOPY DOOR DETAILS, ARRANGEMENTS, LOADS, ATTACHMENTS, SUPPORTS, BRACKETS, HARDWARE, ETC. MUST BE VERIFIED BY THE CONTRACTOR PRIOR TO BIDDING. ANY CONDITIONS THAT REQUIRE CHANGES FROM THE PLANS MUST BE COMMUNICATED TO THE ARCHITECT FOR HIS APPROVAL PRIOR TO BIDDING AND ALL COSTS OF THOSE CHANGES MUST BE INCLUDED IN THE BID PRICE.[78]

The ASBCA found the three-pick-point design for lifting the doors to be a defective design specification. The board then addressed the government's contention that the drawing note, calling for verification of the information, was an exculpatory clause shifting the risk of design defects in the drawings to the contractor. The board rejected the government's defense, stating as follows:

> As the designer of how the door would be installed, the Government warranted that the door load could be evenly distributed to the specified three pick points and corresponding points on the truss if appellant adhered to its design. *See United States v. Spearin*, 248 U.S. 132 (1918). We are not persuaded that the disclaimers contained in the annotations to the design details and the note on drawing S13 shifted the risk of defective specifications to Edsall for a number of reasons.
>
> First, the Government's contention isolates the disclaimers from the written specifications and the other design features it concedes are contained on the drawings. It therefore violates established rules of contract interpretation which require the contract to be read as a whole, with all of its parts harmonized. *E.g., Thanet Corp. v. United States*, [25 CCF ¶ 82,997] 591 F.2d 629, 633 (Ct. Cl. 1979). Next, if the number of pick points and the requirement to distribute the load to three points on the truss were not design specifications because of the disclaimers as the Government asserts, there would be no reason for the note on drawing S13, because bidders would have been free to select the method of performance, and it would not have been

necessary for them to seek the architect's permission to make "changes from the plans."

In any event, it is settled that a contractor is not obligated to inspect the Government's specifications and drawings to ascertain their accuracy and ferret out hidden ambiguities and errors in the documents. *See Blount Bros. Construction Co. v. United States*, [10 CCF ¶ 73,058] 171 Ct. Cl. 478, 496 (1965; *Federal Contracting, Inc.*, ASBCA No. 48280, 95-2 BCA ¶ 27,792. Indeed, "Governmental disclaimers of responsibility for the accuracy of specifications which it authors are viewed with disdain by the courts." *Bromley Contracting Company*, ASBCA Nos. 14884 et al, 72-1 BCA ¶ 9252 at 42,902. In this case, while appellant might be required to verify if the door weighs 21,000 pounds, it had no obligation to ferret out if the Government's three-pick point design would provide the proper load distribution.

The record here established that the Government did not verify the accuracy of the design, a fact which "weighs heavily" against it. *Bromley*, 72-1 BCA at 42,902. It also established that appellant's review of the specifications and drawings for purposes of bidding was entirely reasonable. In addition to the fact that USF's president saw nothing "obviously wrong" with the design, neither USI nor Edsall interpreted the note on drawing S13 to require a full analysis by professional engineers.

* * *

In short, only the drafter of the design was of the view that his note required bidders to undertake a full verification of the design itself.

We are satisfied, therefore, that the detail annotations and the note on drawing S13 did not shift the risk of any design inadequacies to appellant. *See e.g., Radionics, Inc.*, ASBCA No. 22727, 81-1 BCA ¶ 15,011 (Government disclaimer imposing design responsibility on contractor unenforceable where contractor's interpretation of an ambiguous clause was reasonable) and *Essex Electro Engineers Inc.*, ASBCA No. 49915, 99-1 BCA ¶ 30,229, *rev'd on other grounds*, 224 F.3d 1283 (Fed. Cir. 2000) (alleged notice disclaiming warranty and obligating contractor to identify incorrect drawings and to develop new, updated drawings not enforced).[79]

In another case, the ASBCA ruled that a contractor was not obligated to verify specified elevations provided in the detailed design information to the hundredth of a foot for the construction of concrete aprons included in the contract for the renovation of 28 ammunition bunker igloos.[80] The board wrote:

> The government's position—that LEC should have verified the elevations provided in the contract documents, to the hundredth of a foot and thereby avoided the problem—would have required, at the least, a detailed survey of the roadway in front of every igloo. Such a

position negates any possibility of reliance by a reasonable contractor on any information, no matter how detailed, when provided by the government.[81]

In another case, the Federal Circuit confirmed in *E.L. Hamm & Associates, Inc. v. England*[82] that the contract to police and mow certain acreage around naval base housing in Norfolk, Virginia, did not require the contractor during the bidding process to perform "a series of complicated calculations using figures from various sections of the specification" to identify that the specifications understated the acreage. The court wrote in particular:

> Any discrepancy was hidden within the contract documents and therefore latent. Without any reason to reverse engineer the daily mowing and policing acreages . . . [the contractor] had no reason to discern the error. . . . Because the numbers here did not reveal a glaring facial discrepancy such that [the contractor] would have had a duty to inquire into them at the outset, we conclude that the understatement of the yearly policing figure was a latent, not patent, defect.[83]

Not all cases, however, have resulted in rejection of the government's reliance on exculpatory, disclaimer, or other contract clauses to overcome a contractor's allegation of defective specifications. For example, in *Blake Construction Co.*,[84] the government successfully relied on the "omissions and misdescriptions" clause of the contract to rebut the contractor's claim that the contract drawings omitted necessary information for the installation and placement of fire extinguisher cabinets. In essence, the clause provided that the contractor was responsible for performing work necessary to carry out the intent of the drawings and specifications or that is customarily performed even though omitted or misdescribed in the drawings or specifications. The ASBCA agreed the drawings did not contain detailed information concerning the dimensions, elevations, and finish materials for the fire cabinets. Nevertheless, the board found the omission to be minor in nature, concerning work required to be done in furtherance of stated contract requirements so as to come within the purview of the "omissions and misdescriptions" clause of the contract.[85]

To the same effect is *David Boland, Inc.*,[86] in which the ASBCA relied on the "omissions and misdescriptions" clause to deny the contractor's claim for defective specifications because of omitted electrical circuitry details. The contract called for the Army training center to be made complete and ready for use, and the electrical circuitry was deemed manifestly necessary for the installed equipment. The current version of the "omissions and misdescriptions" clause is included in the Department of Defense FAR Supplement (DFARS) at Part 252.236-7001, entitled "Contract drawings and specifications."

In other cases, the boards have considered the effect on claims of defective specifications when the contract terms call for the contractor to coordinate and review the drawings and specifications and through the shop drawing process assure that equipment fits in the space provided, yet the contractor failed to do so. If the contractor would have discovered the defect in the specifications but did not because of its failure to comply with the contract's review and coordination terms, or did not present evidence of coordination and review, the boards have rejected a contractor's claim of defective specifications:

> To the contrary, we conclude that, had appellant submitted the required, detailed, scaled shop drawings and coordination drawings, and performed the necessary coordination of the duct, electrical and ceiling installations in Room LE112, prior to commencing work, it would have discovered the need for more space to accommodate the recessed light fixtures and the parties could have agreed to adjustments to resolve the problem. Appellant is responsible for the consequences of its apparent failure to do so. *See Greenhut Construction Company, Inc.,* ASBCA No. 36912, 90-3 BCA ¶ 23,259.[87]

In a case involving a contract for the renovation of a military hospital, *Conner Bros. Construction Co. v. United States,*[88] the contractor claimed its added costs for demolition of existing above-ceiling piping to be due to the government's defective specifications. The contract included the Site Investigation and Conditions Affecting the Work clause (FAR 52.236-3) requiring all bidders to perform a site investigation before bidding.[89] The contractor contended it was precluded from performing a reasonable inspection of the above-ceiling space by physical obstructions. The COFC rejected the contractor's claim, writing:

> It is this court's opinion that, where the contract's site investigation clause requires all bidders to perform a site inspection and where plaintiff alleges it was precluded from the reasonable conduct of such inspection by physical obstructions, plaintiff was faced with an obvious inconsistency, discrepancy or omission of significance, such as would have required plaintiff to bring the situation to the government's attention before bid submission if it intended to subsequently resolve the issue in its own favor. *See Space Corp. v. United States,* 200 Ct. Cl. 1, 470 F.2d 536, 538 (1972) (citations omitted). Significant policy considerations underpin this rule. Requiring contractors to bring to the government's attention major discrepancies or inconsistencies allows for clarification of ambiguities or correction of problems before a contract is awarded and thereby avoids the need for expensive and complex litigation during contract administration. *See Beacon Constr. Co. v. United States,* 161 Ct. Cl. 1, 314 F.2d 501, 504 (1963). The preaward clarification requirement bolsters the competitive bidding system. *See Monarch Painting Corp v. United States,* 16 Cl. Ct. 280, 287 (1989). The duty to inquire prevents one contractor from exploiting ambiguities

in the contract so as to bid on only a portion of the work solicited and thereby appear to be the low bidder. *Id.* Requiring disclosure of ambiguous provisions or constructions of the contract before bidding ensures that all contractors bid on the basis of identical specifications. *Id.* (citing *Beacon Construction*, 314 F.2d at 504).[90]

3. Design-Build Contracts and the Implied Warranty of Design Specifications

FAR 36.102 defines design-build to mean combining design and construction in a single contract with one contractor. At Subpart 36.3, the FAR prescribes policies and procedures for the use of two-phase design-build selection procedures in federal government construction contracting.[91] The question then arises whether the design-build concept of combining design and construction in one contract with a design-builder transfers the *Spearin* implied warranty of design specifications to the design-builder. In *Donahue Electric, Inc.,*[92] the Veterans Affairs Board of Contract Appeals (VABCA) provided the answer, writing:

> A properly written and administered design-build contract transfers the risk of design insufficiency from the [owner] to the design-builder. The owner is shielded when the design results in cost overruns or does not work. *M.A. Mortenson*, ASBCA No. 39978, 93-3 BCA ¶ 26,189. There are two stages in design/build contracting. First, the design/ build team must determine from the conceptual drawings in the RFP, if any, and, more importantly, from the specifications, if any, what it will cost to design and build the project. Specifications included in a design/build contract, however, to the extent specific requirements, quantities and sizes are set forth in those specifications, place the risk of design deficiencies on the owner.[93]

In *United Excel Corp.,*[94] the design-build contractor contended that because the contract was design-build, the contractor was relieved of inquiring about apparent specification deficiencies in the request for proposals documents. The VABCA replied:

> We also see nothing in the case law, and UEC has provided none, for the proposition that the well settled law relating to the contract interpretation is suspended or abrogated in a design-build contract. To the contrary, the case law indicates that a design build contract shifts risk to a contractor that a final design will be more costly than the bid price to build and that the traditional rules of fixed-price contract interpretation still obtain. UEC was not relieved of its obligation to inquire about the aluminum stainless steel diffuser discrepancy because the Contract was design-build.[95]

In an ASBCA decision, *Strand Hunt Construction, Inc.*,[96] the contract was for the design and construction of a Joint Security Forces Complex in Alaska. The request for proposals (RFP) included thermal and blast resistance performance requirements for the complex's windows. The contractor's subcontractor design firm authored the specifications for the building of the project, including the thermal and blast force performance window requirements. During construction, the contractor contended that no window could be found with the combined thermal and blast resistance performance characteristics; therefore the RFP requirements had to have been defective. The board rejected the design-build contractor's contention, writing:

> There is no dispute that the government authored the RFP performance specification. However, the design/build contract required SHC and its architect of record, Design Alaska, to design the project. That is, they were responsible to create the design specifications in accordance with which the project was to be constructed, making sure those specifications met the various performance characteristics set forth in the RFP. As the design/build contractor of the project, SHC was the party responsible for any defects in the design as set forth in its design specifications. (See findings 3, 4, 6)
>
> SHC argues that it had no choice but to make sure its proposal and design specifications mirrored the RFP requirements. However, if SHC indicated in its proposal and design specification submissions that it would meet the RFP performance requirements without adequate investigation, it did so at its own risk. SHC was obligated to not just say that it would meet requirements, but also to be sure it could actually do so. SHC apparently assumed, even though the RFP made no such representation, that a ready-made window existed or that a compliant custom made window could be acquired within its budget that met the RFP requirements (finding 29).
>
> * * *
>
> The record does not support SHC's argument that there were no windows capable of meeting the RFP performance requirements. The RFP does not require nor promise the availability of ready-made windows. There is evidence that windows meeting all RFP performance requirements could be manufactured given enough time (finding 36).[97]

B. When the Implied Warranty Does Not Apply

Aside from disclaimer or other contract clauses, there are other circumstances when the government's implied warranty of specifications is simply not applicable as a basis for an equitable adjustment. For example:

■   When the contract documents expressly advised the contractor that its obligations would be governed by a contract package containing estimated information, that the contract package was imperfect and the contract package would be supplemented and refined as necessary during construction, the implied warranty was not applicable;[98]

■ When the specifications at issue are performance specifications, the government's implied warranty is not applicable;[99]

■ When the contract specified a particular sandblasting material, without specifying its availability, the government's implied warranty did not encompass that the material would be easily found within set time frames;[100] and

■ When the government uses a purchase type specification, the implied warranty extends to performance of the item but not to price.[101]

1. Performance, Purchase, and Composite Specifications

The scope of the government's implied warranty has its limits.[102] Importantly, the implied warranty applies to the government's design specifications but not to its performance specifications.[103] In *Travelers Casualty & Surety of America v. United States*,[104] the COFC wrote of the implied warranty's application to design specifications, but not to performance specifications, in federal construction contracts:

> In the world of government contracts, a jurisprudential difference exists between what are known as "design specifications" and "performance specifications." *See* JOHN CIBINIC, JR. ET AL., ADMINISTRATION OF GOVERNMENT CONTRACTS 276–86 (4th ed. 2006). This dichotomy is crucial to the warranty doctrine, enunciated in *United States v. Spearin*, 248 U.S. 132, 54 Ct. Cl. 187, 39 S.Ct. 59, 63 L.Ed. 166 (1918), which holds that a contractor is not responsible for defects in the item constructed if he strictly follows the design specifications given to him by the government. *Id.* at 136, 39 S.Ct. 59; *see White v. Edsall Constr. Co., Inc.*, 296 F.3d 1081, 1084–85 (Fed. Cir. 2002); *Hercules, Inc. v. United States*, 24 F.3d 188, 197 (Fed. Cir. 1994); *Ordnance Research, Inc. v. United States*, 221 Ct. Cl. 641, 670, 609 F.2d 462 (1979); *Hol-Gar Mfg. Corp. v. United States*, 175 Ct. Cl. 518, 525, 360 F.2d 634 (1966). As the Supreme Court explained:
>
> > [I]f the contractor is bound to build according to plans and specifications prepared by the owner, the contractor will not be responsible for the consequences of defects in the plans and specifications. This responsibility of the owner is not overcome by the usual clauses requiring builders to visit the site, to check the plans, and to inform themselves of the requirements of the work. . . .
>
> *Spearin*, 248 U.S. at 136, 39 S.Ct. 59 (internal citations omitted). In *Spearin*, the Supreme Court held that contract provisions "prescribing the character, dimensions and location of" a structure to be constructed

"imported a warranty that, if the specifications were complied with, the [structure] would be adequate." *Id.* at 137, 39 S.Ct. 59.

* * *

The *Spearin* implied warranty doctrine has its limits. As was explained above, the warranty does not extend to performance specifications which "merely set forth an objective without specifying the method of obtaining the objective." *White,* 296 F.3d at 1084.[105]

In one case, however, the COFC found the specification, standing alone, to be a performance-type specification with the attendant risk of failure on the contractor, but the acts of the government were grounds for the court to find the government equitably liable, in part, for the contractor's failure to comply with the performance specification.[106] In *George Sollitt Construction Co. v. United States,*[107] a Navy base buildings renovation contract included electrical drawings showing 600 amp service to be wired pursuant to the drawings to power a new chiller to be selected and installed by the contractor in one of the buildings. The court found the electrical service drawing to be a design-type specification but the contract's chiller specification to be "a typical performance specification, where the government specified only the result, that of a particular cooling capacity, and Sollitt was left the discretion to install an appropriate chiller."[108] The contractor submitted its choice of chiller to the Navy for approval, and the Navy approved the selected chiller. Upon installation of the chiller, it was found that the 600 amp electrical service was not adequate to power the chiller. The contractor then had additional work performed to revise the electrical service to 800 amp.

The court found that the evidence did not establish the original design specification for the 600 amp electrical service to be defective and so rejected the contractor's contention of government liability for defective specifications. The court further wrote, however, that it was reluctant to absolve the Navy from all responsibility for costs the contractor incurred to revise the electrical service. The court reasoned:

> Sollitt had a duty to install a functioning chiller that would serve Building 122. The Navy had a duty to provide a workable design, and to approve Sollitt's submittals if Sollitt's submittals would permit that design to work. Both parties neglected these duties to some extent, and extra costs were incurred to rip out the 600 amp service and install an 800 amp service. Because this particular controversy does not fit neatly into liability under a particular contract clause, the court turns to equitable solutions from controlling contract law. As the Court of Claims stated, in some cases one finds "the general proposition that, when a misunderstanding results from carelessness by both parties to a contract, neither should benefit at the expense of the other." *Cover v. United States,* 174 Ct. Cl. 294, 356 F.2d 159, 160 n. 4 (1966).[109]

The court found the Navy liable for half the incurred costs to revise the chiller electrical power service.[110] This is not, however, the typical outcome when the specification at issue is a performance-type specification.

The implied warranty applies to purchase specifications, but the government typically warrants only that the specified purchase item can meet the requirements of the contract; there is no warranty to the item's price.[111] Likewise, if the government's purchase specification designates a particular manufacturer or even a sole source, the government does not warrant that the manufacturer or sole-source supplier will properly manufacture or timely deliver an item.[112] The government only represents "that the requirements of the contract can be met by the use of the item."[113]

With composite specifications, the issue is more complex; while the government's implied warranty will apply to the design components of the specification, it will not apply to the performance components. Each part of the composite specification must stand on its own insofar as contractor responsibility is concerned.[114]

Consequently, whether the specifications or drawings at issue are design or performance or purchase-type specifications becomes critically important in determining whether the government's implied warranty applies and, if so, to what extent.[115] When the specification at issue is a composite specification, then the specific provisions, drawings, or terms at issue must be distinguished and determined whether those terms qualify as design or performance or purchase specifications.[116]

2. Other Risks

In a case involving OSHA violations, *Fabi Construction Co. v. Secretary of Labor*,[117] the D.C. Circuit found that the contractor was not shielded from OSHA liability under the *Spearin* doctrine for the collapse of a parking garage during construction resulting in fatalities. In *Fabi*, the OSHA Commission had affirmed citations of the contractor for failing to place rebar in accordance with industry practice, thereby violating the General Duty clause of OSHA to provide a working environment free from recognized hazards causing or likely to cause death or serious physical harm to employees. In its appeal, the contractor had argued that it had followed the approved shop drawings in placing the rebar. Four trade associations, arguing amici curiae, contended that, because the contractor had followed the approved shop drawings, it was shielded from liability for the OSHA citations under the *Spearin* doctrine. The court disagreed, writing, "*Spearin* does not shield Petitioners from liability here. *Spearin* dealt with contractual obligations—not OSH Act obligations to provide a safe worksite."[118] The court affirmed the OSHA Commission finding that the contractor knew or should have known its rebar placement, even though in accordance with the approved shop drawings, violated industry practice. The court held: ". . . *Spearin* does not extend to the facts before us. Petitioners knew or should have known that the drawings upon which they relied created a serious hazard for their employees."[119]

The result in *Fabi* should be compared with that in *AAB Joint Venture v. United States*.[120] In *AAB*, the COFC held, "Specifications having major safety defects are fully as much in breach of the implied warranty as defects in the feasibility, practicability or commercial possibility of performance as specified."[121]

In another case, *Rick's Mushroom Service, Inc. v. United States*,[122] the Federal Circuit held that the *Spearin* doctrine did not apply to a waste facility operator's claim for equitable indemnification to recoup the costs incurred in defending and settling claims brought by a third party for violation of state and federal environmental laws.[123] Citing the U.S. Supreme Court's ruling in *Hercules Inc. v. United States*,[124] the court wrote that the *Spearin* doctrine does not extend as far as third-party claims.[125]

In addition, as discussed in the next section, the government's implied warranty is not applicable if the defective design specifications did not result in increased costs.[126] Likewise, if the contractor is not "misled" by the defective design specifications, the government's implied warranty will not apply.[127]

III. Recovery for the Government's Defective Specifications

A defective specification within the scope of the government's implied warranty does not mean per se liability for the government. The claimant contractor must still prove all the elements of a contract-based claim for breach of warranty.[128]

A. Elements and Burden of Proof

A contractor claiming an equitable adjustment for the breach of the implied warranty of the government's plans and specifications must demonstrate:

- That the specifications were of the type warranted;
- That reasonable reliance on, and compliance with, the specifications at issue were demonstrated;
- That the specifications were defective; and
- That the defective specifications caused the claimed increased costs.[129]

The burden is on the claimant asserting the breach to establish the fundamental facts of liability, causation, and resultant injury.[130]

For example, a claimant's failure to demonstrate reasonable reliance on, and compliance with, the defective specifications will preclude recovery for breach of the implied warranty of specifications.[131] In *Travelers Casualty & Surety of America v. United States*,[132] the COFC wrote:

> Moreover, the contractor must fully comply with and follow the design specifications, although faulty, to enjoy the protections of the implied warranty, unless the departure from the specifications is "entirely irrelevant to the alleged defect." *Al Johnson Constr. Co. v. United States*,

854 F.2d 467, 470 (Fed. Cir. 1988); *see Gulf Western Precision Eng'g Co. v. United States*, 211 Ct. Cl. 207, 218, 543 F.2d 125 (1976); *Sterling Millwrights, Inc. v. United States*, 26 Cl. Ct. 49, 88 (1992).[133]

Likewise, a failure to demonstrate that the specifications at issue were defective and within the scope of the implied warranty is fatal.[134] In *Caddell Construction Co. v. United States*,[135] the COFC addressed the extent to which a design specification must be defective to be found a breach of the government's implied warranty:

> Although design specifications are meant to give a contractor a very detailed guide on how to complete a project, they "need not be paragons of perfection" but must be "reasonably accurate." *Travelers*, 74 Fed. Cl. at 89 (citations omitted). A defective specification is one that is "so faulty as to prevent or unreasonably delay completion of the contract performance." *Wunderlich v. United States*, 173 Ct. Cl. 180, 351 F.2d 956, 964 (1965). Furthermore, the government's documents must be "substantially deficient or unworkable" in order to be considered a breach of the contract. *Id.* If there are many errors or omissions in the specifications, the government breached the contract if "the cumulative effect or extent of these errors was either unreasonable or abnormal" taking into account the scope and complexity of the project. *Id.*[136]

In *Caddell*, in the words of the court, "[t]he majority of plaintiff's witnesses' testimony at trial focused on the effect of the RFI [request for information] process on the fabrication of steel. Essentially, plaintiff's witnesses tried to show that the number of RFIs and the short time period during which they were generated indicated that the plans were faulty."[137] The court did not find the contractor's evidence sufficient to meet its burden of showing that the specifications were defective, concluding:

> After reviewing the evidence presented at trial under the standard set out above, the court concludes that the plans were not defective. Although plaintiff's witnesses repeatedly testified that the plans were massively flawed, their statements were conclusory, with little or no evidentiary support. At trial, plaintiff's witnesses pointed to the number and nature of the RFIs generated on the project as a basis for the conclusion that the plans were defective. The court, however, does not agree that there was an unusually high number of RFIs or that the RFIs showed that the plans were so riddled with conflicts or missing information that the problems with the contract documents rose to the level of defectiveness.

* * *

What SSC faced in this case was a collection of small errors. A design specification, however, is defective when it contains some fundamental flaw or collection of flaws that requires a major revision that delays the project. All of this leads to the inexorable conclusion that the plans were not defective. The plans were not "unworkable" nor were they "substantially deficient." *See Wunderlich*, 351 F.2d at 964. Furthermore, the plans were not "so faulty as to prevent or unreasonably delay completion of the contract performance." *Id.* Instead, the work stalled because of Caddell's mishandling of the RFIs. Finally, plaintiff provided nothing beyond the self-serving statements of its witnesses that demonstrated that "the cumulative effect or extent of the[ ] errors was either unreasonable or abnormal." *Id.* Therefore, defendant is not responsible for any of plaintiff's claimed damages that it may have suffered as a result of the alleged delay.[138]

Lastly, a failure to show that the defective specifications were the cause of the claimed costs will preclude recovery.[139] In *Caddell*,[140] the COFC wrote:

> Even if plaintiff had been able to present evidence that the government breached the implied warranty of specifications, it would still have had to prove that the government caused SSC's alleged damages.
> * * *
> Defendant, therefore, is not liable for breach of contract, or causes of action that rely upon a defective design claim, or government hindrance of performance, unless plaintiff proved that the alleged defects, changes or hindrances negatively impacted costs and performance of the contract.[141]

B. Government Defenses to Recovery for Breach of the Implied Warranty of Specifications

1. Knowledge of Defects or Deficiencies in the Specifications before Contract Award
A contractor with actual knowledge or constructive knowledge of defects in the specifications prior to contracting with the government cannot subsequently recover damages for breach of the implied warranty of specifications.[142]

In *Wickham Contracting Co. v. United States*,[143] the bid documents included two different scales causing the contractor to underbid the work. Because of the discrepancy between the bids and the government's estimate, the government asked the low bidder to review and verify its bid. During the contractor's review of its bid, the contractor discovered the two different drawing scales. Nevertheless, the contractor verified its initial bid. During performance, the contractor notified the government of discrepancies between the actual field measurements and the measurements as scaled on the contract drawings. The board denied the contractor's claim for the added costs resulting from the erroneous scale on the contract drawing. The then Court of Claims affirmed the board's decision as follows:

If a contractor enters into a contract aware of the fact of defective specifications, it is not entitled to recover on a claim based on these defective specifications. *R.E.D.M. Corp. v. United States*, 428 F.2d 1364, 192 Ct. Cl. 891, 902–03 (1970). At the same time it was requested to verify its bid, plaintiff was aware of the drawing scale error, and whatever duty the government may have had before to advise plaintiff of the error faded. Both sides had equal knowledge of the error at this point. *Donald M. Drake Co. v. United States*, 439 F.2d 169, 172, 194 Ct. Cl. 549, 555 (1971). Plaintiff confirmed its bid after being asked to verify it. This verification request was certainly a warning that the bid might be inadequate. At this time the drawing error was known to both sides. Confirmation of its bid under these circumstances was surely "an unreasonable assumption of risk" by plaintiff. *Highway Products, Inc. v. United States*, 530 F.2d 911, 208 Ct. Cl. 926 (1976); *see also Rixon Electronics, Inc. v. United States*, 536 F.2d 1345, 1351, 210 Ct. Cl. 309, 319 (1976). [Footnote omitted.] If plaintiff embarked on a "ruinous course of action," it was a journey it chose to take with its eyes wide open. It certainly did not act reasonably under the circumstances and the Board rightly denied its claim. *See Ling-Temco-Vought, Inc. v. United States*, 475 F.2d 630, 201 Ct. Cl. 135 (1973); *see also Snyder Corp. v. United States*, 68 Ct. Cl. 667, 676 (1930). Since plaintiff was aware of the drawing error at the time it entered into the contract, refusing an opportunity to withdraw its bid based on the drawing error, it is not entitled to recover additional costs by way of a contract price adjustment based on said error. *See Union Painting Co. v. United States*, 198 F. Supp. 282, 284 (D.C. Alaska 1961).[144]

In *Fox Construction Inc.*[145] the contract was to construct buildings and support facilities at a naval air facility. A dispute arose over whether certain irrigation ditch canal work was part of the contract. The contractor argued that it was reasonable to conclude that the ditch work was not in the contract scope inasmuch as there was no section of the specifications that dealt directly with ditches. The contractor also contended that the Navy did not answer the contractor's questions at the pre-bid conference with respect to the means and methods to install the ditches and what the Navy's expectations were with respect to the ditches. The ASBCA rejected the contractor's claim to construct the ditches:

> We are not persuaded that the specification was defective or misleading simply because it did not put all the ditch work under a separate section or division pertaining specifically and exclusively to ditch work. As the Navy explained, the contract specification for the project was divided into 16 divisions. The specification in this case followed the CSI format "used widely in construction." (Finding 17)

* * *

Fox also appears to argue that it was justified in concluding that ditch work was not a part of the contract because the Navy did not answer its pre-bid conference inquiry with respect to "the means and the method to install the ditch" and what the Navy's "expectations" were (finding 5). Despite the lack of an answer to the questions it found troubling, Fox nonetheless chose to bid the contract, and after it bid the contract, verified its bid (findings 6, 10). Even assuming the specification was defective in not containing sufficient information for a contractor to build the ditches shown on the drawings, Fox cannot recover. It is well established that when a contractor enters into a contract aware of the specification defects, it is not entitled to recover on a claim based on those defects. *Robins Maintenance, Inc. v. United States*, 265 F.3d 1254, 1258 (Fed. Cir. 2001); *Johnson Controls, Inc. v. United States*, 671 F.2d 1312, 1320 (Ct. Cl. 1982); *Wickham Contracting Co., Inc. v. United States*, 546 F.2d 395, 400 (Ct. Cl. 1976).[146]

In *Nielsen-Dillingham Builders J.V. v. United States*,[147] the COFC similarly ruled that the existence of a patent ambiguity in government contract specifications raises a duty on the part of the contractor to inquire as to the correct meaning of the contract prior to submitting a bid. Likewise, more recently in *Travelers Casualty*,[148] the COFC wrote:

A contractor is subject to a "duty to investigate or inquire about a patent ambiguity, inconsistency, or mistake when the contractor recognized or should have recognized an error in the specifications or drawings." *White*, 296 F.3d at 1085; *see Woodcrest Constr. Co. v. United States*, 187 Ct. Cl. 249, 260, 408 F.2d 406 (1960); *Jefferson Constr. Co. v. United States*, 176 Ct. Cl. 1363, 1368–69, 364 F.2d 420 (1966); *PBI Elec. Corp. v. United States*, 17 Cl. Ct. 128, 132–33, 135 (1989).[149]

If a contractor fails to conduct such an inquiry, a patent ambiguity in a contract will be construed against the contractor and preclude recovery for increased costs due to the ambiguous specification.[150] The fact that the contract is design-build does not relieve a contractor with actual knowledge of an ambiguity in the specifications from the duty of inquiry.[151]

An exception to the patent ambiguity rule does exist when the government also has notice of the ambiguity.[152] In that event, even if the contractor failed to inquire, the government has the duty to clarify the ambiguous specification, and, if the government fails to do so, the contractor will be entitled to recover its resultant added costs.[153]

2. During Contract Performance

A contractor will not be allowed recovery for breach of the implied warranty of specifications if it builds to a known erroneous specification or standard and then seeks recovery from the government for correction of the error.[154]

A contractor will not recover for breach of the implied warranty of specifications if, during contract performance, the defective specifications did not cause the contractor's claimed added performance costs.[155] Likewise, if the contractor undertook to perform the defective specifications, but did not incur increased costs as a consequence, the implied warranty of specifications is not applicable as a basis for recovery.[156]

IV. Impossibility and Impracticability of Performance
A. The Doctrines of Impossibility and Impracticability

At common law, the doctrines of impossibility and impracticability of performance excuse a contractor's failure to perform its contract.[157] In federal government contracting, if the government's specifications are found to be impossible or commercially impracticable to meet, then the specifications are defective, and a breach of the government's implied warranty of its specifications.[158] Such a finding can excuse the contractor's performance,[159] and the contractor can recover its costs resulting from relying on those specifications as a constructive change.[160]

The doctrines of impossibility and commercial impracticability are distinct from one another, and both are distinct in application from the implied warranty of specifications under *Spearin*. In *Columbia Engineering Corp*,[161] the ASBCA distinguished between the application of the doctrines of impossibility and impracticability and application of the implied warranty of specifications under *Spearin*:

> In their briefs, both parties misapply standards imposed by the doctrine of commercial impossibility or impracticability to this case. Generally, that doctrine is applicable only to disputes arising from performance specifications, which specify required performance characteristics but leave design responsibility with the contractor. Where, as here, tolerances are specified, the specifications fail to achieve a satisfactory result, then the specifications are deficient. As we stated in Dynalectron Corporation, Pacific Division, ASBCA Nos. 11766, 12771, 69-1 BCA ¶ 7595 at 35,275, these are not the circumstances that create the legal doctrine of commercial impossibility, and: "There is no necessity . . . for a leap into the complicated, turbulent and esoteric seas of 'impossibility.'"[162]

Whether the federal government's specifications are impossible or commercially impracticable to perform is a question of fact, not of law, and the burden of proof lies with the one claiming impossibility or impracticability.[163]

1. Impossibility

While some boards and courts interchange the terms "impossibility" and "impracticability,"[164] the doctrines are separate and distinct. The doctrine of impossibility means actual, objective impossibility for the industry as a whole, not just for the claimant contractor.[165] In contrast, commercial impracticability means the increased cost of performance is so great, as compared to the anticipated cost, that the performance is commercially senseless;[166] nonetheless, performance is not objectively impossible.

In order to establish impossibility, the contractor must show that the government's specifications required performance beyond the state of the art.[167] To conclude that the performance is beyond the state of the art, the evidence must support an objective finding that neither the claimant nor any other contractor could accomplish the specified task.[168]

In *Oak Adec, Inc. v. United States,*[169] the United States Claims Court set forth the factors to be considered in evaluating a claim of impossibility:

> 1) whether any other contractor was able to comply with the specifications; 2) whether the specifications require performance beyond the state of the art; 3) the extent of the contractor's efforts in meeting the specifications; and 4) whether the contractor assumed the risk that the specifications might be defective. [Footnotes omitted.][170]

In regard to the first factor, "whether any other contractor was able to comply with the specifications," the court reaffirmed that the contractor "must prove that the industry as a whole found the specifications impossible."[171] In regard to the second factor going to the "state of the art," the court acknowledged, "[i]t does not necessarily follow, however, that because a specification embraces state-of-the-art technology, that it thereby advances the state-of-the-art."[172] As for the third factor, the extent of the contractor's efforts, the court wrote:

> A third factor to be considered is the extent of the contractor's effort. *Norair Eng'g. Corp.*, 79-1 B.C.A. (CCH) ¶ 13,636 (A.S.B.C.A. 1979) (more than "half-hearted" efforts are necessary to show commercial impossibility).
>
> * * *
>
> As stated by the board in *ASC Sys. Corp.*, 78-1 B.C.A. (CCH) ¶ 13,119 (A.S.B.C.A. 1978): "[p]erformance impossibility must be established on an objective basis, not subjectively. That is, the contractor may not rely solely upon his own inability to accomplish the specified task; he must also negate the possibility of performance by others. . . ." *ASC Sys.*, 78-1 B.C.A. at 64,134. The objective standard is thus not intended to be a "sword" for plaintiffs alleging impossibility, but rather a "shield" to be used by defendants to deflect such charges by a contractor whose own inability was the cause of non-performance. The standard does not operate as *Oak Adec* contends to insulate the contractor from inquiry into its management of the contracts.[173]

In regard to the fourth factor, whether the contractor assumed the risk that the specification may be defective, the court wrote:

> The fourth issue relates to whether the plaintiff assumed the risk of non-performance.
>
> * * *
>
> The appropriate consideration with respect to performance specifications is whether the plaintiff can demonstrate that the defendant should bear the risk that the specifications may be impossible to achieve.
>
> In that regard, certain issues arise. One made relevant here is, which party drafted the specifications? The party that drafts specifications normally bears the risk that those specifications will be possible to perform, *J.A. Maurer*, 485 F.2d at 594–95, but a contractor's participation in the drafting of performance specifications can operate to shift the risk of impossibility to the contractor. *Bethlehem Corp. v. United States*, 199 Ct. Cl. 247, 462 F.2d 1400 (1972).
>
> * * *
>
> A similar degree of contractor participation in the drafting of specifications occurred in *Bethlehem Corp. v. United States*, 199 Ct. Cl. 247, 462 F.2d 1400 (1972). In *Bethlehem*, the Government contacted several contractors, including the plaintiff, for consultation on the development of an environmental test chamber. *Bethlehem*, 199 Ct. Cl. at 250, 462 F.2d at 1401. During the course of the ensuing discussions, the plaintiff assured the Government that it was possible to build a test chamber meeting the Government's proposed specifications. *Id.* The court held that these assurances, coupled with the fact that the plaintiff was an expert in the field, constituted an assumption of the risk of impossibility. *Id.* 199 Ct. Cl. at 254, 462 F.2d at 1404. In support of its holding, the *Bethlehem* court reasoned:

> > Acceptance of [the contractor's] argument would mean that though a purchaser makes his choice because of the attractiveness of a manufacturer's representation and will be bound by it, the manufacturer is free to express what are only aspirations and gamble on mere probabilities of fulfillment without any risk of liability. In the fields of developing technology, the manufacturer would thus enjoy a wide degree of latitude with respect to performance while holding an option to compel the buyer to pay if the gamble should pan out.

> *Id.* at 255, 462 F.2d at 1404 (quoting *United States v. Wegematic Corp.*, 360 F.2d 674 (2d Cir. 1966)).[174]

As noted before, if the contractor can prove that the specifications it relied on were impossible to achieve, then such a finding can excuse the contractor's performance and be ruled a constructive change entitling the contractor to recover its resultant added costs.[175]

2. Impracticability

Commercial impracticability is a subset of the doctrine of legal impossibility.[176] Commercial impracticability ". . . excuses a party's delay in performing or nonperformance when the attendant costs become excessive and unreasonable 'by an unforeseen supervening event not within the contemplation of the parties at the time the contract was formed.'"[177] Commercial impracticability has been said to equate to commercial senselessness.[178] In *Southern Dredging Co.*,[179] the Corps of Engineers Board of Contract Appeals (ENG BCA) wrote: "Commercial senselessness occurs when unreasonable, excessive and unforeseen increases occur in a contract's cost of performance making performance senseless from a business standpoint."[180] A court or board may treat a finding of impracticability as a constructive change to the contract.[181]

To prevail on a defense or claim of commercial impracticability, a party must show that:

(i)   a supervening event, either an Act of God or an act of a third party, made performance impracticable;

(ii) the nonoccurrence of the event was a basic assumption upon which the contract was based;

(iii) the occurrence of the event was not the party's fault; and

(iv) the party did not assume the risk of the event's occurrence.[182]

Whether performance according to a particular contract would be commercially impracticable is a question of fact.[183] The party asserting the doctrine of impracticability bears the burden of showing "that it explored and exhausted alternatives before concluding that the contract was . . . commercially impracticable to perform."[184]

For example, it is not enough to invoke the doctrine of commercial impracticability when a party only shows its actual costs exceeded its forecasted costs. In *Southern Dredging*,[185] the contract was for maintenance dredging of the Atlantic Intracoastal Waterway, where the cost of fuel was a significant portion of the contract price. During the performance of the contract, the Gulf War erupted, spiking the cost of fuel oil. Among other allegations, the contractor contended the escalated fuel costs rendered continued performance commercially impracticable. The ENG BCA disagreed, finding that the fuel cost increase of 6.62 percent of the contract price did not rise to the level of commercial senselessness.[186]

In *Spindler Construction Corp.*,[187] the contractor passed through the claim of its steel fabrication/erector subcontractor for increased costs of prefabricated steel amounting to 23 percent over the estimated costs. The claimant attributed the cost increase to "unforeseen steel cost increases" due to the "global steel crisis."[188] Citing the four elements of proof to establish commercial impracticability, the ASBCA found that the contractor failed to demonstrate three of those elements of proof. In specific regard to the supervening event element of proof, the board wrote:

> As to the first element, the supervening market fluctuation in the price of steel here did not make contract performance impracticable. *See Seaboard Lumber*, 308 F.3d at 1294; *Jennie-O Foods, Inc. v. United States*, 580 F.2d 400, 409 (Ct. Cl. 1978). On the contrary, the 23 percent increase in the cost of steel represents less than a five percent cost overrun of the subcontract price. *See Raytheon Co. v. White*, 305 F.3d 1354, 1368 (Fed. Cir. 2002); *Gulf and Western Industries, Inc.*, ASBCA No. 21090, 87-2 BCA ¶ 19,881 at 100,575 (cost overruns of 57 percent and 70 percent, respectively, did not make performance commercially impracticable).[189]

Regarding the second and fourth elements of proof, the board in *Spindler* held that it was not a basic contract assumption that material cost increases would not occur for the very reason that a fixed-price contract normally assigns the risk of cost increases to the contractor, not the government. The board considered it irrelevant that the subcontractor may have assumed that the steel market would remain within a generally predictable range because the subcontract was for a fixed price as well, and the subcontractor could not therefore unilaterally shift risk away from itself. The board went further to hold that, even if the contractor or the government had shared the subcontractor's assumption that steel prices would stay within a generally predictable range, such an assumption was irrelevant because assumptions as to market shifts do not usually change the basic fixed-price contract risk allocation that, absent an economic adjustment contract clause, the risk of material cost increases rests with the contractor.[190]

Of course, when claiming commercial impracticability, as with the doctrine of impossibility, the contractor must also demonstrate that its failure to perform was not due to its own inadequacy, but must relate its failure to the commercial impracticability imposed by the work specified in the contract. In *GLR Constructors*,[191] the ENG BCA denied the contractor's claim of commercial impracticability, writing:

> Moreover, since some of the end plate assemblies passed the air pressure test, performance was proven to be possible by the Appellant itself. We agree with the Government's experts that performance was commercially practicable if the contractor had properly planned and executed the work.[192]

Despite the difficulty in proving impossibility or commercial impracticability, the doctrines are applicable to defective performance specifications and may apply to excuse a nonperforming contractor or be the basis of recovering costs when the implied warranty of the *Spearin* doctrine would not otherwise apply.[193]

B. Assumption of Risk

The doctrines of impossibility and commercial impracticability will not apply regardless of the type of specification if the contractor has assumed the risk that the specifications may be defective.[194] For example, in *Reflectone, Inc.,*[195] the ASBCA acknowledged the principle that a contractor may assume the risk of defective specifications, and after having done so, cannot claim impossibility:

> Even if the integration of these existing and conventional technologies did go beyond the state of the art as Reflectone asserts, however, a contractor may bind itself to do so. *See United States v. Wegematic Corp.,* 360 F.2d 674, 675–77 (2d Cir. 1966) (evidence of true impossibility far from compelling where contractor proposed a computer it characterized as "a truly revolutionary system utilizing all of the latest technical advances" and what "seemingly did become impossible was on-time performance.") *See also Kinn Electronics Corp.,* ASBCA No. 13526, 69-2 BCA ¶ 8061, at 37,475, where we stated that the Government must prevail where the contractor, by its assurances that it possesses novel and "'revolutionary' technical processes," induces contract award and that, "[h]aving held himself out as capable beyond ordinary measure, he cannot shield himself behind technical impossibility of performance when it fails."[196]

Likewise, a contractor may be deemed to have assumed the risk of defective specifications if it participated in the development and writing of the defective specification.[197] A contractor also may assume the risk of defective specifications by using either its own specification, or its own means and methods of performance.[198]

As discussed above, a contractor will be deemed to have assumed the risk of defective specifications if it had actual or constructive knowledge of the defect before contracting[199] or performed the work knowing the specifications to be erroneous.[200]

Further, an impracticability defense or claim will not apply if the party relying on the doctrine has assumed the risk of the supervening event.[201] For example, in *Spindler Construction Corp.,*[202] the ASBCA wrote:

> [T]he non-occurrence of increased costs was not a basic contract assumption because a fixed-price contract normally assigns the risk of price increases to the contractor. Spindler's contract with the government to design and build a new aircraft depot maintenance hangar was a fixed-price contract that insulated the government from the risks of cost increases. Spindler's subcontract with Sanpete Steel was also fixed-price. While Sanpete Steel assumed that the steel market would remain within a "generally predictable range," this was not a basic, or normal, assumption about the general risk of possible cost increases for a fixed-price contract.[203]

Other decisions have similarly recognized that fixed-price contractors are generally said to have assumed the risk of cost increases in the absence of an economic adjustment clause.[204]

In *L.W. Matteson, Inc. v. United States*,[205] the COFC found that the dredge work contractor assumed the risk that a local county would deny a land use permit for the contractor's disposal operations. Among other claims, the contractor contended that the contract was commercially impracticable to perform because the Corps of Engineers had failed to warn the contractor of the County's hostility toward hydraulic dredging. The court did not agree, finding that the contract placed the risk of complying with applicable laws on the contractor:

> It can hardly be denied that the contract clearly and unequivocally assigned to Matteson the burden of complying with all applicable, federal, state, and local laws, including those of Wabasha County. It therefore follows—in the absence of express provisions to the contrary—that the burden of complying with all applicable laws must include the inchoate obligation to research relevant local legal requirements, as well as any attending problems (such as local opposition) that may hinder the contractor's compliance. Compliance becomes meaningless without the necessary and understood requirement of due diligence. The practical effect of this contractual duty, consequently, is that the contractor bears the risk that local authorities would deny the necessary permission needed to dump on an alternative private property site. The court reiterates that Matteson is a sophisticated contractor.[206]

Consequently, the court rejected the contractor's claim of commercial impracticability based on the county's denial of a permit:

> Here, the court must reject Matteson's commercial impracticality [sic] claim because Matteson fails to allege facts legally sufficient to demonstrate either that the obstruction to the performance of the contract was the cause of the defendant, or that the contract placed the risk the [sic] of the supervening event on the Corps and not Matteson. To be sure, this claim is deficient because as the court previously held, it was Matteson that bore the risk of the County's denial of a conditional-use permit under the contract.[207]

In considering whether assumption of risk will preclude a claim of commercial impracticability, the court in *L.W. Matteson*, quoting *Corbin on Contracts* section 1333, wrote: "[T]he Court must be alert to weigh any evidence to show a custom throwing the risk upon one party rather than the other, or to show that there was a mutual contemplation of it being borne by one of them."[208]

V. Conclusion

Defective drawings and specifications are a reality of construction contracts. In those circumstances, one of the fundamental questions to address is who assumed the risk of those defects. The government's implied warranty of its specifications and the doctrines of impossibility and impracticability often-times provide the answer.

Notes

1. FAR 36.000. Federal A-E contracts are discussed in detail in Chapter 4.
2. FAR 36.192.
3. Metric Constr. Co., Inc. v. United States, 80 Fed. Cl. 178, 181 (2008) (no dispute that the roof of the warehouse was a performance specification and that the structural steel framework to which the roofing system attached was a design specification); Travelers Cas. & Sur. of Am. v. United States, 74 Fed. Cl. 75, 89 (2006); Hawaiian Bitumuls & Paving, a Div. of Dillingham Constr. Pac., Ltd. v. United States, 26 Cl. Ct. 1234, 1240 (1992); Santa Fe Eng'rs, Inc., ASBCA No. 24469, 92-1 BCA ¶ 24,665 at 123,068 (1991).
4. *See, e.g.,* PCL Constr. Servs., Inc. v. United States, 47 Fed. Cl. 745, 796 (2000); Hardwick Bros. Co., II v. United States, 36 Fed. Cl. 347, 411 (1996); Kimmins Con-tracting Corp., ASBCA No. 41615, 91-2 BCA ¶ 23,914 (1991); Monitor Plastics Co., ASBCA No. 14447, 72-2 BCA ¶ 9626 (1972).
5. *PCL Constr. Servs.,* 47 Fed. Cl. at 795.
6. Caddell Constr. Co., Inc. v. United States, 78 Fed. Cl. 406, 411 (2007); Travel-ers Cas., 74 Fed. Cl. at 89–90; George Sollitt Constr. Co. v. United States, 64 Fed. Cl. 229, 296–97 (2005); *Monitor Plastics,* 72-2 BCA ¶ 9626.
7. 188 Ct. Cl. 684, 412 F.2d 1360 (1969).
8. 188 Ct. Cl. at 689, 412 F.2d at 1362.
9. *See, e.g., PCL Constr. Servs.,* 47 Fed. Cl. at 795–96 (2000).
10. *Caddell Constr.,* 78 Fed. Cl. 406.
11. *Id.* at 412.
12. 74 Fed. Cl. 75 (2006).
13. *Id.* at 89.
14. *See, e.g., Caddell Constr.,* 78 Fed. Cl. at 411.
15. George Sollitt Constr. Co. v. United States, 64 Fed. Cl. 229, 297 (2005).
16. *See, e.g., George Sollitt Constr.,* 64 Fed. Cl. 229; J.L. Simmons Company, Inc. v. United States, 188 Ct. Cl. 684, 412 F.2d 1360 (1969); Monitor Plastics Co., ASBCA No. 14447, 72-2 BCA ¶ 9626 (1972).
17. 64 Fed. Cl. 229.
18. *Id.* at 297.
19. GSBCA No. 4270, 79-1 BCA ¶ 13,728 (1979); *see also* Santa Fe Eng'rs Inc., ASBCA No. 24469, 92-1 BCA ¶ 24,665 (1991), (dimensions not shown on the draw-ings do not result in recovery for contractor when the omitted dimensions are performance requirements).

20. ASBCA No. 55251, 08-1 BCA ¶ 33,800 (2008).

21. *Id.* at 167,331.

22. 277 F.3d 1346 (Fed. Cir. 2002).

23. *Id.* at 1357–58; *see also* Hardwick Bros. Co., II v. United States, 36 Fed. Cl. 347, 410–12 (1996).

24. 47 Fed. Cl. 745 (2000).

25. *Id.* at 799.

26. *Id.* at 797.

27. *Id.* at 798.

28. Monitor Plastics Co., ASBCA No. 14447, 72-2 BCA ¶ 9626 (1972); *see also* Frank Briscoe Co., NASA BCA No. 879-15, 80-1 BCA ¶ 14,273 (1980).

29. FAR 36.202.

30. FAR 36.202(c).

31. FAR 11.104, 11.105.

32. *See, e.g.,* Gen. Ship Corp. v. United States, 634 F. Supp. 868 (D. Ma. 1986); Monde Constr. Co., Inc., ASBCA Nos. 44993 et al., 96-2 BCA ¶ 28,400 (1996); Ala. Dry Dock & Shipbuilding Corp., ASBCA No. 39215, 90-2 BCA ¶ 22,855 (1990).

33. *See, e.g.,* Santa Fe Eng'rs, Inc., ASBCA No. 22950, 79-2 BCA ¶ 14,084.

34. ASBCA No. 28731, 89-3 BCA ¶ 22,090 (1989).

35. *Id.* at 111,086.

36. ASBCA No. 14447, 72-2 BCA ¶ 9626 (1972).

37. *Id.* at 44,971.

38. 47 Fed. Cl. 745 (2000).

39. *Id.* at 795.

40. Caddell Constr. Co., Inc. v. United States, 78 Fed. Cl. 406, 411 (2007), citing White v. Edsall Constr., 296 F.3d 1081, 1085 (Fed. Cir. 2002); Hercules, Inc. v. United States, 24 F.3d 188, 197–98 (Fed. Cir. 1994).

41. 42 Fed. Cl. 94 (1998).

42. *Id.* at 96.

43. *Id.*

44. *Id.*

45. 47 Fed. Cl. 745 (2000).

46. *Id.* at 796.

47. 78 Fed. Cl. 406 (2007).

48. *Id.* at 411–12.

49. United States v. Spearin, 248 U.S. 132 (1918).

50. *Id.*

51. *Id.* at 136.

52. PCL Constr. Servs., Inc. v. United States, 47 Fed. Cl. 745, 795 (2000).

53. Blake Constr. Co. v. United States, 987 F.2d 743, 745 (Fed. Cir. 1993).

54. George Sollitt Constr. Co. v. United States, 64 Fed. Cl. 229, 296 (2005); Essex Elec. Eng'rs, Inc. v. Danzig, 224 F.3d 1283, 1289 (Fed. Cir. 2000).

55. *See* La Crosse Garment Mfg. Co. v. United States, 193 Ct. Cl. 168, 180–81, 432 F.2d 1377, 1385 (1970); AEI Pac., Inc., ASBCA No. 53806, 08-1 BCA ¶ 33,972 (2008).

56. Lamb Eng'g & Constr. Co., ASBCA Nos. 53304 et al., 06-1 BCA ¶ 33,178 (2006); Cable & Computer Tech., Inc., ASBCA No. 47420, 48846, 03-1 BCA ¶ 32,237 (2003).

57. *George Sollitt Constr.*, 64 Fed. Cl. at 296; White v. Edsall Constr., 296 F.3d 1081, 1084, 1087 (Fed. Cir. 2002); *Lamb Eng'g*, 06-1 BCA ¶ 33,178 at 164,417.

58. Abcon Assocs., Inc., PSBCA No. 5291, 08-1 BCA ¶ 33,762 at p. 167,136–37 (2008). For further discussion of convenience terminations, see Chapter 13.

59. PCL Constr. Servs., Inc. v. United States, 47 Fed. Cl. 745, 795 (2000).

60. *See, e.g.*, Blake Constr. Co., ASBCA No. 36300, 90-3 BCA ¶ 23,077 (1990); David Boland, Inc., ASBCA No. 48715, 97-2 BCA ¶ 29,166 (1997).

61. *See, e.g.*, Wickham Contracting Co. v. United States, 212 Ct. Cl. 318, 546 F.2d 395 (1976).

62. *See, e.g.*, Rick's Mushroom Serv., Inc. v. United States, 521 F.3d 1338, 1345–46 (2008) (implied warranty under *Spearin* does not extend to indemnity for third-party claims.); Fabi Constr. Co., Inc. v. Sec'y of Labor, 508 F.3d 1077, 1084–85 (2007) (*Spearin* doctrine does not shield employer from failure to comply with OSHA obligations).

63. Swan Wooster Eng'g, AGBCA No. 83-104-1, 87-2 BCA ¶ 19,894 (1987); Brunson Assocs., Inc., ASBCA No. 41201, 94-2 BCA ¶ 26,936 (1994); William Tao & Assocs., Inc., ASBCA No. 32986, 89-2 BCA ¶ 21,588 (1989).

64. FAR 52.236-23.

65. *See, e.g.*, United States v. Rixon Elecs, Inc., 210 Ct. Cl. 309, 536 F.2d 1345 (1976).

66. Bromley Contracting Co., ASBCA Nos. 14884 et al., 72-1 BCA ¶ 9252 (1971).

67. Morrison-Knudsen Co. v. United States, 184 Ct. Cl. 661, 686 (1968).

68. ASBCA Nos. 14884 et al., 72-1 BCA ¶ 9252 (1971).

69. *Id.* at 42,901.

70. *Id.* at 42,902.

71. DOTCAB No. 3012, 97-1 BCA ¶ 28,644 (1996).

72. *Id.* at 143,075.

73. *Id.* at 143,076–77.

74. *Id.* at 143,077.

75. IBCA 2841, 97-2 BCA ¶ 29,072 (1997).

76. *Id.* at 144,677.

77. ASBCA No. 51787, 01-2 BCA ¶ 31,425 (2001), *aff'd*, 296 F.3d 1081 (Fed. Cir. 2002).

78. *Id.* 01-2 BCA at 155,177.

79. *Id.* at 155,180–81.

80. Lamb Eng'g & Constr. Co., ASBCA Nos. 53304 et al., 06-1 BCA ¶ 33,178 (2006).

81. *Id.* at 164,418.

82. 379 F.3d 1334 (Fed. Cir. 2004).

83. *Id.* at 1343.

84. ASBCA No. 36300, 90-3 BCA ¶ 23,077 (1990).

85. *Id.* at 115,876.

86. ASBCA No. 48715, 97-2 BCA ¶ 29,166 (1997); *see also* Elter, S.A., ASBCA No. 52971 02-1 BCA ¶ 31,672 (2002); *compare* Price/CIRI Constr. J.V., ASBCA No. 36988, 89-3 BCA ¶ 22,146 (1989).

87. M.A. Mortenson Co., ASBCA Nos. 53105 et al., 04-2 BCA ¶ 32,713 at 161,847 (2004).

88. 65 Fed. Cl. 657 (2005).

89. For further discussions of this clause, and site inspections generally, see Chapter 10.

90. *Id.* at 676.

91. Design-build and other alternate project delivery systems are discussed in Chapter 5.

92. VABCA No. 6618, 03-1 BCA ¶ 32,129 (2002).

93. *Id.* at 158,826–27.

94. VABCA No. 6937, 04-1 BCA ¶ 32,485 (2003).

95. *Id.* at 160,693.

96. ASBCA No. 55671, 08-2 BCA ¶ 33,868 (2008).

97. *Id.* at 167,642; see also Acquest Gov't Holdings U.S. Geological, LLC v. Gen. Servs. Admin., CBCA No. 439, 07-1 BCA ¶ 33,576 (2007) (GSA's build/lease scheme depicted in Solicitation for Offers not a model of clarity with disclaimers not to rely on conceptual design, and evidence that showed that design-builders did not have unfettered discretion to change initial design to meet performance requirements and to work within design parameters of Solicitation for Offers).

98. PCL Constr. Servs., Inc. v. United States, 47 Fed. Cl. 745 (2000).

99. *Id.*

100. JEM Dev., ASBCA No. 42644, 92-1 BCA ¶ 24,451 (1991).

101. Frank Briscoe Co., NASA BCA No. 879-15, 80-1 BCA ¶ 14,273 (1980).

102. Travelers Cas. & Sur. of Am. v. United States, 74 Fed. Cl. 75, 89 (2006).

103. *See, e.g.,* George Sollitt Constr. Co. v. United States, 64 Fed. Cl. 229, 296–97 (2005).

104. 74 Fed. Cl. 75 (2006).

105. *Id.* at 89.

106. *George Sollitt Constr.,* 64 Fed. Cl. 229.

107. *Id.*

108. *Id.* at 297.

109. *Id.*

110. *Id.* at 298.

111. Frank Briscoe Co., NASA BCA No. 879-15, 80-1 BCA ¶ 14,273 (1980).

112. Demusz Mfg. Co., Inc., ASBCA No. 55310, 07-1 BCA ¶ 33,510 (2007); Demusz Mfg. Co., Inc., ASBCA No. 55311, 07-1 BCA ¶ 33,463 (2006); Monde Constr. Co., Inc., ASBCA Nos. 44993 et al., 96-2 BCA ¶ 28,400 (1996); Ala. Dry Dock & Shipbuilding Corp., ASBCA No. 39215, 90-2 BCA ¶ 22,855 (1990); Cascade Elec. Co., ASBCA No. 28674, 84-1 BCA ¶ 17,210 (1984); Gen. Ship Corp. v. United States, 634 F. Supp. 868, 869–70 (D. Mass. 1986).

113. *Id.;* Interstate Coatings Inc. v. United States, 7 Cl. Ct. 259, 261 (1985).

114. Santa Fe Eng'rs, Inc., ASBCA No. 22950, 79-2 BCA ¶ 14,084 at 69,275–76 (1979); Penguin Indus., Inc. v. United States, 209 Ct. Cl. 121, 530 F.2d 934 (1976).

115. PCL Constr. Servs. Inc. v. United States, 47 Fed. Cl. 745, 795 (2000).

116. Blake Constr. Co. v. United States, 987 F.2d 743 (Fed. Cir. 1993).

117. 508 F.3d 1077 (D.C. Cir. 2007).

118. *Id.* at 1084–85.

119. *Id.* at 1085.

120. 75 Fed. Cl. 414 (2007).

121. *Id.* at 429.

122. 521 F.3d 1338 (Fed. Cir. 2008).

123. *Id.* at 1344–45.

124. 516 U.S. 417, 116 S. Ct. 981, 134 L. Ed. 2d 47 (1996).

125. 521 F.3d at 1346.

126. Franklin Pavkov Constr. Co. v. Roche, 279 F.3d 989 (Fed. Cir. 2002).

127. Robins Maint., Inc. v. United States, 265 F.3d 1254 (Fed. Cir. 2001).

128. M.A. Mortenson Co. v. United States, 40 Fed. Cl. 389, 419 (1998).

129. *Id.; see also* D.F.K. Enters., Inc. d/b/a Am. Coatings v. United States, 45 Fed. Cl. 280, 287 (1999).

130. *M.A. Mortenson Co.,* 40 Fed. Cl. at 419.

131. *Id.* at 419–20; Hardwick Bros. Co., II v. United States, 36 Fed. Cl. 347, 413 (1996).

132. 74 Fed. Cl. 75 (2006).

133. *Id.* at 89–90.

134. Fru-Con Constr. Corp. v. United States, 42 Fed. Cl. 94 (1998).

135. 78 Fed. Cl. 406 (2007).

136. *Id.* at 413.

137. *Id.*

138. *Id.* at 413–16; *see also* Travelers Cas. & Sur. of Am. v. United States, 74 Fed. Cl. 75, 89 (2006) (only if design specifications are so substantially deficient or unworkable as to constitute a breach of the contract may the contractor recover).

139. Franklin Pavkov Constr. Co. v. Roche, 279 F.3d 989, 995 (Fed. Cir. 2002); Robins Maint., Inc. v. United States, 265 F.3d 1254, 1257 (Fed. Cir. 2001).

140. 78 Fed. Cl. 406 (2007).

141. *Id.* at 416.

142. Wickham Contracting Co. v. United States, 212 Ct. Cl. 318, 546 F.2d 395 (1976).

143. *Id.*

144. *Id.* 212 Ct. Cl. at 328–29, 546 F.2d at 400–01.

145. ASBCA Nos. 55265 et al., 08-1 BCA ¶ 33,810 (2008).

146. *Id.* at 167,370.

147. 43 Fed. Cl. 5 (1999).

148. 74 Fed. Cl. 75 (2006).

149. *Id.* at 90.

150. Nielsen-Dillingham Builders J.V. v. United States, 43 Fed. Cl. 5, 10–11 (1999).

151. United Excel Corp., VABCA No. 6937, 04-1 BCA ¶ 32,485 (2003).

152. The Hunt Constr. Group, Inc. v. United States, 48 Fed. Cl. 456 (2001).

153. Price/CIRI Const., J.V., ASBCA No. 37002, 89-3 BCA ¶ 22,059 (1989); Peter Kiewit Sons' Co., ASBCA No. 17709, 74-1 BCA ¶ 10,430 (1974), *aff'd on recon.*, 74-2 BCA ¶ 10,975 (1974).

154. Hunter Ditch Lining, AGBCA No. 87-391-1, 91-2 BCA ¶ 23,673 (1991).

155. Fru-Con Constr. Corp. v. United States, 42 Fed. Cl. 94, 96 (1998) (citing Gulf W. Precision Eng'g. Co. v. United States, 211 Ct. Cl. 207, 218, 543 F.2d 125, 130–31 (1976)).

156. Robins Maint. Inc. v. United States, 265 F.3d 1254, 1257 (Fed. Cir. 2001).

157. Restatement (Second) Of Contracts § 266(1) (1981).

158. Ehlers-Noll v. United States, 34 Fed. Cl. 494, 499 (1995).

159. L.W. Matteson, Inc. v. United States, 61 Fed. Cl. 296, 320 (2004); Raytheon Co. v. White, 305 F.3d 1354, 1367 (Fed. Cir. 2002).

160. Ace Constructors, Inc. v. United States, 499 F.3d 1357, 1364 (Fed. Cir. 2007). Changes are discussed in detail in Chapter 8.

161. ASBCA Nos. 32139 et al., 89-2 BCA ¶ 21,689 (1989).

162. *Id.* at 109,055; *see also* AAB Joint Venture, 75 Fed. Cl. at 429 (citing and quoting Frank J. Baltz & Daniel S. Herzfeld, *Impracticable Specifications*, 34 Procurement Law. 3, 5 (1999): "Generally, a contractor should use the doctrine of commercial impracticability when the appeal involves 'performance specifications,' but should use the *Spearin* Doctrine when the appeal involves 'design specifications.'").

163. *L.W. Matteson*, 61 Fed. Cl. at 320.

164. *See, e.g.,* McElroy Mach. & Mfg. Co., Inc., ASBCA No. 46477, 99-1 BCA ¶ 30,185 (1998) (to come within the limited doctrine of impossibility, which encompasses commercial impracticability, one must show actual impossibility or that the performance could be achieved only at excessive and unreasonable cost); *see also* Reflectone, Inc., ASBCA No. 42363, 98-2 BCA ¶ 29,869 (1998).

165. Seaboard Lumber Co. v. United States, 41 Fed. Cl. 401, 417 (1998); Oak Adec, Inc. v. United States, 24 Ct. Cl. 502 (1991).

166. *L.W. Matteson*, 61 Fed. Cl. at 320; Raytheon Company v. United States, 305 F.3d 1354, 1367–68 (2002); Spindler Constr. Corp., ASBCA No. 55007, 06-2 BCA ¶ 33,376 (2006).

167. *Reflectone*, 98-2 BCA ¶ 29,869 (citing Foster Wheeler Corp. v. United States, 513 F.2d at 594, 598 (Ct. Cl. 1975)).

168. *Id.*; *see also* Tri Indus., Inc., ASBCA No. 47880, 99-2 BCA ¶ 30,529 (1999) (where a specialty contractor performed the specified welds at the subcontract price, that demonstrated that the welds were neither impossible nor commercially impracticable to make, and, while claimant may have lacked the knowledge and experience to perform the welds, such subjective impossibility/commercial impracticability is no basis for relief).

169. 24 Ct. Cl. 502 (1991).

170. *Id.* at 504.

171. *Id.*

172. *Id.* at 505.

173. *Id.* at 505–06.

174. *Id.* at 507–08.

175. Ace Constructors, Inc. v. United States, 499 F.3d 1357, 1364 (Fed. Cir. 2007).

176. L.W. Matteson, Inc. v. United States, 61 Fed. Cl. 296, 320 (2004).

177. *Id.*

178. S. Dredging Co., ENG BCA No. 5843, 92-2 BCA ¶ 24,886 (1992).

179. *Id.*

180. *Id.* at 124,118.

181. Ace Constructors, Inc. v. United States, 499 F.3d 1357, 1364 (Fed. Cir. 2007) (citing Raytheon Co. v. White, 305 F.3d 1354, 1367 (Fed. Cir. 2002)).

182. *L.W. Matteson*, 61 Fed. Cl. at 320 (citing Seaboard Lumber Co. v. United States, 308 F.3d 1283, 1295 (Fed. Cir. 2002); Spindler Constr. Corp., ASBCA No. 55007, 06-2 BCA ¶ 33,376 (2006)).

183. *Raytheon Co.*, 305 F.3d at 1367.

184. *L.W. Matteson*, 61 Fed. Cl. at 320.

185. ENG BCA No. 5843, 92-2 BCA ¶ 24,886 (1992).

186. *Id.; see also Raytheon Co.*, 305 F.3d at 1367–68 (when calculating an overrun for purposes of determining commercial impracticability, it is reasonable to compare the estimated cost of completion with the contract price at the time of termination and, even assuming the cost overrun was 57 percent rather than 24 percent, an overrun of that size does not by itself establish commercial impracticability); Gulf & W. Indus. Inc., ASBCA No. 21090, 87-2 BCA ¶ 19,881, at 100,575 (1987) (finding a contract with a claimed 70 percent overrun not commercially impracticable); Soletanche Rodio Nicholson (JV), ENG BCA No. 5796, 94-1 BCA ¶ 26,472, at 131,774 (1993) (finding commercial impracticability when compliance with the specification would have taken more than 17 years at a cost of more than $400 million, rather than 720 days and $16.92 million); Numax Elecs., Inc., ASBCA No. 29080, 90-1 BCA ¶ 22,280, at 111,916 (1989) (finding commercial impracticability when the contractor obtained a yield of only 300 acceptable units out of 8,000, or 3.75 percent); Whittaker Corp., Power Sources Div., ASBCA No. 14740 et seq., 79-1 BCA ¶ 13,805, at 67,688–89 (1979) (granting relief where what the parties thought would be a seven-month production contract turned into an unsuccessful four-year development effort with a 148 percent cost overrun).

187. ASBCA No. 55007, 06-2 BCA ¶ 33,376 (2006).

188. *Id.* at 165,462.

189. *Id.; see also* Fulton Hauling Corp., PSBCA No. 2778, 92-2 BCA ¶ 24,858 (1992) (board ruled increase in land fill dumping fees from $6.00 to $9.77 per cubic yard alone, even when unforeseen, did not make performance commercially impracticable).

190. ASBCA No. 55007, 06-2 BCA ¶ 33,376 at 165,462–63 (2006).

191. ENG BCA No. 6021, 94-3 BCA ¶ 27,216 (1994).

192. *Id.* at 135,653; *see also* Tri Indus., Inc., ASBCA No. 47880, 99-2 BCA ¶ 30,529 (1999) (specialty contractor performed the specified welds at the subcontract price, demonstrating that the welds were neither impossible nor commercially impracticable to make, and, while claimant may have lacked the knowledge and experience to perform the welds, such subjective impossibility/commercial impracticability is no basis for relief).

193. Guy F. Atkinson Co., ENG BCA No. 4771, 88-2 BCA ¶ 20,714 (1988); dicta, Columbia Eng'g Corp., ASBCA No. 32139, 89-2 BCA ¶ 21,689, *aff'd. on recon.*, 89-3 BCA ¶ 21,999 (1989).

194. Reflectone, Inc., ASBCA No. 42363, 98-2 BCA ¶ 29,869 (1998).

195. *Id.*

196. *Id.* at 147,831 (citing Foster Wheeler Corp. v. United States, 513 F.2d 588, 598–602 (Ct. Cl. 1975)).

197. Austin Co. v. United States, 161 Ct. Cl. 76, 314 F.2d 518 (1963), *cert. denied*, 375 U.S. 830 (1963).

198. Bethlehem Corp. v. United States, 462 F.2d 1400, 1404 (Ct. Cl. 1972); Wallace C. Boldt, General Contractor, Inc., ASBCA Nos. 24862 et al., 83-2 BCA ¶ 16,765 (1983).

199. *See, e.g.,* Wickham Contracting Co. v. United States, 212 Ct. Cl. 318, 546 F.2d 395 (1976).

200. *See, e.g.,* Hunter Ditch Lining, AGBCA No. 87-391-1, 91-2 BCA ¶ 23,673 (1991).

201. Spindler Constr. Corp., ASBCA No. 55007, 06-2 BCA ¶ 33,376 (2006); AGH Indus., Inc., ASBCA Nos. 25848 et al., 85-1 BCA ¶ 17,784 at 88,845 (1984).

202. 06-2 BCA ¶ 33,376.

203. *Id.* at 165,462–63.

204. *See, e.g.,* Chevron, U.S.A Inc., ASBCA No. 32323 90-1 BCA ¶ 22,602 at 113,426 (1989); *AGH Indus.*, 85-1 BCA ¶ 17,784 at 88,845 (1984).

205. 61 Fed. Cl. 296 (2004).

206. *Id.* at 312.

207. *Id.* at 320.

208. *Id.*

CHAPTER 19

Delay, Suspension of Work, Acceleration, and Disruption

ANDREW D. NESS

I. Basic Delay Concepts

"Time is money" is the fundamental premise underlying disputes regarding delays to construction projects. A project that has experienced a delay in its expected completion date, from whatever cause, incurs added costs as a result. A significant portion of the costs of the management and support of an ongoing construction project is directly tied to the overall duration of the construction work. The longer the work goes on, the higher the costs. Because of this direct relationship, disputes regarding the delayed completion of construction work are one of the most common types of construction disputes. When the construction has taken longer to complete than planned or than contractually required, the issue quickly becomes which party—the contractor or the government—must bear the inevitable added costs that result.

II. Critical Path Delays versus Use of Available Float or Slack Time

A "delay" in the construction context can be specifically defined as any failure to complete a specific construction activity within the time planned for it. However, when we think of construction delays that cause added costs and disputes, it is almost always delay to the overall construction contract completion, rather than delay to just a single activity or part of the construction process. Delays to individual work activities typically have relatively little or no distinguishable cost. The damages due to delays are primarily incurred when the overall project duration is extended. So when assessing the costs and consequences of a particular event that delays some aspects of the construction work, the first step is to assess that event's effect on the project's overall completion schedule.

There are some situations in which the overall delaying effect of a particular event is obvious, such as where the government is preventing any construction progress due to a failure to provide access to the work site. Beyond such patent delay situations, however, the basic technique utilized for assessing the effect of individual delaying events on the overall project duration is network analysis, also known as critical path method (CPM) analysis. The longest sequence of required construction activities to complete the project is its critical path. By definition, delay to any of the critical path activities will extend the overall project duration, unless remedial steps can be taken, such as resequencing activities or reducing other critical path activity durations. Other, noncritical activities can be delayed to some extent without affecting the overall project duration and, as such, are said to contain "float" or "slack" to that extent. A delaying event that solely utilizes available float does not affect the critical path, does not delay the project, and so is not the basis for a possible claim for time extension or delay damages (although it may, in some cases, provide the basis for a disruption claim—discussed in Section VIII, below).[1] On the other hand, an event that causes an activity to be delayed by more than the available float time affects the critical path once the float has been exhausted, and so becomes a critical path delay to that extent.

As the U.S. Claims Court (as it was then called) stated in 1992: "Essential to a determination that an activity belongs on the Critical Path of a project is an understanding of how that activity affects other activities."[2] Determining what activities constitute the critical path on a particular project, how much float is associated with the noncritical activities, and what, if any, effect on the critical path resulted from a particular delay event are most typically the subject of detailed expert witness testimony in construction delay litigation, as is discussed further in Section VI below.

III. Categories of Delays

Not all delays that affect the critical path are grounds for a time extension or added compensation for the contractor. Whether the contractor will be entitled to additional time or compensation for a critical path delay depends on which party is responsible for creating the delaying event, or has otherwise taken responsibility for it under the terms of the contract. In simplest terms, the contractor is not entitled to a time extension for delays it has caused itself but will have a basis to obtain relief to some degree for delays caused by either the government or events entirely beyond the contractor's control. In this respect, delay events can be divided into three basic categories: excusable, compensable, and unexcused. An *excusable* delay is a delay for which the contractor is entitled to an extension to the contract time for completion. In other words, the contractor's late completion is excused. *Compensable* delays are a subset of excusable delays for which the contractor is entitled not only to a time extension but also to compensation. The compensation takes the form of an adjustment to the contract price for any added costs that flowed directly from the delay (that is, delay damages).[3] *Unexcused* delays are those for which the contractor has responsibility; they entitle the contractor to neither a time extension nor any added compensation. If the contractor has not completed the work when required and the delay is unexcused, the government will be entitled to its delay damages for the contractor's failure to complete on time. These damages generally take the form of daily liquidated damages. The government may also be entitled to terminate the contract for default for failure to timely complete.[4]

A. Excusable Delay

The terms of the construction contract, which set out the risk allocation between the contractor and the government, define these three categories of delays. The standard form of fixed-price government construction contract describes this risk allocation in several standard form clauses in the Federal Acquisition Regulation (FAR). Each of these clauses has a significant history and thus well-understood meanings. For example, the category of excusable delays is defined, oddly enough, in the Default (Fixed Price Construction) clause,[5] as follows (emphasis added):

  (b) The Contractor's right to proceed shall not be terminated nor the Contractor charged with damages under this clause, if—
   (1) *the delay in completing the work arises from unforeseeable causes beyond the control and without the fault or negligence of the Contractor.* Examples of such causes include—
      (i) Acts of God or of the public enemy,
      (ii) Acts of the Government in either its sovereign or contractual capacity,
      (iii) Acts of another Contractor in the performance of a contract with the Government,
      (iv) Fires,
      (v) Floods,
      (vi) Epidemics,
      (vii) Quarantine restrictions,
      (viii) Strikes,
      (ix) Freight embargoes,
      (x) Unusually severe weather, or
      (xi) Delays of subcontractors or suppliers at any tier arising from unforeseeable causes beyond the control and without the fault or negligence of both the Contractor and the subcontractors or suppliers; and
   (2) The Contractor, within 10 days from the beginning of any delay (unless extended by the Contracting Officer), notifies the Contracting Officer in writing of the causes of the delay. . . .

Note that the keystone requirement is that the delay arise from "unforesee-able causes beyond the control of and without the fault or negligence of the Contractor," and that subsections (i) to (xi) are simply a nonexclusive list of specific examples of delays that fall within this description. As the examples make clear, excusable delays include both delays caused by the government ("acts of the government") and those that are beyond the control of either the contractor or the government (acts of God, fires, floods, strikes, and so forth). In every instance, however, the dual requirements that the delay be unfore-seeable and not arise from the fault or negligence of the contractor still apply.[6] For example, while general economic strikes are excusable delays, strikes that are the result of poor employee treatment and mismanagement by the con-tractor are not.[7] Similarly, a nationwide steel strike that began several months before submission of bids has been held not to be an excusable delay, because this cause of delay was not "unforeseeable."[8]

Note also that while delays caused by *unusually severe weather* are consid-ered excusable, normal weather delays, such as a rainstorm that halts earth-moving work, are *not*. As stated in *Allied Contractors, Inc.*:

> [T]he term "unusually severe weather" does not include any and all weather which prevents work under the contract. The phrase means only that weather surpassing in severity the weather usually encoun-tered or reasonably to be expected in the particular locality during the time of year involved.[9]

The Army Corps of Engineers (Corps) has further specified when weather delays can be classified as unusually severe and has developed frequently used guidelines for how to assess such claims of unusually severe weather.[10]

Another specific rule relates to bankruptcies of subcontractors and suppli-ers. Boards of contract appeals (BCAs) have consistently rejected arguments that delays caused by the bankruptcy or financial inability of a subcontrac-tor or supplier to perform are not the result of either the contractor's or sub-contractor's fault or negligence and so are excusable. There is a quite limited exception where the bankruptcy was directly caused by the government's improper action (such as a chronic failure to pay for that subcontractor's work that causes the financial distress).[11]

Entitlement to a time extension for an excusable delay also does not arise merely because of the occurrence of the delay event, as the contractor must still provide written notice to the government of the cause of delay.[12] It also remains the contractor's burden to demonstrate the actual delaying effect of the event on the work and the project duration. For instance, in *Sauer, Inc. v. Danzig*,[13] the contractor argued that the government's presence at the site and interference with the construction work prevented the contractor from being able to complete the contract on time. The Court of Appeals for the Federal Circuit rejected the contractor's claim on the basis that, although there was a "potential for delaying interference" from the government's actions, "Sauer did not meet its burden of showing such delay actually occurred in this case."[14]

B. Compensable Delay

Compensable delays are generally limited to those situations where the delay flows directly from the actions or inactions of the government as the other contracting party, as opposed to delays that are not due to the fault of either party. Again, however, whether a particular delay can be categorized as compensable is determined by the terms of the contract, and this topic is addressed by several standard FAR clauses. For example, a change to a fixed-price construction contract directed by the government is an action that specifically entitles the contractor to compensation for delay caused by the change, under the Changes clause:[15]

> (d) If any change under this clause causes an increase or decrease in the Contractor's cost of, *or the time required for*, the performance of any part of the work under this contract, whether or not changed by any such order, the Contracting Officer shall make an equitable adjustment and modify the contract in writing. [Emphasis added.]

Any type of constructive change that qualifies for an equitable adjustment under the Changes clause[16] would similarly include compensation for resulting delay, because an equitable adjustment of both the time and the cost of performing the contract is expressly authorized by that clause.[17]

Similarly, a delay to the contract completion stemming from the government's failure to deliver government-furnished property by the required date is a compensable delay, because an equitable adjustment for such delays is expressly provided for by the Government Property clause.[18] A delay due to the government taking over the use and possession of all or a part of the work prior to completion is another specific circumstance that may give rise to a compensable delay, by the terms of the Use and Possession Prior to Completion clause.[19]

While a delay due to the discovery of a differing site condition is not a delay directly stemming from the government's actions or inactions, it nevertheless is made a compensable delay by the terms of the Differing Site Conditions clause, which again expressly provides for an equitable adjustment in both time and cost of performance to the extent they are affected by the differing site condition.[20] Differing site conditions are a situation where the government, for reasons of overall cost efficiency, has allocated to itself the risk of site conditions being materially different than represented, or than reasonably expected in the locality.[21]

Compensable delays can also arise from the government's breach of an express or implied duty under the contract, such as the duty to cooperate with the contractor, or not to hinder or interfere with the contractor's work. In

SIPCO Services & Marine, Inc. v. United States,[22] government orders to adhere to more stringent containment requirements than required by the contract, as well as a pattern of overregulation and excessive supervision, were found to constitute a breach of the implied duty to cooperate. The government was accordingly liable for the delays and added costs that resulted.

On occasion, situations arise in which the government has granted a time extension for at least some of the delay claimed by the contractor, but the parties are unable to reach a final resolution of the total cost and time due the contractor. Formerly, the government's previous granting of a time extension was treated as giving rise to a rebuttable presumption that there was compensable delay to the extent of the time extension granted, a rule known as the "McMullan presumption."[23] In *Sherman R. Smoot Corp. v. England*,[24] the Federal Circuit in 2004 found that this presumption conflicted with the Contract Disputes Act (CDA)[25] and is no longer good law. Accordingly, the contractor must now prove the compensability of delays even where the government has previously provided a time extension via change order.

C. Unexcused Delay

Unexcused delays (i.e., those that do not entitle the contractor to either a time extension or added compensation) are generally not specified in the contract as such. The contractor in a fixed-price construction contract with the government takes responsibility for many types of risks, such as the availability and quality of labor; the availability, delivery, and quality of materials; submission of adequate shop drawings and submittals; the performance of subcontractors and suppliers; site conditions and work restrictions identified in the contract; and safety. To the extent that delays arise out of any of these risks that have been assumed by the contractor, those delays will be considered unexcused. In fact, one can go so far as to say that delays that cannot be brought within the definition of excusable delays (as quoted previously, from FAR 52.249-10) are by definition unexcused.[26]

For instance, if the contractor experiences delays because of difficulties in obtaining the permits and certifications that were contractually required to construct the work, the contractor will not be entitled to a time extension or compensation:

> [T]he responsibility for acquiring the state certification for the facility was [the contractor's] and as the conditions imposed by the state for its granting of the certification were normal and to be expected, [contractor's] Appeal requesting a time extension for meeting those conditions is denied.[27]

Similarly, delay to the project resulting from the government's delay in holding the preconstruction conference has also been found to be an unexcused delay. In *Program & Construction Management Group, Inc. v. General*

Services Administration,[28] the contractor asserted that the government's failure to schedule a preconstruction conference forced the contractor to delay beginning work on the contract. Despite evidence showing preconstruction conferences were standard within the industry, the General Services Board of Contract Appeals (GSBCA) stated:

> [R]egardless of whether there is an industry custom to hold a meeting shortly after issuance of the notice to proceed . . . a preconstruction meeting is not a prerequisite to the start of work . . . the lapse of time between the issuance of the notice to proceed and the convening of a preconstruction conference does not establish a compensable delay to the contractor.[29]

Since the preconstruction conference was not contractually made a precondition to starting work, the delay in holding it did not cause delay to the contractor; the contractor was free to proceed in the meantime. Thus, the resulting delay was unexcused.

When delays to contract completion are unexcused, the government has the right to recover its own delay damages, as typically measured by liquidated damages. Where the government asserts a claim for liquidated delay damages, the government has the overall burden of proof to demonstrate that the contractor is in default, that the government did not prevent performance or contribute to the delay, and that the contractor was the sole cause of the delay.[30] To defeat the claim, the contractor must come forward with evidence to show that the government prevented performance or contributed to the delay or that the delay was otherwise excusable.[31] If the responsibility for days of delay is unclear, or if both parties contributed to the delay, the government must, in order to recover, prove a clear apportionment of the delay attributable to each party, in accordance with the general rule regarding concurrent delay.[32]

Absent a particular clause that defines the delay period differently, the relevant delay period in government claims for liquidated delay damages extends from the contractually required completion date until the work is "substantially complete." Substantial completion is achieved when the completed work is capable of being utilized for its intended purpose.[33] The distinction between items required for substantial completion and those not so required is well illustrated by *Batson-Cook, Inc.*,[34] where the government withheld liquidated damages for late completion of a building until both fire doors and office doors were installed. The Armed Services Board of Contract Appeals (ASBCA) held that the absence of the fire-rated doors precluded substantial completion, because they were a fundamental safety feature of the building, but that the absence of the office doors was a mere inconvenience that did not prevent beneficial use of the building. Accordingly, withholding of liquidated damages was not appropriate once the fire doors were installed, since at that point the work was substantially complete.[35]

IV. Suspension of Work

The Suspension of Work clause[36] is another required clause in fixed-price government construction contracts that addresses a form of delay.[37] The clause provides the government with the ability to order the contractor to suspend, delay, or interrupt all or any part of the contract work for a period determined by the contracting officer (CO).[38] A suspension of work may be invoked expressly, through a *directed suspension*, or implicitly, by means of a *constructive suspension*. In the event of either type of suspension, the contractor is entitled to an adjustment to its contract price for the associated costs and delay to contract completion. In practice, however, contractors rarely invoke their rights under the Suspension clause, because such claims can generally be addressed as easily under the Changes clause, which affords a more liberal remedy to the contractor. Instead, it is the government that is more likely to be contending that a particular work interruption is appropriately treated under the Suspension clause and not the Changes clause.

A. Directed Suspension of Work

When the government issues a written directive to the contractor to cease or interrupt operations, this is the classic case of a *directed* suspension of work. Entitlement to an adjustment to the contract price for a directed suspension involves four elements:

1. The resulting delay was for an "unreasonable period of time";
2. The delay was proximately caused by the government's actions;
3. The delay resulted in some injury to the contractor; and
4. There is no delay concurrent with the suspension that is the fault of the contractor.[39]

The determination as to whether a particular suspension was for an "unreasonable period of time" is made in light of the surrounding circumstances.[40] The relevant surrounding circumstances include the relative length of the work suspension and the reason why the government ordered the suspension. For instance, where the contractor was able to mitigate the effect of the suspension by focusing on other activities, courts have deemed "quite lengthy delays . . . to be reasonable."[41] On the other hand, where the government work suspension impacts not only the targeted portion of the contract but also most other portions of the work, courts will "scrutinize with care" and find a "reasonable time to be quite brief."[42]

That the suspension of work must be proximately caused by the government is a strictly construed requirement. In *Merritt-Chapman Scott Corp. v. United States*,[43] the court of claims ruled that for the government to be deemed the "sole proximate cause" of a suspension, the court must be satisfied that the contractor's conduct, in and of itself, would not have delayed the contract, nor would the contractor have been delayed for any other reason.[44] As discussed in more detail below, however, apportionment of concurrent contractor delays is permissible.[45]

The requirement that the delay caused some harm to the contractor is simply a requirement that the contractor be able to prove some damages—that is, additional costs incurred by the contractor as a result of the work suspension.[46]

B. Constructive Suspension

Even in the absence of a formal order by the government to suspend work, where the conduct of the government nevertheless compelled the contractor to interrupt or cease its work, that conduct may be treated as a *constructive* suspension. The requirements for a claim of constructive suspension are substantially similar to those required for directed suspension, and are also generally stated as a four-part test: "(1) contract performance was delayed; (2) the government directly caused the delay; (3) the delay was for an unreasonable period of time; and (4) the delay injured the contractor in the form of additional expense or loss."[47]

In the absence of a direct suspension order, there must instead be specific incidents, conduct, or behavior by the government that caused the interruption to the contractor's work on the project, in addition to the elements of a directed suspension.

In *W.M. Schlosser, Inc. v. United States,*[48] the Court of Federal Claims (COFC) in 2001 revisited the requirements for a constructive suspension. In that case, the contractor was seeking to recover under the Suspension of Work clause because a specific clause precluded it from recovering extended field overhead costs under the Changes clause. Although the trial record noted evidence that "there were periods of time when virtually nothing was happening on the project," the court also noted that the contractor never fully demobilized.[49] From this, the court reasoned that any suspension did not cause "unreasonable" delays, since "[p]eriodic starting and stopping during a large construction project is to be expected and does not constitute the 'unreasonable period of time' which characterizes a suspension of work."[50] As indicated by this decision, a constructive suspension requires an actual interruption of work; it is not sufficient if the contractor's performance has simply been delayed or hindered.[51] The contractor's performance with regard to a specific contract activity (or in the case of a complete suspension, all of the work under the contract) must be brought to a complete stop.[52]

C. Notice Requirements for Suspension of Work

The Suspension of Work clause[53] requires prompt notice by the contractor, in the absence of a direct suspension order, to preserve entitlement to a constructive suspension of work claim. Under the clause, the claim is disallowed "for any costs incurred more than 20 days before the Contractor shall have notified the Contracting Officer in writing" of the actions considered to constitute a suspension.[54] In the case of either direct or constructive suspension orders, the suspension claim must be asserted in a specific amount, "in writing as soon as practical after the termination of the suspension," but in no event after the date of final payment.[55]

D. Changes versus Suspension of Work

Two notable restrictions in the Suspension of Work clause make it a significantly less utilized remedy for a compensable delay than the Changes clause. First, the clause provides specifically that "no adjustment shall be made under this clause for any suspension, delay, or interruption . . . for which an equitable adjustment is provided for or excluded under any other term or condition of this contract."[56] In effect, this language makes the Suspension of Work clause secondary to other remedy-granting clauses, and applicable only when those other clauses are inapplicable. The expansion over the years of the concept of a "constructive change" under the Changes clause has thus substantially reduced the scope and applicability of the Suspension of Work clause. As a practical matter, almost every situation that could be construed as a constructive suspension of work can also be construed as a constructive change with equal ease.

Second, the Suspension of Work clause defines the adjustment to the contract price that is available under the clause as follows: "[A]n adjustment shall be made for any increase in the cost of performance of this contract (excluding profit) necessarily caused by the unreasonable suspension, delay or interruption."[57] By contrast, the parenthetical "excluding profit" does not appear in the Changes clause, and profit is a generally accepted component of an equitable adjustment under that clause. Since contractors naturally prefer the opportunity to recover profit as part of the equitable adjustment, they generally favor the Changes clause over the Suspension of Work clause. As a result of these two limitations, the Suspension of Work clause has become much less significant in recent years.

V. Additional Delay Concepts and Issues

There are several specific recurring issues that arise frequently enough in delay disputes that substantial case law has developed regarding them. These recurring issues include situations in which the contractor is arguing that it was deprived of the ability to finish the contract earlier than required, the problem of sorting out concurrent delays, and situations where multiple delays lay claim to the use of the available float time.

A. The Right to Finish Early

The right-to-finish-early scenario involves a contractor that is planning and expecting to complete the contract early, which is to say, prior to the completion date required by the contract. However, a government-caused delay is encountered that would, in other circumstances, qualify as a compensable delay, in that it extends the contract's critical path and delays the contract completion from when it otherwise would have occurred. Because the contractor had been working toward an early completion, however, the contractor still completes the contract on or before the original contract completion date. Can the contractor validly claim that it has been delayed and is entitled to an equitable adjustment for the government-caused, critical path delay, even though its performance has not extended beyond the contract completion date?

The general answer under the law of federal government contracts is "yes," the contractor can recover in this situation, even though the contract completion date was met or beaten. The contractor's right to finish early, unless specifically precluded by contract, has long been recognized as a right flowing from the government's obligation to avoid hindering the contractor's progress. In *Metropolitan Paving Co. v. United States*,[58] the court of claims noted that, although "there is not an 'obligation' or 'duty' of [the government] to aid a contractor to complete prior to completion date, from this it does not follow that [the government] may hinder and prevent a contractor's early completion without incurring liability."

The contractor that actually plans to complete early saves on fixed costs due to the shorter performance period and achieves other benefits such as improved cash flow and the opportunity to redeploy its resources on other contracts. Presumably some or all of the contractor's anticipated savings were passed along to the government, because that contractor was able to offer a lower bid price than its competitors. Accordingly, when the contractor is deprived of the ability to complete early, it suffers a real loss. The government, having obtained the benefit of the early completion date via the competitive bidding process, should fairly pay for that loss by way of an equitable adjustment for the compensable delay it caused.

Because of the potential for abuse inherent in this rule, however, there are special requirements imposed before the contractor may recover for delay to a claimed early completion date. Moreover, the boards and courts have not been entirely consistent in stating these added requirements.

At the least, the contractor is required to demonstrate not only that it actually planned an early completion but also that its early completion schedule was realistic and achievable under the circumstances.[59] The mere fact that the government did not object to the early finish schedule, acquiesced in that schedule, or even specifically approved that schedule is not by itself enough to demonstrate reasonableness.[60] Accordingly, in *Emerald Maintenance, Inc.*,[61] the ASBCA denied the contractor's delay claim based on the government's alleged interference with the contractor's planned early finish—because the contractor's early completion schedule was patently unrealistic (although it was approved by the government) and the board was not convinced that the contractor would have completed the project early, even in the absence of government-caused delay.

Beyond the hurdle of demonstrating the reasonableness of the contractor's early completion schedule and its ability to meet its early completion goal,[62] recent decisions have imposed one additional requirement: the contractor must demonstrate that, from the outset, it intended to finish early. The rationale for this added requirement is that, if the rule were otherwise, the government would be subject to contractors taking advantage of government delays to obtain an equitable adjustment when they really had no intention to finish early and were not kept on the project for longer than anticipated.

The Federal Circuit has settled on a three-part test that requires the contractor to demonstrate "that from the outset of the contract it: (1) intended to complete the contract early; (2) had the capability to do so; and (3) actually would have completed early, but for the government's actions."[63] In *Blinderman Construction Co. v. United States*,[64] the COFC went even further, utilizing a four-part test that added the further requirement that the contractor provide notice to the government of its intention to finish early. However, later cases have clarified that such notice is not required but may provide sufficient evidence of the intent requirement.[65] As a result, older cases reflecting the notion that "the proper focus is not whether [the contractor] informed the government, at the inception of the contract, that it *intended* to complete the project [early]; rather, the focus should be on whether [the contractor] *would have* completed the project early, but for the government-caused delay" (emphasis in original)[66] may no longer be valid statements of the law, though not expressly overruled.

Finally, it is important to note that the government likely can, by a proper contract clause, immunize itself from early finish claims. Such a provision would be appropriate, for example, where the government had no use for the project if it was finished early. In *Maron Construction Co. v. General Services Administration*,[67] however, it was held that a contract provision stating that the government would be responsible only for delays that exceed total float did not preclude the contractor from asserting an early finish claim for delay damages. The GSBCA found that the clause upon which the government relied was not specific enough, and the board stated that a clause precluding early completion delay damages must specifically and expressly state that such damages are not recoverable.

B. Concurrent Delays

In performing a complex construction contract, there are often many activities on the project schedule that are not completed when planned and just as many reasons why. As a result, it is by no means unusual to encounter situations in which the overall project has unquestionably been delayed but there are two or more possible explanations of the underlying cause of the delay, one being actions or inactions by the government (compensable delays) and the other being actions or inactions by the contractor (unexcused delays). How are the effects of these different possible sources of project delay to be sorted out? This, in essence, is the problem of concurrent delay.[68]

The basic rules applicable to concurrent delay situations are readily stated, but the reality is that the rules are extremely fact-specific and often difficult to apply in real-world circumstances. Where a contractor is seeking recovery on the basis of a compensable delay, but it is clear that the contractor also caused delay, "there can be no recovery where the [government's] delay is concurrent or intertwined with other delays."[69] The same is true in reverse. Where the government is claiming liquidated delay damages, it is the government's burden to demonstrate the relative effects of its delays as compared to those of its contractor. If the delays are inextricably intertwined, the government cannot recover.[70] Since neither party can recover damages in such an "inextricably intertwined" situation, the net effect is that the period of concurrent delay is effectively treated as if it were excusable but noncompensable delay. That is, the contractor is effectively given a noncompensable time extension because the government is precluded from recovering liquidated damages for the concurrent period.[71]

Most early cases held that the court would not get involved in attempting to apportion delays among the various competing causes,[72] but this is no longer the law. The rule today is that the party claiming delay damages has the burden to segregate or apportion the delays attributable to each party, and recovery will be denied only if there is a failure to do so.[73] As such, in *Blinderman Construction Co. v. United States*,[74] the Federal Circuit stated that, where both parties contribute to the delay, "neither can recover damages, unless there is in the proof a clear apportionment of the delay and the expense attributable to each party."

Generally, courts and boards will deny recovery where the delays are "concurrent or intertwined" and the contractor has not met its burden of separating its delays from those chargeable to the government.[75] While in theory, modern techniques of network analysis such as "windows" or "time impact" analysis (discussed below in Section VI) can be used in most instances to achieve such apportionment, there are many examples of recent cases where recovery has been denied either for failure to meet this burden of proof[76] or where the court was convinced that the contractor's concurrent delay would have delayed the contractor in any event.[77]

Beyond these key principles, there are a number of additional rules that are of assistance in resolving multiple-delay situations. For example, in *Fischbach & Moore International Corp.*,[78] the government's delay due to placing a "hold" on proceeding with a substantial part of the work clearly affected the critical path, whereas the contractor's delays mainly affected only work that was not on the critical path. To the extent that the contractor's delays affected only float, the ASBCA clarified that these were not "delays" at all within the meaning of the concurrent delay rule. Thus, if the contractor can demonstrate that its delays were not on the critical path, there is no concurrent delay. The government's delay to the critical path is then the sole cause of the project delay, and thus compensable. The converse should also be true in situations where the government is claiming liquidated delay damages.

Another important distinction that helps resolve many apparent concurrent delay situations is the difference between two independent causes of delay and an alleged "concurrent delay" that is really just a secondary effect of the primary delay. The leading case on this point is *John Driggs Co.*,[79] in which the contractor claimed an 88-day critical path delay in its ability to drive piles, due to a government-caused problem with providing access to the location where the piles were to be driven. In response, the government alleged that the contractor was in no position to begin pile-driving when the government's access delay started, pointing to the lack of approval of the contractor's shop drawings, incomplete negotiations with the pile-driving subcontractor, and the fact that the contractor had not yet procured the piles themselves. The Corps Board of Contract Appeals (ENG BCA) disagreed, referencing these alleged concurrent delays as "speculative or theoretical," and noting:

> When a significant owner-caused, construction delay . . . occurs, the contractor is not necessarily required to conduct all of his other construction activities exactly according to his pre-delay schedule, and without regard to the changed circumstances resulting from the delay. *The occurrence of a significant delay generally will affect related work, as the contractor's attention turns to overcoming the delay rather than slavishly following its now meaningless schedule.* [The government] is required to demonstrate that, but for the delay caused by [the government], the contractor could not have performed the project in less time, and would necessarily have been delayed to the same extent in any case.[80] [Emphasis added.]

A failure to proceed with the other construction activities as planned, once the government has caused a critical path delay, is not treated as a concurrent delay unless there is an independent cause of delay on the part of the contractor that would have prevented its timely performance even if the government delay did not occur. Once a critical delay has occurred, it is recognized that the contractor's plans and priorities often change in response, and the contractor is not required to "hurry up and wait."[81]

With respect to which party bears the burden of proof in concurrent delay situations, the case law is somewhat confusing. The quote from *John Driggs Co.*,[82] reproduced above, plus *MCI Constructors, Inc.*[83] and *Bechtel Environmental, Inc.*,[84] all indicate that the contractor does not need to affirmatively disprove all possible causes of concurrent delay and that the government has the burden to demonstrate that the contractor could not have avoided the delay and would necessarily have been delayed by the concurrent delay to the same extent, even if the government delay had not occurred.

On the other hand, *PCL Construction Services, Inc. v. United States*[85] can be read to imply otherwise, in that it states that, in order to recover for an alleged compensable delay, "the contractor must show that the government was the 'sole proximate cause' of the delay and that no concurrent cause would have equally delayed the contract, regardless of the government's action or inaction."[86] However, the context of this quote is that the contractor was attempting to assert a "total time" claim, where it simply attributed all delay to the government. In that context, to make it the government's burden to prove concurrent delays would have been to place an effectively impossible burden on the government, since the contractor had never provided a network schedule analysis or otherwise demonstrated the delaying effect caused by the government's actions in the first place. Where the contractor has directly proven the compensable delay by a network analysis, however, it is fair and reasonable to require the government (as the boards did in the three cases cited above) to raise and prove the existence of alleged concurrent delays, rather than require the contractor to explain the lack of significance to each and every activity that was not completed as scheduled.

Finally, it is clear that in some fact situations the effects of multiple causes of delay may be considered by the court as "inherently apportionable." *Essex Electro Engineers, Inc. v. Danzig*[87] involved a series of delays due to multiple submissions, rejections, resubmissions, and further rejections of both engineering change proposals (documenting necessary changes to the drawings) and documentation under the prototype inspection procedure. The Federal Circuit found that both Essex and the government had in part caused the delays in obtaining final approval of these submissions but that "[t]he sequential nature of Essex's submissions and the government's responses renders each party's delays inherently apportionable."[88] In other words, the linear nature of the process (submission, rejection, preparation of the resubmission, and so on) was inherently such that at any particular time it could be clearly determined which party was causing the delay, and so the court could and would undertake this assessment. Clearly not all concurrent delay situations involve such linear delays, but this is a frequently seen scenario in the case of delays to shop-drawing submittals and approvals, answers to requests for information, and similar delays.

C. Apportionment of Project Float (Who Owns the Float?)

Float, in the context of construction scheduling, is "the amount of time any given activity or path of activities may be delayed before it will affect the project completion time."[89] Another aspect of the general problem of assessing the effect of multiple delays involves the problem of apportioning entitlement to the available float time when there are multiple delays, some by each party, that in total exceed the available float time. This issue generally goes by the question, "Who owns the float?" A simple example illustrating this issue is when a given series of activities starts with (for example) 30 days of float, but a 30-day government delay (such as a change directive) utilizes that float, making the subsequent activities critical. Subsequently, the contractor experiences an unexcused delay (such as a poorly performing subcontractor) that further delays the same—now critical—activities by another 30 days. Each party caused a separate 30-day delay, but which party is responsible for the overall 30-day delay in contract completion?

Three potential outcomes are possible in this situation, depending upon whether the contractor is entitled to ("owns") the available float, the government does, or neither does and the float is shared in some manner. If, for example, the contractor is entitled to the benefits of the float in its schedule, then if the government deprives the contractor of that float, and the contractor needs it later, the government should compensate the contractor for the delay.

When early cases on this issue proved unclear and inconsistent, the government response was to develop several variants of contract clauses to address the point. One commonly used provision regarding ownership of float provides: "Float is not time for the exclusive use or benefit of either the Government or the Contractor, but must be used in the best interest of completing the project on time."[90] This provision adopts the neutral position that the schedule float is a shared resource, and not the property of either contractor or government, so when one party causes a delay and needs to utilize the available float to keep the project on schedule, that party is entitled to use it without compensating the other. If the other party then causes *another* delay to that activity (or chain of activities), the float is already gone, and the second party must bear the consequences of extending the project duration due to its critical path delay. Put another way, the party that uses the float first gets the benefit of it.

An alternative formulation used in one decision is that the float should belong to the party performing the schedule activity having the float, which of course will usually (but not always) be the contractor. In *Turner Construction Co.*,[91] decided in 1990, the ASBCA accepted the proposition that "general industry practice dictates that the contractor responsible for performance of [the] activity 'owns' that float and is entitled to consume or use the float as he sees fit or necessary."

Overall, however, the position of most commentators who have addressed the subject is to embrace the "shared resource" approach mentioned previously: "Unless specifically defined in the contract specifications, float is a resource that belongs to the project and is available for all parties to use."[92] This position is also consistent with the above-quoted contract clause in most frequent use on this issue.

VI. Proving Delays

To recover on a delay claim, the contractor is required to prove with "reasonable certainty" that its operations have been delayed.[93] The required elements of liability, causation, and resultant injury must each be proven. More specifically, "the contractor has the burden of showing the extent of the delay, that the delay was proximately caused by government action, and that the delay caused damages to the contractor."[94] Even more specifically, demonstrating damages for delays requires proving: "(1) the number of days of delay attributable to the defendant's wrongful actions; and (2) that these delays were on the project's critical path" (that is, they extended the critical path and delayed completion of the overall project).[95] This last step normally requires utilizing the techniques of network analysis (typically CPM), which is well recognized as the most reliable and trustworthy tool for analyzing delay.[96]

A. Need for Realistic and Accurate CPM Analysis

Use of a CPM schedule analysis to prove a government contract delay claim is not always required, but obtaining a recovery in the absence of a CPM analysis has become quite rare and essentially limited to situations where the impact of the delay is readily apparent or can be determined as a matter of common sense.[97] However, such situations are few, and certainly this will not be possible in most construction delay claims. Additionally, where the contract specifically requires use of a CPM schedule to establish the claimed delay, that requirement will be enforced.[98]

The practical imperative to utilize CPM analysis was well stated in *Hoffman Construction Co. v. United States*,[99] where the court noted:

> Even if a critical path analysis per se is unnecessary, a contractor must supply some form of "specific proof that [its] performance was affected by the Government's undue delays.". . . The mere identification of five potential causes of delay and extended performance time does not establish that the former caused the latter. It is immaterial that some particular event came along which disrupted certain work or delayed its start or completion. It may well have been that that item was not one which would delay the project completion or have any effect on it. We cannot presume that, merely because some extra work was ordered and compensation paid by the contracting officer, there would have been a delay to the completion of the project.[100]

In recent years, the BCAs and the COFC have become quite sophisticated in assessing the adequacy and usefulness of proffered CPM analyses in delay situations. Not infrequently, they will reject inadequate or incomplete efforts. For instance, in *Blinderman Construction Co. v. United States*,[101] the CPM schedule was deemed "utterly useless," as the contractor presented only the first of several pages of the schedule, thereby preventing the court from tracing the critical path through the project. Indeed, the contractor had never designed the schedule so that the critical path could be discerned. Further, although the schedule had been updated on several occasions, the revisions were "indiscernible and [were] unaccompanied by explanatory text."[102] In general, any schedule analysis that fails to clearly identify the critical path of the project is highly likely to be rejected.[103] Additionally, schedules that have never, or only rarely, been updated during the course of the project, particularly where there were changes in circumstances normally significant enough to require such revisions, may preclude the schedule from serving as an acceptable basis for measuring delays.[104]

Similarly, significant variations from the schedule that was utilized contemporaneously on the project require adequate explanation and justification. Otherwise, they will be looked upon with a great deal of skepticism or may even lead to outright rejection of the schedule analysis. In *J.A. Jones Construction Co.*,[105] the contractor's claim was rejected because the work at issue was not shown on either the original critical path nor on any subsequent revision of the project schedule. Not until the contractor prepared its claim in litigation did the activity appear on the critical path. In general, CPM schedules created only after the fact are generally subject to attack and often subject to rejection.[106] For example, in *Blinderman Construction Co.*,[107] the COFC noted that a schedule generated 17 months after project completion "warrants a skeptical reception on the part of the court, because the required nexus between the Government delay and a contractor's performance at some unspecified earlier date cannot be shown merely by hypothetical, after-the-fact projection."

A schedule may also be rejected because the scheduled activities could not reasonably be performed in the manner depicted. In other words, a CPM schedule must be realistic and achievable to be utilized as the basis for a delay analysis. For instance, in *Neal & Co. v. United States*,[108] the CPM schedule presented as the reasonable as-planned schedule depicted simultaneous completion of all of the 30 housing units to be built under the contract. To accomplish this required 30 work crews, one working on each unit, and thus compressed the critical path work at the end of the project. As this was clearly unrealistic, the contractor's schedule analysis was rejected in favor of the government's analysis that included reasonable resource constraints.

The "total time approach" to proving delay is comparable to the "total cost" method of proving damages, in that it attributes all project delay experienced to the actions of the government, without demonstrating cause and effect specifically. Predictably, this method is, like the total cost method, not well received by the courts and boards. In *Morganti National, Inc. v. United States*,[109] the COFC found the total time approach to be of "virtually no value" in meeting the contractor's burden of proof and to be generally unsatisfactory and subject to the same inherent flaws as the total cost method of proving damages. As stated in *Jackson Construction Co. v. United States*,[110] "the contractor must do more than allege its work was delayed by the Government's disruptions or changes—it must present specific evidence of which activities were delayed and how those delays extended the duration of the contract."

Several different varieties of CPM analysis have been utilized in the years since network analysis techniques were first introduced in litigated cases. With accumulated experience in the technique, the courts and boards have also become more discriminating over time as to the specific methodology utilized. The four basic varieties of CPM analysis addressed in most reported construction delay claim decisions are the *impacted as-planned* method, the *collapsed as-built* method (also sometimes called the "but for" analysis), the *time impact* method, and the *windows* method. However, the nomenclature used by the courts and boards, and by the schedule experts themselves, has by no means been close to consistent, so different courts can mean different things when referring to these techniques. Adding to the confusion, each technique has several possible variants.

In June 2007, the first attempt at a formal protocol and set of recommended practices for forensic schedule analysis was published by the Association for the Advancement of Cost Engineers (AACE).[111] This effort, among other advances, identifies and categorizes specifically some eight distinct methods of schedule analysis, while noting the high degree of overlap in common names applied to these techniques. Beyond a brief description of their relative advantages and disadvantages, this protocol does not address the judicial acceptability of any of these methods, but it at least represents a major step forward in providing common terms of reference when discussing schedule analysis techniques in future cases.

B. Impacted As-Planned Method

Classic impacted as-planned analysis is identified as Method 3.6 in the AACE Recommended Practice.[112] The starting point for the impacted as-planned method is the as-planned (or baseline) schedule. The as-planned schedule is a representation of the contractor's intended work schedule as of the outset of the work. Among other things, it indicates the contractor's pre-dispute intended plan for executing and timely completing the contract, including the intended critical path. Using the impacted as-planned method, the delays to the work are depicted by inserting the government-caused delays into the as-planned schedule, generating the impacted as-planned schedule that is supposed to indicate the net effect of the government's delays. The claimed delay period is thus the difference between the adjusted completion date in the impacted as-planned schedule and the completion date in the baseline schedule.

The inherent flaws in the impacted as-planned method are well-known, and generally a delay analysis premised on this method will not be accepted. First, the as-planned schedule, on which the method is entirely premised, often does not reflect the actual course of events on the project. A construction project is by its nature dynamic. Contractors quite typically and properly will adjust their plans after construction begins and follow a somewhat different schedule that better reflects the realities the contractor encounters and facts the contractor learns once construction is under way. As a result, an original baseline schedule often bears no resemblance to the actual construction schedule (even excluding the claimed delay events), and thus does not provide a reasonable foundation for the delay analysis. This was the case in *Chaney & James Construction Co.*,[113] where the contractor's CPM analysis was rejected because it represented neither the schedule that the contractor intended when it bid the project nor the actual sequence of construction.

Similarly, in *Kaco Contracting Co.*,[114] the contractor revised its activity sequencing prior to encountering the delay but never submitted a revised schedule reflecting these changes. The ASBCA noted, in rejecting the claim, that "the record does not contain a schedule which we can use to measure delay as the re-sequencing makes the [prior] schedules not credible for delay analysis."[115]

A second major flaw in the impacted as-planned method is that it fails to consider delays other than those being claimed as caused by the government, namely "excusable delays and delays [caused] by the contractor."[116] For example, in *Galaxy Builders, Inc.*,[117] the contractor's delay analysis was found not to be credible because it failed to take into account *all* of the delays that had occurred prior to the delay for which the contractor was seeking compensation. The ASBCA noted that, if the contractor were not required to account for all changes to the planned schedule, the contractor could potentially recover compensation and receive time extensions for which it was not entitled. These weaknesses have led to a general discrediting of the impacted as-planned method.[118]

It is, of course, possible to modify the impacted as-planned analysis to first insert all delays encountered, other than those caused by the government, in an initial impacted as-planned, and then to generate a second impacted as-planned that also includes the government-caused delays. In theory, the difference should be the net effect of the government-caused delays.[119] However, this variation still suffers from the fundamental shortcoming of the impacted as-planned method, which is that it treats the contractor's planned schedule as essentially fixed and unchangeable. In reality, the contractor's schedule is intended to be a living and flexible management tool, and it changes as the project progresses to reflect changed circumstances and new events. For example, when delays occur, whether excused, compensable, or unexcused, the contractor will typically revise its schedule to mitigate the effects of the delays and work around them to the extent possible. Ultimately, its failure to take the natural evolution of the contractor's plan into account is the most significant weakness of the impacted as-planned method of analysis.

C. Collapsed As-Built Method

The collapsed as-built method of delay analysis (Method 3.8 in the AACE Recommended Practice)[120] is essentially the opposite of the impacted as-planned method. Instead of focusing on the as-planned schedule as the foundation for the analysis, this method utilizes the as-built schedule for this purpose. The as-built schedule simply depicts the sequence of construction as it actually occurred, complete with all delays from all causes. As such it is simply a historical record, created by utilizing the project schedule either as updated during the course of the project or as reconstructed from the contemporaneous records. The identified government-caused delays are then removed from this schedule, allowing the delayed activities to move back to the earlier date when they would have been completed but for the government's delay. This process creates a "collapsed" as-built schedule that, in theory, depicts when the project could have been completed but for the government-caused delays.

The collapsed as-built method has been frequently accepted as an appropriate means of delay analysis, though it is not the preferred method today.[121] While accepted, the collapsed as-built method is nevertheless frequently criticized for its potential inaccuracies and the opportunity for abuse of the method.[122] One such weakness is its potential for disguising possible concurrent delays by the contractor: when activities are "pulled back" after removing a government-caused delay, it is assumed that no other delay was operating that also would have prevented that activity from occurring earlier, which may not be accurate. Another weakness is that it does not take into account ways the contractor would have proceeded differently except for the government's delay. A third criticism is that it fails to consider the as-planned schedule upon which the contractor based its estimate for the project. Finally, limitations in the available as-built information generally require the analyst to make numerous assumptions regarding what the relationships between activities would have been absent the government-caused delays. These assumptions are difficult if not impossible to verify (since they involve inferring relationships in a set of circumstances that did not actually occur), and the outcome of the analysis is usually quite sensitive to the specific assumptions that have been made.

D. Time Impact Analysis and Windows Analysis

Schedule analysis methods described as either time impact analysis or windows analysis are the most commonly used methods today to attempt to demonstrate the effect of individual delays on the project as a whole. The problem is that these terms are applied to a wide variety of schedule analysis techniques, making the use of either term problematic, at least if the intended reference is to one specific technique. The AACE Recommended Practice notes that the name "time impact analysis" has been applied to six of the eight distinct methods it identifies (Methods 3.3 through 3.8). The term "windows analysis" has been applied to Methods 3.2, 3.3, 3.4, 3.5, and 3.7, with all but Method 3.2 also sometimes called "time impact analysis," among other common names.[123] This variation is indicative of the nomenclature confusion that was one of the driving forces in development of the AACE Recommended Practice.

Notwithstanding the confusion in terminology, the underlying concept of the state-of-the-art methods of schedule analysis is to utilize the schedule in effect just prior to a delay as the basis for assessing the impact of the delay. In concept, if the schedule has been carefully and accurately updated, the schedule used as the basis for measuring delay then should already reflect (to the extent practicable) all prior events, problems, and delays that preceded the delay under analysis. The effect of the delay is thus measured by the contractor's current plan for performance, rather than its original plan, as in the impacted as-planned method. This technique is particularly useful for complex projects that have encountered a series of multiple or overlapping changes or other delays. However, its assumption that the schedule has been accurately maintained and updated prior to the delay in question is by no means correct in all instances, since many times contractors may not invest the time and effort required to do this.

Considering the wide range of techniques to which these terms have been applied, it is perhaps unwise to venture too far in generally characterizing the differences between the windows and time impact approaches. Some general differences in the common understanding of these terms do exist, however. Time impact analysis considers the state of the schedule both just before the start and just after the conclusion of each delay encountered, typically relying on monthly schedule updates. The windows method can do the same but is more likely to utilize somewhat longer intervals or "windows" of time, correlating to important interim milestones in the life of the project, as the basic time unit for analysis. By "statusing" the project schedule at each of these interim points, it is possible to identify any gains or losses of time on the critical path activities during the interim period or window. The analyst then must assess the causes of the gain or loss along the critical path during that period and determine whether each cause of delay was unexcused, excused, or compensable. The revised schedule as of the end of each period is then used as the baseline schedule for measuring gain or loss on the critical path during the next period.[124]

Again, time impact analysis may effectively amount to the same thing, but frequently a time impact analysis will instead insert a mini-network of new activities (a "fragnet") into the schedule in effect as of the start of a particular delay. The inserted fragnet activities are those associated with the compensable event, such as (classically) the addition of extra work to the project. The effect of the added fragnet on the project completion date then represents the delay due to that compensable event. In this version, a time impact analysis is much like an impacted as-planned analysis, but it avoids the problem of relying on one static plan for the project generated at its beginning by instead utilizing the current schedule at the time of the compensable event as the baseline for the analysis.[125]

A good example of the successful use of the windows method of analysis is *SAE/American Mid Atlantic, Inc.*,[126] in which the GSBCA explained the technique as follows:

> The methodology used "marches through the project [and] measures where the project stood during certain milestones." Some of the milestones used included completion of caissons, completion of critical concrete work, completion of structural steel, and various milestones applicable to installation of exterior skin. [The expert witness] undertook to determine where the project was both prior to and after an alleged delay or change, and to measure the effect on the project completion date.

As a practical matter, accurate and realistic application of the time impact or windows method generally requires a project schedule that has been regularly updated.[127] As noted in *Blinderman Construction*,[128] "the only way to accurately assess the effect of the delays alleged in [the contractor's claim] is to contrast updated CPM schedules prepared before and immediately after each purported delay ... the CPM mathematical analyses in evidence are not contemporaneous with the alleged delay." However, it is possible in some instances to perform a limited form of windows analysis using just the as-built information in project historical records, without regular schedule updates, or to recreate the schedule updates after the fact.[129]

Bell BCI Co. v. United States[130] is a recent case accepting the use of a time impact analysis consistent with Method 3.7 of the AACE Recommended Practice. The expert there identified some 184 extra work orders that were not in dispute and, after inserting fragnets representing the added work associated with 49 of them, found he had accounted for the entire nine-month period of delay, so went no further. The court accepted this method as reasonable and noted that its use was required by the contract terms when the contractor was requesting a time extension.

Sunshine Construction & Engineering, Inc. v. United States,[131] by contrast, illustrates the potential for confusion due to varying terminology. In this case, the contractor's delay analysis, described as a "time entitlement analysis," was rejected in favor of the government expert's analysis that instead compared the as-planned critical path with the as-built critical path and identified and analyzed the individual work activities causing the differences. The "time entitlement analysis," by contrast, utilized the monthly schedule updates to assess gain or loss on the critical path each month, plus broke the project into fragnets. The court concluded that the contractor failed to demonstrate that the "time entitlement analysis" was a recognized method of CPM analysis, but found the government's method of analysis clear, comprehensive, and reliable. Unfortunately, neither of these methods was really described in sufficient detail to provide much guidance for future claims.

Overall, the time impact and windows techniques of CPM analysis are finding widespread acceptance when appropriately utilized.[132]

VII. Acceleration

The term "acceleration" refers to an increased pace of construction by the contractor, which may be accomplished by any combination of several means, including increasing the number of workers on the project; working overtime; working a second shift; or adding construction equipment such as more cranes, scaffolding, concrete forms, and the like. Accelerating the work almost always involves incurring additional costs, as added workers, overtime with existing staff, or added equipment all cost money and may well also reduce the contractor's efficiency overall due to factors like overcrowding, dilution of supervision, and fatigue.

A contractor may accelerate its construction efforts to make up for its own delays, to make up for government (compensable) or excusable delays, or simply to achieve an earlier completion date. Where the contractor is behind schedule due to unexcused delays, the government is entitled to require acceleration without incurring any corresponding obligation to reimburse the contractor for its acceleration efforts.[133] However, the contractor may be able to recover its acceleration costs in situations where it has been either directed to achieve an earlier completion date than provided by the contract (*directed* acceleration) or *constructively* accelerated by a combination of excusable delays and failure to grant an appropriate time extension.[134]

A. Directed Acceleration

Directed acceleration occurs as the result of an express order to accelerate in situations in which the government determines that it needs the completed work earlier than the contractual completion date and so issues a written directive to speed the pace and complete early. An acceleration directive is expressly recognized as a change to the contract that is compensable under the Changes clause.[135] Additionally, where the contractor is ordered to increase its pace of work without using the word "accelerate" or other language typifying acceleration, and the contractor is not behind schedule, such an order has "uniformly been considered an acceleration order entitling the contractor to compensation."[136]

Under the Changes clause, the compensable costs of directed acceleration would be all costs reasonably incurred in complying with the directive, including both direct costs such as overtime premiums and (subject to the problems of proof discussed in Section VIII) any resulting labor inefficiency due to the acceleration efforts. For example, a program of extended overtime implemented as part of an acceleration effort has been expressly recognized as an accepted cause of reduced labor productivity.[137]

B. Constructive Acceleration

The great majority of acceleration disputes involve claims of constructive acceleration, rather than directed acceleration. In concept, if the contractor is entitled to a time extension until a later completion date because of an excusable delay, but the government refuses to grant the time extension and instead makes it clear that the contractor is still required to meet the original completion date, the contractor's performance is almost as surely being accelerated by the government's actions (or inactions) as if the contractor had received a direct order to accelerate.

Accordingly, it is recognized that in this situation the contractor has been *constructively* ordered to accelerate. Of course, the contractor in this situation also has the option, at least in theory, not to accelerate, to finish by the properly adjusted completion date, and to dispute the assessment of liquidated damages for its "late" completion until it obtains the time extension to which it is justly entitled. In many instances, however, this option is not practically available. Among other problems, it also will expose the contractor to the risk of a default termination because of what the government perceives as its "late" completion. The doctrine of constructive acceleration gives the contractor in this situation another option, in that the contractor can avoid liquidated damages by accelerating to complete on the original completion date, and recover its acceleration costs incurred in doing so.

There are several elements of proof required to recover on a claim for constructive acceleration, which makes the claim a substantial challenge to prove. First, the contractor must demonstrate the existence of a given period of excusable delay.[138] The contractor must then show that it provided notice of the excusable delay along with an appropriately substantiated request for an extension of time,[139] which the government failed to grant within a reasonable period.[140] The government must then, either expressly or implicitly,[141] require the contractor to take steps to overcome the excusable delay, such as by insisting that the contractor must complete by the unadjusted completion date or will be assessed liquidated damages if it fails to do so.[142] Additionally, the contractor must have made reasonable efforts to accelerate its work as a result of these circumstances that resulted in actual and identifiable increased costs.[143] Since a constructive acceleration claim is a form of constructive change order, timely notice to the government of the intended claim for acceleration is another element, although not separately mentioned in most cases listing the elements of constructive acceleration.[144]

An example of how proving these elements can be difficult in practice is *Intermax, Ltd.*,[145] in which the ASBCA held that the government did not accelerate performance where it neither granted nor affirmatively denied the time extension requested by the contractor. The contractor had failed to submit the required justification with its time extension request; the government had promptly requested the missing documentation; and the contractor did not supply it in time for evaluation and action by the government before the work was substantially completed. Since, under these facts, the requirement that the government has failed to grant the time extension due within a reasonable time was not met, there was no recovery. However, in many other cases, contractors have succeeded in demonstrating all the required elements to recover.[146]

A number of cases have addressed the issue of just what constitutes an implied order by the government to accelerate. While a clear threat of assessing liquidated damages or issuing a default termination will generally suffice, it is also clear that there must be something more than mere "impressions, understandings, and desires" on the part of the contractor and government.[147] Quite simply, "[t]here must be something tantamount to a Government order that deprives the contractor of the extended delivery date to which it is entitled."[148] A coupling of a government expectation along with coercive conduct is also sufficient.[149] On the other hand, constructive acceleration requires more than the government merely informing the contractor of the government's right to default,[150] or the government putting reasonable pressure on the contractor to complete work in accordance with the project schedule.[151] Nor is the mere refusal to grant a time extension for a perceived excusable delay, in and of itself, grounds for finding constructive acceleration.[152]

The sequence in which the key events occurred is also important in proving constructive acceleration. In *Solar Foam Insulation*,[153] the contractor's constructive acceleration claim was rejected because there was no evidence that the contractor actually accelerated work as a result of the government's letter requesting speedier progress. The contractor had increased its crew size *before* receiving that letter, but took no further acceleration actions in *response* to that letter. Accordingly, the requirement that the contractor had actually accelerated due to the government's actions was missing.[154]

Subsequently executed change orders providing for time extensions may also frustrate the contractor's recovery of acceleration costs. In *Bart Associates, Inc.*,[155] the contractor was precluded from asserting a constructive acceleration claim where, following a government delay, a bilateral modification was executed extending the completion date into the winter season. In that modification, the contractor accepted the new completion date without negotiating any further adjustment due to the winter working conditions with which it would now have to contend. By contrast, in *Algernon Blair, Inc.*,[156] a claim of constructive acceleration was allowed, notwithstanding bilateral modifications for individual changes that stated they were full and complete equitable adjustments for the change. Prior to executing the modifications, the contractor had made clear that the modifications were not to be construed to preclude an acceleration claim. Additionally, the board reasoned that the claim for constructive acceleration was not associated with the specific changes, "but rather with the procedure by or manner in which the Government recognized appropriate time extensions for performance as a result of the changes," and so was not foreclosed by the bilateral modifications.

VIII. Disruption: Loss of Labor Efficiency

The term "disruption" refers to an interruption, change in the manner of performance, or change in the planned sequence of work. The characteristic, if not defining, effect, of disruption is loss of labor efficiency from that anticipated by the contractor. Because labor costs are generally a substantial fraction of overall construction cost, a decrease in labor efficiency can lead to significant financial losses for the contractor. When the loss of efficiency is a result of government actions or inactions addressable as constructive changes, it is properly the subject of an equitable adjustment. The possible causes of loss of labor efficiency are legion; among other causes, disruption can be the result of more difficult or crowded working conditions, extended overtime or second shift work (the worker fatigue factor),[157] working out of sequence or with frequent interruptions, access problems, and working in less favorable weather or temperature conditions. All of these can be compensable when they are the result of disrupting actions or inactions of the government that entitle the contractor to an equitable adjustment.

However, there are just as many possible noncompensable causes of lower-than-anticipated labor efficiency, including such factors as high labor turnover, poor supervision or planning, inadequate coordination of subcontractors, or simply an overly optimistic estimate. Additionally, all of the previously mentioned causes of inefficiency that may result from compensable actions can just as easily be the result of noncompensable root causes. This overlap, and the possibility that an observed manifestation of inefficiency may be due to multiple root causes, are two reasons why proving compensable loss of efficiency is particularly challenging.

Most typically, loss of labor efficiency is asserted as an element of a contractor's equitable adjustment claim for delay or acceleration, but disruption is actually an independent concept from delay or acceleration and can occur (and be compensable) even in the absence of any compensable delay.[158]

The basis for entitlement to compensation for loss of labor efficiency, as with compensable delay, stems from either a government breach of the contract or one of the contract clauses providing for an equitable adjustment, most commonly the Changes clause. For example, since acceleration is expressly compensable as a change under the Changes clause, to the extent that the contractor can demonstrate that disruption/loss of labor efficiency resulted from its acceleration efforts, the added costs associated with that loss of efficiency are properly included in the equitable adjustment for compensable acceleration.[159] Elements of loss of efficiency or disruption claims that involve unique problems of proof include demonstrating that the loss of labor efficiency was indeed caused by the compensable event and quantifying the amount of disruption so caused.

The boards have acknowledged that determining the amount compensable for disruption/labor inefficiency with exactitude is "essentially impossible,"[160] and that some form of reliable methodology to arrive at an inexact approximation must be utilized, typically based on expert analysis and testimony. The reason is that loss of productivity cannot generally be directly observed, measured, or recorded as it occurs but, instead, can be seen only in the reduced units of completed work installed by the workforce over a period of time.

The method of proof generally recognized as the best available in most instances is the "measured mile," which involves comparing the contractor's productivity (efficiency) on the work that was affected by the compensable event with the productivity actually achieved on similar work that was not so affected.[161] Measured-mile analyses have been successfully used in a number of cases.[162]

P.J. Dick, Inc.[163] illustrates a successful application of the measured-mile method to recover labor inefficiency caused by acceleration of an electrical subcontractor's work. The subcontractor had planned to have separate crews do the rough-in, wire-pulling, and installation of switches and devices for the branch circuits, but, under the accelerated conditions, each crew had to perform all three of these activities, thereby causing a loss of labor efficiency due to loss of the learning-curve effect.[164] Through a qualified expert witness, the subcontractor compared its efficiency on the branch circuit work with its efficiency on the feeder circuit work on the same project. While the branch and feeder circuit work were not identical, the measured-mile analysis was accepted because it compared kinds of electrical work that were reasonably alike. The approximations involved in this comparison were not exact, but nevertheless meaningful, and produced a reliable, though inexact, quantification of the resulting labor inefficiency. This was sufficient to meet the requirement to prove damages with a reasonable degree of certainty.

Considerably more controversial than the measured-mile method of proving loss of labor efficiency is the use of published industry studies containing estimates of productivity loss due to specific causes. The most commonly used of these are studies or guidelines published by the Business Roundtable, the Mechanical Contractors Association of America (MCAA), and the National Electrical Contractors Association (NECA).[165] There are some cases in which the use of such sources has been allowed, when applied with the proper degree of specificity and care.[166]

Hensel Phelps Construction Co. v. General Services Administration[167] offers a good example of how to successfully prove disruption using loss of productivity (inefficiency) factors from the MCAA guidelines. In that case, a mechanical subcontractor was found to have incurred labor inefficiencies stemming from three causes that were each the responsibility of the government: an incomplete design at contract award, a change directive adding vibration dampers late in construction, and a direction to increase the number of workers to accelerate progress. The subcontractor's expert witness utilized the inefficiency percentages associated with six factors found in the MCAA guidelines as a starting point and then applied them to particular portions of the work effort based on his extensive knowledge of the project gained from reviewing the project records, interviewing the personnel involved, and creating a detailed project as-built schedule. The expert used his experience in mechanical work and knowledge of the project to make his own assessments of the percentage impact of the disruption, which were conservative compared to the MCAA percentages. The GSBCA found this approach highly credible and accordingly awarded over $1.5 million for lost labor productivity.

By contrast, other cases have rejected efforts simply to apply loss of productivity factors based on academic studies without first validating the factors through analysis of actual labor costs and production in the field and then relating the loss of productivity to actual project events.[168] For example, in *Community Heating & Plumbing Co.*,[169] the ASBCA rejected the use of loss of efficiency factors taken from published studies that were then applied to all labor costs, stating:

The formula presented by appellant and its expert setting forth factors extracted from textual material applicable to general situations are too vague and disconnected from the specifics of the instant situation to permit a determination of the exact amount of increased costs due to loss of productivity resulting from differing site conditions. We conclude that the conversion of factors to a "multiplier," i.e., an overall factor applicable to appellant's total labor, has no logical support. The mere invocation of an expert's general rationale is insufficient.[170]

Other cases have been more emphatic in rejecting the use of such loss of efficiency studies.[171] Similarly, expert testimony regarding loss of productivity will likely be rejected as intrinsically unpersuasive where the expert has reviewed few of the project records and has not talked with project personnel who were on-site during the alleged disruption period.[172]

The least accepted method of proving loss of labor efficiency is the "total cost" method, which in this application is actually more accurately referred to as the "total labor" method. In the total labor method, the entire difference in labor costs between the actual labor cost and the estimated labor cost is attributed to the decreased labor productivity sought to be recovered. Because this method makes no attempt to identify and segregate the compensable and noncompensable factors that may have adversely affected productivity, it should be viewed as the method of last resort in determining disruption damages. The four elements generally required for application of the total cost method are also applicable to successful use of the total labor method.[173] The "modified" total labor method, in which either the actually incurred labor costs or the estimated labor costs (or both) are adjusted to eliminate added costs not caused by the government or to correct bid inaccuracies, is somewhat more accepted but remains a challenging route to recovery.[174]

On occasion, it is possible to determine by direct evidence an appropriate yardstick for measuring the compensable loss of efficiency, and such opportunities should not be overlooked. In *Batteast Construction Co.*,[175] for example, the contractor's use of a total labor calculation was rejected in favor of a direct measure of lost productivity derived from the testimony of the contractor's superintendent. The superintendent testified that before the change to the masonry specifications that allegedly caused disruption, the masons placed about 140 blocks per day on average, and after the change they placed about 75 blocks per day. The ASBCA then computed the adjustment in labor costs utilizing this ratio, which in effect was a form of measured-mile analysis established without benefit of expert analysis or testimony.

Notes

1. Claims are discussed in detail in Chapter 15.
2. Wilner v. United States, 26 Cl. Ct. 260, 263 (1992).
3. Equitable adjustments are discussed in Chapter 15.

4. Default terminations are discussed in Chapter 14.

5. FAR 52.249-10.

6. Sauer, Inc. v. Danzig, 224 F.3d 1340, 1345 (Fed. Cir. 2000); FAR 52.249-10(b)(1).

7. Sauer, Inc. v. Danzig, 224 F.3d at 1345; Reading Clothing Mfg. Corp., ASBCA No. 3912, 57-1 BCA ¶ 1290.

8. Allied Contractors, Inc., IBCA No. 265, 1962 BCA ¶ 3501.

9. Id.

10. Reg. ER 415-1-15, Construction Time Extensions for Weather (Oct. 31, 1989), *available at* http://140.194.76.129/publications/eng-regs/er415-1-15/entire.pdf. In addition, the *Engineering & Construction Bulletin* (ECB) 2008-23 (July 15, 2008) provides more detailed guidance on the application of ER 415-1-15.

11. Monroe Constr. Corp., PSBCA No. 1518, 87-2 BCA ¶ 19,745; N.S. Meyer, Inc., ASBCA No. 27144, 83-1 BCA ¶ 16,214.

12. FAR 52.249-10(b)(2).

13. Sauer, Inc. v. Danzig, 224 F.3d 1340 (Fed. Cir. 2000).

14. Id. at 1346.

15. FAR 52.243-4. For a detailed discussion of this clause, and changes generally, see Chapter 8.

16. See FAR 52.243-4(b) and the discussion of constructive changes in Chapter 8.

17. A recent decision of the Court of Federal Claims, *George Sollitt Constr. Co. v. United States*, 64 Fed. Cl. 229 (2005), has introduced some possible confusion on this point because it references only the Suspension of Work clause in setting forth the applicable rules associated with recovering for delay. However, it by no means excludes the Changes clause as a source of recovery, nor could it in view of the language of the clause.

18. FAR 52.245-1(d)(2).

19. FAR 52.236-11(b).

20. FAR 52.236-2(b); Comtrol v. United States, 49 Fed. Cl. 294, 299 (2001).

21. See Chapter 9.

22. 41 Fed. Cl. 196, 224 (1998).

23. Robert McMullan & Son, Inc., ASBCA No. 19023, 76-1 BCA ¶ 11,728.

24. 388 F.2d 844, 851–57 (Fed. Cir. 2004).

25. 41 U.S.C. §§ 601–613.

26. Triax-Pac. v. Stone, 958 F.2d 351, 354 (Fed. Cir. 1992).

27. JTL, Inc., ENG BCA No. 6363, 98-2 BCA ¶ 29,873.

28. GSBCA No. 14178 et al., 00-1 BCA ¶ 30,641.

29. Id.

30. Idela Constr. Co., ASBCA No. 45070, 01-2 BCA ¶ 31,437.

31. Id.

32. Sunshine Constr. & Eng'g, Inc. v. United States, 64 Fed. Cl. 346, 371–73 (2005); Sauer, Inc. v. Danzig, 224 F.3d 1340, 1347 (Fed. Cir. 2000). In reaching its holding, the court in Sunshine described the law on apportionment of liquidated damages as still being "unsettled," a residual effect of two conflicting U.S. Supreme Court decisions from 1914 and 1923. The alternative rule, however, that

the existence of some degree of government delay annuls the liquidated damages provision entirely, should no longer be considered good law in light of these cases. An extended discussion of both rules, citing numerous cases, is found in *PCL Construction Services, Inc. v. United States*, 53 Fed. Cl. 479 (2002). *See also* discussion of the concurrent delay problem in Section V.B below.

33. Blinderman Constr. Co. v. United States, 39 Fed. Cl. 529, 573 (1997).

34. ASBCA No. 44902, 97-1 BCA ¶ 28,754.

35. *Id.*

36. FAR 52.242-14.

37. FAR 42.1305(a).

38. FAR 52.242-14(a).

39. FAR 52.242-14(b); CEMS, Inc. v. United States, 59 Fed. Cl. 168, 230 (2003). The fourth factor is a recent addition to the standard recitation of the required elements. As recently as *CJP Contractors, Inc. v. United States*, 45 Fed. Cl. 343, 374 (1999), only the first three elements were listed.

40. Tri-Cor, Inc. v. United States, 458 F.2d 112, 131 (Ct. Cl. 1972).

41. *Id.; accord*, Cont'l Ill. Nat'l Bank & Trust Co. v. United States, 101 F. Supp. 755, 757 (Ct. Cl. 1952).

42. Tri-Cor, 458 F.2d at 131.

43. 528 F.2d 1392, 1397 (Ct. Cl. 1976).

44. *Accord*, Triax-Pac. v. Stone, 958 F.2d 351, 354 (Fed. Cir. 1992) (government's actions must be sole proximate cause of delay).

45. *E.g.*, William F. Klingensmith, Inc. v. United States, 731 F.2d 805, 809 (Fed. Cir. 1984); PCL Constr. Servs., Inc. v. United States, 47 Fed. Cl. 745, 801 (2000), *aff'd*, 96 Fed. App'x 672 (2004); and cases cited in Section V.B below.

46. CJP Contractors, Inc. v. United States, 45 Fed. Cl. 343, 372 (1999).

47. W.M. Schlosser, Inc. v. United States, 50 Fed. Cl. 147, 152 (2001) (quoting Melka Marine, Inc. v. United States, 38 Fed. Cl. 545, 546 (1997)); Commercial Contractors v. United States, 29 Fed. Cl. 654, 661 (1993).

48. 50 Fed. Cl. 147.

49. *Id.* at 152.

50. *Id.*

51. *Accord*, Shams Eng'g & Contracting Co., ASBCA 50618 et al., 98-2 BCA ¶ 30,019.

52. Bay Constr. Co., VABCA No. 5594, 02-1 BCA ¶ 31,795; Fire Sec. Sys., Inc., VABCA No. 3086, 91-2 BCA ¶ 23,743.

53. FAR 52.242-14(c).

54. FAR 52.242-14(c); Potomac Marine & Aviation, Inc., ASBCA No. 42417, 93-2 BCA ¶ 25,865.

55. FAR 52.242-14(c).

56. FAR 52.242-14(b).

57. *Id.*

58. 325 F.2d 241, 242 (Ct. Cl. 1963); *see also* Weaver-Bailey Contractors, Inc. v. United States, 19 Cl. Ct. 474, 479 (1990).

59. *E.g.*, Interstate Gen. Gov't Contractors, Inc. v. West, 12 F.3d 1053, 1060 (Fed. Cir. 1993); Preventive Maint. Servs., Inc., ASBCA No. 44661, 94-3 BCA ¶ 27,115.

60. ASBCA No. 43929, 98-2 BCA ¶ 29,903.

61. *Id.*

62. J.A. Jones Constr. Co., ENG BCA No. 6348 et al., 00-2 BCA ¶ 31,000 (early completion schedule must be reasonable).

63. *Interstate,* 12 F.3d at 1059 (adopting the three-part test as set forth in Elrich Contracting, Inc., GSBCA No. 10936, 93-1 BCA ¶ 25,316, and Frazier-Fleming Co., ASBCA No. 34537, 91-1 BCA ¶ 23,378). The same three-part test was more recently referenced in P.J. Dick, Inc. v. Principi, 324 F.3d 1364, 1373 (Fed. Cir. 2003), and West v. All-State Boiler, Inc., 146 F.3d 1368, 1379 (Fed. Cir. 1998).

64. 39 Fed. Cl. 529, 586–87 (1997).

65. Jackson Constr. Co. v. United States, 62 Fed. Cl. 84, 98, 101 (2004).

66. Weaver-Bailey Contractors, Inc. v. United States, 19 Cl. Ct. 474, 479 (1990).

67. GSBCA No. 13625, 98-1 BCA ¶ 29,685.

68. *See* Peter M. Kutil & Andrew D. Ness, *Concurrent Delay: The Challenge to Unravel Competing Causes of Delay,* CONSTR. LAW., Oct. 1997, at 4.

69. *E.g.,* Gasa, Inc. v. United States, 79 Fed. Cl. 325, 368 (2007); Essex Electro Eng'rs, Inc. v. United States, 24 F.2d 1283, 1292 (2000); Commerce Int'l Co. v. United States, 338 F.2d 81, 90 (Ct. Cl. 1964).

70. C.D. Murray Co., ENG BCA No. 5018, 89-1 BCA ¶ 21,275.

71. Morganti Nat'l, Inc. v. United States, 49 Fed. Cl. 110, 132 (2001).

72. *E.g.,* United States v. United Eng'g & Constr. Co., 234 U.S. 236, 244 (1914).

73. Modestly extending this principle, in *George Sollitt Constr. Co. v. United States,* 64 Fed. Cl. 229, 239–40 (2005), the COFC held that, where concurrent delays can be apportioned, the additional costs of working in winter weather that stem from those concurrent delays can also be apportioned based on the amount of delay attributed to each party.

74. 695 F.2d 552, 559 (Fed. Cir. 1982).

75. *Id.* (quoting Coath & Goss, Inc. v. United States, 1010 Ct. Cl. 702, 714–15 (1944)); *accord,* William F. Klingensmith, Inc. v. United States, 731 F.2d 805, 809 (Fed. Cir. 1984).

76. *E.g.,* Smith v. United States, 34 Fed. Cl. 313, 325 (1995), *appeal dismissed,* 91 F.3d 165 (Fed. Cir. 1996).

77. *E.g.,* Cape Romain Contractors, Inc., ASBCA No. 50557 et al., 00-1 BCA ¶ 30,697; Green v. Gen. Servs. Admin., GSBCA No. 12621, 96-2 BCA ¶ 28,306.

78. ASBCA No. 18146, 77-1 BCA ¶ 12,300.

79. ENG BCA No. 4926 et al., 87-2 BCA ¶ 19,833.

80. *Id.* at 122.

81. *Accord,* MCI Constructors, Inc., DCCAB No. D-924, 1996 WL 331212 (1996); Bechtel Envtl., Inc., ENG BCA No. 6137 et al., 97-1 BCA ¶ 28,640.

82. 87-2 BCA ¶ 19,833.

83. 1996 WL 331212.

84. 97-1 BCA ¶ 28,640.

85. PCL Constr. Servs., Inc. v. United States, 47 Fed. Cl. 745 (2000), *aff'd,* 96 Fed. App'x 672 (Fed. Cir. 2004). Additionally, under the Suspension of Work clause, recent COFC cases have added lack of concurrent delay as a required element of proof in obtaining recovery under that clause. *E.g.,* CEMS, Inc. v. United States, 59 Fed. Cl. 168, 230 (2003).

86. *PCL Constr. Servs.*, 47 Fed. Cl. at 801.

87. 224 F.3d 1283 (Fed. Cir. 2000).

88. *Id.* at 1292.

89. Maron Constr. Co. v. Gen. Servs. Admin., GSBCA No. 13625, 98-1 BCA ¶ 29,685. More technically, this is "total float," specifically defined as the "time between the early start date and the late start date, or the early finish date and the late finish date, of any of the activities in the [given] schedule." Galaxy Builders, Inc., ASBCA No. 50018 et al., 00-2 BCA ¶ 31,040. An activity that is on the critical path is said to have no float. Utley-James, Inc., GSBCA No. 5370, 85-1 BCA ¶ 17,816.

90. *E.g.*, Titan Pac. Constr. Corp. v. United States, 17 Cl. Ct. 630 (1989); Maron Constr., 98-1 BCA ¶ 29,685; The Gassman Corp., ASBCA No. 44975, 00-1 BCA ¶ 30,720; Galaxy Builders, Inc., 00-2 BCA ¶ 31,040; J.A. Jones Constr. Co., ENG BCA No. 6348, 00-2 BCA ¶ 31,000.

91. ASBCA No. 25447 et al., 90-2 BCA ¶ 22,649.

92. Robert F. Cushman, John D. Carter, Paul J. Gorman & Douglas F. Coppi, CONSTRUCTION DISPUTES: REPRESENTING THE CONTRACTOR 519, 538 (3d ed. 2001).

93. G.M. Shupe, Inc. v. United States, 5 Cl. Ct. 662, 737 (1984); Wunderlich Contracting Co. v. United States, 351 F.2d 956, 968 (Ct. Cl. 1965).

94. Bell BCI Co. v. United States, 81 Fed. Cl. 617, 636 (2008), *rev'd in part on other grounds*, 570 F.3d 1337 (Fed. Cir. 2009); *accord*, Wilner v. United States, 24 F.3d 1397, 1401 (Fed. Cir. 1994).

95. Jackson Constr. Co. v. United States, 62 Fed. Cl. 84, 97 (2004); *accord*, Commercial Contractors, Inc. v. United States, 29 Fed. Cl. 654, 662 (1993); Youngdale & Sons Constr. Co. v. United States, 27 Fed. Cl. 516, 550 (1993). While the discussion is solely in the context of the Suspension of Work clause, *George Sollitt Constr. Co. v. United States*, 64 Fed. Cl. 229, 236–43 (2005), contains an extensive collection of cases, most applicable more broadly, regarding the contractor's burden of proof when claiming compensable delay.

96. Haney v. United States, 676 F.2d 584, 595 (Ct. Cl. 1982).

97. Ventas De Equipo, S.A., ENG BCA No. PCC-135 et al., 2000-1 BCA ¶ 30,913.

98. Galaxy Builders, Inc. 2000-2 BCA ¶ 31,040. The Corps' standard schedule specification, for example, specifically requires that both the CPM schedule and the approved project schedule be utilized for assessing any time extension requests. Unified Facilities Guide Specification FGS-01 32 01.00 10, Project Schedule (Aug. 2008).

99. 40 Fed. Cl. 184 (1998).

100. *Id.* at 199 (quoting Commercial Contractors, Inc. v. United States, 29 Fed. Cl. 654, 662 (1993)); Essential Constr. Co., Inc./Hinmount Constructors, Ltd., Joint Venture, ASBCA No. 18706, 89-2 BCA ¶ 21,632.

101. 39 Fed. Cl. 529 (1997), *aff'd*, 1999 WL 780831 (Fed. Cir. 1998).

102. *Id.* at 585.

103. John T. Jones Constr. Co., ASBCA No. 48303 et al., 98-2 BCA ¶ 29,892.

104. Fortec Constructors v. United States, 8 Cl. Ct. 490, 504 (1985); Coffey Constr. Co., VABCA No. 3361 et al., 93-2 BCA ¶ 125,788 (refusing to rely on CPM schedule that had not been updated throughout the contract).

105. ENG BCA No. 6252, 97-1 BCA ¶ 28,918.

106. Pathman Constr. Co., ASBCA No. 23392, 85-2 BCA ¶ 18,096.

107. 39 Fed. Cl. at 585.

108. 36 Fed. Cl. 600, *aff'd,* 121 F.3d 683 (Fed. Cir. 1997).

109. 49 Fed. Cl. 110 (2001).

110. 62 Fed. Cl. 84, 99 (2004).

111. AACE Int'l Recommended Practice No. 29R-03, Forensic Schedule Analysis (June 28, 2007) (hereinafter AACE Recommended Practice).

112. *Id.* at 58.

113. FAACAP No. 67-18, 66-2 BCA ¶ 6066.

114. ASBCA No. 44937, 2001-2 BCA ¶ 31,584.

115. *Id.*

116. Jon M. Wickwire & Stuart Ockman, Use of Critical Path Method on Contract Claims—2000, CONSTR. LAW., Oct. 1999, at 14.

117. ASBCA No. 50018, 00-2 BCA ¶ 31,040.

118. *E.g.,* Titan Pac. Constr. v. United States, 17 Cl. Ct. 630 (1989).

119. Richard Smith & John M. Cook, *Obtaining Time Extensions,* in CONSTRUCTION LAW HANDBOOK 745, 792 (Richard K. Allen & Stanley A. Martin eds., 2d ed. 2009).

120. AACE Recommended Practices, § 3.8, at 68.

121. *E.g.,* Fischbach & Moore Int'l Corp., ASBCA No. 18146, 77-1 BCA ¶ 12,200; CA No. 418, 79-1 BCA ¶ 13,836.

122. Smith & Cook, *supra* note 119, at 854–55.

123. AACE Recommended Practice, tbl.1, at 11.

124. This description of windows analysis corresponds to AACE Recommended Practice Methods 3.3 and 3.4.

125. This description of time impact analysis corresponds to AACE Recommended Practice Method 3.7.

126. GSBCA No. 12294 et al., 98-2 BCA ¶ 30,084.

127. Smith & Cook, *supra* note 119, at 854–55; Cogefar-Impresit U.S.A., Inc., DOTBCA No. 2721, 97-2 BCA ¶ 29,188.

128. 39 Fed. Cl. at 585.

129. The AACE Recommended Practice identifies this as Method 3.5, and notes that it is usually adopted when contemporaneous updates are not available or never existed. AACE Recommended Practice, § 3.5, at 55.

130. 81 Fed. Cl. 617, 640 (2008), *rev'd in part on other grounds,* 570 F.3d 1337 (Fed. Cir. 2009).

131. 64 Fed. Cl. 346, 368–69 (2005).

132. *E.g.,* Donohoe Constr. Co., ASBCA No. 47310 et al., 98-2 BCA ¶ 30,076; John T. Jones Constr. Co., ASBCA No. 48303 et al., 98-2 BCA ¶ 29,892; *Cogefar-Impresit,* 97-2 BCA ¶ 29,188; Gulf Contracting, Inc., ASBCA No. 30195 et al., 89-2 BCA ¶ 21,812.

133. Donald R. Stewart & Assocs., AGBCA No. 84-226-1 et al., 92-1 BCA ¶ 24,705; Carney Gen. Contractors, Inc., NASA BCA No. 375-4 et al., 79-1 BCA ¶ 14,243.

134. A. H. Gaede, Jr., Jonathan B. Head, & David H. Bashford, in CONSTRUCTION LAW HANDBOOK 919, 970 (Richard K. Allen & Stanley A. Martin eds., 2d ed. 2009).

135. FAR 52.243-4.

136. Donald M. Drake Co., ENG BCA No. 1634, 1960 WL 223 (1960).

137. ACE Constructors, Inc. v. United States, 70 Fed. Cl. 253, 281–83 (2006).

138. P.J. Dick Inc., VABCA No. 5597 et al., 2001-2 BCA ¶ 31,647; *Donald R. Stewart*, 92-1 BCA ¶ 24,705. In *ACE Constructors, Inc.*, 70 Fed. Cl. at 280, it was held that the same standards applied where the source of the delay was extra work due to a differing site condition.

139. The request for an extension of time must be supported by "information sufficient to allow the Government to make a reasonable determination." Fermont Div., Dynamics Corp. of Am., ASBCA No. 15806, 75-1 BCA ¶ 11,139, *aff'd*, 216 Cl. Ct. 448 (1978).

140. *Donald R. Stewart*, 92-1 BCA ¶ 24,705. The Government must "be afforded an opportunity to grant or deny a time extension on account of the delay." Greulich, Inc., ENG BCA No. 3832, 78-2 BCA ¶ 13,417, at 65,588. "[I]t is not unusual for parties to negotiate after the fact as to the number of days that are justified under a contract and to incorporate the extensions in contract modifications issued weeks after the fact." Fraser Const. Co. v. United States, 384 F.3d 1354, 1363 (Fed. Cir. 2004) (citing cases).

141. An express or implied order may take the form of coercive acts having the effect of an order, such as the threat of default—Lewis Constr. Co., ASBCA No. 5509, 60-2 BCA ¶ 2732—or threat of assessment of liquidated damages for failure to meet the government's completion date—Unarco Material Handling, PSBCA No. 4100, 00-1 BCA ¶ 30,682; Pathman Constr. Co., ASBCA No. 14285, 71-1 BCA ¶ 8905.

142. *Fermont Div.*, 75-1 BCA ¶ 11,139.

143. *Id.*

144. *Norair Engineering Corp. v. United States*, 666 F.3d 546 (1981), is undoubtedly the most frequently cited source for the required elements of a constructive acceleration claim. In *Fraser Construction Co.*, 384 F.2d at 1361, the Federal Circuit stated the test as comprising five elements, noting that *Norair* compressed them into just three but indicating no intent to make any substantive change.

145. ASBCA No. 41828, 93-2 BCA ¶ 25,699.

146. *ACE Constructors, Inc.*, 70 Fed. Cl. 253, 280–81 (2006), is a well-reasoned recent example.

147. Midwest Bank Note Co., 1998 GPOBCA LEXIS 1.

148. *Id.*

149. Fermont Div., Dynamics Corp. of Am., ASBCA No. 15806, 75-1 BCA ¶ 11,139, *aff'd*, 216 Cl. Ct. 448 (1978).

150. Donald R. Stewart & Assocs., AGBCA No. 84-226-1 et al., 92-1 BCA ¶ 24,705.

151. *Fermont Div.*, 75-1 BCA ¶ 11,139.

152. Fraser Const. Co. v. United States, 384 F.2d 1354, 1363 (Fed. Cir. 2004).

153. ASBCA No. 46278, 94-1 BCA ¶ 26,288.

154. *Cf.* Allen L. Bender, Inc., PSBCA No. 2322 et al., 91-2 BCA ¶ 23,828 (no constructive acceleration where government exhortations to get moving and maintain the schedule preceded any excusable delays).

155. Bart Assocs., Inc., EBCA No. C-9406176, 97-2 BCA ¶ 29,206.

156. Algernon Blair, Inc., ASBCA No. 45369, 94-2 BCA ¶ 26,638.

157. In *ACE Constructors*, 70 Fed. Cl. 253, 281–83 (2006), the COFC expressly accepted use of an interpolation from charts appearing in the Business Round-table report, *Scheduled Overtime Effect on Construction Projects, Report C-2* (November 1980, available at http://www.curt.org/pdf/156.pdf), for calculating the effect of extended overtime on labor productivity.

158. Sauer, Inc. v. Danzig, 224 F.3d 1340, 1348 (Fed. Cir. 2000).

159. *ACE Constructors*, 70 Fed. Cl. at 281–83.

160. VABCA No. 5597 et al., 01-2 BCA ¶ 31,647, *rev'd in part on other grounds sub nom.* P.J. Dick, Inc. v. Principi, 324 F.3d 1364 (Fed. Cir. 2003).

161. Bell BCI Co. v. United States, 81 Fed. Cl. 617, 641 (2008), *rev'd in part on other grounds*, 570 F.3d 1337 (Fed. Cir. 2009); U.S. Industries, Inc. v. Blake Const. Co., 671 F.2d 539, 547 (D.C. Cir. 1982).

162. W.G. Yates & Sons Constr. Co., ASBCA No. 48398, 01-2 BCA ¶ 31,428; Clark Concrete Contractors, Inc., GSBCA No. 14340, 99-1 BCA ¶ 30,280; Danac, Inc., ASBCA No. 33394, 97-2 BCA ¶ 29,184 (1997); Goodwin Contractors, Inc., AGBCA No. 89-148-1, 92-2 BCA ¶ 24,931; Time Contractors, J.V., DOTBCA No. 1669, 87-1 BCA ¶ 19.582; Flex-Y-Plan Indus., GSBCA No. 4117, 76-1 BCA ¶ 11,713; Luria Bros. & Co. v. United States, 369 F.2d 701 (Ct. Cl. 1966).

163. *P.J. Dick*, 01-2 BCA ¶ 31,647.

164. The learning curve effect, sometimes referred to as the production line effect, is premised on the increased efficiency that a given crew will develop over time when it performs essentially the same task repetitively.

165. MECH. CONTRACTORS ASS'N OF AM., MANAGEMENT METHODS BULLETIN PD2 (2005); NAT'L ELEC. CONTRACTORS ASS'N, THE EFFECT OF TEMPERATURE ON PRODUCTIVITY (1974); NAT'L ELEC. CONTRACTORS ASS'N, OVERTIME AND PRODUCTIVITY IN ELECTRICAL CONSTRUCTION (1989); BUS. ROUNDTABLE, SCHEDULED OVERTIME EFFECT ON CONSTRUCTION PROJECTS, REPORT C-2 (1980, *available at* http://www .curt.org/pdf/156.pdf).

166. Stroh Corp. v. Gen. Servs. Admin., GSBCA No. 11029, 96-1 BCA ¶ 28,265 (25 percent labor inefficiency allowed based on methodology that included use of MCAA factors); The Clark Constr. Group, Inc., VABCA No. 5674, 00-1 BCA ¶ 30,870.

167. GSBCA No. 14744, 01-1 BCA ¶ 31,249, *aff'd sub nom.* Perry v. Hensel Phelps Constr. Co., 36 Fed. App'x 649 (Fed. Cir. 2002).

168. Fire Sec. Sys., Inc., VABCA No. 5559-63 et al., 02-2 BCA ¶ 31,977 (most asserted inefficiencies not borne out by daily logs, payrolls, and other records); Cosmic Constr. Co., ASBCA No. 24041 et al., 88-2 BCA ¶ 20,623; Luria Bros. & Co. v. United States, 177 Ct. Cl. 676, 713 (1966).

169. ASBCA No. 37981 et al., 92-2 BCA ¶ 24,870.

170. *Id.*

171. Sunshine Constr. & Eng'g, Inc. v. United States, 64 Fed. Cl. 346 (2005) (attributing delay to MCAA factors rejected as "not an accepted approach by . . . peers or by any trade association"); Herman B. Taylor Constr. Co. v. Gen. Servs. Admin., GSBCA No. 1542, 03-02 BCA ¶ 32,320, at 54–56 (use of MCAA factors was inappropriate because the labor force consisted of laborers, not mechanical workers).

172. Hensel Phelps Constr. Co., ASBCA No. 49270, 99-2 BCA ¶ 30,351; Dravo Corp., EBCA No. 3800, 79-1 BCA ¶ 13,575; Cont'l Consol. Corp., ASBCA No. 14372, 71-1 BCA ¶ 8,742.

173. Youngdale & Sons Constr. Co. v. United States, 27 Fed. Cl. 516, 541 (1993). The four standard elements for use of the total cost method are typically stated as: (1) the nature of the particular losses makes it impossible or highly improbable to determine them with a reasonable degree of accuracy; (2) the contractor's bid estimate was realistic; (3) the contractor's actual costs were reasonable; and (4) the contractor was not responsible for the cost overruns. *See* Servidone Constr. Corp. v. United States, 931 F.2d 860, 861 (Fed. Cir. 1991).

174. Hardrives, Inc., IBCA No. 2319/2514 et al., 94-1 BCA ¶ 26,267; Sovereign Constr. Co., ASBCA No. 17792, 75-1 BCA ¶ 11,251; J&T Constr. Co., DOTCAB No. 73-4, 75-2 BCA ¶ 11,398; WRB Corp. v. United States, 183 Ct. Cl. 409 (1968).

175. ASBCA No. 35818 et al., 92-1 BCA ¶ 24,697.

CHAPTER 20

Pricing of Claims

STEPHEN B. SHAPIRO
BRYAN R. PHILLIPS
DIRK D. HAIRE
CHERYL A. FEELEY

I. Introduction

While the primary focus in construction disputes usually concerns liability, proving and pricing damages are crucial elements of any construction claim. Despite its obvious importance, pricing is frequently taken for granted or ignored. An improperly priced claim can result in the loss of money that the claimant would otherwise be entitled to receive. In some instances, civil or criminal penalties for fraud, false claims, false statements, and other violations result from defective or improper pricing. Therefore, pricing should be approached with the same attention to detail that is used to establish the substantive basis for the claim itself.

A host of complexities make calculating construction damages more challenging than determining ordinary contractual losses. This chapter describes the concepts and approaches for pricing typical construction claims. It also examines the implications of failing to understand and follow appropriate claim pricing methodologies or taking improper liberties in a request for additional costs.

II. Fundamental Principles of Contract Pricing

A. General Rule of Damages

The general goal of contract damages is to place the injured party in the financial position it would have occupied had the other party fully performed under the contract.[1] The basic damage calculation formula provides that damages should be given for the net amount of losses caused and gains prevented, in excess of savings made possible.[2] Therefore, contract damages are limited to losses that are economic in nature.

B. Proof of Adjustment

Perhaps the most difficult element in pricing a construction claim is proving that the claimant is entitled to receive the damages it is requesting. In almost all cases, the party pursuing a claim has the burden of proof in establishing the amount of the adjustment to which it is entitled.[3] To satisfy this burden, the claimant must both prove that the costs being claimed are reasonable and establish a causal connection between those costs and the event(s) upon which the claim is based.[4]

Typically, the claimant's right to recover damages must be established by a "preponderance of the evidence."[5] To meet this burden, the claimant must develop sufficient evidence to establish that the amount it is requesting is reasonable in relation to the work in question. For many years, the rule in federal government contracts was that the amount actually spent in performing the work was presumed to be reasonable unless the government presented evidence to the contrary.[6] The regulations have been changed, and this rule has been modified to place the burden on the claimant to establish that its actual costs represent a reasonable amount for the work being performed.[7]

The contractor should be aware that its actions in incurring a cost will be examined in determining whether a particular cost expenditure was reasonable. Normally, a contractor has great discretion in determining its means and methods of performance.[8] Therefore, unless a contractor abuses its discretion, the costs it incurs will not be deemed unreasonable.[9] Nevertheless, the reasonableness of a contractor's actual costs has been rejected where the contractor failed to use readily available industry practices.[10] Similarly, a contractor can be denied recovery if the government is able to demonstrate that the contractor did not follow the government's directive in performing a change.[11]

In terms of causation, there has been divergent case law addressing the precision with which a claim for damages must be established. The traditional rule was set forth by the U.S. Court of Claims as follows:

> A claimant need not prove his damages with absolute certainty or mathematical exactitude. Dale Construction Co. v. U.S., 168 Ct. Cl. 692 (1964); Houston Ready-Cut House Co. v. U.S., 96 F. Supp. 629 (Ct. Cl. 1951). It is sufficient if he furnishes the court with a reasonable basis for computation, even though the result is only approximate. F.H. McGraw & Co. v. U.S., 130 F. Supp. 384 (Ct. Cl. 1955); Locke v. U.S., 283 F. 3d 521 (Ct. Cl. 1960). Yet this leniency as to the actual mechanics of computation does not relieve the contractor of his essential burden of establishing the fundamental facts of liability, causation and resultant injury. River Constr. Corp. v. U.S., 159 Ct. Cl. 254 (1962); Addison

Miller, Inc. v. U.S., 70 F. Supp. 893 (Ct. Cl. 1947), cert. denied, 332 U.S. 836; J.D. Hedin Construction Co., Inc. v. United States, supra, 171 Ct. Cl. at 86–7, 347 F.2d at 246–47. It was plaintiff's obligation to prove with reasonable certainty the extent of unreasonable delay which resulted from defendant's actions and to provide a basis for making a reasonably correct approximation of the damages which arose there from. Aragona Constr. Corp. v. U.S., 165 Ct. Cl. 382 (1964); Laburnum Construction Corp. v. United States, 163 Ct. Cl. 339, 325 F.2d 451 (1963). Broad generalities and inferences to the effect that defendant must have caused some delay and damage because the contract took 318 days longer to complete than anticipated are not sufficient.[12]

Although some flexibility is permitted, it is important to prove causation using evidence that directly links the applicable losses to the events underlying the claim. The U.S. Court of Federal Claims (formerly U.S. Claims Court, and successor to the original jurisdiction of the Court of Claims) has required some precision in this area, particularly where actual costs are available and a more precise methodology for proving the costs attributable to a claim event can be established.[13] Furthermore, requests for damages that the claimant knows or should know are attributable to events that are not covered by the claim can result in criminal or civil penalties. Therefore, proving a causal nexus between a claim event and the damages being requested is a good practice that avoids problems and increases the likelihood that the claim will be paid.

There are no hard and fast rules for establishing a causal link between a claim event and the damages that may be included in a claim. Nevertheless, it is basic that the cost increase or decrease must be created by the event for which the adjustment is being requested.[14] To meet this standard, the claimant should establish a sequential or chronological relationship between the costs being claimed and the event upon which the claim is being made. The costs also should be logically connected to the underlying claim in a predictable sequence. While this type of causal link is easier to establish for standard change order claims that impact a specific component of the work, it is important to focus on the causal connection in every circumstance in which a claim is made. In complicated situations, expert testimony may be required to satisfy the contractor's burden of proof.

C. False Claims

The federal government has promulgated several powerful statutes that impose criminal or civil penalties for fraud, false claims, false statements, and the knowing receipt of public property with intent to convert it for gain,[15] and the government has become increasingly aggressive in seeking these penalties.[16] Similarly, Congress recently passed the Fraud Enforcement and

Recovery Act (FERA), which amends the False Claims Act by expanding the ability of the government and qui tam relators to pursue false claims against prime contractors, subcontractors, and suppliers.[17] Courts have also dramatically increased the consequences of filing false claims in recent years. For example, in the recent case *Daewoo Engineering & Construction Co. v. United States*,[18] the Court of Appeals for the Federal Circuit upheld the Court of Federal Claims' decision to impose $50.6 million in damages upon a contractor who filed a $64 million false claim against the government. Additional criminal penalties can arise if a claim is prepared or submitted using a fax machine, radio signal, or electronic, overnight, or regular mail.[19] Further sanctions are imposed if more than one person is involved in developing the claim.[20] If a claimant has a pattern or practice of submitting claims that are found to violate criminal statutes, additional penalties, including treble damages, can be imposed under criminal and civil Racketeer Influenced and Corrupt Organizations (RICO) Act statutes.[21] A more detailed discussion of false claims is contained in Chapter 24.

III. Basic Methods of Calculation

A. Actual Cost Method

"Actual costs" are universally preferred for quantifying construction damages.[22] Further, the remedy-granting clause in federal government contracts allows for recovery of the increase or decrease in "costs" resulting from the compensable event.

The preference for actual cost data also is expressed in the Truth in Negotiations Act (TINA), which requires a prime contractor seeking a change or modification over $650,000 to submit detailed pricing data and to certify that such data is "accurate, complete and current" as of the date of agreement on price or, if applicable, an earlier date agreed upon between the parties that is as close as practicable to the date of the agreement on price.[23] Furthermore, the government's contracting officer can require the contractor to segregate costs attributable to a change even when it is less than the $650,000 statutory threshold set forth in TINA. Federal Acquisition Regulation (FAR) 43.205(f) permits contracting agencies to use the following clause in appropriate circumstances:

> The Contracting Officer may require change order accounting whenever the estimated cost of a change or series of related changes exceeds $100,000. The Contractor, for each change or series of related changes, shall maintain separate accounts, by job order or other suitable accounting procedure, of all incurred segregable, direct costs (less allocable credits) of work, both changed and not changed, allocable to the change. The Contractor shall maintain such accounts until the parties agree to an equitable adjustment of the changes ordered

by the Contracting Officer or the matter is conclusively disposed of in accordance with the Disputes clause.[24]

To determine actual costs, the contractor normally establishes cost codes to which it records labor and equipment hours and other data associated with a claim event. Even if costs are not segregated during performance, the contractor should use its records to attempt to establish the actual cost of the change or other compensable event.

If actual cost records are available, the contractor's failure to submit those records may preclude recovery by the contractor.[25] Claims also have been rejected where the contractor was found to be "imprudent" or "irresponsible" for failing to segregate the costs attributable to a change where it should have known to do so.[26] However, the absence of a contract provision requiring cost segregation may help the claimant if its accounting records are not sufficiently detailed to derive costs associated with a specific claim and if another method for deriving those costs can be identified and supported.[27] Cost segregation also may be avoided where a large number of change orders makes it impractical to segregate costs that are attributable to a particular claim.[28] Despite these judicial exceptions, it is always best to rely on actual cost data wherever possible.

B. Cost Estimates

Where actual cost data is unavailable or work covered by a claim is so intermingled with other activities that actual costs cannot be accurately identified or calculated, another method must be utilized to price the claim. Alternative methods also are necessary if forward pricing for a change order is required or a claim involves deleted work. "Estimated costs" are commonly relied upon in these situations.

Like other methods of calculating construction damages, estimates should be substantiated by detailed cost data and clear factual support.[29] Estimates have been rejected where they were supported by vague evidence, uncorroborated calculations, or unpersuasive calculations.[30] Therefore, estimates should be supported by specific facts and verifiable cost data.[31]

To avoid problems where substantiating data is difficult to obtain, it is advisable to retain a consultant or refer to statistical data to corroborate the claimant's estimating techniques. Consultants are particularly useful if a claim is litigated because they can also serve as expert witnesses.[32] Therefore, the consultant should be qualified to present testimony regarding the issue in question and be prepared to offer an opinion that meets the standards set forth in *Daubert v. Merrell Dow Pharmaceuticals* and its progeny.[33]

Care should be taken to assure that the consultant is familiar with the facts underlying a claim, particularly in situations where the consultant may be asked to offer an expert opinion in subsequent judicial proceedings.[34] The best practice is to provide the consultant with access to all relevant facts and information and verify that the expert has thoroughly researched his opinions. In this regard, both the claimant and recipient of a claim should be aware that the facts in a given situation will be considered in ascertaining whether the expert made a proper investigation.[35]

Statistical estimating techniques are useful in supporting estimated claim calculations if backed by actual data, particularly if the statistics are corroborated by an expert consultant.[36] Trade associations such as the National Electrical Contractors Association (NECA), Mechanical Contractors Association of America (MCAA), and American Concrete Institute (ACI) publish data that are commonly used in developing certain types of estimates. Similarly, manuals published by the R.S. Means Company, Inc. (R.S. Means), F.W. Dodge/McGraw Hill (Dodge), and the U.S. Army Corps of Engineers (COE), and The Blue Book of Building and Construction (Blue Book) have been relied upon. The particular manual that is to be used in pricing a claim is sometimes specified by contract or regulation. However, statistical estimating techniques will be rejected if inappropriately utilized or if the statistics do not apply to the instant situation.[37] Given the inherent lack of specificity in general published statistical data, such data should be relied upon only if actual cost data is unavailable and the statistics can be corroborated by the facts of a specific case.

C. Total Cost Method

A controversial but frequently used approach to calculate construction damages is the "total cost" method. In a pure total cost claim, the contractor computes its damages by subtracting its total actual costs from its initial bid estimate to determine the amount of its claim.

While a total cost claim is simple to prepare, it is highly disfavored by the courts because of its inherent flaws.[38] The total cost claim fails to tie the damages claimed to the compensable event.[39] A total cost claim assumes that the claimant's initial estimate was error-free and neither the contractor nor its subcontractors or suppliers was responsible for any part of the cost overruns that were experienced during construction. Obviously, complete responsibility for a cost overrun is difficult to attribute to one party. Bidding inaccuracies can unjustifiably reduce the contractor's estimated costs, while performance inefficiencies can inflate the cost of construction. These types of inaccuracies and inefficiencies can skew the results of the total cost approach.[40]

Despite its inherent flaws, the total cost method is accepted where the appropriate safeguards are established.[41] Typically, the courts require four "indicia of reliability" to be present before the total cost method may be used:

1. The nature of the particular losses must make it impossible or highly impractical to determine the claimant's losses with a reasonable degree of accuracy;

2. The claimant's original estimate or bid must be reasonable and free of material error;

3. The claimant's actual costs must be reasonable and accurately recorded; and

4. The claimant must show that it was not responsible for any cost over-runs.[42]

Even when these criteria are satisfied, the claimant is still obligated to show that the use of a segregated cost system or other more reasonable approach to measuring damages was not feasible, and that the total amount requested is reasonable for the work covered by the claim.[43] In most circumstances, the total cost method will be rejected if the proper safeguards are not satisfied.[44]

D. Modified Total Cost Method

The modified total cost method seeks to overcome shortcomings inherent in the total cost approach by making adjustments to remedy deficiencies in satisfying the four requirements for a pure total cost claim.[45] A modified total cost claim uses a pure total cost calculation as a starting point for pricing the claim, from which a downward adjustment is made to reflect the claimant's inability to satisfy one or more total cost prerequisites. For example, where the contractor is responsible for some of its increased costs, it subtracts those costs from the total costs.[46] Similarly, adjustments are made for errors or inaccuracies in the contractor's bid estimate.

The modified total cost approach has been employed both for "tempering a total cost award" and for "preventing the [government] from obtaining a windfall stemming from the [claimant's] inability to satisfy all the elements of the total cost method."[47] Use of the modified total cost method also bolsters the credibility of a standard total cost claim because the claimant accepts responsibility for its own overruns. While a modified total cost claim more closely represents the additional costs than a pure total cost submission, it still fails to provide a direct link between the costs and the impacts suffered. Therefore, it is not preferred if a more reliable means can be used to calculate losses.

E. Equitable Adjustment

The term "equitable adjustment" is typically used in the context of public contracts as a reference to contract modifications that are based on remedy-granting provisions including the Changes, Differing Site Conditions, Suspension of Work, and Termination for Convenience clauses.[48] Prior to the Contract Disputes Act of 1978 (CDA),[49] this distinction was significant because the judicial and administrative bodies with statutory authority to resolve federal government contract disputes had jurisdiction over construction claims only if the contract contained a provision providing for the remedy requested by the contractor.[50] This limitation meant that breach of contract claims, i.e., claims that did not arise under a contractual remedy-granting clause, could not be resolved under the basic federal government contract disputes procedures. Today, however, virtually all claims against the federal government are covered under the same dispute resolution procedures.

The remedy-granting clauses in a federal government contract provide that the contractor is entitled to an equitable adjustment for its increase or decrease in costs resulting from a change, differing site condition, suspension of work, or termination for convenience. By the explicit terms of the contract, the equitable adjustment is a "cost"-based calculation.

An equitable adjustment is intended to leave the contractor in the same profit or loss position on the contract as a whole that the contractor would have been in had the change or modification not occurred.[51] The adjustment should be equal to the difference between the actual costs incurred in completing the changed or modified work and the reasonable costs of performing the work as originally specified.[52] This calculation should include any price advantages that the contractor obtained with respect to the original work, plus overhead and profit on the work already performed.[53]

To be entitled to an equitable adjustment, a contractor must show liability, causation, and injury.[54] The contractor must prove that the government accelerated, augmented, or complicated the work, and thereby caused the contractor to incur specific additional costs. In addition to proving that the other party specifically caused its increased costs, the contractor also must prove that those costs were "reasonable, allowable, and allocable to the contract."[55]

F. "Jury Verdict" Method

Generally, the contractor must prove its damages by providing evidence that establishes a nexus between specific claim events and the costs being claimed. However, there are circumstances where the contractor can prove it was damaged, but cannot establish the amount of those damages with reasonable certainty. In situations where liability is plain, but incomplete or conflicting evidence is presented regarding the appropriate amount of damages and precise damages are difficult or impossible to derive, courts sometimes employ the "jury verdict" approach to independently determine the compensation that the contractor is entitled to receive.[56]

Generally, jury verdict awards are a last resort and only will be relied upon if (1) clear proof of injury exists; (2) there is no more reliable method for computing damages; and (3) the evidence is sufficient to make a fair and reasonable approximation of the damages.[57] This "methodology" contemplates that the contractor will present the trier of fact with the broadest possible range of information and evidence related to its increased costs.[58] Because proof of actual costs is generally unavailable, other types of evidence may be offered, including accounting records, expert witness testimony, and costs experienced on similar projects.[59] The contractor uses this evidence to ask the trier of fact to formulate its own calculations and award damages using its best judgment.[60]

While jury verdict awards have limited applicability, this approach has been used in cases where the amount claimed was unsupported but the evidence showed that the owner's calculation was too low.[61] Jury verdict damages also have been awarded based on "rough estimates" and for items whose cost was "totally unclear."[62] Judicial bodies have even developed their own computations.[63]

While the jury verdict approach may be equitable, it is not favored.[64] When there is no reasonable excuse for the lack of accurate cost information, damages based on the jury verdict method usually will not be awarded.[65] Furthermore, if insufficient evidence makes a reasonable approximation impossible, the claim may be partially denied.[66] The jury verdict also may reduce the amount claimed if causation is not fully determined.[67]

On a practical level, the jury verdict approach is a high-risk proposition. The contractor has little or no control over the court's or board's decision to formulate damages on its own or ability to guide the manner in which the court or board computes damages after it determines that a jury verdict award is appropriate. Therefore, jury verdict damages should not be relied upon as a substitute for proving a claim through actual costs or estimates.[68]

IV. Elements of Construction Damages

The specific cost elements that may be used in pricing construction damages are dependent on the facts underlying the claim and contract terms under which the claim is presented. While each claim is unique, there are common elements of damage that should be considered in pricing typical construction claims. The most typical pricing elements are described below.

A. Labor Costs

Labor is one of the most expensive and volatile elements in construction. Labor costs are difficult to predict and can be influenced by countless occurrences. Change order directives, differing site conditions, defective specifications, adverse weather conditions, strikes, site obstructions, acceleration orders, and other events can impact labor costs. Given its inherent volatility, labor is a common element in most construction claims.

Like other elements of damage, the claimant has the burden of establishing that the additional labor component of a construction claim is attributable to the events on which the claim is based.[69] The claimant also must demonstrate that the amount claimed for labor is reasonable in relation to the work performed.[70] This burden will typically be met if a reasonable basis for the calculation can be supported.[71] What is reasonable depends on the type of labor claim being presented.

1. Direct Labor Costs

The most basic type of claim for increased labor seeks additional compensation for the direct cost of completing changed or disrupted work. Ideally, the contractor will maintain detailed payroll records and daily logs of workers deployed to the site that can be used to derive costs attributable to extra work orders, differing site conditions, defective specifications, and other specific issues that underlie extra work claims. Properly maintained field records should include daily reports that record the number of hours each person worked on a project and the specific location and type of work each person performed.

The additional hours of work are typically determined by assigning separate cost codes to the changed or additional work. Care must be taken to assure that the field personnel charge their hours to the appropriate accounts or the contractor could face false claim allegations. Another potential pitfall is the establishment of too many cost codes, making it impossible for the field personnel to accurately code their time to the proper accounts without a full-time cost person in the field to assist in the accounting.

Once the hours are determined, the cost of the labor hours can be determined from the contractor's books and records. Obligations imposed by labor, tax, and employment laws require contractors to track hours worked, wage rates, overtime expenditures, health and welfare benefits, social security and unemployment taxes, worker's compensation and other insurance premiums, apprentice fund contributions, and other costs directly attributable to labor hours on construction projects.[72] Furthermore, the Davis-Bacon Act requires contractors to submit certified payroll records.[73] Since actual labor cost data is available in most situations, the difficult aspect of pricing labor costs is proving the amounts attributable to the claim event.

The additional labor costs are calculated by multiplying the number of labor hours expended due to the claim event by the wages paid, with an appropriate markup for taxes, insurance, and other types of labor burden.[74] Given the complex interrelationships between trades and the possibility that an event will affect one work sector while leaving another unaffected, segregated labor records can be invaluable. Since labor records are typically available, claims seeking additional labor costs without them have been rejected, even when they are supported by estimates and expert testimony.[75]

2. Alternate Direct Labor Calculations

If the claimant's records do not segregate individual work items, or the nature of a claim makes reliance on such records difficult or impossible, increased labor costs must be calculated using an alternative methodology. The preferred alternative is the use of estimates that are based on published industry standards and supported by expert testimony and available project data.[76] A nexus must be established between underlying claim events and the damages being requested when estimated labor costs are submitted to support a construction claim.[77] The amount claimed also must be reasonable in proportion to the work performed.

It also may be possible to support additional labor costs using a total cost or modified total cost calculation.[78] Given the general disfavor of the total cost approach, it is best to use a more reliable pricing methodology whenever possible. If no other alternative can be identified, a modified total cost claim is preferable.[79] As a last resort, direct labor costs have been recovered based on a jury verdict award.[80] Whether the claimant relies on estimates or a variation of the total cost method, it is still important to demonstrate a causal connection between the claim event and additional labor costs being requested.[81]

3. Cumulative Impact and Ripple Effect Claims

Even where detailed cost records exist, there are situations where discrete claim events impact labor costs in ways that are difficult to track with precision. For example, it is not uncommon for contractors to encounter multiple change orders, delay events, site obstructions, differing site conditions, and other items that are so intermingled with other work that a precise accounting of additional labor hours is impossible. In these situations, it may be possible to aggregate labor costs attributable to multiple claim events into a single "cumulative impact" claim.[82] Claim events also may create a "ripple effect" that impacts the cost of labor that is needed to perform unchanged work that may be recoverable.[83] To prevail on a ripple effect claim, it is important to demonstrate a causal connection between the claim event and the downstream activity that was impacted to the extent it caused additional costs to be incurred.[84] Cumulative impact and ripple effect claims will be denied where this causal connection is not established.

4. Labor Overhead

Although wages represent the majority of the cost of labor, there are associated labor expenses, sometimes referred to as labor overhead or labor burden, that represent significant additions to employee salaries and hourly wages, including fringe benefits, payroll taxes, workers' compensation insurance, and other costs that are mandated by statute. In the absence of a contractual exclusion, which exists in rare instances, labor overhead may be included in calculating claims for direct labor costs.[85] The common types of labor overhead are described below.

a. Site Supervision

Most construction projects are complicated undertakings that require careful planning and meticulous coordination of labor, material, and equipment. Depending on a project's size and complexity, various levels of supervisory personnel are assigned to oversee the work. Since the events that generate claims can also increase the required supervisory presence on-site, additional site supervision costs are frequently included in construction claims.

If supervisory personnel directly oversee laborers on the job, they may be included in the calculation of direct labor costs.[86] However, most supervisory level personnel above the foreman level are treated as an overhead expense. Therefore, the cost of employing supervisors is typically added to claims through field or home office overhead calculations.[87]

Additional site supervision costs are generally determined through actual cost records or by comparing the supervisory effort anticipated in the contractor's estimate with the supervisory effort actually required.[88] Contractors typically estimate supervision based on a ratio of supervisors to tradespeople. If the ratio increases because of events that support a valid claim, the difference between the estimated and actual cost of supervisory personnel can be calculated and recovered using the basic methodologies described above.

As in all aspects of calculating labor costs, the claimant must demonstrate a causal relationship between the events underlying the claim and the need for increased supervision.[89] The claimant also must establish that the increased supervisory costs are reasonable.

Supervisory personnel above the foreman level are generally paid a fixed salary regardless of the hours worked, and their cost is time-related. As a result, the contractor will incur additional field supervision costs if the project is delayed. A detailed discussion of field overhead/general conditions is contained in Chapter 21.

b. Union Contracts and Fringe Benefits

As noted above, additional labor costs for union contractors are typically calculated using wage rates set forth in collective bargaining agreements. Fringe benefits set forth in union contracts, such as pension contributions and holiday pay, are reimbursable costs.[90] Union fringe benefits are typically included as burden on wages. However, some contractors include these benefits as an overhead cost. Either approach is permissible.

c. Payroll Tax and Insurance

Employer payroll tax contributions and workers' compensation insurance premiums are reimbursable if connected with wages that are properly included in an otherwise valid claim.[91] Payroll taxes imposed by federal, state, and local authorities are mandatory legal obligations and represent an inescapable cost of doing business. As a result, payroll taxes may be included in most claims.[92]

Base workers' compensation insurance rates are set by the state in which the work is performed. This base or "manual" rate is multiplied by a "modifier" that the state establishes for each contractor based upon the contractor's safety record and experience level. Each company is "evaluated, and percentages of [state-determined] premiums [are] charged under a set formula; [thus]

by keeping losses low, an insured may get a return of premiums, but if his losses are high, he is penalized."[93] In most situations, increased workers' compensation premiums may be included in a claim as a markup on direct labor costs.

d. Small Tools and Equipment

It is axiomatic that construction workers must be properly equipped to complete their assignments. While heavy equipment costs are relatively easy to track, small tools are often fungible and difficult to monitor and allocate. Therefore, most contractors account for small tools, such as power tools, ladders, and hammers, in a different manner than large equipment.

Large equipment generally is treated as a reusable asset that is either rented or capitalized across the life of the project, while small tools generally are regarded as temporary items that can be lost, stolen, broken, or otherwise become unsuitable for future use. Given their inherent fungibility, small tools are customarily bundled with another, larger cost item, such as labor or overhead.[94] Thus, labor increases are deemed to carry commensurate increases in small tool usage.

Because the attrition rate for small and frequently replaced items is difficult to predict with precision, small tools are commonly estimated based on a fixed percentage of labor costs or overhead. However, when additional labor costs are incurred, the calculation of the value of small tools in the initial estimate should be verified. Therefore, records showing actual expenditures made on small tools should be maintained to avoid situations where the application of the percentage expressed in a bid is rejected for lack of corroboration.[95]

It also is possible to derive a value for additional small tools costs by dividing the total amount spent on small tools by the total labor costs incurred on a project. The resulting percentage is applied to the additional labor cost, however determined. The resulting amount is the additional small tools cost.[96]

The actual base amount used in this calculation may vary if the cost of direct labor contains significant amounts for site support and supervision, which do not utilize small tools. In such cases, a reduced labor base may be used, incorporating only those costs representing direct field labor.[97] As an alternative, small tools have been calculated as a separate cost item and evaluated based upon the reasonableness of the contractor's estimate.[98]

5. *Labor Escalation*

Another reality in construction is that labor costs increase with the passage of time. Therefore, project delays can cause contractors to incur additional labor costs by virtue of variances between the anticipated and actual time when work is performed. "Labor escalation" refers to increased labor costs that are incurred when a contractor is forced to pay higher labor rates for performing work at a later point in time than anticipated.[99]

a. Escalation Caused by Extended Performance

To recover labor escalation costs for delay, the contractor must prove that the owner is responsible for the delay and that it paid increased labor costs because of the delay. Labor escalation costs on federal government projects are typically established by comparing prevailing wage rates established by the U.S. Department of Labor (DOL) for the anticipated and actual periods when the impacted work was performed.[100] The prevailing wage rates for each time period are multiplied by labor hours expended. The delta between the two calculations represents the amount of the contractor's labor escalation claim. Labor escalation claims on federal government contracts are relatively easy to support because the contractor is required to pay prevailing wage rates established by the DOL by the Davis-Bacon Act.[101]

b. Labor Escalation during the Anticipated Contract Performance Period

The contractor typically bears the risk of escalated wages during the original period of contract performance.[102] However, if a contractor can establish that a portion of work would have been performed at an earlier time within the original contract schedule, it may be able to recover escalation costs incurred prior to the original contract completion date. Establishing entitlement in this situation requires proof of the contractor's original schedule, responsibility for the delay, and increases in the cost of labor. If the project extends through several periods of wage escalation, anticipated or not, this calculation must be performed across all periods.

6. Labor Inefficiency

Contractors also can incur additional labor costs by being forced to perform work in a different sequence or under different conditions than anticipated when construction started. As described throughout this book, construction projects can be extremely complicated. Contractors must plan for and schedule the delivery of labor, material, and equipment in a logical sequence that maximizes efficiency and minimizes expense. If that plan is delayed, disrupted, or impacted, the cost of construction will increase. This section describes the events that can cause labor inefficiency and methods for pricing labor inefficiency costs.

a. Causes of Labor Inefficiency
i. Weather

Disruptive or inclement weather can have a significant impact on labor productivity. Usually, increased costs associated with adverse weather conditions are absorbed by the contractor, especially if the contractor could or should have anticipated weather variations in developing its bid estimate.[103] However, in appropriate situations, contractors may be able to recover for decreased productivity if owner-caused delays force the contractor to work in adverse weather seasons that were not originally expected.[104] The contractor also may recover for other weather-related productivity losses that are attributable to owner-caused delays.[105]

ii. Delays and Overtime

It is well recognized that overtime has an adverse effect on labor productivity.[106] This phenomenon is primarily attributable to increased fatigue and poor morale. Studies indicate that when a crew works 50 hours a week for 10 weeks, its overall productivity can decrease by as much as 35 percent.[107] When the workweek is increased to 60 hours over the same period, these studies indicate that labor productivity can be expected to decrease by as much as 45 percent.[108] Payroll records can be examined to reach an estimate of productivity-associated losses caused by overtime work on a project.

iii. Crew Size Variations

The amount and extent of worker supervision on a project is usually determined based on the anticipated size of the workforce and the number of expected work crews. If the contractor is forced to increase the size of the workforce on a project, productivity may suffer from a dilution of adequate management and supervision. The inefficiency caused by workforce increases operates at all levels. Project superintendents can oversee only a finite number of work crews, and foremen can only supervise a limited number of workers with efficiency. Therefore, increases in the number of crews or crew sizes can have a negative impact on productivity.[109] Also, the expanded work crews may be "watered down" with less-skilled workers who are generally less efficient and require additional supervision.[110]

iv. Manpower Availability/Unavailability

Contractors usually bear the risk of loss associated with labor shortages.[111] In limited circumstances, however, contractors may be permitted to recover for decreased productivity resulting from a lack of qualified laborers. For instance, contractors have recovered additional costs that resulted from labor shortages caused by the lack of skilled labor in remote locations.[112] Valid labor claims in this situation are the exception rather than the rule.

v. Stacking of Trades

Typically, the contractor prefers to work only one trade in a single area. The next trade follows behind when the predecessor trade is finished. This tends to result in optimum efficiency.

The term "trade stacking" refers to the simultaneous operation of multiple trades, e.g., sheet metal workers, plumbers, and electricians, in a limited workspace that can generate disruptions, confusion, and congestion. These conditions frequently decrease labor productivity.[113] The simultaneous operation of trades can result from a host of factors, including owner-caused delay, acceleration, and change directives. Trade stacking is a common and well-recognized factor in productivity losses that accounts for a significant volume of construction claims.[114]

vi. Lack of Site Access

Another common contributing factor to lost labor productivity is restricted site access. Site access can be denied in a number of ways.[115] In addition to actual physical obstruction, access to a work site may be effectively denied by unworkable site conditions or circumstances that make it abnormally difficult and inefficient to perform work on the site.[116] For example, contractors have been permitted to recover for productivity losses resulting from the completion of major renovations while a building was occupied.[117] In at least one case, the mere possibility or threat that a work site would be inaccessible was sufficient to decrease productivity.[118] In that situation, there was a possibility that future work areas on a multisite project would not be available for construction when ongoing activities were complete.[119] The contractor was permitted to recover for decreased productivity because its work crews purposely slowed the pace of construction to avoid layoffs that could have resulted if subsequent work sites were not accessible.[120]

vii. Out-of-Sequence Performance

Productivity also can be decreased if the contractor is forced to perform work in a different sequence than it originally planned. Contractors typically control their means and methods of performance. Therefore, they are usually entitled to adhere to their original work schedule.[121] Productivity losses attributable to material deviations from the anticipated work schedule are a common element in construction claims.[122] The most common problem associated with an out-of-sequence work claim is establishing causation and proving damages.[123]

viii. Material Delivery Delays

Material delivery delays can waste valuable man-hours and undermine labor productivity.[124] In addition to creating idle work crews, late material deliveries can force contractors to stop and start work as materials arrive at the job site. Therefore, contractors may be allowed to recover for lost productivity costs associated with wasted man-hours and disruption resulting from the unavailability of materials.[125]

ix. Excessive Storage and Material Handling

Contractors are generally able to recover costs related to additional handling and processing of material if the additional material handling is attributable to another party.[126] For instance, when materials have been ordered and delivered to the site but cannot be used when contemplated by the contractor's schedule because of owner-caused work interruptions, the contractor can recover additional costs incurred in storing those materials for an extended time period.[127] Labor productivity also can be lost when the contractor is forced to divert workers from other tasks to handle the materials.

b. Calculating Lost Labor Productivity

Quantifying lost labor productivity claims is one of the most difficult tasks in pricing construction damages. The Armed Services Board of Contract Appeals (ASBCA) acknowledged the subjectivity inherent in calculating loss of productivity, stating that "inefficient work is an intangible commodity."[128] Regardless of how productivity losses are measured, the contractor must show that the inefficiencies were caused by the owner or its agents.[129] Simply proving that the contractor did not work as efficiently as it anticipated at the onset of the project is insufficient. In all cases, the claimant must demonstrate that the owner disrupted its work and establish a connection between the owner's actions and its loss of efficiency.

i. Measured Mile through Comparison of Similar Work Activities on the Same Project
The preferred method for proving lost labor productivity is the "measured mile" analysis. A measured-mile inefficiency claim compares the claimant's productivity in performing work during the claimed or impacted period with the productivity achieved during an unimpacted or least-impacted time period, i.e., the measured mile, to quantify how the productivity was influenced by delays, disruptions, acceleration, and changes during the claim period. The strength of a measured-mile analysis is enhanced if the contractor is able to compare impacted and unimpacted work segments on the same project because baseline factors affecting productivity, such as site conditions, design complexity, equipment, and most importantly, workforce and supervision, will not influence the analysis.

In choosing a measured-mile period, the contractor should select a time segment that is of sufficient duration to provide representative work activities and comparable work conditions. For example, a measured-mile baseline that features a two-month period without rain is open to criticism if the region normally receives five days of rain per month.

The measured-mile period also should reflect the same type of work that was impacted by the underlying claim events.[130] For instance, a productivity analysis assessing curtain wall glazing activities will be significantly less credible if the measured-mile period covers interior glazing and the impacted work required exterior glazing. Selecting similar work activities is the most important consideration in developing a measured-mile analysis.

Productivity levels from impacted and unimpacted time periods can be derived from job cost reports, daily logs, inspection reports, and labor and payroll reports.[131] A detailed analysis of such data should yield a viable productivity comparison demonstrating the influence that claim events had on the contractor's activities. This type of analysis can be corroborated by expert testimony or expert analysis.

ii. Measured Mile through Comparison of Similar Work Activities on Different Projects
A measured-mile analysis also may compare impacted and unimpacted work
on different projects having similar activities. This approach is less reliable
than comparing work activities on the same project because productivity on
the baseline project may be impacted by dissimilar circumstances that preclude it from being an appropriate benchmark. For example, the equipment,
labor force, or supervision on the two projects may be dramatically different,
which would impact productivity on each job. Therefore, when selecting a
similar project for this type of measured-mile analysis, the contractor should
try to approximate the geographic location, complexity, means and methods
of construction of the projects, equipment, labor market, and supervision to
the maximum extent possible.[132]

iii. Industry Standards and Manuals

If an appropriate measured mile cannot be identified, it may be possible to
support a lost labor productivity claim using generic productivity rates published by industry organizations such as the Associated General Contractors
of America, NECA, MCAA, and the COE.[133] These organizations have studied the impact of trade stacking, adverse weather, site obstructions, and other
types of disruptions on labor efficiency and developed charts depicting productivity losses depending upon the severity of the condition.

These publications can be useful in supporting a contractor's measured-
mile analysis or its expert's testimony. However, such a comparison is susceptible to attack on the grounds that the industry standards are generic and have
no relation to the unique circumstances of the subject project.[134] Therefore,
industry standards should be corroborated with actual project data wherever
possible.

iv. Expert Testimony

Actual project cost data also may be supported by expert testimony in pricing
lost labor productivity claims.[135] Given the complexity involved in proving
labor inefficiency losses and the difficulty inherent in persuasively presenting
a measured-mile analysis, consultant support and expert testimony is helpful.
Although expert testimony supported by little or no actual project cost data
has been accepted, a claimant should present as much actual project cost data
as possible to demonstrate a lost productivity claim.[136]

B. Equipment Costs

Equipment is another large expense on most construction projects. Heavy
equipment such as cranes, trucks, and earthmovers requires a substantial initial investment and is expensive to operate and maintain. To maximize the
return on this investment, contractors attempt to operate equipment on a
continuous and efficient basis. Therefore, requests for additional or extended
equipment costs are common elements in pricing construction damages.

1. *Identifying and Pricing Equipment Usage*

Additional equipment claims are typically segregated between owned and rented equipment. While most equipment costs are recoverable, it is more difficult to allocate specific costs for owned equipment to a particular project because most equipment ownership costs (e.g., depreciation, state and local taxes, insurance, general repairs and maintenance, and storage) are paid out of general operating accounts and are not specifically tied to a particular job, much less a specific piece of equipment. While most contractors track when a specific piece of equipment is mobilized on a job site, they rarely record actual equipment usage hours. Even when they track the equipment usage hours, actual costs are almost never tied to those hours. Further, most equipment is depreciated on the contractor's books and records based on the accelerated methods that are allowed for tax purposes and not based on the actual useful life of the equipment. This is significant because the actual useful life of equipment is generally far greater than the useful life used for tax purposes. As a result, it is difficult, if not impossible, to track the actual cost needed to operate a specific piece of equipment for one hour.

In contrast, rented equipment is typically leased for a specific project. Therefore, the cost of maintaining rental equipment is relatively easy to track. Allocating rental expenses is more difficult if the lease agreement requires the contractor to pay for fuel and standard operating and maintenance expenses. Although these costs will not appear on the vendor's invoice, they are recoverable, and the contractor should have a method of tracking such costs.[137]

a. Contractor-Owned Equipment

As with most other elements of damage, actual cost records for equipment are the preferred method for pricing additional equipment claims.[138] The FAR specifically requires the use of actual costs for owned equipment unless actual costs cannot be determined:

> Actual cost data shall be used when such data can be determined for both ownership and operations costs for each piece of equipment, or groups of similar serial or series equipment, from the contractor's accounting records. When such costs cannot be so determined, the contracting agency may specify the use of a particular schedule of predetermined rates or any part thereof to determine ownership and operating costs of construction equipment.[139]

Most contractors establish equipment rates for internal use that may or may not have any relationship to actual costs. The contractor uses those rates to bid work and charges each job based on those rates. The company then charges most of the equipment costs, including depreciation (for tax purposes), interest, repair and maintenance, property taxes, storage, and insurance costs to an equipment pool. At the end of the year, it determines whether it has had an over- or under-recovery on its equipment charges and may allocate that over- or under-recovery back to the job or include it in the company overhead.

While the contractor's records are helpful for management purposes, they can be problematic for pricing claims because they do not track the actual cost of each piece of equipment. However, the courts continue to express a preference for using actual costs when they are available.[140] As a result, contractors should attempt to maintain more precise equipment ownership records or derive other allocation methodologies.

b. Rented Equipment

It is typically easier to recover equipment costs if the claimant rents the equipment from an unrelated, third-party equipment supplier.[141] Most costs associated with third-party rentals are easy to track because they are identified on the vendor's invoice. Nevertheless, most rental agreements require the contractor to be responsible for fuel, maintenance, minor repairs, and other consumables. While these ancillary equipment rental costs are recoverable in most instances, the claimant needs to prove that it actually incurred such costs to establish recovery.[142]

c. Related Party Rentals

Many organizations have internal equipment pools or subsidiary corporations from which they rent equipment. The issue therefore arises as to the costs that can be charged to the contract. The FAR has specific regulations regarding the allowability of such rentals.

Generally speaking, when the contractor owns the equipment, it is entitled to recover the reasonable cost of ownership and operation of the equipment.[143] The reasonable cost is determined by the contractor's actual cost data when the data can be determined for each piece of equipment or groups of similar equipment.[144] If the contractor's actual costs cannot be determined, the procuring agency may allow the contractor to use a schedule of predetermined rates.[145] In pricing claims for related party rentals, courts and boards may look to industry or commercial equipment rate guides to determine appropriate rental costs.[146]

Because of concerns regarding reliability and potential abuses associated with pricing equipment that the claimant has rented from a related or wholly owned entity, that type of transaction is more likely to be scrutinized than an arm's-length rental transaction. The federal government allows contractors to rent equipment from related companies, but limits payment for such transactions to the normal cost of ownership.[147] Specifically, federal government contractors are limited to "[c]harges in the nature of rent for property between any divisions, subsidiaries, or organizations under common control, to the extent that they do not exceed the normal costs of ownership, such as depreciation, taxes, insurance, facilities, capital cost of money, and maintenance (excluding interest or other unallowable costs pursuant to Part 31), provided that no part of such costs shall duplicate any other allowed cost."[148] If the related company has an established practice of leasing the same or similar equipment to third parties, however, the contractor is entitled to recover the related company's rental rate as long as the rate is reasonable.[149]

d. Equipment Rate Manuals

Although actual costs are preferred for calculating additional equipment costs, it can be difficult to determine actual equipment costs in many situations. To address this problem, the construction industry has developed several manuals to assist in estimating and pricing equipment claims. The most commonly used equipment pricing manuals are described below.

i. Rental Rate Blue Book

The *Rental Rate Blue Book* (Blue Book) is widely used in the industry and is based on regularly updated independent cost studies. The mechanics of using the Blue Book are relatively simple. The Blue Book consists of three volumes. Volume I lists current equipment rates. Volume II lists rates for equipment that has been discontinued for between five and 10 years. Volume III lists rates for equipment that has been discontinued for between 11 and 20 years. The rates shown in the Blue Book are based on independent cost studies of contractor-owned equipment that evaluates both ownership and operating costs. The "ownership" cost component of the Blue Book includes depreciation, property taxes, storage, licenses, record keeping, insurance, original freight, security, facilities capital, major renovation, and other expenses. The "operating" cost component includes labor, parts, fuel, lubricants, and other routine operating and maintenance costs.

The Blue Book includes monthly, weekly, daily, and hourly rates, adjusted to account for lost productivity when equipment is used for shorter durations. In addition, various rate adjustment tables are included to enable contractors to account for differences in depreciation, geography, type of job, manufacturers' discounts, and other items.

While contractors prefer to use the Blue Book to price claims, it has been criticized for being susceptible to manipulation. Also, there is a perception among some owners that the Blue Book results in inflated ownership costs unless proper safeguards are implemented. Therefore, the Blue Book is frequently precluded by contract or rejected when actual cost data is available.

ii. Cost Reference Guide

The *Cost Reference Guide* (CRG) provides the most comprehensive cost data available, including the estimated hourly costs for more than 15,000 pieces of heavy equipment, major attachments, and small tools. The CRG is a supplement to the Blue Book and was created to provide estimators with an effective tool to price the detailed components of the operating costs listed in the Blue Book. Because of its status as an adjunct to the Blue Book, many CRG expense components are identical to those found in the Blue Book.

The CRG assumes that a particular piece of equipment is operated 2,112 hours per year (12 months at 176 hours per month). By basing the CRG on such a high utilization rate, estimators are better able to make productivity assumptions and pricing adjustments for particular projects. For purposes of pricing claims, the CRG's high utilization rate results in lower rates for owned equipment costs than are reflected in the Blue Book. This makes CRG rates less attractive to contractors for pricing claims but more acceptable to owners in approving contract adjustments.

iii. U.S. Army Corps of Engineers' Manual

The COE publishes an equipment manual that is divided into 12 geographical regions, each with predetermined ownership and operating expense rates for construction and marine equipment. The COE Manual is referenced in many public contracts as the appropriate guideline to use in pricing equipment claims. The FAR states:

> Predetermined schedules of construction use rates (e.g., the Construction Equipment Ownership and Operating Expense Schedule, published by the U.S. Army Corps of Engineers, industry sponsored construction equipment cost guides, or commercially published schedules of construction equipment cost use) provide average ownership and operating rates for construction equipment. The allowance for operating costs may include costs for such items as fuel, filters, oil, and grease; servicing, repairs, and maintenance; and tire wear and repair. Costs of labor, mobilization, demobilization, overhead, and profit are generally not reflected in schedules, and separate consideration may be necessary.[150]

The COE Manual includes sections on operating conditions, equipment selection, equipment values, equipment life, salvage value, ownership cost, standby rates, and various adjustments. It does not include data for mobilization, taxes, insurance, or storage. Therefore, the COE Manual generally produces a lower total equipment cost rate than other equipment manuals and is preferred by owners, who believe that the COE rates are more realistic than the Blue Book or other guidelines. Of course, contractors typically criticize the COE Manual because it fails to account for the full spectrum of ownership expenses.

iv. Associated Equipment Distributors Manual

Associated Equipment Distributors (AED) publishes *Rental Rate Compilation*, which is frequently referred to as the "Green Book." The Green Book includes nationally averaged rental rates (AED rates) that are compiled from thousands of AED member companies. Since many standard American Institute of Architects (AIA) contracts permit contractors to include rental costs of machinery and equipment in pricing contract modifications, many private contractors rely on AED rates for pricing claims.

AED rates do not account for the age and condition of equipment, geography, or the manner in which the owners of rental equipment estimate ownership expenses. Instead, AED rates reflect an averaging of age, condition, and operating efficiency of equipment, which is described in the Green Book by type and capacity. Monthly, weekly, and, in most cases, daily rates are listed, which are calculated based on an 8-hour day, 40-hour week, and 176-hour month. To determine an hourly rate, the weekly rate is divided by 40, or the monthly rate is divided by 176.

AED rates are disfavored by contractors because they fail to account for the age and condition of equipment, geographic considerations, or the manner in which the equipment is used. Interestingly, at least one board of contract appeals has rejected the use of AED rates for contractor-owned equipment because the Green Book is published by dealers engaged in the business of renting equipment.[151] For that reason, the board found that AED rates should be considered rental rates rather than ownership costs.[152] In light of this criticism, caution should be exercised in attempting to use AED rates for pricing or evaluating claims for additional or extended equipment usage.

2. Measuring Additional Equipment Utilization and Calculating Damages

As a prerequisite to recovering additional equipment costs, the claimant must prove that another party was responsible for its additional equipment utilization based on factors that were beyond the claimant's ability to control and for which the claimant is eligible to recover damages.[153] These factors are typically established by contract, although case law can be significant in determining the claimant's rights in this area. Once entitlement is established, the claimant must prove its damages. Several common methods that are used to calculate extended or additional equipment costs are described below.

a. Recording Additional Hours

The claimant must be able to identify the additional time that was expended in completing the project to prepare a viable equipment utilization claim. Three basic scenarios must be considered in deriving this information:

1. Situations where equipment was needed for a longer time period than anticipated;
2. Situations where equipment on a project had to be supplemented by adding new equipment or changing the type of equipment being used; and
3. Situations that required both extended and supplemental equipment usage.

Calculating damages associated with each scenario has unique elements of proof.

It also may be necessary to distinguish the time that equipment was actually used from time during which the equipment was on standby. Fully utilized equipment is typically charged at full ownership rates. In contrast, ownership rates for idle equipment may be reduced to account for decreased wear and tear.[154] Therefore, additional and extended equipment usage should be recorded separately from idle or standby time.

The easiest and most effective way to track additional equipment usage is through a daily log that identifies each piece of equipment on a project, the task performed with such equipment, the operator of such equipment, and the time such equipment was in use. The claimant also should maintain cost records that record depreciation, interest, overhaul, repair, property tax, storage, and insurance costs to supplement or replace equipment manuals. The use of actual cost data is always preferred, and the contractor has the burden of demonstrating that actual costs are not available when seeking to use published rates.[155]

b. Equipment Productivity

Detailed equipment utilization records are frequently deficient or ignored during a project. When actual cost records are unavailable, it may be possible to derive additional equipment utilization information through the type of measured-mile approaches described in the labor productivity section above. As with claims for lost labor productivity, additional equipment costs can be derived by comparing productivity records for the same types of equipment on (1) impacted and unimpacted areas of the same project or (2) impacted and unimpacted areas of similar projects.[156]

c. Additional Contract Time

Measuring additional equipment usage for extended contract performance claims is the most common and straightforward method of calculating equipment damages. This type of calculation assumes that once a piece of equipment is assigned to a job, it is chargeable to the job for the entire duration of the assignment. Therefore, the claimant typically requests the full per diem value of the assigned equipment on the basis of an eight-hour work day, regardless of whether the equipment was actually in use for the full eight hours.[157] This assumption has logical appeal because it is difficult to move equipment between projects to avoid downtime. Nevertheless, the claimant should attempt to avoid problems associated with turning an extended equipment claim into a total cost calculation.

A potential complication in developing an extended equipment usage calculation is that it is often difficult to establish precisely when a piece of equipment was mobilized and removed from a particular project. In such instances, the circumstances relevant to a particular piece of equipment must be examined. Claimants also have been successful in recovering full equipment hours from the date that the equipment was released from its former project and for the period by which the owner was responsible for extending the time period required to complete a contract.[158]

Of course, contractors may not use a job site to store equipment and still recover extended usage costs. Therefore, the claimant must prove that the equipment was on the project for a legitimate purpose and that the equipment was actually being operated.[159] Absent such proof, which typically consists of daily reports, time or payroll records for equipment operators, or maintenance logs, extended equipment usage claims will be rejected.

d. Correlating Equipment Usage to Man-Hours

Another method of pricing additional equipment costs is to correlate equipment hours to the number of additional labor hours expended in operating the equipment. This is a straightforward calculation if a single operator is assigned to a particular piece of equipment.[160] The analysis is more complicated where operators rotate among various pieces of equipment. In this situation, it may be possible to correlate an item of equipment to a particular equipment operation crew on a percentage or pro rata basis.

e. Total and Modified Total Usage Claims

The "total usage" calculation is a variation of a total cost claim that compares anticipated and actual equipment utilization. Under this methodology, the claimant typically focuses on a particular segment of the work and identifies (1) each piece of equipment assigned to the segment and (2) the time period that each piece of equipment was utilized for such segment. If a pure total usage claim is used, the claimant requests additional compensation for the total number of hours by which actual equipment utilization exceeded what was projected.[161] A modified total usage claim makes adjustments for the claimant's estimating or planning errors and the claimant's own operational problems. This methodology is subject to the same criticisms that are inherent in other types of total cost calculations.

3. Standby Time

As indicated above, additional equipment hours may be recovered if the claimant is required to keep its equipment at the project site for an extended time because of a work stoppage.[162] Under these circumstances, the claimant is often entitled to recover its full costs for each piece of idle equipment.[163] Standby costs also may be recovered in situations where the claimant was not delayed beyond the anticipated substantial completion date if the claimant can prove "(1) it intended to complete the contract early; (2) had the capacity to do so; and (3) actually would have completed early but for the [owner's] actions."[164]

Some limits are placed on idle equipment claims. For example, maintenance and repairs are not typically recoverable for standby time.[165] The claimant also must prove that the idle equipment could have been used in a productive capacity if the claimant were not required to maintain the equipment on standby for an uncertain time period.[166] The claimant also may be required to reassign some or all of the idle equipment to another project to mitigate its damages.[167] In this regard, idle equipment claims have been rejected where the contractor failed to demonstrate "that the equipment for which compensation is claimed was reasonably and necessarily set aside and awaiting use in performing the contract."[168]

C. Material Costs

1. Basis for Material Claims

Claims for additional material costs are relatively simple to price. Contractors typically maintain accounting records that show amounts paid for material deliveries and have delivery tickets and installation records that show material quantities incorporated into a project. Therefore, establishing entitlement is frequently more challenging than pricing additional material claims.

The most common types of additional material claims seek reimbursement for materials that are needed to accommodate or overcome change order directives, design errors, or differing site conditions. Changes in the type or quantity of materials generally are compensable and should be easy to calculate.[169] In most situations, these costs are derived by computing the sum of invoices for the materials incorporated into the work and adding markup for overhead and profit.

Additional material costs also may be recoverable in delay and inefficiency claims.[170] In this regard, a compensable delay that forces the claimant to pay escalated material costs can be a valid basis for a request for additional compensation.[171] For example, periodic shortages in the supply of framing lumber and certain gypsum products during the 1990s caused dramatic price fluctuations within very short time periods. If an owner-caused delay forced a contractor to pay increased material costs during one of these fluctuations, the contractor may have been entitled to compensation for the cost overrun.[172] The key to recovering material escalation costs is proving that (1) the claimant based its bid on lower prices that were available at an earlier point in time; (2) it was impossible or impractical for the claimant to purchase the materials prior to the price escalation; and (3) the claimant actually paid more for the materials than it anticipated.[173] Providing evidence of a compensable delay will not be sufficient to establish a material escalation claim if the delay did not preclude the claimant from purchasing materials at an earlier time.[174]

2. Calculating Additional Material Quantities and Costs

As noted above, increased material costs attributable to changes in the work or compensable delays are generally recoverable.[175] Where an error, omission, or design change requires the contractor to obtain additional materials, the claimant is entitled to reimbursement if it can demonstrate the increase in materials over the amount contemplated by the contract. Since additional material costs are based on actual material purchases rather than estimates, proof of the quantity and cost of additional materials should be readily available. If advance pricing is requested for a change order proposal, the additional material quantities should be estimated based on a take-off from the contract documents or calculations using actual field measurements. As with any change order claim, the claimant should develop and maintain accurate and complete files tracking each change. Notably, if actual cost records for procured materials are available, the claimant should avoid claims based on estimated costs.[176] In that situation, a precise accounting of the additional required materials may be needed to satisfy the required burden of proof.[177]

The methods used to calculate additional materials also should comport with accepted industry standards. Beyond the calculation of the actual material quantities, there are several accepted standards for pricing additional quantities. For example, for over half a century, R.S. Means has published "Means Construction Cost Data," which has been recognized as an appropriate benchmark for pricing of construction materials.[178] Means Construction Cost Data includes geographically adjusted pricing data through "City Cost Indexes." The *Dodge Manual* and various estimating guidelines published by the NECA, MCAA, and ACI are other accepted standards for material estimating. Direct quotes from material suppliers also will be accepted in appropriate circumstances.[179] Material pricing rules on federal government contracts are prescribed by regulation:

> When the materials are purchased specifically for and are identifiable solely with performance under a contract, the actual purchase cost of those materials should be charged to the contract. If material is issued from stores, any generally recognized method of pricing such material is acceptable if that method is consistently applied and the results are equitable.[180]

Changes in the project design also can result in reduced material quantities. Although the owner may be entitled to a credit when a change reduces certain material quantities, the burden shifts to the owner to prove the appropriate contract adjustment.[181] A change in the contractor's means or methods resulting in a material quantity reduction does not necessarily entitle the owner to a credit. At least one court has held that the contractor is entitled to select any method of performance that results in compliance with the contract documents.[182]

3. Calculating Escalated Material Costs

As noted above, to recover material escalation costs, the claimant must establish that it was forced to perform work during a period of higher material costs because of a compensable delay.[183] Therefore, the claimant must prove that the delay caused it to purchase specific materials later than originally scheduled.[184] Owner-caused delays, defective specifications, expired purchase orders, storage limitations, and restricted site facilities are the types of situations that can cause a contractor to delay material purchases.[185]

Proof of the actual costs incurred, presented with the actual invoices or quotes showing what the claimant anticipated it would pay for the materials, is the best proof of material escalation costs.[186] If actual cost records are not available, industry-wide averages may be accepted as proof of percentage material escalations.[187]

4. Ancillary Material Costs

It also is possible to recover miscellaneous costs related to the materials, such as additional handling or storage costs.[188] Contractors also may be able to recover the value of purchasing discounts that were lost because of an owner's delay.[189]

D. Field Office Overhead/General Conditions and Home Office Overhead

The contractor is entitled to recover its additional field office overhead (often referred to as general conditions) as well as its additional home office overhead as part of an equitable adjustment. These costs are discussed in detail in Chapter 21.

E. Profit

Profit is generally recoverable as part of an equitable adjustment for changed or extra work,[190] but does not include unearned anticipated profits.[191] However, if the contractor has not performed any additional or changed work but has merely been delayed, profit may not be allowed.[192] In fact, the suspension of work clause specifically excludes the recovery of profit from the equitable adjustment resulting from a suspension of work.[193]

F. Interest

The contractor is entitled to recover interest on "claims" from the date it submits a properly certified claim pursuant to the Contract Disputes Act of 1978, which is discussed in detail in Chapter 15. Interest on unpaid invoices may be recovered under the Prompt Payment Act of 1982,[194] which is discussed in detail in Chapter 11.

G. Additional Bond Premiums

Contractors are entitled to recover additional bond premiums that result from a compensable event. Most construction bonds are priced based on a percentage of the total value of the bonded contract. The percentage is normally determined by the size of the project and the financial and technical ability of the company. If a compensable claim event causes the amount of the contract to increase, the contractor's bond premium also will increase. Therefore, additional bond premiums can be recovered as part of the equitable adjustment. If the increased contract amount causes the bond rate to decrease, the contractor should charge the reduced rate on the increased costs. It is important for contractors to explain the basis for their bond premium and be able to support the premium in accordance with federal cost and pricing requirements. In at least one recent case, a failure to disclose a premium commission splitting arrangement led to a violation of the Anti-Kickback Act.[195]

H. Attorney's Fees and Legal Costs

The FAR makes the cost of pursuing a claim against the government an unallowable cost.[196] Legal fees incurred in pursuing claims against the government are only recoverable if allowed by the Equal Access to Justice Act (EAJA),[197] which is discussed in detail in Chapter 16. But note that under certain circumstances, the reasonable costs of legal and consulting fees attributable to preparing and justifying a request for equitable adjustment (REA) may be reimbursable.[198]

I. Cost Principles

FAR Part 31 contains the cost principles that apply to federal government contract requests for equitable adjustments and claims. FAR 31.205 addresses the allowability of selected costs and should be consulted by the contractor to assure that it does not include items in its claim that are unallowable.

V. Miscellaneous Pricing Issues

A. Deductive Modifications

The purpose of the equitable adjustment is to leave the contractor in the same profit or loss position it would have occupied had the change not occurred.[199] This is true whether the change deletes or adds work.[200] A downward adjustment on a deductive modification is the reasonable cost the contractor would have incurred had it performed the work plus a reasonable markup for overhead and profit.[201] If the contractor's bid estimate was too low, and the contractor would have to perform the deleted work at a loss, the change should not leave the contractor in a better position than it would have occupied if it had performed the deleted work. The equitable adjustment is therefore the cost of performing the work, not a remedy for the contractor's low estimate.[202] Similarly, if the contractor is able to perform the work at a cost far less than the amount included in its bid, it should be able to retain that profit.

B. Variations in Estimated Quantities

Many construction contracts pay for some or all work on a unit price basis. The contract sets out estimated quantities for items such as concrete, earthwork, and pipe installation. The contractor bids a unit price to perform each unit of work and is paid for the actual quantity performed. When there is a major overrun or underrun in quantity, the bid price may not be equitable for the government or the contractor.

As a result, FAR 52.211-18 contains a provision, "Variation in Estimated Quantity," that provides for an equitable adjustment in the unit price if the actual quantity varies more than 15 percent from the estimated quantity. This clause states that the "equitable adjustment in the contract price shall be based upon any increase or decrease in costs due solely to the variation above 115 percent or below 85 percent of the estimated quantity."[203] The courts and boards have concluded that the equitable adjustment is not based on the actual cost of performing the overrun or underrun quantity of work. The unit price is adjusted only to the extent that the increase or decrease in quantity causes the contractor's cost to perform that item to increase or decrease.[204] For example, where there is an overrun in cost, the contractor has recovered all of its fixed costs included in the unit price and must reduce its unit price for the overrun quantity above 115 percent to remove those costs.

Equitable adjustments under other contractual remedy-granting provisions provide for payment of increased costs that are covered by an estimated quantity provision. For example, if a quantity overrun constitutes a differing site condition, the contractor would recover the actual cost of performing the overrun quantity under the Differing Site Conditions clause, which would differ from the adjustment under the Variation in Estimated Quantity clause. Federal decisions have made it clear that the Differing Site Conditions clause overrides the Variation in Estimated Quantity clause.[205] As a result, a contractor typically seeks adjustment under the Differing Site Conditions clause when its unit price is low and adjustment under the Variation in Estimated Quantity clause when its unit price is high.

Notes

1. *See* CORBIN ON CONTRACTS § 992 (2001); Jack Picoult, VACAB 1221, 78-1 BCA ¶ 13,024.

2. RESTATEMENT (SECOND) OF CONTRACTS § 329 (1981); Celesco Indus., Inc., ASBCA 22251, 79-1 BCA ¶ 13,604.

3. G&H Mach. Co. v. United States, 16 Ct. Cl. 568 (1989); Victory Constr. Co. v. United States, 510 F.2d 1379 (Ct. Cl. 1975); Nager Elec. Co., Inc. v. United States, 442 F.2d 936 (Ct. Cl. 1971); Wunderlich Contracting Co. v. United States, 351 F.2d 956 (Ct. Cl. 1965).

4. S.W. Elec. & Mfg. Corp., ASBCA 20698, 77-2 BCA ¶ 12,631, *aff'd*, 655 F.2d 1078 (Ct. Cl. 1981); Se. Servs., Inc. ASBCA 21278, 78-2 BCA ¶ 13,239.

5. Youngdale & Sons Constr. Co. v. United States, 27 Fed. Cl. 516 (1993) (citing Teledyne McCormick Sleph v. United States, 588 F.2d 808 (Ct. Cl. 1978)); E. Contracting Co. v. United States, 97 Ct. Cl. 341 (1942).

6. *See, e.g.*, Bruce Constr. Corp. v. United States, 324 F.2d 516 (Ct. Cl. 1963).

7. FAR 31.201-3(a); Delco Elecs. Corp. v. United States, 17 Cl. Ct. 302 (1989).

8. CRF v. United States, 624 F.2d 1054 (Ct. Cl. 1980); Boeing Aerospace Operations, Inc., ASBCA 46274, 94-2 BCA ¶ 26,802; Teledyne Indus., Geotech Div., ASBCA 18049, 73-2 BCA ¶ 10,088.

9. *Boeing*, 94-2 BCA ¶ 26,802; Bromley Contracting Co., ASBCA 20271, 77-2 BCA ¶ 12,715; DeMauro Constr. Co., ASBCA 12514, 73-1 BCA ¶ 9830.

10. White Buffalo Constr., Inc. v. United States, 52 Fed Cl. 1 (2002); Pa. Drilling Co., IBCA 1187-4-78, 82-1 BCA ¶ 15,697.

11. Norcoast-Beck Constructors, Inc., ASBCA 25261, 83-1 BCA ¶ 16,435.

12. Wunderlich Contracting Co. v. United States, 351 F.2d 956, 968–69 (Ct. Cl. 1965); *see also* Commerce Int'l Co. v. United States, 338 F.2d 81 (Ct. Cl. 1964).

13. Weaver v. United States, 209 Ct. Cl. 714 (1976).

14. Ralph L. Jones Co., Inc. v. United States, 33 Fed. Cl. 327 (1995); Boeing v. United States, 480 F.2d 854 (Ct. Cl. 1973); Elec. & Missile Facilities, Inc. v. United States, 416 F.2d 1345 (Ct. Cl. 1969); *Wunderlich*, 351 F.2d 956.

15. False Statements Act, 18 U.S.C. § 1001; Criminal False Claims Act, 18 U.S.C. § 287; Civil False Claims Act, 31 U.S.C. § 3729; Program Fraud Civil Remedies Act, 31 U.S.C. § 3802; Public Money, Property or Records, 18 U.S.C. § 641.

16. See Chapter 24 for a detailed discussion of recent false claims cases.

17. Fraud Enforcement and Recovery Act, Pub. L. No. 111-21 (May 20, 2009).

18. Daewoo Eng'g and Constr. Co. v. United States, No. 2007-5129, 2009 WL 415490 (Fed. Cir. Feb. 20, 2009).

19. Mail Fraud Act, 18 U.S.C. § 1341; Fraud by Wire, Radio or Television, 18 U.S.C. § 1343.

20. Criminal Conspiracy, 18 U.S.C. § 371.

21. 18 U.S.C. §§ 1961–1968.

22. *See, e.g.*, Baldi Bros. Constructors v. United States, 50 Fed. Cl. 74 (2001) (explaining that the actual cost method is preferred because it provides the court with detailed documentation of underlying expenses, ensuring that the final award will be equitable and not a windfall for either the government or the contractor); Am. Line Builders, Inc. v. United States, 26 Cl. Ct. 1155 (1992); Cen-Vi-Ro of Texas, Inc. v. United States, 210 Ct. Cl. 684 (1976).

23 10 U.S.C. § 2306a; 41 U.S.C. § 254b; FAR 15.403-4(a)(1); FAR 15.403-4(b)(2); FAR 15.406-2.

24. FAR 52.243-6.

25. Weaver v. United States, 209 Ct. Cl. 714 (1976); *but see* Ramar Co., ASBCA 16060, 74-2 BCA ¶ 10,874; Dillon Constr., Inc., ENG BCA PCC-101, 96-1 BCA ¶ 28,113.

26. G. Marine Diesel Corp., 155 B.R. 851 (Bankr. E.D.N.Y. 1993); Baifield Indus., Div. of A-T-O, Inc., ASBCA 13418, 77-1 BCA ¶ 12,308, *aff'd*, 706 F.2d 320 (Fed. Cir. 1983).

27. Neal & Co. v. United States, 17 Cl. Ct. 511 (1989); *Dillon*, 96-1 BCA ¶ 28,113; Cent. Mech., Inc., DOTCAB 1234, 83-2 BCA ¶ 16,642; Keco Indus., Inc., ASBCA 15061, 72-1 BCA ¶ 9450, 72-2 BCA ¶ 9575.

28. Neal & Co. v. United States, 17 Cl. Ct. 511 (1989); Parsons of Cal., ASBCA 20867, 82-1 BCA ¶ 15,659.

29. Rice v. United States, 428 F.2d 1311 (Ct. Cl. 1970); Paccon, Inc., ASBCA 7890, 65-2 BCA ¶ 4996.

30. Rice v. United States, 428 F.2d 1311 (Ct. Cl. 1970); Piracci Corp., GSBCA 6007, 82-2 BCA ¶ 16,047; Md. Painting Co., ENG BCA 3337, 73-2 BCA ¶ 10,223; C.W. Stack & Assoc., DOTCAB 72-4, 72-1 BCA ¶ 9313.

31. Luria Bros. & Co. v. United States, 369 F.2d 701 (Ct. Cl. 1966); Fid. Constr. Co., DOTCAB 1113, 81-2 BCA ¶ 15,345.

32. Turnbull, Inc. v. United States, 389 F.2d 1007 (Ct. Cl. 1967).

33. 516 U.S. 869 (1995). *See also* Kumho Tire Co., Ltd. v. Carmichael, 526 U.S. 137, 119 S. Ct. 1167 (1999) (Daubert's "gatekeeping" obligation, requiring an inquiry into both relevance and reliability, applies not only to "scientific" testimony, but to all expert testimony).

34. B.P.O.A., ASBCA 25276, 82-2 BCA ¶ 15,816; 82-2 BCA ¶ 15,924.

35. A & J Constr. Co., IBCA 11421-2-77, 79-1 BCA ¶ 13,621.

36. Monroe Garment Co., ASBCA 14465, 75-2 BCA ¶ 11,569.

37. Mel Williamson Constr. Co., VACAB 1199, 76-2 BCA ¶ 12,168.

38. United States v. Howard P. Foley Co., 329 U.S. 64 (1946); United States v. Rice, 317 U.S. 61 (1942); Wunderlich Contracting Co. v. United States, 351 F.2d 956 (Ct. Cl. 1965); W. Rds., Inc., AGBCA 84-243-1, 86-3 BCA ¶ 19,333.

39. Boyajian v. United States, 423 F.2d 1231 (Ct. Cl. 1970).

40. Cavalier Clothes, Inc. v. United States, 51 Fed. Cl. 399, 418 (2001) (citing Servidone Constr. Corp. v. United States, 931 F.2d 860, 861 (Fed. Cir. 1991)).

41. Concrete Placing Co., Inc. v. United States, 25 Cl. Ct. 369 (1992); WRB Corp. v. United States, 183 Ct. Cl. 409 (1968).

42. *See, e.g.*, Baldi Bros. Constructors v. United States, 50 Fed. Cl. 74 (2001) (citing *WRB Corp.*, 183 Ct. Cl. at 426).

43. *Cavalier*, 51 Fed. Cl. 399; *Wunderlich*, 351 F.2d 956; Ingalls Shipbuilding Div., Litton Sys., Inc., ASBCA 17579, 78-1 BCA ¶ 13,038.

44. J & T Constr. Co., DOTCAB 73-4, 75-2 BCA ¶ 11,398; Whitmeyer Bros., Inc. v. State, 406 N.Y.S.2d 617 (N.Y. App. Div. 1978).

45. Hewitt Contracting Co., ENG BCA 4596, 83-2 BCA ¶ 16,816.

46. Hardrives, Inc., IBCA 2319, 94-1 BCA ¶ 26,267.

47. Neal & Co. v. United States, 36 Fed. Cl. 600, 638 (1996); *see also Cavalier*, 51 Fed. Cl. 399.

48. FAR 52.243-4; FAR 52.236-2.

49. 41 U.S.C. §§ 601–613.

50. United States v. Utah Constr. & Mining Co., 384 U.S. 394 (1966).

51. VHC Inc. v. Peters, 179 F.3d 1363 (Fed. Cir. 1999) (noting that the aim of an equitable adjustment is to "make the contractor whole" after the contract is modified); Bruce Constr. Corp. v. United States, 324 F.2d 516 (Ct. Cl. 1963), *superseded by rule on other grounds*; Massman Constr. Co., ENG BCA 3660, 81-1 BCA ¶ 15,049.

52. Allen Constr. Co. v. United States, 646 F.2d 487, 484 (Ct. Cl. 1981).

53. *Id.*

54. Morrison Knudsen Corp. v. Fireman's Fund Ins. Co., 175 F.3d 1221 (10th Cir. 1999) (citing Wunderlich Contracting Co. v. United States, 351 F.2d 956, 958 (Ct. Cl. 1965)); McDonnell Douglas Corp. v. United States, 40 Fed. Cl. 529 (1998), *rev'd and vacated in part and remanded on other grounds*, 182 F.3d 1319 (Fed. Cir. 1999).

55. *See, e.g., McDonnell Douglas*, 40 Fed. Cl. at 536 (citing 48 C.F.R. (FAR) §§ 31.201-1 to .201-4 (requiring and defining allowability, reasonableness, and allocability)), *rev'd and vacated in part and remanded on other grounds*, 182 F.3d 1319 (Fed. Cir. 1999).

56. Azure v. United States, 129 F.3d 136 (table, text at 1997 WL 665763) (Fed. Cir. 1997); Joseph Pickard's Sons Co. v. United States, 532 F.2d 739 (Ct. Cl. 1976).

57. Baldi Bros. Constructors v. United States, 50 Fed. Cl. 74 (2001) (citing Dawco Constr. v. United States, 930 F.2d 872 (Cl. Ct. 1991)).

58. *Azure* v. United States, 129 F.3d 136.

59. Harold Benson, AGBCA 384, 77-1 BCA ¶ 12,490; Johnson, Drake & Piper, Inc., ASBCA 9824, 65-2 BCA ¶ 4868.

60. *See Johnson, Drake & Piper*, 65-2 BCA ¶ 4868.

61. Harold Benson, AGBCA 384, 77-1 BCA ¶ 12,490.

62. Rocky Mountain Constr. Co., IBCA 1091-12-75, 77-2 BCA ¶ 12,692; Custom Roofing Co., ASBCA 19164, 74-2 BCA ¶ 10,925.

63. Varo, Inc., ASBCA 15000, 72-2 BCA ¶ 9717.

64. Doninger Metal Prods. Corp. v. United States, 50 Fed. Cl. 110 (Ct. Cl. 2001).

65. Bluebonnet Sav. Bank, F.S.B. v. United States, 266 F.3d 1348, 1357–58 (Fed. Cir. 2001) ("The amount of the recovery can only be approximated in the format of a 'jury verdict' where the claimant can demonstrate a justifiable inability to substantiate the amount of his resultant injury by direct and specific proof").

66. Corbetta Constr. Co., PBCA 817, 82-2 BCA ¶ 15,805.

67. Steven P. Rados, Inc., AGBCA 77-130-4, 82-1 BCA ¶ 15,624.

68. Joseph Pickard's Sons Co. v. United States, 532 F.2d 739 (Ct. Cl. 1976).

69. Nager Elec. Co. v. United States, 442 F.2d 936 (Ct. Cl. 1971).

70. *See* Ralph L. Jones Co., Inc. v. United States, 33 Fed. Cl. 327 (1995); Boeing v. United States, 480 F.2d 854 (Ct. Cl. 1973).

71. *See id.*

72. Clarke Baridon, Inc. v. Merritt Chapman and Scott Corp., 311 F.2d 389 (4th Cir. 1962); J.D. Hedin Constr. Co. v. United States, 347 F.2d 235 (Ct. Cl. 1965), *overruled on other grounds by* Wilner v. United States, 24 F.3d 1397 (Fed. Cir. 1994).

73. *See* 40 U.S.C. § 3141; 29 C.F.R. tit. A, pt. 1-7.

74. White Buffalo Constr., Inc. v. United States, 52 Fed. Cl. 1 (2002).

75. *Id.*

76. *See* Ed Goetz Painting Co., DOTCAB 1168, 83-1 BCA ¶ 16,134.

77. *See* Nager Elec. Co. v. United States, 442 F.2d 936 (Ct. Cl. 1971).

78. H. John Homan Co. v. United States, 418 F.2d 522 (Ct. Cl. 1969).

79. Hewitt Contracting Co., ENG BCA 4596, 83-2 BCA ¶ 16,816.

80. Sovereign Constr. Co., ASBCA 17792, 75-1 BCA ¶ 11,251.

81. Baldi Bros. Constructors v. United States, 50 Fed. C1. 74 (2001).

82. J.D. Hedin Constr. Co. v. United States, 347 F.2d 235 (Ct. Cl. 1965); *overruled on other grounds by* Wilner v. United States, 24 F.3d 1397 (Fed. Cir. 1994).

83. Paul Hardeman, Inc. v. United States, 406 F.2d 1357 (Ct. Cl. 1969); Coley Props. v. United States, 593 F.2d 380 (Ct. Cl. 1979).

84. Hoffman Constr. Co. of Or. v. United States, 40 Fed. Cl. 184 (Fed. Cl. 1998), *rev'd in part on other grounds*, 178 F.3d 1313 (Fed. Cir. 1999).

85. Transtechnology Corp., Space Ordnance Sys. Div. v. United States, 22 Cl. Ct. 349 (1990).

86. Morrison-Knudsen Co. v. United States, 397 F.2d 826 (Ct. Cl. 1949) (contractor only eligible to be reimbursed for project superintendent who supervised crew of nine workers).

87. Altmayer v. Johnson, 79 F.3d 1129 (Fed. Cir. 1996).

88. Mid-West Constr. Co. v. United States, 461 F.2d 794 (Ct. Cl. 1972) (actual cost records); Sovereign Constr. Co., ASBCA 17792, 75-1 BCA ¶ 11,251.

89. L.I. Waldman & Co. v. United States, 106 Ct. Cl. 159 (1946).

90. *See, e.g.*, ITT Fed. Support Servs. Inc. v. United States, 531 F.2d 522 (Ct. Cl. 1976).

91. *See, e.g.*, J.D. Hedin Constr. Co. v. United States, 347 F.2d 235 (Ct. Cl. 1965), *overruled on other grounds in* Wilner v. United States, 24 F.3d 1397 (Fed. Cir. 1994).

92. *Id.*

93. Adami v. Highlands Ins. Co., 512 S.W.2d 737 (Tex. Civ. App. 1974). *See also* Judicial Review of Rates, 82 Am. Jur. 2d *Workers' Compensation* § 467 (2002).

94. Reliance Ins. Co. v. United States, 20 Cl. Ct. 715 (1990).

95. Kit-San-Azusa, J.V. v. United States, 32 Fed. Cl. 647 (1995).

96. Hensel Phelps Constr. Co. v. Gen. Servs. Admin., GSBCA 14744, 01-1 BCA ¶ 31,249.

97. *Id.*

98. In re Stone & Webster, Inc., 279 B.R. 748 (Bankr. D. Del. 2002).

99. Luria Bros. & Co. v. United States, 369 F.2d 701 (Ct. Cl. 1966).

100. *See, e.g.*, FAR 22.404-12.

101. 40 U.S.C. § 3141.

102. J.D. Hedin Constr. Co. v. United States, 347 F.2d 235 (Ct. Cl. 1965), *overruled on other grounds by* Wilner v. United States, 24 F.3d 1397 (Fed. Cir. 1994).

103. *See, e.g.*, Corry Bridge & Supply Co., AGBCA 81-149-1, 82-2 BCA ¶ 16,008.

104. H. John Homan Co. v. U.S., 418 F.2d 522 (Ct. Cl. 1969).

105. *See, e.g.*, Luria Bros. & Co. v. United States, 369 F.2d 701 (Ct. Cl. 1966); *J.D. Hedin Constr.*, 347 F.2d 235; Abbett Elec. Corp. v. United States, 162 F. Supp. 772 (Ct. Cl. 1958).

106. Md. Sanitary Mfg. Corp. v. United States, 119 Ct. Cl. 100 (1951); Casson Constr. Co., GSBCA 4884, 83-1 BCA ¶ 16,523; Cont'l Consol. Corp. v. United States, ENG BCA 2743, 68-1 BCA ¶ 7003, *modified in part*, 200 Ct. Cl. 737 (1972); Lew F. Stillwell, Inc., ASBCA 9423, 1964 BCA ¶ 4128.

107. BUSINESS ROUNDTABLE, SCHEDULED OVERTIME EFFECT ON CONSTRUCTION PROJECTS, REPORT C-2 (1980), *available at* http://www.curt.org/pdf/156.pdf. This method for estimating productivity loss resulting from overtime work is known as the "BART Curve."

108. *Id*. This formula cannot be applied to measure productivity loss due to overtime when there was only "spot or sporadic overtime" worked on the project. *See* J.A. Jones Constr. Co., ENG BCA 6348, 00-2 BCA ¶ 31,000 (noting that "[t]he BART curve is a cumulative curve and applies when the total job is placed on extended overtime of 50–60 hours per workweek").

109. J.D. Hedin Constr. Co. v. United States, 347 F.2d 235 (Ct. Cl. 1965); *overruled on other grounds in* Wilner v. United States, 24 F.3d 1397 (Fed. Cir. 1994); Lew F. Stillwell, Inc., ASBCA 9423, 1964 BCA ¶ 4128.

110. Paccon, Inc., ASBCA 7890, 65-2 BCA ¶ 4996.

111. *See, e.g.*, S&M Traylor Bros., ENG BCA 3942, 82-2 BCA (CCH) ¶ 15,937.

112. Arthur Painting Co., ASBCA 20,267, 76-1 BCA ¶ 11,894.

113. *See, e.g.*, Flex-Y-Plan Indus., Inc., GSBCA 4117, 76-1 BCA ¶ 11,713; Lew F. Stillwell, Inc., ASBCA 9423, 1964 BCA ¶ 4128.

114. *See, e.g.*, Centex Bateson Constr. Co. v. West, 250 F.3d 761 (text, table at 2000 WL 898731) (Fed. Cir. July 6, 2000); Hoffman Constr. Co. v. United States, 40 Fed. Cl. 184 (Fed. Cl. 1998), *rev'd in part on other grounds*, 178 F.3d 1313 (Fed. Cir. 1999); S. Leo Harmonay, Inc. v. Binks Mfg. Co., 597 F. Supp. 1014 (S.D.N.Y. 1984); Flex-Y-Plan Indus., GSBCA 4117, 76-1 BCA ¶ 11,713; Lew F. Stillwell, Inc., ASBCA 9423, 1964 BCA ¶ 4128.

115. *Paccon*, 65-2 BCA ¶ 4996.

116. *See Flex-Y-Plan*, 76-1 BCA ¶ 11,713.

117. *See id.*

118. *See Paccon*, 65-2 BCA ¶ 4996.

119. *Id.*

120. *Id.*

121. *Id.*

122. Fischbach & Moore Int'l Corp., ASBCA 18146, 77-1 BCA ¶ 12,300, *aff'd*, 617 F.2d 223 (Ct. Cl. 1980); Louis M. McMaster, Inc., AGBCA 76-156, 79-1 BCA ¶ 13,701.

123. *See Paccon*, 65-2 BCA ¶ 4996.

124. *See, e.g.*, Smith-Cothran, Inc., DOTBCA 1931, 89-1 BCA ¶ 21,554.

125. H. John Homan v. United States, 418 F.2d 522 (Ct. Cl. 1969).

126. Luria Bros. & Co. v. United States, 369 F.2d 701 (Ct. Cl. 1966).

127. *See id.* (allowing contractor to recover costs incurred in storing, protecting and handling materials on the site for almost a year).

128. Sovereign Constr. Co., ASBCA 17792, 75-1 BCA ¶ 11,251.

129. Nat Harrison & Assocs., Inc. v. Gulf States Utils. Co., 491 F.2d 578 (5th Cir. 1974); McCarty Corp. v. Pullman-Kellogg, 571 F. Supp. 1341 (M.D. La. 1983), *rev'd in part on other grounds*, 751 F.2d 750 (5th Cir. 1985); Luria Bros. & Co. v. United States, 369 F.2d 701 (Ct. Cl. 1966).

130. Clark Concrete Contractors, Inc. v. Gen. Servs. Admin., GSBCA 14340, 99-1 BCA ¶ 30,280 (board would accept "a comparison if it is between kinds of work which are reasonably alike, such that the approximations it involves will be meaningful.").

131. *See, e.g.*, Flex-Y-Plan Indus., Inc., GSBCA 4117, 76-1 BCA ¶ 11,713.

132. *See, e.g.*, Elliott Constr. Co., ASBCA 23483, 81-2 BCA ¶ 15,625; Robert McMullan & Sons, Inc., ASBCA 19129, 76-2 BCA ¶ 12,072.

133. *See, e.g.*, *Clark Concrete*, 99-1 BCA ¶ 30,280; Stroh Corp. v. Gen. Servs. Admin., GSBCA 11029, 96-1 BCA ¶ 28,265 (accepting Mechanical Contractors Association of America's manual to establish percentages of labor inefficiency).

134. *See, e.g.*, Groves-Black, ENG BCA 4557, 85-3 BCA ¶ 18,398; Capital Elec. Co., GSBCA 5316, 83-2 BCA ¶ 16,548, *rev'd and vacated in part and remanded on other grounds*, 729 F.2d 743 (Fed. Cir. 1984).

135. *See, e.g.*, Luria Bros. & Co. v. United States, 369 F.2d 701, 713 (Ct. Cl. 1966) ("almost always [loss of productivity] has to be proven by the opinions of expert witnesses.").

136. *See, e.g., id.*; Warwick Constr., Inc. GSBCA 5070, 82-2 BCA ¶ 16,091 (expert testimony regarding loss of efficiency related to adverse weather accepted by the board with minimal supporting cost data).

137. Tom Shaw Inc., ASBCA 28596, 95-1 BCA ¶ 27,457 (non-AGC estimating manual not permitted to substitute for actual records of fuel usage).

138. Meva Corp. v. United States, 511 F.2d 548 (Ct. Cl. 1975).

139. FAR 31.105(d)(2)(i)(A).

140. *See generally* Reflectone, Inc. v. Dalton, 60 F.3d 1572 (Fed. Cir. 1995) (actual costs are evidence that contractor will receive an equitable adjustment, not a windfall profit); C. Mourer Constr., Inc. v. United States, 23 Cl. Ct. 533 (1991) (citing FAR's expressed preference for actual costs); Cen-Vi-Ro of Texas, Inc. v. United States, 210 Ct. Cl. 684 (1976) (remanded to board for determination of actual costs); Labco Constr., Inc., AGBCA 90-115-1, 94-2 BCA ¶ 26,910 (actual costs must be used if available).

141. Folk Constr. Co. v. United States, 2 Cl. Ct. 681 (1983).

142. FAR 31.105.

143. FAR 31.105(d)(2)(i)(A).

144. *Id.*

145. FAR 31.105(d)(2)(i)(A)–(C); FAR 31.205-36.

146. Degenaars Co. v. United States, 2 Cl. Ct. 482 (1983).

147. FAR 31.205-36(b)(3).

148. *Id.*

149. *Id.*

150. FAR 31.205(d)(2)(i)(B).

151. Blake Constr. Co., GSBCA 1176, 66-1 BCA ¶ 5589.

152. *Id.*

153. *See, e.g.*, Layne-Minnesota P.R., Inc. v. Singer Co., 574 F.2d 429 (8th Cir. 1978) (contractor recovered for additional equipment costs caused by more difficult subsurface conditions than represented by owner).

154. Brand Inv. Co. v. United States, 102 Ct. Cl. 40 (1944); C.L. Fairley Constr. Co., Inc., ASBCA 32581, 90-2 BCA ¶ 22,665.

155. *See* FAR 31.105(d)(2)(i)(A); Meva Corp. v. United States, 511 F.2d 548 (Ct. Cl. 1975); Nolan Bros., Inc. v. United States, 437 F.2d 1371 (Ct. Cl. 1971).

156. *See, e.g.*, N. Slope Technical Ltd., Inc., v. United States, 14 Cl. Ct. 242 (Cl. Ct. 1988).

157. R.C. Nash, Government Contract Changes, 17–21 (2d ed. 1989); *see also* William Langnion, ENG BCA 4287, 82-2 BCA ¶ 15,939 (rejecting attempts to limit recovery to hours depicted on the claimant's logs after finding the logs did not account for all of the claimant's equipment costs—typically only the equipment operating on that day—and allowing recovery from the first to the last date that the equipment first appeared on the claimant's daily log to the last day it is referred to).

158. Woodington Corp., ASBCA 43283, 94-2 BCA ¶ 26,934.

159. C.L. Fairley Constr. Co., ASBCA 32581, 90-2 BCA ¶ 22,665 (no recovery for idle equipment unless it can be shown that but for the delay, the equipment would be in use).

160. *See* River Equip. Co., ENG BCA 6042, 94-3 BCA ¶ 26,996.

161. *See, e.g., N. Slope Technical*, 14 Cl. Ct. 242.

162. Ronald Adams Contractor, Inc., AGBCA 91155, 94-3 BCA ¶ 27,018.

163. Tom Shaw, Inc., DOTCAB 2106, 90-1 BCA ¶ 22,580.

164. Labco Constr., Inc., AGBCA 90-115-1, 94-2 BCA ¶ 26,910. *See also* Interstate Gen. Gov't Contractors, Inc. v. West, 12 F.3d 1053 (Fed Cir. 1993).

165. *Ronald Adams*, 94-3 BCA ¶ 27,018.

166. C.L. Fairley Constr. Co., Inc., ASBCA 32581, 90-2 BCA ¶ 22,665.

167. Hardeman-Monier-Hutcherson, ASBCA 11785, 67-1 BCA ¶ 6210.

168. J.D. Shotwell Co., ASBCA 8961, 65-2 BCA ¶ 5243.

169. *See* FAR 31.205-26(d).

170. *See* Excavation-Constr., Inc., ENG BCA 3858, 82-1 BCA ¶ 15,770; Samuel N. Zarpas, Inc., ASBCA 4722, 59-1 BCA ¶ 2170.

171. *Id.*

172. *Id.*

173. *See* George Hyman Constr. Co., ENG BCA 4541, 85-1 BCA ¶ 17,847.

174. *See* Paccon, Inc., ASBCA 7890, 65-2 BCA ¶ 4996.

175. *See, e.g.*, Am. Line Builders, Inc. v. United States, 26 Cl. Ct. 1155 (1992); Luria Bros. & Co. v. United States, 369 F.2d 701 (Ct. Cl. 1966); Essex Elec. Eng'rs, Inc., ASBCA 49915, 99-2 BCA ¶ 30,418.

176. C.F.I. Constr. Co., DOTCAB 1782, 87-1 BCA ¶ 19,547 (rejecting claims for additional materials where claimant could have made precise counting of the bolts in question, but instead based its claim on an estimate that merely applied a percentage increase to the material quantity).

177. *Id.*

178. *See, e.g.*, C.W. Over & Sons, Inc. v. United States, 48 Fed. Cl. 342 (Fed. Cl. 2000); Williams v. Weyerhaeuser Co., No. A093975, 2002 WL 373578 (Cal. Ct. App. Mar. 8, 2002); Korneffel v. Flynn, Nos. 207234, 208899, 2000 WL 33415012 (Mich. Ct. App. Aug. 1, 2000).

179. *See* Metric Constructors, Inc., ASBCA 48852, 98-1 BCA ¶ 29,384 (accepting the government's pricing of a deductive change order where the contractor failed to provide actual material costs and the government offered proof of the lowest of three quotes from electrical wholesalers), *rev'd on other grounds*, 169 F.3d 747 (Fed. Cir. 1999).

180. FAR 31.205-26(d).

181. *See* Nager Elec. Co. v. United States, 442 F.2d 936 (Ct. Cl. 1971).

182. *See, e.g.,* Hardaway Constructors, Inc. v. N.C. Dep't of Transp., 342 S.E.2d 52 (N.C. Ct. App. 1986) (contractor who bid project based upon certain quantity of concrete was not compelled to credit the owner when the contractor changed its method of performance resulting in a lesser quantity of concrete actually used).

183. Samuel N. Zarpas, Inc., ASBCA 4722, 59-1 BCA ¶ 2170.

184. *See* George Hyman Constr. Co., ENG BCA 4541, 85-1 BCA ¶ 17,847.

185. *See, e.g.,* Luria Bros. & Co. v. United States, 369 F.2d 701 (Ct. Cl. 1966) (delay forced storage of excess materials and forced use of off-site storage).

186. *See* Ingalls Shipbuilding Div., Litton Sys., Inc., ASBCA 17579, 78-1 BCA ¶ 13,038.

187. *See* Excavation-Constr., Inc., ENG BCA 3858, 82-1 BCA ¶ 15,770.

188. *See, e.g., Luria Bros.,* 369 F.2d 701 (allowing costs for "rehandling materials"); Stapleton Constr. Co. v. United States, 92 Ct. Cl. 551 (1940) (allowing recovery of storage costs of structural steel caused by owner's delay); C.F.I. Constr. Co., DOTCAB 1782, 87-1 BCA ¶ 19,547 (allowing recovery of "additional freight costs").

189. *See* Samuel N. Zarpas, Inc., ASBCA 4722, 59-1 BCA ¶ 2170.

190. E.V. Lane Corp., ASBCA No. 9741, 65-2 BCA ¶ 5076.

191. Gen. Builders Supply Co. v. United States, 409 F.2d 246 (Ct. Cl. 1969).

192. *See* Chaney & James Constr. Co. v. United States, 421 F.2d 728 (Ct. Cl. 1970).

193. FAR 52.242-14(b).

194. 41 U.S.C. § 611; 31 U.S.C. §§ 3901–3905, 3907.

195. Morse Diesel v. United States, 66 Fed. Cl. 788 (2005).

196. FAR 31.205-47(f)(1).

197. 5 U.S.C. § 504.

198. FAR 31.205-33.

199. Bruce Constr. Corp. v. United States, 324 F.2d 516 (Ct. Cl. 1963), *superseded by rule on other grounds.*

200. *Id. See also* Varo, Inc. v. United States, 212 Ct. Cl. 432 (1977) *and* Hensel Phelps Co., ASBCA No. 15142, 71-1 BCA ¶ 8796.

201. *Bruce Constr.,* 324 F.2d 516; G&M Elec. Contractors Co., GSBCA No. 4771, 78-2 BCA ¶ 13,452, *recon. denied,* 79-1 BCA ¶ 13,791.

202. Nager Elec. Co. v. United States, 442 F.2d 936 (Ct. Cl. 1971); Keco Indus. v. United States, 364 F.2d 838 (Ct. Cl. 1966), *cert. denied,* 386 U.S. 958 (1967); S.N. Nielsen Co. v. United States, 141 Ct. Cl. 793 (1958); Lofstrand Co., ASBCA No. 4366, 58-2 BCA ¶ 1962.

203. FAR 52.211-18.

204. Foley Co. v. United States, 26 Cl. Ct. 936 (1992); Victory Constr. Co. v. United States, 510 F.2d 1379 (Ct. Cl. 1975); *cf.* Bean Dredging Corp., ENG BCA No. 5507, 89-3 BCA ¶ 22,034, *declined to be followed by* Clement-Mtarri Companies, ASBCA 38170, 92-3 BCA ¶ 25,192.

205. Morrison-Knudsen Co. v. United States, 397 F.2d 826 (Ct. Cl. 1968); Peter Kiewit Sons' Co. v. United States, 109 Ct. Cl. 517 (1947); United Constr. Co., ASBCA No. 8234, 1963 BCA ¶ 3641.

CHAPTER 21

Overhead Costs

RICHARD SMITH
W. STEPHEN DALE

I. Introduction

Changes to a government construction contract, formal or constructive, often extend the time of contract performance. This may require simultaneous performance of the original contract work and the additional work during the original contract time period and/or during a time after the contract was scheduled to be completed. Changes may also result in extra work without an extension of the contract time. More importantly, the change may require additional overhead support in the use of manpower, equipment, facilities, supplies, and utilities.

Where the cause of the change is compensable, the contractor may recover an equitable adjustment that includes the reasonable cost of the work plus appropriate adjustments for overhead and profit.[1] If the cause of delay is controlled by the Suspension of Work clause, the contractor may be due an "adjustment," without profit, but with adjustments that account for overhead costs incurred as a result of the suspension.[2] Regardless of the factual scenario giving rise to the project delay, contractors face the often difficult challenge of recovering overhead costs incurred during and as a result of the delay.

A. Applicable FAR Provisions

Project overhead[3] and home office overhead costs represent an important cost element to a contractor, and a major component in pricing any change order or claim. To ensure proper and adequate overhead recovery, a government contractor must thoroughly review Federal Acquisition Regulation (FAR) Part 31,

The authors would like to thank M. Clay Hamrick for his tireless efforts on this chapter.

Contract Cost Principles and Procedures, and carefully follow the restrictions and guidance provided.[4] FAR Part 31 provides extensive definition and guidance as to the allowability of a cost,[5] the reasonableness of a cost,[6] and the allocability of a cost.[7]

Overhead costs, both project and home office, qualify as an indirect cost under the FAR. More specifically, FAR 31.203, Indirect Costs, defines these types of costs as follows:[8]

> After direct costs have been determined and charged directly to the contract or other work, indirect costs are those remaining to be allocated to intermediate or two or more final cost objectives. No final cost objective shall have allocated to it as an indirect cost any cost, if other costs incurred for the same purpose, in like circumstances, have been included as a direct cost of that or any other final cost objective.

The remainder of FAR 31.203 provides additional basic direction as to the means to account for these costs on federal contracts.[9] For contracts subject to the Cost Accounting Standards (CAS), additional rules for determining indirect costs may apply.[10]

FAR 31.205, Selected Costs, provides an extensive list of costs that merit special treatment in government contracting. This list includes both allowable costs fully compensable[11] under a government contract as well as unallowable costs that cannot be recovered and therefore must not be claimed.[12] A contractor must carefully review submissions to exclude any unallowable costs from equitable adjustment proposals or recovery claims. Moreover, a contractor should pay close attention to FAR 31.201-6, Accounting for Unallowable Costs, and properly identify costs described in that provision. In some extreme cases, failure to adhere to the provisions of FAR 31.201-6 and segregate costs deemed unallowable may result in under-recovery of claimed costs, and in some cases of gross errors, penalties.[13]

B. FAR Supplements

Government agencies may have an independent supplement to the FAR. For example, in contracts with a military organization, a contractor must refer to the Defense Federal Acquisition Regulation Supplement (DFARS) for guidance as well as the specific supplement for the applicable branch of military service. These volumes can contain additional regulatory guidance applicable to the recovery of overhead costs and can present a challenging task for a contractor to review and follow.

C. Contract Limitation of Overhead Clauses

Many government agencies have limited the amount of overhead a contractor can recover by drafting and including special contract terms. A contractor must review these terms, if present, to determine any specific limitations on overhead calculation or recovery. In general, these clauses may expressly describe the elements of overhead a contractor or subcontractor can recover and set limits as to the maximum percentages that a contractor can recover. These clauses may place additional risks on the contractor in the event of delay and should be considered as an important factor pre-bid. For example, the Public Buildings Service of the General Services Administration (GSA), the U.S. Postal Service, and the Department of Veterans Affairs have all used clauses limiting the overhead rate a contractor may charge in connection with additional work. More specifically, clause 23 of GSA Form 1139 limits overhead as follows:

Overhead, Profit, and Commission

Maximum allowable overhead, profit and commission percentage given in this paragraph shall be considered to include, but are not limited to, job-site office expense, incidental job burdens, small tools and general office overhead allocation. The percentages for overhead, profit and commission shall be negotiated and may vary according to the nature, extent and complexity of the work involved, but in no case shall exceed the following:

	Overhead	Profit	Commission
To contractor on work performed by other than his own forces	—	—	10%
To first tier subcontractor on work performed by his subcontractors	—	—	10%
To contractor and/or the subcontractors for the portion of the work performed with their respective forces	10%	10%	—

Not more than four percentages, not to exceed the maximum percentages shown above, will be allowed regardless of the number of tier subcontractors. The contractor shall not be allowed a commission on the commission received by a first tier subcontractor. Equitable adjustments for deleted work shall include credits for Overhead, Profit, and Commission. On proposals covering both increases and decreases in the amount of the contract, the application of overhead and profit shall be on the net change in direct costs for the Contractor or subcontractor performing the work.

These types of limitation of overhead clauses, when incorporated, have successfully restricted the amount of overhead recovery in an equitable adjustment. Contractors have attempted to directly attack or restrict the application of specific government overhead limitation clauses, but without great success.[14] In most cases, where the government can demonstrate that the costs incurred and claimed by the contractor qualify as overhead costs subject to the limitation clause, courts and boards will enforce the limitations as written.[15] By contrast, where the contractor can demonstrate that the costs claimed fall outside the scope of the limitations clause, boards have allowed the cost. For example, in *AMEC Construction Management, Inc.*,[16] the General Services Board of Contract Appeals (GSBCA) considered whether an overhead limitation clause could exclude certain costs labeled by both the government and the contractor as "general condition costs" when the contractor charged those costs directly. In that decision, the board focused on how the contractor accounted for these costs, and refused to grant the government's motion for summary judgment seeking to exclude them as overhead costs by another name. In doing so, the board noted:

> Although it is true that general conditions costs are sometimes called job site or field overhead, overhead is actually an expense "that cannot be allocated to a particular product or service." The percentage limitations in the contract clause in question relate to markups for indirect costs (and profit). But as we noted in our previous decision, the general conditions costs which AMEC seeks to recover are direct costs—"costs that are directly attributable to the performance of a specific contract and can be traced specifically to that contract." The Federal Acquisition Regulation permits "costs incurred at the job site incident to performing the work" to be treated as direct costs, "provided the accounting practice used is in accordance with the contractor's established and consistently followed cost accounting practices for all work." If they are so treated, as AMEC says it treated them, the costs are not truly overhead and thus are not subject to a percentage limitation for markups. "General conditions costs" turns out to be a far more accurate label for these costs than "job site" or "field" "overhead" would be.[17]

Accordingly, the board held that these direct costs would fall outside the scope of the limitation on overhead clause.

D. Other Sources

Other sources of information concerning overhead claims and allowable costs appear in the extensive number of cases decided and published by the several boards of contract appeals and federal courts, such as the U.S. Court of Federal

Claims and the Court of Appeals for the Federal Circuit. Additionally, government publications, such as the Defense Contract Audit Agency (DCAA) Contract Audit Manual and the Cost Accounting Standards, published at 48 C.F.R. Chapter 99, cover specific accounting requirements for government contracts. The DCAA manual, in particular, advises DCAA auditors of current policies, regulations, and cases for audit guidance that can guide a contractor equally well. Understanding the unique aspects of the regulatory framework of government construction contracts, the pertinent contract clauses, and available resource materials is essential to effectively obtain full recovery.

II. Recovery of Project Overhead Costs on Equitable Adjustments and Claims

As noted by the court of claims:

> Equitable adjustments in this context are simply corrective measures utilized to keep a contractor whole when the Government modifies a contract. Since the purpose underlying such adjustments is to safeguard the contractor against increased costs engendered by the modification, it appears patent that the measure of damages cannot be the value received by the Government, but must be more closely related to and contingent upon the altered position in which the contractor finds himself by reason of the modification.[18]

When a change order or government-caused delay necessitates an equitable adjustment to the contract price, courts and boards have traditionally accepted overhead as an appropriate element of recovery.[19] In order to become whole in the event of a change, contractors must generally account for two types of overhead costs on change orders and claims: project overhead and home office overhead.

Project overhead refers to those costs incurred at the site of the project that relate to the overall supervision and administration of the specific contract. These costs generally are not identifiable to any particular construction activity, but inure to the benefit of the project as a whole. The following examples illustrate typical project overhead costs:

- Project management and clerical salaries
- Engineering support and quality control personnel
- Job-site trailer, facilities, or office costs
- On-site equipment, furniture, office supplies, telephone, computers, copiers, fax machines, etc.
- Nondirect labor including warehousemen, janitorial support, job-site security
- Utilities, such as temporary electrical power, water

- Small tools and consumables
- Support equipment, such as forklifts and service cranes[20]

These project overhead costs are allocable to several construction activities within the contract and therefore generally serve as indirect costs within the project work. A contractor cannot reasonably track office supplies to individual construction activities or determine office rent chargeable to a particular construction activity or event. Accordingly, the FAR allows project overhead costs to generally be treated as indirect costs within each individual contract or project.[21]

Project overhead costs typically increase or decrease as a function of the construction work performed and/or the duration of the project. Moreover, project overhead costs typically follow a bell curve, increasing as the project progresses, and tapering off as the project reaches completion. As a result, change order work that increases the overall amount of work to perform under the contract may force the contractor to purchase additional tools and consumables, employ additional supervisory personnel, or purchase additional drinking water and electric power. As the project duration increases, the contractor may also incur additional project overhead costs for the extended period in the form of additional supervisory and administrative salaries, trailer rentals, and security services. In both cases, project overhead costs generally increase in a manner not subject to segregation or identification with any particular contract activity.

The cost of equipment often plays a significant part in any claim based on changed work or government-caused delay.[22] Changed or delayed work may require additional equipment, additional equipment hours, or the original equipment to remain longer at the site, either working, idle, or both, to complete the new work or to perform activities for a period beyond the original completion date. If contractor-owned equipment remains idle during periods of delay, courts recognize a corresponding reduction in the cost of equipment.[23] For third-party rental, a contractor may claim the full rental value if the contractor cannot mitigate the costs by using the equipment on other jobs.[24] Special rules apply where a contractor rents equipment from a source under common control.[25] Depending on the nature of the changes or delays, the nature of the equipment at issue, and the contractor's accounting system, equipment costs may qualify as recoverable overhead.

A. Methods to Price Project Overhead

Since project overhead costs generally cannot be attributed to any single construction activity, additional overhead costs attributable to a specific change cannot be discretely quantified in the same manner as direct costs. As a result, a contractor generally must quantify overhead costs and associate them with particular activities based upon an allocation method. The two methods of allocation most commonly used in pricing project overhead are a percentage markup and a daily rate.

1. *Project Overhead as a Percentage Markup*

A contractor can determine project overhead as a percentage of its direct costs, and in turn apply that percentage to changed or delayed work.[26] For example, if a contractor incurs total project overhead costs of $500,000, and total direct costs of $5 million, project overhead represents 10 percent of direct costs. The contractor can then use the percentage factor to allocate indirect costs to all direct costs incurred on the contract, including changed or delayed work. Accordingly, if the contractor incurred $100,000 in direct costs for changed work, it would allocate $10,000 in project overhead to the changed work. Allocating project overhead as a percentage markup most commonly arises in determining the cost of added work due to change orders. In fact, one federal board has acknowledged this method and held that a percentage markup for project overhead "is customarily recognized and allowed" on changed work.[27]

2. *Project Overhead as a Daily Rate*

If a change order results in significant project delay without significant additional direct costs, the percentage markup method may not recover substantial portions of incurred overhead costs. To address that situation, a contractor may calculate project overhead using a daily rate. When using the daily rate approach, contractors normally identify the time-related project overhead costs incurred, aggregate these costs into a pool, and divide this pool of costs by the number of days of project performance.[28] This procedure results in an average daily overhead rate, which a contractor may then multiply by the number of days of compensable delay to determine the project overhead cost of the delay.[29] A contractor may also calculate a daily rate by dividing the total overhead costs incurred during the identified delay period by the length of the delay.

Calculating a daily overhead rate can pose theoretical problems. Contractors must maintain consistency within the calculation. That consistency applies to the use of total calendar days, including weekends and holidays, as opposed to only workdays, when determining the daily overhead rate, the duration of the project, and the number of delay days. Failure to do so may distort the rate calculation and either under- or overcompensate a contractor for the delay. Moreover, as an allocation method, a daily rate can effectively ignore the total increase in project overhead related to added work or added direct costs performed in the base contract period. Disputes also arise over identifying the pool of overhead costs needed to calculate the daily rate, or more generally, whether the delay occurred during the project or at the end of the project. Since overhead normally peaks during the middle of performance, and then generally tapers off as contract performance reaches completion, the determination of when the delay occurred will impact the amount of project overhead allocated to the change. Picking the slice of the overhead pie can alter the overhead rate used by the contractor. Taking a slice from the peak of the overhead curve may overcompensate a contractor if the delay occurred in the waning days of the project where overhead costs are diminished. By contrast, taking a slice from the tail of the curve may undercompensate the contractor that incurred a substantial delay in the middle of a project. Accordingly, matching the delay with the appropriate overhead cost pool becomes an important part of the compensation exercise.

The court of claims in *Laburnum Construction Corp. v. United States*[30] examined issues related to the timing of delay and the recovery of overhead in relation to that delay. In *Laburnum,* the contractor sought delay damages for work disrupted by defects in the government's design, including project overhead. The government argued that the overhead expenses should be measured at the end of the project period since design defects merely extended contract performance. This measurement of overhead costs would have occurred when the contractor had already "ramped down" and decreased its recoverable project overhead. The court rejected the government's contention, holding instead that, where a contractor can clearly identify the relevant delay periods, recovery of field and home office expenses related to that period of disruption is appropriate.

Significant disputes can also arise in connection with properly accounting for and distinguishing time-related overhead costs from fixed overhead costs. Contractors should thoroughly review their overhead cost pool to identify costs that vary based either on time or extra work, and those that do not vary based on time or extra work. In the context of delays to the work, the contractor should eliminate overhead costs that are not time or extra-work related, along with costs that are generally inappropriate by their nature, and those that may be claimed elsewhere. Non-time-related costs might include mobilization and demobilization costs, and other one-time costs such as connecting temporary power or installation of site telephones. Failing to segregate these costs from the time-related cost pool may result in a prohibited double recovery.

Certain project overhead costs that increase with delay or disruption remain difficult to determine or quantify. For instance, project delay may impact warranty costs, since the warranty period will start at a later date due to the delay, and the shift in time may result in increased costs. Additionally, in order to mitigate costs in the event of a delay, a project may experience partial demobilization and remobilization. These remobilization costs carry additional overhead expenses that do not always readily appear. Moreover, a delay may push the work into periods of adverse weather, requiring, among other things, temporary enclosures, additional heating equipment, or added utility costs. A contractor must carefully scrutinize its records to identify and include in any equitable adjustment proposal or claim the full measure of these extra overhead costs.

B. Full Recovery of Project Overhead

In the past, the majority of government construction contractors typically used a combination of the percentage markup and daily rate method to quantify project overhead related to changed work. For projects or discrete change orders that result in added work without delays, contractors have often calculated allocable project overhead based on a percentage markup on direct costs. On projects or discrete change orders with significant delays, contractors have often quantified project overhead based upon a daily rate. This dual approach was effective for a change involving only added project overhead or only extended project overhead.

When a project experiences a combination of added work as well as excusable delays, a contractor typically cannot attain full recovery of overhead using either of these methods alone. In general, where a project involves a significant amount of both added work and delay, the contractor incurs both added and extended project overhead costs. Since each traditional method of allocating overhead primarily captures either extended costs or added costs, but not both, using only one method may ignore at least some portion of overhead costs incurred. For example, substantial changes can result in significant delays and require a wide variety of additional site overhead costs, including additional supervisory and engineering personnel; added office trailers, furniture and equipment; added tools purchased to perform the new work; added support equipment rentals; and added utility costs. The costs of the changes may also include extended project overhead, such as the project manager's and project secretary's salary, and the original job trailer rental for an extended period. In this instance, use of only one method, i.e., markup percentage or daily rate, generally will undercompensate the contractor for its added site overhead costs since a single method may disregard a segment of overhead costs. As a result, under the circumstances described above, use of only one method to calculate overhead costs would result in an under-recovery of the added site overhead costs attributable to the changes.

To fully recover overhead costs, contractors might consider (a) use of both methods; (b) use of segregated cost pools; or (c) use of a daily rate to quantify delay costs and added costs.[31] The first, and easy, solution might be for the contractor to use both methods in estimating the project overhead costs related to its changes or claims.[32] The contractor could assert a percentage markup on claimed direct costs to quantify added project overhead allocable to the added work and a daily rate to quantify extended project overhead.[33] Since this approach may result in duplication of costs between the two calculations, a contractor must make appropriate adjustments to eliminate any duplication.

Under the second method, a contractor could attempt to segregate project overhead costs into two pools, a fixed (time-related) cost pool and a variable (activity-related) cost pool. Fixed costs would be those costs that increase with the passage of time and variable costs would be those costs that vary with the amount of work performed. In the event of claims or change order work, the contractor could use the recorded variable costs as the pool for determining its percentage rate on a base of direct costs. The recorded fixed costs could serve as the pool for determining the daily rate for the quantification of any delay costs. Such an approach would be consistent with the economic reality of the increased project overhead costs since the recovery method incorporates the forces that actually caused the increased costs to be incurred. However, the precise segregation of project overhead costs into discrete pools often proves very difficult in practice.

As a third method, a contractor could discretely identify and claim overhead costs as "added fixed costs" directly attributable to a particular change. This approach is sanctioned in principle by FAR 31.105(d)(3), which states: "Costs incurred at the job site incident to performing the work, such as the cost of superintendence, timekeeping and clerical work, engineering, utility costs, supplies, material handling, restoration and cleanup, etc., are allowable as direct or indirect costs, provided the accounting practice used is in accordance with the contractor's established and consistently followed cost accounting practices for all work."[34] However, the interaction of many changes on the project may render this method impractical in certain situations. In many instances where a contractor incurs "added fixed costs" for project overhead, these costs result from the accumulation of many changes, rather than a particular change. Accordingly, these increased overhead costs may prove difficult to present and negotiate except as part of a consolidated omnibus claim presentation. Thus, since the ability to identify and segregate such costs to the satisfaction of the government is, at best, a difficult task, this method may not succeed in all situations.

III. FAR Restriction on Project Overhead Costs Recovery— The Requirement to Select and Use a Single Method

A. M.A. Mortenson Co.

Decisions of the Armed Services Board of Contract Appeals (ASBCA) in *M.A. Mortenson Co.*[35] and *Caddell Construction Co., Inc.*[36] apply the FAR cost regulations in a manner that may lead to incomplete recovery of project overhead costs on federal construction contracts impacted by both added work and extended project duration. Although revisions to the FAR provisions relied upon by the ASBCA in *Mortenson* and *Caddell* may mitigate the impact of these decisions, the basic rule remains the same, namely, that a contractor may only use a single allocation method to recover project overhead. This direction compels a contractor to elect a means of accounting for overhead regardless of its impact on the contractor's ability to recover costs or to become whole in the event of a change to the contract.

The original *Mortenson* decision, decided in October 1996, reviewed performance under three separate contracts for construction work at Whiteman

Air Force Base in Missouri. During performance of the contract, the Air Force issued several change orders that did not increase the time of contract performance. The government denied project overhead claimed on these change orders. The government argued that an earlier audit concluded that all of Mortenson's project overhead costs were fixed, and further concluded that no additional costs were incurred on change orders that did not extend the contract performance time.

Mortenson did not specifically identify fixed and variable overhead costs, but instead argued that once the direct costs of the change were determined, then project overhead and home office overhead were recoverable based on Mortenson's standard accounting practices, without having to specifically identify how the project overhead costs were directly impacted by the change.[37] Mortenson stipulated that its project overhead costs were fixed and offered no credible evidence of increased project overhead costs as a result of the change. In a split decision, Mortenson's claim was denied, and the board held that Mortenson was not entitled to include fixed costs in its claimed project overhead rate. More specifically, the Board focused on causation and held:

> Since change orders/modifications are reimbursed on the basis of an increase in cost, it follows that there must be some showing that an increase in cost has actually occurred. If fixed costs in a contractor's overhead have not increased because of a change order/modification, the contractor is not entitled to include the fixed costs in its overhead rate.[38]

The decision in *Mortenson* significantly impacted the long-standing construction industry practice used to recover field overhead as a percentage markup without a causal connection.[39] Generally, fixed costs such as supervisory and staff personnel make up a significant part of a typical project overhead pool. Even though Mortenson's supervisory and staff personnel spent time processing the changes, reviewing and evaluating the changed work performed, and justifying the costs incurred, due to the fixed nature of their compensation and other incurred costs, the *Mortenson* board denied recovery for the time spent since additional costs were not proven. Thus, under this initial iteration of *Mortenson,* assigning fixed salaried personnel to process change order work, without adding more staff or specifically identifying fixed overhead costs resulting from the change, effectively provides an uncompensated service to the government.

On reconsideration, the board's Senior Deciding Group affirmed the *Mortenson* decision, but added a critical rule. The Senior Deciding Group identified two items at issue: (1) whether a contractor could recover costs consistently charged indirectly without proof of an increase in costs from the change, and (2) whether the use of two different bases to claim indirect costs, depending upon whether a time extension was involved, violated the requirements of FAR 31.203, Indirect Costs, governing the selection of a distribution base. The first issue addressed the board's original causation finding, while the second examined a more fundamental understanding of how to calculate the damages.

With respect to the second issue, Mortenson's standard practice with regard to government contracts was to use different allocation methods depending on the nature of the change. Mortenson charged project overhead costs indirectly on a daily basis for changes that extended the time of performance, and used a percentage markup on the direct costs of changes that did not extend the time of performance.[40] The board found that Mortenson's dual charging practice served as a different method of allocating, through use of a different distribution base, the same project overhead pool to multiple changes. This practice, the Senior Deciding Group concluded, violated FAR Part 31, Cost Principles, requiring the use of one distribution base for allocating a given indirect cost pool. Citing FAR 31.203(b), the Senior Deciding Group concluded that the requirement to use "a distribution base common to all cost objectives to which the grouping is to be allocated" requires the use of a *single* distribution base for allocating a given overhead pool to cost objectives. Since this FAR provision applied to changes, the board concluded that Mortenson could have only one distribution base for allocating its project overhead. Accordingly, switching between a daily rate and a percentage markup, depending upon the time impacts involved, violated FAR 31.203. While the decision stated that Mortenson could choose any reasonable distribution base for allocating its project overhead pool to particular cost objectives, it could have no more than one. Since Mortenson had previously selected the use of a daily rate, it could not later switch to a percentage rate.

B. *Caddell Construction Co., Inc.*

Shortly after the *Mortenson* decision, the ASBCA reaffirmed the single-method principle in granting the government summary judgment in the appeal of *Caddell Construction Co., Inc.* Like Mortenson, Caddell customarily used a daily overhead rate on time-related changes, and a percentage markup on changes where time was not an issue but direct costs were. Indeed, during performance of the work the government made it clear to Caddell that "field overhead costs are not allowed when contract time is not extended."[41] Accordingly, Caddell used a daily rate for changes involving time, while reserving its right to a percentage markup on modifications that did not involve an extension of time. During performance of the work, Caddell identified various categories of project overhead costs that increased as a result of contract changes, but had not kept track of its actual increased costs. As a result, Caddell could not offer these cost categories as direct costs of the change, but was forced to recover them indirectly through an allocation method.

Caddell filed an appeal with regard to the denial of field overhead on changes without time impacts. The board, citing *Mortenson*, rejected the argument that FAR 31.105(d)(3) allowed the use of both a daily and a percentage basis overhead claim. Although the provision allows a contractor to consider overhead costs as direct costs, the board concluded that this provision did not negate the requirements of FAR 31.203 that a contractor be consistent in how it accounted for those costs. Indeed, the provision requires that the "accounting practice used is in accordance with the contractor's established and consistently followed cost accounting practices for all work."[42] Switching allocation methods, according to the board, violated FAR 31.203(b) and was controlled by the *Mortenson* decision.[43]

On reconsideration, the board again denied the appeal. The board, however, did respond to Caddell's argument that the Federal Circuit allowed the use of two allocation methods for home office overhead, a percentage basis and an *Eichleay* daily rate calculation, under limited circumstances.[44] Citing *Capital Electric Co. v. United States*,[45] the board distinguished application of the *Eichleay* formula as an essential part of the government contract claims process, subject to modification only by statute, not regulation. The board also noted the special conditions placed on *Eichleay* recovery, while remarking that these special requirements do not cover project overhead. The board, however, did not address Caddell's argument under FAR 31.201-5(c), namely, that only costs in excess of the amounts using an inconsistent practice should be disallowed. That approach would have allowed Caddell to reprice its claims for equitable adjustments using a percentage basis for project overhead. The board, however, held that it lacked jurisdiction to address this aspect of Caddell's argument, deferring the issue to a later matter.

C. Impacts of *Mortenson* and *Caddell*

The federal government moved quickly to implement the decisions of *Mortenson* and *Caddell* and the board's direction to use a single accounting method for project overhead. The DCAA incorporated the *Mortenson* rationale into the January 2000 edition of the DCAA Contract Audit Manual for audits of indirect costs. In doing so, the DCAA advised its auditors that "[j]ob site/field overhead costs are allowable as direct or indirect costs provided the costs are charged in accordance with the contractor's established accounting system and consistently applied for all contracts (FAR 31.105(d)(3))."[46] The agency continued its direction by providing a summary of the *Mortenson* and *Caddell* decisions. This provision remains in the DCAA manual as does its reliance on *Mortenson* and *Caddell*.

Complicating the matter, however, FAR 31.203 was revised in 2004, which may have implications for application of the *Mortenson* rule. In 2004, the Civilian Agency Acquisition Council and the Defense Acquisition Regulation Council drafted and published a revision to FAR 31.203. The revision slightly modified the language expressly relied upon in *Mortenson* for the proposition that "The contractor shall determine each grouping so as to permit use of an allocation base that is common to all cost objectives to which the grouping is to be allocated."[47] While this language change does not impact the thrust of the *Mortenson* decision, the addition of subsection (e) arguably provides flexibility to the contractor beyond that recognized in *Mortenson* and *Caddell*. The new FAR 31.203(e) states in part: "The method of allocating indirect costs may require revision when there is a significant change in the nature of the business, the extent of subcontracting, fixed-asset improvement programs, inventories, the volume of sales and production, manufacturing processes, the contractor's products, or other relevant circumstances." This provision allows the contractor to argue that certain significant changes in the nature of the work may qualify as "other relevant circumstances" that would allow multiple allocation methods. Such an interpretation does not contradict the standing interpretation of FAR 31.203(c) relied upon in *Mortenson* and *Caddell*, but simply provides relief where circumstances warrant a departure from the existing allocation methods. Such an approach would coincide with the remedial goal of an equitable adjustment, namely to "keep a contractor whole when the Government modifies a contract."[48] Accordingly, where a contractor plans to appeal a contracting officer's final decision to the ASBCA, it must be prepared to justify a single method to charge project overhead or to identify "other relevant circumstances" that would allow a departure from the single method. Of equal importance, a contractor may need to make certain to use this method on all government construction contracts, and possibly on all construction contracts, to support its claims for overhead costs.

The decisions in *Mortenson* and *Caddell* have not specified or limited the time at which the election to use a percentage or daily rate has to be made, or whether that election applies only to government contracts or to all of the contractor's construction contracts. Furthermore, as noted above in the context of 31.203(e), no panel has addressed the circumstances under which a contractor can modify its election, or whether the requirement for consistency extends to the method used to determine project overhead in the contractor's bid. Additionally, these decisions raise issues that remain unaddressed related to the government's entitlement to overhead costs on a deductive change. Regardless of the outstanding issues raised by these decisions, for the present, the clear result remains that the ASBCA and the DCAA Contract Audit Manual only allow contractors to employ a single method to allocate overhead.[49]

IV. *Recovery of Home Office Overhead*

The second category of overhead costs necessary to make a contractor whole in the event of a government change is home office overhead. With respect to home office overhead, the Court of Federal Claims (COFC) and boards of contract appeals are required by a series of decisions of the Federal Circuit to use a single method to allocate increases in home office overhead expenses resulting from delay or disruption. Since a contractor must often allocate home office overhead, like project overhead, indirectly, these formulas can create disputes over accuracy and propriety.

A. Home Office Overhead Increases Resulting from Delay

FAR 31.001 defines "home office" as follows:

> Home office means an office responsible for directing or managing two or more, but not necessarily all, segments of an organization. It typically establishes policy for, and provides guidance to, the segments in their operations. It usually performs management, supervisory, or administrative functions, and may also perform service functions in support of the operations of the various segments. An organization which has intermediate levels, such as groups, may have several home offices which report to a common home office. An intermediate organization may be both a segment and a home office.

Thus, home office overhead deals with "those costs which are expended for the benefit of the business as a whole and which usually accrue over time."[50] Project changes and delays extend the duration of activities at the project site and force administrative costs and general overhead expense associated with the project to continue longer than expected. A contractor may normally account for home office overhead costs such as salaried personnel, computers, and furniture as general and administrative (G&A) expenses not directly attributable to any particular project. These types of G&A expenses may also include costs such as executive salaries, clerical staff, professional fees, home office rent, some travel costs, bidding costs, advertising, home office utilities, and liability insurance. Generally, these costs are fixed and accumulate with time independent of any particular project. A contractor normally plans to recoup these home office expenses by allocating a certain proportion of these costs to revenue from individual projects.[51]

When a particular project is delayed or suspended, a contractor continues to incur these G&A expenses. The delay or suspension prevents the contractor from collecting revenue under the contract and therefore prevents it from recouping the home office overhead costs allocated to that contract. As a result, the corporate or home office overhead rate increases due to these revenue shortfalls attributable to the delays to the particular project. If a contractor cannot generate additional revenue to compensate for the reduction of revenue caused by the delay, or conversely eliminate some of its overhead expenses, a contractor's overall revenue will continue to underabsorb the fixed overhead costs.

In most instances of suspension or delay, the contractor's workforce must remain on standby and be prepared to resume work on the suspended or delayed project activity.[52] As a result, the contractor often cannot take on additional work to mitigate the effects of the suspension or delay on home office expenses. Furthermore, a contractor may have reached the limits of its bonding capacity, further exacerbating its ability to take on additional work to mitigate the effects of the delay on home office expenses.[53] Moreover, while billing to the delayed project may have essentially ceased, home office expenses will continue to accrue. As a result, the delayed or suspended contract cannot absorb its anticipated share of the home office expenses. Thus, "a contractor is entitled to recover as damages the amount of overhead on a daily basis allocable to the period of overrun for which the Government is responsible."[54]

B. Evaluating Home Office Overhead

Estimating the amount of home office overhead attributable to any single project and its delay or disruption becomes an intricate and contentious task. Essentially, a contractor must reallocate the amount of indirect costs that it would have otherwise theoretically recouped from a delayed or disrupted contract.[55]

Courts primarily use formulas to evaluate home office overhead in delay damage claims. Courts and boards have, on occasion, used percentage factors or rates to estimate home office overhead.[56] These percentage rates calculate the value of extended home office overhead as a fraction of the direct costs the contractor incurred during the delay or disruption period. Percentage rates may depend on several factors, including restrictive contract provisions, the customary rate for the type of work involved, and the contractor's experience in the marketplace.[57] A percentage rate may vary from as low as 3 percent[58] to as high as 50 percent[59] depending on the factors noted above. The U.S. Claims Court in *C.B.C. Enterprises, Inc. v. United States*[60] stated a preference for percentage rates over other formulas in situations involving simple contract extensions, saying, "When a contract period is extended for additional work, rather than a suspension of work, home office overhead generally can be calculated more accurately by applying a percentage overhead markup to direct costs rather than by use of the *Eichleay* formula."[61] As a general rule, however, the plain arithmetic of a percentage markup may fail to capture home office overhead costs and tends to under-compensate a contractor for these extended costs due to delay. Carried to the extremes, delay may reduce a contractor's direct costs to zero, causing a percentage rate likewise to estimate a home office overhead at zero despite the actual continuation of home office expenses.[62]

The formula most widely employed by courts and boards to evaluate extended or unabsorbed home office overhead arose before the ASBCA in *Eichleay Corp.*[63] That case produced the following formula to calculate extended and unabsorbed home office overhead due to delay:

(a) $\dfrac{\text{Contract Billings}}{\text{Total Billings for Contract Period}} \times \begin{array}{l}\text{Total Home Office} \\ \text{Overhead for full} \\ \text{Contract Period}\end{array} = \begin{array}{l}\text{Overhead Allocable} \\ \text{to the Contract}\end{array}$

(b) $\dfrac{\text{Allocable Overhead}}{\begin{array}{l}\text{Total Days of Performance} \\ \text{(including delay)}\end{array}} = \text{Daily Contract Overhead}$

(c) Daily Contract Overhead **X** No. Days Delay = Amount Claimed

(a) Divide the total contract billings for the delayed project by the total corporate billings for the entire contract period (including delay). Multiply the result by the total overhead for the contract period (including delay). The result is the overhead allocable to the contract in question.
(b) Divide total days of actual performance (including delay) into the allocable overhead identified in step (a). The result is the daily overhead allocable to the contract.
(c) Multiply the daily overhead allocable to the contract by the number of days of compensable delay.

C. Evolution of the *Eichleay* Formula

In 1945, the U.S. Court of Claims in *Fred R. Comb Co. v. United States*[64] outlined a precursor to the *Eichleay* formula.[65] In that case the court recognized that a contractor could recover unabsorbed home office overhead in situations that rendered personnel reductions unreasonable. It stated what has become the underlying rationale for recovery of home office overhead:

> [I]t is, ordinarily not practicable to lay off main office employees during a short and indefinite period of delay such as occurred here. So the contractor, instead of saving the salary of that proportion of main office staff which is attributable to this contract, is obliged, in effect, to waste it, and to spend a similar amount at the end of the contract for the extra time made necessary by the delay.[66]

The court thereafter apportioned the amount of home office overhead using the formula proposed by the contractor. That formula created a multiplier from the ratio of contract revenues to actual revenues. When multiplied by the total home office overhead for the relevant period, this multiplier generated a value for home office overhead attributable to the relevant impacted period. The technique used by the *Fred R. Comb* court mirrored the primary components of the *Eichleay* formula, with significant differences lying in the measurement of delay duration using months rather than days.[67]

Fifteen years later, the ASBCA affirmed the underlying logic of *Fred R. Comb* in *Eichleay Corp.* In that case, the contractor experienced intermittent delays in three construction contracts that resulted in an overall prolonged performance period. The changes and suspensions that delayed the work arose sporadically, resulting in disjointed prosecution of the work. As a result, the board held that "[u]nder these circumstances it would not have been prudent or practical for appellant to risk the layoff of Home Office personnel or facilities, or, on the other hand, to absorb personnel and facilities so made idle by taking on new commitments."[68] Moreover, the board recognized the need to distribute home office overhead among various contracts, saying:

Overhead costs, including the main office expenses involved in this
case, cannot ordinarily be charged to a particular contract. They represent the cost of general facilities and administration necessary to the
performance of all contracts. It is therefore necessary to allocate them
to specific contracts on some fair basis of proration.

The ASBCA also held that, once a contractor has demonstrated delay,
the government must bear the burden of showing that no damages arose as
a result.[69] Furthermore, the board rejected the government's assertion that
actual decreases in home office overhead due to the delay should preclude
recovery.

Initially, the formula developed in *Eichleay* was not universally adopted
as the exclusive method for recovering home office overhead expenses. In
fact, as late as 1982, the GSBCA took issue with the formula developed in
Eichleay. In *Capital Electric Co.,*[70] the contractor sought damages for extended
overhead arising from government-caused delays and disruptions to work
on a contract for electrical, mechanical, and plumbing work in the federal
building in Miami, Florida. The GSBCA criticized the *Eichleay* formula,
stating:

The daily rate concept of recovery of extended overhead that *Eichleay*
represents comports with neither the pervasive principles nor the
broad operating principles that encompass generally accepted accounting principles. It neither associates cause with effect nor allocates costs
that cannot be so associated to a specific accounting period or periods.
It does not assign indirect costs to an appropriate cost objective during
the period in which those indirect costs were incurred. . . . The concept is but a species of the *damnum absequae injuria* as suggested by the
New York Court of Appeals in *Berley,*[71] and that has led to its limitation and ultimate rejection in many forums.[72]

The Federal Circuit, however, overruled that decision, and refused to "jet-
tison this settled line of authority" in favor of other methods.[73] In rejecting
the GSBCA approach, the Federal Circuit emphasized the historical use of
Eichleay-style formulas to recover home office overhead, saying:

[W]e do not believe these precedents should be overruled. They are of
such long standing and have been followed in so many decisions of
the various boards of contract appeals that such action should more
properly be taken by Congress. Nor are we persuaded that this would
be an appropriate case for breaking precedent.[74]

Thus, the Federal Circuit appeared to value certainty and predictability over
causation.

In 1997, in *Wickham Contracting Co., Inc. v. Fischer,*[75] the Federal Circuit confirmed the *Eichleay* formula as "the exclusive means for compensating a contractor for unabsorbed overhead."[76] There, the contractor argued that the *Eichleay* formula only applied when no other better method could estimate the unabsorbed overhead. The contractor argued that it devoted 80 percent of its home office activities to the delayed contract, and should be reimbursed accordingly. The court rejected this argument and justified the *Eichleay* approach, saying, "Allocation based on a pro-rata share is necessary because overhead cannot be traced to any particular contract since overhead consists of expenses which benefit and are necessary to every contract."[77]

Despite the clear language of *Wickham,* the Federal Circuit has indicated that it will allow a departure from the *Eichleay* formula under certain limited circumstances. In *Nicon v. United States,*[78] the government awarded a contract for dormitory repair, but, as a result of protests, never issued a notice to proceed. When the government terminated the contract, having issued no notice to proceed, the awardee sought recovery of its termination costs, including home office overhead. The COFC previously held that Nicon could not avail itself of the *Eichleay* formula because it could not show any contract billings as called for by the formula. The Federal Circuit rejected that position, holding:

> When contract performance has not begun, but all of the other strict requirements for entitlement to home office overhead are met, there is no bar to the award of home office overhead in a termination for convenience settlement, provided a reasonable method of allocation is available on the particular facts of the case.[79]

The court construed those cases, like *Wickham,* which name *Eichleay* as the sole method for recovery of home office overhead, and found that those cases apply only where contract performance had begun and was interrupted by government delay. Where performance had not begun, by contrast, the contractor may utilize another method to recover unabsorbed home office overhead expenses. Accordingly, *Nicon* highlights at least one factual circumstance under which an alternative to the *Eichleay* formula exists.

D. The *Eichleay* Elements in Practice

In order to recover damages using the *Eichleay* formula, a contractor must first demonstrate three elements:[80]

1. A government-caused delay of inestimable length occurred;
2. The delay increased the original time of performance; and
3. The contractor was on standby and unable to take on other work during the period of delay.[81]

Following this preliminary showing, the government has the burden of showing that the contractor could have taken on replacement work and mitigated its damages.[82] If the government makes this showing, then the contractor must demonstrate a fourth element, rebutting the government's proof by showing that replacement work was not a practical possibility.[83] As a final matter, the contractor must provide some, yet not precise, evidence of actual damages suffered due to extended home office overhead before the court will allow use of the *Eichleay* formula.[84]

1. Government Delay

A contractor must demonstrate government responsibility for the delay or extended cost period to recover any damages for extended overhead. Further, a contractor must also show that the government has delayed overall contract performance, not simply delayed some component.[85] The source of the delay or disruption may arise from an actual or constructive suspension, an actual or constructive change, or a differing site condition.[86] A court or board may deny a contractor damages for home office overhead in situations of concurrent delay. In these situations, courts and boards have generally refused to allow recovery of home office overhead expenses where the government successfully demonstrated concurrent delay. A court or board may deny a contractor damages for home office overhead in situations of concurrent delay.[87] In these situations, courts and boards have generally refused to allow recovery of home office overhead expenses where the government successfully demonstrated concurrent delay.[88] Additionally, where the government has knowledge of the performance obligations of a subcontractor, the prime contractor may recover *Eichleay* damages on behalf of the subcontractor in a pass-through suit.[89]

The contractor must also show that the government's delay caused the performance to extend beyond what the contractor originally anticipated. Typically, the contractor may meet this prong by simply demonstrating that the government-caused delay extended the period of performance beyond that specified in the contract.[90] Where a contractor anticipated an early completion, however, it may still recover under the *Eichleay* formula, provided it makes an appropriate showing of its intent to complete early.[91] In such a case, a contractor must prove three things: first, that the contractor intended to finish early; second, that the contractor had the capability to finish early; and third, that the contractor actually would have finished early had the government not caused the delay or disruption.[92]

2. Standby

The standby prong of the *Eichleay* test centers on the suspension of work, and the corresponding decrease in contract billings, rather than the idleness of the contractor's workforce.[93] In particular, this prong focuses on the uncertain duration of the suspension, disruption, or delay caused by the government.[94] Courts have held that the *Eichleay* formula best applies in situations in which "at least some element of uncertainty arising from suspension, disruption, or delay of contract performance" affects the calculation of overhead damages.[95] Where a contractor knows a definitive date on which work will resume, boards may presume that the contractor can reallocate its resources.[96] In *Oak Environmental Consultants, Inc. v. United States*,[97] the COFC noted the following regarding the uncertainty requirement:

> [I]ndefiniteness is essential in determining whether a contractor has been forced into standby mode. Indefiniteness is the very characteristic that distinguishes a period during which resources can be allocated elsewhere, versus one in which the uncertainty of resumption creates a risk of breach if a contractor does not stand at the ready. The indefiniteness, as related to home office overhead, restricts a contractor from making organizational changes that could otherwise be made, such as "laying off" employees and shifting resources elsewhere.[98]

The Federal Circuit evaluated this issue in *Interstate General Government Contractors, Inc. v. West*. The ASBCA found that the contractor's workforce had been either reassigned to other contracts or discharged, and as a result failed the "standby" prong of the *Eichleay* test. The Federal Circuit disagreed, stating:

> Suspension or delay of contract performance results in interruption or reduction of the contractor's stream of income from direct costs incurred. Home office overhead costs continue to accrue during such periods, however, regardless of direct contract activity. Consequently, this decrease in direct costs necessary to support the continuing overhead creates unabsorbed overhead, unless office workers are laid off or given additional work during such suspension or delay periods. *Even then, fixed overhead costs usually remain.*[99]

The Court continued to define this requirement in *C.B.C. Enterprises, Inc. v. United States*.[100] In that case, the contractor performing modifications to a Navy air station sought home office overhead damages for contract modifications that extended contract performance. In upholding the award of extended overhead damages, the court articulated the "standby" test, saying that "the proper standby test focuses on the delay or suspension of contract performance for an uncertain duration, during which a contractor is required to remain ready to perform."[101] The court also articulated the limits of the *Eichleay* formula, namely that:

> [t]he raison d'etre of *Eichleay* requires at least some element of uncertainty arising from suspension, disruption or delay of contract performance. Such delays are sudden, sporadic and of uncertain duration. As a result, it is impractical for the contractor to take on other work during these delays. By contrast, CBC negotiated a change order with the

government which extended contract performance for a brief known period of time. CBC experienced no suspension of work, no idle time and no uncertain periods of delay during the agreed upon extended contract performance period. Where no element of uncertainty is imposed on the contractor, use of the *Eichleay* formula to calculate extended home office overhead is not permissible. This approach has been echoed by other courts and boards that have rejected the application of *Eichleay* to simple additions to the contract scope that extend the time of performance.[102]

In 2003, the Federal Circuit in *P.J. Dick, Inc. v. Principi*[103] attempted to clarify the standby requirement further, articulating what a contractor must show to prove that it remained on standby in the absence of a written instruction.[104] Without such direct proof, the contractor must prove the standby element by showing:

1. "[T]he government caused delay "was not only substantial but was of indefinite duration;"
2. That "during the delay the contractor was required to be ready to resume work on the contract, at full speed as well as immediately;" and,
3. "The effective suspension of much, if not all, of the work on the contract was suspended."[105]

The first two elements of this test restated existing case law, and reinforced the notion that home office overhead should only be recovered when the contractor cannot predict the end of the delay and thereby mitigate its costs. The court attempted to illuminate element two further, noting:

Our case law has not elaborated on this requirement, but it is clear that once the suspension period is over, the contractor must be required to be ready to "resume full work immediately." Thus, where the government gives the contractor a reasonable amount of time to remobilize its work force once the suspension is lifted, the contractor cannot be on standby. Presumably, the same result would follow if the government required immediate resumption of the work, but only with a reduced work force and allowed the contractor to gradually increase its work force over some reasonable amount of time. In addition, satisfaction of this element of standby clearly requires something more than an uncertain delay as this is a separate requirement of the case law; the implication is that the contractor must be required to keep at least some of its workers and necessary equipment at the site, even if idle, ready to resume work on the contract (i.e., doing nothing or working on something elsewhere that allows them to get back to the contract site on short notice).[106]

The court's real work began with its explanation of element three. Before *P.J. Dick*, the ASBCA and the Federal Circuit had not clearly established the meaning of "suspension of much, if not all, of the work."[107] Indeed, in earlier decisions, the ASBCA sent mixed messages, suggesting that a contractor need not remain completely idle during the pendency of the delay or disruption to recover *Eichleay* damages. In *Roy McGinnis & Co.*,[108] the contractor managed to perform some contract work during the suspension period, rendering the suspension less than complete. Moreover, the contractor bid on additional work in the same area during the pendency of the suspension. The ASBCA, relying on *Interstate General*, held that *Eichleay* does not require a completely idle workforce, rather "[I]t is enough 'that overhead be unabsorbed because performance of the contract has been suspended *or significantly interrupted.*'"[109] The question of how much work a contractor could perform in a period of delay, however, remained unanswered by *Roy McGinnis*. On this issue the court in *P.J. Dick* held, "Thus, even though it is the typical scenario, formal suspension is not an absolute prerequisite. Contract performance also could be stopped or significantly slowed by government inaction, such as failure to vacate spaces in which the contract was to be performed."[110] At what point "significantly slowed" transforms into suspension remains undefined and subject to additional inquiry.[111]

3. Unable to Take on Other Work

This prong of the *Eichleay* test addresses the contractor's efforts to mitigate its damages in the event of government-caused delays or disruptions. To prevail on this element, a contractor must demonstrate an inability or impracticability of obtaining new work or shifting its workforces to other contracts.[112] Although the government must first show that the contractor could have performed replacement work, a contractor must rebut that showing to recover under the *Eichleay* formula.[113] The ASBCA has articulated a set of subjective criteria to examine in reviewing a contractor's ability to take on additional work, including:

1. The amount of notice and certainty as to the length of the delay period;
2. Bidding, mobilization, geographic, and submittal restraints attendant upon starting other work;
3. The size, resources, capabilities, and expertise of the contractor; and,
4. The size and degree of completion of the job and the amount of work planned for the delay period.[114]

To the extent that resources may be shifted to other activities, some boards have imposed a corresponding adjustment to the overhead recovery based on the additional work.[115] Once the contractor makes a prima facie showing that the government-imposed delay was of an uncertain duration, and that the government required the contractor to remain on standby, then the burden shifts to the government. To meet its burden the government must show:

1. That it was not impractical for the contractor to obtain replacement work during the delay or disruption, or
2. That the contractor's inability to obtain such work, or to perform it, was not caused by the government.[116]

In *Altmayer v. Johnson*,[117] the Federal Circuit held that where a contractor has "met the standby test and established that the government-imposed delay was for an uncertain duration, it presented a prima facie case of entitlement to Eichleay formula damages."[118]

The Federal Circuit has also clearly stated that additional work must be replacement work for the delayed contract and not merely other work performed in the contractor's ordinary course of business.[119] In *Melka Marine v. United States*,[120] the contractor for a dredging project shifted its resources to work on other existing contracts and submitted bids for future contracts during the pendency of government-caused delays on the contract at issue. The court rejected the government's argument that this resource-shifting and prospective bidding constituted "replacement" work for *Eichleay* purposes. More specifically it stated:

> [I]t would be inconsistent with the purpose behind *Eichleay* recovery to require a contractor to cease all normal, on-going operations during a government-caused suspension on one contract in order to guarantee its recovery of unabsorbed overhead costs. A healthy contractor may well be simultaneously engaged in multiple contracts, at different phases of performance. A government-imposed suspension during performance of one contract will not necessarily affect a contractor's ability to obtain and perform others.[121]

Accordingly, the continuation of normal operations by a contractor should not prevent recovery of home office overhead.[122]

E. Criticism of the *Eichleay* Method

Critics have attacked the *Eichleay* formula as producing awards for home office overhead expenses that far exceed the actual costs.[123] In fact, the GSBCA in *Wickham Contracting* identified the academic complications underlying the *Eichleay* formula, stating: "The idea is entirely theoretical. It assumes that the contractor, at the moment one of its projects is delayed, is engaged in exactly the optimum number of projects for the equipment, talent, people and financing then on hand. This of course is almost never the case."[124] Principally, commentators have argued that the *Eichleay* formula fails to accurately capture planned allocation of home office overhead expenses to the particular contract in question. More specifically:

> The *Eichleay* formula does not take into consideration the first and most important factor—how much fixed overhead would have been allocated to the contract. In addition, it calls for a determination of the total overhead incurred during the contract period. This too is defective since fixed overhead (the subject of unabsorbed overhead) is incurred for accounting periods . . . , not contract periods.[125]

Critics have also attacked the disappearing distinction between extended and unabsorbed home office overhead, and courts' and boards' election to treat both under the *Eichleay* formula.[126] While critics argue that the chosen method may fail to accurately capture these types of overhead costs, the fact remains that the *Eichleay* formula predominates as the preferred method of recovering home office overhead on federal contracts.

V. Cost-Reimbursement Contracts

Courts and boards have not addressed overhead recovery in cost reimbursement contracts in the construction arena as frequently as in other areas of government contracting. By their nature, cost-reimbursement contracts do not require the completion of a task, such as construction of a facility, but instead compensate the contractor for a specified amount of work, regardless of the degree of completion.[127] Accordingly, the recovery of overhead costs in that environment presents a distinctly different exercise governed by different regulations. Nevertheless, those construction situations that invoke these issues tend to reflect the same principles developed elsewhere in government contracting.

A. Types of Cost-Reimbursement Contracts

Four basic types of cost-reimbursement contracts prevail in government contracting. Those are (1) cost-sharing, (2) cost-plus-incentive-fee, (3) cost-plus-award-fee, and (4) cost-plus-fixed-fee.[128] Under a cost-sharing contract, the contractor receives no fee and receives reimbursement only for agreed-upon portions of its allowable costs.[129] A contractor may agree to a cost-sharing contract and absorb a portion of the costs where it expects to share in substantial future benefits.[130] A cost-plus-incentive-fee contract typically provides for adjustments to an initially negotiated fee based on the relationship of total allowable costs to total target costs.[131] A cost-plus-award-fee contract provides for a fee consisting of a base amount determined at the time of contracting and an award amount determined by the government and designed to provide a performance incentive.[132] A cost-plus-fixed-fee contract provides for payment to the contractor of a negotiated fee, fixed at the inception of the contract and unrelated to actual costs, but designed to compensate for changes to the contract scope.[133]

B. Allowable, Reasonable, and Allocable

Regardless of the type of cost-reimbursement contract at issue, the FAR imposes a multi-pronged requirement that governs a contractor's ability to recover indirect costs under a cost-reimbursement contract. The three items most commonly addressed include the allowability of those costs under the contract, the reasonableness of those costs, and the allocability of costs to the particular contract. To recover overhead under a cost-reimbursement type contract, a contractor must demonstrate that the claimed costs are allowable, reasonable, and allocable, and bears the burden of accounting for the costs properly and maintaining adequate records of those costs.[134] As demonstrated below, however, although reasonableness and allocability may be defined as subunits of allowability, these factors are generally treated as stand-alone issues.

1. Allowable

The issue of allowability "is addressed to the question whether a particular item of cost should be recoverable as a matter of public 'policy.'"[135] FAR 31.201-2 provides, in part, the following guidelines for determining allowability:

> 31.201-2 Determining allowability.
>   (a) A cost is allowable only when the cost complies with all of the following requirements:
>   (1) Reasonableness.
>   (2) Allocability.
>   (3) Standards promulgated by the CAS Board, if applicable; otherwise, generally accepted accounting principles and practices appropriate to the circumstances.
>   (4) Terms of the contract.
>   (5) Any limitations set forth in this subpart.

FAR 31.201-2(a)(2) highlights the overlap and interaction between allocability and allowability, making allocability a factor in determining allowability.[136] According to the Federal Circuit, the interaction is simply a common sense touchstone:

> [T]his provision [FAR 31.201-2 (a)(2)] merely codifies the general principle that a cost is not allowable if the cost cannot be allocated to a government contract. For example, a contractor's cost of certain materials (e.g., nuts and bolts) may be allowable when the materials are used to perform a government contract (and the costs therefore can be allocated to the contract), but not when the materials are used to perform a contract with a private party. Thus, cost allowability may turn on whether the cost is allocable. On the other hand, even when a cost

is allocable, it is not necessarily allowable. For example, a contractor's cost of materials (e.g., nuts and bolts) used to perform a government contract may be allocable to the contract, but the cost may be unallowable if it is unreasonable (e.g., if the contractor overcharges the government for the materials).[137]

In many situations, contract terms play an important role in determining whether certain indirect costs qualify as allowable, and may place additional restrictions on recovery of overhead.[138] To properly determine allowability of overhead costs, a contractor must review both the regulatory guidelines and the contract terms specific to the project.

2. Reasonable

The second element required to recover indirect costs is reasonableness. FAR 31.201-3 describes a reasonable cost as follows:

> (a) A cost is reasonable if, in its nature and amount, it does not exceed that which would be incurred by a prudent person in the conduct of competitive business. Reasonableness of specific costs must be examined with particular care in connection with firms or their separate divisions that may not be subject to effective competitive restraints. No presumption of reasonableness shall be attached to the incurrence of costs by a contractor. If an initial review of the facts results in a challenge of a specific cost by the contracting officer or the contracting officer's representative, the burden of proof shall be upon the contractor to establish that such cost is reasonable.
> (b) What is reasonable depends upon a variety of considerations and circumstances, including—
> (1) Whether it is the type of cost generally recognized as ordinary and necessary for the conduct of the contractor's business or the contract performance;
> (2) Generally accepted sound business practices, arm's length bargaining, and Federal and State laws and regulations;
> (3) The contractor's responsibilities to the Government, other customers, the owners of the business, employees, and the public at large; and,
> (4) Any significant deviations from the contractor's established practices.

This provision provides a contractor and the government common ground to determine whether submitted indirect costs are reasonable. Courts and boards traditionally held that costs actually incurred were reasonable.[139] However, for those contracts subject to FAR 31.201-3, revisions instituted in 1987 shifted the burden to the contractor to "establish that such cost is reasonable."[140] Nevertheless, the contractor's burden remains to demonstrate that the expense comports with reasonably prudent business practice and is consistent with the contractor's overall business routine.[141]

3. *Allocable*

Allocability relates to accounting principles that dictate the proper relationship between incurred costs and the contracts to which a contractor charges those costs. "In summary, the concept of allocability is addressed to the question of whether a sufficient 'nexus' exists between the cost and a government contract."[142] A contractor must pay particular attention to properly segregate costs among several government contracts or between government and nongovernment activities.[143] FAR 31.201-4 provides guidelines for determining allocability as follows:

31.201-4 Determining allocability.

A cost is allocable if it is assignable or chargeable to one or more cost objectives on the basis of relative benefits received or other equitable relationship. Subject to the foregoing, a cost is allocable to a Government contract if it—

(a) Is incurred specifically for the contract;

(b) Benefits both the contract and other work, and can be distributed to them in reasonable proportion to the benefits received; or

(c) Is necessary to the overall operation of the business, although a direct relationship to any particular cost objective cannot be shown.

Certain Cost Accounting Standards, including CAS 403 and 410, may also apply if the contract falls under the CAS. Types of general and administrative costs that have been found allocable to government contracts under some of these guidelines include legal expenses,[144] certain idle facilities,[145] and officers' salaries.[146]

C. Cost Limitation Provisions

After demonstrating that the claimed indirect costs are allowable, reasonable, and allocable, a contractor must also comply with other limitations provisions contained in the contract. Cost-reimbursement type contracts typically contain either a Limitation of Cost provision or a Limitation of Funds provision to enable the government to control a contractor's expenditures.[147] Both provisions allow a contractor to stop work, or conversely, relieve the government of the obligation to pay for work whose cost may exceed the amount provisionally funded to the contract. The section below briefly addresses the role of these clauses in recovering overhead expenses in government contracting.

1. *Limitation of Cost*

FAR 32.705-2(b) requires that a contracting officer (CO) insert the Limitation of Cost clause codified at FAR 52.232-20 in solicitations and contracts for consolidated facilities and facilities acquisition.[148] That clause provides in pertinent part as follows:

(a) The parties estimate that performance of this contract will not cost the Government more than the estimated cost specified in the Schedule. The Contractor agrees to use its best efforts to perform the work specified in the Schedule within the estimated cost.

This clause is designed to assist the government in monitoring costs, including overhead costs, on cost type contracts, and imposes on the contractor an obligation to advise the government when costs may exceed the contract estimates. As noted by the Federal Circuit:

There is sound reason for the notice requirement of the Limitation of Costs provision. It protects the contractor by either providing assurance of reimbursement or permitting the contractor to cease performance. It protects the government from paying more than it had expected for the project. The choice as to whether to incur additional costs is the government's, not the contractor's.[149]

The clause and its notice requirements are consistent with the Anti-Deficiency Act, 31 U.S.C. §§ 1341. The clause imposes the following reciprocal obligations:

(d) Except as required by other provisions of this contract, specifically citing and stated to be an exception to this clause—

(1) The Government is not obligated to reimburse the contractor for costs incurred in excess of the estimated cost specified in the Schedule; and

(2) The Contractor is not obligated to continue performance under this contract (including actions under the Termination clause of this contract) or otherwise incur costs in excess of the estimated cost specified in the Schedule, until the Contracting Officer (i) notifies the Contractor in writing that the estimated cost has been increased and (ii) provides a revised estimated total cost of performing this contract.

In certain circumstances, actual overhead costs may exceed the estimates used to prepare the contract schedule. Where overhead costs cause the contract price to exceed the estimated cost specified in the contract schedule, the contractor must provide notice to the government of the expected overrun.[150] In that instance, the clause contemplates that the contractor will:

1. Notify the government in writing that the actual costs and estimated costs to be incurred will exceed the estimated costs in the contract schedule (i.e., the funded amount);

2. Receive a written notification from the CO acknowledging that the costs in the contract schedule (i.e., the funded amount) have been increased; and

3. Not continue performance or incur additional performance costs absent the notification from the CO.[151]

Most litigation surrounding the Limitation of Cost clause centers on the notice required of the contractor. Unless the increased overhead costs were unforeseeable,[152] a contractor's failure to provide the government with notice that costs are exceeding estimates may result in the contractor's inability to recover those costs.[153]

However, additional situations exist in which a contractor may be excused for lack of notice pursuant to the Limitations of Costs clause. Those include situations where:

1. The contractor "could not have reasonably foreseen the cost overrun during the time of performance of the contract";

2. The costs were not avoidable by the contractor through stoppage of work;

3. The government was not prejudiced by lack of notice of the potential overrun;

4. The CO "effectively exercised his discretion in favor of allowing over-run costs to the contractor"; or

5. Under all the circumstances, "it would be inequitable for the Government to refuse additional funding . . . ."[154]

In limited circumstances, in the case of indirect costs such as overhead, the courts and boards have recognized the difficulty in accurately determining these expenses in advance. In *George W. Sturm Associates v. United States*,[155] the court of claims, in reviewing a cost-plus-fixed-fee contract, held that overhead is "hard to calculate in the early stages of performance" and determined that the contractor's lack of notice was not fatal to its claim.[156]

2. Limitation of Funds

Where the government chooses to fund a cost-reimbursement contract on an incremental basis, the CO must insert FAR 52.232-22, the Limitation of Funds clause, in the contract.[157] That clause states its purpose in part as follows:

(b) The Schedule specifies the amount presently available for payment by the Government and allotted to this contract, the items covered, the Government's share of the cost if this is a cost-sharing contract, and the period of performance it is estimated the allotted amount will cover. The parties contemplate that the Government will allot additional funds incrementally to the contract up to the full estimated cost to the Government specified in the Schedule, exclusive of any fee. The Contractor agrees to perform, or have performed, work on the contract up to the point at which the total amount paid and payable by the Government under the contract approximates but does not exceed the total amount actually allotted by the Government to the contract.

Many of the provisions of the Limitation of Funds clause mirror those of the Limitation of Cost clause, giving rise to many of the same notice issues.[158] Contractors faced with this clause and a lack of funds have frequently attacked this provision using equitable arguments such as waiver and estoppel.[159] Although often unsuccessful, courts have left room for contractors to prevail on these equitable theories under the Limitation of Funds clause.[160]

D. Other Cost Principles

As discussed in Section A above, FAR 31.205, Selected Costs, provides specific guidance on the allowability of certain specific costs. This section addresses over fifty different types of costs and covers a wide range of issues. The treatment of each topic varies from a simple statement to detailed description of the limits of allowability in a particular context. Some costs expressly allowed under this regulation include:

- Public relations and advertising costs;[161]
- Certain compensation for personal services;[162]
- Certain costs of money;[163]
- Normal depreciation;[164]
- Costs of materials;[165]
- Professional and consultant services;[166]
- Recruitment costs;[167]
- Transportation;[168] and
- Travel.[169]

On the other hand, the following list highlights some costs generally not allowed by the FAR:

- Bad debts;[170]
- Bonding costs;[171]
- Contingencies;[172]
- Contributions or donations;[173]
- Entertainment costs;[174]
- Fines, penalties, or mischarging costs;[175]
- Lobbying and political action costs;[176] and
- Alcoholic beverages.[177]

These few examples briefly touch on the issues raised in the Selected Costs provision. Notwithstanding approval as an allowable cost, to qualify as recoverable costs, each of these costs must also be allocable and reasonable.

Considering the complexity and detail included in FAR 31.205, a prudent contractor should consult this provision in determining what items to include in a government contract claim.

VI. Conclusion

Construction contractors working with the federal government must pay careful attention to the estimation and recovery of overhead costs on government contracts. While a full and equitable recovery for changes, delays, disruption, or suspension often includes an overhead component, contractors must approach these claims with caution. Federal regulations and contract provisions may render certain costs unallowable, and contractors must consider these restrictions to avoid submitting claims for unallowable costs. A contractor must also implement appropriate cost accounting measures to capture and allocate additional overhead on the project and support any claims it may make to recover these costs.

In the case of project overhead, a contractor must follow the instructions contained in the *Mortenson* decision and determine whether it will use a daily rate or a percentage to capture these costs. Until the courts and boards clarify that decision or reverse it, the issues raised in *Mortenson* regarding the election of recovery may deny government contractors full compensation for additional overhead expenses. While similar complaints regarding the effects of the *Eichleay* formula on home office overhead remain, federal courts have adopted that method with rare exception and have shown little willingness to reverse the trend. These particular issues, along with the provisions contained in the FAR, must guide a contractor's efforts to recover overhead of all varieties on government contracts.

Notes

1. *See, e.g.,* Ocean Tech., ASBCA No. 21363, 78-1 BCA ¶ 13,204.
2. *See* FAR 52.242-14.
3. Sometimes referred to as site support costs or general conditions costs.
4. *See* 48 CFR pt. 31.
5. *See* FAR 31.201-2. *See also infra* notes 135 to 138 and accompanying text.
6. *See* FAR 31.201-3. *See also infra* notes 139 to 141 and accompanying text.
7. *See* FAR 31.201-4. *See also infra* notes 142 to 146 and accompanying text.
8. FAR 2.101 provides a similar definition of indirect costs, stating, "'Indirect cost' means any cost not directly identified with a single final cost objective, but identified with two or more final cost objectives or with at least one intermediate cost objective."
9. Changes to FAR 31.203 effective in 2005 attempted to clarify these rules as they relate to the Cost Accounting Standards (CAS). *See* FAR 31.203—Indirect Costs (Public Comments), 69 Fed. Reg. 17,766 (Apr. 5, 2004).

10. For rules on CAS coverage *see* FAR 30.201, Contract Requirements ("Title 48 CFR 9903.201-1 (FAR Appendix) describes the rules for determining whether a proposed contract or subcontract is exempt from CAS. Negotiated contracts not exempt in accordance with 48 CFR 9903.201-1(b) shall be subject to CAS. A CAS-covered contract may be subject to either full or modified coverage. The rules for determining whether full or modified coverage applies are in 48 CFR 9903.201-2 (FAR Appendix).").

11. *See, e.g.,* FAR 31.205-1 (certain public relations and advertising); FAR 31.205-4 (bonding costs); 31.205-21 (labor relations costs); 31.205-33 (professional and consultant costs).

12. *See, e.g.,* FAR 31.205-3 (bad debts); FAR 31.205-8 (certain donations and contributions); FAR 31.205-14 (certain entertainment costs); FAR 31.205-15 (fines and penalties in certain situations); FAR 31.205-22 (certain lobbying and political costs).

13. *See, e.g.,* Ryan-Walsh, Inc. v. United States, 37 Fed. Cl. 639, 649 (1997) (Where contractor blatantly failed to comply with FAR 31.201-6, court "is persuaded that, while actual allowable costs exceed this figure, RWI should bear the difference as a penalty for its approach to proclaiming costs.").

14. The seminal decisions addressing limitations on overhead recovery reviewed the "G-10" clause inserted in Veterans' Administration contracts, but the rationale applies equally to newer approaches such as the GSA provision noted above. *See* Reliance Ins. Co. v. United States, 20 Cl. Ct. 715 (1990), *aff'd*, 932 F.2d 981 (Fed Cir. 1991); Sante Fe Eng'rs Inc. v. United States, 801 F.2d 379 (Fed Cir. 1986).

15. *See* Lawrence Constr. Co., GSBCA 1450, 65-2 BCA ¶ 4963; Norair Eng. Corp., GSBCA No. 1178, 66-1 BCA ¶ 5312, *recon. denied* 66-1, BCA ¶ 5702; Blake Constr. Co., GSBCA No. 1834, 66-2 BCA ¶ 5741; Jack Picoult, GSBCA 3516, 72-2 BCA ¶ 9621, recons. denied, 73-1 BCA ¶ 9971.

16. GSBCA No. 16223, 2006-1 BCA ¶ 33,177.

17. 2006-1 BCA ¶ 33,177 at 164,384–85.

18. Bruce Constr. Corp. v. United States, 163 Ct. Cl. 97, 100, 324 F.2d 516, 518 (1963).

19. *See, e.g.,* Ordnance Materials, Inc., ASBCA 32371, 88-3 BCA ¶ 20,910 ("A contractor who proves the direct costs of a Government directed change is entitled to his overhead, computed in accordance with the applicable regulations and generally accepted accounting principles."). *Cf.* Algernon Blair, Inc., ASBCA No. 10738, 65-2 BCA ¶ 5127 (including overhead in deductive change order).

20. *See, e.g.,* FAR 31.105(d)(3).

21. However, FAR 31.105(d)(3) allows the contractor to charge project overhead costs as either indirect or direct costs "provided the accounting practice used is in accordance with the contractor's established and consistently followed cost accounting practices for all work." FAR 31.105(d)(3). This approach was supported by the GSBCA in *AMEC Construction Management, Inc.,* GSBCA No. 16223, 2006-1 BCA ¶ 33,177, which allowed overhead-type costs to be charged as direct costs where the contractor demonstrates that those costs are "'costs that are directly attributable to the performance of a specific contract and can be traced specifically to that contract.'"

22. For a detailed review of equipment costs, see Chapter 19 and FAR 31.105(d)(2).

23. *See* Luria Bros. & Co. v. United States, 177 Ct. Cl. 676, 369 F.2d 701 (1966) (allowing one-half fair rental value for equipment idled due to government delay); Cyrus Contracting, IBCA No. 3233, 98-2 BCA ¶ 30,036 (idle equipment at 50 percent of rental rate); Tom Shaw, Inc., DOTBCA 2106, 90-1 BCA ¶ 22,580.

24. *See, e.g.,* Folk Constr. Co. v. United States, 2 Cl. Ct. 681 (1983); Issac Degenaars Co., ASBCA No. 11045, 72-2 BCA ¶ 9764.

25. *See, e.g.,* Degenaars Co. v. U.S., 2 Cl. Ct. 482 (1983) (arms-length transactions); Weaver-Bailey Contractors, Inc. v. United States, 19 Cl. Ct. 474, *recon. denied,* 20 Cl. Ct. 158 (1990) (examining reasonableness under FAR 31.205-36(b)). *See also* FAR 31.205-36, Rental Costs.

26. In many instances, the terms of the contract itself will dictate the percentage markup allowed for project overhead. *See, e.g.,* W.M. Schlosser, Inc. v. United States, 50 Fed. Cl. 147 (2001); George Hyman Constr. Co., ENG BCA No. 11085, 85-1 BCA ¶ 17,847.

27. *See* Gen. Ry. Signal Co., ENG BCA No. 4250, 85-2 BCA ¶ 17,959.

28. This method closely resembles the *Eichleay* formula designed to capture home office overhead. *See infra* note 67, and accompanying text.

29. *See* Shirley Contracting Corp., ASBCA No. 29,848, 85-1 BCA ¶ 17,858 (1995); Kemmons-Wilson, Inc. ASBCA No. 16,167, 72-2 BCA ¶ 9689 (1972).

30. 325 F.2d 451 (Ct. Cl. 1963).

31. *See* Richard F. Smith & Scott D. Gray, *Recovery of Project Overhead on Changed Work: A Significant Dilemma for Government Contractors,* CONSTR. LAW., Fall 2001, at 8.

32. *But see supra* notes 40–54 and accompanying text.

33. Claims for both a percentage overhead markup and a daily rate for additional project overhead have been recovered in jury awards. *Accord* Techdyne Sys. Corp. v. Whittaker Corp., 427 S.E. 2d 334 (Va. 1993).

34. FAR 31.105(d)(3).

35. ASBCA No. 40750 et al., 97-1 BCA ¶ 28,623, *aff'd on recon.,* 98-1 BCA ¶ 29,658 (Sr. Deciding Group).

36. ASBCA No. 49333, 00-1 BCA ¶ 30,702; *aff'd on recon.,* 00-1 BCA ¶ 30,859.

37. *See* FAR 31.203(a).

38. 97-1 BCA ¶ 28,623 at 142,918.

39. *See* Granite Constr. Co., ENG BCA No. 5849, 93-1 BCA ¶ 25,450 (awarding percentage markup on a change performed by a subcontractor consistent with the contractor's past and present accounting practices). *See also* Atlas R.R. Constr. Co., ENG BCA No. 5972, 94-3 BCA ¶ 26,977; Ordnance Materials, Inc., ASBCA No. 32371, 88-3 BCA ¶ 20,910; Gen. Ry. Signal Co., ENG BCA No. 4250 et al., 85-2 BCA ¶ 17,959.

40. Since *Mortenson* uses direct costs as its basis for distributing indirect costs, it is questionable whether the majority opinion in *Mortenson* would comply with the requirements of FAR 31.203(d), as removing change order work from the base would create fragmented bases.

41. *See* ASBCA No. 49333, 00-1 BCA ¶ 30,702 at 151,688, n. 3.

42. FAR 31.105(d)(3).

43. *But see, e.g.,* DANAC, Inc., ASBCA No. 33394, 97-2 BCA ¶ 29,184; *aff'd on recon.,* 98-1 BCA ¶ 29,454.

44. *See* C.B.C. Enters. v. United States, 978 F.2d 669 (Fed. Cir. 1992); Cmty. Heating & Plumbing Co. v. Kelso, 987 F.2d 1575 (Fed. Cir. 1973).

45. 729 F.2d 743 (Fed. Cir. 1984).

46. *See* DEFENSE CONTRACT AUDIT AGENCY, CONTRACT AUDIT MANUAL at ¶ 12-802.4b (2008).

47. FAR 31.203(b) (2005).

48. Bruce Constr. Corp. v. United States, 163 Ct. Cl. 97, 100, 324 F.2d 516, 518 (1963).

49. A contractor has not successfully challenged *Mortenson* and *Caddell* in a published decision; thus, the dictate that a contractor must elect its method to recover project overhead remains. Moreover, the 2004 changes to FAR 31.203(d) were "not modified to reflect the outcome of any court case." *See* FAR 31.203—Indirect Costs (Public Comments), 69 Fed. Reg. 17,766 (Apr. 5, 2004).

50. Wickham Contracting Co. v. Fischer, 12 F.3d 1574, 1578 (Fed. Cir. 1994).

51. *See* Paccon, Inc. ASBCA No. 7890, 65-2 BCA ¶ 4996 (home office overhead expenses are "continuing expense[s] which is allocated to the contractor's various contracts on a total job basis.").

52. *See* Capital Elec. Co. v. United States, 729 F.2d 743, 748 (Fed. Cir. 1984) (Friedman, J., concurring) (recognizing the impracticality of laying off home office staff during periods of delay).

53. *See generally,* Michael W. Kaufman & Craig A. Holman, *The Eichleay Formula: A Resilient Means for Recovering Unabsorbed Overhead,* 24 PUB. CONT. L.J. 319, 320–21 (1995). *See also* Interstate Gen. Gov't. Contractors, Inc. v. West, 12 F.3d 1053 (Fed Cir. 1993).

54. J.D. Hedin Constr. Co. v. United States, 347 F.2d 235, 259 (Ct. Cl. 1965).

55. *See generally* Michael K. Love, *Theoretical Delay and Overhead Damages,* 30 PUB. CONT. L.J. 33 (2000).

56. *See, e.g.,* Reliance Ins. Co. v. United States, 20 Cl. Ct. 715 (1990) (imposing 10 percent overhead limit under standard VA contract).

57. *See generally* RICHARD J. BEDNAR, ET AL., CONSTRUCTION CONTRACTING 752 (1991).

58. *See* Meva Corp. v. United States, 511 F.2d 548 (Ct. Cl. 1975).

59. *See* Arthur Painting Co., ASBCA No. 20267, 76-1 BCA ¶ 11,894.

60. C.B.C. Enters., Inc. v. United States, 24 Cl. Ct. 187 (1991) *aff'd* 978 F.2d 669 (Fed. Cir. 1992).

61. 24 Cl. Ct. at 190. *See also* Program & Constr. Mgmt. Group, Inc., GSBCA No. 14149, 99-2 BCA ¶ 30,579.

62. *See, e.g.,* Capital Elec. Co. v. United States, 729 F.2d 743, 746 (Fed. Cir. 1984) (rejecting application of percentage markup in favor of *Eichleay* variation).

63. ASBCA No. 5183, 60-2 BCA ¶ 2688 (1960), *aff'd on recon.,* 61-1 BCA ¶ 2894 (1961).

64. 103 Ct. Cl. 174 (1945).

65. Of historic interest, *Brand Inv. Co. v. United States,* 102 Ct. Cl. 40 (1944), preceded *Fred R. Comb* and gave essentially the same formula, but receives less academic attention.

66. *Id.* at 183–84. *See also* West v. All State Boiler, Inc., 146 F.3d 1368, 1372 (Fed. Cir. 1998) ("a government contractor incurs indirect costs which are not attributable to one contract in particular but arise because of its general operations.").

67. British courts have flirted with the Hudson formula named after the book *Hudson's Building and Engineering Contracts.* That formula follows:

$$\frac{\text{Contract amount}}{\text{Contract period (weeks)}} \times \frac{\text{Home office overhead in dollars}}{100} \times \text{Period of delay}$$

The Hudson formula differs from *Eichleay* by determining overhead as a percentage of prime cost from the original bid and including profit as well as overhead. This formula has been attacked in much the same manner as the *Eichleay* formula in American courts. *See generally* BARRY B. BRAMBLE & MICHAEL T. CALLAHAN, CONSTRUCTION DELAY CLAIMS 379–81 (2d ed. 1992). The ASBCA has formulated an additional method to calculate overhead costs known as the *Allegheny* formula. That formula has been superseded by the *Eichleay*, but appeared as follows:

$$\text{Actual Overhead Rate} = \frac{\text{Corporate Overhead Rate for Total Performance Period}}{\text{Total Direct Labor Costs for Total Performance Period}}$$

$$\text{Planned Overhead Rate} = \frac{\text{Corporate Overhead for Planned Period}}{\text{Total Direct Labor Costs for Planned Period}}$$

$$\text{Actual Overhead Rate} - \text{Planned Overhead Rate} = \text{Increased Overhead Rate}$$

$$\text{Increased Overhead Rate} \times \text{Contract's Direct Labor Costs} = \text{Excess Overhead Amount}$$

See Allegheny Sportswear, ASBCA No. 4163, 58-1 BCA ¶ 1684 (1958).

68. *Eichleay Corp.,* 61-1 BCA ¶ 2894 at 15,117.

69. The *Eichleay* formula applies to subcontractor delays as well. *See* E.R. Mitchell Constr. Co. v. Danzig, 175 F.3d 1369 (Fed. Cir. 1999) (subcontractor delayed by government may recover even though prime contractor was not delayed).

70. Capital Elec. Co., GSBCA Nos. 5316, 5317, 82-2 BCA ¶ 16,548, *rev'd,* Capital Elec. Co. v. United States, 729 F.2d 743 (Fed. Cir. 1984).

71. The *Berley* court in New York, as well as others, echoed many of the same causation problems previously expressed by federal courts. *See, e.g.,* Guy James Constr. Co. v. Trinity Indus., Inc., 644 F.2d 525 (5th Cir. 1981) (contractor must prove proximate causation between delay and overhead costs); W.G. Cornell Co. v. Ceramic Coating Co., 626 F.2d 990 (D.C. Cir. 1980) (contractor must demonstrate impracticality of additional work); Berley Indus., Inc. v. New York, 385 N.E.2d 281 (N.Y. 1978).

72. *Capital Elec.*, 82-2 BCA ¶ 16,548 at 82,313.

73. *Capital Elec.*, 729 F.2d at 747.

74. *Id.* at 744.

75. 12 F.3d 1574 (Fed. Cir. 1997).

76. *Id.* at 1580. *See also* Nicon, Inc. v. United States, 51 Fed. Cl. 324, 327 (2001) (where contractor "seeks unabsorbed overhead under a formula that is not recognized by the Federal Circuit, its claim must be rejected."). *But see* Clark Concrete Contractors, Inc., GSBCA No. 14340, 99-1 BCA ¶ 30,280 (contractor allowed alternative method for calculating overhead where accounting of method provided).

77. Wickham Contracting Co. v. Fischer, 12 F.3d 1574, 1578 (Fed. Cir. 1994).

78. 331 F.3d 878 (Fed. Cir. 2003).

79. *Id.* at 885.

80. *See* West v. All State Boiler, 146 F.3d 1368 (Fed. Cir. 1998).

81. P.J. Dick, Inc. v. Principi, 324 F.3d 1364, 1370 (Fed. Cir. 2003). *See also* Satellite Elec. Co. v. United States, 105 F.3d 1418, 1421 (Fed. Cir. 1997); Altmayer v. Johnson, 79 F.3d 1129, 1133 (Fed. Cir. 1996); Mech-Con Corp. v. West, 61 F.3d 883, 885 (Fed. Cir. 1995).

82. Melka Marine, Inc. v. United States, 187 F.3d 1370, 1376 (1999).

83. *Id. See also* Pete Vicari, Gen. Contractor, Inc. v. United States, 53 Fed. Cl. 357 (2002) (the government bears the initial burden of production, but the contractor bears the ultimate burden of persuasion).

84. *See* Capital Elec. Co. v. United States, 729 F.2d 743 (Fed. Cir. 1984), quoting Story Parchment Co. v. Paterson Parchment Paper Co., 282 U.S. 555 (1931) ("the extent of those damages need not be quantified to a mathematical certainty."); Allen Ballew Gen. Contractor, Inc., VABCA Nos. 6987, 7042, 7043, 2007-1 BCA ¶ 33,465 (refusal to allow home office overhead costs where "Appellant failed to provide sufficient evidence of its home office overhead costs.").

85. *See, e.g.,* Nicon, Inc. v. United States, 51 Fed. Cl. 324, 327–28 (2001) (delay in notice to proceed must delay actual contract performance); Gavosto Assocs., Inc., PSBCA Nos. 4058, 4131, 4144, 4333, 01-1 BCA ¶ 31,389 (must show government delayed activities on critical path and delayed overall completion); Precision Dynamics, Inc., ASBCA No. 50519, 05-2 BCA ¶ 33,071 at 163,920 ("in order for the contractor to recover, it must establish that completion of the entire project was delayed by reason of the delay to the segment").

86. *See generally* Adrian L. Bastianelli & Lori Ann Lange, *Recovering Delay Damages for Home Office Overhead/Edition III*, 20001-5 CONSTR. BRIEFINGS 1, 5 (2001).

87. *See, e.g.,* Sauer, Inc. v. United States, 224 F.3d 1340, 1347–48 (Fed. Cir. 2000) (concurrent delay fails to establish required showing of compensable delay for recovery of home office overhead); Kato Corp., ASBCA No. 51462, 2006-2 BCA ¶ 33,293.

88. *See* Young Enters. of Ga., Inc., GSBCA Nos. 14437, 14603, 00-2 BCA ¶ 31,148 (2000) (concurrent delays may preclude recovery under the *Eichleay* formula); S-Tron, ASBCA No. 45893, 96-2 BCA ¶ 28,319 (no recovery where contractor-caused delays present); Gaffney Corp., ASBCA No. 36497, 92-1 BCA ¶ 23,811 (no recovery where government delays not sole cause for suspension); Aydin Corp., EBCA No. 355-5-86, 89-3 BCA ¶ 22,044 (contractor must pinpoint "unjustified delay period of delay attributable" to government alone).

89. *See* E.R. Mitchell v. Danzig, 175 F.3d 1369 (Fed. Cir. 1999); Ace Constructors, Inc. v. United States, 70 Fed. Cl. 253 (2006).

90. *See, e.g.,* Orlosky v. United States, 68 Fed. Cl. 296, 314–15 (2005).

91. Interstate Gen. Gov't Contractors, Inc. v. West, 12 F.3d 1053, 1059 (Fed. Cir. 1993); Jackson Constr. Co., Inc. v. U.S., 62 Fed. Cl. 84 (2004) (discussion of proof required to prevail on overhead claim in the context of early completion).

92. *Id.* (citing Elrich Contracting, GSBCA No. 10936, 93-1 BCA ¶ 25,316 and Frazier-Fleming Co., ASBCA No. 34,537, 91-1 BCA ¶ 23,378).

93. *See* Interstate Gen. Gov't Contractors, Inc. v. West, 12 F.3d 1053, 1057 (Fed. Cir. 1993). Some commentators have considered this element "the most difficult to understand and apply." JOHN CIBINIC, JR., RALPH C. NASH, JR. & JAMES F. NAGLE, ADMINISTRATION OF GOVERNMENT CONTRACTS 721 (4th ed. 2006).

94. *See* C.B.C. Enters., Inc. v. United States, 978 F.2d 669 (Fed. Cir. 1992).

95. *Id.* at 675. *See also* Cmty. Heating and Plumbing Co., ASBCA Nos. 37981, 38166, 38167, 38467, 40151, 92-1 BCA ¶ 24,870.

96. *See* Richmond Am. Homes of Colo., Inc. v. United States, 80 Fed. Cl. 656 (2008) (a "final response plan" that outlined tasks required for lifting of suspension did not create a definitive duration for the suspension); Jackson Elec. Co., Inc., ENG BCA No. 6238, 96-2 BCA ¶ 28,431. *But cf.* DCO Constr., Inc., ASBCA Nos. 52701, 52746, 02-1 BCA ¶ 31,851 (contractor not required to plead standby status in complaint).

97. 77 Fed. Cl. 688 (2007).

98. *Id.* at 698–99.

99. *Interstate,* 12 F.3d at 1057 (emphasis added).

100. 978 F.2d 669 (Fed. Cir. 1992).

101. *Id.* at 1058.

102. *See, e.g.,* Program & Constr. Mgmt. Group, Inc. GSBCA No. 14149, 99-2 BCA ¶ 30,579 (agreed extensions to contract performance necessitate fixed rate compensation).

103. P.J. Dick, Inc. v. Principi, 324 F.3d 1364, 1371 (Fed. Cir. 2003).

104. Of note, the court reminded the parties that the standby test it articulated only applied where the contracting officer had not already acted, stating: "In making that [standby] inquiry, the court should first determine whether the CO has issued a written order that suspends all the work on the contract for an uncertain duration and requires the contractor to remain ready to resume work immediately or on short notice. See *Interstate,* 12 F.3d at 1055, 1057 n.4. In such a case, the contractor need not offer further proof of standby." 324 F.3d at 1371.

105. *Id.*

106. *Id.* (citations omitted).

107. *Id.*

108. ASBCA No. 49867, 01-2 BCA ¶ 31,622.

109. *Roy McGinnis,* 01-2 BCA ¶ 31,622, quoting *Interstate,* 12 F.3d at 1057 (emphasis added).

110. 324 F.3d at 1372–73.

111. *See* Oak Envtl. Consultants, Inc. v. United States, 77 Fed. Cl. 688 (2007) (performing $43,000 in work items on a $2 million contract did not qualify as resumption of work; suspension element met).

112. *See* R.G. Beer Corp., ASBCA No. 4885, 86-3 BCA ¶ 19,012.

113. *See* Satellite Elec. Co. v. Dalton, 105 F.3d 1418 (Fed. Cir. 1997) ("despite the shift in the burden of production, the contractor must nevertheless establish that it was unable to take on other work").

114. *See id.* at 96,027.

115. *See* E.C. Morris & Son, Inc., ASBCA No. 36706, 91-2 BCA ¶ 23,778.

116. *See* Melka Marine v. United States, 187 F.3d 1370, 1375 (Fed. Cir. 1999).

117. 79 F.3d 1129 (Fed. Cir. 1996).

118. *Id.* at 1134.

119. *See* Satellite Elec. Co. v. United States, 105 F.3d 1418 (Fed. Cir. 1997).

120. 187 F.3d 1370 (Fed. Cir. 1999).

121. *Id.* at 1379, quoting West v. All State Boiler, 146 F.3d 1368, 1376 (Fed. Cir. 1998).

122. *See, e.g.,* Indus. Steel, Inc., ASBCA No. 50754, 98-2 BCA ¶ 30,020.

123. For additional criticisms of *Eichleay,* see Lisa Lieberman Thatch, *Eichleay Jurisprudence: Its Strengths, Weaknesses and Inappropriately Broad Application,* 12 Fed. Cir. B.J. 107 (2002).

124. Wickham Contracting Co., GSBCA No. 8675, 92-3 BCA ¶ 25,040 at 124,817.

125. John Cibinic, *The Eichleay Formula: Does It Spell Relief?,* 5 Nash & Cibinic Report 62 (1991) (citations omitted).

126. *See* R.G. Beer Corp., ASBCA No. 4885, 86-3 BCA ¶ 19,012. It is important to note that while the extended versus unabsorbed distinction has diminished, courts still recognize a distinction for *Eichleay* purposes between home office overhead and project overhead. The *Eichleay* formula does not apply to extended project overhead. *See, e.g.,* K.L. Cornwell Corp., ASBCA Nos. 35489, 35490, 90-1 BCA ¶ 22,487.

127. *See generally* John Cibinic, Jr. & Ralph C. Nash, Jr., Cost-Reimbursement Contracting 1 (2d ed. 1993) (hereinafter Cibinic & Nash).

128. *See generally* FAR Subpart 16.3, Cost Reimbursement Contracts; Cibinic & Nash, supra note 127, at 68–71.

129. *See, e.g.,* FAR 16.303; City & County of San Francisco v. United States, 130 F.3d 873 (9th Cir. 1997).

130. *See, e.g.,* Jacobs Eng'g Group, Inc. v. United States, 434 F.3d 1378 (Fed. Cir. 2006).

131. *See* FAR 16.304. *See also* FAR Subpart 16.4, Incentive Control.

132. *See* FAR 16.305. *See also* FAR 16.405.

133. *See* FAR 16.306.

134. *See* FAR 31.201-2(d) ("A contractor is responsible for accounting for costs appropriately and for maintaining records, including supporting documentation, adequate to demonstrate that costs claimed have been incurred, are allocable to the contract, and comply with applicable cost principles in this subpart and agency supplements. The contracting officer may disallow all or part of a claimed cost that is inadequately supported.").

135. Rice v. Martin Marietta Corp., 13 F.3d 1563, 1569 (Fed. Cir. 1993).

136. *See also* United States v. Boeing Co., 802 F.2d 1390 (Fed. Cir. 1986) (discussing allowability and allocability).

137. Boeing N. Am., Inc. v. Roche, 283 F.3d 1320, 1326–27 (Fed. Cir. 2002).

138. *See, e.g.,* Ferguson-Williams, Inc., ENG BCA No. 6482, 00-1 BCA ¶ 30,731 (express contract provision excluded certain G&A expenses).

139. *See* Bruce Constr. Corp. v. United States, 163 Ct. Cl. 97, 324 F.2d 516, 519 (1963) ("'Since the presumption is that a contractor's claimed cost is reasonable, the Government must carry the very heavy burden of showing that the claimed cost was of such a nature that it should not have been expended, or that the contractors' costs were more than were justified in the particular circumstance'").

140. FAR 31.201-3 was amended in 1987 with respect to indirect costs only. *See* 52 Fed. Reg. 19,804 (1987); CIBINIC & NASH, supra note 127, at 729. *See also* Morrison Knudsen Corp. v. Fireman's Fund Ins. Co., 175 F.3d 1221 (10th Cir. 1999); *Herman B. Taylor Constr. Co.,* GSBCA No. 12915, 96-2 BCA ¶ 28,547.

141. *See, e.g.,* McDonnell Douglas Corp. v. United States, 40 Fed. Cl. 529, 544 (1998) ("reasonableness is judged by considering 'all of the relevant circumstances existing at the time of the incurrence of the costs'"); Stanley Aviation Corp., ASBCA No. 12292, 68-2 BCA ¶ 7081 (overhead rates in excess of predictions were reasonable).

142. Lockheed Aircraft Corp. v. United States, 179 Ct. Cl. 545, 375 F.2d 786, 794 (1967).

143. *See generally* Boeing N. Am., Inc. v. Roche, 283 F.3d 1320, 1327 (2002).

144. *See* Caldera v. Northrop Worldwide Aircraft Servs., 192 F.3d 962 (Fed. Cir. 1999).

145. *See* Aerojet Gen. Corp., ASBCA No. 12292, 68-2 BCA ¶ 9932.

146. *See* Vare Indus., Inc., ASBCA No. 12126, 68-2 BCA ¶ 7120.

147. *See* Scan-Tech Sec. v. United States, 46 Fed. Cl. 326 (2000).

148. FAR 45.301 defines "Facilities" and Facilities Contract as:

> Facilities, as used in this subpart and when used in other than a facilities contract, means property used for production, maintenance, research, development, or testing. It includes plant equipment and real property (see 45.101). It does not include material, special test equipment, special tooling, or agency-peculiar property.
>
> Facilities contract, as used in this subpart, means a contract under which Government facilities are provided to a contractor or subcontractor by the Government for use in connection with performing one or more related contracts for supplies or services. It is used occasionally to provide special tooling or special test equipment. Facilities contracts may take any of the following forms:
>
> (a) A facilities acquisition contract providing for the acquisition, construction, and installation of facilities.
>
> (b) A facilities use contract providing for the use, maintenance, accountability, and disposition of facilities.
>
> (c) A consolidated facilities contract, which is a combination of a facilities acquisition and a facilities use contract.

149. *See* Titan Corp. v. West, 129 F.3d 1479 (Fed. Cir. 1997).

150. *See* FAR 52.232-20(b), Limitation of Cost.

151. *See Titan,* 129 F.3d 1479.

152. *See* RMI, Inc. v. United States, 800 F.2d 246 (Fed. Cir. 1986).

153. *See Titan,* 129 F.3d 1479; City of Wichita Falls, HUDBCA No. 76-13, 77-2 BCA ¶ 12,764. *But see* Gen. Elec. Co. v. United States, 440 F.2d 420, 423 (Ct. Cl. 1971)

("if the contractor has no reason to believe that an overrun is imminent, he is not required to give notice.").

154. *See* Johnson Controls World Servs., Inc. v. United States, 48 Fed. Cl. 479, 486–87 (2001). *See also* Int'l Sci. & Tech. Inst., Inc. v. United States, 58 Fed Cl. 798 (2002).

155. 421 F.2d 723 (1970).

156. *Id.* at 727.

157. *See* FAR 32.705-2(c).

158. *See, e.g., Int'l Sci. & Tech. Inst.,* 2002 U.S. Claims LEXIS 118; Am. Elec. Labs., Inc. v. United States, 774 F.2d 1110 (Fed. Cir. 1985).

159. *See, e.g.,* McDonnell Douglas Corp. v. United States, 37 Fed. Cl. 295 (1997) (estoppel) *rev'd on other grounds,* 182 F.3d 1319 (Fed. Cir. 1999); Dynamics Concepts, Inc., ASBCA No. 44738, 93-2 BCA ¶ 25,689 (waiver).

160. *See, e.g., Am. Elec. Labs.,* 774 F.2d 1110 (equitably estopping government from relying on Limitation of Fund clause); Ebasco Servs., Inc. v. United States, 37 Fed. Cl. 370 (1997) (contracting officer may waive Limitation of Funds).

161. *See* FAR 31.205-1.

162. *See* FAR 31.205-6.

163. *See* FAR 31.205-10.

164. *See* FAR 31.205-11.

165. *See* FAR 31.205-26.

166. *See* FAR 31.205-33.

167. *See* FAR 31.205-34.

168. *See* FAR 31.205-45.

169. *See* FAR 31.205-46.

170. *See* FAR 31.205-3.

171. *See* FAR 31.205-4.

172. *See* FAR 31.205-7.

173. *See* FAR 31.205-8.

174. *See* FAR 31.205-14.

175. *See* FAR 31.205-15.

176. *See* FAR 31.205-22.

177. *See* FAR 31.205-51.

CHAPTER 22

Subcontracting

LAWRENCE M. PROSEN
JOEL S. RUBINSTEIN

I. *Subcontractors*

Subcontractors perform a large portion (and often the vast majority) of the work on government construction projects. This places them in the unique position of often having a profound and direct impact on the ability of a prime contractor to perform its government contract. Courts recognize the special position of subcontractors on large construction projects:

> [E]xcept in the middle of a battlefield, nowhere must men coordinate the movement of other men and all materials in the midst of such chaos and with such limited certainty of present facts and future occurrences as in a huge construction project such as the building of this 100 million dollar hospital. Even the most painstaking planning frequently turns out to be mere conjecture and accommodation to changes must necessarily be of the rough, quick and ad hoc sort, analogous to ever-changing commands on the battlefield. Further, it is a difficult task for a court to be able to examine testimony and evidence in the quiet of a courtroom several years later concerning such confusion and then extract from them a determination of precisely when the disorder and constant readjustment, which is to be expected by any subcontractor on a job site, become so extreme, so debilitating and so unreasonable as to constitute a breach of contract between a contractor and a subcontractor.[1]

This chapter discusses the more significant, relevant issues involving subcontractors on federal government construction projects, including not only those issues directly impacting subcontractors but also those impacting other parties such as the government, the prime contractor, other subcontractors, and suppliers.

A. Definition of "Subcontractor"

In the field of construction, the term "subcontractor" sometimes is used interchangeably with other terms, including supplier, materialman, contractor, laborer, and "sub." A subcontractor generally is defined as "One who is awarded a portion of an existing contract by a contractor, esp. a general contractor."[2] The Federal Acquisition Regulation (FAR) defines the term "subcontractor" (at least in the context of subcontractor kickbacks) as "(1) . . . any person, other than the prime contractor, who offers to furnish or furnishes any supplies, materials, equipment, or services of any kind under a prime contract or a subcontract entered into in connection with such prime contract; and (2) includes any person who offers to furnish or furnishes general supplies to the prime contractor or a higher tier subcontractor."[3]

The FAR defines the term "subcontract" as "any agreement (other than one involving an employer-employee relationship) entered into by a government prime contractor or subcontractor calling for supplies and/or services required for performance of the contract, contract modification, or subcontract."[4] "Subcontract" is further defined in the FAR Subcontracts clause as "any contract, as defined in FAR Subpart 2.1, entered into by a subcontractor to furnish supplies or services for performance of the prime contract or a subcontract. It includes, but is not limited to, purchase orders, and changes and modifications to purchase orders."[5]

The Supreme Court has provided significant discussion of the term "subcontractor" in a number of contexts. For example, in discussing the issue of who is considered a subcontractor on a federal government project under the Miller Act,[6] the Court stated that, in a generic sense, a subcontractor includes *anyone* who has a contract to furnish labor or material to the prime contractor.[7] In addition, several boards of contract appeals (BCAs) have held that the term "subcontractor" includes suppliers of material.[8]

B. What Law Controls?

Generally speaking, federal law controls many significant aspects of subcontracts involving federal government construction projects. While a subcontractor is not expressly a party to the prime contract, as discussed below, a number of the prime contract's FAR clauses typically flow down to the subcontractor.[9] As relates to disputes strictly between the prime contractor and subcontractor, there is nothing explicitly in the FAR or applicable statutory law that governs disputes between those parties. For example, in a dispute strictly between the prime and subcontractor where their subcontract contains an arbitration clause, that arbitration clause is enforceable under the Federal Arbitration Act or the applicable state arbitration acts.[10] Similarly, if a state court has jurisdiction over a dispute between a prime and subcontractor on a federal project, either party may pursue an action in state court or, if jurisdiction allows, federal court.

Where a subcontractor's claim arises from the actions or inactions of the government or its agents (other than the prime), such as the architect, then the prime must pursue the government on its subcontractor's behalf. Such claims, known as pass-through claims, must be pursued under the Disputes clause of the prime contract in accordance with the *Severin* doctrine.[11]

II. *Government Control over Subcontracting*

A. Government Consent to Subcontractors/Subcontracts

In firm fixed-price contracts and fixed-price contracts with economic price adjustments, the prime contractor need not obtain the government's consent to enter into subcontracts.[12] Consent, however, may be required for unpriced actions (that is, modifications and delivery orders) under fixed-price contracts that exceed the simplified acquisition threshold of $100,000 when (1) the subcontract is one of either cost reimbursement, time and materials, or labor hour, or, alternatively, (2) the subcontract is for a fixed price but exceeds a specified threshold.[13] The threshold for Department of Defense, U.S. Coast Guard, and NASA contracts is the greater of the simplified acquisition threshold or five percent of the total estimated cost of the contract.[14] For other agencies, the threshold is either the simplified acquisition threshold or 5 percent of the total estimated cost of the contract.[15] The FAR does not, however, state when each of these two thresholds is to be utilized. Presumably, that is up to the agency to determine in its reasonable discretion.

Agency consent generally also is required when the prime contract is a cost reimbursement, time and materials, labor hour, or letter contract.[16] Similarly, consent may be required for subcontracts under architect-engineer (A-E) prime contracts.[17] As the reader can discern, the situations that require government consent are those in which cost is not fixed in a predetermined format. Requiring consent in those situations thus allows the government to have an initial check to protect against cost overruns up front.

Consent of the contracting officer (CO) to a given subcontractor is not to be construed as a governmental determination as to the acceptability of the subcontract's terms, price, cost allowability, or other related contract items.[18] To the contrary, COs expressly are prohibited from giving their consent to subcontract terms, including:

1. Providing for cost-reimbursement fees in excess of the limitations in FAR 16.301-3;
2. Providing for payment on a cost-plus-a-percentage-of-cost basis;
3. Obligating the CO to deal directly with the subcontractor;
4. Making the results of arbitration, judicial determination, or voluntary settlement between the prime contractor and the subcontractor binding on the government; or

5.   Containing repetitive or unduly protracted use of cost reimbursement, time and materials, or labor hour subcontracts.[19]

The FAR also advises COs that they should not withhold their consent to a subcontract merely because the subcontract grants the subcontractor a right of indirect appeal to a BCA where the subcontractor is affected by a dispute between the prime contractor and the government.[20]

Another exception to the need to obtain agency consent is the situation where the prime contractor has an approved purchasing system. Where this situation exists, the prime contractor generally is not required to obtain government consent to subcontracts that are cost reimbursement, time and materials, or labor hour, or that exceed the specified threshold.[21] An "approved purchasing system" is defined simply as a contractor's purchasing system that has been reviewed and approved in accordance with FAR Part 44.[22] Even when the prime contractor has an approved purchasing system, the CO may require consent to a subcontract if the CO determines that consent is necessary to protect the government because of the subcontract type, complexity, or value, or because the subcontract otherwise requires special or unique surveillance.[23] This usually occurs when the subcontract in question involves expensive or complex work. For the agency to effectuate such a consent requirement, however, the agency must specifically identify those subcontracts requiring consent in the FAR Subcontracts clause in the prime contract.[24]

When the prime contract is a cost-reimbursement contract, the contractor is required to notify the agency prior to the award of a cost-plus-fixed-fee subcontract or a fixed-price subcontract that exceeds a specified threshold.[25] The subcontract advance notice requirement applies even if the contractor has an approved purchasing system and consent is not required under FAR 44.201-1.

An administrative contracting officer (ACO) is responsible for consent to the subcontracts required under FAR Subpart 44.2. The sole exception to this is where the CO retains contract administration rights or otherwise withholds subcontract consent authority from the ACO.[26] Regardless, the CO responsible for consent must consider the risks associated with such consent and whether such consent is "consistent with current policy and sound business judgment."[27] In making such consent determinations, the FAR requires the CO to consider the following:

(1) Is the decision to subcontract consistent with the contractor's approved make-or-buy program, if any (see [FAR] 15.407-2)?
(2) Is the subcontract for special test equipment, equipment, or real property that are available from Government sources?
(3) Is the selection of the particular supplies, equipment, or services technically justified?
(4) Has the contractor complied with the prime contract requirements regarding—

(i) Small business subcontracting, including, if applicable, its plan for subcontracting with small, veteran-owned, service-disabled veteran-owned, HUBZone, small disadvantaged and woman-owned small business concerns (see [FAR] Part 19[28]); and

(ii) Purchased from nonprofit agencies designated by the Committee for Purchase from People Who Are Blind or Severely Disabled . . . (see [FAR] Part 8)?

(5) Was adequate price competition obtained or its absence properly justified?

(6) Did the contractor adequately assess and dispose of subcontractors' alternate proposals, if offered?

(7) Does the contractor have a sound basis for selecting and determining the responsibility of the particular subcontractor?

(8) Has the contractor performed adequate cost or price analysis or price comparisons and obtained accurate, complete, and current cost or pricing data, including any required certifications?

(9) Is the proposed subcontract type appropriate for the risks involved and consistent with current policy?

(10) Has adequate consideration been obtained for any proposed subcontract that will involve the use of Government-provided equipment and real property?

(11) Has the contractor adequately and reasonably translated prime contract technical requirements into subcontract requirements?

(12) Does the prime contractor comply with applicable cost accounting standards for awarding the subcontract?

(13) Is the proposed subcontractor in the Excluded Parties List System (see [FAR] Subpart 9.4)?[29]

The FAR further warns that such an analysis warrants particular consideration where, among other things, the prime contractor possesses an inadequate purchasing system; close affiliation, ownership, or other working relationship exists between the subcontractor and prime contractor such that it precludes fee competition; the prime contractor proposes to make award of subcontracts on an other than competitive basis or where the prices are either unreasonably high or higher than those provided to the government under comparable circumstances; or, again, where there is a cost reimbursement, time and materials, or labor hour basis for awarding such subcontracts.[30]

B. Award of Subcontracts

The FAR's Competition in Subcontracting clause requires the prime contractor to select subcontractors and suppliers on a competitive basis to the maximum practical extent consistent with the objectives and requirements of the prime contract.[31] The clause, however, has limited application. It does not apply to prime contracts below the simplified acquisition threshold, firm fixed-price contracts awarded on the basis of adequate price competition, firm fixed-price contracts whose prices are set by law or regulation, time and materials contracts, labor hour contracts, and A-E contracts.[32]

C. Government-Directed Subcontractors

On occasion, a government contract may require the prime contractor to use a government-designated subcontractor. Notwithstanding this designation, the risk of delay or nonperformance by that designated subcontractor generally does not shift from the prime contractor to the government.[33]

A similar rule applies when the government directs the use of a sole-source product. In such cases, the government does not warrant that the product will be properly manufactured by the sole-source supplier or delivered in a timely manner.[34] The government merely warrants that the product satisfies the performance requirements of the contract.[35]

D. Subcontractor Listing Requirements

Many government solicitations require a prime contractor to list each of its proposed subcontractors for a given procurement, or at least those performing more than a specified percentage of the prime contract work (typically, one-half or 1 percent). This listing allows the government to verify that each subcontractor has adequate resources to perform its scope of work and to confirm that the subcontractor is not suspended or debarred. This also makes it more difficult for a prime contractor to bid-shop after award of the prime contract.

E. Subcontractor Responsibility Determinations

In *Omneco, Inc.*, the U.S. Government Accountability Office (GAO) held that, while a prospective contractor must demonstrate both its responsibility and responsiveness, "there are times where that contractor must also affirmatively demonstrate . . . the responsibility of its proposed subcontractors."[36] The GAO found that, where the prime contract required a significant amount of subcontracting, "the contracting officer [could] directly determine a prospective subcontractor's responsibility using the same standards to determine a prime contractor's responsibility."[37]

Likewise, the GAO consistently has held that, after evaluating a prime contractor's proposal, the agency is permitted to consider a subcontractor's past performance as part of the overall prime contractor's proposal. This review of a subcontractor is permitted, however, only where the request for proposal (RFP) allows the use of a subcontractor to perform contract work and does not otherwise expressly prohibit consideration of a subcontractor's past performance experience in evaluating a given proposal.[38]

F. Small and Disadvantaged Business Requirements[39]

To effectuate the regulatory policies of full and open competition among prime contractors and subcontractors and give small and disadvantaged businesses reasonable opportunities to participate in federal work, prime contractors generally are required to develop a small business and small disadvantaged business subcontracting plan. This plan must be included in negotiated procurements, modifications, or sealed-bid acquisitions expected to exceed $1 million for construction when the resulting prime contract has subcontracting possibilities.[40] The plan must include a number of items, such as participation percentages by small and disadvantaged businesses, types and quantities of materials and labor to be supplied by these businesses, and other related information.[41]

There are a limited number of groups excepted from these requirements. Contracts awarded to small businesses, personal service contracts, contract modifications to be wholly performed outside the United States, and other limited contracts and modifications are excluded from this small business subcontracting plan requirement.[42]

The FAR also provides detailed requirements and regulations for subcontracting to small businesses. In particular, the government often provides significant incentives for inclusion of small businesses in contract performance. For example, all contractors receiving a contract that exceeds the simplified acquisition threshold must agree that small businesses[43] are given "the maximum practicable opportunity to participate in contract performance consistent with its efficient performance."[44] Failure to make good-faith efforts to comply with this mandate is a material breach of contract and may subject the breaching contractor to liquidated damages liability.[45]

In addition to the requirements set forth above, where a contractor is required by FAR 19.702(a)(1) and (2) to provide a subcontracting plan, the plan must include other information, including percentage goals for using and spending contract monies on small and disadvantaged businesses.

III. Formation of Subcontracts

Disputes in subcontract formation arise primarily from the following causes: (1) the development of the bid price; (2) the use by the prime contractor of a subcontractor's bid in its bid only to try to use a different subcontractor during contract performance; and (3) an effort by a subcontractor to withdraw or change its bid price. These issues are discussed in this section.

A. Offer and Acceptance of Subcontractor's Bid

Contract law long has recognized that any contract, including a subcontract, requires an offer, acceptance, and consideration.[46] A contract is defined as "a

CHAPTER 22: SUBCONTRACTING

promise or a set of promises for the breach of which the law gives a remedy, or the performance of which the law in some way recognizes as a duty."[47] In order to form a contract, there must be a "manifestation of mutual assent to the exchange and the consideration."[48]

This raises the question of whether a subcontractor who provides a bid or price quote to a prime contractor is bound to perform the work at the quoted price when the prime contractor includes the subcontractor's bid in its bid to the government. One of the early cases discussing the formation of a subcontract and the binding effect of subcontractor bids is *James Baird Co. v. Gimbel Brothers*.[49] In that case, Baird, a prime contractor, sued Gimbel, a linoleum supplier, for breach of contract because of Gimbel's refusal to perform based upon its bid.

The court concluded, "Unless there are circumstances to take it out of the ordinary doctrine, since the offer was withdrawn before it was accepted, the acceptance was too late."[50] To meet its burden, Baird relied on the doctrine of promissory estoppel, arguing that Gimbel knew that Baird might use Gimbel's bid in its bid to the owner and that to allow a subcontractor to withdraw its bid after the prime contractor relied on it and used it would put the prime contractor in an untenable predicament. The court rejected application of the doctrine and found no enforceable contract because the supplier timely withdrew its bid before acceptance by the prime contractor.

Other jurisdictions have reached a different result, holding that a subcontractor is estopped from withdrawing its bid when the prime contractor relies on that bid.[51] For example, in *Drennan v. Star Paving*,[52] a paving subcontractor submitted a bid to the prime contractor that the prime contractor relied upon in computing its own bid price. The court held that the prime contractor's use of the bid in its subsequently successful bid bound the contractor to use the subcontractor's bid and made the subcontractor's bid irrevocable. Thus, the question of whether a subcontractor is bound to its bid revolves around whether a definitive offer was made and whether that offer was withdrawn in a timely fashion prior to use by the prime contractor in its subsequently winning bid.[53] This is called promissory estoppel or detrimental reliance. See note 54, *infra*.

As a general statement of the law, if a prime contractor wants to hold a subcontractor to its bid, the prime contractor must make a clear showing that the subcontractor offered to perform the work at a specific price and that the prime contractor relied upon and accepted the subcontractor's bid (presumably by using that bid in its prime contract bid to the owner). A failure to prove all of these elements eviscerates both a subcontractor's ability to force the prime contractor to award it a subcontract and a prime contractor's ability to force a subcontractor to perform based on its bid.[54]

B. Effect of Subcontractor Mistakes

A subcontractor sometimes may attempt to alter its obligations based on a claim of mistake in its bid. The courts have recognized two types of mistake:

(1) a unilateral mistake by the subcontractor that the prime contractor should have known the subcontractor had made in its bid, and (2) a mutual mistake in which both parties made a mistake based upon improper assumptions on which the subcontract bid was prepared.

The defense of unilateral mistake is rarely successful. For example, the courts have held that subcontractors cannot avoid liability for unilateral mistakes in their bid that arise from defects in the plans and specifications prepared by the government.[55] Where, however, a prime contractor knows of a mistake made by a subcontractor in its proposed bid, the prime contractor cannot rely on that bid, and the subcontractor may avoid its liability for the mistake.[56]

Where the mistake is mutual, i.e., the prime contractor and subcontractor both had reason to know that they each had a different interpretation of the same contract documents at the time of bid, the typical remedy is rescission of the subcontract.[57]

Additionally, in limited situations, the government may be responsible for knowing or having reason to know of a subcontractor's improper or mistaken bid. For example, in the realm of sealed bidding, COs are required to examine all bids for mistakes. Where there is an apparent or obvious mistake, the CO is required to verify the prime contractor's proposed price by "calling attention to the suspected mistake."[58] The CO's failure to timely raise the possible mistake with the contractor may subject the government to liability in the form of a change order or modification. Likewise, if the CO observes a line item for a given item of work that appears irregular, the CO is obligated to inquire of the prime contractor whether its subcontractor's bid is accurate.

C. Failure to Comply with All Material Terms of the Solicitation

A sealed bid on a federal contract that fails to comply with all material terms of an invitation for bids (IFB) is considered nonresponsive and cannot be considered for award.[59] Although not couched in terms of responsiveness, a proposal that fails to comply with a material term of a request for proposals (RFP) under a negotiated or best value–type procurement is considered technically unacceptable and also cannot be accepted by the government.[60] With limited exception, a party must bid or propose/offer on all items within the defined scope of work without taking any exceptions thereto. Likewise, if a subcontractor's bid fails to provide for full performance of its work scope, the prime contractor may preclude the subcontractor from consideration for award.

D. Bid Expiration

A prime contractor must keep its bid open for award for a specified period of time (usually 60 days but sometimes longer depending on the situation).[61] This time period permits the CO to review all bids and to make award to the lowest responsible and responsive bidder. Prior to the expiration of this open period, the CO can request that each bidder keep its bid "open" for an additional period of time if the government requires additional time.[62] Likewise, the prime contractor often will require its bidding subcontractors to keep their bids open for an identical period of time to guarantee that the price is fixed.

Some case law recognizes, however, that where a prime contractor fails to accept a subcontractor's bid for an unreasonable period of time, the doctrine of promissory estoppel[63] cannot be used to enforce that bid.[64]

IV. Bid Protests by Subcontractors

The bid protest process for handling disputes regarding the award of prime contracts is described in Chapter 6. Because of their heavy involvement in federal procurements, one might expect that subcontractors would be able to protest failures to receive subcontract awards. With limited exceptions, however, subcontractors are precluded from filing bid protests before the GAO, the Court of Federal Claims (COFC), or other venues. GAO case law is replete with examples in which the GAO has held that a subcontractor or supplier is not an actual or potential bidder and therefore not an interested party.[65]

The GAO's bid protest regulations allow only an "interested party" to file a bid protest.[66] The regulations define an "interested party" as "an actual or prospective bidder or offeror whose direct economic interest would be affected by the award of a contract or by the failure to award a contract."[67] The GAO consistently has interpreted the term "interested party" as being a prospective bidder or offeror with a direct economic interest in the bid/award outcome. Therefore, the GAO will not consider subcontractor protests except where the subcontract is by or for the government (discussed below).[68]

Section 21.5 of the GAO's bid protest regulations, entitled "Protest Issues Not for Consideration," states that the GAO will not consider a protest of the award or proposed award of a subcontract except where the agency awarding the prime contract has requested in writing that subcontract protests be decided pursuant to 4 C.F.R. Section 21.13.[69]

There are, however, limited circumstances in which the GAO will take jurisdiction over protests filed by subcontractors. These circumstances include awards to subcontractors by a prime contractor where the subcontract in effect was awarded on behalf of the government as a result of the government's involvement in the award process or the contractual relationship between the prime contractor and the government.[70] Where the prime contractor's role is predominantly ministerial, the prime contractor effectively is acting as the government's procurement agent, and the GAO will maintain jurisdiction over a subcontractor's protest.[71]

V. Flow-Down Clauses

Flow-down clauses are used widely in both private and public construction subcontracts. These clauses provide that the subcontractor is liable to the prime contractor, as the prime contractor is liable to the owner, and that all (or specifically referenced) relevant clauses set forth in the prime contract are incorporated by reference into the subcontract. Flow-down clauses provide continuity in contract obligations and performance. In the context of government contracts, however, the FAR, issues of privity of contract, and the *Severin* doctrine (explained below in subsection VII) sometimes can have a dramatic effect on the use and enforceability of such flow-down provisions.

A. Impact of Flow-Down Clauses on Subcontractor's Rights against the Government

Flow-down clauses cannot give the subcontractor direct access to the government. The FAR states that COs are not permitted to consent to subcontracts that require that the CO (government) deal directly with the subcontractor.[72] The CO, however, may consent to a subcontract that gives the subcontractor a right of indirect appeal to a BCA, so long as the appeal involves a claim that arises from the subcontractor being impacted by a dispute between the prime contractor and government.[73] Both the boards and courts recognize that they lack jurisdiction to adjudicate disputes between a prime contractor and its subcontractors on federal contracts.[74]

In order for the prime contractor to force the subcontractor to use the prime contract disputes process to resolve disputes involving the government, the courts consistently have held that the subcontract must specifically incorporate or flow down the prime contract Disputes clause.[75] A general flow-down clause incorporating all terms of the prime contract will not suffice.

B. Mandatory Flow-Down Clauses

Most prime contractors will flow down all prime contract clauses to their subcontractors. The FAR, however, specifies that certain clauses *must* be flowed down to subcontracts.

As evidenced by the discussion throughout this treatise, the FAR plays a significant role in the formation and administration of government construction contracts. Of critical import is the fact that such regulations have the force and effect of law.[76] Tied directly to this long-standing proposition is the fact that under the *Christian* doctrine, federal courts have long held that "a mandatory contract clause that expresses a significant or deeply engrained strand of public procurement policy is considered to be included in a contract by operation of law."[77] Given the fact that certain clauses are mandatory and are effectively read into the contract as a matter of law, the question comes down to which of these clauses actually are governed by this doctrine and which are not.

With regard to subcontracts in particular, the FAR explicitly states which clauses are to be flowed down from the prime contract to the subcontract. Such clauses include, to name a few, the labor standards required by the FAR[78] and the Payrolls and Basic Records clause.[79] In fact, the CO is required by the FAR to insert the Subcontracts clause at FAR 52.244-2 into various types of contracts including time and materials, labor hour, and other like contracts.[80]

Without providing a detailed discussion, mandatory flow-down provisions include the Anti-Kickback clause,[81] the Certification and Disclosure Regarding Payments to Influence Certain Federal Transactions clause,[82] the Limitation on Payments to Influence Certain Federal Transactions clause,[83] the Security Requirements clause,[84] the Audit and Records clause,[85] the Price Reduction for Defective Cost or Pricing Data clause,[86] the Subcontractor Cost and Pricing Data clause,[87] the Davis-Bacon Act clause,[88] the Contract Work Hours and Safety Standards Act—Overtime Compensation clause,[89] the Apprentices and Trainees clause,[90] the Payrolls and Basic Records clause,[91] the Compliance with Copeland Act Requirements clause,[92] the Withholding of Funds clause,[93] the Subcontracts (Labor Standards) clause,[94] the Contract Termination—Debarment clause,[95] the Disputes Concerning Labor Standards clause,[96] the Compliance with Davis-Bacon and Related Act Regulations clause,[97] the Certification of Eligibility clause,[98] the Equal Opportunity clause,[99] the Affirmative Action Compliance clause,[100] the Affirmative Action for Workers with Disabilities clause,[101] the Service Contract Act clause,[102] the Cost Accounting Standards clause,[103] the Disclosure and Consistency of Cost Accounting Practices clause,[104] and the Prompt Payment for Construction Contracts clause.[105]

As for whether the *Christian* doctrine is applicable to subcontracts, there is little in the way of decisional law on this point. The few cases that do appear to have addressed this issue, however, appear to hold that absent an explicit flow-down provision, the *Christian* doctrine generally does not apply to subcontracts.[106]

VI. Payment

There is little doubt that payment is the single most important element of any contract to a subcontractor. After all, the reason subcontractors agree to perform is to make money. It is also, however, one of the most contentious areas of construction litigation, whether on private or public projects. The following subsections discuss a number of tools and strategies afforded subcontractors when it comes to receipt of payment and, if necessary, prosecution of litigation to pursue recovery thereof.

A. Progress Payments

Under most construction contracts, subcontractors are paid through "progress payments," usually on a monthly basis. The subcontractor normally is paid based upon the percentage of completion or quantity of work performed by that subcontractor (i.e., its progress). Prime contractors often withhold retainage, generally between 5 and 10 percent, to ensure completion by the subcontractor and to protect themselves in the event of the subcontractor's default. Such retention is specified in the subcontract language.

The FAR provides for monthly or more frequent progress payments based on estimates of the amount of work satisfactorily completed on government construction contracts.[107] While the CO is authorized to retain up to 10 percent of the amount due, he may only do so if the contractor does not make satisfactory progress.[108] The FAR specifically states that retainage is not to be used as a substitute for good contract administration.[109]

The federal construction clauses contain many aids for subcontractors with regard to payment issues. The contractor must list amounts included in each payment estimate for subcontractors and amounts previously paid to subcontractors.[110] The prime contractor also must certify that payments to subcontractors have been made from previous payments and that timely payments will be made from the proceeds of the payment covered by the certification.[111] Other mechanisms regarding payment follow.

B. Prompt Payment Act

The Prompt Payment Act (PPA) was created as a means to avoid government delay in processing and making payments to prime contractors on federal projects. The PPA requires the procuring agency to pay for services or supplies rendered "before the sixteenth day after the required payment date."[112] If payment is not promptly made, interest accrues at the rate established by the Secretary of the Treasury under Section 12 of the Contract Disputes Act (CDA).[113] That interest begins to run the day after the required payment is due and ends on the date that payment actually is made.[114]

The FAR requires that all government construction subcontracts contain a prompt payment clause requiring (1) payment to the subcontractor within seven days from receipt by the contractor of payment from the government, (2) an interest penalty in favor of the subcontractor for late payments by the prime contractor, and (3) a requirement that first-tier subcontractors flow down the prompt payment clause to their lower-tier subcontractors.[115]

The prompt payment provisions of the FAR specifically allow the contractor to withhold retainage from the subcontractor without cause.[116] The FAR also allows the prime contractor to withhold payment in accordance with the subcontract agreement and incur no interest penalty *if* the contractor gives timely notice to the subcontractor of the withholding, gives notice as to the reason for the withholding, and provides a copy of the notice to the CO.[117]

The prompt payment provisions may only be enforced in proceedings between the prime contractor and subcontractor, and the government cannot be made a party to the lawsuit.[118] The interest penalty is not an obligation of the government but, rather, solely that of the prime contractor.[119] There is no private cause of action by a subcontractor against the government for a violation of the Prompt Payment Act.[120]

The Prompt Payment Act does, however, have limitations. Perhaps the most significant limitation arises when some portion of payment is in dispute. Where there is a dispute between the government and prime contractor as to an amount claimed (either by the prime contractor or arguably by a subcontractor), the Act does not apply to the disputed amount and no interest under the Act accrues.[121] Under the FAR, if there is a dispute as to amounts due, the government is not required to pay interest penalties if the delay in payment is due to that dispute or it relates to amounts temporarily withheld regarding that dispute.[122] Interest thereon would be calculated in accordance with the Prompt Pay Act regulations. In order to recover interest on disputed amounts, the contractor must turn the unpaid invoice into a claim and pursue recovery under the CDA.

C. Mechanic's Liens

A mechanic's lien generally is defined as a statutory lien that secures payment for labor or materials supplied in improving, repairing, or maintaining real or personal property.[123] Liens do not, however, apply on federal government contracts. Sovereign immunity prevents the placement of liens on federal property. As a result, subcontractors and prime contractors on federal projects are not afforded the protections that the mechanic's lien offers on private construction projects. This does not leave the subcontractor without any protection or remedy. While a mechanic's lien is not available to the subcontractor on a government contract, as described below, the subcontractor may still pursue its rights either under a payment bond as set forth in the Miller Act or for a breach of contract action for nonpayment.

D. The Miller Act

Perhaps the single most useful "tool" that a subcontractor has on a federal construction project to recover monies owed by the prime contractor is the Miller Act.[124] The Miller Act generally requires that prior to award of a federal construction contract, the proposed awardee/prime contractor furnish a performance bond and a payment bond for the total contract price.[125]

As a result of the Miller Act's broad language, a subcontractor on a federal project is permitted to sue both the prime contractor and its Miller Act surety for payments not made by the contractor, even if the contractor is without blame and the sole reason for nonpayment is nonpayment by the government.[126]

The cases recognize that the Miller Act is intended to protect subcontractors, suppliers, and labor/materialmen on government contracts in lieu of mechanic's liens.[127] In interpreting and enforcing the Miller Act, the courts view the Act as remedial in nature and therefore give the Act's language a broad interpretation and construction, generally to the subcontractor's benefit or preference.[128]

Payment bonds effectively act as guarantees that subcontractors and suppliers covered by the Miller Act will receive payment for the work that they timely and properly perform.[129] It is important to note, however, that a surety's liability under the Miller Act bond is limited to the penal sum set forth on the bond's face.[130] Therefore, if the total amount of claims on a bond exceeds this penal amount, then the total pool of successful bond claimants will take a pro rata share of the penal sum. All other damages incurred in excess of this amount have to be pursued in a private breach of contract action against the prime or higher-tiered subcontractor, which is often judgment-proof or insolvent by the time this type of action arises.

The subcontractor's Miller Act rights do not accrue until 90 days after the date on which the labor or materials were last performed or furnished on the project.[131] Additionally, the Miller Act's protection extends to suppliers and subcontractors so long as the labor or material supplied by the supplier or subcontractor is considered "within the prosecution of the work."[132]

Where the Miller Act bond claimant has a direct contractual relationship only with a subcontractor but not with the prime contractor furnishing the payment bond (that is, it is a subcontractor or supplier at or below the second tier), then the bond claimant must provide written notice to the prime contractor within 90 days from the date on which the claimant last performed labor or last supplied materials.[133] The written notice must state with substantial accuracy the amount claimed and the name of the party to whom the material was supplied or for whom the labor was performed.[134] The courts consistently have construed this 90-day written notice as a condition precedent to a lower-tier party's right to sue under the Miller Act.[135]

Additionally, all lawsuits brought by subcontractors and lower-tier subcontractors/suppliers must be brought in the name of the United States "for the use and benefit of" the party actually suing in the federal district court where the contract work was performed. Suit must be brought no later than one year after the date on which the plaintiff last performed labor or provided materials to the project.[136] Failure to meet either this one-year period or, if applicable, the 90-day period for lower-tier subcontractors will act to bar suit on the bond.

Lastly, Congress has precluded a generic waiver of subcontractor or supplier payment bond rights under the Miller Act. The Miller Act expressly provides that "any waiver of the right to sue on the payment bond required [by the Miller Act] shall be void unless it is in writing, signed by the person whose right is waived, and executed after such person has first furnished labor or material for use in the performance of the contract."[137] This last requirement will render unenforceable any purported waiver of Miller Act rights in the subcontract itself.[138]

E. Material Breach of Contract for Nonpayment

Generally, a subcontractor is in the same contractual position regardless of whether the project is public or private. The only privity of contract the subcontractor has is with the prime contractor. With some exceptions, there is no privity of contract directly with the government. Therefore, a subcontractor generally is limited to pursuing a private contractual cause of action against a prime contractor for nonpayment. As an aside, under the doctrine of "first breach" or "clean hands," if the subcontractor has breached its subcontract first, such as a breach of the disputes provision, the subcontractor may be found to have "unclean hands," thereby preventing the subcontractor from pursuing a breach claim for nonpayment by the prime. This is because, under typical subcontract disputes provisions, the subcontractor must continue performance without delay during resolution of the dispute.

As discussed above, a subcontractor has a right under the Miller Act payment bond to bring suit against both the prime contractor and its surety.[139] Additionally, where a prime contractor fails to remit full payment for all work performed by the subcontractor, the contractor has failed to complete its contract performance and is in material breach of contract. In such an instance, the nonbreaching party, in this case the unpaid subcontractor, has additional remedies at law, including a breach of contract action. The merits of such an action turn upon a determination of whether that nonpayment breach is a material or minor breach of contract.[140]

Courts routinely distinguish between material and minor (that is, nonmaterial) breaches as follows:

> A minor breach may allow the aggrieved party to recover damages or a set-off against the breaching party, but it does not excuse the aggrieved party from performing. A material breach, on the other hand, does entirely discharge the aggrieved party's obligation to perform.
>
> A material, as a opposed to a minor, breach occurs when the non-breaching party did not receive the substantial benefit of its bargain. [Citation omitted.] In making this determination, courts assess the totality of the circumstances, including: 1) the extent to which a party's breach deprived the injured party of the benefit it reasonably expected; 2) the extent of the contract that the breaching party did perform; 3) the likelihood that the breaching party will cure its failure to perform, including any reasonable assurances of performance it did or did not give; and 4) the extent to which the behavior of the party failing to perform comported with standards of good faith and fair dealing.[141]

Among the various types of breaches that have been found to be "material" is nonpayment.[142] In such instances, where the sole material breach of contract is due to the owner's payment or lack thereof (or, as in the context of this chapter, a prime contractor's nonpayment of a subcontractor), that material breach of contract may permit a subcontractor to cease performance of its remaining work.

There are, however, limitations to this general holding. For example, a series of bankruptcy cases recognize that delay in making payment on disputed work alone is not a breach of contract.[143] These cases generally arise from situations where a subcontractor terminates performance for nonpayment of disputed amounts.

Most of these decisions conclude that in such a situation, nonpayment of disputed amounts did not constitute a material breach. The subcontractor's improper termination of its performance for these "nonmaterial" nonpayments, however, was a material breach. The material breaches by the subcontractor in turn precluded the subcontractor from seeking enforcement (that is, payment) of its subcontracts.[144] The courts concluded that, where a party fails to comply with the applicable disputes and claims clauses and walks off the job, such action constitutes a material breach of contract.[145]

Where the unpaid amount is an insignificant portion of the total contract price, nonpayment does not constitute a material breach.[146] While nonpayment of a progress payment may justify a subcontractor's suspension of performance, the amount of the deficiency is important in determining whether there was a material breach.[147]

VII. Claims and Disputes

A. Lack of Privity between the Subcontractor and Government— Enter the *Severin* Doctrine

As the definitions of subcontractor and subcontract establish, there typically is no direct contractual relationship between a subcontractor and the government. As a result of this lack of privity, a subcontractor usually is not permitted to directly pursue a contract claim it has against the government.[148] The prime contractor must submit any subcontractor claim.[149]

The *Severin* doctrine generally provides that a subcontractor is not permitted to seek and obtain recovery of a pass-through claim by the prime contractor against the government where the subcontractor has released the prime contractor, by contract or otherwise, from any liability to the subcontractor.[150] An extensive line of cases has narrowed this otherwise broad rule by finding that the government has the burden of proving that a sponsoring prime contractor has no liability to its subcontractor making the claim.[151]

In fact, this case law now limits the *Severin* doctrine to require an ironclad release or contract provision completely immunizing the prime contractor from liability to the subcontractor.[152] As a result, a prime contractor may "pass through" its subcontractor's claim to the government so long as the prime remains liable to the subcontractor. The mechanism to assert a subcontractor claim against the government often is referred to as a "liquidation," "sponsorship," or "pass-through" agreement. This agreement normally states that the prime contractor remains liable to its subcontractor to the same extent that the government is liable to the prime contractor. By use of this language, the parties establish that the sponsoring prime contractor remains and is liable to its subcontractor, thereby overcoming the *Severin* doctrine's limitations on government liability to the subcontractor.

Where a subcontractor signs a standard form release of liens and claims to obtain its monthly progress payment from the prime contractor, it releases the government from a claim in the prime contractor's name against the government if the effect of the release is to extinguish any rights it has against the prime contractor.[153]

B. Implied Contracts between the Government and the Subcontractor

In some rare instances an implied-in-fact contract may arise directly between the subcontractor and the government. An implied-in-fact government contract only will be found where both (1) the usual contract requirements of a mutually intended offer and acceptance, consideration, meeting of the minds, and definite terms exist and (2) the government official had the requisite contracting authority.[154]

In contrast to implied-in-fact contracts, implied-in-law contracts arise from instances in which equity demands that some relief be granted regardless of what the parties actually intended.[155] The courts lack authority, however, to consider and make awards under an implied-in-law contract against the government.[156]

Under limited circumstances, subcontractors have been able to recover for unjust enrichment or quantum meruit for work performed for which the government received a benefit even though no "formal" contract exists (that is, such as where a subcontractor performs change order work but no written change order was issued).[157] The courts have permitted subcontractors to recover their reasonable damages under the implied-in-fact contract doctrine. For such a recovery to occur, however, the subcontractor must establish some degree of contractual privity between itself and the government.[158] Such privity may be established through an assignment of a subcontract by a prime contractor to the government or where the subcontractor and government enter into direct claim-settlement discussions.[159]

C. The Small Business Act 8(a) Program

Under the Small Business Act 8(a) program,[160] the general contractor performing the work literally is a subcontractor. The Small Business Administration (SBA) acts as the prime contractor for purposes of contract award. In such instances, the boards view set-aside contracts as establishing an intent both by the procuring agency and the SBA that the subcontractor is the party that will perform the contract work. Therefore, the 8(a) contractor is the real party in interest. This is sufficient to establish the necessary privity of contract with the agency.[161]

VIII. Conclusion

As the foregoing readily establishes, the typical complexities associated with private construction subcontracting are amplified significantly when one enters the federal construction contracting realm. While the overarching themes associated with both public and private construction contracts are the same, by entering into a federal construction contract, a subcontractor subjects itself to a number of additional requirements such as prevailing wage requirements, waiver of mechanic's liens, the *Severin* doctrine, and other issues unique to federal contracting. While posing a daunting task, the opportunities presented on federal construction contracts to subcontractors are significant, often represented by some of the largest construction contracts, and in turn subcontracts, in a given region.

Critical to the success of a project are both the subcontractor's continuous and ongoing dialogue with the prime contractor and the subcontractor's intimate knowledge of, and compliance with, the subcontract and prime contract documents. Project success depends on these types of issues and knowledge. Of course, the same can, and should, be said about private construction projects. With involvement of the federal government, however, these basic keys to success are amplified several times over.

Notes

1. Blake Constr. Co., Inc. v. C.J. Coakley Co., Inc., 431 A.2d 569, 575 (D.C. 1981).

2. BLACK'S LAW DICTIONARY 1437 (7th ed. 1999).

3. FAR 3.502-1; *accord* 48 C.F.R. § 44.101 ("'subcontractor' means any supplier, distributor, vendor or firm that furnishes supplies or services to or for a prime contractor or another subcontractor").

4. FAR 44.101; *see also* FAR 19.701.

5. FAR 52.244-2(a).

6. *See* discussion in subsection VI.D of this chapter, and Chapter 23 for a detailed discussion of the Miller Act.

7. Clifford F. MacEvoy Co. v. United States *ex rel.* Calvin Tomkins Co., 64 S. Ct. 890, 894 (1944) (discussing difference between materialman and subcontractor for purposes of Miller Act coverage); *see also* J.W. Bateson Co. v. United States *ex rel.* Bd. of Trs. of Nat'l Automatic Sprinkler Indus. Pension Fund, 98 S. Ct. 873, 434 U.S. 586 (1978); F.D. Rich Co. v. United States ex rel. Indus. Lumber Co., Inc., 417 U.S. 116 (1974).

8. Richmond Eng'g Co., IBCA No. 426, Oct. 1, 1964, 64 BCA ¶ 4465.

9. *Id.*

10. *See, e.g.,* FAR 44.203(b)(4) (precluding a CO from consenting to a subcontract that requires that any judgment in an arbitration, judicial dispute, or other ADR type of matter or voluntary settlement subject the government to be bound thereto).

11. For discussion of the *Severin* doctrine, see Section VII, *infra.*

12. FAR 44.201-1; FAR 52.244-2. Consent is needed for cost-reimbursement, time-and-materials, labor-hour, and letter contracts. FAR 44-201-1(b).

13. *Id.*

14. FAR 44.201-1(b)(2)(i).

15. FAR 44.201-1(b)(2)(ii).

16. FAR 44.201-1(b).

17. FAR 44.201-1(c). For further discussion of government A-E contracts, see Chapter 4.

18. FAR 44.203(a).

19. FAR 44.203.

20. FAR 44.203(c) (such a clause may not, however, seek to "obligate the contracting officer or the appeals board to decide questions that do not arise between the Government and the prime contract or that are not cognizable under the clause at § 52.233-1, Disputes").

21. FAR 44.201-1(a).

22. FAR 44.101.

23. *Id.,* such as in the case of critical, complex, secure or similar subcontract scopes of work.

24. FAR 44.201-1(a).

25. FAR 44.201-2.

26. FAR 44.202-1(a). For further discussion of debarments from contracting with the government, *see* Chapter 7.

27. *Id.* at (b).

28. For further discussion of requirements under these government socio-economic programs, *see* Chapter 12.

29. FAR 44.202-2(a).

30. *Id.* at (b).

31. FAR 52.244-5.

32. FAR 44.204(c).

33. Stephenson Assocs., Inc., GSBCA Nos. 6573, 6815, May 22, 1986, 86-3 BCA ¶ 19,071.

34. Ala. Dry Dock & Shipbuilding Co., ASBCA No. 39215, Apr. 13, 1990, 90-2 BCA ¶ 22,855; Cascade Elec. Co., ASBCA No. 28674, Feb. 28, 1984, 84-1 BCA ¶ 17,210.

35. *Id.* For further discussion of the government's warranty of specifications under the *Spearin* doctrine, *see* Chapter 18.

36. Omneco, Inc., B-218343 et al., 85-1 CPD ¶ 660 (citing 48 C.F.R. § 9.103(c)); for further discussion of responsibility and responsiveness, *see* Chapter 6.

37. *Id.*, citing 48 C.F.R. § 9.104-4(b).

38. Myers Investigative Serv., B-286971.2 et seq., 2001 CPD ¶ 59 at 4, citing Premiere Cleaning Sys., Inc., B-249179.2, 92-2 CPD ¶ 298 at 4; Commercial Bldg. Servs., Inc., B-237865.3, 90-1 CPD ¶ 473 at 6.

39. For further discussion of government small and disadvantaged business programs applicable to federal construction, see Chapter 12.

40. FAR 19.702(a).

41. FAR 19.704. The U.S. Small Business Administration's various small disadvantaged business programs, such as the Section 8(a), HUBZone, and others, are located in Title 13 of the Code of Federal Regulations.

42. FAR 19.702(b).

43. The relevant FAR section defines small business as including small business, veteran-owned small business, service-disabled veteran-owned small business, HUBZone small business, small disadvantaged business, and women-owned small business concerns. FAR 19.702.

44. FAR 19.702.

45. FAR 19.702(c).

46. *See generally* CORBIN ON CONTRACTS, §§ 1.2 *et seq.* (2001 Supp.).

47. RESTATEMENT (SECOND) OF CONTRACTS § 1.

48. *Id.* at § 17.

49. 64 F.2d 344 (2d Cir. 1933); Allied Comm'c'ns Corp. v. Cont'l Cellular Corp., 821 F.2d 69 (1st Cir. 1987).

50. *Id.* (citing RESTATEMENT (SECOND) OF CONTRACTS § 35).

51. *See generally* John Price Assocs., Inc. v. Warner Elec., Inc., 723 F.2d 755 (10th Cir. 1983); Alaska Bussell Elec. v. Vern Hickel Constr., 688 P.2d 576 (Alaska 1984); Haselden Langley Constructors, Inc. v. D.E. Farr Assocs., 676 P.2d 709 (Colo. App. 1983); *but see* Home Elec. v. Hall & Underdown Heating & Air Conditioning, 358 S.E.2d 539 (N.C. App. 1987).

52. 333 P.2d 757 (Cal. 1958).

53. *Id.; see generally* Preload Tech., Inc. v. A.B.& J. Constr. Co., 696 F. 2d 1080 (5th Cir. 1983); *John Price*, 723 F.2d 755.

54. For an excellent discussion of the "firm bid rule," which is now being called "detrimental reliance" in some jurisdictions, albeit in nonfederal contract situations, *see* Pavel Enters., Inc. v. A.S. Johnson Co., Inc., 342 Md. 143, 674 A.2d 521 (Md. 1996); Citiroof Corp. v. Tech Contracting Co., 159 Md. App. 578, 588, 860 A.2d 425, 432 (Md. App. 2004).

55. *John Price*, 723 F.2d 755.

56. *Drennan v. Star Paving*, 333 P.2d 757 (Cal. 1958).

57. *Id.*

58. FAR 14.407-1.

59. FAR 14.301(a) ("To be considered for award, a bid must comply in all material respects with the invitation for bids. Such compliance enables bidders to stand on equal footing and maintaining integrity of the sealed bidding system."); FAR 14.404-2(A), (B) ("Any bid that does not conform to the applicable specifications shall be rejected.").

60. Nextira Fed., LLC, B-290820, B-290820.2, 2002 U.S. Comp. Gen. LEXIS 147 (Oct. 4, 2002), citing Techseco, Inc., B-284949, 2000 CPD ¶ 105; ASOMA Instruments, Inc., B-251674, 93-1 CPD ¶ 317.

61. *See* FAR 52.214-15, 52.214-16.

62. *Id.*

63. Discussed earlier in this subsection.

64. *See, e.g.,* Preload Tech., Inc. v. A.B.& J. Constr. Co., 696 F.2d 1080 (5th Cir. 1983).

65. IKG, 85-2 CPD ¶ 159, citing AME Matext Corp., B-218588.2, 85-1 CPD ¶ 704; PolyCon Corp., B-218162, 85-1 CPD ¶ 593; PolyCon Corp., B-218304 et al., 64 Comp. Gen. 523, 85-1 CPD ¶ 567; *see also* Gulie Research Labs., Inc., B-219370, 85-2 CPD ¶ 185.

66. 4 C.F.R. § 21.1(a).

67. 4 C.F.R. § 21.0(a).

68. Omneco, Inc., B-218343 et al., 85-1 CPD ¶ 660, citing 4 C.F.R. § 21.3(f)(10).

69. 4 C.F.R. § 21.5(h).

70. RGB Display Corp., B-284699, 2000 CPD ¶ 80, citing St. Mary's Hosp. & Med. Ctr. of S.F., Cal., B-243061, 91-1 CPD ¶ 597; Univ. of Mich. and Indus. Training Sys. Corp., B-225756 et al., 87-1 CPD ¶ 643 at 5–6.

71. *Id.*

72. FAR 44.203(b)(3).

73. FAR 44.203(c).

74. *See, e.g.,* Boeing Co., ASBCA No. 30404, Aug. 18, 1986, 86-3 BCA ¶ 19,314 (board lacked jurisdiction to consider subcontractor's claim that prime contractor improperly withheld progress payment). For a more thorough discussion of a subcontractor's rights against the government, see subsection VII below.

75. Wash. Metro. Area Transit Auth. *ex rel.* Noralco Corp. v. Norair Eng'g Corp. and Employees Commercial Union Ins. Co., 553 F.2d 233, 23 CCF ¶ 81,164 (D.C. Cir. 1977) (citing John W. Johnson, Inc. v. Basic Constr. Co., 429 F.2d 764, 774–76 (D.C. Cir. 1970) (subcontractor not bound by prime contract disputes clause because subcontract failed to adequately incorporate said disputes clause by reference); Beacon Constr. Co. of Mass., Inc. v. Prepakt Concrete Co., 375 F.2d 977, 12 CCF ¶ 81,043 (1st Cir. 1967) (subcontractor's mere notice of prime contract disputes/arbitration clause under prime contract with U.S. Postal Service inadequate to force subcontractor to submit to same clause).

76. *See* G. Chrysler Corp. v. Brown, 441 U.S. 281, 295–96, 99 S. Ct. 1705 (1997); *Splane v. West*, 216 F.3d 1058, 1065 (Fed. Cir. 2000).

77. S.J. Amoroso Constr. Co., Inc. v. United States, 12 F.3d 1072, 1075 (Fed. Cir. 1993) (citing G.L. Christian & Assocs. v. United States, 312 F.2d 418, *aff'd on reh'g,* 320 F.2d 345, 160 Ct. Cl. 58 (1963); Gen. Eng'g & Mach. Works v. O'Keefe, 991 F.2d 775, 779 (Fed. Cir. 1993); SCM Corp. v. United States, 645 F.2d 893, 903–04, 227 Ct. Cl. 12 (1981); Chris Berg, Inc. v. United States, 426 F.2d 314, 317, 192 Ct. Cl. 176 (1970). For further discussion of the *Christian* doctrine, see Chapter 7.

78. FAR 52.222-11 as required by FAR 22.406-5.

79. As required by FAR 22.406-6 and found at FAR 52.222-8.

80. *See* FAR 44.204.

81. FAR 52.203-7.
82. FAR 52.203-11.
83. FAR 52.203-12.
84. FAR 52.204-2.
85. FAR 52.214-26.
86. FAR 52.214-27.
87. FAR 52.214-28.
88. FAR 52.222-6.
89. FAR 52.222-4.
90. FAR 52.222-9.
91. FAR 52.228-8.
92. FAR 52.222-10.
93. FAR 52.222-7.
94. FAR 52.222-11.
95. FAR 52.222-12.
96. FAR 52.222-14.
97. FAR 52.222-13.
98. FAR 52.222-15.
99. FAR 52.222-26.
100. FAR 52.222-25.
101. FAR 52.222-36.
102. FAR 52.222-41.
103. FAR 52.230-2.
104. FAR 52.230-3.
105. FAR 52.232-27.
106. *See, e.g.,* K.L. Conwell Corp., EBCA No. 399-10-87, Apr. 6, 1988, 88-2 BCA ¶ 20,712.
107. *See* FAR 32.102, 32.103, 32.111, 52.232-5.
108. FAR 32.103, 52.232-5(e).
109. FAR 32.103.
110. FAR 52.232-5(b).
111. FAR 52.232-5(c)(2).
112. 31 U.S.C. § 3902(d)(3).
113. *Id.; see also* 41 U.S.C. § 611.
114. *Id.;* FAR 52.232-27(d)(2).
115. FAR 52.232-27(c).
116. FAR 52.232-27(d)(1).
117. FAR 52.232-27(d)(2).
118. FAR 52.232-27(i).
119. FAR 52.232-27(k).
120. United States *ex rel.* Maris Equip. Co., Inc. v. Morganti, Inc., 163 F. Supp. 2d 174, 197 (E.D. N.Y. 2001); 31 U.S.C. § 3905(b), (g).
121. 31 U.S.C. § 3907(c); L & A Jackson Enters. v. United States, 38 Fed. Cl. 22, 44–45 (1997); *see also,* FAR 32.904(d)(iii)(B)(i).
122. FAR 52.232-27. Under the circumstances where a dispute exists, while partial payments may be made on the undisputed amounts, the government is

permitted to withhold payments on the disputed portions and the contractor, or if applicable, subcontractor, would be required to file as a dispute under the FAR. FAR 52.233-1.

123. BLACK'S LAW DICTIONARY 935 (7th ed. 1999).

124. 40 U.S.C. §§ 3131 *et seq.* (formerly at 40 U.S.C. § 270a–f).

125. *Id.*

126. Maris Equip. Co., 163 F. Supp. 2d 174; United States *ex rel.* Va. Beach Mech. Servs., Inc. v. Samco Constr. Co., 39 F. Supp. 2d 661, 670 (E.D.Va. 1999) ("Under the Miller Act, a subcontractor in a Federal government project may bring actions against the project's general contractor and surety for claims surety in contract or quasi-contract").

127. Skip Kirchdorfer, Inc. v. United States, 6 F.3d 1573 (Fed. Cir. 1993); United States *ex rel.* Wulff v. CMA, Inc., 890 F.2d 1070 (9th Cir. 1989); United Bonding Ins. Co. v. Catalytic Constr. Co., 533 F.2d 469 (9th Cir. 1976).

128. *See, e.g.,* United States *ex rel.* Maddux Supply Co. v. St. Paul Fire & Marine Ins. Co., 86 F.3d 332 (4th Cir. 1996); United States ex rel. Bryant Elec. Co. v. Aetna Cas. & Sur. Co., 297 F.2d 665 (2d. Cir. 1962); United States *ex rel.* DDC Interiors, Inc. v. Dawson Constr. Co., Inc., 895 F. Supp. 270 (D. Col. 1995), *ed.* 82 F.3d 427 (10th Cir. 1996).

129. Aetna Cas. & Sur. Co. v. United States, 526 F.2d 1127 (Ct. Cl. 1975), *cert. denied,* 425 U.S. 973 (1976); *DDC Interiors,* 895 F. Supp. 270.

130. Transamerica Premier Ins. Co. v. Ober, 894 F. Supp. 471, 474 n.4 (D. Maine 1995).

131. 40 U.S.C. § 3133(a) (formerly 40 U.S.C. § 270b(a)).

132. *Id.;* United States *ex rel.* Lab. Furniture Co. v. Reliance Ins. Co., 274 F. Supp. 377 (D. Mass. 1967); United States f/u/b/o Luis A. Cabrera, S.E. v. Sun Eng'g Enters., Inc., 817 F. Supp. 1009 (D.P.R. 1993).

133. 40 U.S.C. § 3133(a) (formerly 40 U.S.C. § 270b(a)).

134. *Id.*

135. Fleischer Eng'g & Constr. Co. v. United States *ex rel.* Hallenback, 311 U.S. 15 (1940); United States *ex rel.* Am. Builders & Contractors Supply Co. v. Bradley Constr. Co., 960 F. Supp. 145 (N.D. Ill. 1997) (liberal interpretation of Miller Act goes to remedy, not 90-day notice of this period, and such written notice to general contractor and performance payment bond surety is a strict condition precedent).

136. 40 U.S.C. § 3133(b)(4).

137. 40 U.S.C. § 3133(c); *see also* 40 U.S.C. § 3134 (discussion of certain limited waivers of the Miller Act under various scenarios by various department secretaries).

138. *See, e.g.,* United States of America f/u/b/o DDC Interiors, Inc. v. Dawson Constr. Co., Inc., 895 F. Supp. 270, 272 et seq. (D. Co. 1995); *but see* United States f/u/b/o Trans Coastal Roofing Co., Inc. v. David Boland, Inc., 922 F. Supp. 597, 598 (S.D. Fl. 1996).

139. United States *ex rel.* Coastal Steel Erectors, Inc. v. Algernon Blair, Inc., 479 F.2d 638, 640 (4th Cir. 1973); United States *ex rel.* B&M Roofing of Colo., Inc. v. AKM Assocs., Inc., 961 F. Supp. 1441, 1443–45 (D. Colo. 1997).

140. United States *ex rel.* Va. Beach Mech. Servs., Inc. v. Samco Constr. Co., 39 F. Supp. 2d 661 (E.D. Va. 1999).

141. Samco Constr., 39 F. Supp. 2d at 970, citing Stone Forest Indus., Inc. v. United States, 973 F.2d 1548, 1552 (Fed. Cir. 1992); United States *ex rel.* Cortolano & Barone, Inc. v. Morano Constr. Corp., 724 F. Supp. 88, 98–99 (S.D.N.Y. 1989).

142. *Id.* at 670.

143. *In re* Fordson Eng'g Corp. v. Gen. Motors Corp., 25 B.R. 506, 510 (E.D. Mich. 1982), citing In re Northup-Johnson, Inc., 15 B.R. 767 (D. Md. 1981); *In re* 1616 Remind L.P., 14 B.R. 484 (E.D. Va. 1981); *see also In re* Cornell & Co., 229 B.R. 97, 103 (E.D. Pa. 1999); *In re* Jandous Elec. Constr. Corp, v. Slattery Assocs., Inc., 1989 WL 81139 (S.D.N.Y. Mar. 24, 1989).

144. *Id.; see also* McGee Constr. Co. v. Neshobe Dev., Inc., 594 A.2d 415, 418 (Vt. 1991).

145. Stewart v. C&C Excavating & Constr. Co., 877 F.2d 711, 714 (8th Cir. 1989); citing United States *ex rel.* Aucoin Elec. Supply Co. v. Safeco Ins. Co. of Am., 555 F.2d 535, 541 (5th Cir. 1977).

146. *Id.*

147. *Stewart*, 877 F.2d at 714; *see also* United States v. S. Constr. Co., 293 F.2d 493, 428 (6th Cir. 1961) (breach relied upon must be a breach that is so substantial and fundamental as to go to the very root of the contract).

148. Severin v. United States, 99 Ct. Cl. 435 (1943), *cert. denied*, 322 U.S. 733 (1944).

149. Beacon Constr. Co. of Mass., Inc. v. Prepakt Concrete Co., 375 F.2d 977 (1st Cir. 1967); quoting Cent. Steel Erection Co. v. Will, 304 F.2d 548, 551 (9th Cir. 1962); Fanderlik-Locke Co. v. United States for the Use of Morgan, 285 F.2d 939, 942 (10th Cir.), *cert. denied*, 365 U.S. 860 (1961); Raytheon Co. Elec. Sys. Div., ASBCA No. 50221, Dec. 6, 1996, 97-1 BCA ¶ 28,663.

150. *Id.*

151. S. Constr. Co. v. United States, 364 F.2d 439 (Ct. Cl. 1966); Folk Constr. Co. v. United States, 2 Cl. Ct. 681 (1983); Strand Hunt Constr. v. Gen. Servs. Admin., GSBCA No. 12860, Jan. 30, 1996, 96-1 BCA ¶ 28,185; Planning Research Corp. v. Dept. of Commerce, GSBCA No. 11286-COM, Nov. 14, 1993, 94-1 BCA ¶ 26,566, *on further proceedings*, Sept. 26, 1995, 96-1 BCA ¶ 27,954; Systemhouse Fed. Sys., Inc., GSBCA No. 9446-C, Apr. 26, 1989, 89-2 BCA ¶ 21,773.

152. Planning Research Corp., 94-1 BCA ¶ 26,566, quoting Cross Constr. Co. v. United States, 225 Ct. Cl. 616, 618 (1980).

153. George Hyman Constr. Co. v. United States, 30 Fed. Cl. 170 (1993), *aff'd*, 39 F.3d 1197 (Fed. Cir. 1994).

154. E. Trans-Waste of Md., Inc. v. United States, 27 Fed. Cl. 146 (1992), citing H. Landau & Co. v. United States, 886 F.2d 322, 324 (Fed. Cir. 1989); Cal. Sand & Gravel, Inc. v. United States, 22 Cl. Ct. 19, 26–27 (1990), *aff'd*, 937 F.2d 624 (Fed. Cir. 1991), *cert. denied*, 112 S. Ct. 934 (1992).

155. *Id.*

156. *Id.*, citing Merritt v. United States, 267 U.S. 338, 341 (1925); Lopez v. A.C. & S., Inc., 858 F.2d 712, 714 (Fed. Cir. 1988).

157. For a general discussion of this issue, see *E. Trans-Waste*, 27 Fed. Cl. 159.

158. Kaiser Marquardt, ASBCA Nos. 49800, 50177, Jan. 11, 1999, 99-1 BCA ¶ 30,216 (subcontractor unsuccessfully alleged existence of three-way oral agreement between government, prime contractor, and subcontractor).

159. *See* Sandblasting & Coating, Inc., ASBCA Nos. 30873, 35517, Mar. 9, 1989, 89-3 BCA ¶ 21,892; McMillin Bros. Constructors, Inc., EBCA No. 328-10-84, July 11, 1986, 86-3 BCA ¶ 19,179.

160. For further discussion of government small business programs, including the 8(a) program, see Chapter 12.

161. *See, e.g.,* Decorama Painting, Inc., ASBCA No. 25299, Feb. 11, 1981, 81-1 BCA ¶ 14,992; Kyle Eng'g Co., ASBCA No. 25168, Feb. 19, 1981, 81-1 BCA ¶ 14,990.

CHAPTER 23

Surety Bonds in Federal Contracting

DANIEL E. TOOMEY
TAMARA M. McNULTY

I. *Legal Requirements for Bid, Payment, and Performance Bonds*

Since 1847 and the passage of the Heard Act, the government has required contractors on federal projects to be bonded. Much later, during the administration of Franklin Delano Roosevelt, Congress enacted the Miller Act.[1] This Act governs performance and payment bonds, and it has been amended a number of times. The most recent amendments, the Construction Industry Payment Protection Act of 1999,[2] made two major changes in the Miller Act.

First, Miller Act payment bonds must equal the contract amount unless the contracting officer (CO) determines in writing that this is impractical, in which case the amount of the bond may be less, but not less than the amount of the performance bond. Second, prime contractors cannot require subcontractors to waive their rights to sue under the payment bond *prior to* their supplying labor and materials. Finally, in rules associated with this legislation, the Federal Acquisition Regulation (FAR) requires that, in contracts between $30,000 and $100,000, payment bonds must be in an amount equal to 100 percent of the contract amount.[3]

The current version of the Miller Act provides:

Bonds of contractors of public buildings or works.
    (a) Definition.—In this subchapter, the term "contractor" means a person awarded a contract described in subsection (b).
    (b) Type of bonds required.—Before any contract of more than $100,000 is awarded for the construction, alteration or repair of any public building or public work of the Federal Government, a person must furnish to the Government the following bonds, which become binding when the contract is awarded:
        (1) Performance bond.—A performance bond with a surety satisfactory to the officer awarding the contract, and in an amount as the officer considers adequate, for the protection of the Government.

(2) Payment bond.—A payment bond with a surety satisfactory to the officer for the protection of all persons supplying labor and material in carrying out the work provided for in the contract for the use of each person. The amount of the payment bond shall equal the total amount payable by the terms of the contract unless the officer awarding the contract determines, in a writing supported by specific findings, that a payment bond in that amount is impractical, in which case the contracting officer shall set the amount of the payment bond. The amount of the payment bond shall not be less than the amount of the performance bond.[4]

A. Bid Bonds

1. Law

The Miller Act does not address bid bonds, but they are required by the FAR. FAR 28.101 specifies when a bid guarantee is required:

28.101-1 **Policy on use**.

(a) A contracting officer shall not require a bid guarantee unless a performance bond or a performance and payment bond is also required (see 28.102 and 28.103). Except as provided in paragraph (c) of this subsection, bid guarantees shall be required whenever a performance bond or a performance and payment bond is required.

(b) All types of bid guarantees are acceptable for supply or service contracts (see annual bid bonds and annual performance bonds coverage in 28.001). Only separate bid guarantees are acceptable in connection with construction contracts. Agencies may specify that only separate bid bonds are acceptable in connection with construction contracts.

(c) The chief of the contracting office may waive the requirement to obtain a bid guarantee when a performance bond or a performance and payment bond is required if it is determined that a bid guarantee is not in the best interest of the Government for a specific acquisition (e.g., overseas construction, emergency acquisitions, sole-source contracts). Class waivers may be authorized by the agency head or designee.[5]

B. Performance Bonds

1. Law

For most federal construction projects, the Miller Act requires "[a] performance bond with a surety satisfactory to the officer awarding the contract in an amount the officer considers adequate, for the protection of the Government."[6]

2. *Regulations*

FAR 28.102 governs performance bonds and discusses what types of security the CO may require for performance. For example, the FAR provides:

28.102-1 **General**.

   (a) The Miller Act (40 U.S.C. 3131 *et seq.*) requires performance and payment bonds for any construction contract exceeding $100,000, except that this requirement may be waived—

   (1) *By the contracting officer for as much of the work as is to be performed in a foreign country upon finding that it is impracticable for the contractor to furnish such bond; or*

   (2) As otherwise authorized by the Miller Act or other law.[7] [Emphasis added.]

The FAR also contains requirements regarding performance bond amounts. It states:

   (a) **Definition**. As used in this subsection—
   "Original contract price" means the award price of the contract; or, for requirements contracts, the price payable for the estimated total quantity; or, for indefinite-quantity contracts, the price payable for the specified minimum quantity. Original contract price does not include the price of any options, except those options exercised at the time of contract award.
   (b) Contracts exceeding $100,000 (Miller Act)—
   (1) Performance bonds. Unless the contracting officer determines that a lesser amount is adequate for the protection of the Government, the penal amount of performance bonds must equal—
   (i) 100 percent of the original contract price; and
   (ii) If the contract price increases, an additional amount equal to 100 percent of the increase.[8]

Finally, FAR Part 49 sets out some of the rights of sureties in the event of a default of its principal.

C. Payment Bonds

1. *Law*

Section 3131 of the Miller Act requires for federal construction:

[a] payment bond with a surety satisfactory to the officer for the protection of all persons supplying labor and material in carrying out the work provided for in the contract for the use of each person. The amount of the payment bond shall equal the total amount payable by the terms of the contract unless the officer awarding the contract determines, in writing supported by specific findings, that a payment

bond in that amount is impractical, in which case the contracting officer shall set the amount of the payment bond. The amount of the payment bond shall not be less than the amount of the performance bond.[9]

2. *Regulations*

The FAR addresses waiver of the payment bond as follows:

(b) . . . .

(1) Pursuant to 40 U.S.C. 3132, for construction contracts greater than $30,000, but not greater than $100,000, the contracting officer shall select two or more of the following payment protections, giving particular consideration to inclusion of an irrevocable letter of credit as one of the selected alternatives:

(i) A payment bond.

(ii) An irrevocable letter of credit (ILC).

(iii) A tripartite escrow agreement. The prime contractor establishes an escrow account in a federally insured financial institution and enters into a tripartite escrow agreement with the financial institution, as escrow agent, and all of the suppliers of labor and material. The escrow agreement shall establish the terms of payment under the contract and of resolution of disputes among the parties. The Government makes payments to the contractor's escrow account, and the escrow agent distributes the payments in accordance with the agreement, or triggers the disputes resolution procedures if required.

(iv) Certificates of deposit. The contractor deposits certificates of deposit from a federally insured financial institution with the contracting officer, in an acceptable form, executable by the contracting officer.

(v) A deposit of the types of security listed in 28.204-1 and 28.204-2.

(2) The contractor shall submit to the Government one of the payment protections selected by the contracting officer.

(c) The contractor shall furnish all bonds or alternative payment protection, including any necessary reinsurance agreements, before receiving a notice to proceed with the work or being allowed to start work.[10]

Additionally, the FAR contains requirements regarding payment bond amounts and when those amounts must be increased:

(2) Payment bonds.

(i) Unless the contracting officer makes a written determination supported by specific findings that a payment bond in this amount is impractical, the amount of the payment bond must equal—

(A) 100 percent of the original contract price; and

(B) If the contract price increases, an additional amount equal to 100 percent of the increase.

(ii) The amount of the payment bond must be no less than the amount of the performance bond.

(c) Contracts exceeding $30,000 but not exceeding $100,000. Unless the contracting officer determines that a lesser amount is adequate for the protection of the Government, the penal amount of the payment bond or the amount of alternative payment protection must equal—

(1) 100 percent of the original contract price; and

(2) If the contract price increases, an additional amount equal to 100 percent of the increase.[11]

II. *Operation of Bid Bonds under Federal Law*

A. Technical Requirements: Form over Substance

Bidders on federal projects (and the sureties that supply the bid forms) must be particularly cognizant of the FAR requirements and case law interpreting those regulations, particularly the decisions of the Government Accountability Office (GAO) relating to the adequacy of the form of bid bonds that are submitted, as they are required to be, at the time of bidding.

1. *Need for Surety to Be Approved*

There are two kinds of security acceptable to the government: (1) individual or corporate sureties; or (2) security of the type prescribed by the FAR.[12] While the government will accept individual sureties who can demonstrate the financial wherewithal and balance sheet to adequately protect the government's financial interests, by far the most common form of bonds are those provided by corporate sureties. Companies that write such surety bonds must qualify to do so, and evidence of their qualification can be found in the so-called "Treasury List." The Treasury List, also known as Department of Treasury 570,[13] is constantly updated. In the wake of recent failures of a number of sureties, contractors need to be aware that government COs have the right to require contractors whose sureties have failed to secure bonds from other acceptable sureties.[14]

2. *Need for Bid Bonds to Meet All Technical Requirements to Avoid Having Bid Held Nonresponsive*

Most authority interpreting the validity of bid bonds and the effect of defects in these forms may be found in GAO decisions. These decisions state that failure to meet one or more of the numerous literal requirements of the form of bond documents will result in having an otherwise acceptable bid declared nonresponsive, and, where an agency attempted to waive such a failure, these decisions have nullified what otherwise would have been legitimate awards.

The test for the validity of the bond is whether, in looking at the bond documents *alone*, there is any doubt as to whether they are enforceable against the surety.[15] There are innumerable ways in which bond documents can be found defective. Some examples are: (1) the failure of the surety's authorized representative to sign the bond despite the fact that it is sealed, has the proper name of the surety and the proper amount, and is accompanied by a proper power of attorney form authorizing the agent to execute the bond;[16] (2) any alteration to the bid bond without evidence thereon of the surety's authorization;[17] and (3) use of facsimiles without original signatures in lieu of an original bond form.[18]

One case illustrates the point. In *Adrian L. Merton v. United States*,[19] the low bidder on an Army Corps of Engineers project had its bid declared nonresponsive merely because the power of attorney document accompanying the actual bond was photocopied and not an original or facsimile copy. The purpose of this document is to demonstrate that the surety has authorized its "attorney-in-fact," usually certain employees of a bond broker, to sign and issue bonds on its behalf. There was no question that the "attorneys-in-fact" listed in the power were authorized to act on behalf of the surety, nor was there any defect in the bid bond itself, which was signed with a "wet signature" by an agent-in-fact of the surety. While the court's decision was primarily based on the fact that the project was of a short duration and had already been awarded and that requiring a change in the contractor was not in the best interests of the public, the court upheld the Corps' finding that Merton's bid was nonresponsive, stating:

> The contracting agency's actions here in determining plaintiff's bids nonresponsive were not without a rational basis. A clear line of decisions by the General Accounting Office (GAO) holds that photocopied powers of attorney are unacceptable in support of bid bonds. When actions of procurement officials have been expressly validated by considered decisions of the Comptroller General or are in compliance with a reasonably consistent pattern of Comptroller General's determinations, the courts should be extremely reluctant to overturn such actions. *Kinnett Dairies, Inc. v. Farrow*, 580 F.2d 1260, 1272 (5th Cir. 1978) (citing *M. Steinthal & Co.*, 455 F.2d at 1300). *See also Shoals American Industries, Inc. v. United States*, 877 F.2d 883 (11th Cir. 1989). Thus, "[i]t would be entirely justifiable for the contracting officer to follow the general policy of acceding to the views of the [GAO] in [the area of competitive bidding] even though he had another position on the particular issue of legality or propriety." *John Riener & Co. v. United States*, 325 F.2d 438, 442 (Ct. Cl. 1963), *cert. denied*, 377 U.S. 931 (1964).[20]

Since *Merton*, however, the FAR has addressed some of these "traps." For example, the FAR now states:

Authority of an attorney-in-fact for a bid bond.

(a) Any person signing a bid bond as an attorney-in-fact shall include with the bid bond evidence of authority to bind the surety.

(b) An original, *or a photocopy or facsimile of an original, power of attorney is sufficient evidence of such authority*.

(c) For purposes of this section, *electronic, mechanically-applied and printed signatures, seals and dates on the power of attorney shall be considered original signatures, seals and dates, without regard to the order in which they were affixed*.

(d) The contracting officer shall—

(1) Treat the failure to provide a signed and dated power of attorney at the time of bid opening as a matter of responsiveness; and

(2) Treat questions regarding the authenticity and enforceability of the power of attorney at the time of bid opening as a matter of responsibility. These questions are handled after bid opening.

(e)

(1) If the contracting officer contacts the surety to validate the power of attorney, the contracting officer shall document the file providing, at a minimum, the following information:

(i) Name of person contacted.

(ii) Date and time of contact.

(iii) Response of the surety.

(2) If, upon investigation, the surety declares the power of attorney to have been valid at the time of bid opening, the contracting officer may require correction of any technical error.

(3) If the surety declares the power of attorney to have been invalid, the contracting officer shall not allow the bidder to substitute a replacement power of attorney or a replacement surety.

(f) Determinations of non-responsibility based on the unacceptability of a power of attorney are not subject to the Certificate of Competency process of subpart 19.6 if the surety has disavowed the validity of the power of attorney.[21] [Emphasis added.]

In short, while the wet signature problem is no longer an issue, the failure to sign the bond, as in the *Kennedy Electric Company* case, still renders the bid nonresponsive.[22]

3. *Commercial Form Valid Only if It Provides the Same Protection as the Standard Form*

Standard Form (SF) 24, set forth in the FAR, is the acceptable form of Miller Act bid bond.[23] This standard does not necessarily mean that use of another commercial form will invalidate the bid. Rather, the test is whether the bond form provides essentially the same protection to the government as the SF 24 and/or in the solicitation.[24] Needless to say, bidders and sureties are well advised to use the standard form.

4. Waiver of Defects

The FAR does provide certain instances when form defects may be waived. Such instances include (1) where there is only one bidder; (2) a bid guarantee, though less than called for, is in an amount that is greater than the next-highest bid; (3) lateness in filing of the bid bond in certain instances; (4) erroneous or blank date on the bid bond form; and (5) failure to list the United States as obligee, though correctly listing the project solicitation.[25]

Both bidders and sureties owe it to themselves to carefully review the bid bond documents at the time they are submitted for any possible defects, correcting them before they are submitted. Correspondingly, unsuccessful bidders are well advised to ask for and review the bid documents submitted by successful bidders since defects may render awards to those bidders invalid.

B. Responsibility of the Bid Bond Surety in Event of Default

The responsibilities of the surety under a bid bond are separate and distinct from those under performance and payment bonds. The purpose of a bid bond is to ensure that, in the event that a principal refuses to enter into the contract on which it successfully bid, the government is protected for the additional costs of reprocurement.

As with all bonds, the bid bond surety's obligation is limited to the penal sum of the bond. Typically, such bonds are limited to a fraction of the total bid amount, such as 5 or 10 percent, but this may be higher depending on the terms of the procurement.

It is important to note that, while a surety issuing a bid bond agrees to provide a payment and performance bond, the bid bond does not guarantee performance of the contract. Rather, failure of the surety to provide a performance and payment bond subjects the surety to liability up to the penal limits of the bid bond, as measured by the difference between the lowest and second lowest responsible, responsive bids.[26]

C. Defenses of Surety to Claims under Bid Bond

Bid bond sureties may invoke numerous defenses against attempts to hold them liable, depending on the facts of the particular situation.

1. Mistake

If a surety's principal has made a material mistake in its bid (that is, it failed to include items in its bid or incorrectly added elements of its bid), the contractor *and the surety* may be relieved of responsibility. Typically, however, (1) the mistake must be so substantial as to make enforcement of the obligation to perform unconscionable; and (2) the contractor must give notice before the government has changed its position to its prejudice.[27]

Courts in various jurisdictions take different views on whether to allow a contractor to withdraw from its bid for mistake. Some require a showing of "reasonable care" in the preparation of the bid before excusing the bidder.[28] Other courts allow withdrawal of a bid so long as the mistake was not due to gross negligence.[29] Others eschew a negligence/gross negligence standard and look to see whether it would be equitable or inequitable to require performance.[30]

2. Solicitation Irregularities

If the government fails to follow its own solicitation procedures, theoretically, this can be grounds for both the contractor and the surety to avoid performance.[31]

3. Material Changes of Contractual Risk

If the government materially alters the terms of the solicitation after bid opening, this also would provide grounds for a surety's avoidance of liability on the bond.[32]

4. Bond Form Discrepancies

While sureties may endeavor to avoid responsibility by pointing out discrepancies in the bond documents, such as a failure of the principal to sign a bond, this defense generally has proven to be ineffective.[33] Given the strictness with which the government interprets the form requirements for bid bonds, however, this may be a legitimate defense in the federal arena, with the bidder or surety arguing the nonresponsiveness of the bid.

5. Fraud or Misrepresentation by Principal

The defense of fraud or misrepresentation by a surety's principal has been similarly unsuccessful in allowing a surety to avoid its liability to an obligee.[34]

6. Failure to Mitigate

The defense of failure to mitigate is always available to a bid bond surety in defense against the government's claim against its bond. In order for the government to avoid this defense, it is required to show three things: (1) that it has procured a contract identical to or sufficiently similar to the terminated contract; (2) that it has suffered damages that are certain in amount; and (3) that it has acted reasonably to minimize its additional costs arising from the failure of the contractor and its surety to enter into the contract.[35]

7. Limitation of Liability

As a last resort, the surety can always argue that its liability is limited to the penal limit of its bond, irrespective of the amount of losses sustained by the government. In other words, if the difference in the bid price between the first and the second bidder exceeds the penal amount of the defaulting bidder's surety bond, the surety's liability is nevertheless limited to its penal sum.[36]

III. Operation of Performance Bonds in Event of Default

A. Need for Declaration of Default

While there is much controversy over whether an obligee needs to declare formally and in writing a contractor in default under a performance bond,[37] the formality of government contracting practice rarely makes this an issue. This is because the FAR makes written notice mandatory. For instance, FAR 49.402-3(e)(2) provides: "When a termination for default appears imminent, the contracting officer *shall provide a written notification* to the surety. If the contractor is subsequently default terminated for default, a copy of the notice of default *shall be sent* to the surety." (Emphasis added.) As one commentator has indicated, this "reflects the actual practice of contracting officers providing notice to the sureties on most occasions."[38]

B. Obligations and Options of Surety in Event of Declaration of Default

Under federal law and regulation, the surety has a number of options once there has been a written declaration of default. These are covered by FAR 49.404(b) and (c):

> (b) Because of the surety's liability for damages resulting from the contractor's default, the surety has certain rights and interests in the completion of the contract work and application of any undisbursed funds. Accordingly, the contracting officer shall carefully consider proposals by the surety concerning completion of the work. The contracting officer shall take action on the basis of the Government's interest, including the possible effect of the action upon the Government's rights against the surety.
>
> (c) If the surety offers to complete the contract work, this should normally be permitted unless the contracting officer has reason to believe that the persons or firms proposed by the surety to complete the work are not competent and qualified and the interests of the Government would be substantially prejudiced.

While the FAR provides that sureties should be permitted to take over the defaulted contractor's work, the surety has several options upon the default of its principal.

1. Tendering Penal Sum of Bond

The Miller Act prescribes that the penal sum of a performance bond shall be adequate for the protection of the government.[39] Perhaps the simplest option of the surety is to tender the penal sum of the bond when a default is declared. The only advantage of this option is that the surety avoids the exposure of liability beyond the penal sum. As will be discussed *infra*, if a surety determines to enter into a takeover agreement with the government without a clause limiting the surety's obligation to the penal sum, the surety exposes itself to potentially limitless liability for its own breaches.[40] What the surety relinquishes in tendering the penal sum, however, is the right to control its destiny and possibly mitigate its losses below the penal sum. As will be seen, entering into a takeover agreement enables the surety to potentially avoid the loss of the entire penal sum of the bond by retaining the services of a bonded completion contractor for a fixed price under the penal limits of the bond.

If a surety wishes to cap its risk by tendering the penal sum, it is important that it perform a sufficient investigation such that the indemnitors on the surety's General Indemnity Agreement cannot later claim that the surety has acted as a "volunteer," or that it settled with the government in bad faith. If indemnitors were to successfully prove either, the surety could potentially lose its rights to be indemnified.[41]

2. Takeover and Completion

Under the FAR, the surety does not have the absolute right to complete the work of its principal after a default termination, although refusal to allow a surety to do so would be fairly exceptional.[42] Indeed, when a surety is permitted to complete the work it assumes the rights and obligations of its principal under the contract, as if it were the contractor.[43]

a. Takeover Agreement

The means by which a surety assumes the contract of its defaulted principal is through a "takeover agreement." As is the case with most of the relationships between the government and its contractors, this too is governed by federal regulation. Indeed, the FAR prescribes mandatory clauses that every takeover agreement must contain, as follows:

> (e) Any takeover agreement must require the surety to complete the contract and the Government to pay the surety's costs and expenses up to the balance of the contract price unpaid at the time of default, subject to the following conditions:
>
> (1) Any unpaid earnings of the defaulting contractor, including retained percentages and progress estimates for work accomplished before termination, must be subject to debts due the Government by the contractor, except to the extent that the unpaid earnings may be used to pay the completing surety its actual costs and expenses incurred in the completion of the work, but not including its payments and obligations under the payment bond given in connection with the contract.

(2) The surety is bound by contract terms governing liquidated damages for delays in completion of the work, unless the delays are excusable under the contract.

(3) If the contract proceeds have been assigned to a financing institution, the surety must not be paid from unpaid earnings, unless the assignee provides written consent.

(4) The contracting officer must not pay the surety more than the amount it expended completing the work and discharging its liabilities under the defaulting contractor's payment bond. Payments to the surety to reimburse it for discharging its liabilities under the payment bond of the defaulting contractor must be only on authority of—

(i) Mutual agreement among the Government, the defaulting contractor, and the surety;

(ii) Determination of the Comptroller General as to payee and amount; or

(iii) Order of a court of competent jurisdiction.[44]

In addition to allowing a surety to control, to some extent, the likely cost of completion, one of the more important reasons for a surety to enter into a takeover agreement is to protect its right to claim for extras and change orders, as will be discussed *infra*, in the event of a later suit for wrongful termination of the surety's principal, or in connection with the performance of the surety's completion contractor. This is because, unless a surety enters into such an agreement, the surety is limited to the balance owed to the defaulted principal and may not make claims for changes or extra work required by the government.[45] The surety should also be careful to reserve the right of both it and its principal to pursue claims against the government for wrongful termination.[46]

In *Travelers Insurance Co. v. United States*,[47] the fact that the surety had entered into a takeover agreement prevented the Court of Federal Claims (COFC) from dismissing its claims against the government for breach of its performance bond in making unauthorized payments to the surety's defaulted principal. The court held that the surety acquired the rights of a "contractor" under the Contracts Disputes Act (CDA)[48] and, as such, stood in the shoes of its defaulted principal as a party to the contract with the government.[49]

Of course, the principal risk in entering a takeover agreement with the government is the exposure to potential damages beyond the penal sum of the bond.[50] Although sureties are often able to persuade obligees to agree to limit their liability to the penal sum in a takeover agreement, the government will rarely agree to such a limitation.

b. Completion Contract

The two most important considerations for the takeover surety in retaining and contracting with a completion contractor is to ensure (1) that the contractor agrees to all of the terms of the defaulted contractor's contract, including warranty obligations; and (2) that the contractor is bonded by a reputable bonding company. In addition, in light of the surety's desire to avoid the administrative hassle of contract administration, it is not uncommon for sureties to allow the completion contractor to deal directly with the government's contract administrative staff without surety participation.

While a completion contract may have its benefits, it is imperative for the surety to require that it be copied on all communications between the government and its completion contractor. On occasion, the government has relieved a completion contractor of certain contractual obligations, which obligations may have been the cause of the defaulted contractor's inability to perform in the first place. In such instance, these decisions may assist the surety in proving its wrongful termination claims, or, at a minimum, entitle the surety to a deductive change of its contract amount with the completion contractor.[51]

c. Need to Preserve the Surety's Rights to Recover for Wrongful Termination and Other Pre-termination Claims

A contractor has a direct right to pursue claims for wrongful termination by virtue of its privity of contract with the government.[52] Unfortunately for the surety, very often its principal does not survive the default termination and ceases to exist. Sureties often are reluctant to enter into a takeover agreement, particularly if the government refuses to agree to limit the surety's liability to the penal sum of the bond or is uncooperative in approving the company (such as the surety's principal) the surety wants to complete the project.

The dilemma that the surety faces is that without a formal executed takeover agreement with the government, the surety's rights are limited under the equitable doctrine of subrogation[53] to recovery of the balance of the contract due its principal, assuming the balance is not totally consumed in the completion of the contract. The rationale behind this view is that surety bonds are not express contracts with the government since the government has not signed them and does not undertake any obligations to the surety by accepting the bond.[54] In addition, the surety is not considered a party to the contract between the government and the contractor.[55]

The right of recovery under the doctrine of equitable subrogation has application in a number of contexts. If a surety has met all of its payment obligations, then it is equitably subrogated to the contract balance held by the government.[56] Additionally, where a surety either finances a principal to completion[57] or tenders a completion contractor to the government while making payments to the completing contractor for the excess costs of completion, the surety has a right to claim any contract balance under this doctrine.[58] An assignment to a surety of a contractor's claim alone, however, is not sufficient to confer jurisdiction to hear such claims on the COFC since such assignment would be invalid under the Assignment of Claims Act.[59]

On the other hand, if the surety enters into a takeover agreement, and properly reserves its rights, it will be able both to pursue the wrongful termination claim (and any other pre-termination claims its principal might have had) and to recover the excess procurement costs, as well as the claims for changes and extras of the principal that were denied to the principal during contract performance.[60]

Correspondingly, where the surety has entered into a takeover agreement and enters into a release agreement with the government in order to obtain final payment from the contracting authority, this will have the effect of releasing the defaulted contractor's claims for wrongful termination.[61]

It is important to note that actions against the government under the Tucker Act[62] must be brought within six years of their accrual.[63] On the other hand, the Contract Disputes Act,[64] which governs those claims that sureties may have who have entered takeover agreements,[65] have a much shorter time limitation. These typically involve appeals from CO decisions and, in the case of appeals to boards of contract appeals (BCAs), must be made within 90 days of receipt of the final decision[66] or, in the case of appeals to the COFC, within 12 months of the final decision.[67]

3. Tendering a New Contractor

Tendering a new contractor means that the surety has found a contractor that is ready, willing, and able to enter into an agreement with the government to complete the principal's contract, while the surety agrees with the government to make up the difference between the contract balance available and the tendered contractor's price to complete. Typically, the surety will also ensure that another surety bonds the tendered contractor. This is a means by which a surety pays a certain sum to "buy back" its bond. In return, the surety is released from its bond obligations, capping its losses at the amount paid to the government.

Unfortunately for sureties, the government will rarely agree to such a tender, and there is no FAR provision requiring a CO to do so. On the other hand, there does not appear to be a legal prohibition unless, straining one's interpretation, there is a contention that to do so violates fundamental procurement law. Arguably, it should not matter to the contracting authority whether (a) the completion is performed pursuant to a tender of a new, *bonded* and acceptable completion contractor with whom the authority directly contracts or, instead, (b) a surety entering into a takeover agreement presents a contractor acceptable to the government with whom the surety will enter into a completion agreement. Indeed, it could easily be argued that in accepting a tendered bonded contractor, the government is gaining more control of its project because it can deal directly with the contractor without interference by the surety. Moreover, few sureties have either the capability or desire to supervise the work of their completion contractors.

Sometimes completion contractors are unwilling to warrant the work that has been previously performed by the defaulted principal. In such instances, the surety can either provide a warranty bond or allow the original bond to remain in place to cover warranty work. Getting the government to agree to accept a tender, however, is a continuing difficulty.

4. Financing the Surety's Principal

Another option available to the surety is to finance its principal through the completion of the project. This typically occurs when the surety's principal has financial problems unrelated to its performance and/or the surety determines that the most cost-effective way to complete the project is to keep the principal in place. In such cases, a surety will provide infusions of cash to enable its principal to meet its payroll and pay subcontractors and suppliers. The surety may also be trying to prevent a domino effect where its principal's default will trigger defaults on other bonded jobs, potentially exacerbating the economic and administrative burdens on the surety.

The major danger to the surety in undertaking the financing of its principal is that, if the principal nonetheless defaults or is forced into bankruptcy, none of the payments made by the surety to keep its principal afloat are debited against the penal sum of the bond(s).[68] Thus, the surety is exposed (as in the case of a takeover agreement without a Limitation of Liability clause) to payments well in excess of the bonded amount. Moreover, a surety can unwittingly subject itself to the tax liability of its principal for failure to withhold payroll taxes when advancing payroll to its principal.[69]

Generally, if a surety elects to finance its principal, it will demand in writing that the government refrain from making contractual payments directly to the contractor and instead make payments directly to the surety. Unfortunately, the government is not required to comply with the surety's request and may continue to make payments to the principal provided the contractor's progress is satisfactory and there is a reasonable expectation that the contractor will complete the project.[70] As a practical matter, this should not become an issue, since the surety is unlikely to finance its principal without an assurance that it has control of future payments by the government. Absent financing, it is likely that the principal, unable to meet its obligations, will be forced to default.[71]

One advantage to the surety in financing its principal is that it may acquire the status of a completing surety.[72] As completing surety, it has a priority over certain claimants to any unpaid contract balances.[73] This higher status, however, does not rise to the level of a takeover surety, because the takeover agreement is the sine qua non of submitting claims for extras and changes.

5. Doing Nothing

The final option of the surety is to do nothing. This is a strategy employed when the surety believes it is on solid ground and that its principal has been wrongfully terminated. In such an instance, the government may enforce its rights under the bond against the surety in an appropriate federal district court, because merely providing a bond is not consent to the jurisdiction of either a BCA or the COFC. Because the surety must be sued in district court, it has the right to a jury trial, whereas no such right is afforded before either a board or the COFC.

The risks to the surety in choosing to do nothing are: (1) an inability to control how the government chooses to reprocure and the costs associated with reprocurement (leaving only the defense of the government's failure to mitigate); (2) a risk that the government will try to seek recovery of amounts in excess of the penal sum of the bond on the ground that the surety, in failing to meet its obligations, acted in bad faith;[74] (3) a waiver of the right of the surety to assert affirmative counterclaims against the government for claims of its principal (as it could have if it had entered into a takeover agreement); and (4) the risk that it will be stricken from the Treasury's List of Approved Sureties for failure to meet its obligations.[75]

C. Claims of Surety Against the Government under a Takeover Agreement

A surety that has entered into a takeover agreement can assert a variety of claims against the government under the CDA.[76]

1. Progress Payments and Retainage

The most fundamental claim a takeover surety has is the right to recover progress payments and retainage, which derives from the doctrine of equitable subrogation.[77] It should be noted, however, that the government's duty to withhold payments from its principal to protect the surety under this doctrine generally arises only *after* the surety had given written notice to the government to do so.[78] If the government fails to heed this notice, it may be liable to the surety for any overpayments.[79]

2. Surety Claims for Delay, Acceleration, Wrongful Termination, Return of Liquidated Damages, and Improper Payments by the Government

A takeover surety also has the right to assert claims arising from delay and acceleration under either the original defaulted contract or under its takeover agreement.[80] Furthermore, the takeover surety has the right to contest its principal's default termination as wrongful.[81] Similarly, the surety may also assert that liquidated damages were improperly assessed, thereby increasing the contract balance available to it for completion or to reduce its ultimate takeover expenses. Finally, where the government has previously made payments improperly to the defaulted contractor, such as overpayments[82] or improper payments of retainage, the takeover surety is entitled to claim these amounts as well.[83]

3. Contract Balances from Other Jobs

Where a surety has taken over a contract from which it has incurred a loss, and its defaulted principal has another contract with the government where it has completed successfully at a profit, the surety has some recourse to its principal's funds on the profitable contract. If the surety gives timely notice to the government of its rights to such monies, the surety has a cause of action against the government to recover the funds if its notice is ignored.[84]

4. Interest

A surety is entitled to collect interest as a takeover surety under the Prompt Payments Act of 1982.[85] In addition, it is entitled to recover interest under the CDA on disputed claims from the time of certification of the claims.[86]

D. Government's Right of Setoff Against the Surety

The government does not have a right to set off contract funds against a performance bond surety to meet other obligations, such as costs it incurs on another contract or an obligation to another governmental entity, such as the Internal Revenue Service.[87] Where the government's obligation arises under the contract at issue (for example, to pay subcontractors that were underpaid in violation of Davis-Bacon Act wage requirements),[88] the results are somewhat different.[89] In addition, in the case of a payment bond surety's claim of equitable subrogation, the government's right of setoff is superior.[90] It is well established, however, that the surety has a superior right to contract funds to that of a bankruptcy trustee for the defaulted principal, and the government subjects itself to liability to the surety (either performance or payment) for improperly making payments to the trustee or, for that matter, a debtor in possession.[91]

E. Damages Recoverable under Performance Bonds

1. Costs of Contract Completion

Assuming that the surety has not entered into a takeover agreement and has chosen to do nothing, it will be subject to the cost of completion incurred by the government not in excess of the penal limits of the bond, even though the total cost may exceed the original contract price. The surety will nonetheless be entitled to any setoff defenses available.[92] Additionally, the surety will be entitled to assert the defense of failure to mitigate damages, particularly if the government has not made an effort to reprocure using competitive procedures.

2. Damages for Delay and Liquidated Damages[93]

A surety is liable to the government for delay damages resulting from the late completion of its principal. As a practical matter, such a claim is subsumed into the Liquidated Damages clause, if there is one in the contract, because the government does not ordinarily incur identifiable costs for delay, such as those expressed in *Eichleay Corp.*[94] Virtually every federal contract contains a Liquidated Damages clause. Such clauses are regularly recognized by the courts and boards, unless they are found to be penalties.[95]

3. Attorney's Fees

In light of the fact that federal contracts do not contain attorney's fees clauses, this is generally not an issue. The Equal Access to Justice Act,[96] however, may subject the surety to damages if the surety raises defenses found to be frivolous.[97]

4. Warranty Claims

Since the surety bond form incorporates the terms and conditions of the construction contract, the surety is required to honor any warranty obligations contained in the contract. Similarly, a surety may be expected to be responsible for the costs of correcting its principal's defective work.

5. Interest

A decision of the COFC holds that, under the Debt Collection Act of 1982,[98] the government has the right to collect interest on what is owed, beginning from the time that the CO renders a decision that the surety is obligated to the government.[99]

F. Defenses of Surety to Declaration of Default

Not surprisingly, the defenses available to sureties against owners in the private sector are generally available against the government as well.

1. Overpayment

Where the government has made improper payments to a surety's principal, it will likely be responsible to the surety for the overpayments. The U.S. Claims Court (now the COFC) has determined that, where a surety had entered into a takeover agreement, it was entitled to recover the balance of the unpaid contract amount as well as any payments that were made to the contractor not in conformance with the contract.[100] The court reasoned, "In other words, Transamerica's [the surety's] right to recover should not be reduced because the government mistakenly made an excessive payment to George S. Rush, d/b/a Rush Engineers [the contractor] before the default."[101] Indeed, the Court of Appeals for the Federal Circuit has recognized the right of the surety to recover improperly paid retainage paid to the defaulted contractor.[102] The court determined that the "retainage provision in a bonded construction contract serves to protect the surety as well as the government . . . ."[103]

2. Plan Defects

A surety that either enters into a takeover agreement and seeks to recover damages for the wrongful termination of its principal or defends against suit by the government under the performance bond has the same right as its contractor to assert the defense that the plans and specifications were defective under the *Spearin* doctrine,[104] wherein the government is held to impliedly warrant that the plans and specifications are free from material defects.[105]

3. *Impossibility of Performance/Commercial Impracticability*
Likewise, the surety has the right to assert either affirmatively as a takeover surety or defensively that its principal's contract was impossible to perform, thereby relieving it of responsibility under the bond.[106]

4. *Wrongful Termination of Contractor*
In the non-federal setting, sureties have successfully asserted claims that their principals were wrongfully terminated because of failures of the obligees to give proper notices or opportunities to cure defaults.[107] This is not likely to be a viable defense in federal contracting in light of the fact that the FAR outlines clear procedures for notification to the contractor of the possibility of a default termination.[108] In addition, the FAR mandates that the surety be given written notice if the default of its principal is imminent and when the principal is actually default terminated.[109]

5. *Material Alteration/Cardinal Change*
A surety may assert the defenses of material alteration or cardinal change.[110] As will be discussed elsewhere, these theoretical defenses are rarely, if ever, successful either as affirmative claims or as defenses, for either principals or their sureties.[111]

IV. Operation of Payment Bonds in Event of Default

Courts often state that the purpose of a Miller Act payment bond is to protect persons supplying labor and materials on a government construction project where they are not afforded the protection they might receive under state lien laws.[112] In addition, courts are instructed to construe the Miller Act liberally so as to ensure that it protects its intended beneficiaries.[113]

A. Parties Covered by Payment Bonds

The parties that are covered under the Miller Act payment bond are well defined. The applicable provision of the statute provides:

> (b) Right to Bring a Civil Action.—
>   (1) In general.—Every person that has furnished labor or material in carrying out work provided for in a contract for which a payment bond is furnished under section 3131 of this title and that has not been paid in full within 90 days after the day on which the person did or performed the last of the labor or furnished or supplied the

material for which the claim is made may bring a civil action on the payment bond for the amount unpaid at the time the civil action is brought and may prosecute the action to final execution and judgment for the amount due.[114]

The government must be named as a party to a Miller Act suit, even though it has no financial interest in the suit.[115] There is no requirement, on the other hand, that the prime contractor be named,[116] and when a completion contractor is retained by a takeover surety, the proper party to a Miller Act suit is not the completion contractor but the surety.[117] Making certain that the surety is named may have important ramifications when the suit is brought at or near the limitations period on the bond, since a suit against the contractor alone does not toll the running of the limitations period against the surety. Thus, it is prudent in Miller Act suits to name both the surety and the prime contractor.

1. First-Tier Subcontractors and Suppliers

Unquestionably, a Miller Act bond covers the first-tier subcontractors and suppliers, that is, subcontractors and suppliers in a direct privity with the prime contractor. As discussed below, the status of first-tier subcontractor and supplier relieves that party of certain notice requirements.

2. Second-Tier Subcontractors and Suppliers

Those subcontractors and suppliers who are in privity of contract with first-tier *subcontractors* will have the right to maintain an action under the payment bond, although these second-tier parties have important notice obligations.

3. Parties Not Covered

The Supreme Court has ruled that subcontractors and suppliers below the second tier are not covered by the Miller Act.[118] Some third-tier subcontractors have tried to establish a second-tier relationship by creating joint-check arrangements with a second-tier subcontractor or supplier. These attempts, however, have proven ineffectual.[119]

In addition, unless second-tier subcontractors and suppliers provide those services to first-tier subcontractors (as opposed to suppliers), they are not considered eligible for coverage under the Miller Act.[120] Thus a second-tier supplier to a first-tier supplier is not eligible under the Act. In order for an entity to be considered a subcontractor, that entity must supply both materials and labor.[121]

There is precedent holding that even though a subcontractor's work was not performed on-site, if the equipment or material is manufactured specifically for a contract (that is, not "off the shelf" items), then the subcontractor is a proper claimant.[122] Some courts have listed the following factors to determine whether a provider of material or equipment is to be considered a subcontractor for purposes of the Miller Act:

- Custom manufacture of the product;
- Product supplied in a complex integrated system;
- Close financial relations between the parties;
- Continuous relationship between parties including shop drawings and/or requirement of representative of supplier on-site;
- Supplier performance on-site;
- Contract for labor and for materials;
- Use of term "subcontractor" in contract;
- Materials provided not out of existing inventory;
- Contract of supplier is significant part;
- Requirement that the supplier provide all material of a particular type;
- Requirement of a surety bond;
- Contractual provision for back-charging for correction of mistakes; and
- Periodic payments to supplier.[123]

These factors in favor of subcontractor status are weighed against such other factors as:

- Whether a purchase order form is used;
- Materials from pre-existing inventory;
- Simplicity of product provided;
- Small proportion of contract amount; and
- Whether sales tax is included in price.[124]

By the same token, where a prime contractor seeks to create a sham subcontractor, thereby creating an additional tier in an effort to insulate the prime contractor, the court will not hesitate to ignore the sham.[125]

Joint-venturers or partners of the prime contractor are not considered eligible under the Miller Act, even where there appears to be an agreement reflecting a contractor-subcontractor relationship.[126] Additionally, where the government employs a contractor to operate one of its facilities and requires that the contractor obtain a bond covering them both, they are considered to be joint obligees, and a construction subcontractor is considered to be the *contractor* for the purpose of determining the tiers.[127]

B. Notice Requirements

1. Statutory Language

The Miller Act has specific notice requirements set forth as follows:

> A person having a direct contractual relationship with a subcontractor but no contractual relationship, express or implied, with the contractor furnishing the payment bond may bring a civil action on the payment bond on giving written notice to the contractor within 90 days from the date on which the person did or performed the last of the

labor or furnished or supplied the last of the material for which the claim is made. The action must state with substantial accuracy the amount claimed and the name of the party to whom the material was furnished or supplied or for whom the labor was done or performed. The notice shall be served—

(A) by any means that provides written, third-party verification of delivery to the contractor at any place the contractor maintains an office or conducts business or at the contractor's residence; or

(B) in any manner in which the United States marshal of the district in which the public improvement is situated by law may serve summons.[128]

2. Ninety-Day Notice Requirement

It is clear that second-tier eligible subcontractors and suppliers, not in direct privity with the prime contractor, are required to give written notice to the prime contractor of their claims within 90 days of the last date on which either materials were supplied or work was performed. Importantly, this notice need not be sent to the surety, though it is hard to understand why a claimant would not do so, unless the surety was unknown.[129]

Courts strictly construe this notice requirement, viewing it as a condition precedent to the right of recovery.[130] If a subcontractor or supplier can prove an express or implied contract directly with the prime contractor, however, such notice is not required. For instance, if a prime contractor assures an otherwise second-tier claimant that it will continue to be paid if it continues to work or make deliveries, an express or implied contract arises.[131] The creation of a joint-check arrangement by the prime contractor with a subcontractor and sub-subcontractor will not, however, create an express or implied contract.[132]

There have been recent legislative changes affecting the technical aspects of the notice requirement. In 1999 when the Miller Act was amended, the requirement that the notice be sent by registered mail was eliminated. Now, "any means which provides written, third party verification of delivery" is acceptable.[133] Courts interpreting the earlier version of the Miller Act had been liberal in their interpretation, excusing literal compliance.[134] It is anticipated that this trend will continue and that the courts will rely on the prior precedents. Nevertheless, the courts have continued to be relatively strict in requiring that the notice be in *writing*.[135]

Some courts have recognized effective notice where the subcontractor showed a writing to the prime contractor but did not leave it with the prime contractor, followed by an oral demand for payment.[136] At least one court was willing to recognize an oral notice where the prime contractor responded in writing acknowledging the notice.[137] Another case has held that a lawsuit filed by a second-tier subcontractor against a prime contractor in a state court was sufficient notice to meet the 90-day notice requirement, since it would have prompted a reasonable contractor to withhold payment from the first-tier subcontractor.[138]

There is a split of authority over whether a notice sent before the last date upon which services were performed, or when material was delivered, is effective. For example, the Fourth and Tenth Circuits hold that such service is effective.[139] The Fifth Circuit holds that such service is ineffective.[140] If the last day on which the 90-day period falls is a weekend or holiday, then Rule 6(a) of the Federal Rules of Civil Procedure provides that the period must be extended to the next business day.[141] Nonetheless, it has been held that notice mailed within the 90-day period but not received within that time is inadequate.[142]

In addition, the courts are in conflict over the adequacy of the contents of the notice. Failure of the second-tier subcontractor to inform the prime contractor (either expressly or implicitly) of the intent to look to the prime contractor for the payment of its bills, according to some courts, constitutes inadequate notice.[143] Other courts are more lenient regarding the contents of the notice, provided the contractor is put on written notice of the second-tier's claims.[144] In some instances, courts have gone so far as to approve notice where the notice was not originally sent to the prime contractor but to the surety directly, which then forwarded it to the prime contractor.[145] Other courts have held that, if a notice appears to have "plainly brought home to the prime contractor the nature and state of the indebtedness," that will be sufficient.[146]

The best practice for a second-tier claimant to ensure effective notice is to send written notice of a claim (1) *after* the last day when services were performed or materials were delivered but *within* 90 days thereof; (2) by a means that will allow the claimant to prove both when the notice was sent and that it was received by the prime contractor; and (3) stating that the claimant is looking to the prime for payment. While not necessary, it is probably prudent that the claimant copy the surety on the notice. Notice to the surety may often be a greater stimulus to action than a notice to the prime contractor.

C. Limitations of Actions

1. Filing of Lawsuit

Under the Miller Act, suit must be filed within one year of the last day on which services were performed or on which materials were delivered *but not before* 90 days from those dates.[147] Though technically a suit that is filed within the initial 90-day period is premature, courts have generally permitted plaintiffs to amend their pleadings to cure this defect.[148]

On the other hand, failing to file within a year's time has been held to be a jurisdictional defect by many courts and a condition precedent to recovery under the Act.[149] The modern trend, however, is to treat the failure to file timely as a restriction on the *right to recover* without affecting the court's ability to *exercise subject matter jurisdiction* over the case.[150] Importantly, this allows the court to deal with issues of tolling, waiver, and estoppel relative to the statute of limitations.[151]

Filing a timely action in the wrong federal district court will normally toll the statute of limitations, but filing in a state court will not.[152] Filing a demand for arbitration alone will not toll the statute.[153] Again, it is not enough to merely file the suit against the prime contractor; rather, a claimant must file against the *surety* within one year, or else the claim against the surety is lost.[154]

2. Commencement of Ninety-Day Notice and Limitation Periods

Determining the last day that services have been provided or materials have been delivered can be very tricky.[155] Generally, as far as labor is concerned, the courts construe this as actual physical labor as opposed to general administrative work, such as requisitioning.[156]

Also, it is important to note that, unless the work is that which is called for under the original contract or change orders, as opposed to incidental, warranty, or repair work, the statute will not toll.[157] This interpretation of labor is not always rigidly enforced. For example, some courts have considered punch list items "work" because such work is part of the required work of the contract.[158] Similarly, where the subcontractor can prove that repairs related to an integral part of the contract and that the subcontractor was not responsible for the original defects requiring the repairs, the completion period will be included in the calculation of time.[159]

The courts treat delivery of materials and issues relating to equipment differently. While the date of the last shipment to the site is generally the start of notice and limitations periods, the inadvertent exclusion and later delivery of parts of a shipment will extend the time.[160] In connection with a rental agreement relating to pilings, however, a court determined that the date when the pilings were returned to the lessor—not the date when they were first removed from the site—was the date that started the running of the limitations period. In so holding, the court also determined that a notice that was sent earlier than that time would have been premature and therefore insufficient.[161]

Where the issue concerns repair work, the courts consider a number of factors. In attempting to draw what is admittedly a hazy line, the considerations are: (1) whether the subcontractor properly performed its work; (2) the value of the work or material; (3) whether the nature of this work was unexpected; and (4) how significant the work or materials were to the systems or construction involved.[162]

There are some courts that take the position that work accomplished after substantial completion, or after the subcontractor's work becomes workable, should not be used in the computation of time.[163] Generally, this is not considered an appropriate way to calculate time.[164]

Another conundrum for courts is how to determine proper notice by a supplier of multiple orders or a single supplier to multiple projects. The better rule with regard to multiple orders on a single project is one notice for all deliveries.[165] One line of cases holds, however, that where there is a lapse of 90 days between last delivery and its predecessors, it will not save the prior deliveries.[166] As to orders on multiple projects, the courts are divided. The Fourth

Circuit has held that only a single notice was required, running from the last delivery where the supplier had delivered materials to the same subcontractor, where the prime contractor and surety were the same on both projects.[167] Other courts take the view that where deliveries are made to separate projects with separate payment bonds, they must all be treated separately.[168]

Equipment rentals are also subject to the notice and limitation requirements, and compliance is measured in two ways. If the lease agreement is still in effect, the time begins to run from the last date that the equipment was on-site and available for use in the project.[169] If, however, the agreement has expired, is abandoned, or has terminated, even if the leasing company is unaware of that fact, the date runs from that time.[170] There is some support for the view that the time does not begin to run until the date the equipment is returned.[171]

3. Waiver and Estoppel as Defenses to Lack of Notice and Late Filing of Suit

It is generally agreed that waiver and estoppel are viable defenses to contentions of failure to give timely notice or to file with the limitations period, despite earlier Supreme Court decisions speaking of these requirements as "conditions precedent."[172]

If it can be shown that someone on behalf of a surety lulls a claimant into a false sense of security—by words, actions, or conduct—such that the claimant fails to serve a timely notice, or lets the limitations period on the bond expire, a court will deny the defense on the basis of estoppel.[173] Usually the claimant has the burden of showing "representation, reliance, change of position, and detriment" to prove an estoppel.[174] Importantly, it is a representative of the *surety* who must make the representation, and a representation made by the surety's principal is insufficient to prove estoppel.[175]

Neither the surety's failure to inform a claimant of the existence of the limitation period nor the acknowledgement, investigation, or attempt to settle a claim, in and of itself, will constitute estoppel.[176] Typically, the way a surety effectively avoids a finding of estoppel is by declaring in writing its intent to reserve and not waive any of its rights.[177] Those who have dealt with sureties and their attorneys will note this reservation on almost all correspondence.

Theoretically, sureties can waive the defense of limitations by not asserting the defense in their pleadings, yet most sureties will be looking for this defense at the inception of a claim. Moreover, a surety's mere silence about the limitations defense before suit is not construed to be a waiver.[178]

D. Venue and Jurisdiction Issues

1. Jurisdiction

Federal courts have exclusive subject-matter jurisdiction under the Miller Act,[179] notwithstanding that the Supreme Court has referred to this part of the statute as one that merely governs venue and is not a reference to jurisdiction.[180] The Miller Act, however, has no applicability unless the project is a project of the U.S. government (rather than one that merely has financing from the federal government).[181] The Supreme Court has defined the term "public work" as "projects of the character heretofore constructed or carried on directly by public authority or with the public and to serve the interests of the general public."[182]

Miller Act claimants are not required to meet the monetary requirements of diversity jurisdiction, nor is their citizenship important, because the jurisdiction is based on a federal question.[183]

2. Venue

The Miller Act provides that venue for payment bond claims is in the district court "in which the contract is to be performed and executed."[184] This is an inflexible provision. For example, where the prime contractor has filed for bankruptcy, the proper place to bring the action is not in the district where the bankruptcy case is pending (unless it is the same district in which the project is located).

The Supreme Court has determined, however, that proper venue is not a jurisdictional necessity,[185] and, as such, a valid venue selection clause may result in a waiver of the statutory venue requirements.[186] Courts have also held that a valid arbitration clause with a provision is enforceable despite the venue requirements of the Miller Act.[187] Sureties may successfully invoke such clauses unless the claimants can prove that the forum selection clause is so inconvenient as to be unenforceable.[188]

If contract performance is in another country, the general venue statute applies.[189] Thus, courts have held in this instance that the proper venue of a Miller Act claim on an overseas project is the principal place of business of the defendant.[190]

Moreover, as in any venue contest, the proper remedy if a case is filed in the wrong district court is to transfer it to the right court, not to dismiss the action.[191]

3. Joinder of Other Claims

Where state common law contract claims are closely related and do not predominate over the Miller Act claims, they may be joined under the doctrine of pendant jurisdiction.[192] Similarly, related claims based on diversity of jurisdiction may also be joined to Miller Act claims.[193]

E. Impact of Arbitration Clauses on Miller Act Payment Bond Claims

The question of whether arbitration clauses in subcontracts are enforceable against sureties or against subcontractors is still evolving.[194] In one case, a court granted the motion of the prime contractor and the surety to stay the Miller Act proceeding pending the arbitration of the claim, as required by the claimant's subcontract, reasoning that nothing in the law provided that Miller

Act claims were not arbitrable and that the requirement to arbitrate does not constitute a waiver of Miller Act rights.[195]

In another case, however, where the subcontract contained a valid arbitration clause, the surety and prime contractor were denied a stay of the Miller Act claimant's suit against them pending arbitration. In that case, the court reasoned that, absent an express waiver in the arbitration clause of the claimant's right to proceed in federal court, there was no basis to stay the Miller Act case.[196]

If the surety enters into a takeover agreement on the project that incorporates its principal's contract and that contract contains an arbitration clause, the requirements of the Miller Act do not apply, however.

Where claimants in arbitration seek damages not covered by the Miller Act and where other common law claims are brought, sureties can take advantage of the multiplicity of claims and request that the arbitration panel issue an award categorizing the damage.[197]

The mere making of a demand for arbitration is not sufficient, in and of itself, to toll the running of the Miller Act's one-year statute of limitations. Failure to file a claim within the time period, even when a timely demand for arbitration has been made, will result in dismissal of the action against the sureties under the Act.[198] The prudent action is to file the Miller Act suit in the district court where the project is located and simultaneously move to stay the proceedings pending the outcome of the arbitration.

Because of an apparent conflict between the Federal Arbitration Act (FAA),[199] which discusses arbitration clauses specifying a state for enforcement, and the Miller Act, which mandates venue at the location of the project,[200] problems had previously arisen in determining where an arbitration award could be enforced. The Supreme Court's ruling in *Cortez Byrd Chips, Inc. v. Bill Harbert Construction Co.* resolved the conflict by ruling that a Miller Act arbitration award could be enforced under the venue provisions of the FAA or the Miller Act.[201]

F. Claims Generally Covered

1. Labor

It is axiomatic that a Miller Act payment bond covers labor performed on the project, since it is specifically called for under the express terms of the Act.[202] Architectural, engineering, and other professional services that involve superintendence, supervision, or inspection and are performed at the job site are covered.[203] On the other hand, a court has held that a subcontractor did not furnish "labor" when it provided on-site project management, supervision, administrative, and oversight "services" to the contractor, because the work involved was administrative and clerical, rather than manual labor or physical work.[204] Beyond the actual payment of labor, numerous related costs are also covered, including the costs discussed below.

a. Withholding Taxes

Since the Miller Act's 1966 amendments, sureties have been responsible for the payment of withholding taxes that their principals or second-tier subcontractors have failed to pay.[205] The government is required to give notice to the surety, however, within 90 days after the contractor files a return or within 180 days after a return was due, and it must file suit against the surety within one year after the notice is given.[206]

b. ERISA

Originally, there had been a fair amount of controversy as to whether ERISA claims were covered under the Miller Act.[207] It is now generally accepted that such claims are covered.[208]

c. Union Dues

It is not seriously questioned that unpaid union dues are recoverable, and normally the party bringing the action is the labor union.[209]

d. Food and Lodging

Only where food and lodging are a necessary part of the project, as where work is being performed in a remote or wilderness area, are such items recoverable under a Miller Act payment bond.[210]

2. Materials

Without question, the costs of materials that are incorporated into the project are recoverable under the Miller Act bond.[211] Additionally, costs of materials that are not incorporated may also be recovered if there is a reasonable expectation that the materials will be substantially consumed or incorporated into the project.[212] Thus, the payment bond will cover materials such as fuel and tires[213] and specially fabricated materials for a project so long as they are integral to, and a substantial part of, the project.[214]

Generally speaking, however, suppliers are only entitled to recover under the bond for materials that are incorporated into the project. There are numerous exceptions. If a supplier is able to demonstrate that it supplied the materials to the project in good faith, the cost of these materials will be covered,[215] despite the fact that the materials were ultimately used on other jobs.[216] The costs of materials wrongfully rejected because they allegedly failed to conform to plans and specifications are recoverable upon a showing that they were conforming.[217]

The supplier bears the burden of demonstrating to the satisfaction of the trier of fact that the materials were supplied for a particular bonded project and not a number of projects.[218] Suppliers of fungible materials, such as concrete, lumber, etc., would do well to keep careful records of where they are delivering materials by creating delivery tickets that clearly identify the project to which their materials are being delivered and in what amounts.

3. Equipment
a. Rental

One of the more significant costs incurred on a project can be the cost of rental fees for equipment utilized in the performance of the work.[219] An equipment-leasing company may recover the fair rental value for equipment, even though the equipment may be idle for periods of time during the project.[220] More-over, even where the equipment is returned before the conclusion of the lease, unless the lessee is able to demonstrate that the leasing company could re-let the equipment, the lessor may be able to recover the balance of the lease.[221]

b. Repairs

Courts dealing with the issue of rental costs must decide whether repairs to the equipment involved in the project (1) are for the normal wear and tear experienced in the use of equipment on the project or (2) were to permit the equipment to be used for other jobs. In the latter case, courts will consider these as capital improvements and not allow coverage.[222]

c. Capital Goods

Typically, unless equipment is expected in the ordinary course to be con-sumed in the performance of the contract, there will be no coverage for capital expenditures for equipment.[223] Authority exists for the proposition that where an equipment rental agreement contained a clause shifting the risk of loss to the lessee for loss thereof, and that equipment is later lost, the value of the equipment may be covered under the Miller Act.[224]

4. Services

The determination of whether the payment bond covers professional services turns on whether or not the subcontractor performed such services on-site and whether they were more than mere administrative tasks.[225] With respect to such services as transportation and freight charges, the courts differ. Gen-erally, the costs of transporting materials to the site are recoverable.[226]

5. Loans

Unless a lender obtains a valid assignment of a claim under the bond from its debtor,[227] loans, even if used to pay for labor and materials on the job, are generally not included within the payment bond coverage.[228] Moreover, even in the case of a proper assignment, the assignee will have to comply with the Miller Act notice requirements.[229]

6. Delay Damages

As with other areas of claims that formerly have not been covered by Miller Act bonds,[230] the modern trend appears to allow payment bond claimants to recover delay damages.[231] Indeed, one court awarded delay damages despite the existence of a "no damages for delay" clause in the contract, where the court determined that the prime contractor had actively interfered with the claimant's progress on the project.[232] Another court reasoned that, without recovery of delay damages, the subcontractor's options for recovery of said damages would otherwise be inadequate.[233]

7. Attorney's Fees

The Supreme Court ruled in 1974 that there could be no recovery for attorney's fees under a Miller Act payment bond despite a state statute allowing recovery on state payment bond cases.[234] Since that ruling, a surety was successful in having *state* law claims allowing the claimant to obtain interest and attorney's fees dismissed from a Miller Act suit on the grounds that the Miller Act preempted such laws.[235] Currently, unless there is a federal statute authorizing payment or a valid contractual provision, there is no right to recover attorney's fees under the Miller Act.[236] Importantly, the contractual provision must be directly between the prime contractor and the first-tier subcontractor to state a valid claim for attorney's fees against the Miller Act surety.[237]

8. Interest

A number of courts have permitted the recovery of interest if state law so provides.[238]

9. Bad Faith Claims

There is a substantial amount of dispute as to whether Miller Act claimants can recover extra-contractual damages, such as punitive damages. Reasoning that the Supreme Court had already held that a Miller Act surety is not subject to a state law requiring the award of attorney's fees to a prevailing party, one court has held that a state law authorizing recovery from a surety for bad faith is inapplicable in a Miller Act case.[239] Other courts have not agreed, holding that there is no preemption of state bad faith claims.[240] Recently, a court held that the Miller Act preempts state law claims for enhanced damages for failure to investigate or pay a Miller Act claim under the Massachusetts Unfair and Deceptive Trade Practices Law.[241]

G. Claims Generally Not Covered

1. Lost Profits

The Miller Act generally does not cover lost profits for work not yet performed.[242] The Act may cover claimed profits as part of the work that has already been performed by the claimant.[243]

2. Insurance Premiums

Brokers or insurance companies that have advanced premiums for defaulting subcontractors are generally unsuccessful in recovering those advances under a Miller Act bond, since the provision of insurance is not deemed either labor or material.[244]

3. Damages Relating to Termination

Attempts by claimants to recover for damages attributable to their terminations, other than for materials and/or labor supplied, have also largely been unsuccessful.[245]

4. Contracts Implied in Law or in Fact and Quantum Meruit

While the Miller Act speaks of contracts "express or implied,"[246] courts have held that the Act does not permit recovery under an implied-in-law or quasi-contract claim.[247] *Quantum meruit* claims, however, are recognized under the Miller Act.[248] "A subcontractor may recover in quantum meruit from the prime contractor and surety in at least two instances,"[249] where (1) "there is a substantial breach of the subcontract, [in which case] the subcontractor 'may forego any suit on the contract and sue for the reasonable value of his performance,'";[250] or where (2) "the subcontractor . . . has performed work outside the terms of the contract that benefits the prime contractor."[251] In the latter case, "[f]ailure to make . . . a payment [for work done under the terms of sub-contract] when due is a substantial breach of the contract entitling the sub-contractor to recover in *quantum meruit* for the reasonable value of the work performed."[252] This ruling somewhat begs the question of a claim where there is an existing written contract, since in order to make such a claim, the claimant must declare the contract at an end for material breach and not continue accepting payments under the contract.[253]

In any event, claimants that perform extra work under the contract are entitled to recover under the Miller Act bond, including their profits on the extra work.[254] In that regard, where there is neither a written contract nor an agreement on the contract price for such work, the *quantum meruit* measure of damages (i.e., the reasonable value of the work) is the standard by which such damages will be calculated.[255]

H. Other Payment Bond Defenses

Issues relating to eligibility to file a claim, notice, and limitations have been discussed above. This section will address the substantive and contractual defenses available to the Miller Act surety against payment bond claimants.

1. General Right to Assert All of Principal's Defenses

The defenses of the principal (prime contractor) against the claims of subcontractors and suppliers generally are available to the surety.[256] This is because the surety is only derivatively liable to claimants under the Miller Act.[257] Moreover, in cases in which the principal and the surety are sued jointly, the surety may assert the principal's counterclaims against the claimant as defenses.[258]

2. Payment

Perhaps the best defense is that of payment to the claimant, since it is a defense against both first- and second-tier claimants and it does not matter who has made the payment.[259] The difficulty, however, is that payment to the first-tier subcontractor does not insulate the prime contractor from claims by the second-tier subcontractors and suppliers who have provided labor and materials but have not been paid,[260] except in circumstances as shown in the various defenses set forth below.

3. Payment Misapplication

Often suppliers provide materials to the same contractor on both bonded and nonbonded jobs. There may be a great temptation to apply all funds received to the nonbonded debts regardless of the source of the funds. Suppliers, however, are not free to apply their payments to any project they wish. If the supplier has actual knowledge of the source of payments or reasonably should know of their source, the supplier is bound to apply those payments to that contract.[261]

Courts have set out several guiding principles when determining whether there has been a misapplication. Examples are: (1) whether the payment made by the subcontractor was applied to the project the way the subcontractor intended and/or whether the subcontractor made this known to the creditor; (2) where the subcontractor has not indicated the source of its payment, whether the creditor nonetheless independently determined the source within a reasonable time after payment; (3) where no indication is made, whether the creditor applied it to the oldest debt as would be fairest to the debtor (as opposed to a more recent but nonbonded debt); and, (4) if the subcontractor knows that the payments are to discharge the debt of a particular prime contractor, but directs the creditor to apply the payment to another debt, the creditor must, nonetheless, apply it to the appropriate debt.[262] Where the supplier does not know the source of the subcontractor's payments, however, it was not a misapplication to follow the subcontractor's instructions as to its application.[263] In any event, the surety has the burden of demonstrating that the supplier has misapplied the funds.[264]

4. Joint-Check Arrangements

In an effort to ensure that a prime contractor's payments to its subcontractors are properly applied to the claims of that subcontractor's second-tier subcontractors and suppliers, a prime contractor will often write joint checks to the subcontractor and its subcontractor or supplier. Naturally, the prime contractor wants to assert as a defense the fact that payments have been made by joint check. Unfortunately, in the Miller Act context, this fact, by itself, has not proven to constitute a waiver of the supplier's claims to recover under the bond.[265] Some courts have refused to deem the joint check as a direction to apply the payments in a particular way and have refused to interpret release language on the check because there was no specific reference in the release to bond rights.[266] Other courts have been extremely critical of the rationale of the federal cases, arguing that payment by joint check, in and of itself, should be a viable defense.[267]

5. *Setoff and Recoupment*

While the terms "recoupment" and "setoff" are often used interchangeably, the two actually differ. The right of setoff refers to the application of *unrelated* obligations of a claimant as a "setoff" against its current claim—for example, back-charges for one contract are set off against claims on a second contract. Recoupment results from withholding monies in the same contract on which the subcontractor is asserting a claim, such as a back-charge for resolving defective work against a claim for extra work. As with many of the defenses that a surety can assert, privity of contract will be an important issue.[268]

6. *Compromise and Settlement*

Both the surety and prime contractor should insist on clear language in any settlement agreement between the latter and a claimant stating that the claimant does not have the right to seek payment in full for its claim against the surety and that the surety has the right to fully rely upon the agreement reached between its principal and claimant, provided there is some consideration and it is clear what claims have been compromised.[269] Where an agreement so states, it will provide a defense the surety may assert against claims on the bond.

7. *Accord and Satisfaction*

A similar defense that a surety may rely on is the defense of accord and satisfaction. This most commonly occurs when the prime contractor tenders a "full payment" check with either an endorsement as such or a letter accompanying the check that the payment is "in full" or in "accord and satisfaction of all outstanding claims."[270] The fact that the claimant objects at the time of the tender is not normally a defense, and negotiating the check will extinguish the right to a greater sum.[271]

8. *Pay-When-Paid Clauses*

While the "pay-when-paid" defense is generally recognized on private bonds,[272] the same cannot be said with respect to Miller Act payment bond sureties. As early as 1914, federal courts have determined that a pay-when-paid clause is not a defense to a Miller Act payment bond claim because it contravenes federal public policy.[273]

Other federal courts have similarly held that sureties may not enforce such clauses on the basis that either the clause was an ineffective waiver of Miller Act rights[274] or the Act does not contain a condition precedent, requiring the government to pay the prime contractor before the Miller Act surety's obligation arises.[275] Thus, while the law in many places is not finally determined on this issue, there is ample case law upon which the claimant can rely in an effort to defeat such a defense to a Miller Act claim.

9. Pass-Through or Disputes Clause Requirements

So-called "pass-through" clauses, that is, clauses that seek to have the dispute provisions of the prime contract "flow down" to the subcontractor and obligate the subcontractor to await the outcome of that process, have also shown little success in protecting sureties against subcontractor claims. For example, in *Fanderlik-Locke Co. v. United States ex rel. Morgan*, subcontract language provided (a) that a subcontractor is bound to the "disputes" clause of the prime contract; (b) that the government's approval of payment was a condition precedent to its being paid; (c) that the subcontractor assumed the same duties to the prime that the prime contractor agreed to with the owner; and (d) that the prime's duty to the subcontractor was the same as the government owed to the prime. The Tenth Circuit did not allow the surety to delay payment of its obligation to the claimant based on this language.[276]

Other federal courts have similarly refused to require Miller Act claimants to wait for the outcome of the disputes process between the prime contractor and the government before they could recover from the Miller Act surety.[277]

Arguably, the amendments to the Miller Act voiding waivers of Miller Act rights before the actual performance of labor or delivery of materials[278] only strengthened the argument against pass-through claims as viable defenses against Miller Act claims. It should be understood, however, that sureties are not thereby prevented from making other substantive defenses, such as that the work or materials are defective. It only means that it will be incumbent upon the surety to prove those defenses.

10. Changes Clauses

Sureties' reliance upon the changes clauses, either in a flow-down context or those contained in the actual subcontract, have been successful as defenses against claimants. For instance, where the changes clause requires that the claimants provide notice in order to recover for extra work, a surety has successfully argued that failure to provide the requisite notice bars the claim against the surety.[279] Additionally, failure of a claimant to obtain prior written approval before undertaking work may also preclude recovery under the Act.[280]

11. No Damage for Delay Clauses

Many subcontracts contain no damage for delay clauses. As its name implies, a no damage for delay clause provides that the subcontractor is entitled to no money damages for delay, with a time extension its only remedy. As such, the subcontractor would have no right to extended field overhead, extended home office overhead, or other delay-related costs, even if caused by the prime contractor. Courts have generally held that the no damage for delay clause defense is available to the surety, on the theory that the surety's obligation can be no greater than that of its principal.[281] Many jurisdictions, however, recognize exceptions to the enforcement of the no damage for delay clause, such as (a) where the delay involves deceit or fraud;[282] (b) where there was active interference by the party asserting the defense;[283] (c) where the delay constitutes a cardinal change justifying abandonment; (d) where the delay was caused by events that neither party anticipated; (e) where the party asserting the defense caused the delay by its gross negligence;[284] or (f) where there was a failure to grant a warranted time extension.[285]

12. Waiver and Estoppel

The 1999 amendments to the Miller Act extinguished the ability of the subcontractor to waive its rights under the Act *before the performance* of labor or the actual providing of materials.[286] Thus, while there is a great deal of case law relating to waiver and estoppel, caution must be taken in using these cases because most preceded the change in the statute.

In any event, the defense of waiver and estoppel is frequently raised. Assuming that the contentions are timely in that the acts of either estoppel or waiver have occurred after the labor has been performed or the materials provided, these are generally not difficult defenses to prove in the case of first-tier subcontractors. The difficulty of proof arises in connection with second-tier subcontractors and suppliers.[287] In either case, in order to prove the waiver of a Miller Act right, the evidence must be specific as to the waiver or release of their rights.[288] One court has gone so far as to say that this requirement of specificity requires the actual mention of the Miller Act.[289] Generally, only where the actions of the second-tier claimant caused the prime contractor to detrimentally rely on those actions will the defenses lie against the second-tier claimant.[290]

Requests for joint-check arrangements are not considered waivers by themselves because they merely are requests for additional security.[291] Additionally, where a surety sought to assert waiver and estoppel against a claimant that colluded with its principal by signing a letter that it had been paid in full, the defense was held not sufficient since the surety had not relied on the letter to its detriment.[292]

In order for a surety to prevail against a claimant on the theory of estoppel, it must show (a) words, conduct, or acts leading the prime contractor or surety to think that a certain situation existed; (b) negligence or willfulness on the part of the claimant with regard to the words, conduct, or actions; and (c) reasonable and detrimental reliance thereon.[293] If a prime contractor pays a subcontractor in reliance on a statement by the subcontractor's supplier that it has been paid, the supplier will be estopped from its claim against the prime because of the detrimental reliance by the prime on the supplier's statement.[294]

Sometimes mere silence by the claimant can lead to an estoppel. An example is a supplier that, in the presence of the prime contractor and the subcontractor, stood mute regarding outstanding obligations to it by the subcontractor, and the prime released substantial payment to it in reliance on that silence.[295]

13. Licensing Requirements

In many jurisdictions, entities performing construction activity are required to be licensed by a state authority, and the failure to be licensed may render the contractor seeking compensation ineligible because of the lack of a license. Because federal projects are not subject to state licensing requirements, however, and prime contractors and subcontractors need not be licensed in the state where the project is being performed, this is not a viable defense for a Miller Act surety.[296]

14. Failure to Mitigate

Nothing in the Miller Act, or in the case law interpreting it, precludes the defense by a surety that the claimant has failed to mitigate its damages. The issue is whether the claimant acted in a "wholly unreasonable way in avoiding damages created by a party's breach of its contract."[297]

The doctrine of mitigation, however, is always a two-way street. Though claimants are subject to the defense of mitigation,[298] prime contractors or their Miller Act sureties that seek to reduce or avoid payment of claims via recoupment and setoff are equally subject to the requirement of mitigation in connection with the losses that support their recoupment and setoff defenses.[299]

An example of the application of the defense of mitigation against claimants in the Miller Act context is the failure of equipment renters to retrieve and attempt to re-let equipment within a reasonable time after the party renting the equipment is defaulted.[300]

15. Penal Sum Limitation

It is a fundamental precept of surety law that the surety's liability does not exceed the penal sum of its bond, that is, the amount for which the bond is written. Under the Miller Act, the penal sum is generally the contract amount.

A problem arises when the Miller Act surety is faced with subcontractor claims that exceed the penal limits of the bond. This issue was more acute prior to the 1999 amendments to the Miller Act, since under the former law, the prime contractor was not required to provide a bond for the full amount of the contract. It is not inconceivable, however, that a surety will face a situation where the claims exceed the bond principal, as in a case where significant change orders were agreed to or the prime contractor's actions cause significant delays.

Where a surety is faced with multiple payment bond claims, it may be prudent, before negotiating immediate settlements with early claimants, to wait and get a "lay of the land." Then, if there is a significant potential for the claims to exceed the penal sum, the surety should give careful consideration to filing an interpleader action.[301] Failure to do so could subject the surety to liability for an amount in excess of the penal sum of the bond.[302]

Where a surety is subject to the payment of interest on the claim, normally it expects that the total of the claims and interest should not exceed the penal sum of the bond. Authority exists, however, that if the surety's default on its obligations under the bond caused delay in payment, it may be subject to exposure exceeding the penal sum.[303]

16. Material Breach by Claimant

One of the more fundamental defenses of any prime contractor to a claim by a subcontractor or supplier is the material breach by that subcontractor or supplier. Indeed, material breaches often form the bases for the defenses of offset and recoupment. Thus, if a subcontractor or supplier, without justification, abandons the project, prevents substantial completion, or performs materially defective work, the defense of material breach is available to both the prime contractor and the surety. Indeed, such breaches may give rise to counterclaims by the prime and its surety against a Miller Act claimant.[304]

Notes

1. 40 U.S.C. §§ 3131–3134 (2006) (formerly 40 U.S.C. § 270a et seq.).
2. Pub. L. No. 106-49 (1999).
3. See revisions to FAR 28.102 and 52.228-13, -15, and -16, implementing the changes.
4. 40 U.S.C. § 3131(a)–(b) (formerly 40 U.S.C. § 270a).
5. FAR 28.101-1.
6. 40 U.S.C. § 3131(b)(1) (formerly 40 U.S.C. § 270a(a)(1)).
7. FAR 28.102-1(a) (emphasis added).
8. FAR 28.102-2(a)–(b)(1).
9. 40 U.S.C. § 3131(b)(2) (formerly 40 U.S.C. § 270a(a)(2)).
10. FAR 28.102-1(b)–(c).
11. FAR 28.102-2(b)(2)–(c).
12. *See* FAR 28.201 and 28.204.
13. "Companies Holding Certificates of Authority as Acceptable Sureties on Federal Bonds and as Acceptable Reinsuring Companies." FAR 28.202(a)(1).
14. FAR 28.203.
15. *See* Adrian L. Merton, Inc. v. United States, No. 02-237C at 8 (Ct. Cl. 2002).
16. *See* Kennedy Elec. Co., B-239687, May 24, 1990, 90-1 CPD ¶ 499.
17. *See* Ameron, Inc., B-2118262, Apr. 29, 1985, 85-1 CPD ¶ 485.
18. *See* G&A Gen., B-236181, Oct. 4, 1989, 89-2 CPD ¶ 308; Sundance Helicopters, B-208435, Nov. 15, 1982, 82-2 CPD ¶ 442.
19. *See Adrian L. Merton* No. 02-237C at 8.
20. *Id.*

21. FAR 28.101-3.

22. 90-1 CPD ¶ 499.

23. FAR 53.301-24.

24. *See* W&M Constr., Inc., Sept. 18, 1990, B-239847, 90-2 CPD ¶ 227.

25. 48 C.F.R. § 28.101-4(c).

26. *See* Chas. H. Tompkins Co. v. Lumbermans Mut. Cas. Co., 732 F. Supp. 1368 (E.D. Va. 1990). Note that there have been some cases, however, where federal courts, on rather questionable reasoning, determined that the failure of a bid bond surety to provide a payment bond exposed the surety to possible payment bond claims. *See, e.g.,* United States *ex rel.* Empire Plastics Corp v. W. Cas. & Sur. Co., 429 F.2d 905 (10th Cir. 1970); United States *ex rel.* Victory Elec. Corp. v. Md. Cas. Corp., 213 F. Supp. 800 (E.D.N.Y. 1963). *See also* Frederic E. Cann, *What to Do When There Is No Miller Act Payment Bond*, 23 Pᴜʙ. Cᴏɴᴛ. L.J. 413 (1994).

27. *See* Nat'l Fire Ins. Co. v. Brown & Martin Co., 726 F. Supp. 1036 (D.S.C. 1989).

28. *See* M.J. McGough Co. v. Jane Lamb Mem'l Hosp., 302 F. Supp. 482 (S.D. Iowa 1969).

29. *See* Marana Unified Sch. Dist. No. 6 v. Aetna Cas. & Sur. Co., 696 P.2d 711 (Ariz. 1984).

30. See Reg'l Sch. Dist. No. 4 v. United Pac. Ins. Co., 493 A.2d 895 (Conn. App. Ct. 1985).

31. In *Hanover Area School District v. Sarkisian Bros., Inc.*, 514 F. Supp. 697 (M.D. Pa. 1981), the court held that the owner's failure to strictly adhere to the "Instructions to Bidders" violated a "material part of the contract," thereby freeing both the contractor and the surety of liability.

32. Nw. Constr. Co. v. Winston-Salem, 83 F.2d 57 (4th Cir. 1936) (reducing scope of work by 15 percent after bid opening); Commr's of Sewerage v. Nat'l Sur. Co., 140 S.W. 62 (Ky. 1911) (refusal to accept steel when contractor's bid was accepted on basis of use of type steel).

33. *E.g.,* Bolivar Reorganized Sch. Dist. No. 1 v. Am. Sur. Co., 307 S.W.2d 405 (Mo. 1957); State v. Integon Indem. Corp., 735 P.2d 528 (N.M. 1987).

34. Peerless Ins. Co. v. United States, 674 F. Supp. 1202 (E.D. Va. 1987).

35. *Id.* at 1207.

36. Chas. H. Tompkins Co. v. Lumbermans Mut. Cas. Co., 732 F. Supp. 1368 (E.D. Va. 1990).

37. *Compare* L&A Contracting Co. v. S. Concrete Servs., Inc., 17 F.3d 106 (5th Cir. 1994), *with* Moore Bros. Co. v. Brown & Root, Inc., 207 F.3d 717 (4th Cir. 2000).

38. Bedner, et al., *Chapter 17: Bonds*, Cᴏɴsᴛʀᴜᴄᴛɪᴏɴ Cᴏɴᴛʀᴀᴄᴛ Lᴀᴡ at 1321 (1990). *See also* Andrew W. Stephenson, Douglas L. Patin & George J. Mitchell, *Surety's Role in Default Termination*, Cᴏɴsᴛʀᴜᴄᴛɪᴏɴ Bʀɪᴇғɪɴɢ No. 90-4 (Fed. Pubs. 1990).

39. 40 U.S.C. § 3131(b)(1).

40. Caron v. Andrew, 284 P.2d 544 (Cal. 1955). *See also* Employers Mut. Cas. Co. v. United Fire & Cas. Co., 682 N.W.2d 452 (Iowa Ct. App. 2004).

41. *E.g.,* Arntz Contracting Co. v. St. Paul Fire & Marine Ins. Co., 54 Cal. Rptr. 2d 888 (Cal. Ct. App. 1996).

42. *See generally* FAR 49.404(b)–(c).

43. Carchia v. United States, 485 F.2d 622 (Ct. Cl. 1973).

44. FAR 49.404(e).

45. *Compare* Westech Corp. v. United States, 20 Cl. Ct. 745 (1990), *with* Transamerica Ins. v. United States, 6 Cl. Ct. 367 (1984).

46. Essex Constr. Co. v. United States, GSBCA 6784, 85-1 BCA ¶ 17,818.

47. 16 Ct. Cl. 142 (1988).

48. 41 U.S.C. §§ 601–617.

49. *Id.*

50. McWaters & Bartlett v. United States *ex rel.* Wilson, 272 F.2d 291 (10th Cir. 1959).

51. *See generally* Daniel E. Toomey et al., *Good Faith and Fair Dealing: The Well-Nigh Irrefragable Need for a New Standard in Public Contract Law*, 20 Pub. Cont. L.J. 87 (1990).

52. Wolfe Constr. Co., ENG BCA 5309, 88-3 BCA ¶ 21,122.

53. Pearlman v. Reliance Ins. Co., 371 U.S. 132 (1962); Int'l Fid. Ins. Co. v. United States, 41 Fed. Cl. 706 (1998).

54. Ransom v. United States, 900 F.2d 242 (Fed. Cir. 1990).

55. *See* Fireman's Fund Ins. Co. v. United States, 909 F.2d 495, 500 (Fed. Cir. 1990); *see also* Nova Cas. Co. v. United States, 69 Fed. Cl. 284 (2006).

56. *See Int'l Fid. Ins. Co.*, 41 Fed. Cl. 706; *but see* Admiralty Constr., Inc. by Nat'l Am. Ins. Co. v. Dalton, 156 F.3d 1217 (Fed. Cir. 1998), where the court speaks only of a takeover agreement or a surety's financing its principal as creating a right to claim directly against the government. It may be that this is simply an oversight by that court.

57. *See* Universal Sur. Co. v. United States, 10 Cl. Ct. 794 (1986); *see also* Westchester Fire Ins. Co. v. United States, 52 Fed. Cl. 567 (2002).

58. *See* Transamerica Ins. Co. v. United States, 6 Cl. Ct. 367 (1984).

59. 31 U.S.C. § 3727; 41 U.S.C. § 15; *see also* Intercargo Ins. Co. v. United States, 41 Fed. Cl. 449 (1998).

60. *See Transamerica Ins. Co.*, 6 Cl. Ct. 367.

61. *See* Essex Constr. Co., GSBCA No. 6784, 85-1 BCA ¶ 17,818.

62. 28 U.S.C. § 1491.

63. 28 U.S.C. § 2501.

64. 41 U.S.C. §§ 601–617.

65. Note that it is a prerequisite that a surety enter into a takeover agreement in order to sue the government under the CDA. Reliance Ins. Co. v. United States, 27 Fed. Cl. 815 (1993); Universal Sur. Co., 10 Cl. Ct. 794. A completing contractor retained by a surety does not have a right to sue the government because it is not in contractual privity. George W. Kane., Inc. v. United States, 26 Cl. Ct. 655 (1992).

66. 41 U.S.C. § 606.

67. 41 U.S.C. § 609(a).

68. *See generally* Richard E. Tasker, G. Wayne Murphy Sr. & William Schwartzkopf, *Chapter 9: Financing the Principal*, in Practical Guide to Construction Contract Surety Claims (1997).

69. *See* 26 U.S.C. § 3505(b).

70. *See* Royal Indem. Co. v. United States, 529 F.2d 1312, 1319–21, 208 Ct. Cl. 809 (1976).

71. For a discussion of whether a surety has any obligation to finance its surety, see Daniel E. Toomey, *Is a Surety Obligated to Investigate Financing a Contractor Who Requests Financial Assistance?* CONSTR. LAW., Nov. 1992, at 11.

72. *See* Aetna Cas. & Sur. Co. v. United States, 845 F.2d 971 (Fed. Cir. 1988).

73. *Id.*

74. *See* United States v. Seaboard Sur. Co., 622 F. Supp. 882 (E.D.N.Y. 1985) (court rejected the government's bad faith claims holding that the surety's actions amounted to nothing more than assertion of the right to have the issue of wrongful termination litigated).

75. *Cf.* Fid. & Deposit of Md. v. Bristol Steel & Iron Works, Inc., 722 F.2d 1160 (4th Cir. 1983).

76. 41 U.S.C. §§ 601–617.

77. *See* Prairie State Nat'l Bank v. United States, 164 U.S. 227, 240, 17 S. Ct. 142, 41 L. Ed. 412, 32 Ct. Cl. 614 (1896); Ins. Co. of the W. v. United States, 243 F.3d 1367 (Fed. Cir. 2001).

78. *See* Westchester Fire Ins. Co. v. United States, 52 Fed. Cl. 567 (2002); Fireman's Fund Ins. Co. v. United States, 909 F.2d 495 (Fed. Cir. 1990); Am. Ins. Co. v. United States, 62 Fed. Cl. 151 (2004).

79. *See, e.g.,* Nat'l Sur. Corp. v. United States, 118 F.3d 1542 (Fed. Cir. 1997); John T. Callahan & Sons, Inc. v. Dykeman Elec. Co., 266 F. Supp. 2d 208 (D. Mass. 2003); Nova Cas. Co. v. United States, 72 Fed. Cl. 755 (2006).

80. *See* Carchia v. United States, 485 F.2d 622 (Ct. Cl. 1973).

81. *See* Westech Corp. v. United States, 20 Cl. Ct. 745 (1990).

82. *See* Transamerica Ins. Co. v. United States, 6 Cl. Ct. 367 (1984).

83. *See Nat'l Surety Corp.,* 118 F.3d 1542; *but see* Great Am. Ins. Co. v. Norwin Sch. Dist., 544 F.3d 229 (3d Cir. 2008) (where payment of retainage did not violate the language of the contract, the payment was not improper and therefore could not be claimed by the surety).

84. *See* Transamerica Ins. Company v. United States, 989 F.2d 1188 (Fed. Cir. 1993).

85. 31 U.S.C. §§ 3901–3907.

86. 41 U.S.C. § 611. For a recent discussion of prompt payment issues, see John W. Hays, *Prompt Payment Acts; Recent Developments and Trends,* CONSTR. LAW., Summer 2002, at 29.

87. *See* Aetna Cas. & Sur. Co. v. United States, 845 F.2d 971 (Fed. Cir. 1988). *See generally* Edward G. Gallagher, *Chapter 10: Entitlement to Contract Proceeds, in* THE LAW OF PAYMENT BONDS (Kevin L. Lybeck & H. Bruce Shreves eds., 1998).

88. 40 U.S.C. §§ 3141–3148. For further discussion of the Davis-Bacon Act, see Chapter 12.

89. *See* Westchester Fire Ins. Co. v. United States, 52 Fed. Cl. 567 (2002); *see also* Nat'l Fire Ins. Co. of Hartford v. Fortune Constr. Co., 320 F.3d 1260 (11th Cir. 2003) (on a private bond, the general's Davis-Bacon Act offsets were superior to the subcontractor's surety's subrogation rights).

90. *See* United States v. Munsey Trust Co., 332 U.S. 234 (1947); Sentry Ins. & Mut. Co. v. United States, 12 Cl. Ct. 320 (1987).

91. *See* Pearlman v. Reliance Ins. Co., 371 U.S. 132 (1962); Cont'l Cas. Co. v. United States, 164 Ct. Cl. 160 (1964); *see also* Commercial Cas. Ins. Co. v. United States, 71 Fed. Cl. 104 (2006).

92. *See* Dale Benz, Inc., Contractors v. Am. Cas. Co., 303 F.2d 80 (9th Cir. 1962).

93. Claims and delays are discussed in detail in Chapters 15 and 19, respectively.

94. ASBCA No. 5183, 60-2 BCA ¶ 2688 (1960), *aff'd on recons.*, 61-1 BCA ¶ 2894 (1961).

95. *See* United States *ex rel*. United Elec. Corp. v. Nat'l Bonding & Acc. Ins. Co., 711 F.2d 131 (9th Cir. 1983).

96. 5 U.S.C. § 504.

97. In recent years, courts have taken a more evenhanded approach in invoking the provisions of the Equal Access to Justice Act (EAJA) to both the government and private litigants for transgressions. The normal invocation of EAJA is by small business contractors (and perhaps their sureties) against the government for asserting and pursuing defenses that lacked "substantial justification."

98. Pub. L. No. 97-365, 96 Stat. 1749; *see* 31 U.S.C. §§ 3701, 3711, 3716–3719, and FAR 52.232-17.

99. *See* Westchester Fire Ins. Co. v. United States, 52 Fed. Cl. 567 (2002).

100. *See* Transamerica Sur. Co. v. United States, 6 Cl. Ct. 367 (1984).

101. *Id*. at 371.

102. *See* Nat'l Sur. Co. v. United States, 118 F.3d 1542 (Fed. Cir. 1997).

103. *Id*. at 1545. *See also* Balboa Ins. v. United States, 775 F.2d 1158, 1161 (Fed. Cir. 1985); *but see* Fireman's Fund Ins. Co. v. United States, 909 F.2d 495 (Fed. Cir. 1990) (contract specifically authorized government in its discretion to release retainage).

104. *See* Spearin v. United States, 248 U.S. 132 (1918).

105. For further discussion of defective specifications, see Chapter 18.

106. *See, e.g.*, H.B. Zachary Corp. v. Travelers Indem. Co., 391 F.2d 43 (5th Cir. 1968).

107. *See, e.g.*, L&A Contracting v. S. Concrete Servs., Inc., 17 F.3d 106 (5th Cir. 1994); Sch. Bd. of Escambia County, Fla. v. TIG Premier Ins. Co., 110 F. Supp. 2d 1351 (N.D. Fla. 2000).

108. FAR § 49.402-3(e)(1).

109. FAR § 49.402-3(e)(2).

110. The doctrine of cardinal change is discussed in Chapter 13.

111. *See, e.g.*, PCL Constr. Servs., Inc. v. United States, 47 Fed. Cl. 745 (2000).

112. *See* GE Supply Co. v. C & G Enters., Inc., 212 F.3d 14 (1st Cir. 2000).

113. *See* United States *ex rel*. Maddux Supply Co. v. St. Paul Fire & Marine Ins. Co., 86 F.3d 332 (4th Cir. 1996).

114. 40 U.S.C. § 3133(b)(1).

115. *See* United States *ex rel*. Haycock v. Hughes Aircraft Co., 98 F.3d 1100 (9th Cir. 1996), *supp'd*, 99 F.3d 1148, *cert. denied*, 520 U.S. 1211 (1997).

116. *See* United States *ex rel.* Apex Roofing & Insulation, Inc. v. Union Indem. Ins. Co. of N.Y., 865 F.2d 1226 (11th Cir. 1989); Faeber Elec. Co. v. Atlanta Tri-Com., Inc., 795 F. Supp. 240 (N.D. Ill. 1992).

117. *See* United States *ex rel.* Aldridge Elec. Co. v. Pickus Constr. & Equip. Co., 249 F.3d 664 (7th Cir. 2001).

118. *See* J.W. Bateson Co. v. United States *ex rel.* Bd. of Trs. of the Nat'l Automatic Sprinkler Indus. Pension Fund, 434 U.S. 586 (1978).

119. *See* United States *ex rel.* Metal Mfg. Inc. v. Fed. Ins. Co., 656 F. Supp. 1194 (D. Ariz. 1987).

120. *See* Clifford F. MacAvoy Co. v. United States *ex rel.* Calvin Tompkins Co., 322 U.S. 102 (1944).

121. *See* F.D. Rich Co. v. United States *ex rel.* Indus. Lumber Co., 417 U.S. 116 (1974).

122. *See* United States *ex rel.* Wellman Eng'g Co. v. MSI Corp., 350 F.2d 285 (2nd Cir. 1965); United States *ex rel.* Parker-Hannifin Corp. v. Lane Constr. Corp., 477 F. Supp. 400 (M.D. Pa. 1979).

123. *See* United States *ex rel.* Conveyor Rental & Sales Co. v. Aetna Cas. & Sur. Co., 981 F.2d 448 (9th Cir. 1992).

124. *Id.*

125. *See* United States *ex rel.* M.A. Bruder & Sons, Inc., v. Aetna Cas. & Sur. Co., 480 F. Supp. 659 (D.D.C. 1979).

126. *See* United States *ex rel.* PCC Constr., Inc. v. Star Ins. Co., 90 F. Supp. 2d 512 (D.N.J. 2000); United States *ex rel.* Briggs v. Grubb, 358 F.2d 508 (9th Cir. 1966).

127. *See* United States *ex rel.* Hillsdale Rock Co. v. Cortelyou & Cole, Inc., 581 F.2d 239 (9th Cir. 1978).

128. 40 U.S.C. § 3133(b)(2) (formerly 40 U.S.C. § 270b).

129. *See* United States *ex rel.* Greenwald-Supon, Inc. v. Gramercy Contractors, Inc., 433 F. Supp. 156 (S.D.N.Y. 1977).

130. Bowden v. United States *ex rel.* Malloy, 239 F.2d 572 (9th Cir. 1956); *see also* United States *ex rel.* Irving Equip., Inc. v. James N. Gray Co., Civil No. 05-215-p-s, 2006 U.S. Dist. LEXIS 567 (D. Me., Jan. 5, 2006).

131. *See* United States *ex rel.* Billows Elec. Supply Co., v. E.J.T. Constr. Co., 517 F. Supp. 1178 (E.D. Pa. 1981); United States *ex rel.* Keener Gravel Co. v. Thacker Constr. Co., 478 F. Supp. 299 (E.D. Mo. 1979).

132. *See* United States *ex rel.* Light & Power Utils. Corp. v. Liles Constr. Co., 440 F.2d 474 (5th Cir. 1971).

133. 40 U.S.C. § 3133(b)(2)(A) (formerly 40 U.S.C. § 270b(a)).

134. *See, e.g.,* United States *ex rel.* Moody v. Am. Ins. Co., 835 F.2d 745 (10th Cir. 1987) (timely receipt of unregistered letter satisfies the notice requirement); *but see* Pepper Burns Insulation, Inc. v. Artco Corp. 970 F.2d 1340 (4th Cir. 1992) (notice mailed within 90-day period but received after 90 days is insufficient notice).

135. *See* United States *ex rel.* Excavation Constr., Inc. v. Glenn-Stewart-Pinckney Builders & Developers, Inc., 388 F. Supp. 289 (D. Del. 1975).

136. *See* United States *ex rel.* Consol. Elec. Distribs., Inc. v. Altech, Inc., 929 F.2d 1089 (5th Cir. 1991).

137. *See* Houston Fire & Cas. Ins. Co. v. United States *ex rel.* Trane Co., 217 F.2d 727 (5th Cir. 1954).

138. *See* United States *ex rel.* NEW Interstate Concrete, Inc. v. EUI Corp., 93 F. Supp. 2d 974 (S.D. Ind. 2000).

139. *See* United States *ex rel.* Honeywell, Inc. v. A&L Mech. Contractors, Inc., 677 F.2d 383 (4th Cir. 1982); United States *ex rel.* Moody v. Am. Ins. Co., 835 F.2d 745 (10th Cir. 1987).

140. *See* United States *ex rel.* Kinlau Sheet Metal Works., Inc. v. Great Am. Ins. Co., 537 F.2d 222 (5th Cir. 1976).

141. *See NEW Interstate Concrete*, 93 F. Supp. 2d 974.

142. *See* United States *ex rel.* B&R, Inc. v. Donald Lane Constr., 19 F. Supp. 2d 217 (D. Del. 1998).

143. *See, e.g.,* United States *ex rel.* S&G Excavating, Inc. v. Seaboard Sur. Corp., 236 F.3d 883 (7th Cir. 2001); McWaters & Bartlett v. United States *ex rel.* Wilson, 272 F.2d 291, 295 (10th Cir. 1959); United States *ex rel.* Old Dominion Iron & Steel Corp. v. Mass. Bonding & Ins. Co., 272 F.2d 73 (3rd Cir. 1959); Bowden v. United States *ex rel.* Malloy, 239 F.2d 572 (9th Cir. 1956).

144. *See, e.g.,* United States *ex rel.* Water Works Supply Corp. v. George Hyman Constr. Co., 131 F.3d 28 (1st Cir. 1997); Maccaferri Gabions, Inc. v. Dynateria Inc., 91 F.3d 1431, 1437 (11th Cir. 1996); Consol. Elec. Distribs., 929 F.2d 1089; United States *ex rel.* Jinks Lumber Co. v. Fed. Ins. Co., 452 F.2d 485 (5th Cir. 1971).

145. *See* United States *ex rel.* Joseph T. Richardson, Inc. v. E.J.T. Constr. Co., 453 F. Supp. 435 (D. Del. 1978).

146. *See* J.D. Fields & Co. v. Gottfried Corp., 272 F.3d 692 (5th Cir. 2001).

147. 40 U.S.C. § 3133(b)(1) (formerly 40 U.S.C. § 270b).

148. *See* United States *ex rel.* McAmis Indus. of Or., Inc. v. M. Cutter Co., 130 F.3d 440 (9th Cir. 1997); United States *ex rel.* Capitol Elec. Supply Co. v. C.J. Elec. Contractors, Inc., 535 F.2d 1326 (1st Cir. 1976).

149. *See* United States *ex rel.* Harvey Gulf Int'l Marine, Inc. v. Md. Cas. Co., 573 F.2d 245 (5th Cir. 1978).

150. *See* United States *ex rel.* Texas Bitulithic Co. v. Fid. & Deposit Co. of Md., 813 F.2d 697 (5th Cir. 1987).

151. *Id.*

152. *Compare* United States *ex rel.* Angell Bros., Inc. v. Cave Constr., Inc., 250 F. Supp. 873 (D. Mont. 1966), *with* United States *ex rel.* Casablanca v. Cont'l Cas. Co., 354 F.2d 1353 (D.P.R. 1972).

153. United States *ex rel.* Portland Constr. Co. v. Weiss Pollution Control Corp., 532 F.2d 1009 (5th Cir. 1976); United States *ex rel.* Lab. Furniture Co. v. Reliance Ins. Co., 274 F. Supp. 377 (D. Mass. 1967).

154. *See* United States *ex rel.* Statham Instruments, Inc. v. W. Cas. & Sur. Co., 359 F.2d 521 (6th Cir. 1966).

155. *See, e.g.,* United States *ex rel.* J&A Landscape Co. v. Reza, Inc., No. 00-55082, 2002 WL 123567 (9th Cir. 2002).

156. *See* Gen. Ins. Co. of Am. v. United States *ex rel.* Audley Moore & Son, 409 F.2d 1326 (5th Cir. 1969), *cert. denied*, 396 U.S. 902 (1969); Polu Kai Servs., LLC v. Ins.

Co. of Pa., Civil Action No. 06-10708 Section "L"(1), 2007 U.S. Dist. LEXIS 93143 (E.D. La., Dec. 19, 2007).

157. United States *ex rel.* Mod-Form, Inc. v. Barton & Barton Co., 769 F. Supp. 235 (E.D. Mich. 1991); Trinity Universal Ins. Co. v. Girdner, 379 F.2d 317 (5th Cir. 1967); *see also* United States *ex rel.* E. Coast Contracting, Inc. v. USF&G, Civil No. CCB-03-3200, 2004 U.S. Dist. LEXIS 14441 (D. Md., July 23, 2004), *aff'd,* 133 Fed. Appx. 58, 2005 U.S. App. LEXIS 9683 (4th Cir. Md. 2005).

158. See United States *ex rel.* Monahan Insulation Co. v. Acme Missiles & Constr. Corp., 221 F. Supp. 733 (W.D. Pa. 1963).

159. *See* S. Steel Co. v. United Pac. Ins. Co., 935 F.2d 1201 (11th Cir. 1991).

160. *See* United States *ex rel.* Noland Co. v. Andrews, 406 F.2d 790 (4th Cir. 1969).

161. *See* J.D. Fields & Co. v. Gottfried Corp., 272 F.3d 692 (5th Cir. 2001).

162. United States *ex rel.* Ga. Elec. Supply Co. v. United States Fid. & Guar. Co., 656 F.2d 993 (5th Cir. 1981); *see also* United States v. Int'l Fid. Ins. Co., 200 F.3d 456 (6th Cir. 2000) (repair work will not toll limitations period).

163. *See* S. *Steel,* 935 F.2d 1201.

164. Robert F. Carney, *Chapter 6: Suit Limitations, in* THE LAW OF PAYMENT BONDS, *supra* note 87, at 119.

165. *See* United States *ex rel.* J.A. Edwards & Co. v. Bregman Constr. Corp., 172 F. Supp. 517 (E.D.N.Y. 1959); United States *ex rel.* GE Supply v. G & C Enters., Inc., 29 F. Supp. 2d 49 (D.P.R. 1998).

166. *See* United States *ex rel.* Trane Co. v. Raymar Contracting Corp., 295 F. Supp. 234 (S.D.N.Y. 1968); United States *ex rel.* Robert DeFilippis Crane Serv., Inc. v. William Crow Constr. Co., 826 F. Supp. 647 (E.D.N.Y. 1993).

167. *See* Noland Co. v. Allied Contractors, Inc., 273 F.2d 917 (4th Cir 1959).

168. *See* United States *ex rel.* Grotnes Mach. Works, Inc. v. Henry B. Byors & Son, Inc., 454 F. Supp. 203 (D.N.H. 1978).

169. *See* United States *ex rel.* Pippin v. J.R. Youngdale Constr. Co., 923 F.2d 146 (9th Cir. 1991).

170. See United States *ex rel.* SGB Universal Builders Supply, Inc. v. Fid. & Deposit Co. of Md., 475 F. Supp. 672 (E.D.N.Y. 1979).

171. *See* United States Dep't of the Navy *ex rel.* Andrews v. Delta Contractors Corp., 893 F. Supp. 125 (D.P.R. 1995).

172. *See* United States *ex rel.* Texas Bitulithic Co. v. Fid. & Deposit Co. of Md., 813 F.2d 697 (5th Cir. 1987).

173. *See* United States *ex rel.* Nelson v. Reliance Ins. Co., 436 F.2d 1366 (10th Cir. 1971).

174. United States *ex rel.* Humble Oil & Ref. Co. v. Fid. & Cas. Co. of N.Y., 402 F.2d 893, 898 (4th Cir. 1968).

175. *See* United States *ex rel.* Witt v. JP, Inc., 655 F. Supp. 480 (D. Alaska 1987).

176. Visor Builders, Inc. v. Devon E. Tranter, Inc., 470 F. Supp. 911 (M.D. Pa. 1978).

177. *See* United States *ex rel.* B & B Welding, Inc. v. Reliance Ins. Co. of N.Y., 743 F. Supp. 129 (E.D.N.Y. 1990).

178. Alaska *ex rel.* Palmer Supply Co. v. Walsh & Co., 575 P.2d 1213 (Alaska 1978).

179. 40 U.S.C. § 270b.

180. *See* F.D. Rich v. United States *ex rel.* Indust. Lumber Co., 417 U.S. 116, 125 (1974); United States. *ex rel.* Expedia, Inc. v. Altex Enters., Inc., 734 F. Supp. 972, 973 (M.D. Fla. 1990).

181. *See* United States *ex rel.* Miller v. Mattingly Bridge Co., 344 F. Supp. 459 (W.D. Ky. 1972).

182. United States *ex rel.* Noland Co. v. Irwin, 316 U.S. 23 (1942).

183. United States *ex rel.* Sligh v. Fullerton Constr. Co., 296 F. Supp. 518 (D.S.C. 1968), *aff'd*, 407 F.2d 1339 (4th Cir. 1969).

184. 40 U.S.C. § 3133(b)(3) (formerly 40 U.S.C. § 270b(b)).

185. *See* F.D. Rich, 417 U.S. 116.

186. *See* United States *ex rel.* Giannola Masonry Co. v. P.J. Dick, Inc., 79 F. Supp. 2d 803 (E.D. Mich. 2000).

187. *See* United States *ex rel.* DeLay & Daniels, Inc. v. Am. Employers Ins. Co., 290 F. Supp. 139 (D.S.C. 1968).

188. Arrow Plumbing & Heating, Inc. v. N. Am. Mech. Servs. Corp., 810 F. Supp. 369 (D.R.I. 1993).

189. 28 U.S.C. § 1391. *See also* United States *ex rel.* Skip Kirchdorfer, Inc. v. M.J. Kelley Corp., 995 F.2d 656 (6th Cir. 1993).

190. *See* United States *ex rel.* Expedia, Inc. v. Altex Enters., Inc., 734 F. Supp. 972 (M.D. Fla. 1990). For discussion of federal government construction contracting outside the United States, see Chapter 25.

191. *See* United States *ex rel.* Coffey v. William R. Austin Constr. Co., 436 F. Supp. 626 (W.D. Okla. 1977).

192. *See* United States *ex rel.* M.G. Astleford Co. v. S.J. Groves & Sons Co., 53 F.R.D. 656 (D. Minn. 1971).

193. *See* United States *ex rel.* A.C. Garrett v. Midwest Constr. Co., 619 F.2d 349 (5th Cir. 1980).

194. *See generally* Laurence J. Zielke, *Arbitrating Miller Act Claims and Problems in Enforcing an Award Under the Federal Arbitration Act*, 24 PUB. CONT. L.J. 410 (1995). For discussion of alternative dispute resolution in federal government construction, see Chapter 17.

195. *See* United States *ex rel.* Humbarger v. Law Co., Case No. 01-4156-SAC, 2002 U.S. Dist. LEXIS 4702 (D. Kan., Feb. 20, 2002); *see also* Elec. & Missile Facilities, Inc. v. United States *ex rel.* Moseley, 306 F.2d 554 (5th Cir. 1962), *rev'd on other grounds sub nom.* Moseley v. Elec. & Missile Facilities, Inc., 374 U.S. 167 (1963).

196. *See* United States *ex rel.* Tanner v. Daco Constr., Inc., 38 F. Supp. 2d 1299 (N.D. Okla. 1999).

197. *Id.*

198. *See* United States *ex rel.* Portland Constr. Co. v. Weiss Pollution Control Corp., 532 F.2d 1009 (5th Cir. 1976).

199. 7 U.S.C. § 9.

200. *Compare* Sunshine Beauty Suppliers, Inc. v. United States Dist. Court for the Cent. Dist. of Cal., 872 F.2d 310 (9th Cir. 1959), *with* Sigma v. Dean Witter Reynolds, Inc., 766 F.2d 898 (2d Cir. 1985), *cert. denied*, 105 S. Ct. 1381 (1986).

201. *See* Cortez Byrd Chips, Inc. v. Bill Harbert Constr. Co., 529 U.S. 193, 202–03, 120 S. Ct. 1331, 146 L. Ed. 2d 171 (2000); *see also* Union Water Power Co. v. Local Union No. 42, IBEW, Docket No. 99-14-P-H, 2000 U.S. Dist. LEXIS 2469, 2000 WL 761632 (D. Me., Feb. 16, 2000), at 7 n.3.

202. 40 U.S.C. § 270a(a)(2); United States *ex rel.* Olsen v. W.H. Cates Constr. Co., 972 F.2d 987 (8th Cir. 1992).

203. *See* United States *ex rel.* Naberhaus-Burke v. Butt & Head, Inc., 535 F. Supp. 1155 (S.D. Ohio 1982).

204. *See* United States *ex rel.* Constructors, Inc. v. Gulf Ins. Co., 313 F. Supp. 2d 593 (E.D. Va. 2004).

205. 40 U.S.C. § 3131 (formerly 40 U.S.C. § 270(a)–(d)); United States v. Am. Mfrs. Mut. Cas. Co., 901 F.2d 370 (4th Cir. 1990).

206. 40 U.S.C. § 3131 (formerly 40 U.S.C. § 270(d)).

207. *See* Williams v. Ashland Eng'g Co., 863 F. Supp. 46 (D. Mass. 1994); *aff'd*, 45 F.3d 588 (1st Cir. 1995), *cert denied*, 516 U.S. 807 (1995).

208. *See* Int'l Union of Operating Eng'rs Local 57 v. Seaboard Sur. Co., 946 F. Supp. 141 (D.R.I. 1996).

209. *See* United States *ex rel.* Sherman v. Carter, 353 U.S. 210 (1957).

210. Equitable Cas. & Sur. Co. v. Helena Wholesale Grocery Co., 60 F.2d 380 (8th Cir. 1932).

211. *See* U.S. Martin Steel Constructors, Inc. v. Avanti Constructors, Inc., 750 F.2d 759 (9th Cir. 1984), *cert. denied*, 474 U.S. 817, 106 S. Ct. 60, 88 L. Ed. 2d 49 (1985).

212. *See* United States *ex rel.* Dragone Bros., Inc. v. Moniaros Contracting Corp., 882 F. Supp. 1267 (E.D.N.Y. 1995).

213. *See* United States *ex rel.* T.M.S. Mech. Contractors, Inc. v. Millers Mut. Fire Ins. Co. of Texas, 942 F.2d 946 (5th Cir. 1991).

214. *See* United States *ex rel.* Conveyor Rental & Sales Co. v. Aetna Cas. & Sur. Co., 981 F.2d 448 (9th Cir. 1992).

215. *See* United States *ex rel.* Int'l Bus. Machs. Corp. v. Hartford Fire Ins. Co., 112 F. Supp. 2d 1023 (D. Haw. 2000); United States *ex rel.* Hawaiian Rock Prods. Corp. v. A. E. Lopez Enters., Ltd., 74 F.3d 972 (9th Cir. 1996).

216. *See* United States *ex rel.* Westinghouse Elec. Supply Co. v. Endebrock-White Co., 275 F.2d 57 (4th Cir. 1960).

217. *See* United States *ex rel.* Purity Paint Prods. v. Aetna Cas. & Sur. Co., 56 F. Supp. 431 (D. Conn. 1944).

218. *See* United States *ex rel.* Balzer Pac. Equip. Co. v. Fid. & Deposit Co. of Md., 895 F.2d 546 (9th Cir. 1990).

219. *See* United States *ex rel.* D & P Corp. v. Transamerica Ins. Co., 881 F. Supp. 1505 (D. Kan. 1995).

220. *See* United States *ex rel.* F & R. Constr. Co. v. Guy H. James Constr. Co., 390 F. Supp. 1193 (M.D. Tenn. 1972), *aff'd*, 489 F.2d 756 (6th Cir. 1974).

221. *See* United States *ex rel.* Carlisle Constr. Co. v. Coastal Structures, Inc., 689 F. Supp. 1092 (M.D. Fla. 1988).

222. *See* United States *ex rel.* Rent It Co. v. Aetna Cas. & Sur. Co., 988 F.2d 88 (10th Cir. 1993).

223. *See* Transamerica Premier Ins. Co. v. Ober, 894 F. Supp. 471 (D. Me. 1995).

224. *Compare* Morgan Towing Corp. v. M.A. Gammino Constr. Co., 363 F.2d 108 (1st Cir. 1966), *with* Transamerica Premier Ins., 894 F. Supp. 471.

225. *See* Naberhaus-Burke Inc., 535 F. Supp. 1155; United States *ex rel.* Constructors, Inc. v. Gulf Ins. Co., 313 F. Supp. 2d 593 (E.D. Va. 2004).

226. *See* United States *ex rel.* Carlisle Constr. Co. v. Coastal Structures, Inc., 689 F. Supp 1092 (M.D. Fla. 1988).

227. *See* U.S.F&G. v. United States *ex rel.* Bartlett, 231 U.S. 237 (1913).

228. *See* Psychiatric Inst. of Wash., D.C. v. Conn. Gen. Life Ins., Co., 780 F. Supp. 24 (D.D.C. 1992).

229. *See* Nickell v. United States *ex rel.* D.W. Falls, Inc., 355 F.2d 73 (10th Cir. 1966).

230. *See* United States *ex rel.* Mobile Premix Concrete, Inc. v. Santa Fe Eng'rs, Inc., 515 F. Supp. 512 (D. Colo. 1981).

231. *See* Consol. Elec. & Mechs., Inc. v. Biggs Gen. Contracting, Inc., 167 F.3d 432 (8th Cir. 1999); United States *ex rel.* Heller Elec. Co. v. William F. Klinensmith, Inc., 670 F.2d 1227 (D.C. Cir. 1982); Mai Steel Serv., Inc. v. Blake Constr. Co., 981 F.2d 414 (9th Cir. 1992); United States *ex rel.* Aldridge Elec. Co. v. Pinkus Constr. Equip. Co., 249 F.3d 664 (7th Cir. 2001); Lighting & Power Serv., Inc. v. Roberts, 354 F.3d 817 (8th Cir. 2004).

232. *See* United States *ex rel.* Wallace v. Flintco., Inc., 143 F.3d 955 (5th Cir. 1998).

233. *See* United States *ex rel.* Walton Tech., Inc. v. Westar Eng'g, Inc., 290 F.3d 1190 (9th Cir. 2002).

234. *See* F.D. Rich Co. v. United States *ex rel.* Indus. Lumber Co., 417 U.S. 116 (1974).

235. *See* United States *ex rel.* Great Wall Constr., Inc. v. Mattie & O'Brien Mech. Contracting Co., Civil No. 00-080-P-H, 2001 WL 127663 (D. Me. 2001); United States. *ex rel.* Cal's A/C & Elec. v. Famous Constr. Corp., 220 F.3d 326 (5th Cir. 2000).

236. *See* United States *ex rel.* Maddux Supply Co. v. St. Paul Fire & Marine Ins. Co., 86 F.3d 332 (4th Cir. 1996); *see also* United States *ex rel.* Varco Pruden Bldgs. v. Reid & Gary Strickland Co., 161 F.3d 915 (5th Cir. 1998); United States *ex rel.* Casa Redimix Concrete Corp. v. Luvin Constr. Corp., 00 Civ. 7552 (HB)(HBP), 2001 U.S. Dist. LEXIS 6184 (S.D.N.Y., May 11, 2001).

237. *See* United States *ex rel.* L.K.L. Assoc. v. Crockett & Wells Constr. Inc., 730 F. Supp. 1066 (D. Utah. 1990).

238. *See* Int'l Bus. Machs., 112 F. Supp. 2d 1023; United States *ex rel.* Canion v. Randall & Blake, 817 F.2d 1188 (5th Cir. 1987); *but see Great Wall Constr.*, 2001 WL 127663; United States *ex rel.* Rocking D&R Ranch, Inc. v. James Talcott Constr., Inc., Civil No. 00-180-P-H, 2007 U.S. Dist. LEXIS 90378 (D. Mont., Dec. 6, 2007).

239. *See* United States *ex rel.* Gen. Elec. Supply Co. v. Minority Elec. Co., 537 F. Supp. 1018 (S.D. Ga. 1982), *relying on* F.D. Rich Co. United States *ex rel.* Indus. Lumber Co., 417 U.S. 116 (1974).

240. *See* K-W Indus. v. Nat'l Sur. Co., 855 F.2d 640 (9th Cir. 1988); United States *ex rel.* Sunworks Div. of Sun Collector Corp. v. Ins. Co. of N. Am., 695 F.2d 455 (10th Cir. 1982). Note that two other courts have severely criticized the holding in *K-W Industries*. *See, e.g.*, Tacon Metal Contractors, Inc. v. Aetna Cas. & Sur. Co., 860 F. Supp. 386 (S.D. Tex 1994).

241. *See* United States *ex rel.* Metric Elec. v. EnviroServe, Inc., 301 F. Supp. 2d 56 (D. Mass. 2003).

242. Consol. Elec. & Mech., Inc. v. Biggs Gen. Contracting, 167 F.3d 432 (8th Cir. 1999); *but see* United States *ex rel.* D & P Corp. v. Transamerica Ins. Co., 881 F. Supp. 1505 (D. Kan. 1995).

243. *See* United States *ex rel.* Morgan & Son Earth Moving, Inc. v. Timberland Paving & Constr. Co., 745 F.2d 595 (9th Cir. 1984).

244. *See* United States *ex rel.* Cobb-Strecker-Dumphy & Zimmerman, Inc. v. M.A. Mortenson Co., 894 F.2d 311 (8th Cir. 1990).

245. *See* United States *ex rel.* T.M.S. Mech. Contractors, Inc. v. Millers Mut. Fire Ins. Co. of Texas, 942 F.2d 946, 953 (5th Cir. 1991).

246. 40 U.S.C. § 3133 (formerly 40 U.S.C. § 270b).

247. *See* Fid. & Deposit Co. of Md. v. Harris, 360 F.2d 402 (9th Cir. 1966).

248. *See* United States *ex rel.* B & M Roofing of Colo., Inc. v. AKM Assocs., Inc., 961 F. Supp. 1441 (D. Colo. 1997); *see also* United States ex. rel. D & P Corp. v. Transamerica Ins. Co., 881 F. Supp. 1505 (D. Kan. 1995).

249. United States v. W. States Mech. Contractors, 834 F.2d 1533, 1550 (10th Cir. 1987); United States *ex rel.* D & P Corp. v. Transamerica Ins. Co., 881 F. Supp. 1505, 1508 (D. Kan. 1995); United States *ex rel.* Joseph Stowers Painting, Inc. v. Harmon Constr. Co., Civil Action No. 87-2523-0, 1989 U.S. Dist. LEXIS 3492, 1989 WL 32195, at *4 (D. Kan. 1989).

250. *W. States*, 834 F.2d at 1550 (quoting St. Paul-Mercury Indem. Co. v. United States *ex rel.* H.C. Jones, 238 F.2d 917, 922 (10th Cir. 1957)).

251. *Id.* (citations omitted).

252. *Id.* at 1551 (citation omitted).

253. *See* United States *ex rel.* E. Gulf, Inc. v. Metzger Towing, Inc., 910 F.2d 775, 780 (11th Cir. 1990).

254. *See* Arthur N. Olive Co. v. United States *ex rel.* Marino, 297 F.2d 70 (D. Mass. 1961).

255. *See* United States *ex rel.* D & P Corp. v. Transamerica Ins. Co., 881 F. Supp. 1505 (D. Kan. 1995).

256. *See* Painters Local Union No. 171 v. Willliams & Kelly, Inc., 605 F.2d 535 (10th Cir. 1979); *see also* Concrete Indus. v. Dobson Bros. Constr. Co., Case No. 06-1325-WEB, 2007 U.S. Dist. LEXIS 36417, 63 U.C.C. Rep. Serv. 2d (CBC) 173 (D. Kan. 2007).

257. *See* Am. Cas. Co. v. Arrow Rd. Constr., 309 F.2d 923 (9th Cir. 1962).

258. *See* Barnard Curtiss Co. v. United States. *ex rel.* D.W. Falls Constr. Co., 244 F.2d 565 (10th Cir. 1957).

259. *See* City *ex rel.* Charles Equip. Co. v. United States Fid. & Guar. Co., 491 N.E.2d 1269 (Ill. App. Ct. 1986).

260. *See* United States *ex rel.* Raymond A. Bergen, Inc. v. DeMatteo Constr. Co., 467 F. Supp. 22 (D. Conn. 1979).

261. United States *ex rel.* C.H. Benton, Inc. v. Roelof Constr. Co., 418 F.2d 1328 (9th Cir. 1969).

262. St. Paul Fire & Marine Ins. Co. v. United States *ex rel.* Dakota Elec. Supply Co., 309 F.2d 22 (8th Cir. 1962).

263. United States *ex rel.* Maddux Supply Co. v. St. Paul Fire & Marine Ins. Co., 86 F.3d 332 (4th Cir. 1996).

264. United States ex. rel. Gen. Elec. Supply Co. v. Wiring, Inc., 646 F.2d 1037 (5th Cir. 1981).

265. *See* United States *ex rel.* Clark-Fontana Paint Co. v. Glassman Constr. Co., 397 F.2d 8 (4th Cir. 1968).

266. *See* United States *ex rel.* Marmet Corp. v. Becon Servs. Corp., 794 F. Supp. 428 (D.D.C. 1992).

267. *See* Iowa Supply Co. v. Grooms & Co. Constr., Inc., 428 N.W.2d 666 (Iowa 1988).

268. *See* United States *ex rel.* A & W Concrete & Bldg. Materials v. A.P. Johnson Contractor, Inc., 225 F. Supp. 727 (E.D. La. 1964).

269. *See* State *ex rel.* Gus T. Handge & Sons Painting Co. v. Tri-State Constr. Co, 662 S.W.2d 928 (Mo. Ct. App 1983).

270. *See* Valley Asphalt, Inc. v. Stimpel Wiebelhaus Assoc., 3 Fed. Appx. 838, 2001 U.S. App. LEXIS 1707 (10th Cir. 2001); Hoeppner Constr. Co. v. United States, 273 F.2d 835 (10th Cir. 1960); J.F. White Eng'g Corp. v. United States, 311 F.2d 410 (10th Cir. 1962).

271. *See* United States *ex rel.* Glickfeld v. Krendel, 136 F. Supp. 276 (D.N.J. 1955). Note, one of the distinctions between accord and satisfaction and compromise and settlement is that the former can succeed in extinguishing an undisputed claim for less than the liquidated claim amount, while the latter may fail because of lack of consideration. *Cf.* All In One Maintenance Serv. v. Beech Mountain Constr. Co., 318 S.E.2d 856 (N.C. Ct. App. 1984).

272. *See, e.g.,* Pac. Lining Co. v. Algernon-Blair Constr. Co., 812 F.2d 237 (5th Cir. 1987); Crown Plastering Corp. v. Elite Assocs., 560 N.Y.S.2d 694 (N.Y. App. Div. 1990).

273. *See* United States *ex rel.* Ackerman v. Holloway Co., 126 F. Supp. 347 (D.N.M. 1914).

274. *See* United States *ex rel.* DDC Interiors v. Dawson Constr. Co., 895 F. Supp. 270 (D. Colo. 1995), *aff'd*, 82 F.3d 427 (10th Cir. 1996).

275. *See* Walton Tech., 290 F.3d 1199; United States *ex rel.* T.M.S. Mech. Contractors, Inc. v. Millers Mut. Fire Ins. Co. of Texas, 942 F.2d 946 (5th Cir. 1991); United States *ex rel.* Straightline Corp. v. Am. Cas. Co., Civil Action No. 5:06-00011, 2007 U.S. Dist. LEXIS 50688 (N.D. W. Va., July 12, 2007).

276. 285 F.2d 939 (10th Cir. 1960).

277. *See, e.g.,* United States *ex rel.* N.U., Inc. v. Gulf Ins. Co., 650 F. Supp. 557 (S.D. Fla. 1986); United States *ex rel.* Pembroke Steel Co. v. Phoenix Gen. Constr. Co., 462 F.2d 1098 (4th Cir. 1972); United States *ex rel.* T/N Plumbing & Heating v. Fryd Constr. Corp., 423 F.2d 980 (5th Cir. 1970).

278. *See* Construction Industry Payment Protection Act of 1999, Pub. L. No. 106-49 (1999), 40 U.S.C. § 270c: "(c) Nonwaiver of rights: Any waiver of the right to sue on the payment bond required by sections 270a to 270d-1 of this title *shall be void unless* it is in *writing*, signed by the person whose right is waived, and executed *after* such person has first furnished labor or material for use in the performance of the contract." (Emphasis added.)

279. *See* United States *ex rel.* Warren Painting Co. v. J.C. Boespflug Constr. Co., 325 F.2d 54 (9th Cir. 1963).

280. *See* Austin Elcon Corp. v. Avco Corp., 590 F. Supp. 507 (W.D. Tex. 1984).

281. *See* Lamparter Acoustical Prods. Ltd. v. Md. Cas. Co., 407 N.Y.S.2d 579 (N.Y. App. Div. 1978).

282. *See* United States *ex rel.* Seminole Sheet Metal Co. v. SCI, Inc., 828 F.2d 671 (11th Cir. 1987).

283. *See* United States *ex rel.* Wallace v. Flintco, Inc., 143 F.3d 955 (5th Cir. 1998).

284. *See* United States *ex rel.* Falco Constr. Corp. v. Summit Gen. Contracting Corp., 760 F. Supp. 1004 (E.D.N.Y. 1991).

285. *See* United States *ex rel.* Pertun Constr. Co. v. Harvesters Group, Inc., 918 F.2d 915 (11th Cir 1990).

286. Pub. L. No. 106-49 (1999). See revisions to FAR 28.102 and 52.228-13, 15, and 16, implementing the changes.

287. *See* United States *ex rel.* Material Serv. Co. v. Wolfson, 362 F. Supp. 454 (E.D. Mo. 1973).

288. *Id.*

289. *See* United States *ex rel.* DDC Interiors v. Dawson Constr. Co., 895 F. Supp. 270 (D. Colo. 1995), *aff'd*, 82 F.3d 427 (10th Cir. 1996).

290. *See* United States *ex rel.* Westinghouse Elec. Co. v. James Stewart Co., 336 F.2d 777 (9th Cir. 1964).

291. *See* United States *ex rel.* Youngstown Welding & Eng'g Co. v. Travelers Indem. Co., 802 F.2d 1164 (9th Cir. 1986).

292. *See* United States *ex rel.* J.W. Smith & Co. v. Aetna Cas. & Sur. Co., 480 F.2d 1095 (8th Cir. 1973).

293. United States *ex rel.* Krupp Steel Prods. v. Aetna Ins. Co., 923 F.2d 1521 (11th Cir. 1991).

294. *See* Moyer v. United States *ex rel.* Trane Co., 206 F.2d 57 (4th Cir. 1953).

295. *See* United States *ex rel.* Robert DeFilippis Crane Serv. Inc. v. William L. Crow Constr. Co., 826 F. Supp. 647 (E.D.N.Y. 1993).

296. *See* Leslie Miller, Inc. v. Arkansas, 352 U.S. 187 (1956).

297. Kaiser Indus. Corp. v. United States, 340 F.2d 322 (Ct. Cl. 1965).

298. *See* United States *ex rel.* Apex Roofing & Insulation, Inc. v. Union Indem. Ins. Co. of N.Y., 865 F.2d 1226 (11th Cir. 1989).

299. *See* Spang Indus., Inc. v. Aetna Cas. & Sur. Co., 512 F.2d 365 (2d Cir. 1975).

300. *See* United States *ex rel.* SGB Universal Builders Supply, Inc. v. Fid. & Deposit Co., 475 F.2d 672 (E.D.N.Y. 1972).

301. *See, e.g.,* Penn. Fire Ins. Co. v. Am. Airlines, 180 F. Supp. 239 (E.D.N.Y. 1960).

302. *See* United States *ex rel.* Howard P. Foley Co. v. Home Indem. Co., 265 F. Supp. 943 (M.D. Fla. 1966).

303. *See, e.g.,* Appley Brothers v. United States, 924 F. Supp. 935 (D.S.D. 1966), *aff'd*, 107 F.3d 876 (8th Cir. 1997).

304. *See* Evergreen Pipeline Constr. Co. v. Merritt Meridian Constr. Corp., 95 F.3d 153 (2nd Cir. 1996).

CHAPTER 24

The False Claims Act
and Other Fraud Statutes

KRISTA L. PAGES

I. Introduction

The trend over the past decade in construction contract claims is to throw in all claims—actual, potential, and sometimes even overstated—and see what falls out. This approach has been encouraged by the use of alternative dispute resolution procedures to resolve claims. The tendency is to make the claims high enough so that, when they are compromised in negotiation, the contractor is still in a profitable position. This philosophy, if applied in the public contract claims arena, can have serious consequences. The government may consider an overstated claim, however a contractor may justify it, to be fraud and in violation of several federal statutes. The recent Court of Federal Claims (COFC) case *Daewoo Engineering & Construction Co. v. United States*, discussed below, is a good example.[1]

The federal government has several powerful weapons to investigate and punish contractors for presenting overstated, "fraudulent" claims. Among the most powerful are the civil False Claims Act (FCA) and the criminal False Claims Act. Other statutes include (1) the Forfeiture Statute,[2] which is often used along with the FCA; (2) the antifraud provision of the Contract Disputes Act (CDA);[3] (3) the Program Fraud Civil Remedies Act of 1986 (PFCRA)[4] for smaller fraud claims; (4) the False Statements Act;[5] and (5) the Truth in Negotiations Act (TINA),[6] which can be brought for the defective pricing of claims. The federal government also added new regulations in the Federal Acquisition Regulation (FAR) that make it mandatory for contractors to notify the Office of the Inspector General (OIG) and the contracting officer (CO) whenever they have credible evidence that a principal, employee, agent, or subcontractor of the contractor has committed a violation of federal criminal law

727

involving fraud, conflict of interest, bribery or gratuity violations, or violation of the civil FCA.[7]

This chapter focuses on contractor claims that allegedly violate the civil FCA.[8] In the past 20 years, the government has used the FCA with increasing frequency against contractors in a variety of situations where false, inflated, or overstated claims have been presented to the government. To understand how and when the FCA and other antifraud statutes come into play, the practitioner must study not only the elements of the statutes but also how courts apply such statutes in construction cases. In an effort to provide practical guidance, this chapter concludes by discussing how a contractor can avoid FCA suits.

II. *History of the Civil and Criminal False Claims Act*

The FCA was enacted during the Civil War at the urging of President Abraham Lincoln to punish unscrupulous businessmen and wartime profiteers who sold broken rifles, lame horses, and useless ammunition to the Union Army.[9] Signed into law in 1863, the first federal FCA provided criminal and civil penalties and contained qui tam provisions, allowing a private citizen to bring suit to enforce the statute.[10] The FCA was significantly revised only once between the time of its enactment and 1986. Those revisions, added in 1943, and the restrictive court opinions that followed, thwarted the effectiveness of the FCA by narrowing a private citizen's ability to bring a qui tam suit and decreasing any potential bounty the private citizen was eligible to receive.[11]

Forty-three years later, in 1986, Congress reinvigorated the FCA because of real and perceived concerns over fraud in government contracting. Allegations of fraud within the U.S. Department of Defense (DoD) served to galvanize support for more stringent antifraud statutes. For instance, during this period, the DoD reported that 45 of the largest 100 (and nine of the largest ten) defense contractors were under investigation for "multiple fraud offenses."[12] Against this backdrop, Congress expanded the FCA's qui tam provisions and stripped away the many legislative and judicial bars that had developed since the 1943 amendments.[13] Since the amendments were passed in 1986, the government has recovered more than $20 billion in settlements and judgments for fraud against the United States.[14]

Today, there is even more effort being made to strengthen the FCA through new legislative amendments and increased enforcement. In October 2006, the U.S. Department of Justice (DoJ) created a National Procurement Fraud Task Force "to promote the early detection, prevention and prosecution of procurement fraud associated with increased contracting activity for national security and other government programs."[15] Senator Charles (Chuck) Grassley (R-Iowa), the leading congressional champion of the FCA, proposed the "False Claims Act Correction Act of 2007."[16] This proposed legislation was intended to "clarify" the FCA legislation and to overturn the controversial decisions in *United States ex rel. Totten v. Bombardier Corp.*[17] and *Rockwell International Corp. v. United States ex rel. Stone.*[18] Most importantly, the Fraud Enforcement and Recovery Act (FERA)[19] was signed into law on May 20, 2009. Although FERA was aimed at enforcement of mortgage fraud, securities and commodities fraud, financial institution fraud, and "other frauds related to Federal assistance and relief programs," Section 4 of the Act significantly revised or "clarified" the civil FCA in several important ways that will impact public procurement as discussed herein. The clarifications also impact certain aspects of the criminal FCA.

III. *Criminal and Civil Liability under the False Claims Act*

A. Criminal False Claims Statute[20]

Under the criminal FCA, presentation of a false, fictitious, or fraudulent claim to the federal government can result in both a fine and imprisonment of up to five years.[21] Fines may be as high as $250,000 for a felony conviction that does not involve fraud against the DoD;[22] for a false claim related to a contract with the DoD, the statute mandates a fine of up to $1 million.[23] Because prison sentences and fines are assessed per violation, a contractor can face substantial fines and the possibility of lengthy imprisonment if the contractor makes multiple false submittals.

In order for a contractor to be convicted of violating the criminal FCA, the prosecution must prove three essential elements for each offense: (1) the contractor presented a claim to the federal government; (2) the claim was false, fictitious, or fraudulent; and (3) the contractor knew the claim was false, fictitious, or fraudulent.[24]

1. *What Is a Claim and When Is It Presented to the Government?*

For criminal FCA liability to attach, the contractor must have presented a claim to the government. Although the criminal FCA statute does not define "claim," FERA amends the FCA to provide that a "'claim' (A) means any request or demand, whether under a contract or otherwise, for money or property and whether or not the United States has title to the money or property, that (i) is presented to an officer, employee, or agent of the United States; or (ii) is made to a contractor, grantee, or other recipient, if the money or property is to be spent or used on the Government's behalf or to advance a Government program or interest" and the government provides or provided any portion of the money or will reimburse the contractor for any portion of the money or property.[25]

Prior to FERA, in order for the presentation element to be satisfied, a contractor had to physically submit the claim to the government; intent to submit a claim was not sufficient. It is not clear from language in FERA whether the criminal FCA will still be interpreted to require the contractor to physically submit the claim to the government. However, nothing in the new amendments should impact the court's finding that the government need not pay the claim for the violation to have occurred; presentation alone is sufficient.[26]

Contractors are not the only individuals potentially liable under the criminal FCA. A subcontractor can be liable for the submission of a false claim even if it did not directly present the claim to the government. Thus, a submittal presented by a subcontractor to a prime contractor that then seeks compensation from the federal government can constitute a claim.[27] The subcontractor is fully liable under the statute as if it had presented its claim directly to the government.[28]

Under the new definition in FERA, the claim may be presented to a "department or agency" of the government or to a recipient of federal funds if the money or property provided to the recipient will be used on the government's behalf or to advance a government program or interest.[29] Departments include all executive branch departments.[30] Claims submitted to agencies, which include any "independent establishment, commission, administration, authority, board or bureau of the United States or any corporation in which the United States has a proprietary interest," also violate the statute.[31] The provision requiring presentation to the federal government has been interpreted expansively by the courts so that it also includes any state contract where federal funds are used, even if the contractor is unaware that the federal government was involved in the project.[32] With the new FERA clarifications, there should be no doubt that state contracts are included. Ignorance of the government's involvement is not a defense to a criminal false claims prosecution.[33]

2. What Makes a Claim False, Fraudulent, or Fictitious?

Because the criminal FCA statute requires that a submitted claim be false, fraudulent, or fictitious, it is not necessary for a claim to "be fraudulent so long as it is false or fictitious."[34] In examining allegations of FCA violations, courts have found exaggerated charges, which are characterized by the contractor as estimates, to be false for purposes of the FCA.[35] In such cases, the courts look to the circumstances surrounding the presentation of the claim when determining its falsity; consequently, a hastily prepared estimate or imprecise claim can be found false or fictitious.

3. When Does a Contractor Act "Knowingly" under the FCA?

For a claim to violate the criminal FCA, the contractor must have known the claim was false, fraudulent, or fictitious when the claim was presented for payment or credit.[36] Once again, the contractor must look to the civil statute for a definition of this element, since the criminal statute does not specify what is considered "knowledge." The civil FCA defines the terms "knowing" and "knowingly" to mean that a contractor "(1) has actual knowledge of the information; (2) acts in deliberate ignorance of the truth or falsity of the information; or (3) acts in reckless disregard of the truth or falsity of the information. Under the civil Act, no proof of specific intent to defraud is required."[37]

While the civil false claims statute clearly indicates that a showing of specific intent is not required, courts interpreting the criminal FCA vary on the degree to which mens rea is necessary for the claim presentation to violate the statute. Some courts equate knowledge of the false claim with intent to deceive and require specific proof of that intent as an essential element of the crime.[38] Other courts, however, hold that simple knowledge of the falsity is sufficient and "do not find that the statute specifies an intent to defraud as an element to be proved."[39]

In these jurisdictions, a contractor who submits claims to the government "knowing them to be false or fictitious or fraudulent . . . act[s] 'willfully,' that is, with . . . a consciousness that he was doing something wrong."[40] Still other courts have determined that a contractor acts with knowledge if it acts with "indifference to the truth"[41] or with reckless disregard for the truth. Reckless disregard can include lack of supervision of the contractor's employees, even if the contractor had no knowledge of and was not personally involved in the alleged fraudulent acts.[42]

B. Civil False Claims Statute[43]

In addition to the criminal FCA, false contractor claims may violate the civil FCA, which provides for trebled damages and civil penalties for the submission of false or fraudulent claims. Under the civil FCA, a contractor may be liable if it: (1) knowingly presents or causes to be presented false or fraudulent claims for payment or approval (this is the most common FCA violation); (2) knowingly makes, uses, or causes to be made or used a false record or statement material to a false or fraudulent claim; (3) conspires to commit a violation of the civil FCA; (4) "has possession, custody, or control of property or money used, or to be used, by the Government and knowingly delivers or causes to be delivered, less than all of that money or property"; (5) fails to verify the truth of the delivery receipt; (6) knowingly exchanges property with a government agent who may not lawfully exchange said property; or (7) knowingly makes, uses, or causes to be used a false record or statement material to an obligation to pay or transmit money or property to the Government, or knowingly conceals or knowingly and improperly avoids or decreases an obligation to pay or transmit money or property to the government.[44]

To establish a civil FCA violation, the government or a qui tam relator must show: (1) the contractor presented or caused to be presented a claim for payment to an agent of the United States; (2) the claim submitted was false or fraudulent; and (3) the contractor knew that the claim was false or fraudulent.[45]

1. What Is a Claim and When Is It Presented to the Government?

As discussed above, the civil FCA definition of a claim has been expanded to include any request for money or property submitted for payment not only to the government or a contractor but also to any recipient of federal funds if the government will reimburse the contractor for any portion of the money or property requested or if the money or property provided to the recipient will "be spent or used on the government's behalf or to advance a government program or interest."[46] For example, liability can occur where a subcontractor submits a false claim to the general contractor, which in turn requests reimbursement from the federal government. In such instances, the subcontractor is liable as if it had submitted the claim directly to the government. As the Supreme Court stated in *United States ex rel. Marcus v. Hess*, "any person who knowingly assisted in causing the Government to pay claims which were grounded in fraud, without regard to whether that person had direct contractual relations with the Government," is liable.[47] The innocent prime contractor that is unaware of the fraudulent claim is not liable under the civil FCA.

Similar to the criminal FCA, the civil Act will apply even if the contractor did not present claims directly to the federal government. Courts routinely find that a claim is presented for purposes of the FCA when it is made through a third party; the provisions of the FCA are "broad enough to include any request for money that was originally obtained from the United States Government."[48] Consequently, the claim may reach the government after passing through a state or local government,[49] an insurance company,[50] or a billing service.[51] Furthermore, as with the criminal FCA, each separate submission to the government seeking payment constitutes a claim under the FCA.[52] Thus, it is possible to have many FCA "claims" that arise from a single contract.[53] Because the statute requires only that the claim be submitted for payment, it is not necessary for the claim to be paid for the violation to occur.[54]

If the project is funded only in part with federal funds and the contractor presents claims to a state or local government or other entity administering the project, the FCA applies. Passage of FERA overturned the Supreme Court decision in *Allison Engine Co. v. United States ex rel. Sanders*, which held that the contractor must intend that the false statement be material to the government's decision to pay or approve the false claim.[55] Now FCA liability may attach simply where a false statement is "material to" a government payment of the claim and not merely where the statement was made "to get" the claim paid.[56] The FCA definition of "material" provided in the FERA amendments is "having a natural tendency to influence, or be capable of influencing, the payment or receipt of money or property."[57]

The contractor may face an FCA allegation even if it does not submit a claim seeking payment. In such cases, termed "reverse false claims," the contractor made a material misrepresentation or omitted certain information to avoid paying money owed to the government.[58] Reverse false claims arise when an obligation owed to the government is concealed, avoided, or decreased through the use of a false statement.[59] Although early decisions on reverse false claims found liability for a contractor's failure to report statutory or regulatory violations in order to avoid potential fines, later courts have declined to embrace such a sweeping interpretation of the FCA.[60] The majority of courts now hold that reverse false claims liability is limited to false statements made to avoid existing, rather than potential, obligations to pay the government.[61]

Under this current line of reasoning, any action by the contractor that has the purpose and effect of causing the government to pay out money it was not obligated to pay, or any action that intentionally deprives the government of money it is lawfully due, may be a claim within the meaning of the FCA.[62] However, contingent obligations (such as obligations arising from civil or criminal monetary penalties imposed after a finding of wrongdoing) cannot constitute reverse false claims because the pertinent obligations do not attach before the contractor makes or uses the false record or statement.[63] Reverse false claims may also arise where the contractor owes no direct obligation to the government.[64] In these cases of indirect reverse false claims, the lack of a contractual relationship between the contractor and the government is immaterial; as long as the contractor is aware of the loss to the government as a result of its fraudulent acts, the FCA applies.[65] In those cases where courts have found a reverse false claim, the government or a qui tam relator demonstrated that (1) the false statement caused the government to incur a loss[66] and (2) the contractor owed an existing obligation (created by statute, regulation, or contract) to the government from which the fraudulent conduct stemmed.[67]

2. What Makes a Claim False or Fraudulent?

While the FCA does not define what gives rise to a false or fraudulent claim, some claims are obviously false on their face. Examples include claims that contain inflated labor costs (where labor hours charged differ from the employees' actual time cards), requests for payment for materials or services that were not provided, or other erroneous information.[68] In such cases, the falsity is indisputable. However, it is possible for an invoice that contains no false information on its face to be a false claim. Thus, a claim can be false under the FCA if it is submitted for work not completed, for defective work, for the use of lesser-quality materials than required in the contract specifications (product substitution), for defective pricing, or for work for which the contractor is not entitled to be paid under the contract (such as a claim for reimbursement for unallowable costs). Because all payment invoices submitted to the federal government implicitly incorporate the terms and conditions of the underlying contract, such an invoice can be a false claim (despite the fact that it contains accurate information), since its submission implies that it conforms to contract specifications.[69]

Several FCA cases involved fraud allegations based on express or implied certifications. In these cases, the contractor's or subcontractor's submission of an invoice or progress report was held to be either an express or implied certification of compliance with laws, and a knowing violation of those laws created liability under the FCA. Even in the absence of an express or implied certification, the government or a relator may argue that the government did not bargain for performance that failed to comply with applicable statutory, regulatory, or contractual requirements and that a claim for such performance is a false claim. Furthermore, an implied certification can violate the FCA, even if it is not expressly false when filed.[70]

For example, recent court cases have found that liability can attach if the claimant violates its continuing duty to comply with the regulations on which payment is conditioned.[71] In *Shaw v. AAA Engineering and Drafting, Inc.*,[72] the Tenth Circuit held that work orders submitted by a contractor constituted false claims because the work orders contained an implied certification that the contractor complied with the environmental requirements set forth in the contract.[73] In this case, the court rejected the contractor's argument that, because the work orders were not linked to any payment from the government, they could not be the basis for FCA liability.[74] In contrast, another recent court opinion noted that a false certification of compliance with a statute or regulation cannot serve as the basis for an action under the FCA unless payment is conditioned on the false or fraudulent certification.[75]

3. When Does a Contractor Act "Knowingly" under the FCA?

In order for a contractor to violate the civil FCA, the government must show that the contractor knew the claim it submitted was false. Specifically, the statute requires that the contractor (1) had actual knowledge of the information, (2) was deliberately ignorant of the truth or falsity of the information, or (3) recklessly disregarded the truth or falsity of the information.[76] These provisions "are to be read broadly and . . . [are] violated not only by a person who makes a false statement or a false record to get the Government to pay the claim, but also by one who engages in a fraudulent course of conduct."[77] An individual must "make simple inquiries which would alert him that false claims are being submitted . . . the inquiry need only be 'reasonable and prudent under the circumstances,' which clearly recognizes a limited duty to inquire as opposed to a burdensome obligation."[78] There is no requirement of a specific intent to defraud.[79]

4. Can a Contractor Be Vicariously Liable for an FCA Violation?

Under general rules of agency law, an employer can be liable for the acts of its employee when its employee acts with apparent authority, even if the employee acts solely for self-benefit.[80] Similarly, in FCA cases, an employer may be vicariously liable for the fraudulent actions of its employee. Liability in these FCA cases is premised upon the belief that the employee's position facilitated the consummation of the fraud, because the actions appeared to be within the employee's scope of authority and the transaction seemed regular on its face to a third person. The issue of whether vicarious liability will be imposed upon a contractor for an employee's FCA violations typically turns on whether the employee acted with apparent authority.

Federal courts disagree on the elements necessary for vicarious liability to attach to the employer in an FCA case. In *United States v. Southern Maryland*

Home Health Services, a Maryland district court refused to hold a company liable for the acts of a nonmanagerial employee because the company was unaware of the acts, did not ratify the acts, and was not reckless in its hiring or supervision of the employee.[81] However, in *United States ex rel. Bryant v. Williams Building Corp.*, a district court in South Dakota held otherwise in a case involving a contract for the remodeling of housing units on a U.S. military installation. In that case, the government alleged that the project superintendent knew of the existence of asbestos but failed to report the discovery to the government when he submitted his daily reports, progress reports, and payment invoices.[82] The court imputed the project superintendent's concealment to the contractor and found that a contractor is liable whenever its employee acts within the scope of employment or with apparent authority, regardless of the contractor's knowledge, culpability, policies, or efforts to restrain the employee's unlawful acts.

The *Bryant* court observed that vicarious liability applies, regardless of the effect of the employee's violation on the contractor, unless there is statutory language or purpose that precludes its imposition. Looking to the legislative language of the FCA and its intent, the court determined that the statute imposes liability upon employers for the fraud of their employees.[83] Thus, even though (1) the corporation received no benefit from the employee's fraud, (2) the employee acted solely for his personal benefit, and (3) the employee acted in a manner adverse to the interests of the contractor, the contractor remained liable because, under the court's analysis, apparent authority, not corporate benefit, is the proper standard to be used in the determination of an employer's vicarious liability under the FCA.[84]

5. What Types of Penalties Can Be Imposed?

Once the court establishes liability, the next issue in an FCA case is the amount of damages. The FCA provides for civil penalties of $5,000 to $10,000 for each violation, plus treble the amount of damages the government suffers as a result of the violations.[85] Effective September 29, 1999, the civil penalty limits were increased by the DoJ to $5,500 to $11,000, respectively.[86] Under the FERA revisions to the FCA, the civil penalties will be adjusted by the Federal Civil Penalties Inflation Adjustment Act of 1990.[87] Damages are calculated as the difference between what the government actually paid and what the government either received or should have been paid had the claim or statement not been false, plus "3 times the amount of damages which the Government sustains because of the act."[88] In assessing damages, the court has discretion to determine whether to award either double or treble damages.[89] If the individual responsible for committing the violations furnishes government investigators with all information known by that person within 30 days after the date on which the person first obtained the information, fully cooperates with the government investigation, and had no knowledge of a previous investigation into this violation, the court at its discretion may assess only double damages.[90]

The new FERA amendments provide that a person violating the civil FCA also "shall be liable to the United States Government for the costs of the civil action brought to recover" FCA penalties and damages.[91]

6. What Is Qui Tam Litigation?

The civil FCA also provides a mechanism by which private citizens can bring actions for fraud on behalf of the United States—the qui tam suit.[92] The phrase "qui tam" is derived from the Latin *qui tam pro domino rege quam pro sic ipso in hac parte sequitur*, meaning one "who brings the action for the king as well as himself." The plaintiff in a qui tam proceeding is referred to as the "relator" and files the complaint on behalf of the United States against a party who the relator believes violated the FCA. While relators are not popular among contractors, the government encourages such suits by private citizens.[93] In order to bring a qui tam suit against a contractor, the relator must possess "direct and independent knowledge of the information on which the allegations are based and [must have] voluntarily provided the information to the government before filing an action."[94] The qui tam provisions of the FCA encourage insiders who might be privy to fraud to act as whistleblowers.[95]

The "classic" qui tam case involves a company insider who brings an action against his employer. Although the disgruntled employee is the most common qui tam relator, the statute allows any person who has knowledge of the fraud to bring a civil action.[96] Circuits differ on the circumstances in which government employees may file actions as qui tam relators.[97]

While qui tam cases are filed as civil actions, the DoJ refers all cases to its Criminal Division for review, and criminal charges may be brought against the contractor as a result of the qui tam suit. However, the government need not take part in the relator's qui tam suit; a private plaintiff can go forward with the action even where the government chooses not to intervene.[98] A court has discretion to allow the government to intervene at a later time if the government can show good cause, so long as the rights of the relator are not compromised by the intervention.[99]

If the government declines to intervene in the relator's case, and the relator then settles with the defendant, the government may still object to the terms of the settlement even though the government did not join as a party to the claim; however, any government objections to a proposed settlement between the relator and the defendant "are not dispositive in such a case."[100] Government refusal to intervene may insulate the government from intrusive discovery requests by the defendant.[101]

The FCA provides for a statute of limitations for the filing of a qui tam suit by a relator. The relator must bring the suit within six years of the date of the infraction or within three years of when the facts become known to the relator.[102] Further, the statute bars any alleged infraction brought more than 10 years from the date of the offense.[103] If the government elects to intervene in the action, the qui tam relator will receive a bounty of between 15 and 25 percent of the penalty or settlement.[104] In a case where the government decides not to take part, the qui tam relator will receive "an amount which the court decides is reasonable for collecting the civil penalty or damages," between 25 and 30 percent of the proceeds.[105]

The Supreme Court has ruled that qui tam relators have the constitutional standing to sue, even though they themselves have not suffered an injury in fact. Under Supreme Court precedent, a plaintiff must possess an "irreducible constitutional minimum of standing" in order to bring suit.[106] Industries that are hard hit by the qui tam provisions of the FCA have long argued that the standing requirement was not met by relators and that the provision was therefore unconstitutional. However, the Court found that the FCA can be regarded as effecting a partial assignment of the government's damages claims, thereby providing a relator standing to assert the injury in fact suffered by the assignor, which in this case is the government.[107]

7. Special Concerns in Qui Tam Litigation

If an employee, contract hire, or agent of the contractor has instituted a qui tam lawsuit under the FCA, the contractor must be very careful in how it deals with that person. The FERA has strengthened the whistleblower protection provisions of the FCA that prohibit employers from harassing, threatening, suspending, discharging, demoting, discriminating, or otherwise retaliating in any way against persons who engage in lawful conduct in furtherance of a qui tam action, including initiating or assisting in an FCA suit.[108]

If an employer violates this provision, the FCA provides the whistleblower with "all relief necessary to make the employee, contract hire or agent whole."[109] Under the statute, the court can force the employer to (1) reinstate the whistleblower with the same seniority status; (2) pay the whistleblower double the amount of back pay plus interest on the back pay; and (3) provide compensation for any special damages sustained as a result of the employer's discriminatory actions, including litigation costs and reasonable attorney's fees.

An important aspect of an FCA retaliation claim requires a showing that the employer acted against the employee because of the employee's lawful acts in furtherance of an FCA suit. Thus, to establish a prima facie case of retaliation, the whistleblower must prove that (1) he engaged in lawful acts in furtherance of a qui tam suit that was filed or is to be filed; (2) the employer was aware that the employee engaged in such protected conduct; and (3) the employer discharged, demoted, suspended, threatened, harassed, or in any other manner discriminated against the employee because of such protected conduct.[110] Some courts have allowed employees to maintain a retaliation claim under the FCA even though an FCA claim ultimately was not filed against the employer.[111]

In the absence of a definition for employee and employer under the FCA, courts have relied on common law agency principles to determine who can sue for unlawful discharge or retaliation under the statute. Thus, courts have found that the rights provided under the FCA can be asserted only if the evidence shows that the whistleblower has or had a direct employment relationship with the defendant.[112] Applying these principles of agency law, several courts have held that independent contractors and partners are not protected employees for purposes of the FCA.[113] However, the FCA revisions provided by FERA have expanded protection to contract employees and agents, so some of these legal precedents will likely no longer apply.

In addition to the FCA, an employee may be able to utilize state law bases for a suit against the employer. For example, some states allow wrongfully discharged employees to assert a cause of action based upon a judicially recognized public policy exception to the employment-at-will doctrine.[114] Other states have enacted whistleblower protection statutes that are intended to protect employees who report an employer's violation of state or federal law.[115] In many cases, employers have defended such suits on the grounds that the sought-for state law remedy is preempted by federal legislation. Although it does not appear that any court to date has found that the FCA preempts state law claims, courts have held that other federal whistleblower acts preempt state law claims for wrongful discharge by protected employees.[116]

IV. The Government's Use of the FCA in Construction Cases

Some of the construction cases that have been prosecuted by the government involve obvious, deliberate fraudulent actions on the part of the contractor. Others involve the gray area in the statute, "deliberate ignorance and reckless disregard of the truth or falsity of information pertaining to the claim."[117] Each FCA case involving a construction contractor is different because of the unique claims asserted on each project. The following cases illustrate some of the ways contractors have become subject to criminal or civil FCA liability. In many of the cases, liability under the FCA came as a surprise when the government brought counterclaims after suit had been brought by the contractor.

A. *Daewoo Engineering and Construction Co. v. United States*[118]

Daewoo won a contract from the government, as low bidder, to construct a 53-mile road on the Pacific island nation of Palau. Daewoo submitted a certified claim for additional compensation under the CDA[119] for $65 million, claiming that unexpectedly adverse weather made the soil impossible to compact to the contract specifications. The claim was denied by the CO, and Daewoo appealed to the COFC.[120] The government filed counterclaims under the FCA and other antifraud statutes,[121] based on the testimony of Daewoo's project management and experts.

The court found Daewoo liable under the FCA because it had added $50 million to its legitimate claim of about $13 million when it "submitted false records and made false statements in preparing, certifying, and pursuing its claim" against the government.[122] The court assessed a $10,000 penalty for the claim and found that Daewoo made 762 other misrepresentations to the government—for example, stating compaction requirements were impossible when its records showed otherwise, alleging that earthwork could not be done in the allotted time when a subcontractor had been successful in so doing, and overstating the cost of 700 items of equipment in its claim. The court stated that these misrepresentations and false claims could result in an additional $7,620,000 in FCA penalties.

The court found that because of its fraudulent conduct, Daewoo had forfeited its legitimate claims of $13 million and was liable to pay an additional forfeiture of $50 million (plus costs) under the antifraud provision of the CDA.[123] On appeal, the Federal Circuit upheld the decision.[124]

B. *United States ex rel. Plumbers & Steamfitters Local Union No. 38 v. C.W. Roen Construction*[125]

In 1994, C.W. Roen Construction Company (Roen) entered into a contract with the City of Santa Rosa, California, to make improvements to the Laguna Wastewater Treatment Plant. As a federally funded project, the plant was subject to the prevailing wage and reporting requirements set forth in the Davis-Bacon Act.[126] The act requires contractors to pay workers prevailing wages and to submit weekly statements reflecting the wages paid.[127] Regulations governing federal construction contracts also dictate that the contractor certify its payments.[128]

Roen submitted the certifications as required. However, the government alleged that the certifications amounted to false statements because the company paid employees performing certain types of work less than the prevailing wage rate. Roen classified workers who performed the piping work as laborers and paid them at the laborer wage rate. It then certified that it was paying these workers the appropriate wage rates under the Davis-Bacon Act. The government claimed that Roen misclassified these workers since all workers who perform this type of piping work on wastewater treatment plants in Northern California must, under Davis-Bacon, be classified as "plumbers and steamfitters," not laborers. Thus, the government alleged that the misclassification and underpayment of the workers violated the FCA.

The court held that the FCA extends to false statements regarding the payment of prevailing wages. Relying on *United States ex rel. Hopper v. Anton*,[129] the court found that "it is the false certification of compliance which creates liability when certification is a prerequisite to obtaining a Government benefit."[130] Thus a false certification that workers have been paid at the legally required wage rate may give rise to liability under the FCA. Furthermore, the court found that a showing of "knowing" false presentation does not require proof of specific intent to defraud.

C. *Commercial Contractors, Inc. v. United States*[131]

In this case, the Army Corps of Engineers (Corps) awarded a contract to Commercial Contractors, Inc. (CCI) to construct several segments of the Telegraph Canyon Channel in Chula Vista, California, as part of a flood control project. The contract required CCI to excavate the areas in which the channel segments were to be built, to build the channel segments by setting up forms and pouring concrete into the forms, and to backfill the excavated areas surrounding the channel segments. The contract contained detailed specifications that governed all aspects of the work, including drawings indicating the areas CCI was required to excavate, quality control standards specifying the hardness of the poured concrete before the supporting forms were removed, and miscellaneous other provisions specifying such factors as the composition and compaction of the backfill materials.

Shortly before CCI completed the contract, and again soon after contract completion, CCI's subcontracted surveyor contacted the Corps to bring attention to CCI's performance. The Corps forwarded this information to the Army's Criminal Investigation Division (CID), which conducted an investigation. CCI sued for additional payment under the contract, and the government asserted counterclaims based on the FCA,[132] the antifraud provision of the CDA,[133] and the Forfeiture of Fraudulent Claims Act (FFCA).[134] The COFC entered a final judgment against CCI for approximately $14.2 million under the FCA and the CDA.

On appeal, the Federal Circuit upheld the trial court's findings based on its conclusion that CCI's payment requests for excavation, backfill shoring, channel length, and concrete testing were false and fraudulent. The most significant false claims involved overstated quantities resulting from its interpretation of the contract. CCI's surveyor testified for the government, stating that he had repeatedly questioned CCI's interpretation and requested that the matter be brought to the Corps' attention. CCI refused. The surveyor went to the Corps after the certified payment requests overstating quantities were submitted, and CCI quickly became the subject of an FCA investigation.

CCI also made the mistake of underestimating the government's ability during discovery to uncover additional evidence of false claims. CCI's contract permitted it to use certain excavated materials as backfill, but the contract prohibited CCI from filling the excavated areas with construction debris. The trial court found that CCI improperly buried debris under and alongside the channel and that it knowingly submitted false claims for properly filling the excavated areas and for clearing the excess fill and debris. A CCI heavy equipment operator testified that CCI's owner and president told him not to haul any debris off the project site and that, pursuant to that direction, CCI employees buried debris and other unsuitable material in numerous locations.

D. *United States v. Montoya*[135]

Ricardo Montoya, doing business as the Ram Corporation, contracted with the Governor's Office of Community Affairs of the State of New Mexico to "weatherize" homes of elderly, low-income residents. Unbeknownst to Montoya, the project was funded by the U.S. Department of Energy. Montoya submitted invoices to the State of New Mexico. None of the work, however, had actually been completed. The United States indicted him for violations of the criminal FCA. At his trial, a jury convicted Montoya on six counts of presenting false claims to the government, and he was sentenced to five years in prison on each count.

On appeal, Montoya asserted that he did not present a claim to the United States since he had contracted with and received payment from the State of New Mexico. Montoya also argued that his actions, if criminal, were only in violation of state law, since the federal government had not supervised the project. Rejecting both arguments, the Court of Appeals affirmed the conviction on the grounds that the state had received the funds from the federal government and a "defendant's awareness that the funds would ultimately be provided by a federal agency" was irrelevant.[136] The court further stated that "claims presented to an intermediary . . . come within [the criminal FCA] as long as payment comes ultimately from the Federal Government."[137]

E. *Ab-Tech Construction, Inc. v. United States*[138]

In this case, Ab-Tech was awarded a subcontract by the Small Business Administration (SBA) for construction of a facility for the Army Corps of Engineers. Ab-Tech requested an equitable adjustment for its costs for extra work allegedly caused by defective government specifications. The claim was denied, and Ab-Tech appealed to the Court of Federal Claims. Court proceedings were stayed while a grand jury investigated Ab-Tech's "business affairs."[139]

The investigation resulted in the criminal indictment and conviction of Ab-Tech's president on two counts of making false statements to the government in violation of the False Statement Act. Ab-Tech's president was sentenced to a term of imprisonment and a $5,000 fine.

The government then amended its answer at the COFC and asserted counterclaims against Ab-Tech for damages and civil penalties under the FCA and forfeiture of Ab-Tech's pending claim under the Forfeiture Statute. The government's criminal investigation, conducted by the SBA and the DoD OIG, revealed that Ab-Tech had not complied with the SBA's 8(a) minority contracting requirements. In fact, Ab-Tech had a silent partner, Pyramid Construction, a nonminority owned enterprise. Pyramid and Ab-Tech had entered into an "Indemnification Agreement" prior to commencing performance of construction of the Corps facility. The SBA had not been apprised of this relationship (although disclosure was required), and Ab-Tech had signed certifications that it was in compliance with the 8(a) program requirements.

In the civil FCA action, the government successfully argued that, because Ab-Tech falsely claimed compliance with SBA 8(a) program requirements, every progress payment voucher submitted to the government was a false claim. The court agreed, stating that "the False Claims Act reaches beyond demands for money that fraudulently overstate an amount otherwise due; it extends 'to all fraudulent attempts to cause the Government to pay out sums of money.'"[140]

Although Ab-Tech successfully performed its obligation under the contract with Pyramid as a subcontractor and probably had incurred some cost for extra work, the court dismissed Ab-Tech's claim as rooted in fraud and thereby forfeited under the Forfeiture Statute. Although the government sought treble the progress payments to Ab-Tech ($4.2 million plus interest) and statutory penalties, the court denied the treble damages because it found no proof that the government suffered any detriment to its contract interest because of Ab-Tech's falsehood. However, the court awarded penalties of $10,000 for each of the 21 payment vouchers submitted by Ab-Tech during the course of contract performance.

F. *United States v. Safe Environment Corp.*[141]

An Amtrak project manager initiated a kickback scheme by instructing contractor Safe Environment Corporation to inflate its bid proposal on an asbestos abatement project for Amtrak to include a "consultant fee" for the project manager. The contractor included the fee in its bid, won the contract, and paid the fee. It included the fee in its invoice, which it presented to Amtrak for payment. The project was paid in part by federal funding and therefore the company and its president were convicted of criminal FCA violations for knowingly submitting false claims.[142]

G. *Morse Diesel Int'l. Inc. v. United States*[143]

Morse Diesel International, Inc. d/b/a AMEC Construction Management, Inc. (AMEC) performed work on three General Services Administration (GSA) construction projects. It obtained performance and payment bonds for these projects through its parent company. AMEC entered into a "commercial partnership" with its bond broker under which the bonding company remitted a portion of the commissions it received on the AMEC bonds to the parent company in the form of a discount or rebate. The court found that this partnership violated the Anti-Kickback Act.[144]

AMEC also requested reimbursement from the government for the bond premiums as part of progress payments before AMEC had paid the premiums. After inquiries from the GSA auditor about the timing of payments, an AMEC employee obtained falsely stamped invoice forms marked paid by its bond broker and submitted them to the GSA. The government brought actions under the civil FCA and the Forfeiture of False Claims Act.[145]

The COFC ruled that AMEC had forfeited more than $53 million in pending claims against GSA, and the court assessed fines in excess of $7 million.

V. *Defending an FCA Claim*

Faced with a charge that it violated the FCA, the contractor and its counsel should consider the following arguments: (1) that the submitted claim was not false or fraudulent; (2) that the alleged falsity was, in fact, an innocent mistake; (3) that the government was already aware of all the facts underlying the allegedly false or fraudulent claim; (4) that the alleged fraud was immaterial; (5) that the alleged false claim was not a demand for payment; (6) that the claim was based on a valid interpretation of the contract; or (7) that the contract's arbitration clause must be complied with before the government can assert an FCA claim.

A defense based on the lack of falsity will typically be a question of fact for the judge or jury. In such fact-intensive cases, expert witnesses (such as auditors or accountants) likely will play a key role in both the prosecution's and defense's case. In addition to arguing that the claim was not false, a contractor accused of violating the FCA may be able to argue that the action was not "knowing." Innocent mistakes and negligence are valid, and sometimes successful, defenses to FCA claims.[146] Additionally, the fact that the contractor acted on advice of counsel may prove that the contractor did not *knowingly* submit any false claims.[147]

Contractors also may be able to argue that the government's knowledge of the facts underlying the false claim negates any fraudulent intent a contractor had in presenting the claim.[148] This defense is premised upon the argument that a claim cannot be false within the meaning of the FCA if the government was told the "truth" about its basis. The most successful cases utilizing this defense have been those in which the defendant was able to prove that: (1) it fully disclosed all of the relevant facts;[149] (2) the disclosure was made before the claim was filed;[150] and (3) the disclosure was made to the government official responsible for the contract.[151]

As stated above, the government is vigorously pursuing potential fraudulent behavior in areas such as interim progress reports and certifications on the basis that misrepresentation has the effect of causing the government to pay outstanding invoices, even though the reports themselves do not request payment. In response to such claims, some contractors have argued successfully that the FCA is inapplicable because their submission to the government was not a demand for money as required by the FCA.[152] For example, the government requires the submission of many forms that do not contain a demand for payment, such as information that is used to tabulate data. The fact that these forms contain false information may violate other antifraud statutes (most notably, the False Statements Act), but it does not mean that they are false claims under the FCA.

A contractor also may be able to defeat an FCA claim by showing that the alleged fraud was not material. While the FCA does not expressly require a showing of materiality, some courts have construed the FCA as requiring a showing that the false claim or statement had a tendency to influence agency action.[153] Several courts, including the Fourth, Fifth, Eighth, and District of Columbia Circuits, as well as the COFC, have held that the materiality of the falsehood is an essential element of a false claim offense. In these cases, the courts found the materiality requirement met if the falsification would have a tendency to influence agency action or would be capable of influencing agency action.[154] In contrast, other courts have refused to read a materiality requirement into the statute.[155] Because the determination of materiality is a mixed question of law and fact, that determination generally is a question for the judge, not the jury.[156] For example, a district court recently dismissed an FCA suit against a contractor who allegedly submitted padded waste costs in a request for progress payments on a Corps fixed-price drilling contract.[157] The court found that the claim lacked materiality given the fixed-price nature of the bid and contract and the absence of any required or actual disclosure of cost information.[158]

Another defense to an allegation of an FCA violation is that the claim was based on the contractor's valid interpretation of the contract. If a contract is ambiguous, and the contractor has relied on one of several reasonable interpretations, it should not be liable for making a false claim.[159] In *Race v. United States*, for example, the court found that a defendant "cannot be convicted . . . for a statement or billing which may be said to be accurate within a reasonable construction of the contract. This is so because one cannot be found guilty of a false statement under a contract when his statement is within a reasonable construction of the contract."[160] If the court concludes that the interpretation is reasonable or plausible, the court cannot find a violation of the FCA unless it also finds "some specific evidence of knowledge that the claim is false or of intent to deceive."[161] However, if the court finds the contractor's interpretation unreasonable or frivolous, it could find FCA liability.[162] At least one court has warned that, if a contractor does not verify certain contract interpretations with the relevant government contracting authority before submitting a claim, then it risks FCA liability.[163] This is particularly true when the interpretation is coupled with deceitful conduct on the part of the contractor.[164]

As a practical matter, courts often are reluctant to dismiss a case on the basis that the contractor's interpretation was reasonable, because the underlying issue is the contractor's state of mind. This forces the contractor to go through the expense of trial and risk an unfavorable verdict. Additionally, some courts have relied on the "reckless disregard" or "deliberate ignorance" standard to impose upon the contractor a duty to inquire as to the meaning of an ambiguous contract term.[165]

In at least one case, the inclusion of an arbitration clause in the contract has precluded the government from initiating a civil suit based on the FCA until it participated in arbitration. In a Fourth Circuit case, the government brought suit against the contractor alleging breach of contract and violation of the FCA.[166] In response, the contractor filed a motion to stay the proceedings pending arbitration, as provided by the contract.[167]

In a split decision, the court ruled that the FCA claim did not preclude arbitration of the contract dispute and that the FCA claim did not interfere with nonbinding arbitration proceedings that were called for by the contract.[168] Because all outstanding claims, even the FCA claim, could be resolved by the arbitration provision of the contract, the court ruled that the arbitration provision must be enforced.[169] Because this is the only case that has forced the government to engage in arbitration before pursuing a civil FCA suit, the success of this argument beyond its facts or outside the Fourth Circuit is uncertain.

While a contractor may wish to argue that its actions did not harm the government, such a defense is meritless under the FCA. The FCA clearly indicates that a showing of government reliance or injury is not a necessary element of an FCA action.[170] Consequently, a contractor cannot assert as a defense that the government did not sustain any actual damages. The mere fact that a contractor submitted a false claim, regardless of the impact to the government, is enough to establish a prima facie case against the contractor.

VI. Other Antifraud Statutes Commonly Used

An FCA suit is often accompanied by other fraud remedies. Actions can arise under multiple statutes. While recovery of civil damages can occur only under one statute, civil penalties are available under other laws.

A. Forfeiture Statute[171]

As demonstrated by the *Daewoo* case discussed above, a contractor can forfeit its entire contract payment if the government uncovers evidence that a portion of the contract is fraudulent. The Forfeiture Statute requires that a "claim against the United States . . . be forfeited to the United States by any person who corruptly practices or attempts to practice fraud against the United States in proof, statement, establishment, or allowance thereof."[172] The statute requires the forfeiture of all claims arising under a contract tainted by fraud against the government.[173] The statute applies only to claims against the United States brought in the COFC. The government has the burden of proving, by clear and convincing evidence, that the defendant knew the claim was false and that, when the claim was submitted, the contractor intended to deceive the CO.[174]

B. The Truth in Negotiations Act (TINA)[175]

Where it applies, TINA requires a contractor to certify that, "to the best of the person's knowledge and belief, the cost or pricing data submitted are accurate, complete and current."[176] Contracts with the DoD fall under the TINA pro-visions at 10 U.S.C. § 2306(a); all other procurements are subject to 41 U.S.C. § 254(d). TINA applies in claims because the statutes require the contractor to submit cost or pricing data before the pricing of a change or modification. In the construction setting, this may arise where unit prices are submitted for excavation of materials. For example, if a differing site condition arises and one price is used for a deductive modification and another is used for the claim, this discrepancy can result in a TINA claim being asserted by the government.

Both TINA statutes provide that if the government makes an overpay-ment to a contractor because of the submission of defective cost or pricing data, the contractor is liable for interest on the amount overpaid and, if it was a knowing submission, for an additional amount equal to the amount of the overpayment.[177]

C. The Contract Disputes Act (CDA)[178]

Under the CDA, a contractor that cannot support any part of its claim because of misrepresentation of fact or fraud is liable to the government for an amount equal to the unsupported part of the claim in addition to all costs the govern-ment incurs in reviewing that portion of the claim.[179] The government has effectively relied on this portion of the CDA as an affirmative defense to valid claims in order to avoid payment of those claims. The CDA defines "misrep-resentation of fact" as a "false statement of substantive fact, or any conduct which leads to a belief of a substantive fact material to proper understanding of the matter in hand, made with intent to deceive or mislead."[180]

To recover under the CDA, the government must demonstrate that the contractor made false or fraudulent statements in the claim that was submit-ted, with the requisite intent to deceive or mislead the government.[181] Without such intent, the contractor cannot be found liable under the anti-fraud provi-sions of the CDA.[182]

D. The Major Fraud Act[183]

The Major Fraud Act of 1988[184] created a new criminal offense, "Major Fraud Against the United States." To establish a case of major fraud, the govern-ment must prove that (1) the contractor knowingly executed or attempted to execute a scheme or artifice; (2) the contractor had the intent to defraud the United States, or to obtain money or property by means of false or fraudulent pretenses, representations, or promises; (3) the alleged fraud must involve the procurement of property or services where the contractor was a prime con-tractor, subcontractor, or supplier on a contract with the United States; and (4) the value of the contract, subcontract, or any constituent part thereof for property and services was $1 million or more.[185]

The maximum penalty for a violation of the Major Fraud Act is a $1 mil-lion fine and 10 years' imprisonment. The fine may increase to $5 million if (1) the gross loss to the government or the gross gain to a defendant is $500,000 or greater, or (2) the offense involves a conscious or reckless risk of serious personal injury.[186] The maximum fine for a single prosecution, including multiple counts, may not exceed $10 million.[187]

E. The Program Fraud Civil Remedies Act[188]

The Program Fraud Civil Remedies Act of 1986 (PFCRA) imposes civil penalties against persons who make false, fictitious, or fraudulent claims to obtain money (including money representing grants, loans, or other benefits) from the federal government.[189] The Act authorizes agencies to establish administrative procedures to prosecute small cases—where the claim or group of claims is less than $150,000 (larger claims are addressed under the FCA)—involving false claims.[190] Recoveries are limited to double damages plus $5,000 per false claim. Cases under the PFCRA are brought before the agency's enforcement division, which obtains approval from the DoJ before initiating proceedings.[191]

F. The False Statements Act[192]

The False Statements Act is another criminal statute that prohibits the knowing or willful falsification, concealment, or cover-up of a material fact as well as knowingly making a false statement, representation, or writing. Similar to the penalties set forth in the criminal FCA, those convicted of making material false statements in a government matter may face up to $10,000 in fines and five years in prison.[193] The false statement may be written or oral, signed or unsigned, sworn or unsworn, voluntary or required by law.[194] Furthermore, the false statement need not be made directly to the government. Statements made to programs partly financed by federal funds[195] and statements that are never submitted to the government but are made in records that may be subject to federal inspection[196] fall under the Act.

G. Corruption and Conspiracy Statutes

Other criminal offenses in the contracting arena include corruption, improper access to government property and documents, obstruction of justice, conspiracy, Racketeer Influenced and Corrupt Organizations (RICO) Act violations, and antitrust violations. The government has several statutory avenues to prosecute corruption: the Anti-Kickback Act,[197] the bribery statute,[198] the Ethics in Government Act,[199] the Procurement Integrity Provisions,[200] and the Byrd Amendment.[201] Obstruction of justice is also governed by several statutes.[202] Finally, the Sherman Antitrust Act[203] imposes criminal penalties for collusive bidding and other anticompetitive conduct.

Conspiracy is prosecuted under 18 U.S.C. Section 371, which makes it a crime to "conspire to commit any offense against the United States, or to defraud the United States."[204] To establish a conspiracy, the government must show (1) that an agreement existed between two or more persons to commit an offense against the United States or to defraud the United States, and (2) the commission of an overt act by one conspirator in furtherance of the conspiracy.

RICO imposes additional criminal and civil penalties on persons who engage in a "pattern of racketeering activity."[205] The violation of two or more of over 30 specified state and federal criminal statutes during a 10-year period constitutes a "pattern of racketeering activity." Mail fraud, wire fraud, bribery, and illegal gratuities are included as RICO predicates. False claims and false statements alone cannot serve as the bases of a RICO violation.

VII. Contractor Code of Business Ethics and Conduct— Avoiding Criminal and Civil FCA and Other Fraud Actions

As a means of making sure federal contractors comply with the procurement statutes and to instill procurement integrity in contracting, the Civilian Agency Acquisition Council and the Defense Acquisition Regulations Council published a Final Rule amending the FAR to require contractors to have a written code of business ethics and conduct.[206] Contractors must also promote compliance with the code of business ethics and conduct and establish a formal training program covering business ethics and conduct awareness and an internal control system. Finally, the new regulations also contain mandatory disclosure provisions.[207]

Every contractor, no matter the size of its government contracts business, should consider having a code of conduct, promoting compliance, and instituting an internal control system to reduce the risk of FCA or other fraud allegations being raised by either the government or qui tam relators.

In addition to its compliance program, the best way for a contractor to avoid investigation and possible FCA or other fraud charges is to create and maintain a good working relationship with the CO and the government's project team from the beginning of the project. The strength of this relationship will go a long way in discussions that may arise later in the project regarding payment requests and pricing of equitable adjustments. The contractor must also create and maintain good project management, including good project documentation. For example, a contractor must be careful to inform the government of its understanding of the contractual obligations and its interpretation of the contract terms. All such communications with the government should be documented. A contractor may even wish to include explanatory information on the face of invoices or requests for payment to avoid misinterpretation. Additionally, contractors should be careful to document any changes made to time sheets or change order logs at the time the changes are made so that, if an investigation of records is done during either discovery or a routine audit, FCA or other charges cannot be brought against the contractor for falsifying documents.

Since even the appearance of fraud can result in a costly investigation and possible litigation, a contractor may wish to develop an early warning system to alert it to possible problems before they come to the attention of the government. By examining the indicators of fraud on which government auditors rely, a contractor may be able to avoid possible fraud allegations. The Defense Contract Audit Agency (DCAA) has published guidelines and manuals for its auditors conducting labor, material, or pricing audits. Such guides are designed to assist auditors in ferreting out potential fraud by advising the DCAA auditors how to recognize and pursue the most common indicators of fraud.

A contractor's vigilance in this area will help avoid allegations of FCA violations. Such compliance and monitoring programs may uncover an employee's illegal acts, of which the contractor is unaware. If a contractor uncovers fraudulent claims submitted by its employees, it could, as stated above, be held vicariously liable for those unlawful acts. In such cases, it may be advisable for the contractor to voluntarily disclose the fraud. In order to ensure that its legal rights are protected, a contractor should always consult with its counsel before any disclosure to the government.

While voluntary disclosure programs offer few definite assurances of leniency in return for the disclosure of possible fraudulent acts, the programs can prevent the filing of a criminal FCA claim and debarment,[208] as well as reduce the assessment of damages and penalties if the government chooses to pursue civil remedies. Furthermore, disclosure is now a FAR requirement as discussed herein. Under the FCA, a contractor who voluntarily discloses possible fraudulent acts can limit its civil liability if it furnishes all information it has about the violation within 30 days of receiving the information, cooperates fully with any government investigation, and did not have actual knowledge of the existence of an investigation into the alleged violation.[209] Additionally, public disclosure may bar civil suits by qui tam relators who do not have direct and independent knowledge of the acts upon which an action is based.[210]

Notes

1. 73 Fed. Cl. 547 (2006).
2. 28 U.S.C. § 2514.
3. 41 U.S.C. §§ 601–613.
4. 31 U.S.C. §§ 3801–3812.
5. 18 U.S.C. § 1001.
6. 41 U.S.C. § 254(d); 10 U.S.C. § 2306(a).
7. FAR § 52.203-13(b)(3). These new FAR requirements are discussed in detail in Chapter 7.
8. *See* CHARLES M. SINK & KRISTA L. PAGES, FALSE CLAIMS IN CONSTRUCTION CONTRACTS: FEDERAL, STATE AND LOCAL (2007) (extensively discussing the government's use of fraud statutes as affirmative defenses to contractors' claims).
9. *See* 132 CONG. REC. H22339 (daily ed. Sept. 9, 1986) (statement of Rep. Berman).

10. The False Claims Act was originally the Act of Mar. 2, 1863, ch. 67, 12 Stat. 696 (1863).

11. *See* S. Rep. No. 99-345, at 8 (1986), *reprinted in* 1986 U.S.C.C.A.N. 5266.

12. *See id.* at 5267.

13. *See id.* at 23–30.

14. Press Release, DOJ No. 07-873, Justice Department Recovers $2 Billion for Fraud Against the Government in FY 2007: More than $20 Billion Since 1986 (Nov. 1, 2007).

15. Press Release, DOJ No. 06-688, Deputy Attorney General Paul J. McNulty Announces Formation of National Procurement Fraud Task Force (Oct. 10, 2006).

16. S. 2041

17. 380 F.3d 488 (D.C. Cir. 2004).

18. 127 S. Ct. 1397 (2007) (the legislation as written would, inter alia, expand liability to situations where companies and individuals submit claims for payment to government grants or other entities that have received government money; permit current and former government employees to file qui tam actions; virtually eliminate the FCA's "public disclosure bar"; and extend the statute of limitations from six to ten years).

19. Pub. L. No. 111-21, 123 Stat. 1617 (2009).

20. 18 U.S.C. § 287.

21. *Id.* For a comprehensive examination of the criminal false claims statute, including legislative amendments not addressed here, see SINK & PAGES, *supra* note 8, at ch. 9.

22. 18 U.S.C. § 3571.

23. Pub. L. No. 99-145, § 931, 99 Stat. 699 (1985).

24. 18 U.S.C. § 287.

25. 31 U.S.C. § 3729(a)(1).

26. United States v. Coachman, 727 F.2d 1293, 1302 (D.C. Cir. 1984).

27. United States *ex rel.* Luther v. Consol. Indus., Inc., 720 F. Supp. 919, 920 (N.D. Ala. 1989).

28. United States v. Montoya, 716 F.2d 1340, 1344–45 (10th Cir. 1983).

29. 18 U.S.C. § 287; 31 U.S.C. § 3729(a)(1).

30. 18 U.S.C. § 6. For a listing of all executive branch departments, see 5 U.S.C. § 101.

31. 18 U.S.C. § 6.

32. *See* the discussion, *infra*, of *Montoya*, 716 F.2d 1340.

33. *See id.*

34. United States v. Irwin, 654 F.2d 671, 683 (10th Cir. 1981).

35. *See* United States v. White, 765 F.2d 1469, 1482 (11th Cir. 1985).

36. 18 U.S.C. § 287.

37. 31 U.S.C. § 3729(b).

38. United States v. Martin, 772 F.2d 1442, 1444 (8th Cir. 1985).

39. United States v. Maher, 582 F.2d 842, 847 (4th Cir. 1978).

40. *Id.* at 845.

41. United States v. Nazon, 940 F.2d 255, 259 (7th Cir. 1991).

42. Abdelkhalik v. United States, No. 94 C 5809, 1996 WL 41234, at *6 (N.D. Ill. 1996).

43. 31 U.S.C. § 3729 (2009). For a thorough investigation of civil false claims, including an in-depth discussion of the issue of qui tam litigation, see JOHN T. BOESE, CIVIL FALSE CLAIMS AND *QUI TAM* ACTIONS (1994).

44. *Id.*

45. Wilkins *ex rel.* United States v. Ohio, 885 F. Supp. 1055, 1059 (S.D. Ohio 1995).

46. 31 U.S.C. § 3729(b)(2).

47. 317 U.S. 537, 544–45 (1943) (emphasis added).

48. *Wilkins*, 885 F. Supp. at 1063.

49. United States v. Littlefield, 840 F.2d 143, 151 (1st Cir. 1988); United States v. Montoya, 716 F.2d 1340, 1342–43 (10th Cir. 1983).

50. United States v. Hooshmand, 931 F.2d 725, 734–45 (11th Cir. 1991); United States v. Catena, 500 F.2d 1319, 1321–23 (3d Cir. 1974).

51. *See* United States v. Mackby, 261 F.3d 821, 824 (9th Cir. 2001).

52. Ab-Tech Constr. Inc. v. United States, 31 Fed. Cl. 429, 435 (1994).

53. *See id.*

54. United States v. Killough, 848 F.2d 1523, 1533–34 (11th Cir. 1988).

55. 128 S. Ct. 2123 (2008).

56. 31 U.S.C. § 3729(a)(1).

57. 31 U.S.C. § 3729(b)(4).

58. 31 U.S.C. § 3730(h).

59. 31 U.S.C. § 3729(a)(7).

60. Pickens v. Kanawha River Towing, 916 F. Supp. 702, 705 (S.D. Ohio 1996); United States *ex rel.* Stevens v. McGinnis, Inc., 1994 WL 799421, at *6 (S.D. Ohio Oct. 26, 1994) (unpublished). These cases were overruled *sub silentio* by Am. Textile Mfrs. Inst., Inc. v. Limited, Inc., 190 F.3d 729, 735 (6th Cir. 1999).

61. *Am. Textile Mfrs.*, 190 F.3d at 738 (6th Cir. 1999); United States v. Q Int'l Courier, Inc., 131 F.3d 770, 773 (8th Cir. 1997); United States *ex rel.* Capella v. Norden Sys., Inc., 2000 WL 1336487, at *11 (D. Conn. Aug. 24, 2000); United States *ex rel.* Lamers v. City of Green Bay, 998 F. Supp. 971 (E.D. Wis. 1998), *aff'd*, 168 F.3d 1013 (7th Cir. 1999).

62. *See* United States v. Rivera, 55 F.3d 703, 708 (1st Cir. 1995); United States v. Richard Dattner Architects, 972 F. Supp. 738, 746–47 (S.D.N.Y. 1997).

63. *See Rivera*, 55 F.3d at 734; *Q Int'l Courier*, 131 F.3d at 773; *Am. Textile Mfrs.*, 190 F.3d at 738.

64. *See* United States *ex rel.* Koch v. Koch Indus., Inc., 57 F. Supp. 2d 1122, 1128–29 (N.D. Okla. 1999).

65. *See id.* at 1129.

66. *See Q Int'l Courier*, 131 F.3d 770, 772–74 (8th Cir. 1997); *Koch*, 57 F. Supp. 2d at 1128; United States *ex rel.* Thompson v. Columbia/HCA Healthcare Corp., 20 F. Supp. 2d 1017, 1047 (S.D. Tex. 1998); *Lamers*, 998 F. Supp. at 997–98; United States *ex rel.* S. Prawer & Co. v. Verrill & Dana, 962 F. Supp. 206, 209 (D. Me. 1997); United States *ex rel.* S. Prawer & Co. v. Verrill & Dana, 946 F. Supp. 87, 95, 94 n.12 (D. Me. 1996).

67. *See* United States v. Raymond & Whitcomb Co., 53 F. Supp. 2d 436, 445 (S.D.N.Y. 1999); United States *ex rel.* Dunleavy v. County of Del., 1998 WL 151030, at *3 (E.D. Pa. Mar. 31, 1998); United States v. Pemco Aeroplex, Inc., 195 F.3d 1234, 1236–38 (11th Cir. 1999).

68. United States v. White, 765 F.2d 1469, 1473–75 (11th Cir. 1985).

69. BOESE, *supra* note 43, at 2:40–2:57.

70. United States *ex rel.* Augustine v. Century Health Servs., Inc., 289 F.3d 409, 414–16 (6th Cir. 2002).

71. *See id.*

72. 213 F.3d 519 (10th Cir. 2000).

73. *See id.* at 533.

74. *See id.* at 530–33.

75. United States *ex rel.* Siewick v. Jamieson Sci. & Eng'g, Inc., 214 F.3d 1372, 1376 (D.C. Cir. 2000); *see also* United States *ex rel.* Thompson v. Columbia/HCA Healthcare, 20 F. Supp. 2d 1017, 1046–47 (S.D. Tex. 1998).

76. 31 U.S.C. § 3729(b).

77. United States v. Inc. Vill. of Island Park, 888 F. Supp. 419, 439 (E.D.N.Y. 1995).

78. *See* S. Rep. No. 99-345 at 21 (1986), *reprinted in* 1986 U.S.C.C.A.N. 5266, 5286.

79. 31 U.S.C. § 3729(b)(1).

80. RESTATEMENT (SECOND) OF AGENCY § 27 (1958).

81. 95 F. Supp. 2d 465, 467–68 (D. Md. 2000).

82. 158 F. Supp. 2d 1001, 1006 (D.S.D. 2001).

83. *See id.* at 1008.

84. *See id.* Other courts have reached similar conclusions. *See, e.g.,* United States v. O'Connell, 890 F.2d 563, 568–69 (1st Cir. 1989).

85. 1 U.S.C. § 3729(a).

86. 28 C.F.R. § 85.3(a).

87. 28 U.S.C. § 2461 note; Pub. L. No. 104-410 (1990).

88. 31 U.S.C. § 2729(a)(2).

89. *Id.; see also, e.g.,* Ab-Tech Constr., Inc. v. United States, 31 Fed. Cl. 429, 434–35 (Fed. Cl. 1994).

90. 31 U.S.C. § 3729(a)(2).

91. *Id.*

92. 31 U.S.C. § 3730(b).

93. 132 CONG. REC. S11238-04 (daily ed. Aug. 11, 1986) (statement of Sen. Grassley) ("The Government needs help—lots of help—to adequately protect the Treasury against growing and increasingly sophisticated fraud").

94. 31 U.S.C. § 3730(e)(4)(B).

95. United States *ex rel.* Fine v. Chevron, U.S.A., 72 F.3d 740, 742 (9th Cir. 1995).

96. 31 U.S.C. § 3730(b)(1).

97. *See* United States *ex rel.* LeBlanc v. Raytheon Co., 913 F.2d 17 (1st Cir. 1990), *cert. denied*, 499 U.S. 921 (1991) (government employee whose job is to investigate fraud is not a proper qui tam relator); United States *ex rel.* Hagood v. Sonoma County Water Agency, 929 F.2d 1416 (9th Cir. 1991) (lawyer for the government qualified as an original source for qui tam purposes even though he gained the original information by representing a government client); United States *ex rel.* Williams v. NEC Corp., 931 F.2d 1493 (11th Cir. 1991) (military lawyer qualified as original source in qui tam suit despite the fact that he obtained the information while conducting an investigation on behalf of the government); United States *ex rel.* Fine v. MK-Ferguson Co., 861 F. Supp. 1544, 1548 (D.N.M. 1994), *aff'd*, 99 F.3d

1538 (10th Cir. 1996) (recognizing that an employee of a government Inspector General's office can be a qui tam relator if he is an original source).

98. 31 U.S.C. § 3730(c)(3).

99. *Id.*

100. Procurement Fraud Comm., Am. Bar Ass'n, Qui Tam Litigation Under the False Claims Act 32 (1994).

101. In *United States ex rel. Farrell v. SKF USA, Inc.*, 1998 WL 265242 (W.D.N.Y. May 18, 1998), a defendant contractor in a qui tam action demanded that the United States reply to its discovery requests pursuant to Federal Rules of Civil Procedure 26 and 34. A magistrate judge held that "while the United States may be the real party in interest in a *qui tam* action, the United States is not an actual party to the action in a case where it has declined to intervene." *Id.* at *3. Therefore, in such a case "the United States is not a party plaintiff and thus is not bound by the Federal Rules of Civil Procedure as they relate to party discovery." *Id.; see also* Procurement Fraud Comm., *supra* note 100, at 41 ("The institution of a *qui tam* action . . . will not expose the government to unlimited discovery, particularly if the suit is being prosecuted by a private plaintiff").

102. 31 U.S.C. § 3731(b).

103. *Id.*

104. A recent court of appeals decision may greatly increase the valuation of a relator's claim where settlements occur. In *United States v. United States ex rel. Thornton*, 207 F.3d 769, 771–72 (5th Cir. 2000), the decision by the court would allow a relator to be eligible for a percentage of the value of counterclaims dropped by contractors in the settlement if the claims were released in exchange for the settlement of the FCA suit.

105. 31 U.S.C. § 3730(d)(2).

106. Lujan v. Defenders of Wildlife, 504 U.S. 555, 560–61 (1992).

107. *See* Vt. Agency of Natural Res. v. United States *ex rel.* Stevens, 529 U.S. 765, 773–74 (2000). This case is most notable for the holding that the FCA does not subject a state or state agency to liability because a state is not a "person" for purposes of qui tam liability under the statute. *See id.* at 787. However, a district court, declining to follow the Eighth Circuit's contrary position, recently found that a state official is a person and thus may be sued in an individual capacity in an FCA qui tam action regardless of whether the official's actions were within the scope of his official duties.

108. 31 U.S.C. § 3730(h).

109. *Id.*

110. McKenzie v. BellSouth Telecomms., Inc., 219 F.3d 508, 513–14 (6th Cir. 2000).

111. United States *ex rel.* Ramseyer v. Century Healthcare Corp., 90 F.3d 1514, 1522 (10th Cir. 1996); Neal v. Honeywell Inc., 33 F.3d 860, 863–64 (7th Cir. 1994).

112. *See* Mruz v. Caring, Inc., 991 F. Supp. 701, 709–10 (D.N.J. 1998).

113. *See* Vessell v. DPS Assocs., 148 F.3d 407, 412 (4th Cir. 1998); Godwin v. Visiting Nurse Ass'n Home Health Servs., 831 F. Supp. 449, 453–54 (E.D. Pa. 1993); Hardin v. DuPont Scandinavia, 731 F. Supp. 1202, 1205 (S.D.N.Y. 1990).

114. In such states, an employee is protected from wrongful discharge, despite the employment-at-will doctrine, for engaging in an act that enhances, or refusing

to do an act that runs counter to, public policy. *See* Wagner v. Globe, 722 P.2d 250, 256–57 (Ariz. 1986).

    115. *See, e.g.,* Conn. Gen. Stat. Ann. §§ 31–51m(b), (c) (1997) (creating civil remedy for employees who are discriminated against for reporting the violation of federal or state law); Fla. Stat. Ann. § 448.103(a) (1997); Mich. Comp. Laws Ann. § 15.362 (2001).

    116. *See* Snow v. Bechtel Constr., Inc., 647 F. Supp. 1514, 1517 (C.D. Cal. 1986) (finding that Atomic Energy Act preempted employee's state wrongful termination action); Olguin v. Inspiration Consol. Copper Co., 740 F.2d 1468, 1473–74 (9th Cir. 1984) (finding that the Federal Mine Safety and Health Act preempted employee's state wrongful discharge action); Rayner v. Smirl, 873 F.2d 60, 63–66 (4th Cir. 1989) (finding that the Federal Railroad Safety Act preempted employee's state wrongful discharge action).

    117. 31 U.S.C. § 3729(b).

    118. Fed. Cir. No. 2007-5129 (Feb. 20, 2009); 73 Fed. Cl. 547 (2006).

    119. 41 U.S.C. §§ 601–613.

    120. 73 Fed. Cl. at 560–81.

    121. *Id.* at 582–91.

    122. *Id.* at 596.

    123. *Id.*

    124. Fed. Cir., No. 2007-5129 (Feb. 20, 2009).

    125. 183 F.3d 1088 (9th Cir. 1999).

    126. 40 U.S.C. § 276a. For further discussion of the Davis-Bacon Act, see Chapter 12.

    127. 40 U.S.C. §§ 276a, 276c.

    128. 29 C.F.R. § 5.5(a)(3)(ii).

    129. 91 F.3d 1261 (9th Cir. 1996).

    130. *Plumbers & Steamfitters,* 183 F.3d at 1092.

    131. 154 F.3d 1357 (Fed. Cir. 1998).

    132. 31 U.S.C. §§ 3729–3731.

    133. 41 U.S.C. § 604.

    134. 28 U.S.C. § 2514.

    135. 716 F.2d 1340 (10th Cir. 1983).

    136. *Id.* at 1344.

    137. *Id.* at 1342.

    138. 31 Fed. Cl. 429 (Fed. Cl. 1994).

    139. *Id.* at 431.

    140. *Id.* at 433 (citing United States v. Neifert-White Co., 390 U.S. 228, 233 (1968)).

    141. 2002 U.S. Dist LEXIS 8421 (N.D. Ill. May 10, 2002).

    142. *Id.* at 9–10.

    143. 72 Fed. Cl. 116 (2007); *see also* Morse Diesel Int'l, Inc. v. United States, 66 Fed. Cl. 788 (2005), 69 Fed. Cl. 558 (2006), and 74 Fed. Cl. 601 (2007). For further discussion of federal anti-kickback laws, see Chapter 7.

    144. 41 U.S.C. §§ 51–54.

    145. 28 U.S.C. § 2514.

146. *See* United States *ex rel.* Hagood v. Sonoma County Water Agency, 929 F.2d 1416, 1421 (9th Cir. 1991).

147. *See* United States *ex rel.* Bidani v. Lewis, 2001 WL 32868, at *6 (N.D. Ill. Jan. 12, 2001) (*vacated on other grounds by* United States *ex rel.* Bidani v. Lewis, 2001 WL 747524 (N.D. Ill. June 29, 2001). Although this case was vacated on other grounds, it appears that the court did embrace the proposition that, if a contractor relies on advice from its lawyers, such reliance may negate the knowledge requirement of the FCA. Note, however, that invoking this defense will usually effect a waiver of the attorney-client privilege.

148. *See* United States *ex rel.* Butler v. Hughes Helicopters, 71 F.3d 321, 326 (9th Cir. 1995); *Hagood*, 929 F.2d at 1421; *but see* United States *ex rel.* Kreindler & Kreindler v. United Tech. Corp., 777 F. Supp. 195, 201–02 (N.D.N.Y 1991), *aff'd on different grounds*, 985 F.2d 1148, 1156 (2d Cir. 1993). The government knowledge defense has also been used to negate the requisite intent. *See* Chemray Coatings Corp. v. United States, 29 Fed. Cl. 278, 285 (Fed. Cl. 1993). In other words, where the government is aware of the false information at the time the claim is made, there can be no liability under the CDA because the contractor did not possess an intent to deceive. *See id.*

149. *See* Chen-Cheng Wang *ex rel.* United States v. FMC Corp., 975 F.2d 1412, 1421 (9th Cir. 1992); United States *ex rel.* Butler v. Hughes Helicopter, Inc., 71 F.3d 321, 326 (9th Cir. 1995).

150. *But see* United States *ex rel.* Lamers v. City of Green Bay, 998 F. Supp. 971, 988 (E.D. Wis. 1998), *aff'd*, 168 F.3d 1013 (7th Cir. 1999) (finding that government knowledge precluded an FCA claim even though the disclosure to the government occurred after the claims had been made).

151. *See* United States *ex rel.* Durcholz v. FKW, Inc., 997 F. Supp. 1159, 1171 (S.D. Ind. 1998).

152. *See* United States v. Nat'l Wholesalers, 236 F.2d 944, 950 (9th Cir. 1956); United States v. Grannis, 172 F.2d 507, 515–16 (4th Cir. 1949); Miller v. United States, 550 F.2d 17, 24 (Ct. Cl. 1977).

153. United States *ex rel.* Berge v. Bd. of Trs., 104 F.3d 1453, 1459–60 (4th Cir. 1997).

154. *See id.* at 1459 (stating that "if previously unclear, we now make explicit that the current civil False Claims Act imposes a materiality requirement"); United States v. Southland Mgmt. Corp., 288 F.3d 665, 675–81 (5th Cir. 2002) ("While this court has indicated that the [False Claims] Act contains a materiality element, we have not yet clarified the exact nature of this requirement," but a false certification submitted to HUD indicating compliance with a condition that is a prerequisite to receiving government funds is a false claim as a matter of law); United States v. TDC Mgmt. Corp., 24 F.3d 292, 296, 298 (D.C. Cir. 1994) (requiring proof that defendant "actually knew it had omitted material information from its monthly progress reports or that it recklessly disregarded or deliberately ignored that possibility"); United States v. Pruitt, 702 F.2d 152, 155 (8th Cir. 1983); Tyger Constr. Co. v. United States, 28 Fed. Cl. 35, 55 (1993); *see also Lamers*, 998 F. Supp. at 991–92; *Durcholz*, 997 F. Supp. at 1167; Luckey v. Baxter Healthcare Corp., 2 F. Supp. 2d 1034, 1045 (N.D. Ill. 1998).

155. *See* United States v. Parsons, 967 F.2d 452, 455 (10th Cir. 1992) (materiality not an element for § 287 violation); United States v. Elkin, 731 F.2d 1005, 1009 (2d Cir. 1984) (proof of materiality not required); United States *ex rel.* Roby v. Boeing Co., 184 F.R.D. 107, 112 (S.D. Ohio 1998).

156. *Berge*, 104 F.3d at 1459–60.

157. *See* United States *ex rel.* Wilkins v. N. Am. Constr. Corp., 101 F. Supp. 2d 500, 520–22 (S.D. Tex. 2000).

158. *See id.* at 522–23.

159. *See* United States v. Mead, 426 F.2d 118, 123 (9th Cir. 1970).

160. United States v. Race, 632 F.2d 1114, 1120 (4th Cir. 1980); *see also* United States *ex rel.* Butler v. Hughes Helicopter Co., 71 F.3d 321, 329 (9th Cir. 1995); Basin Elec., 248 F.3d at 794 (8th Cir. 2001).

161. Commercial Contractors, Inc. v. United States, 154 F.3d 1357, 1366 (Fed. Cir. 1998); *see also* United States v. Basin Elec. Power Coop., 248 F.3d 781, 805 (8th Cir. 2001).

162. *See Commercial Contractors*, 154 F.3d at 1366; *see also* United States *ex rel.* Compton v. Midwest Specialities, 142 F.3d 296, 303 (6th Cir. 1998).

163. *See Commercial Contractors*, 154 F.3d at 1366.

164. Daff v. United States, 78 F.3d 1566, 1573–74 (Fed. Cir. 1996).

165. *See* United States v. Littlefield, 840 F.2d 143, 147–48 (1st Cir. 1988); United States v. Aerodex, Inc., 469 F.2d 1003, 1008–09 (5th Cir. 1972); United States v. Entin, 750 F. Supp. 512, 518 (S.D. Fla. 1990).

166. *See* United States v. Bankers Ins. Co., 245 F.3d 315, 317 (4th Cir. 2001).

167. *See id.* The contract contained the following arbitration provision: "If any misunderstanding or dispute arises between the Company and the FIA [Federal Insurance Administration] with reference to any factual issue under any provisions of this Arrangement . . . such misunderstanding or dispute may be submitted to arbitration for a determination [that] shall be binding upon approval by the FIA." *Id.* at 318.

168. *See id.* at 323, 325.

169. *See id.* at 325.

170. United States *ex rel.* Wilkins v. N. Am. Constr. Corp., 101 F. Supp. 2d 500, 516–17 (S.D. Tex. 2000).

171. 28 U.S.C. § 2514.

172. *Id.*

173. Ab-Tech Constr. Inc., 31 Fed. Cl. 429, 435–36 (Fed. Cl. 1994).

174. Ingalls Shipbuilding v. United States, 21 Cl. Ct. 117, 122 (Fed. Cl. 1990).

175. 41 U.S.C. § 254(d); 10 U.S.C. § 2306a.

176. *Id.*

177. Interest if computed under 41 U.S.C. § 254(f)(1)(A) and 10 U.S.C. § 2306a(f)-(1)(A) is as follows:

> (i) for the period beginning on the date the overpayment was made to the contractor and ending on the date the contractor repays the amount of such overpayment to the United States; and

> (ii) at the current rate prescribed by the Secretary of the Treasury under 6621 Title 26.

178. 41 U.S.C. §§ 601–613.

179. 41 U.S.C. § 604.

180. 41 U.S.C. § 601(7).

181. 41 U.S.C. § 604; Crane Helicopter Servs., Inc. v. United States, 45 Fed. Cl. 410, 435 (Fed. Cl. 1999).

182. *See id.*

183. 18 U.S.C. § 1031.

184. *Id.*

185. 18 U.S.C. § 1031(a).

186. 18 U.S.C. § 1031(b).

187. 18 U.S.C. § 1031(c).

188. 31 U.S.C. §§ 3801–3812.

189. 31 U.S.C. § 3807.

190. The statute defines "authority head" as the head, or the official employee designated as the head, of any executive or military department, the U.S. Postal Service, or an establishment, as defined in Section 11(2) of the Inspector General Act of 1978, that is not an executive department. 31 U.S.C. § 3809.

191. 31 U.S.C. § 3803.

192. *See* 18 U.S.C. § 1001.

193. *See id.*

194. *See* United States v. Beacon Brass Co., 344 U.S. 43, 44–46 (1952).

195. It does not matter if the defendant is unaware of the federal involvement for the False Statements Act to apply. *See* United States v. Stanford, 589 F.2d 285, 297 (7th Cir. 1978).

196. *See* United States v. Kraude, 467 F.2d 37, 38 (9th Cir. 1972); United States v. Hooper, 596 F.2d 219, 223 (7th Cir. 1979).

197. 41 U.S.C. §§ 51–54.

198. 18 U.S.C. § 201(b).

199. 17 U.S.C. § 207.

200. 41 U.S.C. § 423.

201. 31 U.S.C. § 1352.

202. *See* 18 U.S.C. § 1516 (making it a crime to influence, obstruct, or impede the performance of a federal audit); 18 U.S.C. § 1503 (making it a crime to attempt to influence any grand or trial juror or court official or to otherwise impede the due administration of justice); 18 U.S.C. § 1512 (prohibiting witness tampering); 18 U.S.C. § 1505 (prohibiting obstruction of pending administrative proceedings before any department or agency of the United States). Perjury and false declarations under oath are prohibited by 18 U.S.C. §§ 1621, 1623.

203. 15 U.S.C. § 1.

204. 18 U.S.C. § 371.

205. 18 U.S.C. §§ 1961–1968.

206. FAR 52.203-13.

207. FAR 52.203-13(b)(3). For further discussion of the FAR's mandatory disclosure requirements, see Chapter 7.

208. For further discussion of debarment, see Chapter 7.

209. 31 U.S.C. § 3729(a)(7).

210. *See* 31 U.S.C. § 3720(e)(4)(A).

CHAPTER 25

Funding and Related Issues Regarding Federal Government Construction Outside the United States

JOHN S. VENTO

I. Introduction

As the preface to the first edition of this book stated: "The U.S. government has been and, for the foreseeable future, will continue to be the largest purchaser of construction services in the world." Given the events since the first edition was published in 2003, it might now be more accurately stated as: "The U.S. military has been and, for the foreseeable future, will continue to be the largest purchaser of construction services in the world." And since September 11, 2001, with wars in Iraq and Afghanistan, and a "Global War on Terror" (GWOT), a significant amount of U.S. military (and State Department–authorized military) construction has been outside of the continental United States (OCONUS). For that reason, this new chapter is being added to the second edition of this book so that construction lawyers will have a fundamental understanding of the fiscal, statutory, and regulatory legal regime applicable to federal government and, particularly, military construction OCONUS.

Lawyers always like to start with definitions. And quite apart from the philosophical question of "What is construction?" there is actually a definition of "military construction": "Any construction, development, conversion, or extension at any time carried out with respect to a military installation, whether to satisfy temporary or current requirements, is construction."[1] Further, a "military installation" is defined as a "base, camp, post, station, yard, center, or other activity under [military jurisdiction] or, in the case of an activity in a foreign country, under [military] operational control, without regard to the duration of operational control."[2] And quite apart from the breadth of these definitions, each military organization has its own definition of "construction."[3] For example, the Air Force definition states: "to build, develop, convert, or extend real property and real property systems or components."[4] "Components" is not otherwise defined. Moreover, maintenance and repair are not considered construction. Just to ensure that everyone understands the distinction between *construction* and *maintenance and repair*, each of the armed services has its own definitions. For example, the Air Force's is as follows:

What is *not* construction?

> 4.1.1. **Maintenance (EEIC 521)**. Maintenance is work required to preserve real property and real property systems or components and prevent premature failure or wearing out of the same. Maintenance includes work to prevent and arrest component deterioration, as well as landscaping or planting work done in conjunction with a facility repair project . . .
>
> 4.1.2. **Repair (EEIC 522 and 524)**. Repair means to restore real property and real property systems or components to such condition that they may effectively be used for their designated functional purposes. . . . [the] systems or components need not have failed to permit a repair project. . . .[5]

One might ask why it is important to distinguish between construction on the one hand and maintenance and repair on the other in connection with the legal regime of overseas military construction. The answer is simple: it is the "color of money" question that has to be answered when military construction projects are funded. In short, it is a fiscal law issue that has its basis in the U.S. Constitution. The next logical question then is "Why is that important?"

It is important for the lawyer representing contractors that are doing or want to do business with the government for overseas military construction to identify the proper appropriation and/or authorization (or two) that allows the Department of Defense (DoD) to legally fund the construction to ensure that proper funding is being used so that the project is not curtailed, or, worse, terminated for convenience, when the correct funds are not used. Proactive involvement with the sourcing authority by the contractor or prospective contractor and lawyer will be essential to ensuring a smooth and uneventful construction project, to the extent that is possible in a dangerous environment. For this reason, fiscal law authority is at the basis of all OCONUS government construction.

II. The Constitutional Authority to Fund U.S. Military Operations, Including Construction

A. The President's Commander-in-Chief Powers

As you may recall from civics class, Article II of the Constitution appoints the President as the Commander-in-Chief: "The President shall be Commander in Chief of the Army and Navy of the United States . . . ."[6] Notwithstanding, under the Constitution's system of checks and balances, it is Congress that has the power of the purse, that is, the sole fiscal authority to fund military operations, including construction.

B. The Congressional Power of the Purse

Because the U.S. Constitution states that "no Money shall be drawn from the Treasury, but in Consequence of Appropriations made by Law . . . ,"[7] Congress can constrain or expand the appropriated funds available for foreign affairs activities conducted by the executive agencies, including the DoD, and thereby indirectly affect the conduct of foreign affairs, including wars. The Constitution also provides: "The Congress shall have Power to dispose of and make all needful Rules and Regulations respecting the Territory or other Property belonging to the United States . . . ."[8] And while it may seem clear to us now, it has been the subject of decisional law from the founding of our nation to the present to determine the extent of Congress's authority to approve funds before they may be expended. For example, in *United States v. MacCollom*,[9] the Supreme Court held: "The established rule is that the expenditure of public funds is proper only when authorized by Congress, not that public funds may be expended unless prohibited by Congress." To implement its authority, Congress has created an extensive legislative scheme that directly affects the funding of military construction, both contingency and exercise construction (this chapter will address the differences between the two), by controlling appropriations through what is called the fiscal law principles of "purpose, time and amount." Thus, an understanding of the fiscal law constraints on U.S. military construction is critical in understanding DoD overseas construction. And while the majority of Americans might believe that it is the executive power of the President that effectively controls foreign policy—and you would firmly believe so from observing presidential elections—it is actually Congress that controls foreign policy through the implementation of the power of its purse. While the executive branch of government may conceptualize and articulate foreign policy, an effective foreign policy requires "institutions, agencies, people and money, and Congress controls them all."[10] Through the authorization and appropriation process, Congress sets the terms of commerce; it provides military forces and intelligence capabilities; and it establishes the conditions for development assistance, security support programs, and U.S. participation in international organizations. "Hardly any important executive branch decision is taken without consideration of the reaction in Congress."[11]

Understanding funding limitations in military construction is an important recurring issue when contracting with the government. For example, with the influx of funds under the American Recovery and Reinvestment Act of 2009 (ARRA),[12] a number of issues have been addressed in DoD regarding the use of Organization and Maintenance (O&M), as opposed to Military Construction Appropriated Funds, for minor construction. Currently, for projects within the United States, the O&M funding limit is $750,000. In very limited circumstances involving contracting for deficiencies that are life-threatening, health-threatening, or safety-threatening, the limit is $1.5 million. Air Force Instruction AFI 32-1032, Chapter 5, defines "force protection" as constituting one of those deficiencies. Force protection has been broadly defined to include any construction project necessary for the safety of U.S. forces, which may include berms, barriers, or hardened shelters for billeting personnel and storing vehicles, equipment, and the like. For projects outside the United States in support of combat and contingency operations, the Secretary of Defense has temporary authority to use O&M funds for construction projects not carried out at a military installation, where (1) the United States is reasonably expected to have a long-term presence; (2) the United States has no intention of using the construction after the operational requirements have been satisfied; and (3) the level of construction is the minimum necessary to meet the temporary operational requirements. It is important to remember that, even though the Federal Acquisition Regulation (FAR) mandates the procurement of construction services on the basis of sealed bidding, there is an exception to the sealed, firm, fixed-price bidding requirements for construction contracts to be performed outside of the United States, its possessions, and Puerto Rico.[13] For fiscal year 2009, the aggregate limit on O&M funding of such projects is $200 million, but the authority does not extend to projects in Afghanistan. For Joint Chiefs of Staff military exercise-related construction projects outside the United States, O&M funds are not permitted to be used at all, as will be discussed later in this chapter.

Another significant recurring funding issue involves socioeconomic limitations on construction. For example, the Civilian Agency Acquisition and Defense Acquisition Regulations Council has published Federal Acquisition Circular (FAC) 2005-32, which contains six interim rules amending the FAR. The first three implement ARRA. The six rules are:

Item I—Buy American Requirements for Construction Materials (FAR Case 2009-008);
Item II—Whistleblower Protections (FAR Case 2009-012);
Item III—Publicizing Contract Actions (FAR Case 2009-010));
Item IV—Reporting Requirements (FAR Case 2009-009);
Item V—GAO/IG Access (FAR Case 2009-011); and
Item VI—Government Accountability Office (GAO) Access to Contractor Employees (FAR Case 2008-026).

All of these interim rules had a comment period extending to June 1, 2009.

The interim rule associated with FAR Case 2009-008 amends the FAR to implement Section 1605 of ARRA with regard to Buy American Act (BAA) requirements for construction materials.[14] Section 1605 prohibits the use of funds appropriated, or otherwise made available by ARRA, for any project for the construction, alteration, maintenance, or repair of a public building or public work unless *all* of the iron, steel, and manufactured goods used in the project are produced in the United States. The law however requires this prohibition be applied in a manner consistent with obligations under international agreements, and it provides for waiver under three circumstances: (1) iron, steel, or manufactured goods are not produced in the United States in sufficient and reasonably available quantities and of a satisfactory quality; (2) inclusion of iron, steel, or manufactured goods produced in the United States will increase the cost of the contract more than 25 percent; or (3) applying the domestic preference would be inconsistent with the public interest. The rule adds FAR Subpart 25.6,[15] entitled "American Recovery and Reinvestment Act—Buy American Construction Materials," and also adds new contract clauses at FAR 52.225-21 through FAR 52.225.24, with consistent changes to FAR 1.106, 5.207, 25.001, 25.002, and 25.1102. The rule applies to solicitations and contracts awarded on or after March 31, 2009. Contracting officers (COs) must modify, on a bilateral basis in accordance with FAR 1.108(D)(3), existing contracts to include the FAR clauses for future orders, if ARRA funds will be used. If a contractor refuses to accept the modification, the contractor will not be eligible to receive ARRA funds.

In short, knowledge of fiscal law and how that law may be adapted or modified to accommodate U.S. security interests is integral to understanding the complexities of securing and performing overseas military construction.

III. *Legislative Framework Regulating the Funding of U.S. Military Operations, Including Construction*

A. Fiscal Legislative Controls

An agency, whether civilian, such as the Department of State (DoS), or military, may obligate and expend appropriations only for a proper purpose as authorized by Congress. An agency may obligate funds only within the time limits applicable to the appropriation. Finally, an agency must obligate funds within the amounts established by Congress. In a nutshell, there are three limitations on the expenditure of funds: purpose, time, and amount. Most importantly, there is no "deployment exception" to the fiscal law of the United States. In other words, whether deployed for an exercise or an actual real-world contingency—such as in Iraq and Afghanistan—or building an embassy in some part of the world, the fiscal law of the United States applies, without exceptions. Understanding that law is, therefore, important to construction contracting outside the United States.

1. *The Purpose Statute*

The Purpose statute[16] prohibits the application of appropriations to objects or subjects other than those for which the appropriations were made by Congress. This is also known as the "color of money" concept. A three-part test is commonly used in military construction cases, following a Comptroller General decision and response to a request for an opinion by former Congressman

Bill Alexander (D.-Arkansas) relating to funding issues arising out of military exercises conducted in Honduras. In the decision, the Comptroller General described the three parts of the Necessary Expense doctrine used to determine whether or not an appropriation is being used for its proper statutory purpose:

1. The expenditure must be reasonably related to the purposes for which the appropriation was made. (That is, it must be for the particular statutory purpose of the appropriation, or necessary and incident to the proper execution of the general purpose of the appropriation.)
2. The expenditure must not be prohibited by law.
3. The expenditure must not fall specifically within the scope of some other category of appropriations. (This last requirement would prohibit expenditure of funds if another funding source for the particular purpose was enacted by Congress, even if that funding source was expended and no longer available.)[17]

2. Amount

The Anti-Deficiency Act[18] prohibits obligating or spending money before it is appropriated, or in amounts in excess of the amounts appropriated. The Act makes it a crime to knowingly enter into or authorize government contracts in the absence of government funds to pay for such contracts.

3. Time

Appropriations are available only to support bona fide needs during the period of availability. Typically, the period of availability for an annual appropriation is one fiscal year. Funds are presumed to be available for obligation only during the fiscal year in which they are appropriated.[19] Thus, annually appropriated funds are available for obligation, and a contract obligating those funds must be entered into, during the period October 1 through September 30. The Bona Fide Need statute[20] provides that the balance of an appropriation or fund limited for obligation to a definite period is available only for payment of expenses properly incurred during the period of availability or to complete contracts properly made within that period of availability. Thus, the bona fide need is determined when the government obligates the appropriated funds and not at the time of performance. This means that the bona fide need must exist at the time the award is made, regardless of when performance is scheduled to begin or conclude.

IV. Department of State Authorizations and Appropriations

A. Introduction

Title 22 of the U.S. Code codifies the authorizations and appropriations of the DoS. Construction, temporary or permanent, can be authorized and funded under a myriad of different statutory authorizations that sometimes may not, at first glance, be thought of as a source for overseas construction. There are three main categories of DoS authorizations and appropriations that could involve construction:

1. Reimbursable security assistance
2. U.S.-financed security assistance
3. Development assistance

The first category, reimbursable security assistance, falls under the Foreign Military Sales (FMS) program,[21] under which the U.S. government can contract for construction for the benefit of a foreign government pursuant to an FMS (Case Act) authorization pursuant to a letter of agreement (LOA) between the U.S. government and the foreign government. The construction is taxpayer-funded under the authorization. Construction under reimbursable security assistance also may be authorized under the Commodities and Services (C&S) program,[22] whereby the government would provide, for example, construction management services to a foreign government, again funded under the Case Act appropriation.

Although the second category, U.S.-financed security assistance, is most commonly used to provide grants to developing nations to buy U.S. military equipment from U.S. suppliers, it may also be used to authorize construction and appropriate funds under the International Military Education and Training (IMET) program.[23]

The broadest category in which construction can be authorized and funded is development assistance. Under this main category, construction could be funded under the Economic Support Fund (ESF).[24] This fund may be used for security, political and economic projects, and targeted development programs, e.g., for orphanages in Afghanistan. Another category of authorization could be under the International Narcotics and Law Enforcement Affairs Act and its appropriations,[25] under which money may be appropriated and authorized for fighting foreign drug wars or for other anticrime purposes. This would also include international security police training. Also under development assistance are categories such as disaster relief, health, education, energy, and environment, all of which do involve construction programs under DoS authorizations and appropriations.

And, as can be seen from these broad categories, the U.S. military is involved in many of these projects and, therefore, the general rule in funding U.S. military operations abroad is that the DoS has the primary responsibility, authority, and funding to conduct foreign assistance on behalf of the U.S. government.[26] Most of these development assistance programs are financed with direct grants or loans from the DoS or the U.S. Agency for International Development (USAID) to the developing country. These are typically large-scale projects, some of which involve DoD, such as the Economic Support Fund (ESF) and the Bureau of International Narcotics and Criminal Law Enforcement (INCLE), both of which provide DoS funds to provincial reconstruction teams (PRTs) in Iraq and Afghanistan. PRTs are civil-military organizations that are staffed by U.S. government civilian and military personnel to assist foreign provincial governments in their efforts toward reconstruction, the establishing of security and rule of law, and political and economic development. PRTs were first deployed in Afghanistan in 2002. They are funded under the ESF.[27] The success of the PRTs in Afghanistan led the U.S. government to incorporate this concept into its stability and reconstruction strategy in Iraq in 2005.[28] PRTs in both Iraq and Afghanistan provide their own force protection and those in Iraq are generally co-located with large coalition forward operating bases (FOBs) that provide some of the needed force protection.[29]

B. Support for Iraq and Afghanistan

There has been special congressional authorization for funding security assistance in Iraq and Afghanistan. The Iraq Liberation Act of 1998,[30] authorized the President to direct not only the drawdown of defense articles from the stocks of DoD but also the processing of defense services of DoD, including military education and training for Iraqi democratic opposition organizations.

Currently, the Afghanistan Freedom Support Act of 2002[31] authorizes defense articles, defense services, and military education and training for the government of Afghanistan and eligible foreign countries on its borders, such as Pakistan, and even eligible international organizations. This authority is carried out under a section of the Foreign Assistance Act.[32] Additional funding for this Act has come by way of the emergency supplemental appropriations discussed earlier in this chapter. Generally, the ESF has a two-year period of availability and is appropriated annually in the Foreign Operations Assistance Act (FOAA), the DoS equivalent to the annual DoD Appropriations Act. The two most recent FOAA appropriations for the ESF were $2.5 billion appropriated in the fiscal year 2007 FOAA[33] and $2.99 billion appropriated in the fiscal year 2008 Consolidated Appropriations Act.[34]

C. Accessing DoS Appropriations and Authorizations

The U.S. Embassy is going to be the source of access to DoS appropriations and authorizations for construction in foreign countries, including military construction, as we have just noted. However, U.S. military units also may access DoS funds through their deployed DoS political advisors (POLADs), located at the combined joint task force (CJTF) or division level. DoD-deployed units may also coordinate with DoS foreign officers located at the local PRTs. Contractors wishing to do business in overseas deployed environments have to pay attention to the sources of both business and oversight, which are not always located at the U.S. Embassy.

There are also two exceptions to the general rule that foreign assistance is funded with DoS authorizations and appropriations. The first exception is based on historical GAO opinions that allow for the use of DoD O&M funds to train foreign forces, as long as the purpose of the training is interoperability, safety, and familiarization of the foreign element with U.S. forces. This "interoperability training" is known as "little t" training in the Pentagon, as opposed to the "big t" training for security assistance purposes that must be funded with DoS authorizations and appropriations. Construction can emanate from both the "big t" and "little t" training.[35] The second exception requires a specific authorization from Congress for DoD to provide foreign assistance outside of Title 22 DoS appropriations and authorizations.

V. Department of Defense Appropriations and Authorizations

A. Introduction

DoS has the primary responsibility, authority, and funding to conduct foreign assistance on behalf of the U.S. government. The DoS, therefore, is the primary source of the funding of OCONUS U.S. military operations, with the two noted exceptions above. Knowing how military construction projects are authorized and appropriated is also important in assisting construction clients interested in doing business with the DoD. The congressional appropriations committees—the House and Senate Appropriations Committees, respectively—are charged with the responsibility of drafting the federal appropriations acts for consideration and passage by Congress. The congressional authorization committees—the House and Senate Armed Services Committees, respectively—draft the DoD authorization acts for consideration and passage by Congress. Congress traditionally appropriates funds, and authorizes additional purposes for those funds, in three annual public laws: (a) Department of Defense Appropriations Act (DoDAA); (b) Veterans Affairs and Military Construction Appropriations Act (VA/MILCONAA); and (c) National Defense Authorization Act (NDAA).

In the DoDAA, funds are appropriated for the yearly expenses and investment activities of DoD, not including the funding of contingency operations; this is important to the understanding of overseas military construction and how it is funded. The activities falling under the DoDAA are usually referred to as "baseline operations," funded with "baseline funds."

In the VA/MILCONAA, funds are appropriated for DoD construction, domestic and overseas. Division A appropriates funds for military construction, also known as MILCON, for unspecified minor military construction, or UMMC. Division B appropriates funds for specified military construction. VA funds are separately appropriated now that the VA Administration is its own agency. NDAA provides the maximum amounts that the DoDAA may appropriate, and it provides additional authorizations (purposes) for which the appropriated funds drawn by the DoDAA may be used.

It is generally the rule that Congress appropriates DoD funds in the DoDAA and the MILCONAA, and provides authorizations through the NDAA. However, Congress will provide additional appropriations in the

NDAA and additional authorizations in one of the DoD appropriations acts. This has occurred more frequently since September 11, 2001, because of DoD requests for high-priority authorization or appropriation provisions that might be politically controversial in one or other of the congressional committees. DoD Legislative Affairs officers typically request provisions in both Appropriations Committees and the Armed Services Committees when they are unsure if a high-priority appropriation or authorization will survive the respective committee compromises. It is only on rare occasions that the same appropriation and/or authorization provision appears in multiple appropriations and/or authorization acts. The President, on the other hand, has made the policy decision to request funds for the GWOT through separate appropriations acts, other than the DoDAA or the VA/MILCONAA, to avoid delays possible in politically controversial requests in the DoD or MIL-CON appropriations. Since September 11, 2001, GWOT operations have been funded through wartime supplemental appropriations acts. If DoD expends the appropriated funds enacted by Congress prior to the end of the fiscal year (as has been the case in recent years), Congress may provide additional funds with additional appropriations through "emergency supplemental" appropriations acts. This has occurred with both Iraq and Afghanistan. The federal appropriations acts for 2009 are as follows: (a) 2009 DoDAA, enacted by the President on September 30, 2008;[36] (b) 2009 MILCONAA, enacted by the President on September 30, 2008;[37] and, (c) 2009 NDAA, enacted by the President on October 14, 2008.[38]

Generally, DoD may obligate appropriated O&M (and procurement) funds only when the primary purpose is to benefit the U.S. military directly. The general rule with respect to the funding of U.S. military operations is that the DoS, and not the DoD, funds foreign assistance to foreign nations. For that reason, Part IV of this chapter addresses foreign assistance, including security assistance to a foreign military or government, development assistance for major infrastructure projects, and humanitarian assistance directly to a foreign population.

B. Funding Sources of U.S. Military Operations

This subsection discusses the specific statutes under which construction could be authorized and funded in connection with U.S. military operations in the deployed environment or for exercise-related and other training of foreign allies.

There are three main categories of defense authorizations and appropriations that may involve the sourcing and funding of OCONUS military construction. They are:

1. Assistance to allies: coalition support authorizations and appropriations
2. Assistance to allies: Title 10 training authorizations and appropriations
3. Humanitarian assistance authorizations and appropriations

Under the first category, coalition support authorizations, construction may be authorized either under the Emergency and Extraordinary Expenses Authorization,[39] which covers programs carried out through the DoD and includes the use of O&M funds or the COCOM (Combatant Command) Initiative Fund[40] (CCIF), in which the individual combatant commands, such as the new African Command or the better known Central Command, may initiate programs in their respective areas of responsibility (AORs).

Another source of funding comes under the GWOT coalition support authorizations, which would include logistical support for coalition allies, security assistance, including training and equipment, and support to special operations forces for the GWOT. The GWOT has been funded at a level of $35 million a year in O&M funding through fiscal year 2013.[41]

Under training authorizations and appropriations, there is also an additional special operations forces training authorization to fund something known as "incremental expenses" of developing countries in training with U.S. special operations forces. This could also include use of funds for construction.[42] Temporary or non-permanent construction under this category can also be authorized under the developing countries' combined exercise program, which allows funds to be obligated across one fiscal year.[43] There is also separate funding under this broad category for Partnership for Peace Initiatives,[44] which funded $110 million to the Joint Staff until expended and $44 million to the office of the Secretary of Defense until expended, in addition to those funds authorized under the GWOT funding discussed above.

Finally, under the broad area of humanitarian assistance, money has been appropriated to DoD for Humanitarian and Civic Assistance (HCA), which allows the use of DoD O&M funds for such purposes, including construction.[45] Also under this broad category, there has been both funding for Operation Iraqi Freedom/Operation Emerging Freedom for Iraq and separate funding for Afghanistan. Commander's Emergency Response Program (CERP) funds can also be used for HCA.[46]

C. Exercise-Related Construction

Construction that is necessary for use by U.S. military forces, such as base camp construction, a bridge to the base camp, or other construction necessary for operations during a Joint Chiefs of Staff–directed exercise, are funded with military construction, or MILCON, appropriations,[47] and not through funding for the exercise itself or other humanitarian appropriations.[48] Congress has also prohibited the funding of exercise-related construction under $750,000 as O&M funded construction.[49] A decision in 1984[50] found that the construction of four base camps for U.S. troops, roads, runways, and other buildings were permanent structures that were intended to be left behind for the use of Honduran troops after the exercise and, therefore, should have been funded through DoS funds and not with DoD O&M funds.

One caveat is that a military construction project must include all work necessary to produce either a complete and usable facility or a complete and usable improvement to an existing facility.[51] Construction includes the erection, installation, or assembly of a new facility or the addition, expansion, extension, alteration, conversion, or replacement of an existing facility. Also, funded costs include out-of-pocket expenses such as materials, supplies, and services applicable to the project (including property owned by the government), installed capital equipment, transportation costs, civilian labor costs, overhead and support costs, temporary duty (TDY) expenses, the cost of fuel to operate equipment, and site preparation costs. They do not include "sunk costs" (i.e., "expenses charged to a different appropriation"), such as military pay and labor, government equipment depreciation, some licenses/permits/fees, and excess materials, supplies, and equipment obtained on a nonreimbursable basis from another federal entity. Finally, maintenance and repair projects are not construction and therefore are not subject to the $750,000 O&M limitation on construction. Thus, as a practical matter, MILCON funds, those appropriated in the MCAA, may be available for exercise-related construction, but will not be available for contingency operations unless actually specified in combat- and contingency-related construction authorized under the annual military construction authorization acts.

D. Commander's Emergency Response Program (CERP) in Iraq and Afghanistan

One of the greatest sources of military construction in the world in recent years has been the Iraq-funded CERP (I-CERP), which is the use of CERP funds for Iraq reconstruction. The Secretary of Defense has waived the Competition in Contracting Act of 1984 (CICA)[52] requirements for coalition force CERP–funded projects, and, therefore, those projects need not follow the competition requirements of the FAR. In Iraq, these programs were originally funded with seized Iraqi assets.[53] Originally, the Coalition Provisional Authority (CPA) accounted for the seized Iraqi funds, and it administered and distributed the funds to U.S. commanders for "reconstruction assistance" to the Iraqi people. The CPA initially defined reconstruction assistance as the "building, repair, reconstruction, and reestablishment of the social and material infrastructure in Iraq."[54] I-CERP is now an Iraq-funded program based on a Memorandum of Understanding between the Iraq Supreme Reconstruction Council and the Multinational Force Iraq.[55] Thus, U.S.-appropriated funds are not used in this program. The primary purpose of I-CERP is for commanders to execute urgent reconstruction projects, unlike the longer term and planned humanitarian assistance in the coalition force-appropriated funding for the benefit of the Iraqi people. Commanders may use I-CERP for four basic purposes:

1. Water purification plants;
2. Schools;

3. Health clinics; and/or
4. City planning facilities (government of Iraq, provincial and local government offices).

By exception with the approval of the multinational force commander, I-CERP may also be used for roads, sewers, irrigation, and/or non-reconstruction projects that promote small business development. Coalition force CERP has not been replaced by this project. In fact, the 2009 NDAA authorized the use of over $1.5 billion from DoD O&M in fiscal year 2009 for the coalition force CERP.[56] The 2009 NDA also added significant restrictions to the maximum size of I-CERP projects in Iraq, but specifically exempted these restrictions from applicability in Afghanistan. The restrictions include a reporting requirement to Congress for all projects greater than $500,000; the unavailability of funds for projects greater than $1 million unless the funds are certified by the Secretary or Deputy Secretary of Defense; and the unavailability of appropriated funds for projects greater than $2 million, unless the Defense Secretary waives the restriction and provides notice to Congress for the waiver and justification.[57]

E. Conclusion

Between the DoS and DoD appropriations and authorizations discussed in the two previous sections, Congress has provided the funds necessary for DoD to fund the vast world of contingency operations, including GWOT and combat-related construction.

VI. Contracting Forms and Procedures OCONUS

Contractors doing business with the government are most familiar with Standard Form 1442 as the primary contracting form used for construction contracting. However, in a contingency contracting environment where timeliness and effectiveness are most important, this is not the only contracting form that can be used. Other instruments provide the flexibility needed to satisfy most requirements in that environment. For example, during Operations Desert Shield/Desert Storm, the following contracting purchasing instruments were also used, some to purchase construction, such as painting, tile work, etc., as a commodity: SF 44; DD Form 1155; Blanket Purchase Orders (BPOs); and even imprest fund accounts.

As previously stated, in declared contingencies, the simplified acquisition threshold has been raised to $1.5 million outside the continental United States. For these limits, the simplified acquisition procedure in FAR Part 13 can be used. Simplified acquisitions may be accomplished by use of purchase orders (FAR 13.302), BPOs,[58] imprest funds,[59] and government-wide commercial purchase cards.[60] A purchase order is a government offer to buy certain supplies, services, or construction from commercial sources upon specified terms and conditions. The requirements are set out in FAR Part 13.302. COs use an SF 1449 (Solicitation/Contract/Order for Commercial Items) for purchases of commercial items, including some construction items as a commodity. Per Defense Federal Acquisition Regulation Supplement (DFARS) 213.307, if SF 1449 is not used, DD Form 1155 (Order for Supplies or Services) should be used.

COs may also award contracts through sealed bidding procedures or by contracting by negotiation.[61] Competitive proposals may be used, however, under FAR Part 15 when sealed bids are not appropriate[62] or required, as in OCONUS construction contracting. However, it is generally necessary to conduct discussions with offerors in contracts to be made and performed outside the United States.[63] Moreover, by holding negotiations, COs will be able to conduct discussions with offerors to ensure they understand the government's requirements and have a plan and the capability to perform the contract. In addition, COs will be able to advise offerors through discussions of certain deficiencies in the offers, so that deficiencies can be corrected in an offeror's best and final offer.[64]

In a contingency contracting environment, one of the most difficult determinations for a CO is whether the prices offered by offerors considered for award are fair and reasonable. This determination may be extremely difficult because of a limited number of offerors, the urgency of the government's requirements, and the lack of any record of a lengthy history of the government's purchases for these goods or services at the contingency site. However, in *L. White Construction Co./Ansley-Sheppard-Burgess Co.*,[65] the Comptroller General explains that, before awarding any contract, the CO must determine that the price at which the contract would be awarded is reasonable.[66] A determination of price reasonableness is within the discretion of the contracting agency and will not be disturbed unless the determination is unreasonable or the record shows that it is the result of fraud or bad faith on the part of the contracting officials.[67] *A determination that the price is unreasonable may be based on a comparison of the bid price with the government estimate.*[68] If the price is so low as to be unreasonable and there is evidence the prospective contractor cannot perform for the price bid, the contract should not be awarded.

Whether offers are received by sealed bids or as a result of negotiations, COs must still determine that all offered prices are fair and reasonable.[69]

The award of any negotiated contract or modification of any contract in excess of $500,000 will require the contractor to submit certified cost or pricing data, unless one of the exemptions to the requirement for certified cost or pricing data applies.[70] A waiver can be sought by the CO if there exists a reasonable basis for it. Because most local contractors supporting a foreign contingency operation will not have an understanding of the government's legal requirements for certified cost or pricing data, U.S. contractors working with local contractors should expect to spend time explaining this requirement to potential joint venture partners or subcontractors.

FAR Part 9 also requires that contracts may be awarded only to "responsible" prospective contractors.[71] Thus, the contingency CO (CCO) will examine a firm's general ability to perform the work and not rely on a technical evaluation of the proposal (which would be done by the end user).[72] However, FAR 9.102 does recognize an exception outside the United States in those limited instances where the laws of a foreign country would preclude this, i.e., where "application of the subpart would be inconsistent with the laws or customs where the contractor is located." This issue is not further addressed in the FAR or any of its supplements, the FACs, Comptroller General opinions, Board of Contract Appeals (BCA) decisions, court decisions, or the Federal Register. Thus, host nation laws may have to be consulted.

Finally, other practical issues, such as payment and alternate contracting authority, are synopsized in the attached chart.

SYNOPSIS:

Current Legal Authorities for Contingency Operations

The FAR and DFARS spell out exceptions for OCONUS contingency operations as follows:

FAR/DFARS REFERENCE	SUBJECT	EXCEPTIONS ALLOWED
FAR 5.202(a) (2)	Synopsis	Not applicable if unusual and compelling urgency exists.
FAR 5.202(a) (3)	Synopsis	International agreement specifies the source of supply.
FAR 5.202(a) (12)	Synopsis	Does not apply overseas.
FAR 6.001(a)	Competition Requirements	Does not apply to simplified acquisition procedures (SAPs) (i.e., contingencies).
FAR 25.102(a) (1)	Buy American Act	Not applicable for items used outside United States.
FAR 25.302(b)	International Balance of Payments Program	Government may buy foreign if under simplified acquisition threshold ($1 million).
FAR 25.501	Payment in Local Currency	Foreign contracts should be priced and paid in local currency, unless CCO determines it to be inappropriate.
FAR 25.702 and DFARS	Restrictions on Certain Foreign Purchases	Authorized to buy items from North Korea, Vietnam, Cambodia, and Cuba in emergencies.

FAR/DFARS REFERENCE	SUBJECT	EXCEPTIONS ALLOWED
FAR 28.102-1(a)	Bonds	Can be waived by CCO.
FAR 37.104(b) and DFARS	Personal Services	Allowed under 5 U.S.C. § 3109, if advantageous to the national defense. Requires Determination and Findings.
DFARS 5301.601-93	Contracting Support in Emergency Situations	Expedited acquisition procedures allowed to the extent authorized by law. Also, CCO appointment valid in any theater of deployment.
DFARS 5301.602-3	Ratifications	Can be delegated to others by the HCA (for instance, the "on-scene commander").
DFARS 5301-690	Contracting Authority of Personnel Other than CCOs	Imprest fund purchases in accordance with FAR Part 13.4; SF 44 purchases. Individual must be trained and get written authorization from the CO.

Notes

1. 10 U.S.C. § 2801.

2. *Id.*

3. U.S. Army Regs. AR 415-15, AR 420-10; Dept. of the Army pamphlet DA PAM 420-11; Dept. of the Army Repair Memorandum (Aug. 4, 1997): "A facility must exist and be in a failed or failing condition in order to be considered for a repair project." DA PAM 420-11: "facility or facility component has failed, or is in the incipient stages of failing or is no longer performing the functions for which it was designated"; and Office of the Chief of Naval Operations, OPNAV INST 11010.20G: Facilities Projects Instruction (Oct. 14, 2005).

4. *See* Air Force Instruction AFI 32-1032 (Oct. 15, 2003), *available at* http://www.af.mil/shared/media/epubs/AFI32-1032.pdf.

5. *Id.*

6. U.S. CONST. art. II, § 2, cl. 1.

7. U.S. CONST. art. I, § 9, cl. 7.

8. U.S. CONST. art. IV, § 3, cl. 2.

9. 426 U.S. 317 (1976).

10. WILLIAM C. BANKS & PETER RAVEN-HANSEN, NATIONAL SECURITY LAW AND THE POWER OF THE PURSE 166 (1994).

11. 83 AM. J. INT'L LAW, 750 (1989).

12. Pub. L. No. 11-5 (2009).

13. FAR 36.103(a).

14. 41 U.S.C. §§ 10a–10d. For further discussion of the BAA, see Chapter 12.

15. FAR 25.600-25.607.

16. 31 U.S.C. § 1301(a).

17. The Honorable Bill Alexander, B-213137, 63 Comp. Gen. 422 (1984).

18. 31 U.S.C. § 1341(a).

19. 31 U.S.C. § 9502; DEFAS-DER 7004, ¶ 7; DEFAS-INERG 37-1, ¶ 99-5c(2).

20. 31 U.S.C. § 1502(a).

21. 22 U.S.C. § 2761.

22. 22 U.S.C. § 2356.

23. 22 U.S.C. § 2347.

24. 22 U.S.C § 2346.

25. 22 U.S.C. § 2291.

26. The Hon. Bill Alexander, B-213137, 63 Comp. Gen. 422 (1984).

27. 22 U.S.C. § 2346.

28. Press Release, U.S. Embassy, Iraq, Provincial Reconstruction Team (PRT) Fact Sheet (Mar. 20, 2008), *available at* http://iraq.usembassy.gov/root/pdfs/factsheetapril2008.pdf.

29. *See generally* Timothy Austin Furin, *Legally Funding Military Support to Stability, Security, Transition, and Reconstruction Operations*, ARMY LAW., Oct. 2008, at 41–43.

30. Pub. L. No. 105-338, 112 Stat. 3178 (Oct. 31, 1998).

31. Pub. L. No. 107-327, 116 Stat. 2797 (codified at 22 U.S.C. § 7532 (2002)).

32. *See* section 506 of the Foreign Assistance Act, 22 U.S.C. § 2318(a)(1).

33. Pub. L. 110-28, available through Sept. 30, 2008.

34. Pub. L. 110-161, available through Sept. 30, 2009.

35. The Hon. Bill Alexander, House of Representatives, B-213137, Jan. 30, 1986 (unpublished GAO opinion).

36. *See* Consolidated Security, Disastrous Systems, and Continuing Appropriations Act, 2009, Pub. L. No. 110-329, div. C.

37. *Id.* at div. E.

38. Duncan Hunter National Defense Authorization Act for Fiscal Year 2009 [NDAA 2009], Pub. L. No. 110-417.

39. 10 U.S.C. § 127.

40. 10 U.S.C. § 166(a).

41. NDAA 2009, Pub. L. No. 110-417, § 1208.

42. 10 U.S.C. § 2011.

43. 10 U.S.C. § 2010.

44. Pub. L. No. 103-337, § 1307.

45. 10 U.S.C. § 401.

46. NDAA 2009, Pub. L. No. 110-417, § 1214.

47. 10 U.S.C. § 2805(a)(2).

48. 10 U.S.C. § 401.

49. 10 U.S.C. § 2805. *See also* Ahuas Tara II, B-213137, June 22, 1984, 63 Comp. Gen. 422 (1984); this decision arose out of the funding of large numbers of Honduran troops with O&M funds, which was a violation of U.S. security assistance laws. All such funding should come from DoS funds.

50. Ahaus Tara II, B-213137, June 22, 1984, 63 Comp. Gen. 422 (1984); 1984 U.S. Comp. Gen. LEXIS 972.

51. 10 U.S.C. § 2801(b).

52. 41 U.S.C. § 253 *see also* 10 U.S.C. § 2304.

53. Memorandum from President George W. Bush to Secretary of Defense Rumsfeld on Certain State- or Regime-Owned Property in Iraq (Apr. 30, 2003).

54. *See* Combined Joint Task Force Fragmentary Order CJTF-7 FRAGO 89 (June 2003).

55. *See* Multi-National Force–Iraq Fragmentary Order MNF-I FRAGO 08-166, Apr. 17, 2008 and Multi-National Corps–Iraq Fragmentary Order MNC-I FRAGO 08-322, Apr. 19, 2008.

56. *See* National Defense Authorization Act for Fiscal Year 2008, Pub. L. No. 110-181, § 1205.

57. Memorandum from Tina W. Jonas, Under Secretary of Defense Comptroller, the Secretaries of the Military Departments, et al., on Commander's Emergency Response Program (CERP) Guidance (June 6, 2008).

58. FAR 13.303.

59. FAR 13.305.

60. FAR 13.301. The SAP is currently $200,000 for UCONUS contingency operations. *See* 41 U.S.C. § 403(11) as amended by 10 U.S.C. § 2302(7).

61. FAR 6.401. Sealed bidding and competitive negotiation are discussed in Chapters 2 and 3, respectively.

62. FAR 6.401(b)(1).

63. FAR 6.401(b)(2).

64. FAR 15.610.

65. B-245916, Feb. 3, 1992, 92-1 CPD ¶ 138.

66. FAR 14.407-2(f); Adrian Supply Co., B-240871, B -240872, Dec. 21, 1990, 90-2 CPD ¶ 515.

67. G. Marine Diesel Corp., B-238703, B-238704, May 31, 1990, 90-1 CPD ¶ 515.

68. Hawkins Builders, Inc., B-237680, Feb. 5, 1990, 90-1 CPD ¶ 154.

69. *See* FAR 14.407-2(a) and 15.802(b)(1).

70. FAR 15.403-4(a)(i).

71. 10 U.S.C. § 2305(b)(4)(B).

72. *See* Data Preparation Inc., B-233569, Mar. 24, 1989, 89-1 CPD ¶ 300.

CHAPTER 26

Federal Grants

STEPHEN M. SEEGER
ANGELINE R. NELSON

I. Introduction

Grants are a vehicle for the federal government to provide financial assistance when the primary purpose of funding is to accomplish some public support or stimulation.[1] Assistance programs are designed to serve various public purposes, many impacting the construction industry, such as construction of highways, hospitals, housing, sewers, wastewater treatment facilities, state-directed mass transit projects, and airport programs.[2]

Federal government ("grantor") agencies or departments may issue grants to state or local governments or private entities.[3] To award grants, grantor agencies must have specific statutory authority.[4] Grant programs are generally governed by detailed legislation and regulations. As a result, much of the case law involving judicial and administrative grants is not broadly applicable to the extent those cases "hinge on specific statutory or regulatory provisions."[5]

The past four decades have seen a dramatic growth in federal grants.[6] However, beginning with the enactment of the American Recovery and Reinvestment Act of 2009 (ARRA) in February 2009, the United States is witnessing an exponential increase in the use of federal grants.[7] For example, ARRA aims to undertake large-scale weatherization and modernization projects of 75 percent of federal building space and more than 1 million homes, as well as investing heavily in infrastructure by improving roads, bridges, and mass transit systems.[8]

Federal agencies and departments using ARRA grant funds are subject to specific policy objectives, such as job creation and preservation.[9] Further, use of any grant funds dispersed under ARRA will be scrutinized under increased transparency and accountability, creating stricter auditing and reporting procedures.[10] The White House Office of Management and Budget

(OMB) issued guidance for implementing ARRA.[11] In April 2009, a new part 176 was added to Title 2 of the Code of Federal Regulations to provide interim final guidance and standard award terms for grants and cooperative agreements funded with ARRA funds, discussed below.[12]

A. Grants and Cooperative Agreements

Grants and cooperative agreements are related forms of federal assistance agreements, distinguished by the degree of the government's involvement in carrying out the underlying activity.[13] Grants are to be used when substantial involvement is not expected between the executive agency and the state, local government, or other recipient when carrying out the activity contemplated in the agreement.[14] Conversely, cooperative agreements are appropriate where a higher degree of government involvement is expected.

B. The Relationship Between Assistance Agreements and Contracts

Grants are not procurements contracts, and, as such, are not governed by the Federal Acquisition Regulation (FAR), the Contract Disputes Act, or federal procurement case law.[15] There are, however, contractual aspects to the grant relationship, and the rights and obligations of the parties to a federal grant will be determined by general contract law principles.[16] Additionally, Congress may impose conditions upon recipients of federal grants if the grantee is sufficiently aware of the condition.[17]

Grant programs will not "be viewed in the same manner as a bilateral contract governing a discrete transaction. . . . Unlike normal contractual undertakings, federal grant programs originate in and remain governed by statutory provisions expressing the judgment of Congress concerning desirable public policy."[18] As a result, various rules and principles of contract law will not automatically be applied to grants, such as the doctrines of impossibility of performance, quantum meruit, and implied-in-fact contracts.[19]

II. Federal Control of State and Local Contracting

There is a tension between state and local government autonomy and federal interest in assuring that its funds for state and local government projects are not misused. State and local governments that use federal grants or funds must do so pursuant to the terms of the grant and applicable federal law and regulations.[20] The federal government defines and prescribes the permissible uses of federal grant funds primarily through OMB Circular A-102, through the Federal Grants and Cooperative Agreement Act of 1977, discussed at length below, and through other documents frequently incorporated into the grant agreements. Also, all grants utilizing ARRA funds are subject to additional executive guidance regarding transparency, accountability, and objectives of the grant.[21] The FAR does not apply to federal grants or cooperative agreements.

A. OMB Circular A-102 and the Common Rule

The OMB provided the first uniform regulations governing federal grant-in-aid programs in 1971 when it issued Circular A-102, titled "Uniform Administrative Requirements for Grants and Cooperative Agreements to State and Local Governments."[22] Circular A-102 also established standards and guidelines for the procurement of supplies, equipment, construction, and services for federal grant programs.[23]

In 1988, the OMB revised Circular A-102 to create the Grants Management Common Rule (Common Rule), "governing the administration of federal grants by state and local government."[24] The Common Rule standardized and codified the fiscal and administrative requirements federal agencies impose on grant recipients.[25] Most agencies and departments have codified the Common Rule.[26] A chart of departments and agencies and their corresponding regulations codifying the Common Rule appears at the end of this chapter.

The Common Rule contains important elements for procurement procedures, including the following:

- Grantees must apply the same policies and procedures as for procurements with non-federal funds.
- Contracts let by a state that uses grant funds must include any clauses required by federal statutes and executive orders.
- Grantees and subgrantees other than states will use their own procurement procedures that reflect applicable state and local laws and regulations and that comply with applicable federal law.[27]

B. The Federal Grant and Cooperative Agreement Act of 1977

In an effort to minimize misuse of grants versus procurement contracts and to prescribe the distinct uses of grants and cooperative agreements, Congress enacted the Federal Grant and Cooperative Agreement Act of 1977[28] (FGCAA).[29] The FGCAA's primary goals were to achieve uniformity, clearly define the various assistance agreement relationships, and clarify the grantor and grantee responsibilities.[30]

The FGCAA distinguished procurement contracts and the forms of assistance agreements, and defined their appropriate uses, as follows:

> Agencies must use a procurement contract:
> ... as the legal instrument reflecting a relationship between the United States government and a State, a local government, or other recipient when—(1) the principal purpose of the instrument is to acquire ... property or services for the *direct benefit or use of*

the United States Government; or (2) the agency decides in a specific instance that the use of a procurement contract is appropriate.[31]

Grant Agreements are the appropriate instrument when:

> (1) the principal purpose of the relationship is to transfer a thing of *value to the State or local government* or other recipient to carry out a *public purpose of support or stimulation* authorized by a law of the United States instead of acquiring . . . property or services for the direct benefit or use of the United States Government; and (2) substantial involvement is not expected between the executive agency and the [grantee] when carrying out the activity contemplated in the agreement.[32]

Finally, cooperative agreements must be used when "substantial involvement is expected between the executive agency and the [recipient]. . . ."[33] It is thus the basic purpose of the underlying activity that determines whether a procurement contract or assistance agreement should be used.[34]

In choosing between forms of assistance agreements involving the same general public-use purpose, the determining factor is the anticipated level of federal government involvement.[35] It may be difficult to determine whether the government's involvement is substantial enough to require a cooperative agreement instead of a grant. Cooperative agreements are likely the appropriate vehicle in the following situations:

- Federal project management or federal program or administrative assistance would be helpful because of novelty or complexity involved;
- Federal/recipient collaboration is desirable;
- Federal monitoring is desirable to permit certain direction or redirection of the work because of interrelationships among projects in areas such as applied research; and
- Federal involvement is desirable in the early stages of ongoing programs, such as welfare or law enforcement programs, where standards are being developed or the application of standards requires a period of adjustment until recipient capability has been developed.[36]

Agencies must choose the type of agreement with care. Procurement contracts are subject to a variety of statutory and regulatory requirements that are not generally applicable to grants or cooperative agreements.[37] When grants are appropriate, they should not be unnecessarily burdened by the additional requirements of a procurement contract.[38] Likewise, where a procurement contract is appropriate, assistance agreements should not be improperly used to evade competition or other procurement regulation requirements.[39]

The grantor agency's decision to use a particular instrument will have a major effect on that project's administration. Further, choosing the wrong instrument invites potential protests. For example, in *Chem Service, Inc. v.*

Environmental Monitoring Systems Laboratory, a plaintiff properly challenged whether a cooperative research and development agreement between the Environmental Protection Agency (EPA) and a private laboratory actually constituted a procurement contract subject to competition requirements under federal law.[40]

III. State and Local Procurement under Federal Grants

Grantee state and local governments often obtain grants for construction of infrastructure, housing, and other public use projects. Grantees generally use their own procurement procedures, which must comply with certain minimum requirements.[41] Under the new 2 C.F.R. Part 176, effective April 23, 2009, the OMB issued additional guidance for implementing ARRA provisions. Under the new regulations:

> Under the interim final guidance, agencies would use the standard award terms in their financial assistance awards to require recipients and subrecipients (first-tier that are not individuals) to maintain current registrations in the Central Contractor Registration (CCR[located at http://www.ccr.gov]) database; to require recipients to report quarterly on project or activity status, subgrant and subcontract information; to notify recipients of the domestic sourcing ("Buy American") requirements that apply to certain iron, steel and manufactured goods; to notify recipients of the wage rate requirements that apply to certain projects; and to ensure proper accounting and reporting of Recovery Act expenditures in single audits.[42]

A. Procurement Procedure

A grantee may adopt American Bar Association (ABA)-approved procurement procedures. The ABA sections of Public Contract Law and State and Local Government Law developed a Model Procurement Code for state and local governments, which was adopted by the ABA in 1979 and revised in 2000 (the 2000 MPC).[43] The 2000 MPC added new materials to guide state and local jurisdictions wishing to use new methods of public-private collaboration.[44] For public infrastructure projects with state and local governments in particular, the ABA developed the Model Code for Public Infrastructure Procurement (the 2007 MC PIP).[45] The 2007 MC PIP provides "basic policies for the procurement of supplies, services and construction; disposal of supplies"; legal remedies; socioeconomic policies; and "ethical standards for public officials and contractors in connection with governmental procurement."[46]

The MPC Model Codes are not mandatory. However, a grantee must submit for agency approval a procedure that incorporates the Common Rule procurement standards.[47]

B. Competition

Grants are classified as either mandatory or discretionary. In a mandatory grant program, Congress directs awards to one or more classes of prospective recipients who meet specific criteria for eligibility, in specified amounts. Sometimes called "entitlement" or "formula" grants, these are often awarded on the basis of statutory formulas.[48] The grantor agency has no basis to refuse to make the award.[49] Because of the lack of discretion in making the award, competition is not required.

On the other hand, discretionary grants require full and open competition to determine the best possible projects to achieve the agency program objectives.[50] The FGCAA does not offer competition guidelines, and there is no equivalent to the Competition in Contracting Act.[51] Competition guidelines therefore vary by agency, and generally these guidelines do not set forth particular competition procedures.[52] Because of the broad range of competition requirements, the U.S. Government Accountability Office (GAO) does not hear bid protests or complaints that a certain grant should have been awarded to a different entity.[53] GAO will not question funding decisions in discretionary federal assistance programs.[54]

Procurement transactions undertaken by grant recipients must provide full and open competition and not unduly restrict or eliminate competition.[55] In the context of construction projects, competitive sealed bidding is preferred under the Common Rule.[56] The other permissible procurement methods are competitive negotiation, noncompetitive proposals, and small purchase procedures.[57]

When using sealed bids, the grantee issues an invitation for bids (IFB) that must "(1) be publicly advertised with sufficient time to bid before the date set for bid opening; (2) include specifications that define the items or services subject to the procurement to allow the bidder to properly respond; (3) mandate that all bids be opened publicly at the time and place published in the IFB; and (4) provide that any and all bids may be rejected for a reasonable and documented cause."[58] The grantee must award the contract to the lowest responsive and responsible bidder.[59]

Competitive negotiation may be used when sealed bidding is not appropriate.[60] The grantee must issue a request for proposals (RFP) that identifies procurement requirements and the relative importance of the significant evaluation factors.[61] As opposed to sealed bidding, the grantee may conduct negotiations with the offerors. The grantee may make the contract award to that bidder "whose proposal will be most advantageous to the procuring party, price and other factors considered." Thus, the grantee has more flexibility in selecting the winning bidder.[62]

Although generally discouraged, noncompetitive negotiation may be used when award of a contract is not feasible through the other permitted procurement methods. This is generally the case where the goods and services are available only from a sole source, or where there is public exigency or emergency that does not permit the time necessary to use competitive procedures.[63]

C. Construction Contract Provisions

Local grantees procuring construction services must ensure that certain contract provisions are utilized.[64] The Common Rule requires local grantees to include certain contract provisions, and agencies such as the EPA, the Federal Emergency Management Agency (FEMA), Housing and Urban Development (HUD), and the Department of Transportation (DOT), require or permit certain specific contract clauses.[65] The contract clauses that are typically required for construction projects include the Changes, Differing Site Conditions, Suspension of Work, Termination for Default, Termination for Convenience, Davis-Bacon, Work Hours and Safety, and Equal Employment Opportunity clauses.[66] Other socioeconomic requirements, such as disadvantaged business entity participation, are also frequently incorporated.

The EPA administers numerous construction and engineering grant projects, such as construction of water treatment plants.[67] EPA regulations employ model contracts for construction and engineering projects and require use of certain contract clauses.[68] Required EPA model clauses include the Suspension of Work, Termination for Cause, Termination for Convenience, Remedies, Audit and Access to Records, the Covenant Against Contingent Fees, Gratuities, and Buy American clauses.[69]

D. Other Contract Requirements

Before bidding begins, some owners use a prequalification procedure for major equipment items. The grantee must issue a public notice clearly stating the required evaluation criteria.[70] Passing the prequalification is a prerequisite to submitting a bid for that IFB.[71] Grantee solicitations may impose experience requirements on businesses, but such requirements may not be enforced where they act to reduce competition.[72]

In construction and improvement contracts exceeding $100,000, the grantor agency may not require a bond if the agency reviews the grantee's procedures and determines that they are sufficient to protect the agency's interest.[73] If the agency does not make that determination, the grantee must require a bid guarantee (5 percent of the bid price), performance bond (100 percent of contract price), and payment bond (100 percent of contract price).[74]

E. Funding

Grant funds in the hands of the grantee are not subject to all the same restrictions and limitations imposed on direct expenditures by the federal government, but grantees do not have unfettered use of grant funds.[75] Grantees must spend the grant funds for the purposes and objectives of the grant and in accordance with applicable statutory or other conditions for use of the grant funds.[76] This includes the implied requirement of adhering to basic principles of full and open competition, and statutory requirements such as compliance with

Title VI of the Civil Rights Act of 1964,[77] the Rehabilitation Act of 1973,[78] the Age Discrimination Act, and Title IX of the Education Amendments of 1972.[79]

Grant recipients "are expected to use the assistance funds for the purposes for which they are awarded, subject to any conditions that may attach to the award."[80] What constitutes an "allowable cost" is determined by "relevant program legislation, regulations, including OMB circulars and the common rules, and the terms of the grant agreement."[81]

IV. Power of Agencies to Enter into Assistance Agreements

Agencies have no inherent right to enter into assistance agreements. While agencies may enter into contracts for the agency's own use, they cannot enter into a transaction to benefit an entity other than the government without independent statutory authority.[82] The FGCAA itself is not a grant of agency authority to enter into assistance relationships;[83] it merely defines the relationships the agency can enter into.[84]

A. Agency Regulations

The primary source of agency authority to enter into assistance agreements is the agency's regulations.[85] Such regulations must fall within the agency's broader statutory authority, be issued in compliance with applicable procedural requirements, and not be arbitrary or capricious.[86] The Administrative Procedure Act, the law under which federal agencies create rules and regulations to implement and enforce legislation, does not apply to agency grant regulations.[87] Grant regulations are subject to congressional oversight and review by the GAO and Congress under the Congressional Review Act.[88]

B. Other Sources

In addition, there are statutes that themselves provide agency authority to enter into assistance agreements.[89] Pursuant to the powers granted by the Homeland Security Act, the Department of Homeland Security manages federal grants for state and local governments for port security, critical infrastructure protection activities such as freight rail security, regional and local mass transit systems, equipment and training for first responders, and Homeland Security.[90] Also, the Stafford Act gives FEMA the authority to administer grants relating to disaster assistance.[91]

V. Liability under the False Claims Act: Allison Engine Co. and the FERA

A contractor obtaining payment pursuant to a federal grant may be exposed to civil or criminal liability under the False Claims Act (FCA).[92] For an extensive discussion of the FCA, see Chapter 24. The scope of and potential for liability for a recipient of grant funds under the FCA has changed substantially over the past year with the Supreme Court's decision in *Allison Engine Co.*[93] and the recent enactment of the Fraud Enforcement and Recovery Act (FERA).[94] Until *Allison Engine* and FERA, there was debate about whether recipients of grant funds from a private entity grantee could be liable under the FCA. In other words, it was not clear whether project participants on contracts with state and local governments that utilized federal funds were subject to the FCA.

Prior to the enactment of FERA, the FCA provided, in part, for liability for any person who:

(1) Knowingly *presents*, or causes to be presented, to an officer or employee of the United States Government or a member of the Armed Forces of the United States a false or fraudulent claim for payment or approval;

(2) knowingly makes, uses, or causes to be made or used, a false record or statement to get a false or fraudulent claim paid or approved by the government;

(3) conspires to defraud the Government by getting a false or fraudulent claim allowed or paid; . . . .[95]

The debate centered on whether subsections (a)(2) and (a)(3) had a "presentment" requirement, requiring intent that the *government itself* pay or approve payment of funds based on the false or fraudulent representation.[96] The Supreme Court's decision "precluded liability under the FCA when fraudulent claims induce private entities to disburse federal funds over which the private entity has complete control."[97] However, the *Allison Engine* decision "left open the possibility that, if the federal government is somehow involved in the grantee's disbursement of federal money, FCA liability may exist."[98]

On May 20, 2009, Congress passed the FERA to strengthen the FCA and close certain FCA loopholes, arguably in direct response to the *Allison Engine* decision. Specifically, FERA was designed to enforce Congress's initial intent that a "false claim to the *recipient of a grant* . . . under a program financed in part by the United States, [is] to be *considered a false claim to the United States.*"[99]

Congress addressed some of the particular reasons for enacting the FERA, stating:

the Supreme Court ruled in the Allison Engine decision that, absent the "Government itself" inking the check or approving a false claim, the Act does not impose liability for false claims on Government funds disbursed for a Government purpose by a Government contractor or other recipient of Government funds, even if such fraud damages the Government or its programs. Because so many inherently governmental functions are carried out by government contractors these days, including contracting and program management functions, this ruling severely limits the reach of the law. *The primary impetus for the current corrective legislation is to reverse these unacceptable limitations and restore the [FCA] to its original status as the protector of all Government funds or property.* . . . the following conduct clearly violates the [FCA]: . . . submitting a fraudulent application for a grant of Government funds . . . submitting a false application in a multi-staged grant application process, where the second stage of the application would not have been granted had the applicant been truthful in the first stage. . . .[100]

The FERA amends the FCA to no longer require presentment "to an officer or employee of the United States Government or a member of the Armed Forces of the United States" to incur liability. The definition of "claim" was also changed, in part, as follows:

[A]ny request or demand, whether under a contract or otherwise, for money or property, *whether or not the United States has title to the money or property*, that—
(i) is presented to an officer, employee, or agent of the United States; or
(ii) is made to a contractor, *grantee, or other recipient*, if the money or property is to be spent or used on the Government's behalf or to advance a Government program or interest . . . .[101]

Congress has made it clear that the FCA standards will be broadly applied, affecting the rapidly growing number of grant fund recipients.

VI. *Grants and Construction*

A. Where to Find Construction Project Grants

Some of the federal agencies and departments most commonly involved in construction contracting through grants are DOT, EPA, FEMA, and HUD. At least through 2009 and 2010, the majority of new grants for construction are funded through ARRA. For example, HUD has awarded ARRA funds for projects under the Public Housing Capital Fund, Neighborhood Stabilization Program, etc.[102] As of December 7, 2009, HUD has awarded approximately $11.38 billion available under ARRA, and has paid out approximately $2.1 billion.[103] DOT has approximately $31.77 billion available funds and has paid out $683.163 million.[104] More information on particular grant opportunities is available at Grants.gov.

B. Codification of the Common Rule by Department and Agency[105]

Department	Common Rule
Agriculture	7 C.F.R. 3016
Commerce	15 C.F.R. 24
Defense	32 C.F.R. 33

Agency	Common Rule
EPA	40 C.F.R. 31
FEMA	40 C.F.R. 13
GSA	41 C.F.R. 105-71

Education	34 C.F.R. 80
Energy	10 C.F.R. 600
Health & Human Services	45 C.F.R. 92
Housing & Urban Development	24 C.F.R. 85
Interior	43 C.F.R. 12
Justice	28 C.F.R. 66
Labor	29 C.F.R. 97
State	22 C.F.R. 135
Transportation	49 C.F.R. 18
Veterans Affairs	38 C.F.R. 43

NASA	14 C.F.R. 1207
NARA*	36 C.F.R. 1207
NEA**	45 C.F.R. 1157
NEH***	45 C.F.R. 1174
NSF	45 C.F.R. 602
ONDCP****	21 C.F.R. 1403
SBA	13 C.F.R. 143

*National Archives & Records Admin.
** National Endowment for the Arts
***National Endowment for the Humanities
**** Office of National Drug Control Policy

Notes

1. Federal Grant and Cooperative Agreement Act of 1977, 31 U.S.C. § 6304.

2. SMITH CURRIE & HANCOCK'S FEDERAL GOVERNMENT CONSTRUCTION CONTRACTS 531 (Thomas J. Kelleher, Jr., Thomas E. Abernathy, IV & Hubert J. Bell, Jr. eds., 2008).

3. JOHN CIBINIC, JR. & RALPH C. NASH, JR., FORMATION OF GOVERNMENT CONTRACTS 15 (3d ed. 1998).

4. *Id*. at 16.

5. GAO, 2 PRINCIPLES OF FEDERAL APPROPRIATIONS LAW 10-3 (3d ed. 2006) (hereinafter GAO REDBOOK) at 10-5, 10-6, *available at* http://www.gao.gov/special. pubs/d04261sp.pdf.

6. *Id*. at 10-4.

7. The American Recovery and Reinvestment Act of 2009, 123 Stat. 115 (2009), *available at* www.recovery.gov/about/pages/the_act.aspx.

8. *Id*.

9. Memorandum from Peter R. Orszag, Director, Office of Management and Budget, to the Heads of Departments and Agencies, on Initial Implementing Guidance for the American Recovery and Reinvestment Act of 2009 (Feb. 18, 2009), *available at* http://www.whitehouse.gov/omb/assets/memoranda_fy2009/m09-10 .pdf (hereinafter Initial Guidance OMB Memorandum).

10. *Id*.

11. *See* Initial Guidance OMB Memorandum, *supra* note 9; *see also* Memorandum from Peter R. Orszag, Director, Office of Management and Budget, to the

Heads of Departments and Agencies, on Updated Implementing Guidance for the American Recovery and Reinvestment Act of 2009 (Apr. 3, 2009), *available at* http://www.whitehouse.gov/omb/assets/memoranda_fy2009/m09-15.pdf (hereinafter Updated Guidance OMB Memorandum).

12. Updated Guidance OMB Memorandum at 3; 74 Fed. Reg. 18,449.

13. GAO Redbook, *supra* note 5, at 10-5; *see also* Kellehes, *supra* note 2, at 531.

14. 31 U.S.C. § 6304.

15. *See* Cibinic & Nash, *supra* note 3, at 16 (citing Trauma Serv. Group, Ltd. v. United States, 33 Fed. Cl. 426 (1995)).

16. GAO Redbook, *supra* note 5, at 10-6 (citing McGee v. Mathis, 71 U.S. 143, 155 (1866) (determining that contractual elements existed and thus the grant constituted a contract)).

17. Pennhurst State Sch. & Hosp. v. Halderman, 451 U.S. 1, 17 (1981), stating:

> [L]egislation enacted pursuant to the spending power is much in the nature of a contract: in return for federal funds, the States agree to comply with federally imposed conditions. The legitimacy of Congress' power to legislate under the spending power thus rests on whether the State voluntarily and knowingly accepts the terms of the "contract." . . . There can, of course, be no knowing acceptance if a State is unaware of the conditions or is unable to ascertain what is expected of it. Accordingly, if Congress intends to impose a condition on the grant of federal moneys, it must do so unambiguously.

18. Bennett v. Ky. Dep't of Educ., 470 U.S. 656, 669 (1985).

19. GAO Redbook, *supra* note 5, at 10-10.

20. Michael T. Callahan, Procurement of Construction and Design Contracts § 3.01, 243 (2005).

21. Initial Guidance OMB Memorandum, *supra* note 9.

22. Callahan, *supra* note 20, at 243.

23. *Id.*; Office of Mgmt. & Budget, Circular No. A-102 Revised, Attachment 0, 1997 WL 563108 (O.M.B.) (hereinafter OMB Circular A-102).

24. 60 Fed. Reg. 19,638-01; Callahan, *supra* note 20, at 243 (citing A-G-E Corp. v. United States, 753 F. Supp. 836, 844 (D.S.D. 1990), *aff'd*, 968 F.2d 650 (8th Cir. 1992)).

25. Kelleher et al., *supra* note 2, at 533; OMB Circular A-102.

26. Office of Mgmt. & Budget, Grants Management, Codification of Governmentwide Grants Requirements by Department and Agency, http://www.whitehouse.gov/omb/grants_chart/.

27. Kelleher et al., *supra* note 2, at 538–39.

28. Pub. L. No. 95-224, 92 Stat. 3 (1978), codified at 31 U.S.C. §§ 6301–6308.

29. GAO Redbook, *supra* note 5, at 10-13.

30. 31 U.S.C. § 6301 (2); GAO Redbook, *supra* note 5, at 10-14.

31. 31 U.S.C. § 6303 (emphasis added).

32. 31 U.S.C. § 6304 (emphasis added).

33. 31 U.S.C. § 6305 (emphasis added).

34. GAO Redbook, *supra* note 5, at 10-15.

35. *Id.*

36. S. Rep. No. 95-449, at 9–10 (1978); GAO Redbook, *supra* note 5, at 10-15, 10-16.

37. GAO REDBOOK, *supra* note 5, at 10-18.

38. *Id.*

39. *Id.*

40. Chem Serv., Inc. v. Envtl. Monitoring Sys. Lab. of Cincinnati, 12 F.3d 1256 (3rd Cir. 1993).

41. CALLAHAN, *supra* note 20, at 246.

42. 74 Fed. Reg. 18,449; 2 C.F.R. 176.

43. Kelleher et al., *supra* note 2, at 536.

44. AM. BAR ASS'N, 2007 MODEL CODE FOR PUBLIC INFRASTRUCTURE PROCURE-MENT, at ii (2008) (*available at* http://www.abanet.org/dch/committee.cfm?com=PC500500).

45. *Id.* at iii.

46. *Id.* at iv.

47. Kelleher et al., *supra* note 2, at 537.

48. GAO REDBOOK, *supra* note 5, at 10-25.

49. *Id.*

50. *Id.* at 10-26; 31 U.S.C. § 6301(3).

51. Kurt M. Rylander, *Scanwell Plus: Challenging the Propriety of a Federal Agency's Decision to Use a Federal Grant and Cooperative Agreement*, 28 PUB. CONT. L.J. 1, 72 (1998) (citing 41 U.S.C. § 253(a) ("Competition in Contracting Act")).

52. GAO REDBOOK, *supra* note 5, at 10-26.

53. *Id.*; *see, e.g.*, Baltimore Council for Equal Business Opportunity, B-203096, May 20, 1981, 81-1 CPD ¶ 396 ("GAO will not review propriety of federal agency's individual grant award")

54. GAO REDBOOK, *supra* note 5, at 10-27.

55. CALLAHAN, *supra* note 20, at 245.

56. *Id.*

57. *Id.*

58. *Id.*

59. *Id.*

60. *Id.*

61. *Id.*

62. *Id.* at 245–46.

63. *Id.* at 246.

64. Kelleher et al., *supra* note 2, at 534 (citing 40 C.F.R. § 31.36(a)(2008) (EPA Regulations for Post-Award Requirements)); 43 C.F.R. § 12.76(a) (2008) (Department of the Interior Regulations for Post-Award Requirements)).

65. CALLAHAN, *supra* note 20, at 252; Kelleher et al., *supra* note 2, at 535.

66. Kelleher et al., *supra* note 2, at 535.

67. CALLAHAN, *supra* note 20, at 253.

68. *Id.*

69. *Id.*

70. *Id.* at 249.

71. *Id.*

72. *Id.* at 247.

73. *Id.* at 250.

74. *Id.*; *see, e.g.*, 40 C.F.R. § 31.36(h)(3).

75. GAO Redbook, *supra* note 5, at 10-68.

76. *Id.* at 10-71.

77. 42 U.S.C. § 2000d.

78. 29 U.S.C. § 794.

79. 20 U.S.C. § 1681.

80. GAO Redbook, *supra* note 5, at 10-111.

81. *Id.*

82. GAO Redbook, *supra* note 5, at 10-17; Matter of Council on Environmental Quality and Office of Environmental Quality—Cooperative Agreement with National Academy of Sciences, B-218816, June 2, 1986, 65 Comp. Gen. 605, 607.

83. *Id.*

84. *Id.* at 10-18 (citing S. Rep. No. 97-180, at 4 (1981)).

85. *Id.* at 10-45.

86. *Id.* at 10-46.

87. *Id.* at 10-46; 5 U.S.C. § 553(a)(2).

88. Pub. L. No 104–121, tit. II, subtit. E, § 251, 110 Stat. 847, 868 (Mar. 29, 1996), codified at 5 U.S.C. §§ 801–808.

89. Kelleher et al., *supra* note 2, at 537.

90. Department of Homeland Security, http://www.dhs.gov/xopnbiz/grants/ (last visited Nov. 18, 2009).

91. *Id.; see also* Kelleher et al., *supra* note 2, at 538.

92. 31 U.S.C. § 3729 (2008); 18 U.S.C. § 287.

93. Allison Engine Co. v. United States *ex rel.* Sanders, 128 S. Ct. 2123 (2008).

94. Pub. L. No. 111-21, § 386, 123 Stat. 1617 (May 20, 2009).

95. 31 U.S.C. § 3729(a) (emphasis added).

96. *Id.* at 2128.

97. United States Dep't of Transp. v. CMC Eng'g, 564 F.3d 673, 677 (3d Cir. 2009) (citing *Allison Engine*, 128 S. Ct. 2123).

98. *Id.* at 678.

99. 155 Cong. Rec. E1295-03, E1296 (daily ed. May 18, 2009) (statement of Rep. Berman).

100. *Id.*

101. Pub. L. No. 111-21, § 386, 123 Stat. 1617, 1622–23; 31 U.S.C. § 3729 (b)(2) (as amended May 20, 2009).

102. Recovery.gov, Press Release, Dep't of Hous. & Urban Dev., Secretary Donovan Awards Over $1 Billion in Recovery Act Funds to Jump-Start Affordable Housing Construction in 26 States (June 30, 2009), http://portal.hud.gov/portal/page/portal/RECOVERY/news/NO_09-103.

103. Recovery.gov (Recovery.gov/Transparency/agency/reporting/default2 .aspx, then select "Agency" from menu) (last visited Dec. 7, 2009).

104. *Id.*

105. The White House, Office of Mgmt. & Budget, Grants Management, Codification of Governmentwide Grants Requirements by Department and Agency, http://www.whitehouse.gov/omb/grants_chart/.

TABLE OF CASES

INDEX

Use and Possession Prior to
Completion clause, 537

value engineering, 238–240. *See
also* best value concept
Variation in Estimated
Quantity clause, 600
veterans
Service-Disabled Veteran
Business programs, 339–341
Veterans Affairs and Military
Construction Appropriations
Act, 767–768
Veterans Affairs Board of
Contract Appeals, 507
Veterans Benefit Act (2003), 339
Veterans' Entrepreneurship and Small
Business Development Act (1999), 339
vicarious liability, 734–735
violations. *See also* fraud and
misrepresentation
conflicts of interest and,
113–114, 191–196
debarment/suspension for,
180–181, 197–200
disclosure of, 179–185
false claims as (*See* false claims)
improper pricing practices, 185–188
in interaction with federal
employees, 188–191
whistle-blower protection, 183–185,
737–738

Wage and Hour Division, 10, 347, 348–349
wages, 10, 346–349, 579–588, 585, 703–704
waiver
of completion/delivery
date, 396, 413–415
of defects, 684

payment bond claims and, 701, 711
of subcontractor payment rights, 665
Walsh Bros. v. United States, 400–401
warranties. *See also* inspections,
acceptance and warranties
breach of implied, 512–517
implied warranty of
specifications, 230, 499–517
performance bonds and, 694
quality control and, 297–299
*Watson, Watson, Rutland/Architects,
Inc. v. Montgomery County
Board of Education*, 89
weather, 257–258, 400–401,
405, 536, 584–585
*Weeks Dredging & Contracting, Inc.
v. United States*, 258–259, 265
Weeks Marine Inc. v. United States, 15
whistle-blower protection,
183–185, 737–738
White v. Edsall Construction Co.,
111–112
*Wickham Contracting Co. v.
Fischer*, 627
*Wickham Contracting Co. v.
United States*, 514–515
*W.M. Schlosser, Inc. v. United
States*, 541
women
affirmative action for, 329–333
as disadvantaged business
enterprise owners, 334–337
8(a) Business Development
program for, 337–338
Equity in Contracting for Women
Act, 329, 333, 338–339
as small business owners, 333
work method, 231–232
wrongful termination, 689–690, 695